Natural Disasters as Interactive Components of Global Ecodynamics

Kirill Ya. Kondratyev, Vladimir F. Krapivin and
Costas A. Varotsos

Natural Disasters as Interactive Components of Global Ecodynamics

Professor Kirill Ya. Kondratyev
Counsellor of the Russian Academy of Sciences
Scientific Research Centre for Ecological Safety
Nansen Foundation for Environment and Remote Sensing
St Petersburg
Russia

Professor Dr Vladimir F. Krapivin
Institute of Radioengineering and Electronics
Russian Academy of Sciences
Moscow
Russia

Associate Professor Dr Costas A. Varotsos
University of Athens
Department of Physics
Laboratory of Meteorology
Athens
Greece

SPRINGER–PRAXIS BOOKS IN ENVIRONMENTAL SCIENCES
SUBJECT *ADVISORY EDITOR*: John Mason B.Sc., M.Sc., Ph.D.

ISBN 10: 3-540-31344-3 Springer-Verlag Berlin Heidelberg New York
ISBN 13: 978-3-540-31344-1 Springer-Verlag Berlin Heidelberg New York

Springer is part of Springer-Science + Business Media (springeronline.com)

Bibliographic information published by Die Deutsche Bibliothek

Die Deutsche Bibliothek lists this publication in the Deutsche Nationalbibliografie; detailed bibliographic data are available from the Internet at http://dnb.ddb.de

Library of Congress Control Number: 2005938672

Printed in Germany

Cover design: Jim Wilkie
Project management: Originator Publishing Services, Gt Yarmouth, Norfolk, UK

Printed on acid-free paper

Contents

Preface

With the development of civilization the urgency of the problem of predicting the scale of expected changes in the environment, including the climate, has grown. First of all, it is a question of the appearance and propagation of undesirable natural phenomena which are dangerous (even lethal) for living beings, causing large-scale economic losses. Such phenomena have been called natural disasters. In the course of long-term evolution, natural anomalies of various spatial and temporal scales are known to have played an important role in ecodynamics, causing and activating the regulation mechanisms for natural systems. With the development of industry and the growth of the population, such mechanisms have changed considerably and in some cases have reached a life threatening level. First of all, this is connected with the growth and propagation of the amplitude of anthropogenic environmental changes.

Numerous studies of the last decades have shown that in some cases the frequency of occurrence of natural disasters, and their scale, have been growing, leading to an enhancement of the risk of large economic and human losses as well as to destruction of social infrastructure. So, for instance, in 2001, about 650 natural catastrophes happened over the globe, with victim numbers totaling more than 25,000 and economic losses exceeding US$35 bln. In 2002, only 11,000 people perished, but economic damage reached US$55 bln. In 2003, more than 50,000 people died because of natural disasters, with economic damage reaching about US$60 bln. Over the territory of Russia, the number of natural disasters during the last decade increased from 60 to 280. The USA in 2004 was hit by a record number of tornados (562), whereas in 1995 they numbered 399. The early part of 2004 was characterized by an increase of emergency situations, mainly of weather origin, and this year ended with a catastrophic tsunami on 26 December with enormous losses for the countries in the Indian Ocean basin. In Sri Lanka alone the damage reached US$3.5 bln. At 07:59 local time (00:59 UTC) on 26 December 2004, an *M* 9.3 megathrust earthquake occurred along 1,300 km of the oceanic subduction zone located 100 km west of Sumatra and the Nicobar and Andaman

Islands in the eastern Indian Ocean (Titov *et al.*, 2005). The tsunami waves generated on 26 December 2004 transported seismic energy over distances of thousands of kilometers throughout the World Ocean.

The amounts of damage caused by natural disasters depends strongly on the readiness of the territory to reduce the risk of losses and changes substantially over time. Maximum losses are caused by floods and hurricanes. The spatial distribution of disasters is rather non-uniform. On the whole, the relative distribution of the number of catastrophes, by type and over continents, is characterized by the following indicators: tropical storms – 32%, floods – 32%, earthquakes – 12%, droughts – 10%, and the share of other disasters – 14% (including Asia – 38%, America – 26%, Africa – 14%, Europe – 14%, and Oceania – 8%).

Natural disasters can be classified into various categories. Large-scale catastrophes are environmental phenomena leading to human victims and destruction of houses with substantial economic damage over a given region. Hence, the scale of natural disasters depends on the level of economic development of a region which determines the degree of protection for the population. Therefore, studies of phenomena connected with natural disasters should be accompanied by analysis of the level of poverty for a given region. The results of studies for the last 25 years show that in developing countries the dependence of losses due to natural disasters is much stronger than in economically developed regions. Bearing in mind that during the last decade the number and scale of natural disasters increased, on average by 5 times, it becomes clear, what levels of danger may be expected for the population of these countries in the near future. There is no doubt that prediction of, and warning about, critical phenomena over the globe should be of concern for all countries independent of their economic development.

In mathematics, the theory of catastrophes has been well developed. However, its application to the description of events and processes in the real environment requires the use of the methods of system analysis to substantiate the global model of the Nature–Society System (NSS) with the use of the technical facilities of satellite monitoring. Solution of these problems lies in the sphere of ecoinformatics, which provides a combination of analytical simple, semi-empirical, and complex non-linear models of ecosystems with renewable global data sets. Many international and national programs to study the environment could have recently raised the level of coordination in order to reach a required level of efficiency. This coordination can be exemplified by the Global Carbon Project (GCP) and Earth Observing System (EOS), within which the most efficient information and technical facilities to evaluate and predict the NSS dynamic characteristics have been concentrated.

The development of constructive methods of prediction of natural disasters requires the solution of a number of problems, including:

- Adaptation of methods of ecoinformatics as applied to the problem of diagnostics and prediction of natural disasters in all their diversity and scales.
- Formation of statistical characteristics of natural disasters historically, separation of categories and determination of spatial and temporal scales of catastrophic changes in the environment. An analysis of the history of

disasters is important for understanding the present dependences between crises in nature and society. Statistical characteristics of natural disasters in their dynamics enable one to formulate the basic principles of the mathematical theory of catastrophes and to determine the directions of first priority studies.
- Development of the concept and synthesis of the survivability model to use it for evaluation of the impact of natural disasters on the human habitat.
- Study of the laws of interactions between various elements and processes in the global NSS in correlation with the notion of biological complexity (biocomplexity) of ecosystems, considering it as a function of biological, physical, chemical, social, and behavioral interactions of subsystems of the environment, including living organisms and their communities. The notion of biocomplexity correlates with the laws of biospheric functioning as a unity of the ecosystems of different scales (from local to global). In this connection, it is necessary to give a combined formalized description of biological, geochemical, geophysical, and anthropogenic factors and processes taking place at different levels of the spatial–temporal hierarchy of scales. It is important to assess the possibility to use biocomplexity as a form of indicator of an approaching natural disaster.
- Study of correlations between survivability, biocomplexity, and evolution of the NSS with the use of global modeling technology. Creation of the units for the global model which describes the laws and trends in the environment that lead to an appearance of stress situations and are initiated by man's economic and political activity.
- Consideration of demographic conditions for the origin of natural disasters and an identification of mechanisms for regulation of the environment which hamper the formation of these conditions.
- Assessment of the information content of the existing technical means of collection of data on the state of the NSS subsystems and available global databases in order to allocate them for solving the problems of the evaluation of conditions for the origin of stress situations in the environment.

The main objective of this book is to discuss and develop methods of ecoinformatics as applied to the problem of diagnostics and forecast of natural disasters in all their diversities and scales. The book contains nine chapters.

Chapter 1 considers the statistical characteristics of natural disasters in their historical aspect, selecting categories and determining the spatial and temporal scales of catastrophic changes to living beings habitats. An analysis of the history of catastrophes is important for understanding the present dependences between crises both in nature and in society. Statistical characteristics of natural disasters in their dynamics enable one to formulate the basic principles of the mathematical theory of catastrophes and to determine the directions of first priority studies.

Chapter 2 considers the notion of ecosystem survivability. The concept is given and a model of survivability is developed which is used to assess the impact of natural disasters on the human habitat. The nature–society interaction is parameterized in terms of mathematical game theory. Taking into account a conditional division of nature and society, the notion has been introduced of their strategy, objectives, and behavior. Depending on the relationships between these categories, various versions of the model of survivability have been studied, and critical characteristics have been determined for man's survival as an NSS component.

Chapter 3 discusses the problem of interaction between various elements and processes in the global NSS in correlation with the notion of biocomplexity. Biocomplexity is evaluated using the scale characterizing the structural and functional complexity of direct and indirect connections in the system. It is shown that biocomplexity is a possible indicator of an approaching natural catastrophe.

Relationships between survivability, biocomplexity, and evolution of the NSS have been studied in Chapter 4 using simulation modeling. The global model units have been developed which describe the laws and trends in the environment that lead to stress situations and are initiated by man's economic and political activity. Demographic conditions for the origin of natural disasters are considered and mechanisms for environmental regulation are analyzed which hamper the formation of these conditions. Globalization of anthropogenic processes connected with the accelerated development of megalopolises and other crisis-threatening formations has been considered as the main cause destabilizing the NSS dynamics. The ecological state of such territories is evaluated with the use of the indices of population density and economic development.

Chapter 5 describes the modern information facilities that provide data to study the state of environmental systems. A brief analysis has been made of the available global databases, and their role in the assessment of conditions for the origin of stress situations in the environment has been shown. The important problem of synthesis of the information searching monitoring systems has been considered which detect and warn about anomalies in natural systems at an expert level of accomplishment.

Chapter 6 develops methods, algorithms, models, and other elements of information technologies to predict natural disasters. With this aim in view, a global spatially heterogeneous model of the NSS functioning has been substantiated, which integrates the accumulated knowledge of, and data on, the environment and provides computer facilities to predict the moments of the origin and trends of development of stress situations in their various manifestations. The emphasis is firmly on the development and propagation of catastrophic processes on global biogeochemical cycles, greenhouse gas (GHG) dynamics, and the water balance of the biosphere.

Chapter 7 demonstrates the usefulness of ecoinformatics methods in analysis of the state and prediction of one of the quasi-global natural disasters in the Aral and Caspian Seas region. A strategic scenario is exemplified, the accomplishment of which can prevent this catastrophe from further development.

Chapter 8 contains an analysis of relationships between some types of natural phenomena and processes of formation of global changes in the biosphere and in the climatic system. As an illustration, an analysis has been made of the estimates of the role of forest fires and lightning strikes in changes in the atmospheric composition. Various hypothetical scenarios of potential impacts on the environment have been considered, which lead to a drastic change in the trends of global ecodynamics.

The problem of global ecodynamics is closely correlated with the problem of changes within the Earth's climate system. Understanding this correlation is possible in the context of interactivity of the climate and other environmental processes.

The concluding Chapter 9 completes the book's theme with the consideration of interactivity between the climate and natural catastrophes. Anomalous situations in the environment and climate change are considered as interactive processes.

The book continues the development of approaches to studies of the global NSS dynamics proposed earlier by the authors, with an emphasis on the problems of evaluation, detection, prevention, and prediction of disasters both of natural and anthropogenic origin. The proposed approach consists mainly in the combined use of numerical simulation and global monitoring, with the creation of a system of knowledge from different sciences that determines the NSS functioning. The book will be useful for specialists in the sphere of global modeling, studies of climate change, nature–society relationships, geopolitics, international relations, and methodology of interdisciplinary studies. It is of special interest for designers and users of information technologies in the field of population protection from natural disasters.

Figures

Tables

Abbreviations and acronyms

AAHIS	Advanced Airborne Hyperspectral Imaging Spectrometer
AARGOS	Arctic Atmosphere Research Global Observing System
AARS	Asian Association on Remote Sensing
ABL	atmospheric boundary layer
ACE-2	Advanced Composition Explorer-2
ACIA	Arctic Council and International Association
ACR	atmospheric counter-radiation
ACRS	Asian Conference on Remote Sensing
ACSYS	Arctic Climate System Study
ADEOS	Advanced Earth Observation Satellite
AERONET	Aerosol Robotic Network
AGCM	atmospheric general circulation model
AI	aerosol index
AIDS	Acquired Immune Deficiency Syndrome
AIMES	Analysis, Integration and Modeling of the Earth System
AIMR	Airborne Imaging Microwave Radiometer
AIRS	Atmospheric Infrared Sounder
AIS	Airbone Imaging Spectrometer
AMIP	Atmospheric Model Intercomparison Project
AMMA	African Monsoon Multidisciplinary Analysis
AMMR	Airborne Multichannel Microwave Radiometer
AMPR	Advanced Microwave Precipitation Radiometer
AMSR	Advanced Microwave Scanning Radiometer
AMSU	Advanced Microwave Sensing Unit
ANSS	Advanced National Seismic System
AOCI	Airborne Ocean Color Imager Spectrometer
AOT	aerosol optical thickness
APT	Advanced Picture Transmission

AR	autotraphic respiration
ARF	aerosol radiative forcing
ARISTI	All-Russian Institute for Scientific and Technical Information
ASI	Advanced Study Institutes
ASTER	Advanced Spaceborne Thermal Emission and reflection Radiometer
AVHRR	Advanced Very-High Resolution Radiometer
AVNIR	advanced radiometer of visible and near-IR
AWSR	Airborne Water Substrate Radiometer
BID	Banco Interamericano de Desarrollo
BWBM	Biosphere Water Balance Model
C4MIP	Coupled Climate and Carbon Cycle Model Intercomparison Project
CA	Climate Agenda
CAENEX	Complete Atmospheric Energetics Experiment
CAGL	Central Aero-Geophysical Laboratory
CALIPSO	Cloud–Aerosol Lidar and Infrared Pathfinder Satellite Observation
CASA	Carnegie–Ames–Stanford Approach
CASI	Compact Airborne Spectrographic Imager
CATCH	Computer And Technology Crime High
CCM1-Oz	NCAR Community Climate Model 1-Oz
CCM3.6	NCAR Community Climate Model 3.6
CCN	cloud condensation nuclei
CCSP	Climate Change Science Program
CCSR	Center for Climate System Research
CCSS	carbon–climate–society system
CENEX	Complex Energetic Experiment
CEOP	Coordinated Enhanced Observing Period
CEOS	Committee on Earth Observation Satellites
CERES	Clouds and Earth's Radiant Energy System
CETA	Carbon Emissions Trajectory Assessment
CF	cloud feedback
CFC	chlorofluorocarbon compounds
CG	cloud–ground
CGCM	Canadian Global Climate Model
CHRISS	Compact High-resolution Imaging Spectrograph Sensor
CID	Center for International Development
CliC	Climate and Cryosphere
CLIMCYC	CLIMate CYCle project
CLIVAR	CLImate VARiability and predictability
CMDL	Climate Monitoring and Diagnostics Laboratory
COADS	Comprehensive Ocean–Atmosphere Data Set
COLA	Centre for Ocean–Land–Atmosphere studies
COP	Conference of Parties

COPES	Coordinated Observation and Prediction of the Earth System
CR	Club of Rome
CS	cloud sensitivity
CSIRO	Commonwealth Scientific and Industrial Organization
CSRP	Carbon Sequestration Research Program
DEKLIM	Deutsches Klimnaforschungsprogramm
DEMETER	Development of a European Multimodel Ensemble system for seasonal to inTERannual prediction
DIVERSITAS	International Biodiversity Programme
DMS	Disaster Management System
DMUU	Decision Making Under Uncertainty
DOE	Department of Energy
DOM	dissolved organic matter
DPSIR	Driving forces, Pressures, State, Impact, Response
D-SELF	Double Self-Organization
DU	Dobson Units
ECDR	effective cloud droplet radius
ECHAM4.L39(DLR)	European Centre Hamburg Coupled Chemistry–Climate Model
ECSib	External Climate of Siberia
ECYI	Earth Charter Youth Initiative
EEC	European Economic Community
EMIRAD	Electromagnetic Institute Radiometer
EMISAR	Electromagnetic Institute SAR
ENSO	El Niño/Southern Oscillation
EOF	empirical orthogonal function
EOS	Earth Observing System
EPA	US Environmental Protection Agency
EPIC	Environmental Policies and Institutions for Central Asia
ERB	Earth's Radiation Balance
ERBE	Earth Radiation Budget Experiment
ERCD	efficient radius of cloud droplets
EROC	Ecological Rates of Change
ERS	Earth Radiation Satellite
ESRI	Economic and Social Research Institute
ESS	effective scattering surface
ESTAR	Electronically Steered Thinned Array Radiometer
ESSP	Earth System Science Pathfinder
EWT	equivalent water thickness
FAO	Food and Agriculture Organization
FCCC	Framework Convention in Climate Change
FEMA	Federal Emergency Management Agency
FMC	fuel moisture content
GACP	Global Aerosol Climatology Project
GAIM	Global Analysis, Integration, and Modeling

GAME	GEWEX-related Asian Monsoon Experiment
GARP	Global Atmospheric Research Programme
GCC	global carbon cycle
GCM	global circulation model
GCOS	Global Climate Observing System
GCP	Global Carbon Project
GDP	Gross Domestic Product
GEF	Global Ecological Fund
GEOS	GEOstationary satellite
GEOSS	GEOhydrological and Spatial Solutions
GEP	global ecological perspective
GEWEX	Global Energy and Water Cycle Experiment
GHG	greenhouse gas
GHOST	Global HOlocene Spatial and Temporal climate variability
GGD	Greenhouse Gas Dynamics
GIMS	geoinformation monitoring system
GIS	geographic information system
GLOBEC	Global Ocean Ecosystems Dynamics
GMNSS	global model of the nature–society system
GMS-5	Geosynchronous Meteorological Satellite-5
GOES	Geostationary Observational Environment Satellite
GOOS	Global Ocean Observing System
GPM	Global Precipitation Measurement
GRACE	Gravity Recovery And Climate Experiment
GSFC	Goddard Space Flight Center
GTOS	Global Terrestrial Observing System
GWEM	global wildland fire emission model
GWP	Gross World Product (Chapters 4 and 8)
GWP	Global Warming Potential (Chapter 6)
HDI	Human Development Index
HGP	home gross production
HIV	Human Immunodeficiency Virus
HRPT	High-Resolution Picture Transmission
HSB	Humidity Sounder for Brazil
IA	Integrated Assessment
IACCA	Interdisciplinary Commitee on Climate Agenda
IAHS	International Association of Hydrological Sciences
IAMAS	International Association of Meteorology and Atmospheric Sciences
IASI	Infrared Atmospheric Sounder Interferometer
ICESat	The Ice, Cloud and land Elevation Satellite
ICGGM	International Centre for Global Geoinformation Monitoring
ICLRT	International Center for Lightning Research and Testing
ICSU	International Council of Scientific Unions
IDB	Inter-American Development Bank

IDEA	Instituto de Estudios Ambientales
IEA	International Energy Agency
IEEE	Institute of Electrical and Electronics Engineers
IFRC	International Federation of Red Cross and Red Crescent Societies
IG	Institute of Geography
IGAC	International Global Atmospheric Chemistry Project
IGBP	International Geosphere–Biosphere Programme
IGOS	Integrated Global Observing Strategy
IHDP	International Human Dimensions Programme
IKONOS	Remote sensing satellite launched by ESRI Business partner space imaging
INDOEX	Indian Ocean Experiment
INSAR	Interferometric Synthetic Aperture Radar
INSAT	Indian Satellites
IPAB	International Programme for Antarctic Buoys
IPCC	Intergovernmental Panel on Climate Change
IPY	International Polar Year
IR	infrared
IRE	Institute of Radioengineering and Electronics
IRS	Indian Remote Sensing Satellite
ISCCP	International Satellite Cloud Climatology Project
ITCT	Intercontinental Transport and Chemical Transformation
JGOFS	Joint Global Ocean Flux Study
KBGG	Kara–Bogaz–Gol Gulf
KP	Kyoto Protocol
LEA	Leaf Area Index
LANDSAT	Land (Remote Sensing) Satellite
LP	logico–probabilistic
LSM	land surface model
LSU	Leningrad State University
LTER	Long-term Ecological Research
LWR	long-wave radiation
LWRF	long-wave radiative forcing
MAPSS	Mapped Atmosphere–Plant–Soil System
MEA	Millenium Ecosystems Assessment
MERGE	Model for Evaluating Trajectory Regional and Global Effects
METOP	METeorological Operational Polar
MGC	minor gas component
MIDC	minor island developing countries
MISR	Multi-angle Imaging Spectrometer
MLS	Microwave Limb Sounder
MMIA	Methods and Models for Integrated Assessment
MOBY	Marine Optical BuoY
MODIS	Moderate-resolution Imaging Spectroradiometer
MOZART	Model for Ozone and Related chemical Tracers

MPC	minor pollution components
MSR	Microwave Scanning Radiometer
MSSA	multi-channel singular spectral analysis
MSU	Moscow State University
MTLE	Model of Transmission Line with Exponential current decay
NACP	North American Carbon Programme
NAO	North Atlantic Oscillation
NAS	The National Association of Scholars
NASA	National Aeronautical Space Agency
NASDA	National Space Development Agency of Japan
NATO	North Atlantic Treaty Organisation
NCAR	National Center for Atmospheric Research
NCEP–NCAR	National Center for Environmental Protection–National Center for Atmospheric Research
NDVI	Normalized Differential Vegetation Index
NDWI	Normalized Difference Water Index
NEHRP	National Earthquake Hazards Reduction Program
NEP	New Economics Papers
NMHC	Non-Methane Hydrocarbons
NOAA	National Oceanic and Atmospheric Administration
NPOESS	The National Polar-orbiting Operational Environmental Satellite System
NPP	Net Primary Production
NPPP	nuclear power plants production
NSF	National Scientific Fund
NSS	Nature–Society System
NTSLF	National Tidal and Sea Level Facility
OACES	Ocean–Atmosphere Carbon Exchange Study
OCCC	Ocean- Carbon Cycle and Climate studies
OCO	Orbiting Carbon Observatory
OCTS	ocean color and temperature scanner
OSEM	Okhotsk Sea ecosystem model
OSWR	outgoing short-wave radiation
P/B	Production/Biomass
PAGES	Pilot Analysis of the Global Ecosystems
PAR	photosynthetically active radiation
PAUR	Photochemical Activity and Ultraviolet Radiation
PCE	Peruvian Current ecosystem
PCL	permissible concentration level
PDUS	Primary Data User Station
PIK	Potsdam Institut für Klimafolgenforschung e.V.
PNA	Pacific–North America
PNAS	Proceedings of the National Academy of Sciences of the USA
POLDER	Polarization and Directionality of the Earth's Reflectances
RAL	Rutherford Appleton Laboratory

RANS	Russian Academy of Natural Sciences
RAS	Russian Academy of Sciences
RC	Roma Club
RegCM2	Regional Climate Model-2
REMO	REgional MOdel
RF	radiative forcing
RFBR	Russian Fund for Basic Research
RGS	Russian Geographical Society
RIV	Resources, Innovation, and Values
SALLJEX	South American Low-Level Jet Experiment
SAR	Synthetic Aperture Radar
SAT	surface atmospheric temperature
SC	solar constant
SCAR	smoke/sulfate clouds and radiation
SCIMASACHY	Scanning Imaging Absorption Spectrometer for Atmospheric CartograpHY
SDS	Scott Data System
SeaWiFS	Sea-viewing Wide Field-of-view Sensor
SFSO	Swiss Federal Statistical Office
SiB2	Simple Biosphere model-2
SIDC	small island developing countries
SMASHF	simulation model of the Aral Sea hydrophysical fields
SMMR	Scanning Multichannel Microwave Radiometer
SMOS	The Soil Moisture and Ocean Salinity
SOI	State Oceanographic Institute
SP	strategic plan
SPARC	Stratospheric Processes And their Role in Climate
SPE	system "planet Earth"
SPOT	Systéme Probatoire d'Observation de la Terre (French Earth-observing satellite)
SRES	Special Report on Emissions Scenarios
SSM/I	Special Sensor Microwave/Imager
SSROC	simulation system for regional ozonosphere control
SST	sea surface temperature
STEDI	Student Explorer Demonstration Initiative
SWR	short-wave radiation
SWRF	short-wave radiation forcing
TARFOX	Tropospheric Aerosol Radiative Forcing Observational Experiment
TERRAPUB	Terra Scientific Publishing Company
THORPEX	A Global Atmospheric Research Programme
TIROS-N	Television Infrared Observational Satelline-Next
TL	transmission line
TMI	TRMM Microwave Imager
TO	tropospheric ozone

TOGA	Tropical Ocean Global Atmosphere
TOMS	Total Ozone Mapping Spectrometer
TRMM	Tropical Rainfall Measuring Mission
TROPEX	A global atmospheric research programme (TROPospheric Experiment)
TSFP	Turbulence and Shear Flow Phenomena
UAE	United Arab Emirates
UABL	upper atmospheric boundary layer
UAL	upper atmospheric layer
UEM	upwelling ecosystem model
UHF	ultra-high frequencies
UNDP	United Nations Development Programme
UNEP	United Nations Environment Programme
UNESCO	United Nations Educational, Scientific and Cultural Organization
UNFCCC	UN Framework Convention on Climate Change
US GLOBEC	US Global Ocean Ecosystems Dynamics
USGS	US Geological Survey
USSR	Union of Soviet Socialistic Republic
UV	ultraviolet
VAMOS	Variability of the American Monsoon Systems
VAST	Vietnamese Academy of Science and Technology
VDI	Vegetation Dryness Index
VISSR	Visible Infrared Spin-Scan Radiometer
WB	World Bank
WCDR	World Conference on Disaster Reduction
WCP	World Climate Programme
WCRP	World Climate Research Programme
WDI	Water Deficit Index
WEFAX	Weather Faximile
WHIRL	Wide-angle High-resolution Line-imager
WHO	World Health Organization
WI	Wellbeing Index
WMO	World Meteorological Organization
WOCE	The World Ocean Circulation Experiment
WSOC	water-soluble organic compounds
WSSD	World Summit on Sustainable Development
WTF	wet tropical forest
WWRP	World Weather Research Programme
ZEUS	Zentren Europaischen Supercomputings

About the authors

Kirill Ya. Kondratyev received his Ph.D. in 1956 at the University of Leningrad where he also became a Professor in 1956 and was appointed Head of the Department of Atmospheric Physics in 1958. From 1978–1982 he was Head of the Department of the main Geophysical Observatory and from 1982–1991 Head of the Institute for Lake Research in Leningrad. Since 1992 he has been Councillor of the Russian Academy of Sciences. Kirill has published more than 100 books in the fields of atmospheric physics and chemistry, remote sensing, planetary atmospheres, and global change.

Vladimir F. Krapivin was educated at the Moscow State University as a mathematician. He received his Ph.D. in geophysics from the Moscow Institute of Oceanology in 1973. He became Professor of Radiophysics in 1987 and Head of the Applied Mathematics Department at the Moscow Institute of Radioengineering and Electronics in 1972. He is a full member of the Russian Academy of Natural Sciences and has specialized in investigating global environmental change by the application of modeling technology. Vladimir has published 14 books in the fields of ecoinformatics, the theory of games, and global modeling.

Costas A. Varotsos received his BSc in Physics at Athens University in 1980 and his PhD in Atmospheric Physics in 1984. He was appointed Assistant Professor in 1989 at the Laboratory of Meteorology in the Physics Department of Athens University, where he also established the Laboratory of the Middle and Upper Atmosphere. In 1999 he became Associate Professor of the Department of Applied Physics at Athens University. Costas has published more than 200 papers and 15 books in the fields of atmospheric physics, atmospheric chemistry, and environmental change.

Summary

The problems of natural disaster prediction and also substantiation of environmental monitoring systems receiving, storing, and process the necessary information for the solution of relevant problems has been analyzed. A three-level procedure is proposed for decision-making with the use of natural disaster indicators. It is based on the assessment of relative indicators and the use of a mathematical model for environmental dynamics. Some specific natural disasters (with special emphasis on the Aral Sea problems) have been studied. The global Nature–Society System (NSS) is considered as an interactive natural–anthropogenic mechanism driving environmental dynamics. A search for a control strategy to weaken or prevent crisis situations in the NSS has been undertaken.

1

Statistics of natural disasters

1.1 HISTORY OF CATASTROPHIC EVENTS

Natural disasters have always been understood as one of the elements of global ecodynamics. Natural disasters and various natural cataclysms in the past took place together with the development of natural trends, and starting from the 19th century, their dynamics have been influenced by anthropogenic factors. The engineering activity in the 20th century, and formation of a complicated global socio-economic structure, have not only increased the share of anthropogenic disasters but also changed the environmental characteristics toward deteriorating the habitat of living beings, including human beings. According to Schneider (1995), the climatic seasonal variations in the historical past had been characterized by a higher stability. The seasonal shift over 344 years starting since 1651 did not exceed 24 hours per century. Beginning from 1940, in the northern hemisphere, a clearly expressed anomaly of the seasonal shift has been taking place. For instance, the winter of 1994 in the USA was characterized by very low temperatures in the eastern states, and in July of that year there was a record high temperature in the southwest of the country, when the temperature reached 48.8°C. In the summer of 1994, thousands of people died as a result of excessive heat in India. On the contrary, the second half of 1991 was characterized by reduced temperatures, apparently, due to the Mount Pinatubo eruption on the Philippines in June 1991, when huge masses of ash were ejected into the atmosphere. On the whole, along with the processes of climatic destabilization, the catastrophic events grow in number. Tables 1.1–1.10 and Figure 1.1 give some idea about the dynamics of the number of natural disasters and associated calamities (Goudsouzian, 2004).

Historically, there is much information about natural disasters over different parts of the planet. Of course, it is impossible to reproduce a complete pattern of such disasters, since descriptions, chronologies, and characteristics of the 154 anomalous natural phenomena recorded by scientists in chronicles of the period

Table 1.1. Large-scale volcanic eruptions throughout humankind's history (Grigoryev and Kondratyev, 2001).

Year	Volcano	Country	Number of deaths
1500 BC	Santorini	Greece	End of the Minosoic civilization
79 BC	Vesuvius	Italy	30,000
1586	Kelut	Indonesia	10,000
1631	Vesuvius	Italy	18,000
1669	Etna	Italy	10,000
1783	Paradajan	Indonesia	9,340
1792	Unsen	Japan	15,190
1815	Tambor	Indonesia	92,000
1815	Kumbava	Indonesia	100,000
1883	Krakatau	Indonesia	36,420
1902	Bonpele	Martinica (France)	29,500
1902	Santa-Maria	Guatemala	6,000
1919	Kelut	Indonesia	5,050
1937	Maturi	New Guinea	500
1985	Nevado-del-Ruiz	Columbia	24,740
1997	Suffrier Hills	Montserrat in Caribbean Sea	19
2000	Popokatepetl	Mexico	Evacuation of 15,000
2002	El Reventador	Ecuador	Evacuation of 3,000

Table 1.2. Distribution of the number of victims by type of natural disaster for the 50-year period, 1947–1997.

Type of natural disaster	Number of deaths
Cyclones, typhoons, storms on the coast	1,500,000
Earthquakes	400,000
Floods	360,000
Thunderstorms	40,000
Tsunamis	30,000
Volcanic eruptions	15,000
Sudden heat	10,000
Fog	7,000
Sudden cold	7,000
Avalanches	7,000
Landslides, mudflows	6,000
Flash floods	2,000

from the 1st century BC to the 7th century AD differ from the present ones both in the level of their perception and in their descriptions without referencing to formalized scales. For instance, in present conditions, a natural phenomenon like a solar eclipse is not seen as a natural disaster, in contrast to the period of the old

Table 1.3. Continental distribution of natural disasters and resulting damage (Ruck, 2002; Phelan, 2004; Munich Re, 2005a,b).

Region	Number of loss events			Number of fatalities			Economic losses (US$ mln). Insurance payments are given in brackets		
	2002	2003	2004	2002	2003	2004	2002	2003	2004
Africa	51	57	36	661	2,778	1,322	308 (158)	5,158 (0)	444 (0)
America	181	206	185	825	946	4,830	13,933 (6,259)	21,969 (13,247)	68,183 (34,585)
Asia	261	245	245	8,570	53,921	170,254	13,965 (385)	18,230 (600)	72,706 (7,887)
Australia and Oceania	69	65	52	61	47	67	2,192 (11)	628 (246)	343 (124)
Europe	136	126	124	459	20,194	371	24,246 (5,897)	18,619 (1,690)	3,765 (1,218)
Globe	*698*	*699*	*642*	*10,576*	*77,886*	*176,844*	*54,644 (12,710)*	*64,604 (15,783)*	*145,441 (43,814)*

Table 1.4. Statistics of the most powerful natural disasters (Ruck, 2002; Munich Re, 2005a,b).

Years	1950–1959	1960–1969	1970–1979	1980–1989	1990–1999	2000–2002	2003–2005
Number of natural disasters	20	27	47	63	91	70	27
Economic losses (US$bln)	42.1	75.5	138.4	213.9	659.9	550.9	764.6

and early Middle Ages. Nevertheless, the type and character of natural disasters have not practically changed since then. There are various hypotheses among specialists who studied the history of the Earth as a planet, on the origin of disasters taking place on it. So, Walker (2003) discusses the "snowy Earth" hypothesis believing that the observed natural anomalies are weak noises against a background of historical signals, with an amplitude ranging from complete covering of the planet, with a 1-km layer of ice and snow, to the greenhouse state.

By the estimates of Norton (2002), in the 18th century over the territory of England there were seven serious natural disasters, fragmentary information about which does not allow one to reliably assess their scale, but, nevertheless, it is known that over the territory of England there have been several powerful hurricanes with

Table 1.5. Most significant natural disasters in 2003.

Date	Country/region	Event	Losses	
			Human lives	Economic (US$ mln)
2–3 January	Germany Sweden Switzerland	Winter hurricane Calvann, floods. Wind speed reached 200 km h^{-1}.	6	1,000
January	India Bangladesh Pakistan	Sudden drop in temperature. Large losses in agriculture.	1,800	
13–25 January	USA	Winter hurricanes.	2	650
25–27 January	Italy	Floods. Thousands evacuated.	1	150
27 January	Cyprus	Tornado, hurricanes. Buildings and cars damaged.	–	10
January–February	Australia	Forest fires. 1,500 houses destroyed.	4	300
24–26 February	China	Earthquakes of magnitude 6.4. About 70,000 buildings ruined or damaged. More than 4,000 people wounded.	268	150
4–8 April	USA, Texas	Unfavorable weather with hail.	13	2,100
14–15 April	Oman	Floods: the most powerful of the last 10 years. Washed-out roads.	30	1
21–24 April	Bangladesh	Unfavorable weather with hail.	230	
28 April–16 May	Argentina	Flood. 30,000 buildings ruined or damaged.	26	200
2–11 May	USA, Midwest	Tornados, heat, thunderstorms.	42	3,200
May, June	India Bangladesh Pakistan	Heat and drought. The temperature reached 50°C. Heavy losses to agriculture.	2,000	400
18–19 May	Sri Lanka	Floods.	300	
21 May	Algeria	Earthquake of magnitude 6.8, tsunami. About 100,000 buildings ruined or damaged; 200,000 people homeless; more than 10,000 people wounded.	2,200	5,000
May–July	China	Floods. Heavy losses of yield. More than 2 million buildings ruined or damaged.	800	8,000
29 June–1 July	USA (Louisiana, Massachusetts)	Tropical hurricane Bill, wind speed 90 km h^{-1}.	4	30
21–23 July	USA	Unfavorable weather.	7	800
23–25 July	China Philippines	Typhoon Imbudo. Wind speed 230 km h^{-1}. Thousands of buildings ruined.	41	125
28–31 July	Italy	Powerful thunderstorms with coarse hail (diameter up to 10 cm).	1	20

Date	Country/region	Event	Losses	
			Human lives	Economic (US$ mln)
July, August	Europe	Heat and drought. Large-scale forest fires. Record temperatures. Rivers shallowing.	27,000	13,000
8–10 August	Japan	Typhoon Etau No. 10. Record rains.	13	55
22–27 August	China Vietnam Philippines	Typhoon Krovanh. Heavy rains. Dams and hundreds of buildings are damaged. Electric power and water supplies are broken.	7	145
23–25 August	Australia	Thunderstorms. Electric power supply is broken.	1	20
28–31 August	Austria France, Italy Switzerland	Thunderstorms, landslides. Buildings and urban infrastructures are damaged.	5	500
June–September	India Bangladesh Pakistan Nepal	Floods.	1,400	
August–October	China	Floods, thunderstorms. 73,000 buildings ruined or damaged, agricultural infrastructure damaged.	123	2,200
1–3 September	China Taiwan	Typhoon Dujuan. Record rains. Thousands of buildings damaged or ruined. Heavy losses to agriculture.	42	320
5–6 September	Bermudas	Hurricane Fabian, wind speed 235 km h^{-1}.	8	500
11–13 September	Japan South Korea	Typhoon Maemi.	118	4,150
18–20 September	USA Canada (west coasts)	Hurricane Isabel, wind speed 260 km h^{-1}.	40	5,000
25–28 September	Japan, Hokkaido	Earthquakes of magnitudes 7.4 and 8.3. Several buildings ruined.	1	180
October–November	USA, California	Forest fires, drought.	20	3,500
12–14 November	USA	Unfavorable weather.	8	600
2–4 December	France	Flood. Buildings ruined and agricultural infrastructure damaged.	7	1,500
26 December	Iran	Earthquake of magnitude 6.6. The city of Bam is destroyed. 30,000 people are wounded and 100,000 people made homeless.	>40,000	

Table 1.6. The most powerful earthquakes in the history of humankind (Grigoryev and Kondratyev, 2001; Binenko *et al.*, 2004; Hanson, 2005).

Year	Region	Fatalities, thousands of people
365	The east Mediterranean region, Syria	50
844	Syria (Damascus)	50
893	Armenia	100
893	India	180
1138	Syria	100
1268	Turkey (Semjiya)	60
1290	China (Province of Djili)	100
1456	Italy (Naples)	60
1556	China (Province of Shengshi)	830
1626	Italy (Naples)	70
1667	Azerbaijan (Shemakha)	80
1668	China (Shandung)	50
1693	Italy (Sicily)	60
1727	Iran (Tebriz)	77
1730	Japan (Hokkaido)	137
1737	India (Bengaluru, Calcutta)	300
1739	China (Province of Ninaya)	50
1755	Portugal (Lisbon)	60
1783	Italy (Calabria)	50
1868	Ecuador (Ibarra)	70
1908	Italy (Messina)	120
1920	China (Province of Hanshu)	180
1920	China (Province of Ningsha)	200
1923	Japan (Canto)	140
1923	Japan (Tokyo)	99.3
1932	China (Province of Hanshu)	70
1935	Pakistan (Kuetta)	60
1948	Turkmenistan (Ashkhabad)	110
1957	Gobi-Altai (Mongolia, Buriatiya, Irkutsk and Chita Regions)	0
1970	Peru (Chimbote)	66
1970	Iran	50
1972	Nicaragua	5
1972	Iran	5
1974	Pakistan	5.3
1974	Guatemala	22.084
1976	Indonesia	5
1976	Turkey (Van)	4
1976	China (Province of Tien Shan)	255
1976	Guatemala	22.085
1976	Philippines	3.739
1978	Iran	25
1985	Mexico	9.5
1988	Armenia	25
1988	Turkey	25
1990	Iran (western part)	40
1993	India	9.475
1994	USA, California	0.061
1995	Japan, Kobe	6.43
1998	Afganistan (Takhar)	4
1999	Turkey	19.118
1999	Earthquake in Neftegorsk, Russia	2
2001	India, Pakistan	15
2002	USA, Alaska	0
2002	Afghanistan	1
2003	Iran (south-east)	26.271
2004	South-eastern Asia (Sumatra, Thailand, India, Sri Lanka)	283
2004	Japan, Niigata Prefecture	0.04
2005	Indonesia, Malaysia	2
2005	Nias Island, Indonesia, Northern Sumatra	2.5

Table 1.7. Earthquakes and consequences from 1980.

Year	Country	Magnitude (Richter scale)	Fatalities
1980	Algeria	7.4	5,000
1980	Italy	6.9	3,105
1981	Iran	6.9	3,000
1981	Iran	7.3	1,500
1982	Yemen	6.0	2,800
1982	Italy	4.5	0
1983	USA	6.5	0
1983	Japan	7.8	104
1983	Turkey	6.9	1,400
1983	Belgium	4.7	2
1984	Mexico	8.1	10
1984	USA	6.1	0
1984	Italy	5.6	0
1985	Mexico	8.1	10,000
1986	El Salvador	5.4	1,100
1987	USA	5.9	8
1987	Ecuador	6.9	1,000
1988	Burma	6.1	730
1988	China	7.6	730
1988	Turkey	7.6	25,000
1988	Armenia	6.9	25,000
1988	India, Nepal	6.7	1,052
1989	USA	7.1	63
1989	Tadjikistan	6.0	284
1989	Indonesia	5.6	120
1989	China	6.1	29
1989	Australia	5.5	10
1990	Iran	7.4	50,000
1990	China	6.9	126
1990	Peru	6.2	200
1990	Philippines	7.7	1,600
1990	Italy	4.7	20
1991	Pakistan, Afghanistan	6.8	1,200
1991	India	6.1	1,600
1992	Indonesia	6.8	2,200
1993	India	6.4	22,000
1995	Japan	7.2	6,430
1995	Russia (Far East)	7.5	1,989
1997	Iran	5.5	1,000
1997	Iran	7.1	1,560
1998	Afghanistan	6.1	4,000
2000	Indonesia	7.9	103
2000	Indonesia	6.5	48
2000	Sumatra	7.9	103
2000	Japan	7.1	30
2001	India, Pakistan	8.0	15,000
2001	El Salvador	7.7	844
2001	India	7.7	20,005
2002	Afghanistan	6.1	1,000
2003	Algeria	6.8	2,266
2003	Iran	6.5	26,271
2004	Morocco	6.4	628
2004	Indian Ocean	9.0	283,000
2005	Pakistan, Afghanistan, India	7.6	50,000
2005	Nias Island	8.7	2,500

Table 1.8. Form and size of tornados.

Characteristic	Type of tornado		
	Weak	Powerful	Fierce
Share of all tornados, %	69	29	2
Share of mortal issues, %	5	28	67
Life time, minutes	1–10	≈20	≥60
Wind speed, $km\,h^{-1}$	<200	200–380	>380

downpours and thunderstorms, that caused heavy destruction and death. For instance, in 1703 a powerful hurricane occurred over London on the morning of 26 November and continued to the next morning, heavily destroying everything. About 40 barges and other ships sank in the River Thames. There were not many human victims because the citizens of London had time to leave their homes and find safe shelter. The hurricane abated and started anew during the next two days. A similar natural disaster took place twice, in July and December 1725, in Yorkshire and London, respectively. From preserved descriptions, a "fire ship" appeared on 31 July from the ocean, which sent fire and water onto people and their homes. Many people who were in boats or on bridges perished in the storm. The chronicles also contain descriptions of powerful floods over England. The 19 June 1725 downpour caused heavy river flooding practically over all of England, immersing the ground floors of buildings and causing considerable material damage.

Many hurricanes and storms are followed by thunderstorms. The thunderstorm activity is observed almost at all latitudes. Damage caused by thunderstorms is manifested through forest fires and direct destruction of various communication networks as well as human victims. For instance, during 3–4 April 1972, over the territory of the USA, 320 people perished, and economic damage was estimated at 1.7 billion dollars. During the next years, there were two instances of large human losses in May 1999 and 2000, when 42 and 51 people died, respectively. Thunderstorms following in quick succession over the states of Kansas, Oklahoma, and Tennessee from 2–11 May caused material damage estimated at 3.205 billion dollars. On the whole, over the last decade, natural disasters were distributed with the following percentages of type: floods – 33%, storms – 23%, epidemics – 15.2%, droughts – 15%, earthquakes and tsunamis – 7%, landslides – 4.5%, volcanic eruptions – 1.4%, and avalanches – 0.7%.

1.1.1 Earthquakes

Over the history of humankind, considerable damage has been caused by earthquakes being the most powerful natural phenomena and the least well predicted. During earthquakes a large-scale center of damage forms, whose territory is characterized by vast damage to buildings. The majority of people die under such ruins. The number of victims depends on the seismic stability of the buildings, time of day,

Table 1.9. Examples of tornados and characteristics of their consequences.

Date and place of tornado	Consequences
26 April 1896. Bangladesh.	1,300 people perished, several villages were ruined.
15 May 1896. USA: Kansas, Texas, Oklahoma, and Kentucky.	A series of tornados. In Texas, 22 houses were ruined and two people perished. The tornado was most active in the region of Red River, where 50 houses were ruined and 73 people perished. Their bodies were taken away by the tornado to a distance of >1 km.
27 May 1896. USA: Missouri and Illinois.	A series of tornados. 255 people perished, 137 of them died in St. Louis under ruined buildings.
28 October 1896. USA: Texas, Oklahoma, and Louisiana.	Several tornados in different regions of these states. Several houses were ruined, 5 people perished.
11 April 1965. USA: Indiana, Oklahoma, and Michigan.	A series of tornados with heavy rains and hail. 271 people perished. Economic damage was estimated at 1,070 million dollars.
5 May 1995. USA: Louisiana, Oklahoma, Massachusetts, and Texas.	A series of tornados with hail and flood. 32 people perished. Economic damage was estimated at 7,020 million dollars.
21 April 1996. USA: Arkansas.	Two people perished, 50 people were admitted to hospital with different traumas. 35 houses and 78 supermarkets were ruined. 853 buildings were seriously damaged.
13 May 1996. Bangladesh.	More than 80 villages with 10,000 edifices were ruined. More than 700 people perished and 30,000 were wounded.
19 July 1996. Bulgaria.	Sudden tornado-like vortex on the Black Sea shore raised pebbles in the air hitting a man on the beach who died instantly.
6 November 1996. South-east of Sweden.	Tornado tore the roofs off several buildings, turned over several cars, and pulled up many trees.
3 May 1999. USA: Kansas, Oklahoma, Tennessee, and Texas.	Tornado. Heavy coarse hail. 51 people perished. Economic damage is estimated at 2,950 million dollars.
30 March 2001. Pakistan: Province Punjab.	Powerful tornado destroyed about 100 buildings, 4 people perished.
18 June 2001. USA: Wisconsin (north-west).	A powerful tornado covering a band about 900 m wide travelling at a rate of 170 km h^{-1}, for a distance of ~60 km, caused destruction, ruining 120 houses and killing 5 people.
23 June 2003. USA: Nebraska.	11 farms were destroyed, 20 units of cattle were lost, one person died of related traumas.

Table 1.10. Hurricanes on the American continent during the period 1528–1803 (Ludlum, 1963).

Date	Place	Date	Place
2 October 1528	Florida, USA	24–25 October 1716	Massachusetts, USA
29 August 1559	Alabama, USA	9 August 1723	New York, USA
23 September 1565	Florida, USA	10 November 1723	Isl. Rhode
26 September 1566	East Florida coast, USA	23 August 1724	Virginia, USA
23–26 June 1586	Rounoke Island, USA	28–30 August 1724	Virginia and Pennsylvania, USA
31 August 1587	Rounoke Island, USA		
26 August 1591	Rounoke Island, USA	27 September 1727	East of New England, USA
25 August 1635	South-east New England, USA	1–2 November 1743	Eastern coast of USA
13 August 1638	East of New England, USA	18–19 October 1749	Eastern coast of USA
5 October 1638	East of New England, USA	23–24 September 1757	New England, USA
6 September 1667	North Carolina, Virginia, and New York, USA	23 October 1761	South-east New England, USA
		11 September 1766	Virginia, USA
7–8 September 1675	South of New England, USA	7–8 September 1769	Eastern coast of the USA
23 August 1683	Virginia and south of New England, USA	20 October 1770	South-east New England, USA
		26 August 1773	Virginia, USA
4–5 September 1686	South Carolina, USA	2–3 September 1775	Eastern coast of the USA
29 October 1693	Virginia and Delaver, USA	12–13 August 1778	New England, USA
3–4 October 1696	Eastern coast of Florida, USA	7–8 October 1783	Eastern coast of USA
16 September 1700	South Carolina, USA	23–25 September 1785	Virginia, USA
18–19 October 1703	Virginia and South England, USA	19 August 1788	New Jersey, USA
		2–3 August 1795	North Carolina, USA
14–15 October 1706	East of New York and west of New England, USA	12–13 August 1795	North Carolina, USA
		2–3 October 1803	Virginia, USA

and warnings about an impending earthquake. So, the earthquake of magnitude 6.6 (on the Richter scale) in Iran on 26 December 2003 resulted in about 25,000 deaths and about 30,000 wounded. At the same time, a sufficiently strong earthquake (of magnitude 6.5) in California was without victims and noticeable destruction. Such a contrast is largely explained by the different protection afforded to people in these territories and the adjustability of the area's infrastructure to overcoming the consequences. Iran is located in a seismically active zone between the Arabian and Eurasian Plates. Here a high density of population and construction is observed – construction which is poorly adjusted to earthquakes. During the last century, six large-scale earthquakes were recorded with about 5,000 victims. In 1999, the earthquake on the coast of the Caspian Sea claimed the lives of 40,000 people.

Grigoryev and Kondratyev (2001) give a sufficiently detailed chronology of earthquakes for the history of civilization beginning from biblical evidence and legends, which describe the destruction of many cities. From fragmentary information and excavations, numerous scientists have reproduced the past geological processes connected with rock movements and resulting earthquakes. There had been many cases of populated areas sinking to the bottom of the sea in coastal areas of the Black and Mediterranean Seas. Up to now, scientists are debating the real causes of the loss of two cities into the ocean near the coast of India (Delgado, 1998).

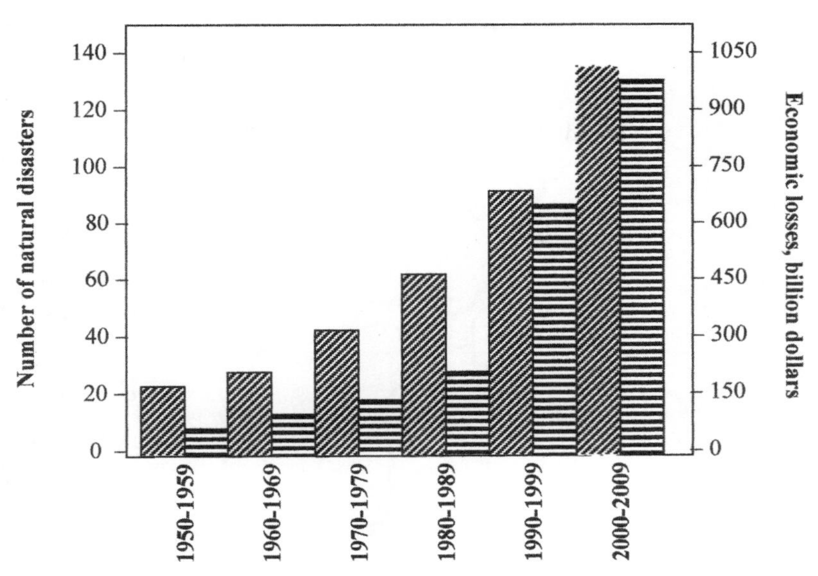

Figure 1.1. Dynamics of the number of the largest natural disasters (Munich, 2004). Notation: diagonal hatching = natural disasters, horizontal hatching = economic losses.

Table 1.6 contains a list of the most destructive earthquakes in the history of humankind. These (catastrophic) and other smaller scale earthquakes are followed, as a rule, by serious economic and human losses. During the period 1900–1989, earthquakes resulted in about 1.2 million human victims globally. The temporal distribution of earthquakes and their consequences cannot be reliably parameterized. So, whereas during the 1980s the earthquakes in the world claimed the lives of 57,500 people, in 1990 alone there were 52,000 victims. On the whole, the earthquakes statistics are very inhomogeneous. The 10-year period 1980–1990, for instance, is characterized by a temporal high density of destructive earthquakes. They totaled 170 and covered 43 states. During this period, the number of human victims was greatest in Asia (86,212 people), followed by the USSR (25,392 people), North and Central America (11,313 people), Africa (5,070 people), Europe (3,253 people), South America (2,506 people), and Australia and Oceania (459 people). Table 1.7 gives some more detailed information.

The number of earthquakes and their consequences differ from year to year. So, in 2002 the global number was about 70 – resulting in human victims and destruction. In particular, several earthquakes took place over the territory of Afghanistan, among which the March earthquake in the Hindu Kush mountains led to more than 2,000 victims. The earthquake of moderate power on 31 October 2002 in Italy at Molise ruined a school and caused serious concern within the population of the central part of the country about the structural quality of buildings within that region. The most powerful earthquake of magnitude 7.9 took place on 3 November 2001 in Alaska – a thinly populated area. In November 2004 alone, there were six strong earthquakes: 11 November in Indonesia (Kepilauan Alor),

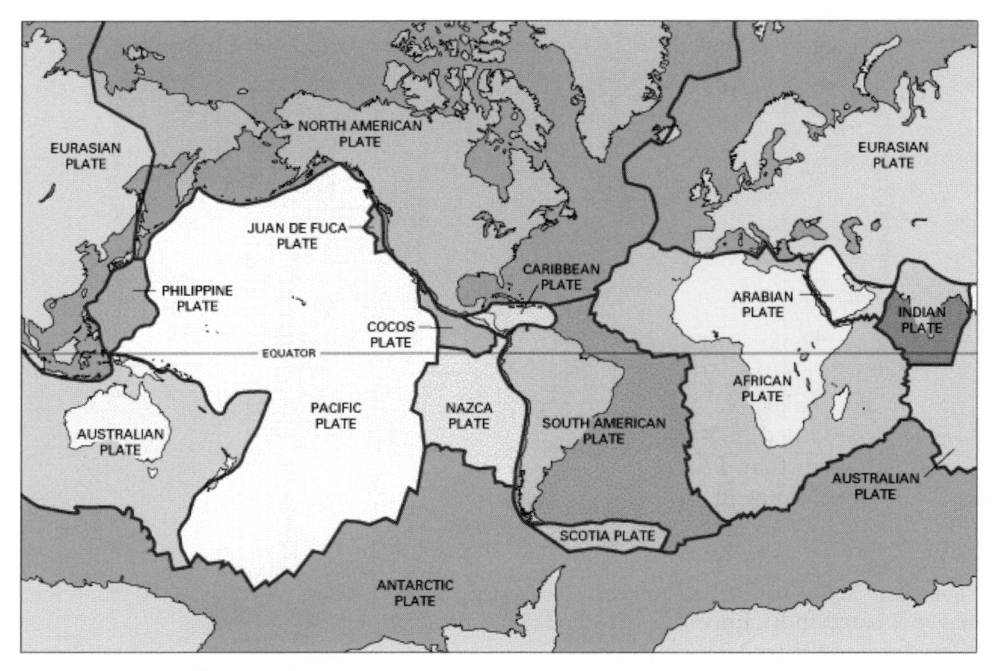

Figure 1.2. Distribution of tectonic plates, whose motion determines the structure of seismic and volcanic activity (*www.ofspiritandsoul.com/earth%20vortices/tectonic.html*).

magnitude 7.5; 15 November on the western coast of Columbia, magnitude 7.2; 20 November at Costa Rica, magnitude 6.4; 22 November on the western coast of New Zealand, magnitude 7.1; and on 26 November in Indonesia (Papua), magnitude 7.2. On the whole, 19 earthquakes of magnitude >5 were recorded in 2004. The most powerful earthquake of magnitude 9.7 took place on 26 December 2004 in the eastern sector of the Indian Ocean. The resulting tsunami led to more than 250,000 victims and huge destruction. Three months later, on 28 March 2005, the western coast of Indonesia, the Sumatra Island, and Malaysia, again, suffered an earthquake of magnitude 8.7 with the center located 350 km south-east of the center of the 26 December 2004 earthquake. There were about 2,000 victims with huge damage to these territories.

The seismically dangerous zones are known, though an earthquake can take place in any region. Nevertheless, one of the functions of national systems of nature monitoring in many countries is to record the oscillations of subsoil rocks. The amplitude of these oscillations determines the level of danger of an earthquake occurrence or another natural disaster, such as mud flows or landslides. In seismically dangerous regions the houses, and industrial and administrative buildings are built with earthquakes taken into account. The coordinates of the seismically active regions of the globe are well known. Earthquakes are caused by sudden movements and transformations of tectonic plates (Figure 1.2). Tensions in the upper layer of land affect these plates and cause changes of energy fluxes, which manifest themselves via the Earth's surface oscillations, and in some cases lead to volcanic eruptions. So, for instance, in California, two plates collide – the Pacific Plate and the North American Plate. Their boundaries rather accurately determine the seismic zones in this region. Numerous geophysical theories of earthquakes are based on studies of the motion of tectonic plates. The laser sounding of the Earth's surface and detection of landscape changes in fractures between the plates is one of the promising methods of the earthquakes prediction.

1.1.2 Floods

A retrieval of the history of floods over the globe is only possible from fragments of available information. Without making it our aim, we shall mention here only some facts. In particular, we have records of floods in Moscow. The first Moscow river flood was recorded in 1496, when, after a cold and snowy winter a "great freshet" happened. After that, floods in Moscow were frequent, which led to the construction of a drainage canal in the 18th century to protect the region of the present-day Bakhrushin, Novokuznetskaya, Piatnitskaya, and Yakimanskikh streets. In April 1908, huge flooding of the Moskva and Yauza Rivers, and the drainage canal, occurred, rising up to 9 m and covering 20% of the urban territory. The buildings of the Dorogomiliv Embankment and Zamoskvorechye, as well as the Kremlin Wall were flooded up to 2.3 m above pavement level.

In the 20th century, floods occurred frequently over the USA, happening quite unexpectedly and leading to large economic losses and numerous deaths. As a rule, water fluxes exhibited a huge force destroying buildings, bridges, and other constructions. In the 20th century, the destructive floods in the USA totaled 32. By type they were different. They could be caused by river ice piling up, high storm waves or dam destruction, as well as by choking up of the rivers due to landslips or landslides. Some of the floods were regional in character. They regularly occurred in high latitudes in spring due to rapid snow melting or downpours. Since in the spring the infiltration is poor because of frozen soil, the excess water flows into rivers, the level of which may rise by several meters during several hours. Such events are also frequent in Russia, and especially in Siberia. For instance, a catastrophic flood happened in England in March 1936, which claimed the lives of 150 people. Of huge destructive force was the flood in 1993 of the Mississippi River basin that caused damage estimated at 20 million dollars. In Chile there were two large floods, in 1891 and 2000, when 24-hour rainfall reached 17.59 mm and 5.84 mm, respectively. On 14 June 2000, the water level in Santiago rose by 1 m, which led to a blocking of 14 main highways in the country.

Short-term floods occur due to a combination of such factors as intensive precipitation and a certain relief of locality. They last for not more than several hours. Conditions for such floods can be observed in cities and highlands. In a city, during a heavy shower, the storm's water cannot rapidly be discharged, and in highlands, with narrow canyons, water has not time enough to infiltrate. In both cases accumulated water can rapidly flow over free spaces, destroying buildings or causing landslides and other unfavourable events. So, in 1972, in the city of Rapid

Figure 1.3. A powerful wave several meters high comes from the sea. This wave could result from an earthquake with an epicenter on the sea bottom or from an unfavorable combination of various geophysical and climatic parameters (volcanic eruption, hurricane, storm, etc.).

(South Dakota, USA), rainfall continuing for 5 hours and reaching 37.5 cm, caused a wave over 9 m high, which moved for many kilometers from it's origin, with the result of 237 deaths.

Floods caused by river ice-jams are very dangerous, since they happen instantly, and their energy, with a growing force, destroys buildings, dams, plants, constructions, bridges, transport links, etc. The floods of the Yukon River in Alaska in the spring of 1992 and the Lena River in May 2001 are examples of such floods. In the latter case the city of Lensk was flooded, and the level of the river rose by 20.13 m.

Unexpected natural disasters include floods due to high sea waves coming onto land (Figures 1.3 and 1.4). In this case, depending on combinations of wind speed and air resistance to moving water, the water can rise to over 6 m. For instance, in September 1900 in Texas (USA) such a high wave formed as a result of a simultaneous storm and hurricane, causing 6,000 deaths.

Floods brought in the past, and are bringing now, huge trouble in many global regions. Many of them being initiated by typhoons. On 12 September 2000, the flood in Japan resulting from downpours following typhoon Saomai, paralysed the functioning of the Toyota automobile factory and caused landslides that led to 5 deaths and an evacuation of more than 500,000 people. The city of Nagoya was flooded. On Honshu Island, rainfall lasted for more than two days, practically closing the entire road network in central Japan. The insurance payments reached 25 million dollars. A sufficiently strong flood was initiated on Taiwan in late October 2000 by typhoon

Figure 1.4. The episode of flooding in the Czech Republic during the summer season of 2002 when catastrophic floods badly affected many regions of Europe and caused significant economic losses.
Photo IFRC (*www.munichre.com*).

Xangsane, resulting in 53 deaths and destruction to agricultural lands over an area of 48,000 ha. The damage reached 77 million dollars. The cyclone Leon–Eline on 4–7 February 2000 caused downpours over Madagascar, South Africa, Mozambique, and Botswana, which ended in an unprecedented flood leaving 500,000 homeless and causing material damage estimated at 160 million dollars. In South Korea, on 22–23 May 2000, a 400-mm downpour caused a powerful flood that covered a territory over 27,500 km^2, claimed the lives of 10 people, and destroyed more than 1,000 buildings, costing 27 million dollars.

On the whole, the year 2000 became the year of powerful floods in Europe. They happened in Italy, France, Germany, Austria, Great Britain, Belgium, and over many territories of other countries. The scale of the floods could be judged by the resulting destruction. So, in Germany, in December, the day's rainfall reached 50 l per m^2, which led to the flooding of many highways and populated areas. As a result, more than 350,000 suffered losses, and the damage exceeded 9.1 billion Euros. In Italy, the October flood due to downpours caused mudflows and rivers to overflow, and as a result, 13 people perished, several bridges were destroyed, and the railway network between several industrial centers was broken. In 2000, the number of floods totaled 18, covering practically all continents. On 11 September, on the Japanese islands Shao-Mai and Kyushu there was flooding, the largest for the last 100 years. From 10 October to 5 November, floods covered Great Britain, Sweden, Italy, and France, with the level of water rising by 2 m or more, which had not happened during the past 30 years. In the south of England the water rose by 5 m. During three days, from 9–11 October 2000 in England, rainfall reached 103.4 mm, with hurricane wind speeds of 140 km h^{-1}. It was the largest flood since 1773.

In Australia, on 20 November, 12 rivers overflowed their banks simultaneously, with damage estimated at 250 million Australian dollars. Nowadays floods have become more frequent in Russia, reaching 19% of its total number of natural disasters. During recent years, not less than 50,000 km^2 are flooded every year, covering more than 300 populated areas. For instance, in the summer of 2002, at the Northern Caucasus (the Rivers Kuban, Terek, Kuma, Podkumok, etc.) there was observed an anomalous hydrological regime, resulting in the largest scale and destructive flood over 345 km^2 resulting in 104 deaths. Floods happen regularly in the center of Europe – Hungary. The River Tisza flows across rich agricultural regions in Hungary. Most recently, these regions suffered large-scale floods in 1993, 1995, 1998, 2000, and 2001.

Dams are being built to protect from floods. Their height and location are calculated with due regard to flood statistics over a concrete territory and with provision of a certain level of risk. One of the characteristic examples, when consideration of the statistics of water rising during floods in a given territory is of vital importance, is the situation in the Netherlands on 1 February 1953. Before this, hydraulic engineers could not decide upon the height of a protective dam on the coast. At first, it was decided that the dam should be 3.9 m high – the level which had never been overflown. However, bearing in mind an absolute maximum recorded over 25 years and taking into account the economic aspects, it was decided to build the dam 3.4 m high, which corresponded to the probability of reaching this level once in 70 years. A tragedy happened on 1 February 1953, when the flood exceeded this level, claiming the lives of 2,000 people, and caused huge destruction. As a result, it was decided to raise the dam up to 5 m, this level can be reached only once in 10,000 years. But, nevertheless, there is a probability that the wave will rise above the dam. In this case the engineering calculations during construction works foresee the protection of the dam itself from destruction. In any case, the dam could be destroyed, resulting in huge masses of water sweeping away everything in its path. Flood situations shown in Figures 1.5 and 1.6 illustrate the level of damage caused by floods. Figure 1.5 characterizes the situation which took place during Typhoon Songda as it swept over Japan at the beginning of September 2004 with wind speed reaching 200 $km\ h^{-1}$. Songda is the second most expensive windstorm in Japan after Typhoon Mireille in 1991.

1.1.3 Tornados, storms, hurricanes, and typhoons

History contains rich information about natural disasters which are now called tropical cyclones and which are mainly formed over the oceans in the tropics, regularly falling onto the eastern and near-equatorial regions of the continents (Figure 1.7). Tropical cyclones are hurricanes and typhoons, happening in the northern and southern Pacific, in the Bay of Bengal, the Arabian Sea, in the southern sector of the Indian Ocean, near the coast of Madagascar, and the north-western coast of Australia. The Atlantic hurricanes form mainly in the zone confined to the coordinates (10°N – 20°N, 20°W – 6°W) and (7.5°N – 17.5°N, 30°W – 100°W). Usually, tropical cyclones are given names (Figures 1.7 and 1.8).

A whirlwind (tornado) is one of the insidious and sudden natural formations in the atmosphere (Table 1.8, p. 8). It is a rotating funnel-shaped cloud which stretches from the bottom of the thundercloud to the Earth's surface. Characteristic tornado

Figure 1.5. Flood in Japan at the beginning of September 2004 during Typhoon Songda (Munich Re, 2005).

wind speeds are 65–120 km h^{-1}, but sometimes this magnitude reaches 320 km h^{-1} or more (Hoinka and de Castro, 2005; Hamill *et al.*, 2005; Krishnamurti *et al.*, 2005; Bernard, 2005). An external indicator of an approaching whirlwind is a noise similar to the thunder of a moving train. The appearance of the tornado is connected to a combination of natural processes, but from the times of Egyptian Pharaohs, artificial tornados had been created above the tops of pyramids and marked an ascention of the Pharaoh's soul to the skies, to the Sun god Ra. The pictures of tornados in Egyptian hieroglyphs do not explain how these artificial tornados were formed.

The USA is the most characteristic region where tornados are a rather frequent phenomena, though tornados happen all over the world. During the period 1961–2004, over the USA territory, tornados claimed the lives of 83 people per year, on average. Most often, tornados happen in the eastern states adjacent to the Gulf of Mexico, in February and March their frequency of occurrence is at a maximum. In

Figure 1.6. Floods paralyse vast territories. This picture characterizes a site in Switzerland during rainfall in 2002 (Roch, 2002).

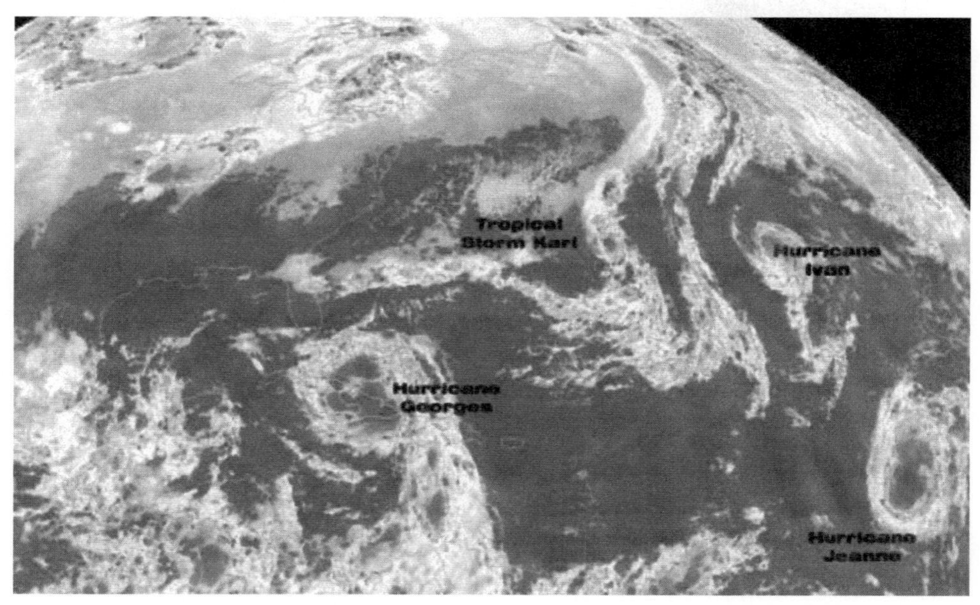

Figure 1.7. The Atlantic sector of hurricane formation observed in the IR region from GOES (Geostationary Operational Environmental Satellite) which has been in operation for more than 25 years providing data on the Earth's climate system. Data of GOES-8 with 8 km resolution, 24 September 1998, 13:45 UTC (Holweg, 2000).

Figure 1.8. Hurricanes Madeline and Lester recorded from the GOES satellite on 17 October 1998 (*http://rsd.gsfc.nasa.gov/goes/pub/goes/981017.lester.jpg*)..

the states of Iowa and Kansas, tornados are most frequent in May–June. The average amount of tornados over the USA is estimated at $\sim 800\,\mathrm{yr}^{-1}$, 50% of which occur during April–June. The territorial heterogeneity of the frequency of occurrence of tornados in the USA has stable characteristics: in Texas – 120 tornados per year, and in the north-eastern and western states – 1 tornado per year. For instance, only in April and November 2002 there were more than 100 tornados that caused numerous destruction with more than 600 cases of insurance payments. Tornados occur in other countries, too. Table 1.9 (p. 9) gives some ideas about the character of destruction and the level of damage from tornados in different parts of the world. It is seen that each tornado causes destruction and death.

Another example of destruction is Hurricane Charlie in the USA in 2004, where over Florida, Cuba, and Jamaica, there were 19, 2, and 1 deaths, respectively. More than 150,000 people remained without telecommunication links, and the damage exceeded 11 billion dollars. Hurricane Katrina caused extensive wind-related property damage that caused flooding and pollution in New Orleans, the disturbance of oil rigs in the Gulf of Mexico, and outages of power. Hurricane Katrina brought destruction in many areas along the Mississippi and Louisiana coast as well as in Gulf Coast towns. The satellite pictures show how the monster Hurricane Katrina, traveling through the Gulf of Mexico, moved from the Gulf into Lake Pontchartrain and flooded the New Orleans area. The city, about 80% of which was below sea level, filled with water (Travis, 2005).

Table 1.10 (p. 10) characterizes the historical chronology of hurricanes happening over the American continent. It is seen that during one year the frequency of occurrence of hurricanes is non-uniform: 92% of events fell between August–October with an approximately uniform distribution.

Typhoons can cause a local rise of the ocean's level by several meters and, if they take place near the coastline, up to a 10-m wave may form which could flood lowlands at an alarming rate. For example, numerous typhoons took place in 2000 over Taiwan, Japan, South Korea, and other global regions. Typhoon "Billis", falling on 20 August (at a rate of 360 km h^{-1}) onto Taiwan, during several hours ruined 25 buildings (banks, schools, offices, and houses), causing damage costing 48 million dollars. In late August, the typhoon "Prapiroon" stormed over the East China Sea with a wind speed of 43–181 km h^{-1}. The international airport in Shanghai was closed for some time. In the World Ocean there are seven regions where typhoons originate. All of them are located near the equator. The main cause of a typhoons appearance is the heating of several tens of meters of the ocean water above a critical level (26.8°C). The satellite monitoring of typhoons has made it possible to trace their routes. A trade-wind-caused typhoon moves mainly westward, gradually turning to the north-east and east under the influence of the Coriolis force. Up to 47% of typhoons move along this parabolic trajectory. The remaining 53% move along other trajectories which are not strictly regular. But still, a great group of typhoons (21%) move, as a rule, along half the parabolic territory mentioned above, and then move from south-east to north-west. Seven per cent of typhoons move east-to-west and from the south-west to north-east, and four per cent of typhoons move south-to-north (Baibakov and Martynov, 1976).

1.1.4 Volcanic eruptions

Volcanic eruptions cause heavy destruction due to propagation of lava flows and ash deposition. Human victims of volcanic eruptions are connected with many factors: lava, mud, and pyroclastic flows, avalanches, ejections of tephra and ballistic bombs, and diseases and hunger. The scale of the disaster in a zone of volcanic eruption depends on the infrastructure and population of the territory adjacent to the volcano (Schmincke, 2004). For instance, in the period 1900–1986, during the catastrophic eruptions of the planet, 85.8% (65,200 people) of all the victims (75,000 people) died because of pyroclastic flows and avalanches as well as mudflows and floods. Japan has 108 active volcanoes, which erupt repeatedly. This brings regular economic losses and deaths. The present means of monitoring from space enables pre-calculation of the directions of these flows and thus enables the determination of the safe areas in the zone of influence of a given volcano (Robock and Oppenheimer, 2003).

Despite the high level of danger in zones of volcanic activity, settlements and towns have been built there because volcanic rocks become covered with fertile soils. In due course, feeling their powerlessness before volcanic activity, people try to find a means to resist this formidable element. During the last century one could explain many aspects of volcanic activity and describe the specific structure of the Earth's crust. Rich information on the formation of magma in the Earth's crust and mantle has made it possible to formulate the laws of volcanic activity and to find external indicators of an approaching volcanic eruption. These laws depend on the type of volcano, its size, and many other parameters. The Mauna Loa Volcano on Hawaii is the highest. The largest crater is located in Alaska. The Mayon Volcano in the Philippines, 350 km south-east of Manila, is one of the most active volcanoes. It erupts every 10 years. The 1814 eruption led to 1,200 deaths. During the 1993 eruption more than 70 people perished.

Historically, a volcanic eruption had been perceived in various fantastic formats depending on a given territory and, as a rule, had been connected with religious interpretation. For instance, in Ancient Greece and Rome, volcanos were considered the place inhabited by the God of fire "Vulkan", and the moment of eruption was taken to represent some confusion in his domain. Numerous legends about the causes of volcanic eruptions were explained by the fact that in the historical past there had been no possibility to look into the depths of the Earth's crust and, furthermore, explain the physics of these processes. Even now, when geophysics and volcanology have powerful theoretical and technical possibilities at their disposal, the problem of reliable prediction of volcanic eruptions and earthquakes is far from being resolved. This is, first of all, connected with the existence of several hypotheses regarding the formation of volcanic centres. One of them is connected with the cooling of the Earth's flaming core. Another hypothesis associates the volcanic activity with processes of compression and heating taking place with the disintegration of radiative elements at great depths. There are also other hypotheses which try to connect the geophysical and climatic processes and to explain a non-uniform distribution over the surface of the centers of volcanic activity. For instance, why are more than 10 volcanos concentrated within the small territory of Kamchatka, and why are some of them active and others idle? All related problems are resolved by geology, which has established a certain chronology of epochal changes in the image of the Earth's surface and, apparently, the answer to many of the mentioned problems of volcanology lies in the knowledge of the laws of motion of the continents. Over periods of long geological history the continents have combined and separated. It has been established that about 2.5 billion years ago there were 20 continents, and during the early Proterosoi (2 billion years) they numbered 13.

It is supposed that in 1.5 billion years the Australian, American, African, and Eurasian continents will again combine with Antarctica into a single super-continent (Sorokhtin and Ushakov, 1996; Sidorov, 1999). At the same time, there will be changes in powerful volcanic bands located along the edges of the continents.

Volcanic activity is an integral part of our planet's life. Volcanos are not only dangerous for populations of adjacent territories, but they also fertilize the soils, give

Table 1.11. Average chemical composition of volcanic lava (in weight %).

Oxides	Nepheline basalt	Basalt	Andesite	Dacite	Phonolite	Trachyte	Rhyolite
SiO_2	37.6	48.5	54.1	63.6	56.9	60.2	73.1
Al_2O_3	10.8	14.3	17.2	16.7	20.2	17.8	12.0
Fe_2O_3	5.7	3.1	3.5	2.2	2.3	2.6	2.1
FeO	8.3	8.5	5.5	3.0	1.8	1.8	1.6
MgO	13.1	8.8	4.4	2.1	0.6	1.3	0.2
CaO	13.4	10.4	7.9	5.5	1.9	2.9	0.8
Na_2O	3.8	2.3	3.7	4.0	8.7	5.4	4.3
K_2O	1.0	0.8	1.1	1.4	5.4	6.5	4.8
H_2O	1.5	0.7	0.9	0.6	1.0	0.5	0.6
TiO_2	2.8	2.1	1.3	0.6	0.6	0.6	0.3
P_2O_5	1.0	0.3	0.3	0.2	0.2	0.2	0.1
MnO	0.1	0.2	0.1	0.1	0.2	0.2	0.1

off heat and generate rich supplies of precious metals and minerals (Table 1.11). In particular, volcanos generate supplies of gold, found for instance on Kamchatka. An accumulation of thinly scattered gold in volcanic rocks can reach several kilograms per ton of ore, but its extraction requires the use of other technologies than those required for extraction of gold from traditional deposits. By composition, volcanic rocks are divided into four base groups determined by the content of silicon dioxide.

1.1.5 Landslides, avalanches, landslips, and collapses

Dangerous natural events in highlands include the slope movements of mountain rocks, for instance, mudflows, collapses, avalanches, landslips, and landslides. Landslips and mudflows are most widely spread. Over the period 1945–2004 there were 43 large-scale land motions with more than 50 victims in each case (Negri *et al.*, 2005).

A landslip is a gravitational shifting of soil masses over the wet surface of the rock base (Figure 1.9). A landslide is a falling of rocks from cliffs and relatively dry loose steep slopes. These phenomena appear as a sudden motion of soils in the regions with steep slopes and, as a rule, in the presence of considerable precipitation. A dramatic situation occurs if a landslip, or a falling, takes place near a settlement. In 1903, at Frang, Canada the top of the mountain Tertle collapsed, throwing more than 30 million m^3 of rock over a miners' settlement. Seventy people perished and the trans-Canada railway was covered. In 1959, a landslip caused by the earthquake at Montana led to the formation of a new lake called Earthquake Lake. Twenty-eight people perished. In July 2004 the flood in south-western China caused heavy mudflows and landslips, falling onto the villages of Yun-nan Province, resulting in 11 deaths, and the flooding of more than 2,000 peasants' houses.

Figure 1.9. Landslides change the relief of the land and lead to serious destruction. View of a landslide on 14 October 2000 in the village of Gondo, Canton Valais, Switzerland. (*inset*) A landslide has ruined part of a road (Roch, 2002).

Avalanches are also dangerous natural events in highlands as a result of abundant snowfall and due to a long accumulation of snow on steep slopes. The avalanche hazard regions are well known. For instance, in Russia they are in large areas of the Caucasus, the Urals, Kamchatka, the North-East Territories, Altai, Sayany, and in vast territories east of the Yenissey River. Not all of them have been studied adequately. One of the best studied avalanche hazard regions is in the Byrranga and Putorana mountains (Troshkina, 1992; Voitkovsky and Korolkov, 1998).

The Byrranga mountains are located in the northern part of the Taimyr Peninsula with a maximum height of 1,146 m (Lednikovaya mountain). In the historical past, this region had twice been under the Taimyr glacier which moved from the northern part of the peninsula to the west Siberian lowland. As a result, in the Tertiary epoch the northern part of the Taimyr had dropped below sea level, and the southern part had risen along the edge of the Yenissey–Khatanga depression. The second glaciation had changed the structure of the Byrranga mountains. High terraces and deep river valleys had appeared with the slopes of 20–25° or more.

The climatic parameters of the Byrranga mountains are as follows: in January the temperature averages −32°C to −34°C, winter lasts for 10 months and the annual sum of precipitation is 400–500 mm (about half being solid precipitation). As a result, the duration of snow cover stability is ~270 days, with an average thickness of 70–90 cm. In the windless regions the snow cover thickness can exceed 2 m. Since forest vegetation is absent, and the moss–grass tundras rise only up to 200–250 m, the snowdrifts on mountain slopes are held in place only by hard frost, and the probability of an avalanche occurrence is low until late May–early June, when after the long polar night the solar radiance starts increasing, reaching in April 9 kcal cm^{-2} per month. From this time, the probability of snow avalanches grows. Tareyeva and Seliverstov (2004a,b) give estimates characterizing the levels of risk of avalanches in the Byrranga and Putorana mountains.

The Putorana mountains are located in the north-western part of the Middle-Siberian plateau stretched from Lake Piasino to Lake Essey and Kotuya and Moyeros Rivers (with their tributaries). Maximum height is 1,664 m, and the depth of ravines varies from 200–1,000 m. Steep slopes not less than 35° without vegetation are sources of snow avalanches. The annual sum of precipitation, depending on the height, varies from 400–600 mm to 1,200–1,600 mm, which favors an accumulation of large snow masses. As in the Byrranga mountains, in winter, in some places the height of snow cover reaches 120 cm with a density of 350–400 kg m^{-3}. However, in contrast to the Byrranga mountains, the Putorana mountains have a more complicated vegetative cover. The foot of the mountains and the lower parts of their slopes are covered with mixed and broadleaved forests. The upper boundary of the forest (and sparse forests) reaches 250 m in the north-west, and 800 m in the south-east. Above is the mountain tundra, which at the height of 1,200 m, the mountain becomes a desert. On the one hand, all this favors the formation of a complicated space structure of large snow supplies, but on the other hand, it clearly outlines the avalanche hazard zones. The volumes of avalanches vary from 10 to 10,000 m^3. As a rule, they begin in the 300–500-m altitudinal zone, where the restraining factors of vegetation are small.

Most of the territory of the Khibin's mountains is characterized by a considerable danger of avalanches. In this connection, in the mid-1930s, the anti-avalanche service was organized. Most avalanches occur on slopes angled at 20–45°. The temporal distribution of avalanches varies from 0–120 avalanches per year for one center with maxima in January–February and April. In the period 1936–1982, every year about 200 avalanches were recorded with volumes from 50 m^3 to 1,255,000 m^3. At present, the anti-avalanche protection is a branch of industry, and as a result, preventive measures are taken against avalanche formation.

The next region of Russia with a high risk of avalanches is Altai characterized by a combination of high ridges (>4,000 m) and deep valleys. More than half the Altai area is a forest band up to 350–400 m in the west, and up to 1,600–1,800 m in the south-east. In some places, forests grow at altitudes of 1,800–2,400 m. Of course, forests together with the relief play a substantial role in the formation of avalanche danger. But the main avalanche-forming factor is climate, which at Altai exhibits a clearly expressed vertical banded character. Circulation processes and the thermal regime are climatic elements that strongly affect the activity and regime of avalanche formation. The main snow masses are brought by westerly and south-westerly cyclones followed, as a rule, by a considerable temperature increase and intense snowfall. As a result, conditions are created for avalanche sliding. In the snowy winters of 1965/1966 and 1968/1969, in the western outlying area of Altai, there were mass snowslides. On the whole, the structure of avalanche danger at Altai is formed due to a combination of the temperature and wind regimes. Because of the extended territory and inhomogeneous relief, these regimes are spatially distributed. So, the average January air temperature at the western Altai ranges between −13.0°C and −27.0°C. At the same time, the average annual temperature at the foothills of the western Altai constitutes 1–3°C. For the southern Altai, the annual mean temperature changes from 5°C to 10°C. On the whole, over the whole territory of Altai, January temperatures can reach −57.5°C. The wind conditions are characterized by vertical changes: near the surface, south-westerlies prevail, and at high altitudes, southern winds prevail. The distribution of solid precipitation is also non-uniform: in the inter-mountain depressions the annual sum of solid precipitation reaches 200 mm, and in the mountains of the western Altai it exceeds 1,000 mm. The share of solid precipitation decreases from the edge of the mountain range to deep into the mountain area and from the top of the mountain to its base. The critical level of snowfall intensity for avalanche formation varies within 5–10 mm per day depending on the snow cover condition.

Considerable avalanche activity is characteristic of the Sayan ridge extending from the sources of the Yenissey and Abakan Rivers to the Angara and Kazyr Rivers. The Sayan ridge is divided into the Western Sayan (2,500–2,700 m) and the Eastern Sayan (2,700–3,400 m). The centers of avalanche formation are mainly concentrated on the windward north-western and western slopes of the Sayan ridge, where air temperature averages about −3.8°C. Due to a very non-uniform distribution of precipitation, the snow cover is also non-uniform, sometimes reaching 1.0–2.5 m, which determines the spatial variety of avalanche danger. Together with the vertical distribution of vegetation cover, the duration of the avalanche hazard period is determined by an appearance of the first avalanches in October, and the last ones in June. At the Western Sayan the avalanches first occur in November, and at the Eastern Sayan, from January to April. The peak of avalanche danger falls during March–May, the period of precipitation and temperature growth. Most powerful avalanches (volumes of up to 100,000 m^3) are observed in the basin of the Kazyr River. In the alpine and sub-alpine landscape bands avalanches of 150,000–500,000 m^3 are possible (Bozhinsky and Losev, 1987; Troshkina, 1992).

The map of the avalanche hazard regions of Russia shows the territories of the Urals stretching by a narrow (150 km) chain of low parallel hills for more than 2,000 km from 48°–68°30′N. A deeply divided relief with intense snowfall during a long winter created such favorable conditions for snowslides (also favored by a high repeatability of snowstorms of 140–160 days) and frequent westerlies at speeds from

8–12 $m s^{-1}$ to 4 $m s^{-1}$. Volumes of avalanches vary from 10–2,000 m^3 depending on season and latitude.

The Great Caucasus is the most dangerous region, where in winter, tragic episodes have repeatedly been recorded of people perishing under snowslides. Mass snowslides were recorded in 1846/1847, 1854/1855, 1899/1900, 1931/1932, 1955/1956, 1986/1987, and 1992/1993. In these years, there were numerous victims, buildings were destroyed, large areas were deforested, cattle were killed, and the roads were blocked. For instance, on 27 January 1993, on the Trans-Caucasian Highway, 17 people were buried under a snowslide. Similar to other regions, the conditions for snowslide formation on the slopes of the Great Caucasus are determined by the slope of ridges, the presence of vegetation on them, and climatic conditions. The northern slope of the Great Caucasus is forested up to 1,200–1,500 m in the east and 2,000–2,300 m in the west. The winter period lasts for 78 and 25 days on the northern and southern slopes, respectively. The air temperature decreases with height by 0.6°C on average for each 100 m of the ascent, which determines the duration of winter at different altitudes.

Though the regimes of snowslide formation on the slopes of the Great Caucasus have been well studied and the most dangerous regions have been well determined, the process of initiation and actual sliding down of avalanches remains difficult to parameterize and predicted. The vertical distribution of avalanches, on average, is determined by the following indicators: 2.2% – below 1 km, 10.3% of avalanches within the band 1.0–1.5 km; 1.5–2.0 km – 13.9%; 2.0–2.5 km – 19.5%; 2.5–3.0 km – 31.5%, and above 3 km – 22.6%. Most of avalanches occur in December–February. By type, avalanches are divided into those occuring during snowfall (75%), during the springtime snowmelt (8%), during the thaw (6%), during snow storms (2%), and during thermal snow loosening (9%). The average volume of an avalanche is estimated at 50,000 m^3 (Zalimkhanov, 1981; Pogorelov, 1998).

Snowslides are an integral part of the processes taking place in highlands. The number of victims and related material damage depends on many factors. For instance, as Ramsley (1978) notes, in Norway, beginning in 1836, every year 12 people, on average, die under snowslides. In Switzerland – 25 people annually. In particular, among the most critical years for Norway were 1679, 1755, 1881, 1886, 1906, and 1919, when the number of victims reached 450, 200, 60, 161, 29, and 31, respectively. This level of human loss still remains today. The 6 August 2004 snowslide in Tian-Shang over the territory of Kirghizia claimed 12 lives. Snowslides on 2 February 2004 in Pakistan (Kashmir) killed more than 30 people.

1.1.6 Heat and drought

Drought refers to an extreme man–nature interaction. Under conditions when precipitation is scarce plants suffer a deficit of water, dying when this deficit is prolonged. Droughts are mentioned in biblical history. Prolonged droughts have brought great suffering to man, leading to food exhaustion. In more recent periods, droughts have been recorded in many regions of the globe. For instance, in the late 1960s–early 1970s, the drought in the Sahel, at the southern edge of the Sahara Desert, led to about 100,000 deaths. In particular, for eastern Kenya, the years 1970 and 1971 were characterized by a decrease in rainfall of 50%, which ruined crops, caused cattle plague, and starvation. Serious droughts in Kenya were recorded in 1836, 1850, 1861, 1880, 1899–1901, 1913–1918, 1925, 1936, 1954, and 1961. In general, for many African countries drought is a national problem. So, for instance, Tanzania looses annually, on average, 10% of its primary production because of droughts.

Another region of the globe, where droughts regularly take place, is Australia. Here heavy droughts were recorded in 1864–1966, 1972–1973, and 1991–1995. During arid years 1911–1916, 19 million sheep and 2 million cattle died, and the 1963–1968 drought led to a 40% loss of the harvest. Especially the central regions of the country suffer from droughts.

During the prolonged periods of high temperatures the river-beds dry up and many plants die.

In general, drought is defined as an extreme deviation of precipitation amount from the mean statistical level characteristic of a given territory. There are three types of droughts: meteorological, hydrological, and agricultural. The meteorological drought is a situation when there is a strong temporal delay in rainfall, and as a result, a deficit of moisture occurs over a territory. The hydrological drought is a deficit of water supply in surface layers and a decrease of the level of groundwater because of the lowering water level in rivers, lakes, and other water basins. The agricultural drought occurs when soil moisture lowers below the critical level when the growth of plants becomes limited. It is clear that these types of droughts are closely connected, since they are functions of the characteristics of the regional hydrological process.

Periods of droughts are closely interwoven with forest fires. During the last decade, these two natural events occur more and more frequently in different regions, many of them do not belong to zones of arid climates. So, for instance, the year 2002 was the year of heavy drought in eastern USA, which from April to July covered about 40% of the USA. At the same time, in the Canadian province Quebec, 45 forest fires were recorded. In 2002, drought covered central Russia, causing numerous fires in its European regions, the smoke from which reached more than 100 populated areas, including Moscow. In 2002, a drought covered many African countries (Zimbabwe, Malawi, Zambia, Mozambique, Kenya, Lesotho, Swaziland, and Ethiopia), and as a result, a famine broke in these countries. At the same time, in the east of Africa, there were floods. Such a regional heterogeneity and instability of climatic situations becomes characteristic of the present environment. In 2002, in the west of India, in August, there was a heavy drought, whilst in the east of India, there were floods. In China, in April, a sand storm broke out, the most serious one for the last 40 years, and in July there were heavy floods with human fatalities.

1.1.7 Wildfires

Forest fires can be of natural and anthropogenic origin. Three hundred million ago, when the first forests appeared on the Earth, the anthropogenic factor was absent, and forests themselves were heavily moistened. The marshy thickets of fern, horsetails, and mosses could not burn. However, in time, climate changes have led to situations where forest fires become an everyday occurrence. It is known that the ratio of lightning strikes over land and oceans constitutes 100:1, which leads to forests frequently catching fire. On the average, the density of lightning strikes (Figure 1.10), for instance, in tropical forests and moderate-zone forests constitutes, respectively, 50 and 5 strikes per km^2 per year. The probability of the forest catching fire depends strongly on the degree of soil moisture content. From the available estimates, every year more than 20,000 forest fires occur over the globe. Their geography is determined by the climate, and the distribution and scale are functions of numerous factors of the environment (soil moisture, temperature, density and types of trees, relief, etc.). Historically, forest fires played the role of regulator of the evolution of the Earth's cover and populating animal species. Human interference has led to a drastic change of the age-long laws of natural evolution and has changed the role of forest fires.

As follows from available chronicles, peaks in forest fire occurrence in the past

Figure 1.10. Photo of lightning strike taken by the NOAA Central Laboratory (USA) during observations of the thunderstorm activity (*www.photolib.noaa.gov/nssl/nssl0010.htm*; photo, C. Clark)

fell within periods of droughts. For instance, the Suzdal chronicle informs that in 1223 and 1298 there were large-scale forest and peat-bog fires over the territory of Russia. The Nikon and Novgorod chronicles mention droughts and forest fires in the 14–17th centuries followed by famine among population and large losses of wild animals. Information about droughts in the 18–19th centuries can be found in many historical documents, in the remaining communications of the known people and in the periodicals of that time. As Sofronov and Vakurov (1981) note, the amount of droughts and forest fires mentioned in the Russian chronicles does not exceed 50 cases. In the past and in the early 21st century the statistics of forest fires acquires a regular character, and the problem itself has gained scientific interest. Numerical models of forest fire propagation have been constructed, remote sensing techniques have been developed to detect and control forest fires, conditions are being studied under which forest material catches fire, and technologies are proposed to localize the areas of fire. In particular, an interdependence has been established between forest fires and the state of the ozone layer, since the main constituents of fire risks, such as temperature and humidity, of inflammable materials in the forest are mainly determined by the concentration of stratospheric ozone (Vaganov *et al.*, 1998).

The beginning of the 21st century has been marked by extensive forest fires over the territory of Russia, Europe, America and South-East Asia. So, in the summer of 2004, there were 197 forest fires of different origins in Russia. Forest fires were recorded in the Krasnoyarsky region, republics of Komi and Sakha, regions of Arkhangelsk, Vladimir, Irkutsk, Kirov, Sverdlovsk, and Chita, as well as in the Far East. The scale of these forest fires can be judged by their numbers of outbreaks: the Arkhangelsk Oblast – 112 outbreaks, Yakutia and Komi – 22 and 20 forest fires, respectively. The maximum number of forest fires in Russia were in the Far East (47 outbreaks).

A forest fire is a dangerous natural disaster, since it destroys areas containing material value, animals and birds, as well as, depending on the zone of burning, propagating into populated areas, industrial enterprises, and causing power outages. Forest fire fills large areas with smoke, changing the state of the atmosphere. Large-scale forest fires reduce the sinks of atmospheric CO_2, which enhances the greenhouse effect. Such fires often occurred in the territory of Russia. So, in 1915 in western Siberia, forest fires covered an area of $\sim$14 million hectares. On average, in Russia, 10,000–30,000 forest fires occur annually, covering an area of 0.5–2.0 million hectares in total. Large-scale forest fires occurred in 1972, 1984, and 2002. For instance, in July 2002, a forest fire took place close (2 km) to a liquid gas production plant near Yakutsk, placing it under threat of destruction.

1.2 THE NATURAL DISASTER AS A DYNAMIC CATEGORY OF ENVIRONMENTAL PHENOMENA

As Walker (2003) notes, the definition of a natural disaster is rather vague, and its definition depends on many factors. Grigoryev and Kondratyev (2001) define a natural disaster as an extreme and catastrophic situation in population viability caused by substantial unfavorable changes in the environment or "as a spasmodic change in the system in the form of its sudden response to smooth changes in external conditions". The number of such critical situations in the environment is increasing. For instance, before 1990 during the preceding 30 years, only in 1973 were more than 1,000 tornados recorded in the USA; after 1990 this threshold was exceeded for each and every year.

In common cases, natural disasters can arise from:

- Weather patterns (storms, cyclones, hurricanes, floods, tornados, thunderstorms).
- Other climatic conditions (droughts, bush fires, avalanches, cold snaps, and winterstorms),
- Changes in the Earth's crust (volcanoes, earthquakes, tsunami, and tidal waves).

At present, natural disasters are floods, droughts, hurricanes, storms, tornados, and tsunamis, volcanic eruptions, landslides, landslips, mudflows, snow avalanches, earthquakes, forest fires, dust storms, bitter frosts, heatwaves, epidemics, locust invasions, and many other natural phenomena (Grigoryev and Kondratyev, 2001; Abrahamson, 1989; Braun *et al.*, 1999; Changnon, 1996, 2000, 2001; Field *et al.*, 2002; Gardner, 2002; Monmonier, 1997; Walker, 2003; Satake, 2005). In the future, this list may be extended due to the appearance of new kinds of natural disasters, such as collisions with space bodies and anthropogenic ones – bioterrorism, nuclear catastrophes, sharp change in the Earth's magnetic field, plague, etc. Therefore, it is important to develop effective quantitative technologies and criteria, which would reliably warn about impending catastrophic natural phenomenon.

Many experts associate the notion of a natural disaster with the notion of ecological safety, which appeared in connection with the necessity to evaluate the danger for human health within a given territory resulting from changing environmental parameters. These changes can be both natural and anthropogenic. In the first case, danger is caused by fluctuations in natural processes connected with changes in the synoptic situation, epidemics, or due to a natural disaster. In the second case, the danger occurs as a response of nature to man's activity. For instance, Gardner (2002), analysing changes in the environment in the Himalayas, India, came to the conclusion that such factors as deforestation and changes in vegetation cover, became the cause and intensifier of instability in this region characterized by a degradation of soil resources and increase of the consequences of environmental destruction due to water flow. Field and Raupach (2004) as well as Abrahamson (1989) connect a change in the pattern of the occurrence of natural disasters with the increase in instability of the carbon–climate–man system. According to Field *et al.* (2002), this instability during the most recent two decades can increase due to a change in many characteristics of the World Ocean's ecosystems. Milne (2004), analysing the history of various large-scale disasters, has come to a pessimistic prognosis with respect to the fate of humankind, using the notion of "Doomsday".

In general, the ecological danger over a given territory results from a deviation of the parameters of humans' habitat beyond their natural limits, where after long residence a living organism starts changing but not in the direction corresponding to the natural process of evolution. As a matter of fact, the notions of "ecological danger" or "ecological safety" are connected with the notions of the stability, vitality, and integrity of the biosphere and its elements. Moreover, the Nature–Society System (NSS) being a self-organizing and self-structuring system and developing by the laws of evolution, creates within itself totalities of ecological niches whose degree of usability for a population of a given territory is determined, as a rule, by natural criteria (totality of maximum permissible concentration, religious dogmas, national traditions, etc.).

Nevertheless, when considering the perspectives of life on the Earth, it is necessary to proceed from a common criteria of evaluation of the levels of degradation of the environment, since in time, local and regional changes in the environment transform into global changes. The amplitudes of these changes are determined by mechanisms of the NSS functioning, which ensures the optimal changes of its elements. Humankind deviates more and more from this optimality in its strategy of interaction with the surrounding abiotic and biotic components of the natural environment. However, at the same time, human society as an NSS element, tries to understand the character of large-scale relationships with nature, applying many sciences and studying cause-and-effect connections in this system. One such cause-and-effect connection is a correlation between El Niño and the ozone layer. El Niño refers to a natural phenomena directly related to natural disasters taking place in the equatorial sector of the East Pacific. It is a complex of interrelated variations of chemical and thermobaric parameters of the atmosphere and ocean. The anomalous character of natural processes called El Niño, observed mainly near the shore of Peru and Chile, is manifested via a sharp increase of water temperature, an air pressure decrease in the Pacific, and a substantial change of air flux direction. All this takes place over one of the most active parts of the world for degassing – the East Pacific elevation. The mechanism of El Niño occurrence is as follows. Hydrogen, rising from the ocean bottom of the rift zone, reaches the ocean surface, and due to heat release in the reaction with oxygen warms the water of the upper photic layer. As a result, the CO_2 solubility decreases, and its flux from the ocean to the atmosphere increases. From the estimates of Monin and Shishkov (1991), during the 1982–1983 El Niño the atmosphere gained 6,000 Tg CO_2. In addition, water evaporation increases, which, together with CO_2, enhances the greenhouse effect. The heating of water leads to the origin of typhoons, a decrease of atmospheric pressure, and breaking the standard trade wind scheme of atmospheric dynamics. All these changes trigger feedbacks that restore the equilibrium of natural processes. One of the important regulators here is the ozone layer whose depletion during El Niño leads to an increase of the temperature gradient between the equatorial and southern sectors of the Pacific.

The human habitat is a complicated dynamic system. Its temporal stability is connected with the constancy of the structure, material composition and energy balance, as well as with stability of its response to the same external forcings. The system's stability can be broken by the impact of both passive and active external forces. In other words, under present conditions, the nature N and human society H being a single planetary system and having hierarchical structures $(|N|, |H|)$, interact, each with the aims of its own $(\underline{N}, \underline{H})$. From the formal point of view, this interaction can be considered as a random process $\eta(x, t)$ with the distribution law unknown, representing the level of tension in relationships between subsystems N and H or assessing the state of one of them. Here $x = \{x_1, \ldots, x_n\}$ is a set of identifying characteristics of subsystems N and H, which are the components of a possible indicator of the origin of a natural disaster, that is, a deviation of $\eta(x, t)$ beyond the limits, where the state of the subsystem N threatens subsystem H. It follows from this that the aims and behavior of subsystems N and H are functions of the indicator η depending on if their behavior is antagonistic, indifferent, or cooperative. The main goal of the subsystem H consists in reaching a high standard of living ensuring a long comfortable life. The goal and behavior of subsystem N is determined by objective laws of co-evolution. In this sense, a division between N and H is conditional, and it can be interpreted as a division of a multitude of natural processes into controled and non-controled. It is clear that with growing population density, natural disasters will intensify the feeling of discomfort, affecting the social and cultural conditions in many regions.

Without dwelling upon philosophical aspects of this division, we shall consider the systems H and N symmetrical in the sense of their description given above, and open. Here the system H disposes of technologies, science, economic potential, industrial and agricultural production, sociological arrangement, size of population, etc. The process of interaction of the systems H and N leads to a change of η, whose level affects the structure of vectors $\underline{H}$ and $\overline{H}$. Really, there is a threshold $\eta_{\max}$, beyond which humankind stops existing, and nature survives. The asymmetry of subsystems H and N in this sense causes a change in the goal and strategy of the system H. Apparently, under present conditions, interactions between these systems $\eta \to \eta_{\max}$ take place rather rapidly, and therefore some elements of vector $\underline{H}$ can be attributed to the class of the cooperative. Since the present socio-economic structure of the world is represented with a totality of states, a country should be considered a functional element of the system H. The function $\eta(x, t)$ reflects the result of interaction between countries and nature. A totality of the results of these interactions can be described by the matrix $B = \|b_{ij}\|$, each element of which having a symbolic sense of its own:

$$b_{ij} = \begin{cases} + & \text{cooperative behavior;} \\ - & \text{antagonistic relationships;} \\ 0 & \text{indifferent behavior.} \end{cases}$$

Many theories have been dedicated to studies of the laws of interactions of complicated systems of various origins. In the asymmetrical case considered here, it is a question of the survival of system H and an attempt to find a means to assess the future dynamics of system N. According to Podlazov (2001), the reflexive behavior of H will help humankind, eventually, to find a behavioral pattern "able to weight profits and danger, to understand principal limitations of our capabilities, and to feel new threats in due time". As Chernavsky (2004) notes, a human being is versatile, and knowledge of this synergetic capability will make it possible in the future to describe the system $H \cup N$ bearing in mind all social peculiarities in their variability, observing the boundaries of integral mentality of human society. Mechanisms of self-organization and self-regulation of natural systems determines the complexity of this method (Ivanov-Rostovtsev *et al.*, 2001). Of course, deep semantic and philosophical notions of personal architectonics (which should have been taken into account when forming the model of vitality, at the present level of a formalized description of intellect) remain beyond the feasibility of present ecoinformatics.

1.3 THE RANGE OF NATURAL DISASTERS

The notion of the range of natural disasters includes geographical, spatial, temporal, ecological, economic, and human factors, each with a specific scale of its own. Historically, this notion had suffered numerous changes acquiring in the present epoch the form of a complex function of its components. From the viewpoint of the present understanding of natural disasters, their range in the past was determined by available information or retrieved data about catastrophic events. It is clear that in the historical past the level of unfavorable natural phenomena was higher than at present. As follows from numerous excavations in the past, volcanic eruptions had been one of the most destructive kinds of natural disasters. In the regions of active volcanic activity, geologists find traces of settlements and towns buried under thick layers of ash, pumice, and lava. So, during the archaeological excavations carried out by the scientists of the Koeln University on the eastern territory of Germany, in the region of Lake Laahersee in the Noiwieder hollow, early settlements of people (11,000 years of age) were found under the 15-m layer of lava. Another example of a terrible natural drama is the disappearance of the two Italian towns of Pompeii and Herculanum in 79 AD as a result of the Somma (Vesuvius) Volcano eruption. History has preserved information about many tragic events on all continents, when not only large settlements but also civilizations vanished (Grigoryev and Kondratyev, 2001).

From the objective point of view, the development and scale of a dangerous natural phenomenon depends on the conditions of natural background, which can either prevent or favor the propagation of the event and, hence, reduce or enhance its impact on the environment. The amount of victims depends on the level of development of a particular society, which manifests itself through the developed system of prediction, warning, and prevention of possible natural disasters. In fact, the matter concerns the formation of a multitude of factors which can be considered as natural and social forerunners to natural disasters (Table 1.12). The scale assessment of a natural disaster depends on the human response to this disaster. For instance, the danger of a tropical cyclone is determined by a combined impact of all its elements – wind, rain, storm surges, and waves. The wind speed in a tropical cyclone can exceed 250 km h^{-1}, covering an areal band of 40–800 km wide. At this wind speed, buildings collapse, communication breaks down, and plants die. As a

Table 1.12. Possible precursors of a natural disasters.

Natural disaster	Precursors of natural disaster
Volcanic eruption	Amplitude of tectonic shifts, surface temperature, a change in the composition of gas emissions, SO_2 content.
Earthquake	Groundwater level, amplitude of surface fluctuation.
Flood	River level dynamics, air temperature, precipitation variability, depth of snow cover.
Tropical cyclone	Wind velocity, atmospheric pressure, ocean surface temperature, air temperature variability, wind shear, ozone hole.
Dust storm	Albedo, wind velocity.
Landslide, mudflow	Changes in relief and landscape, rainfall rate, surface and deep porous water pressure.
Forest fire	Temperature, rainfall rate, soil moisture, forest age.

result, a tropical cyclone can either cause human fatalities or injuries. During a tropical cyclone the rainfall can reach 2,500 mm bringing forth a flood. An important factor to consider is the storm surge – the rise of seawater above the average level of the ocean by up to 7 m or more, which leads to a rapid flooding of low areas inshore. Finally, wind–storm surge combinations lead to the propagation of high waves, which destroy beaches, agricultural lands, constructions, and buildings in the coastal zone. The cyclone's body usually moves at speeds of not more than 24 km h^{-1} increasing to 80 km h^{-1} as the cyclone moves from its center. Gigantic waves accompanying the cyclone have great destructive force. The scale and size of damage caused by tropical cyclones can be judged from the data shown in Table 1.13.

It is not always possible to evaluate the damage from tropical cyclones. In many countries, especially with the agrarian sector of an economy, the financial losses in most cases cannot be estimated because of the absence of statistical services and qualified personnel. Therefore, on the whole, there are no global-scale statistics for the consequences of natural disasters, and hence, it is impossible to estimate their scale in many regions. The service of evaluation of the consequences of natural disasters is best of all developed in the USA and other industrial countries, with developed infrastructures of systems for geoinformation monitoring. These systems are comprised of observations of regions known as areas of potential hurricane occurrence. The space-borne images of tropical hurricanes (Figure 1.11) show that their origins can be clearly seen in the optical and IR spectral intervals.

Multiyear observations of tropical cyclones over the USA have supplied information about their parameters and levels of destruction. Here one should mention Hurricane "Camilla" that flew over the USA in 1969 and was one of the most destructive natural disasters. It claimed the lives of 248 people, injuring more than 8,000 people. The economic damage constituted 1.4 billion dollars. The 10-m waves of Hurricane "Long Snake" hitting the Japanese islands in mid-December 2004

Table 1.13. Tropical cyclones and their consequences (White, 1974).

Year	Country	Name of hurricane	Number of victims	Damage (US$ mln)
2004	Japan	Lightnings matrix	100	nd
1970	Australia	Ada	13	12
1970	Bangladesh	Ada	300,000	nd
1970	USA	Selia	11	453
1969	USA	Camilla	256	1,421
1967	USA	Beulah	–	200
1966	Cuba, USA	Alma	6	100, 10
1966	Mexico	Iness	65	100
	Haiti		750	20
	USA		48	5
1965	Bangladesh		19,279	nd
1965	USA	Betsey	75	1,421
1964	USA	Dora	5	250
1964	USA	Hilda	38	125
1964	USA	Cleo	3	129
1963	Bangladesh		11,468	nd
1963	Cuba, Haiti		7,196	nd
1961	USA	Carla	46	408
1960	Bangladesh		5,149	nd
1960	Japan		5,000	nd
1960	USA	Donna	50	426
1954	China	Olga	4,000	nd

Note: nd = no data.

claimed 62 lives, sank many ships berthed at port, and caused numerous destruction within the coastal zone. The Hurricane "Isabelle" in late September 2003 devastated a large part of the eastern coast of the USA, destroying more than 360,000 buildings with damage estimated at 5 billion dollars. Forty people perished. In total, the year of 2003 for the USA was characterized by a sharp increase in numbers of natural disasters. During this year, tropical hurricanes claimed 68 lives, and the economic damage constituted 5.89 billion dollars (the insurance payments constituted 2.43 billion dollars). Numerous spring–summer hurricanes broke records both in meteorological and insurance sectors. The hurricanes caused landslides and landslips in Texas and North Carolina, a fierce snow storm covered the northern states, and forest fires broke out in the southern regions of California. Due to these events the insurance losses from natural disasters exceeded the highest previous level of 1994. Strong snowfalls in the north-eastern cities of the USA caused numerous-discomfort to the population and municipal services. In March 2003, in Alabama, Denver, and Georgia, snowfall reached 81 cm, and in Colorado the snow cover was 220 cm high. Severe frost in late January blocked the river transport systems in

Figure 1.11. Photos showing the origination of cyclones (from NASA and NOAA libraries) taken in different years from American satellites. (a) Photo of a hurricane taken in 1997 from the American Shuttle during the flight STS-82. (b) Space-derived photo of Hurricane Anita over the Mexican coastline in 1977 (*www.space.com/images/ig151_02_02.jpg*). (c) Space-derived photo of Hurricane Elana over the Gulf of Mexico in 1955 *www.krugosvet.ru/articles/04/1000405/0002401G.htm*). (d) Space-derived photo of a tropical cyclone.

(c)

(d)

north-eastern USA, and in Alabama and Georgia, on 16–18 February 2003, a hail storm occurred with hailstones up to 7.5 cm in diameter.

Every year, tropical hurricanes annihilate large forests, which can have global consequences, since forests are sinks for excess CO_2 from the atmosphere. Hurricanes take carbon from forests and affect thereby the global heat balance. The scale of this impact from estimates by McNulty (2002), only for the USA, constitutes 20 Tg C yr^{-1} withdrawn from circulation. Most of all, hurricanes affect the south-eastern coast of the USA, 55% of which is covered with forest. One hurricane can destroy about 10% of the annual wood production of forests, which constitutes about 1 billion dollars if calculated for the price of the timber. McNulty (2002) analyzed the data for 1900–1996 on the losses of carbon due to the impact of hurricanes on the forest systems of the USA territory and came to the conclusion that hurricanes transform huge amounts of living biomass into dead organics, which are then either used for building or burning, or decompose with the participation of micro-organisms. As a result, a considerable volume of carbon returns to the atmosphere. Thus, hurricanes are an important factor in the long-term impact on climate. Of course, hurricanes promote a rejuvenation of forests, but it takes 15–20 years to re-grow a destroyed canopy, and this means that during this period the sink of atmospheric CO_2 is reduced over a given territory.

Large-scale natural disasters also include floods which often follow tropical hurricanes as well as those that occur in rivers' flood plains during prolonged and heavy rains or rapid snowmelt (Figure 1.12). One of the characteristic examples of a powerful flood is the situation that took place in September 2000 in central Japan due to heavy rains caused by Typhoon "Saomai". The resulting flood paralyzed work at the Toyota factory, eight people perished, and about 500,000 people were evacuated. The city of Nagoya was flooded. The central part of Honshu Island remained without electricity for some time. The passenger trains were delayed. Many highways were blocked for a long time. Almost at the same time, Typhoon "Maemi" swept over the Japan and Okhotsk Seas, which, after having weakened, reached Kamchatka with speeds of 20 m s^{-1} causing little damage in the Yelisovsky, Ust′-Bolsheretsky regions and in the city of Petropavlovsk–Kamchatsky. Several events that took place in 2002 are characteristic examples of the consequences of typhoons and storms. In August, on Majorca, rain falling for 3 hours reached 224 mm, which caused several landslides and led to heavy mudflows. On 12 August, in Dresden, rainfall reached 154 mm during one day, and on the night of 8–9 September, in the south of France, a 36-hour period of rainfall reached 670 mm. These examples show that extreme situations are accidental, and their forecasting is a task for systems of prediction of accidental processes. This task becomes complicated when several phenomena are combined, as, for example, happened in July in Tadjikistan. Here after several days of downpour the area was hit by an earthquake and a hurricane. As a result, more than 700 houses and a multitude of administrative buildings were destroyed, and in many populated areas there was no electricity.

As for the spatial coverage, a flood is an extreme natural event that happens in most cases on territories adjacent to flood plains or in the zones of arid climate with prevailing downpours. Floods are characterized by a high level of damage, since independent of protective constructions, such extreme river overflows are possible, when water flows start destroying bridges, buildings, machinery, swamping roads, and changing the environmental relief. Table 1.14 (p. 42) gives an overview of the most powerful floods on the globe.

Floods occur practically in all regions of the globe. In each case they cause much trouble to the population. In India, the July 2004 flood in the north-eastern part of the country, destroyed scores of villages, leaving 35,000 people homeless, and washing away a great multitude of bulldozers working on the banks of the overflowing river. As a result, 40 people perished. The flood in Russia taking place at the same time at Kuban' caused damage to the Krasnoyarsk Region estimated at about 200 million roubles. The flood on 17 August 2004 in Great Britain resulting from a 6-cm rainfall during one single day in the region of Boscastle, created a flash flood which swept away cars and houses. During the July 2004 flood in south-western China, in two districts of the Yun-nan Province 11 people perished, 6 people were seriously injured, and 34 people went missing (presumed dead). The flood caused mudflows and landslides which seriously damaged more than 2,000 peasants' houses, the damage being estimated at US$33.7 million. Torrential rains over 31 July through to 26 August 2002 caused major flooding across Europe affecting areas of Austria, Cezch Republic, Germany, Russian Federation, Romania, Italy, Spain, and Slovakia. As a result of this event the total economic losses were more, EUR 15 billion (including Germany – 9.2, Austria – 2.9, Czech Republic – 2.3), with the loss of 100 lives.

The unpredictability and insidiousness of floods is confirmed by the 9 January 2005 flood caused by the most powerful hurricane in the north of Europe and Scandinavia. In Germany, England, and Sweden 14 people perished and dozens of people were either injured or missing. The hurricane winds ($\sim 30\,\mathrm{m\,s^{-1}}$) resulted in Sweden and Latvia suffering large losses of electric power. In the Estonian city of Piarnu water rose 2.8 m above its normal level, and as a result, almost 25% of the city's urban territory was flooded. The hurricane took place over the Pskov, Leningrad and Kaliningrad Oblasts of Russia. In St. Petersburg on 9 January 2005 several subway stations were closed because of the threat of an impending flood, the boiler works in the Petrograd region was closed, and preventive measures were taken in some other regions of the city. Further studies will show whether this flood was the result of the 26 December 2004 earthquake in the so-called "band of danger" or whether it was an independent event.

A flood as an extreme natural phenomenon can develop slowly, gradually changing the structure of the environment of a limited region. With global climate warming large-scale floods over large territories are possible. As an example one can look at the rise of the Caspian Sea level over the last few decades by almost 3 m. The result of which: more than 400,000 hectares of coastal territories have become useless for agriculture, leading to economic losses of US$6 billion, and affecting 100,000 people.

Earthquakes are accompanied by powerful destructive factors which lead to the horizontal shifting of land surface layers, and cause tsunamis in the sea. Large-scale earthquakes are felt over large territories reaching an area of more than

(a)

(b)

Figure 1.12. Episodes of floods in different regions of the globe. (a) Flood in August 1998 in Australia. Photo is taken in a settlement 209 km from Sydney. (b) Flood in June 2001 as a result of tropical Hurricane Allison White Oak Bayou (Texas, USA) (*www.texasfreeway.com*, Ivey, 2002). (c) Flood in England in 2002. (d) Water flow in the main street of Buffalo Bayou (Texas, USA), 9 June 2001 (*www.nhc.noaa.gov/HAW2/english/images/buffalo2.jpg*).

(c)

(d)

Table 1.14. Most powerful floods over the globe (White, 1974).

Time	Place	Number of victims and material damage
1 November 1570	Holland	50,000 people perished. The city of Frieland was ruined.
1642	China	300,000 people perished.
1824	St. Petersburg, Russia	Flood was 410 cm above ordinary. The consequences were reflected in Pushkin's poem 'Bronze Horseman'.
1887	Henan, China	Over 900,000 people perished. The Yellow River overflowed destroying numerous settlements.
31 May 1889	Georgetown, Pennsylvania, USA	Over 2,000 people perished.
September 1900	Galveston, Texas, USA	About 6,000 victims were registered. A result of a hurricane.
June 14, 1903	Willow Creek, Oregon, USA	225 people perished. City of Heppner, Oregon, was destroyed.
1911	China	The Yangtze River flood took about 100,000 lives.
25–27 March 1913	Ohio and Indiana, USA	467 people perished. Excessive regional rains caused economic losses of US$143 million.
April–May 1927	Mississippi River from Missouri to Louisiana, USA	Record discharge downstream from Cairo, Illinois. Approximate cost equaled US$230 million.
13 March 1928	Santa-Paula, California, USA	450 people perished. The dam in Saint Francis was destroyed.
March 1936	New England, USA	Excessive rainfall on snow. Fatalities of 150. Economic losses of US$300 million.
September 1938	Northeast USA	494 people died. Economic losses of US$306 million. Result of a hurricane.
July–August 1939	Tientsin, China	1,000 people perished. Millions made homeless.
14 August 1950	Province of Anhoi, China	500 people perished. 10 million made homeless. 2 million hectares of agricultural lands were flooded.
2–9 July 1951	Kansas and Missouri, USA	41 people perished. 200,000 made homeless. Damage constituted US$1 billion.
28 August 1951	Manchuria	Above 5,000 people perished.
31 January–1 February 1953	Northern Europe, The Netherlands	1,835 people perished. Coastal regions were devastated. 72,000 people were evacuated. 3,000 houses were destroyed and 40,000 houses were damaged.
1 August 1954	Kazvin region, Iran	Over 2,000 people perished.
4 October 1955	Pakistan and India	1,700 people perished. Damage reached US$63 million and 2.2 million hectares of sown area were flooded.
2 December 1959	Frejuce region, France	412 people perished. The dam at Malpasset was destroyed.
May 1961	Central regions of the western USA	25 people perished.
27 September 1962	Barcelona, Spain	Over 470 people perished. Damages of US$80 million.
31 December 1962	Northern Europe	Over 309 people perished.
9 October 1963	Belluno, Italy	Over 4,000 people perished.
14 November 1963	Haiti	Over 500 people perished.
8–9 June 1964	Northern Montana, USA	36 people perished.
December 1964	Western USA	45 people perished.

Time of event	Place of event	Number of victims and material damage
18–19 June 1965	South-western USA	27 people perished.
3–4 November1966	Arno River valley, Italy	113 people perished. Historical monuments in Florence and other places were destroyed.
12 November1966	Rio de Janeiro, Brazil	300 people perished.
January-March 1967	Rio de Janeiro and Sâo Paolo, Brazil	Over 600 people perished.
26 November 1967	Lisbon, Portugal	457 people perished.
29–31 May 1968	New Jersey, USA	Eight people perished. Damage of US$140 million.
8–14 August 1968	Gudjarat, India	1,000 people perished.
23 August 1969	North Virginia, USA	100 people perished.
4 July 1969	Southern Michigan and northern Ohio, USA	33 people perished.
25–29 January 1969	Southern California, USA	95 people perished.
11–23 May 1970	Oradia, Romania	200 people perished. 225 settlements were damaged.
2 February 1972	Buffalo Creek, West Virginia, USA	125 people died. Economic losses of US$60 million. Dam failure after excessive rainfall.
9–10 June 1972	South Dakota, Rapid City, USA	237 people perished. Damage of US$160 million. 15 inches of rain in 5 hours.
June 1972	Eastern USA	Over 100 people perished. Damage US$2 billion.
31 July 1976	Big Thompson and Cache la Poudre Rivers, Colorado, USA	144 people lost. Economic losses of US$39 million. Flash floods in canyon after excessive rainfall.
May–August 1993	USA	48 people perished, economic losses came to US$21 billion with insured losses of US$1.270 million.
May–August 1996	China	3,048 people perished, economic losses of US$24 billion with insured losses of US$445 million.
May–September 1998	China	Fatalities reached 4,159, economic losses of US$30.7 billion with insured losses of US$1 billion. 30,000 people perished.
1999	Venezuela	164 people perished, 85,000 people were evacuated.
29 February 2000	Mozambique	Seven people perished, 50,000 people suffered, damage of US$200 million.
May 2001	Yakutia, Russia	298 people perished, 294 people were injured. Many buildings ruined, highways destroyed, electricity network posts brought down, trees uprooted, 5,500 families made homeless.
June 2001	Southern Russia	753 people perished. Over 2 million hectares of agricultural land damaged.
12 November 2001	Northern Algeria	114 people perished, 335 people injured, economic damage of US$484 million.
8 June 2002	China	23 people perished. 40,000 hectares of agricultural land damaged.
12–20 August 2002	Europe	Over 200 people perished with 200 people missing.
8 September 2002	France	37 people perished, economic losses of US$16 billion with insured losses of US$3.4 billion.
18 May 2003	Sri Lanka	About 100,000 people made homeless.

4 million km^2. The earthquake intensity and respective destruction are determined by the type of ground surface. Magnitude scales are used to measure earthquakes. In Japan a scale of 7 magnitudes is used, but the Richter scale is most widely used (Richter, 1969). This scale is determined from the formula: $\log E = 11.4 + 1.5\,M$, where E is the total released energy and M is the magnitude corresponding to the amplitude of the horizontal shifting. According to this dependence, each subsequent unit on the Richter scale means that the released energy is 31.6 greater than the preceding unit on the scale. Hence, the most powerful earthquakes are those of magnitude 7 and higher. Examples of which are: the 1906 earthquakes in San Francisco (M 8.25), in Tokyo in 1923 (M 8.1), in Asam in 1950 (M 8.6), in Alaska in 1957 (M 8.4–8.6), and the largest scale earthquake at Gobi-Altai on 4 December 1957 (M 11). In November 2004 there were four earthquakes of M 7 on the Richter scale: two in Indonesia (11 November – M 7.5 and 26 November – M 7.2), on the western coast of New Zealand (22 November – M 7.1) and the western coast of Columbia (15 November – M 7.2). The number of victims depends on the population density and the preventative/reactionary measures taken. The Gobi-Altai earthquake mentioned above was felt over an area more than 5 million km^2, including the whole territory of Mongolia, the south of Buriatiya, the Yakutsk and Chita Oblasts, and the northern provinces of China, but due to a small population density, victims were not numerous.

Globally, from data of the US Geological Survey (USGS), about 20,000 earthquakes are recorded every year, 18 of them being of M 7.0–7.9, and one of M 8.0. In the USA, 39 states are subject to the risk of earthquakes. A perfect and Advanced National Seismic System (ANSS) has been developed in the USA, with 6,000 sensors, providing 5–10-minute warning times. The USGS, within the framework of the National Earthquake Hazards Reduction Programme (NEHRP), beginning in 2001, established more than 300 sensors in many large cities, such as San Francisco, Seattle, Anchorage, Las Vegas, and Memphis, in order to increase the feasibility of recording an earthquake and thus improving the operative warning systems for the general public. As a result a number of earthquakes in the USA in 2003 did not cause any considerable damage. Among them are the 9 April earthquake in north-east Alabama (M 4.6), the 6 June earthquake in the western part of Kentucky (M 4.0), the 9 September earthquake in Virginia (M 4.5), and the 17 November earthquake in the region of the Aleutian Islands, at a distance of 2,220 km from Anchorage.

The Sumatra–Andaman and Nias Island earthquakes on 26 December 2004 and 28 March 2005 with magnitudes M 9.0–9.3 and M 8.2–8.7 respectively were the largest in 40 years. The Sumatra–Andaman earthquake was the second largest earthquake ever recorded instrumentally and the third most fatal earthquake ever. It was equivalent to a 100-gigaton bomb. Its energy (4.3×10^{18} J) generated a tsunami that traveled to the Antarctic, the east and west coasts of the Americas, and the Arctic Ocean (Bilham, 2005).

One of the peculiar features of earthquakes and volcanic eruptions is their ability to cause secondary natural processes, such as landslides, landslips, mudflows, tsunami, and floods. This cascading nature of earthquakes often manifests itself in thickly populated regions in the form of fire, gas explosions, and other indirect factors (floods, damage from electric shocks, etc.). Lava flowing down the snow- or ice-covered mountain slopes causes mudflows and floods. The hot mudflow, growing in power, can gradually transform into an avalanche and then, with further snow and ice melting, transform into a powerful water flow which, bursting out of gorges, acquires a significant destructive force.

One of the most fearful natural disasters are tsunamis. A tsunami is a gigantic wave reaching, near shore, a height of 10–30 m and moving with enormous velocity. The waves appear in the ocean at the epicenter of an earthquake. For instance, let us consider the 5 November 1952 event that took place on the Kuril Islands. A huge wave from the ocean completely covered Paramushir Island flooding the city of Yuzhnokurilsk. It only took a few minutes to kill everybody and to cause mass destruction. The 30-m tsunami that hit many countries of South East Asia on 26 December 2004 claimed about 232,000 lives in Indonesia, Thailand, India, Bangladesh and other coastal countries of the Indian Ocean basin. The tsunami was initiated by an earthquake of *M* 9 on the Richter scale in the eastern part of the Indian Ocean. A warning about the impending danger issued by the Pacific Center of Tsunami Warning located not far from Honolulu (Hawaii) did not reach them for lack of a coordinated warning system in these countries.

The following events exemplify the large-scale destruction and death caused a tsunami. On 1 April 1946 due to an earthquake of magnitude *M* 7.3 on the Richter scale near the Aleutian Islands, the resulting tsunami wave claimed 159 lives on Hawaii and caused damage estimated at more than US$26 million. The earthquake of magnitude *M* 8.2 which struck south-east of Kamchatka on 5 November 1950 caused a huge tsunami wave which with enormous speed dashed over the northern sector of the Pacific Ocean. On 9 March 1957, in the zone of the Aleutian Islands there was an earthquake of magnitude *M* 8.3. The resulting tsunami wave (23 m high) struck the islands of Umnak and Kanai with huge destructive force. The earthquake of magnitude *M* 8.3 on the sea bottom near Chile caused a tsunami wave that reached Hawaii, claiming 61 lives and destroying 537 houses causing damage exceeding US$23 million. On Hawaii the 14-m tsunami wave of 29 November 1975 (following an earthquake of magnitude *M* 7.2) caused damage estimated at US$4.1 million. Finally, the earthquake of magnitude *M* 7 in the northern sector of the Bismarck Sea on 17 July 1998 caused a tsunami wave which killed 2,202 people and destroyed the homes of 10,000 people in coastal regions.

Similar scale destruction can be caused by tornados (whirlwinds) which usually occur suddenly and affect a confined territory for a short time. The length of a tornado run averages about 25 km with a coverage band not exceeding 400 m. The whirlwind occurrence is connected with thunderclouds in the presence of a sharp contrast of temperature, humidity, density, and other parameters of air fluxes. Usually, a whirlwind appears in the zone of contact between cool and dry air masses in the surface air layer. As a result, high winds start blowing in a narrow transition zone leading to a vortex (Figure 1.13).

The destruction caused by tornados is terrific. Tornados destroy buildings, uproot trees, and lift cars and fragments of buildings into the air. To evaluate the

Figure 1.13. Some occurrences of tornados. Photos from the NOAA archive, Phelan (2003), and by Bove and Thráinsson (2003).

Table 1.15. Characteristics of the Fujita–Pearson scale.

Level of scale	Intensity	Wind speed (km h^{-1})	Analysis of danger
F0	Storm	40–72	Some damage to chimneys, broken tree branches, uprooted trees, signs torn off buildings.
F1	Moderate tornado	73–112	Lowest wind speed of the early hurricane. Blows off roofs, pulls houses from foundations or turns them over, moves cars, can destroy separate or attached garages.
F2	Substantial tornado	113–157	High level of danger. Blows off roofs, destroys mobile dwellings, overturns box garages, breaks or uproots big trees, transports light-weight objects.
F3	Strong tornado	158–206	Destroys roofs and walls of stationary constructions, overturns trains, uproots trees.
F4	Devastating tornado	207–260	Destroys solid buildings, takes away houses with poor foundations, cars, and large objects.
F5	Incredible tornado	261–318	Detaches houses with vigor from foundations. Moves cars to distances of up to 100 m, breaks trees, heavily damages steel and concrete-fastened structures.
F6	Inconceivable tornado	319–379	Such winds are improbable. Their destruction capabilities are therefore difficult to distinguish from destructions by tornados of the type F4 and F5. The emergent vortex exhibits amounts of power over a small territory, it can take large volumes of water and transport them great distances.

possible consequences of tornados, the Fujita–Pearson scale is usually used (Table 1.15), which, despite its subjectivity, helps to range tornados by their level of danger. Tornado statistics in the USA shows that over the USA territory about 69% of moderate (weak) tornados occur for durations of 1 to 10 minutes. Fatal outcomes represent no more than 5% of all victims of tornados. More powerful tornados occur in 29% of cases and last for more than 20 minutes, with a 30% fatalities rate. Frantic and fierce tornados occur in 2% of cases with human victims constituting 65% of the victims of all tornados. They destroy everything during lifetimes of 1 hour or longer. From the data for 2003, during the first 10 days of May only, over the USA, there were 412 tornados with wind speeds of

330–420 km h^{-1}. In South Dakota and Mississippi, on 4 May 2003, 94 tornados were recorded. Many of them covered a band up to 450 m wide at a distance of 24 km. Despite the preventive measures taken, the damage reached US$7 billion.

Serious damage is usually caused by forest fires, which can be of both natural and anthropogenic origin. Their spatial scale can be judged by the number of annually recorded wild fires. Figures 1.14–1.16 demonstrate the consequences of forest fires and their danger for the population. During the last few years the following were recorded: 12,300 fires in 2000, 84,079 fires in 2001, 88,458 fires in 2002, and 29,634 fires in 2003. Respectively, during this period, forests were burned over a territory of 3.4 million hectares in 2000, 1.4 million hectares in 2001, 2.8 million - hectares in 2002, and 4.1 million hectares in 2003. Every year, over the territory of Russia, from 12,000 to 37,000 forest fires occur, eliminating 0.4–4.0 million hectares of forest, with economic damage reaching hundreds of millions of dollars. Independent of the causes, a forest fire usually propagates rapidly becoming a large-scale disaster. In January 2002, fierce forest fires in the environs of Sydney (Australia) ruined several national parks and threatened the city. Forests burned over an area of 0.5 million hectares, and the atmosphere over the south-eastern coast of Australia was filled with smoke. In 2003, in the south of California (USA), unusually active forest fires were recorded in late October. On the whole, the number of forest fires over an area 1.62 million hectares totaled 60,000. In 2002, 90,000 forest fires were recorded over an

Figure 1.14. Landscape of the Yellowstone National Park near the Lake Levis after a forest fire in the summer of 2002 (*www.thefurtrapper.com/forest_fires.htm*).

Figure 1.15. The post-fire forest in the Yellowstone National Park in May 2003 (*www.thefur-trapper.com/forest_fires.htm*).

Figure 1.16. A forest fire in Colorado (USA), 2003. (*www.coloradocomments.com/fire6.jpg*).

area of 2.92 million hectares. On average, during 1993–2003 over the USA, 100,000 forest fires occurred every year over an area of 1.74 million hectares. A very powerful forest fire in the USA was recorded in June 2002 south of Denver (Colorado, USA). The fire covered an area of 36,000 hectares.

Extinction of forest fires requires concentration of a great number of special fire-prevention means based on land and in the air. A characteristic example of the resulting situations is demonstrated by the annual forest fires, and the struggle against them, in the territory of Russia. One of the causes of forest fires is increased air temperature leading to a drying of the litter, creating favorable conditions for fire propagation. Most dangerous are periods during arid seasons with a prolonged deficit of soil moisture, and with shallowing rivers and water basins. Such situations often occur in Africa, Australia, Central America, and Asia. For instance, during the 2002 drought in the south of India, more than 1,000 people died. In the Chinese Province of Anhoi, in the summer of 2003, there was recorded a drought, the heaviest over the last 25 years, resulting from combined high temperatures (41.3°C) and reduced rainfall (by 81.7%). In particular, in the vicinity of Huangshan, the deficit of soil moisture was recorded over an area of 43,000 hectares of crops (89% of the total area of crop growing). Drought often destroys rice fields in Indonesia. In higher northern latitudes, droughts were recorded in Bashkiria, where they occur in a combination of several agro-meteorological phenomena, most important being an intrusion of Arctic air masses containing little moisture.

The spatial scale of the impact of forest fires on climate can manifest itself indirectly through a change in the gas exchange in the atmosphere–soil system in the zones of permafrost. Soil heating (by forest fires) can lead to an enhanced respiration and favor thereby an enhancement of post-fire CO_2 release to the atmosphere over time periods of 10s of years (Kondratyev and Grigoryev, 2004). In this connection, it should be noted that the high-latitude ecosystems cover 22% of land surface and contain about 40% of the global soil carbon. A considerable part belongs to the permafrost zone whose dynamics is determined by the cycles of degradation (thermocarst formation) and aggradation. These cycles are closely connected with forest fires which mainly disturb boreal forests. During the last decades, in the boreal and Arctic regions of Canada and Alaska, the soil temperature has increased by 1.5°C, including also the permafrost zones. As a result, there were changes in the rate of nitrogen fixation, moss growth, the depth of the organic layer, and soil drainage.

As has been mentioned above, the scale of natural disasters increases both in frequency of occurrence and in damage caused. For instance, there are data on losses and damage for the last years. Note that, for instance, in 2003, on a global scale, natural disasters claimed more than 50,000 lives, whereas in 2002 these losses constituted 11,000 deaths. Due to heatwaves in Europe and an earthquake in Iran more than 20,000 people perished. During 2002–2003 the economic losses constituted US$55 and US$60 billion, respectively, the main contribution being made by tornados, heatwaves, forest fires, and floods in Asia and Europe. A colossal typhoon ravaged South Korea 12 September 2003. Economic losses amounted to US$3.7 billion (insured – US$0.27 billion). One should also note the unstable character of the time of occurrence of natural disasters alternating with seemingly incompatible phenomena. For instance, in India, Pakistan, and Bangladesh in May–June temperatures reached 50°C, whilst from June–September there was a chain of heavy floods.

Note should be taken that in the integral estimation of the scale of natural disasters and consequences one should take into account the statistics of economic losses. Unfortunately, these statistics are not available for all countries. Knowledge of the distribution of damage by type of natural disaster is important for insurance companies. For instance, for the USA, in 2003 the distribution of economic (insurance) losses was thus: tropical cyclones – 38% (49)%; earthquakes and volcanic eruptions – 19% (16)%; floods – 16% (3)%; thunderstorms – 13% (16)%; snow storms – 11% (10)%; and other events – 3% (6)%. The ratio of economic losses to insurance payments for damage caused by natural disasters during the last decades in the USA decreased from 6 in 1950 to 2.4 in 2003, which demonstrates the efficiency of the insurance strategy.

A new category of possible natural cataclysms has recently appeared whose source is outer space (Goldner, 2002). Specialists studying planets and astrophysicists believe that during the last 6,000 years there were $\sim$30 large-scale natural disasters caused by collisions of our planet with cosmic objects. These statements are based on archaeological and palaeontological data, and here various models and information of morphological character are used. The spatial scale of some similar disasters is estimated at 10^5–10^6 Mt. In this connection, there are hypotheses linking the origin of many formations on the Earth's surface to such disasters. Among them is the Caspian Sea.

The problem of a collision between the Earth and cosmic bodies has been an issue for discussion over the last two centuries in connection with observations of the motion of comets and prediction of their orbits. In the 18th century, astronomers could reliably assess the possibility of such collisions. The paradox was that with a background of this reliability, there were no predictions of large-scale earthquakes, such as the 1 November 1755 earthquake, which completely destroyed the city of Lisbon (Kendrick, 1957). At the same time, mathematicians reliably predicted the motion of the Galley Comet, convinced of their accuracy when the comet appeared in 1835, 1910, and 1986.

At any time in the future of the Earth there may be a cataclysm caused by a collision with a large cosmic body. When it happens, it will represent an ordinary planetary event, which apparently have taken place in the past, creating conditions for the origin and development of life. Such a hypothesis is being discussed by many scientists who believe that the collision of the Earth with a comet or an asteroid will become, in the future, a crucial moment in evolution, perhaps ruining our civilization and thereby initiating a new one, as has happened in the past. Therefore, the study of the history of disasters (Tables 1.16–1.17) and an extension of the base of their knowledge is a necessary element for the solution of the problems such as:

- Development and improvement of the strategy of man–nature interaction ensuring a sustainable development.

Table 1.16. Characteristics of the damage caused by natural disasters in 2002.

Natural disaster	Number of events	Number of victims	Economic losses (US$ bln)	
			Total	Insurance
Earthquake	61	2,625	1.03	–
Volcanic eruption	18	24	–	–
Tropical storm	34	723	9.21	1.36
Winter blizzard, snowstorm	25	183	1.49	0.73
Tornado	22	61	3.00	2.92
Strong hurricane	137	519	4.85	3.03
Heavy hail	18	49	–	–
Sand storm	6	6	–	–
Local storm	30	7	0.22	–
Storm roughness	2	5	–	–
Flood	160	4,300	27.00	4.74
Rainfall flood	40	283	0.25	–
Drought, heatwave	23	1,221	6.63	–
Forest fire	43	1	0.19	0.11
Landslide	25	366	–	–
Avalanche	13	43	–	–
Severe frost	7	171	–	–

- Analysis of knowledge of pre-history natural disasters and a search for technologies that can use this knowledge to assess the future trends in the development of humankind making it possible to reveal the ultimate loads on nature.
- An understanding of the role and place of the NSS human component in space.

These and similar problems can be resolved with the use of data for the last 6,000 years of existence of the present human civilization, available utilising archaeology and palaeontology. The role of informatics should manifest itself here through the overcoming of a high level of uncertainty of the moment of origin of natural disasters using new highly efficient prediction technologies. Particularly important is the problem of predicting the ecological consequences of natural disasters which, gradually increasing, can manifest themselves over decades taking the form of a reduced productivity of ecosystems, a changed territorial structure to the water balance, and broken vital environmental parameters. In other words, when developing a technology to assessing the spatial scale of natural disasters, it is necessary to take into account a complex of criteria: medico-biological, economic, social, botanic, soil, zoological, and geodynamical (Kharkina, 2000). One of the elements of this type of technology is the grouping of natural disasters and application of knowledge from different scientific fields, especially those with prospective technologies for future usage. One of the urgent problems appearing on the verge of future changes in the

Table 1.17. The dynamics of natural disasters and their consequences. Estimates are based on publications.

Year	Number of natural disasters				Total number	Economic losses (US$ bln)
	Floods	Snowstorms, frosts, hurricanes	Earthquakes, volcanic eruptions	Other natural disasters		
1950	–	–	17	–	17	–
1951	53	–	36	–	89	11.4
1952	–	–	–	–	–	–
1953	19	–	16	–	35	23.0
1954	19	16	–	–	35	1.4
1955	19	16	–	–	35	5.4
1956	–	17	18	–	35	–
1957	–	–	35	–	35	–
1958	–	–	–	–	–	4.0
1959	–	70	–	–	70	15.5
1960	–	36	34	–	70	2.1
1961	18	17	–	–	35	3.8
1962	16	35	–	–	51	3.9
1963	–	35	16	–	51	18.9
1964	15	16	36	–	67	21.2
1965	–	34	–	–	34	9.4
1966	–	16	16	–	32	12.1
1967	–	16	–	–	16	1.4
1968	29	26	16	–	71	3.4
1969	17	19	–	–	36	9.4
1970	–	40	15	–	55	5.4
1971	14	16	38	20	88	11.4
1972	16	53	37	–	106	17.5
1973	55	16	–	–	71	8.7
1974	19	55	14	–	88	18.9
1975	22	36	14	–	72	3.9
1976	14	53	84	–	151	47.7
1977	–	17	18	20	55	16.1
1978	–	–	33	–	33	2.3
1979	–	54	34	–	88	22.9
1980	–	18	50	–	68	42.1
1981	–	7	6	–	13	1.4
1982	18	19	34	18	89	24.2
1983	10	55	32	12	109	14.9
1984	–	34	–	–	34	1.7
1985	–	53	54	–	107	15.6
1986	–	34	71	74	179	16.4
1987	105	69	15	21	210	24.2
1988	67	69	33	32	201	63.2
1989	32	17	34	6	89	36.3
1990	–	102	35	–	137	40.9
1991	29	71	52	29	181	51.8
1992	35	69	36	20	160	62.6
1993	139	89	15	22	265	82.0
1994	53	16	14	4	87	92.8
1995	74	105	35	–	214	170.0
1996	36	70	–	16	122	51.1
1997	18	20	13	–	51	24.9
1998	35	126	33	17	211	71.2
1999	34	126	50	–	210	84.7
2000	14	–	–	–	14	0.1
2001	–	34	34	–	68	16.1
2002	16	–	–	–	16	15.1
2003	69	37	18	18	142	18.5
2004	37	59	13	7	116	45.7

environment is to understand the role of climate change in the processes of genetic modification (Hinchliffe *et al.*, 2002). Similar problems are now hotly discussed by scientists in different spheres of knowledge. Morris *et al.* (2003) considered and compared many models describing environmental change, emphasized their limits, and considered an important problem, such as the ratio of temporal and spatial scales, with due regard in these models to the impact on the environment due to anthropogenic applications of natural resources, water, soil, and the atmosphere, as well as trying to assess the role of technologies and the economy in the formation of these impacts. Blowers and Hinchliffe (2003) focused their analysis on revealing correlations between changes in the environment and technical, economic, and political responses to these changes, as well as setting numerous problems concerning the existing uncertainties and risk of those interactions. Hardy (2003) classified the potential issues of climate change for the planet on the whole and for population, in particular, dwelling upon the possible perspectives of the development of the society–nature relationship.

1.4 CLASSIFICATION OF NATURAL DISASTERS

A natural disaster is an elemental calamity leading to human death and large-scale economic damage (Edward, 2005; Nepomniashchy, 2002). Earthquakes, landslides, snow avalanches, landslips, glacier descent, floods, volcanic eruptions, forest fires, thunderstorms, tornados, storms, drought/heatwaves, hurricanes, etc., lead often to such consequences. Diverse causes of these elemental phenomena create certain difficulties in their prediction and, thereby, the possibility to prevent large-scale losses still remains low. The loss of 136 people in northern Osetia during the descent of the Kolka Glacier in September 2002 is an example of such an event. In the USA, despite a developed monitoring system with functions warning about natural disasters, in 2002 thunderstorms caused serious economic losses connected with insurance payments of US$7.7 billion across 39 states. The main causes of this damage were not thunderstorms themselves but the heavy fall of huge hailstones and accompanying tornados. These were the greatest insurance losses in the USA due to thunderstorms during the period from 1950. Nevertheless, a study of conditions of the origin of extreme natural phenomena makes it possible to discover and formulate the laws of interaction between biological and physical subsystems of the environment, ensuring thereby a possibility, if only of statistical prediction, of dangerous natural phenomena.

Under present conditions of NSS functioning, it is impossible to distinguish between natural and anthropogenic processes whose interactions considerably determine the dynamics of many natural disasters. The level of this interaction is determined by the preparedness and readiness of a population to face the dangers connected with an extreme event. An extreme event is any event in the geophysical system, whose characteristics comparatively strongly deviate from average magnitudes. But the level of catastrophic character of these deviations depends on the understanding, and assessment of, the danger of such a natural event by the population of that region, a factor which is determined by a totality of economic, technical, and moral–ethical parameters. It is known that populations living in places where natural disasters occur regularly, and the population of the territories with a low probability of occurrence of extreme situations, perceive them differently, that is, for them, the scales of danger are different.

Now, natural disasters can be divided into two classes: natural and caused by human activity. With due regard to a strong interdependence of anthropogenic and natural factors, this division becomes more and more conditional. Nevertheless, various criteria are being developed to assess the danger of a catastrophic natural event. For instance, the notions of "lesion factor", "number of victims", "level of economic damage", etc. There have been attempts to introduce the scale of catastrophic consequences of a natural event using a totality of such indicators as the number of victims and the level of economic damage. Such scales used in many countries lead to different estimates of natural disasters. In correspondence with the United Nations Environment Programme (UNEP) criterion, such a scale has only two positions:

(1) There are not less than 10 victims.
(2) The damage exceeds US$2 million.

There is no universal classification of natural disasters. Each type is usually classified by its power. For instance, according to the international agreement, tropical cyclones are classified by wind force: tropical depressions at a wind speed of $<63\,\mathrm{km\,h^{-1}}$; tropical storms (64–$119\,\mathrm{km\,h^{-1}}$); and tropical hurricanes (typhoons) ($>120\,\mathrm{km\,h^{-1}}$). Tornados are usually divided into three groups: weak ($\sim 110\,\mathrm{km\,h^{-1}}$), strong ($110$–$205\,\mathrm{km\,h^{-1}}$), and frantic ($>205\,\mathrm{km\,h^{-1}}$). Scales of Fujita (Table 1.15) and Saffir–Simpson (Table 1.18) are most widely used to classify tornados and hurricanes, respectively.

A formalized solution to the problem of classification of natural disasters requires an introduction of quantitative gradations for processes in the environment. Such attempts are made when determining the types of jet air currents, drawing synoptic maps, and assessing the characteristics of numerous atmospheric phenomena. Knowledge of the global structure of air currents and their dynamics

Table 1.18. The Saffir–Simpson scale for measuring hurricane magnitude (Webster *et al.*, 2005).

Hurricane category	Maximum wind speed		Minimum atmospheric pressure (hPa)	Storm surge height (m)
	$\mathrm{m\,s^{-1}}$	$\mathrm{km\,h^{-1}}$		
1	33–43	119–153	≥980	1.0–1.7
2	43–50	154–177	979–965	1.8–2.6
3	50–56	178–209	964–945	2.7–3.8
4	56–67	210–249	944–920	3.9–5.6
5	>67	>250	<920	≥5.7

gives the basis for predicting catastrophic phenomena of atmospheric origin. Of course, many local atmospheric disturbances, such as thunderstorms, require additional information for their prediction. Thunderstorms are fast atmospheric phenomena occurring under conditions of atmospheric instability. It is very difficult to assess the power of lightning strokes from external forerunners of thunderstorms. For this purpose, it is necessary to know the size distribution of liquid and frozen water droplets in clouds over a limited territory. Realization of this requires a large economic expenditure. The technology of this monitoring is based on measurements of polarization characteristics of self- and scattered radiation in the microwave range (Zagorin, 1998).

The problem of assessing thunderstorm activity is closely connected with studies of the random fields of the Earth created by atmospheric–electrical discharges (Remizov, 1985). Results of these studies are important not only for control of fire risks in forested territories but also for radiotechnical applications. Remizov (1985) was the first to publish data on the global distribution of the index of intensity of thunderstorm activity for different seasons.

Classification of natural disasters used in the scientific literature and in everyday life is in fact connected with the types of phenomena taking place in the environment (Kotlyakov, 1993). They have been enumerated in Section 1.1.

1.5 SPATIAL AND TEMPORAL CHARACTERISTICS OF NATURAL DISASTERS

The spatial and temporal distribution of each type of natural disaster is well known and changes with time insignificantly. Table 1.19 and Figure 1.1 characterize to some extent this distribution. Nevertheless, in connection with the recently growing number of extreme natural phenomena, the zones of extreme danger can shift. For example, there was a record increase of extreme weather conditions in 2004 in Europe. In June, in the south of France, the average temperature exceeded the threshold value 40°C, which is above the norm by 5–7°C. In Switzerland, June was the hottest month during the last 250 years. Many regions of India where the temperature exceeded the average norm by 5°C, suffered from unprecedented heat.

In the USA, in 2004 there were a record number of tornados. In May alone there were 562. Downpours flooded Sri Lanka causing large-scale floods and landslides. For instance, the data in Table 1.20 characterizes to some extent the growing number of one type of dangerous natural disasters: earthquakes.

The growing heterogeneity of the occurrence of natural disasters, both in space and in time, is explained by the increasing anthropogenic constituent and the natural trend of the climate system. The problem of their relationship has been repeatedly discussed in the literature (Grigoryev and Kondratyev, 2001). The authors note that the regional features of the consequences of natural disasters are characterized by clear indicators of their territorial density. The lowest density of natural disasters (one disaster per 470,000 km^2) falls in North America. The middle position by value

Table 1.19. The spatial–temporal distribution of most significant natural disasters in 2000.

Region	Months (2000)											
	1	2	3	4	5	6	7	8	9	10	11	12
Australia												1
Austria							1				1	
Bangladesh							1	1				1
Great Britain										1		
Vietnam							1					
Indian Ocean												
India						1	1	2				
Indonesia						1						1
Italy										1		
China								1				1
Kuwait						1						
Malaysia											1	
Mexico								1				
Mozambique		1										
Nigeria							2				1	
The Netherlands					1							
Russia											1	
USA	3	1	3		3					1	1	
Taiwan												
Uganda	1									1		
Philippines							1			1		
Switzerland										1		
Japan					1				1			

of this indicator attributed to Africa, Europe, and South America, where it is one disaster per 270,000, 240,000, and 230,000 km^2, respectively. The regions of Central America and the Caribbean basin, Asia/Australia, and Oceania are characterized by the highest territorial density of natural disasters (one disaster per 150,000, 120,000, and 80,000 km^2, respectively). Thus, during the last 40 years, taking into account this indicator of territorial density of natural disasters, a conclusion can be drawn that the territories of Australia, Asia, the Caribbean basin, Oceania, and Central America suffer most from natural disasters. But this conclusion does not correspond to the territorial distribution of the scale of damage from natural disasters, since it does not reflect their dependence on the state of the economy and other factors of the societal development.

As has been mentioned above, the spectrum and spatial distribution of natural disasters have been established in the process of the Earth's evolution. However, looking ahead, one should note that with development of civilization both the spatial distribution and the character of natural disasters can change in time. Many

Table 1.20. Distribution of earthquakes by magnitude for the period 1995–2004 and their consequences (2005 is a forecast).

Year	Number of earthquakes	Distribution of earthquakes by magnitude (Richter scale)				Number of victims
		1 to 5	5 to 7	7 to 8	8 and higher	
1995	19,162	17,641	1,501	18	2	7,980
1996	17,751	16,365	1,371	14	1	589
1997	16,453	15,204	1,233	16	0	3,069
1998	19,252	18,144	1,096	11	1	9,430
1999	18,721	16,299	1,220	18	0	22,662
2000	19,131	17,613	1,503	14	1	158,483
2001	20,595	19,210	1,369	15	1	33,819
2002	24,507	23,146	1,348	13	0	1,686
2003	27,677	26,319	1,343	14	1	21,357
2004	26,714	25,235	1,464	13	2	232,000
2005	18,291	16,982	1,297	11	1	1,953

scientists believe that principal changes will take place in the future. Here are some of the possible changes:

- *Earthquakes and floods*, even in several decades, will be killing tens of thousands of people in developing countries; developed countries will continue to suffer large-scale economic losses and thus a breaking of the progress in many spheres of life.
- *Epidemics*, despite the development of medicine, will, as usual, prevent establishment of a healthy mode of life due to appearance of new kinds of diseases, which may be caused by genetic engineering.
- *Aggression* of people living on a territory of other people could create a precedent of colonization and principal change to the way of life of a population on the Earth. A reduction of traditional supplies of biological food and mineral resources could initiate a change of species which would be able to feed on solar energy or some chemical elements, of which there are plenty in the World Ocean (e.g., deuterium).
- *The impact of cosmic bodies* on the Earth could cause a sharp global climate change, which would lead to global catastrophe. A comet or asteroid with the diameter of several kilometers would be able to devastate huge areas, either by direct forcing or due to ensuing fires, tsunamis, and other extreme phenomena, in addition to a potential change to the Earth's orbit. The probability of such an event is negligible.
- *An approach of the Earth to a super new star* could destroy every living being on the Earth's surface due to high levels of radiation.
- *Global glaciation* could happen in the nearest 10,000 years as an alternative to the expected climate warming.

- *Change of the Earth's magnetic field.* Changing the poles could eliminate the ozone layer and cause thereby irreversible change to the biosphere.
- *Anthropogenic disasters* may continue to expand due to the appearance of new kinds of impact on the environment and human society. They may include deviations in the social and cultural spheres, in science, and engineering. Bioterrorism will become enhanced and nano-technologies will change the structure of the energy balance of the planet, raising the efficiency of assimilation of solar energy from the present 10% to 50% in the future.

2

Natural disasters and the survivability of ecological systems

2.1 ESTIMATION OF THE DANGER OF NATURAL DISASTERS

The notion of the danger of natural disasters is closely connected with the notion of vulnerability. According to Vogel and O'Brien (2004), the term "vulnerability" is, in a sense, vague, and its use in assessing the consequences of stress situations in the environment leads often to great uncertainties. To characterize the response of a population and the environment of a given territory to external forcings, along with vulnerability, they use many notions, such as stability, adaptability, survivability, etc. Though the relationship between these notions remains vague, nevertheless, they reflect, if only on the intuitive and terminological levels, the notion of the danger of natural disasters. Vulnerability is the ability of man or a group of people to foresee, struggle, resist, and overcome the damage from a natural disaster. The notion of vulnerability is closely connected with the social characteristics of a given territory and can be defined as a function of ecodynamic constituents, such as the levels of urbanization and economic development, the state of environmental protection, and the medical services.

During the last few decades, there has been a trend toward increasing economic losses due to urbanization and climate change. This means that an estimate of the vulnerability of the natural–anthropogenic systems of a given territory requires the development of a versatile approach toward calculations of the risk of natural disasters, reflecting the interaction of biophysical and socio-economic elements. The events in the beginning of the 21st century confirm the need for a search for more adequate technologies for assessing Nature–Society System (NSS) vulnerability. For instance, on 22 January 2005, in the middle-eastern and north-eastern USA there was a 50-cm snowfall with hurricane wind speeds, which led to the cancelation of hundreds of air flights and blocked many highways. Is this event only a regional-scale episode or does it correlate with the processes of global ecodynamics?

Unfortunately, the data available for complex studies of these processes do not permit one to answer this simple question.

In reality, in 2003–2005 alone, there were several natural disasters in the USA, the scale of which was previously unheard of. Here are some of them. On 16 June 2003 there was 160 mm of rainfall which provoked a heavy flood that destroyed more than 200 buildings and caused damage estimated at US$23 million. The day before, in the center of Texas, a hurricane-type wind, at a speed of 135 km h^{-1}, caused destruction with damage of US$9 million. On 8 August, in Florida, a tornado of F1 magnitude (on the Fujita scale) with a wind speed of 160 km h^{-1} caused damage estimated at greater than US$20 million. Similar damage was caused by thunderstorms with downpours and tornados taking place on 19 August in Las Vegas and on 23 September in Pennsylvania. In late 2003, an earthquake of magnitude M 6.5 in Carolina led to losses estimated at US$50 million. On the whole, in 2003 in the USA, 107 cases of natural extreme events were recorded. As a percentage, they were distributed as follows: tropical cyclones – 30%, thunderstorms – 28%, floods – 18%, winter snowstorms – 17%, earthquakes and volcanic eruptions – 7%.

The year 2004 in the USA was characterized by a number of hurricanes (Charley, Frances, Ivan, and Jeanne) that were destructive for many other countries of the Caribbean Basin. The year 2005 was, for the USA, a year of catastrophic hurricanes. Many more than 180,000 people were killed throughout the world as a result of natural catastrophes in 2004. Similar statistics are available for analysis only in developed countries, and therefore it is this information gap for undeveloped countries that prevents us from developing an efficient technology to warn about catastrophic natural events (Dole, 2005).

The Indian Ocean tsunami raises fundamental questions about the mysteries of nature, life, and death. The Indian Plate moved below the Burmese Plate with extraordinary force which caused one of the most powerful earthquakes ever recorded. The movement of tectonic plates is a completely natural process, the control of which is too difficult for present-day science. The paradox of this situation is the apparent helplessness of present-day science when such a catastrophe has arisen. Here it is appropriate to give a comment from *Guardian Weekly* by Martin Kettle (Schröter, 2005): "From at least the time of Aristotle, intelligent people have struggled to make some sense of earthquakes. Earthquakes do not merely kill and destroy. They challenge human beings to explain the world order in which such apparently indiscriminate acts can occur. Europe in the 18th century had the intellectual curiosity and independence to ask and answer such questions. But can we say the same of 21st-century Europe? Or are we too cowed now to even ask if God can exist that can do such things?"

Estimates of vulnerability of a community (social group) or landscape (ecosystem) are possible with an introduction of some indicator scale. Here, various approaches are possible, one of which is the calculation of sensitivity of NSS elements to global changes. Such estimates can be obtained with the use of the respective model. For example, Krapivin and Kondratyev (2002) estimated the consequences of the greenhouse effect; Vogel and O'Brien (2004) described the index of human insecurity as well as the index of agricultural vulnerability on the territory of India with respect to climate change (O'Brien *et al.*, 2004). One should also note the fact that the notion of vulnerability includes the available means of subsistence for the population of a given territory suffering from natural disaster. This aspect is important in order to assess the vulnerability, since it reflects human behavior under conditions of overcoming the consequences of the disaster and represents the internal mechanisms of correlation between the social and physical factors of the formation of the vulnerability index. Another measurement of vulnerability concerns the state of the household and the presence of food. Here the vulnerability is classified by four levels: weak, moderate, high, and extreme. The index of the household state is related to socio-economic groups with selection of different information categories: information about demography, the level of agriculture development, precipitation, and state of market. This enabled Bohle (2001) to construct a 2-level structure of vulnerability that reflected the interaction between political, economic, and ecological factors, assessing the internal and external aspects of the crisis and conflict of a territory under the conditions of natural cataclysms.

Of course, the territories of developing countries are most vulnerable, since the losses due to natural disasters in poor countries, apart from direct material damage and human victims, break their economic structure and limit the rate of their progress towards sustainable development. The high vulnerability of the developing countries is also explained by a high risk of heavy losses from natural disasters, since the houses of poor people are not protected from natural disasters. Besides this, a developing country needs more time to liquidate the consequences of a natural disaster.

The notion of vulnerability can be used to assess seven categories of threats which form human safety:

- Economic safety (secure economic development) – vulnerability with respect to global economic changes.
- Food safety (physical, economic, and social availability of food) – vulnerability with respect to extreme events, agricultural changes, etc.
- Health safety (relative freedom from diseases and infections) – vulnerability with respect to health.
- Environmental safety (access to preservation of sanitary norms of water, clean air, and non-degraded surface systems) – vulnerability with respect to pollution and degradation of land.
- Personal safety (guarantee against physical violence and threat) – vulnerability with respect to conflicts, natural cataclysms, impending "misfortunes" similar in type to AIDS.
- Community safety (guarantee of cultural integration) – vulnerability with respect to cultural globalization.
- Political safety (preservation of human rights and freedoms) – vulnerability in conflicts and war problems, both for scientists and politicians.

In connection with this classification, three main questions arise:

(1) Which technologies can be suggested by present-day science for a better

understanding and assessment of the present-day (complicated) reality in which the Earth's population exists?
(2) Can the present-day science of making political decisions suggest technologies for more realistic assessments of modern life and thus reduce vulnerability?
(3) Are there means of vulnerability conceptualization from the viewpoint of benefits and losses under conditions of forthcoming global changes?

Clearly, answers to these questions can be sought in different situations, from an individual to the whole NSS. Vulnerability will be different in developed and developing countries, in cities and the countryside, in a desert region and in mountainous regions. Ecoinformatics deals with the development of respective methods of vulnerability assessment (Kondratyev *et al.*, 2002a; Krapivin and Potapov, 2002; Dilley *et al.*, 2005). In connection with the necessity to assess the danger for a population of a given territory to suffer damage to health, constructions, or property due to changes of the environmental parameters, the term "ecological safety" is used. These changes can be caused by both natural and anthropogenic factors. In the first case, the danger results from fluctuations in natural processes connected with a change of the synoptic situation, epidemics, or a natural disaster. In the latter case, the danger happens as a response of nature to human activity (Kondratyev, 1991).

It is evident that the problem of survivability arises when environmental fluctuations are irregular. In this case living organisms do not have the possibility to adapt to a fluctuating environment. Kussell and Leibler (2005) developed the model of survivability of a population under a fluctuating environment. This model parameterizes a population's behavior to survive. It is an example of how necessary it is to develop a demographical block of global biospheric models of different scales.

In general, a threat of ecological danger on a given territory results from a deviation of the environmental parameters beyond limits where after a long stay a living organism starts changing in a direction not corresponding to the natural process of evolution. As a matter of fact, the notions of "ecological danger" or "ecological safety" are connected with the notions of stability, vitality, and integrity of the biosphere and its elements (Kondratyev, 1990, 1991). Moreover, the NSS, being a self-organizing and self-structuring system and developing by the laws of evolution, creates within itself ecological niches whose degree of acceptability for a population of a given territory is determined, as a rule, by natural criteria (a totality of pollution criteria, religious dogmas, national traditions, etc.).

Nevertheless, when considering the prospects for life on Earth, one should proceed from the human criteria of assessing the levels of environmental degradation, because in due course, local and regional changes in the environment are developing into global ones. The amplitudes of these changes are determined by the mechanisms of NSS functioning, which provide optimal changes in its elements (Gorshkov *et al.*, 2000, 2002). Humankind deviates more and more from this optimality in its strategy of interaction with the surrounding inert, abiotic, and biotic components of the environment. But at the same time, humankind as an NSS element tries to understand the character of large-scale relationships with nature, directing the efforts of many sciences to this aim and studying the cause-and-effect relationships in this system. Since the structure of human society is divided into countries, the socio-economic component of the NSS is identified with a country.

The national safety of any country under present conditions should be assessed based on numerous criteria of military, economic, ecological, and social character. The development of an efficient technique for an objective analysis of the problem of national safety requires the use of the latest methods of collection and processing of data on various aspects of the global system's functioning. Such methods have been provided by geoinformation monitoring system (GIMS) technology (Kondratyev *et al.*, 2004a). One of the aspects of national safety is the protection against the rapidly growing number of natural disasters, which, under present-day conditions, can be sources of large-scale social upheavals and, eventually, a destabilizing factor for sustainable development. For instance, in Russia alone, an increase in the number of natural disasters during the last decade constituted 27.3%.

Consider the economic–ecological aspect of national safety. From the viewpoint of system analysis, any country can be considered as an object of system analysis functioning in the space of other complicated systems. The interaction of these systems is connected with the controled and non-controled exchange of elements of economic and ecological categories. A problem that appears is that of the search of optimal strategies for each of the interacting systems. It is necessary to take into account the heterogeneous scientific–technical level of these systems and, hence, the different approaches to the choice of criteria assessing national safety. The GIMS technology proposes the following solution to such problems.

The global NSS model is being developed. This model describes the main processes in the NSS with their discretization in space and time. The model is based on the available data and information space. It is inserted into a single national system of ecological monitoring of a territory of a country, and is combined with similar global and national systems, interaction with which is possible. As a result of the combination of the NSS model, the system of collection of data for environmental and economic parameters of the regions of the country, and the system of computer cartography and informatics, a single national system of observation and control for economic–ecological safety has been synthesized. This system has a hierarchical structure of information channels with a respective hierarchy of problems to be resolved. In particular, it can provide the operational information to regions about the state of the ecology and economy at any spot on the globe. The system ensures the information about:

- the current global changes in the environment;
- expected climate changes, and the role of existing or planned changes in the environment of a country in the changes of the climate and the biosphere, in various regions;
- the state of the atmosphere, hydrosphere, and soil–plant formations in the territory of a country;
- availability of data on ecological, climatic, economic, and demographic parameters of any region;

- the level of ecological safety on a given territory;
- the appearance of events that are dangerous to humans and the environment;
- trends in the changes of the states of forests, marshes, pastures, agricultural crops, river and lake systems, and other natural complexes; and
- the risk that any measures may change the environment.

On the national level, the system can solve the following problems:

- long-term and timely planning and control of economic activity with due regard to its ecological expediency and development of a strategy of rational nature use;
- operational notification and warning about processes taking place beyond and within the territory of a country, which can aggravate the ecological situation and cause long-term changes to the environment with an increasing risk to the health of a population of individual regions;
- assessment of the consequences of a realization of anthropogenic projects for a country and other global territories; and
- working out immediate measures to liquidate the causes of the appearance of ecological catastrophes and natural disasters.

The development and realization of an efficient technology to assess ecological safety on global scales will become possible after the organization of the International Centre for Global Geoinformation Monitoring (ICGGM). The creation of the ICGGM will make it possible to identify the mechanisms of co-evolution of man and nature. The main mechanism to achieve this will be new technologies for data processing, based on the progress of evolutionary informatics and global modeling. As a matter of fact, it is a question of realization of an approach developed by several authors (Bukatova *et al.*, 1991; Nitu *et al.*, 2004) to model the processes under conditions of incomplete a priori information about their parameters and the presence of principally unavoidable information gaps.

According to the scheme in Figure 2.1, the ICGGM is characterized by a set of the following functions:

- collection of information from national monitoring systems and international centers of environmental studies;
- sorting out, primary processing and accumulation of data on natural processes;
- formation of a base of knowledge of the processes operating in the environment;
- simulation, numerical and physical modeling of climatic, biospheric, cosmic, social, and economic processes;
- forecast of the environmental state and formation of the constantly renewed bank of scenarios of anthropogenic activity;
- attending to inquiries from national and international environmental protection agencies; and
- issuing recommendations to national and international centers of environmental monitoring.

Indicators characterizing the GIMS as the base subsystem of the ICGGM are

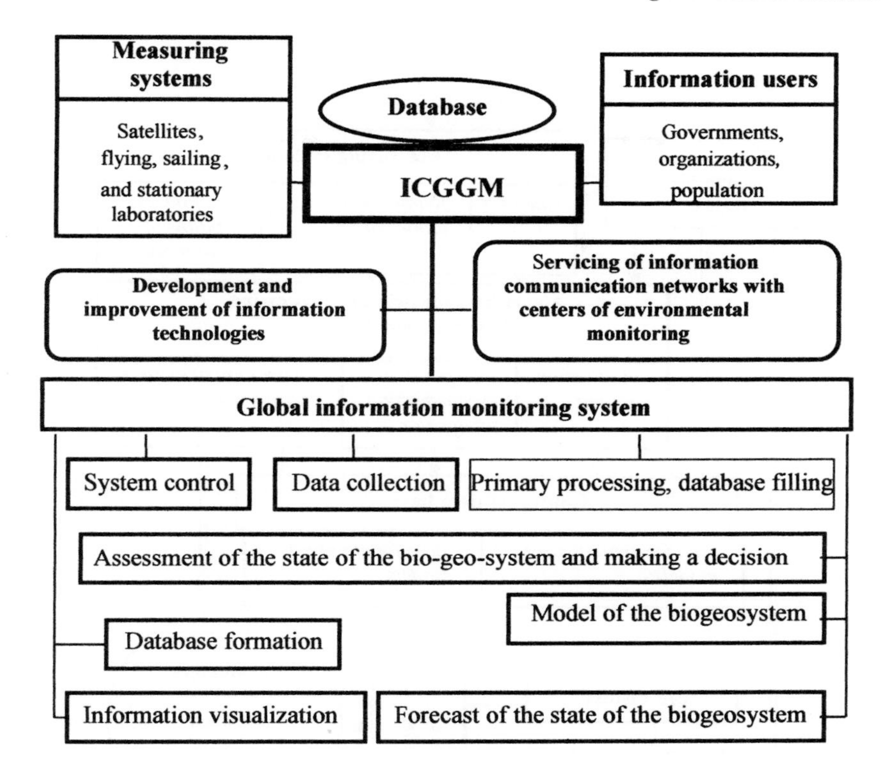

Figure 2.1. The concept of a single center of global geoinformation monitoring.

grouped by the thematic principles of organization of its structure. They are specified in the process of GIMS exploitation, and cover the key characteristics of global topography, synoptic situations in energy active zones, the content of dangerous atmospheric pollutants in characteristic latitudinal belts, and information about catastrophes. An input to GIMS is a multitude of spatially irregular and temporally fragmentary data of measurements of geophysical, geochemical, ecological, biogeocenotic, and synoptic characteristics. *In situ* and remote-sensing measurements are made using devices with different degrees of accuracy. An agreement of obtained measurements with other GIMS units is accomplished by algorithmic procedures from primary data processing. The volume of these data will be reduced in the process of a GIMS functioning. The GIMS input also foresees a possibility to receive signals from scenarios of anthropogenic development from situations under study.

The GIMS model is shown as a conceptual scheme in Figure 2.2. The correlation of input and output parameters is accomplished through a composition of information fluxes mentioned here. The GIMS functions in the adaptive regime, and the final result of the system affects the input characteristics of the measuring unit. The mathematical software of the adaptive GIMS unit is shown in Figure 2.3. Here all biogeochemical and biogeocenotic processes are described by systems of balanced

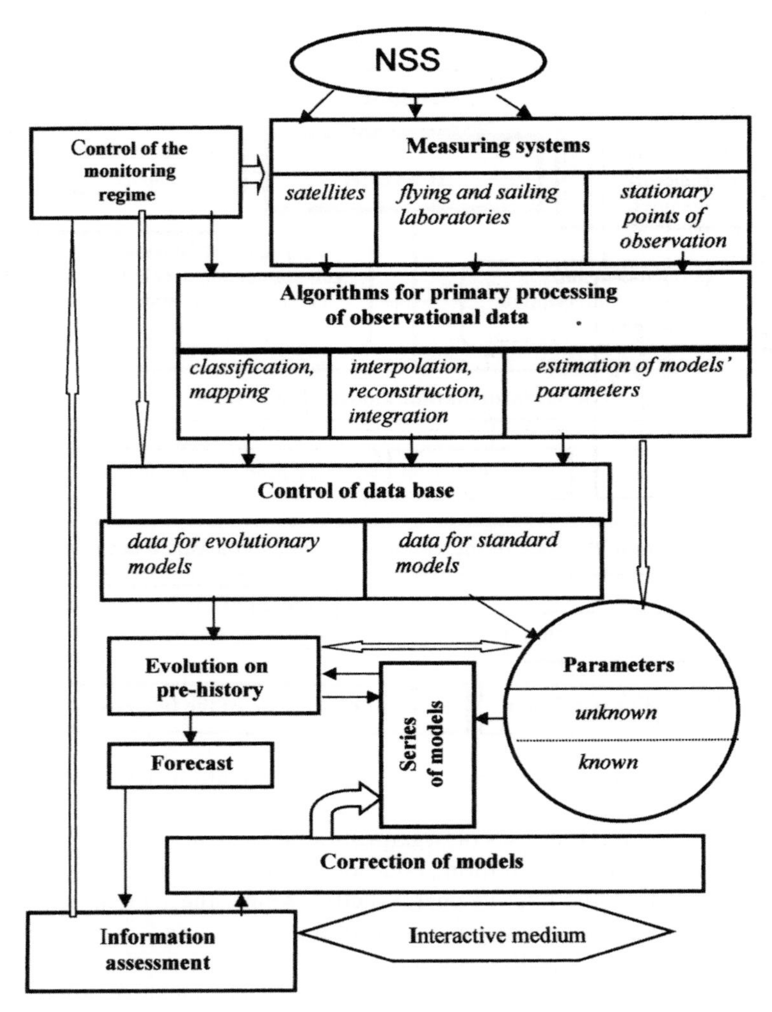

Figure 2.2. The concept of adaptive adjustment of the GMNSS in geoinformation monitoring: GIMS-technology.

equations. However, a considerable part of poorly parameterized processes are described with the use of the method of evolutionary modeling oriented toward the unformulated parameterization of strongly non-stationary processes.

The socio-economic structure of the globe can be divided into m levels. According to Kondratyev *et al.* (1994, 1997), $m \geq 9$, and this structure has three main levels of regional development: developed, developing, and underdeveloped. The realization of the ICGGM project will accelerate the process of leveling of this structure, since an optimization of planning an organized structure of human society will be accelerated, and the purposeful direction of global processes will be provided for the public benefit and without damage to nature, and above all, will favor the

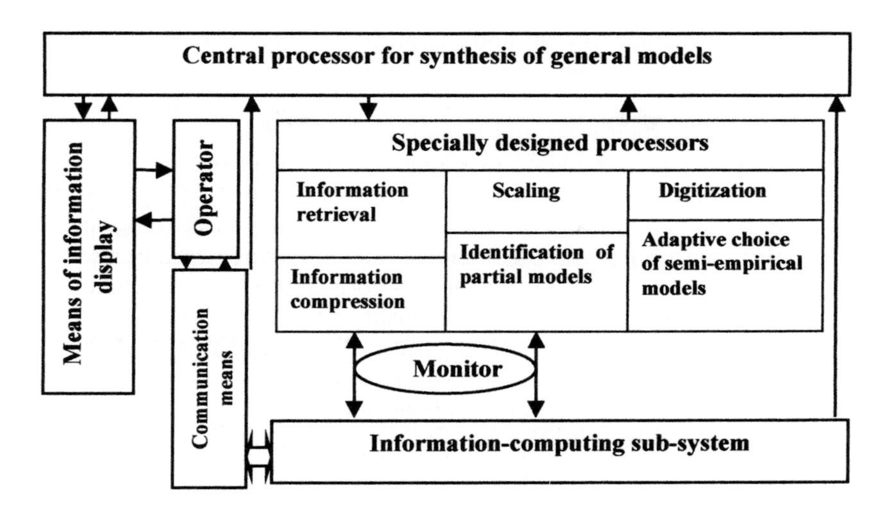

Figure 2.3. The principal structure of information fluxes in the system of geoinformation-monitoring data processing with the use of evolutional modeling technology (Bukatova *et al.*, 1991).

creation of international mechanisms of the coordinated behavior of the global population with respect to nature use. Humankind will profit from the ICGGM in a sense that finances will not be spent in vain on the accomplishment of ecologically unacceptable projects – the equilibrium with nature will be preserved. With progress in science and engineering, this profit will grow, since the transition to new kinds of resources raises no doubts (Gorshkov, 1995).

Let us formulate the problem of a numerical assessment of the ecological safety of a country using the complex systems theory and vitality theory. The national system of the country A, interacts with other similar systems but with different spatial locations. For simplicity, all other systems will be denoted as B. In other words, all other countries will be identified in a first approximation with a single system B. Further, this situation can be complicated by considering many systems with which system A interacts.

The systems A and B have aims, structures, and behaviors (strategies). The aim $\underline{A}(\underline{B})$ of the system $A(B)$ is to reach certain preferable states. These aims can be of diverse hierarchical character. The parametric presentation of the aim is one of the important problems. As a possible suggestion, consider the following components of the system A: $\underline{A}_q$ – integral indicator Q of the quality of the environment for the whole territory of a country should not be below the threshold q; $\underline{A}_L$ – the maximum permissible concentrations $L(i,j)$ $(j = \overline{1,m_i})$ of substances should not be violated at the jth part of the territory of a country in the domain i ($i = 1$ – soil; $i = 2$ – water; $i = 3$ – atmosphere); $\underline{A}_e$ – the economic potential of the country over time Δt should increase by $s\%$.

The aim $\underline{B}$ of system B can refer to A as antagonistic, neutral, or cooperative. This relationship is determined by the type of criterial functions for both sides. An expediency of the structure $|A|(|B|)$ and the purposefulness of behavior $\overline{A}(\overline{B})$ of the system $A(B)$ is estimated by the efficiency with which the system reaches its aim.

The behavior of the systems can either favor or be neutral, or prevent the systems from reaching their aims and the aims of another system. In the first case, a pair of systems can be considered as a single system with the aim in common of interacting with other systems. In other cases it is a question of systems' relationships. In general, the spectrum of interaction of the natural–anthropogenic systems, defined as either quasi-homogeneous regions, individual countries, or their groups, includes a number of factors among which Lomborg (2005) points to climate change, infectious diseases, conflicts, education, financial instability, corruption, migration, inferior food, starvation, trading barriers, and inclination to war. Unfortunately, a formalized consideration of these factors is still difficult.

Since the systems denote national ecological systems, it is natural to introduce some statements with respect to the ways in which they interact. Such systems are open, and their interaction can be presented in the form of exchange with some resources (finances, technical means, natural resources, etc.). It can be formalized by introducing some resource V, spent by the system and the resource W, consumed by the system. As a result, a (V, W)-exchange takes place between the interacting systems. It is clear that each system wants to make this exchange profitable for itself. Hence, there is a possibility of further formalization of the systems' functioning. In other words, let us believe that the aim of each system is the most profitable (V, W)-exchange (i.e., each system, for a minimum of V, tries to get a maximum of W, which is the function of the structure and behavior of the interacting systems):

$$W = W(V, |A|, |B|, \overline{A}, \overline{B}) = W(A, B) \tag{2.1}$$

As a result, the interaction of systems A and B is numerically reduced to the following relationships (models):

$$\underline{W}_a = W_a(V_a, A_0, B_0) = \max_{\{\overline{A},|A|\}} \min_{\{\overline{B},|B|\}} W(V_a, A, B)$$

$$\underline{W}_b = W_b(V_b, A_0, B_0) = \max_{\{\overline{B},|B|\}} \min_{\{\overline{A},|A|\}} W(V_b, A, B)$$

where A_0 and B_0 are optimal systems.

It is seen from these relationships that to determine its aim, each system should decide which is important for itself: to obtain the most profitable (V, W)-exchange or to prevent another system from doing this. In this case, the systems can vary the values of (V, W)-exchanges within some limits $\underline{W}_1 \leq \underline{W}_a \leq \overline{W}_1$, $\underline{W}_2 \leq \underline{W}_b \leq \overline{W}_2$, where $\underline{W}_1$ and $\underline{W}_2$ correspond to maximum aggressive states of the systems, and $\overline{W}_1$ and $\overline{W}_2$ – to the most cautious ones.

With the aims of the systems known, we have a clear situation. But if each system, or one of them, conceals its intentions, we have a game situation with respect to the choice of the aim. Denote as $\underline{A}_i$ and $\underline{B}_j$ $(i = 1, \dots, n; j = 1, \dots, m)$ the sets of aims of the systems A and B, respectively. The aims $\underline{A}_1$ and $\underline{B}_1$ consist in causing maximum damage to another system (maximum aggressiveness), and the aims $\underline{A}_n$ and $\underline{B}_m$ correspond to extreme cautiousness of both systems (maximum favor). All the other aims are scaled in $\{i\}$ and $\{j\}$ in the order of transition from $\underline{A}_1(\underline{B}_1)$ to $\underline{A}_n(\underline{B}_m)$, including the aims A_q, A_L, and A_e. Assuming that in the situation $\{\underline{A}_i, \underline{B}_j\}$ the systems get profits $\underline{W}_a = a_{ij}$ and $\underline{W}_b = b_{ij}$, we obtain a bi-matrix game to determine an optimal aim with pay-off matrices $\|a_{ij}\|$ and $\|b_{ij}\|$. In a special case, at $\underline{W}_a = -\underline{W}_b$ the game becomes antagonistic.

Note that in a general case, the situation should be studied in a probabilistic space (i.e., some probability $P(V, W)$ of reaching its aim by each system). Moreover, it is necessary to consider various manifestations in the systems' behavior: reliability, information content, controlability, learning capability. The systems' elements should have different functions and purposes: protective, vital.

In addition to equations of the (V, W)-exchange, dynamic relationships should be considered which describe the temporal dependence of the systems' parameters. In this case, mathematically the problem of assessing the level of ecological safety of the territory of a country is reduced to a differential game.

On a national level, as has been mentioned above, the criteria are numerous. The State should observe certain sanitary/hygienic and ecological norms in given climatic situations. These situations should be predicted and serve as initial conditions for the system of estimation of ecological safety. The environmental quality is a complex function of: temperature T; wind speed U; total content of heavy metals in water E; air D; soil G; the content of gas of the kth type ($k = 1, \dots, N$) in the atmosphere S_k; biomass of vegetation cover M; and other parameters, $Q = Q(T, M, U, E, D, G, S_1, \dots, S_N)$.

There is a similar situation with prescribed functional presentation of dependences $L(i,j)$ and other environmental characteristics on natural and anthropogenic parameters. Moreover, many of these parameters can be presented as functions of the investment policy of a country. For instance, investments are made into the struggle against pollution, agricultural development, road building, development of new technologies, etc. These parameters are vital for indicators of the environmental quality, and the problem will be reduced to a search for an optimal investment policy. A totality of models of the dynamics of the environmental parameters and optimization relationships mentioned above, determine the problem of synthesis of national policy in the field of nature-protection activities with due regard to respective policies of bordering countries and the whole global community. Of course, the development of a regional strategy to prevent losses from natural disasters should take into account their statistics with the distribution of respective losses. Table 2.1 exemplifies one possible registration of the results of multiyear monitoring of natural disasters.

With due regard to this, the first-priority problem consists in concretization of goal functions and their dependences on parameters, taking into account both internal and external national strategies in the field of ecological monitoring. Finally, the numerical problem is reduced to a boundary value problem for the system of differential equations of parabolic type. The system of equations describes the dynamics of pollution over the territory of a country and the

Table 2.1. Natural disasters and distribution of their consequences for the period 1980–2004. During this period about one million people perished.

Natural disaster	Share of all events (%)	Distribution of deaths (%)	Distribution of economic losses (%)	Distribution of insurance payments (%)
Volcanic activity	4	2	0.1	0
Hail	0.5	4	5	4.1
Heatwaves, droughts	4	4	7	1.2
Earthquakes	27	13	23.8	11
Winter blizzards, snowstorms	1	4	6	17
Avalanches	0.5	2	0.2	0
Forest fires	0.5	5	2.9	2.3
Floods	24	28	30.7	10
Landslides	2	3.9	0.1	0
Severe frosts	1	3	3	5
Tornados	0.3	4	2	4.2
Tropical hurricanes	11	8	16	28
Hurricanes	1.7	18	2	17
Tsunamis	0.5	0.6	0.2	0.1
Storm surges	22	0.5	1	0

boundary value conditions will be determined with due regard to the behavioral strategies of adjacent territories. Solutions of the boundary value problem will be introduced into equations of the (V, W)-exchange which will finally determine ecological safety.

It should be noted that the above mentioned approach to the development of technology for assessing the danger of natural or anthropogenic disasters contain many free elements requiring specification or further development. In particular, as shown by Kondratyev *et al.* (2004b) and Posner (2004), the problem of danger assessment is interdisciplinary, and its solution requires joint efforts from scientific experts, lawyers, economists, psychologists, and sociologists. In principle, at present there is a necessity for a broader understanding of the term "disaster".

Many scientists have recently paid attention to ultimate cases when the danger can threaten most or even the whole of humankind. It is a question of large-scale catastrophes, such as the collision between Earth and a large asteroid, irreversible global climate change, propagation of lethal viruses, nuclear accidents, etc. For instance, Posner (2004), discussing the potential danger for humankind, speaks about the possible transformation of the Earth into a super-dense dwarf with a diameter of 100 m, or its being populated by super-intelligent nano-robots, after which humankind looses control of its own development. Unfortunately, as a rule, such issues are considered alarming, fantastic, or as an object of scientific fiction. However, numerous potential events leading to the extinction of humankind surely contain elements of the possible. It is important that both the psychological and cultural perception of these elements by the public and politicians correspond to a level of certain rationality and understanding.

The problem of sustainable development and human survivability is solved by many countries without consideration of comparative data that characterizes regional correlations between negative anthropogenic impacts on the environment and the possibility of a recovery by nature. Tarko (2005) gives the obvious example of unequal difference between industrial emissions of CO_2 and its sinks for different countries. Russia, Canada, and Brazil have small discrepancies in these CO_2 fluxes but the USA, China, and Japan double the current problem of the greenhouse effect. Tarko says that the survivability problem is to be solved in the globe with optimization of national (V, W)-exchanges.

2.2 THE MODEL OF ECOLOGICAL SYSTEM SURVIVABILITY

The efficiency and stability of complex systems under certain conditions of functioning are evaluated using the theory of potential efficiency of complex systems developed by Fleishman (2003). The temporal stability of the complex system is connected with stability of structure, composition and energy balance of this system, as well as with the stability of its response to the same forcings. The system's stability can be violated due to physical and moral ageing of its elements or by environmental forcings. The impacts of the environment can be passive and active. Therefore, apart from natural ageing, the system's elements can be harmfully affected by some active agents of the environment, such as disasters. These problems are the subject of the theory of survival (Krapivin, 1978), which is a section within the theory of potential efficiency.

The uncertainty and complexity of the interaction of NSS components in extreme situations can be an object of analysis of the theory of survival, since human survival in a catastrophe depends on taking preventive measures to create protective means, as well as considering the possible impact on the environment, to warn about or even prevent a disaster. In this case, the model of survival should take into account the uncertainty of environmental conditions, the variety and instability of the "goals" of interacting processes, and a multitude of behavioral strategies. In the synthesis of the survival model the survival theory recommends performing a conditional division of interacting processes into two sub-sets with limited resources for protection. One set includes elements which determine the structure of the society and its goals. The other set can be conditionally defined as an external medium (i.e., a totality of phenomena taking place in living and non-living nature, and factors capable of violating the normal functioning of the elements of the first set and, maybe, even breaking them down). The system's survivability means its capability to actively resist the external forcing, to preserve its characteristics, with due regard to probable states of the system, in which it goes on functioning, as well as to ensure its functioning under certain conditions of exploitation.

The notion of survival makes it possible to broaden the notion of reliability, since apart from natural ageing of the system's elements, it takes into account the possibility of their malfunctioning as a result of harmful impacts of some active agents in the environment. Such agents of the environment can be natural disasters whose origins are connected with a change of energy processes and re-distribution of matter within some space. Following model (2.1), consider the system N (nature), which interacts with another system H (humankind). Each system can change its state in time. The aim $\underline{N}(\underline{H})$ of the system $N(H)$ is to reach certain preferable conditions. An expediency of the structure $|N|(|H|)$ and purposeful behavior $\overline{N}(\overline{H})$ of the system $N(H)$ is evaluated by the efficiency with which the system reaches its aim. So, without dwelling upon the philosophical aspect of the formulation of the aim of the system N, the system $N(H)$, with its aim known, can be characterized by two factors $N = (|N|, \overline{N}), H = (|H|, \overline{H})$.

The systems' behaviors can either favor or prevent the systems from reaching their aims and the aims of a further system. In the first case, a pair of systems can be considered a single system with a common aim, interacting with the environment. In the second case, the situation is a conflict. If it is neither the first nor the second case, the situation is indifferent, and only in this case will one system be a medium with respect to the other. In the latter case, the medium can occasionally disturb the system. This forcing is usually called noise.

The systems N and H are open systems. Similarly to (2.1), their interaction is described by equations of (V, W)-exchange. In the real situation, the interaction of the systems N and H during the (V, W)-exchange is of stochastic character, and therefore we can only speak about some probability $P(V, W)$ for each system to reach its aim. The value of $P(V, W)$ is the indicator of the system's efficiency. Maximum value of this probability is defined as a threshold efficiency. It is clear that in developed countries this threshold is higher than in poor countries.

The aim of each system can be presented as a totality of four sections: R – reliability (stability), I – information content (knowledge of situation), C – controllability (external activity), and L – self-organization (learning, in particular). In connection with this, the integral system Ξ can be considered as a totality of some systems – "sections" $\Xi = (\Xi_R, \Xi_I, \Xi_C, \Xi_L)$, each being connected only with the respective component of the aim $\underline{\Xi}$. All elements of both systems N and H are divided into three classes: working (vital) n and h elements; protective R_n and R_h elements, and c_n and c_h elements, to affect the environment. The energetic capabilities of the systems are limited by supplies of vital "substrates" $\bar{V}_n = \{V_{nj}, j = 1, \ldots, n_n\}$ and $\overline{V}_h = \{V_{hj}, j = 1, \ldots, n_h\}$, which are spent on the reproduction of vitally important elements in such a way that from $V_{nj}(V_{hj})$ it is possible to produce $N_j(H_j)n_j(h_j)$ – elements of the jth type of values $n_j(h_j)$. The protective and active elements of each system are reproduced (generated) by vital elements. First of all, protective $E^n_{Rm}(E^h_{Rm})$ and active $E^n_{cm}(E^h_{cm})$ "substrates" are created from which R

and c elements of the mth type are generated, with the following limitations:

$$\left.\begin{aligned} E_{Rm}^{n} &= E_{Rm}^{n}(\overline{V}_n, N_1, \ldots, N_{n_n}) = \sum_{j=1}^{n_n} w_{mj}^{n} f_{jR}^{n}(V_{nj}, N_j) \\ E_{Rm}^{h} &= E_{Rm}^{h}(\overline{V}_h, H_1, \ldots, H_{n_h}) = \sum_{j=1}^{n_h} w_{mj}^{h} f_{jR}^{h}(V_{hj}, H_j) \\ E_{cm}^{n} &= E_{cm}^{n}(\overline{V}_n, N_1, \ldots, N_{n_n}) = \sum_{j=1}^{n_n} w_{mj}'^{n} f_{jc}^{n}(V_{nj}, N_j) \\ E_{cm}^{h} &= E_{cm}^{h}(\overline{V}_h, H_1, \ldots, H_{n_h}) = \sum_{j=1}^{n_h} w_{mj}'^{h} f_{jc}^{h}(V_{hj}, H_j) \end{aligned}\right\} \tag{2.2}$$

where $w_{mj}^{n(h)}, w_{mj}'^{n(h)}, f^{n(h)}$ are some prescribed weights and functions, respectively.

Energy supplies (2.2) are spent to form the respective R and c elements. Initial resources $\overline{V}_n = \{V_{nj}, j = 1, \ldots, n_n\}$ and $\overline{V}_h = \{V_{hj}, j = 1, \ldots, n_h\}$ due to their limited character, can reproduce either many low-efficient elements or a few highly efficient elements. Therefore, it is natural to suppose that the functions $f^{n(h)}$ at fixation of the first argument are decreasing functions of the second argument. The point is that "in integration is the force" (splitting the forces is not profitable). Hence, with $N_j(H_j)$ the amount of protective $E_{Rm}^{n(h)}$ and active $E_{cm}^{n(h)}$ energy supplies decreases, and it means that the amount of respective protective and active elements of fixed efficiency decreases or, by the same amount, the efficiency is reduced. Hence, with a fixed amount of vital substrates there are certain critical relationships between the amounts of all the elements, which preserve the survivability of the system at a maximum. For instance, in the USA, to maintain the 3-level monitoring system to warn of hurricanes, billions of dollars are spent annually. It is clear that putting additional resources into this system from the countries of South East Asia would make it possible to broaden the range of its action and to raise its efficiency. The absence of this broadening led to the tragedy on 26 December 2004 caused by a powerful tsunami. Unsolved problems of understanding of the mechanisms for environmental regulation are connected with formalization in the form of the described model of elements of the system N (Kondratyev *et al.*, 2003a; Ivanov-Rostovtsev *et al.*, 2001).

Assume that at the moment of the initiation of interaction, the systems N and H have:

(1) N_j and H_j vital elements of the type $j = 1, \ldots, n_n(n_h)$ and values a_j and b_j, respectively, with:

$$\sum_{j=1}^{n_n} a_j N_j = U_n(0), \sum_{j=1}^{n_h} b_j H_j = U_h(0);$$

(2) r_n and r_h types of protective elements in α_m and β_m in the mth type, with:

$$\sum_{m=1}^{r_n} \alpha_m = U_{R_a}(0), \sum_{m=1}^{r_h} \beta_m = U_{R_h}$$

(3) s_n and s_h types of active elements in ν_m^n and ν_m^h in the mth type, with:

$$\sum_{m=1}^{s_n} \nu_m^n = D_n(0), \sum_{m=1}^{s_h} \nu_m^h = D_h(0)$$

respectively.

The process of substrate formation in real systems is always connected with the hierarchical character of their structure. The form of this transformation is determined by the system's characteristics. For instance, in ecological systems, the energy goes from one trophic level to another. Any impact on the trophic pyramid of the ecosystem can cause either its destruction or transition into another state. In general, the energy fluxes in nature are governed by the mechanisms of self-regulation of natural systems among which are natural disasters. Their role is manifested through the creation of centers of absorption or emission of energy in the form of earthquakes, evaporation, etc. People, through their activity (c elements), interfere with these natural processes, and nature itself affects the distribution of energy by humankind, often making people spend it on prevention or overcoming natural disasters. This interaction takes place during some time periods and consists in exchange with certain portions of c elements.

Assume that R_n and R_h elements of the mth type, when interacting with c_n and c_h elements of the nth type, respectively, have efficiencies d_{mn}^n and d_{mn}^h, and c_n and c_h elements of the mth type in interaction with ω-elements ($\omega = n, h, R, c$) of the nth type have efficiencies $\lambda_{m\omega n}^n$ and $\lambda_{\mathrm{m}\omega n}^h$, respectively. The systems N and H at each time moment t determine their behavior choosing a pair of numbers:

$$\overline{N}(t) = \{\mu_{\mathrm{m}\omega n}^n(t), \sigma_{m\omega n}^n\}, \overline{H}(t) = \{\mu_{m\omega n}^h(t), \sigma_{m\omega n}^h\}$$

where $\mu_{m\omega n}^{n(h)}(t)$ and $\sigma_{m\omega n}^{n(h)}(t)$ are portions of $R_n(R_h)$ and $c_n(c_h)$ elements of the mth type directed to protect and eliminate the ω elements of the nth type, respectively.

In due course, the portions of c_n and c_h elements fill in the systems H and N, respectively, and thus, as time goes by, the vital elements of the systems weaken, if they are not reproduced. It is assumed that one ω element of the nth type of the system $N(H)$ breaks down under the influence of one $c_h(c_n)$ element of the mth type with the probability $p_{m\omega n}^{h(n)}(\lambda_{m\omega n}^{h(n)}; \mu_{1\omega n}^{n(h)}, \ldots, \mu_{r_{n(h)}}^{n(h)}; d_{1m}^{n(h)}, \ldots, d_{r_{n(h)}m}^{n(h)})$.

To obtain a quantitative estimate of the character of interaction of the systems N and H, it is necessary to specify the meaning of (V, W)-exchanges, which both systems try to optimize. Assume that the system's survivability is determined by the presence in it of vital elements. In other words, the system $N(H)$ at a moment t functions normally, provided the following condition is met $Q_n(N_1, \ldots, N_{\mathrm{n}_n}, t) > Q_{n,\min}(t)$; $(Q_h(H_1, \ldots, H_{n_h}, t) > Q_{h,\min})$, where Q_n and Q_h are prescribed functionals of time and structures of the systems N and H, respectively.

In particular, we can assume:

$$Q_n = \sum_{j=1}^{n_n} a_j N_j(t); Q_{n,\min} = \sum_{j=1}^{n_n} \theta_{nj} a_j N_j(0)$$

$$Q_h = \sum_{j=1}^{n_h} b_j H_j(t); Q_{h,\min} = \sum_{j=1}^{n_h} \theta_{hj} b_j H_j(0)$$

where $0 \leq \theta_{nj}$, $\theta_{hj} \leq 1$ are parameters determined by special features of the systems and requirements made on their survivability. These parameters characterize the systems' vitality. The system $N(H)$ at a moment t functions normally if the $(1 - \theta_{nj})$ $((1 - \theta_{hj}))$th share of $n(h)$ elements of the jth type is functioning at this moment.

Processes of the (V, W)-exchange are interrelated. Therefore, the situation of interaction of the systems N and H can be considered as a general conflict, to describe which it is necessary to prescribe the pay-off function, whose form for each system depends on the degree of the conflict (i.e., it depends on how one system affects the goal achievement by another system, as well as on the state of both systems). To measure the degree of the conflict, some scale can be suggested provided the notion of a decision-maker's purposefulness for each system can be expressed in the form of some fixed succession of states preferable from the viewpoint of each system. The magnitude of losses of vital elements by both systems serves the basis for this succession. The significance of these losses in the time interval $[O, T]$ can be expressed as:

$$\Delta Q_n = \sum_{i=1}^{n_n} a_i [N_i(0) - N_i(T)] \qquad \Delta Q_h = \sum_{i=1}^{n_h} b_i [H_i(0) - H_i(T)]$$

for the systems N and H, respectively. Hence, the intensity of the conflict depends on $\Delta Q = \Delta Q_n + \Delta Q_h$.

2.3 A BIOSPHERE SURVIVABILITY MODEL

The NSS can be presented as a totality of nature N, and human society H (*Homo sapiens*), which constitute a single planetary system. Therefore, in the development of global or regional models, their division should be considered conditional. The systems N and H have hierarchical structures $|N|$ and $|H|$, goals $\underline{N}$ and $\underline{H}$, and behaviors $\overline{N}$ and $\overline{H}$, respectively. From the mathematical point of view, the interaction of the systems N and H can be described by a set of relationships (parameterizations) reflecting in a general case an accidental process $\eta(t)$ with an unknown law of distribution and consisting of a composition of partial processes of the interaction of these systems. Therefore, the goals and behaviors of the systems are functions of the indicator η. There are intervals in which the η varies, and in which the behaviors of the systems can be antagonistic, indifferent, or cooperative. Humankind is adapted to some formalized division of space into countries grouped on the principle of economic development, primarily.

A country H_i has m_i possible means to reach goal $\underline{H}_i$, in other words, it uses a set of strategies $\{\overline{H}_i^1, \ldots, \overline{H}_i^{m_i}\}$. The weight of each strategy $\overline{H}_i^j$ is prescribed by the parameter $p_{ij}(\sum_{j=1}^{m_i} p_{ij} = 1)$. The resulting value of the parameter η is a function of the characteristics mentioned above, and on the whole, at each moment it is described by a game theory model.

An objective assessment of the environmental dynamics $N = (N_1, N_2)$ is possible with certain assumptions using the models of the biosphere N_1 and climate N_2. Such models have been developed by many scientists, and an experience gained covers the point, regional, box, coupled, and spatial models (Marchuk and Kondratyev, 1992; Krapivin, 1993; Bartsev *et al.*, 2003; Degermendji and Bartsev, 2003). This experience enables one to synthesize a global model of new type covering the key connections between the hierarchical levels of natural and anthropogenic processes.

In general, the state of the systems H and N can be described with vectors $x_H(t) = \{x_H^1, \ldots, x_H^n\}$ and $x_N(t) = \{x_N^1, \ldots, x_N^m\}$, respectively. The combined trajectory of these systems in the $n + m$ dimensional space is described by the function $\eta(t) = F(x_H, x_N)$, the form of which is determined by solutions of the global model equations. The form of F is determined within the knowledge of the laws of co-evolution, and therefore here we have a wide field to investigate in different spheres of knowledge. The available estimates of F (Krapivin, 1996) indicate an interaction of the notions of "survivability" and "stability". It is evident that the dynamic system is "alive" in the time interval $(t_a\, t_b)$ if the phase coordinates determining it are within "permissible limits" $x_{H,\min}^i \le x_H^i \le x_{H,\max}^i; x_{N,\min}^j \le x_N^j \le x_{N,\max}^j$. Since the systems H and N have the biological bases and limited resources, one of these boundary conditions is unnecessary (i.e., for the components of vector $x = \{x_H, x_N\} = \{x_1, \ldots, x_{n+m}\}$ the condition $x_{\min} \le \eta = \sum_{i=1}^{n+m} x_i$ should be met). This simple scheme includes the requirements to preserve the total energy in the system and its variety of elements.

Of course, the notion of biospheric survivability is more capacious and informative. As many authors believe, in the system ecology, this notion means stability and integrity of the system, which is the ability of the biosphere to withstand an external forcing. In other words, the survivability of the biosphere is measured by its tendency to suppress strong oscillations in its structures and elements, restoring their equilibrium. Thus, biospheric survivability is its ability to actively resist the external forcings, to keep its characteristics for a long time with due regard to the probable states of its subsystems in which it is still functioning, and to continue its functioning in the case of deviations in these external forcings. The development of the so-called strategies of natural and anthropogenic components of the biosphere is connected with a search of life-securing technologies (self-protection technologies) able to help people in crisis situations to adapt to extreme environmental changes. These technologies are genetics, robotics, artificial intelligence, and nanotechnology. One of the strategies for humankind's survival under conditions of large-scale disasters is to create adapted-to-danger transport systems, engineering constructions, water resource protective systems, robotic mechanisms for rescue and reconstruction operations, etc. For instance, the existing transport highways in the zones of possible origins of tsunamis cannot ensure an operative evacuation of the population to secure regions. It is clear that for these purposes, in coastal regions, air, underground, or sub-marine means are needed to rescue people. Similarly, in the zones of extreme natural phenomena like floods, tsunamis, or earthquakes, it is necessary to construct systems to protect drinking water and food resources. An alternative solution to these problems is to build protective dams which can resist tsunamis, lava, and water fluxes, and thus protect people. With this aim in view, for instance, in 1998 in the USA the Federal Emergency Management Agency (FEMA) was organized with the task of developing protective means and constructions which would make it possible to avert extreme situations in the regions where these situations are highly probable. Clearly, it is high time for humankind to think about these problems. Civilization achieved over the last 5,000 years could be annihilated by an accidental meteoritic storm, the onset of glaciation, global-scale land flooding, or the southward turn of the Gulf Stream. All possible scenarios, even the almost improbable, should be studied during this century. Humankind cannot stop further progress, people must use the achievements of cosmonautics in order to start a new stage in the use of the Solar System's resources instead of the resources existing on our planet.

2.4 IMPACT OF NATURAL DISASTERS ON HABITATS

Natural cycles of various timescales are followed by unavoidable destructive forcings on the environment, which, in principle, can throw back the present-day civilization. Of course, present-day civilization's possesses and knowledge might successfully withstand such global natural disasters. But, nevertheless, there is a danger of collisions with large meteorites, powerful volcanic eruptions, and earthquakes, when a drastic climate change may occur with a resulting change to the environment, which will create conditions unfavorable for the existence of living organisms. Among numerous scenarios of the development of such situations in the future, one should study those which admit the possibility of people's survival, if they can find the way to resist possible threats.

The problem of humankind's survival is too complex to be resolved based only on a model. Here, both philosophic and medical aspects are also important, without the understanding of which it is impossible to determine the strategies and behavior of people. For many decades scientists have disputed the character of biospheric evolution, formulating specific features of individuals and their communities, as well defining of the notion of the biosphere. But still there are no constructive solutions to how the biosphere evolves. Along with biotic and physico-chemical processes, it is necessary here to take into account psychological, ethical, religious, linguistic, cultural, and political peculiarities of the population of a given region. In these spheres, the positions of people are often diametrically opposite, and hence, it is difficult to formulate the goal $\underline{H}$ for the system H. Some people want to see the world green and not destroyed by anthropogenic processes, but for others comfortable living conditions is a priority. The compromise between these aspects of the goal $\underline{H}$ is often impossible, and therefore the final choice is made by one of the sides regardless of the desire of the other. But natural disasters meddle with this choice, which can permit a compromise between the aspects of the goal $\underline{H}$.

Natural disasters have been a decisive factor in some periods of the history of life on Earth. Only in the present-day has humankind become one of these decisive factors, though he has not yet managed to overcome many troubles caused by natural disasters and/or eliminate their causes. The role of natural disasters in a change of the social structure and even in political life is not always understood.

Nature gives humans even less opportunities to solve the strategic problems of the improvement of life for all. It demands of them a creation of a planetary strategy of interaction with the environment. In other words, the global problem of NSS sustainable development becomes more and more urgent every year. At present, practically there are no unpopulated territories on the globe without cities and settlements, without energy systems, and without pollution emitted to the atmosphere and oceans. The growing need for natural resources causes concerns about their depletion and the possible annihilation of forests, the fact that land may be unfit for agriculture, and that water resources will be unusable. In this situation people may starve and die as a result of various epidemics. From this point of view, the present notion of a natural disaster changes its meaning.

Speaking about the habitat of living beings, it should be borne in mind that the environment constantly changes as a result of the matter–energy exchange between its components. As a result of these changes, the earlier connection between the elements of this system breaks, which can be the cause of the origin of natural disasters on territories where they have not previously happened. Therefore, the level of interaction between man and the natural–geographic domain has a historical character in many respects. It is clear that the historical aspect of nature–society interaction is strictly referenced geographically and it is connected with special features of both natural and socio-economic conditions of a given region. For instance, the Near Eastern countries are located in the zone of arid climate, and they are characterized by a social and cultural way of thinking that differs from Europe or America, which explains the different perception of extreme natural phenomena. For this region, limited water resources, a nomadic way of life, a desire to build irrigation systems, etc., are normal. As a result, an adaptability of the population of the countries of the Near and Middle East to environmental conditions is characterized by a preserved equilibrium between the possibilities of the environment and the needs of society, which has been achieved by establishing a certain social order.

Speaking about relationships between natural and anthropogenic processes under conditions of a probable appearance of an extreme environmental situation, it is necessary to take into account that a human, in contrast to an animal, is a unity of natural and social elements, and this means that his exchange with the environment is socially conditional and realized non-biologically. Therefore, the human habitat is determined by a strategy of expedient activity and perception is subjective in many respects. The main supply of Earth's resources crucial to the maintenance of life in its present form are limited, and therefore they are a common property and should not be changed by an individual region.

3

Biocomplexity as a predictor of natural disasters

3.1 THE NOTION OF BIOCOMPLEXITY FOR ECOLOGICAL SYSTEMS

A natural catastrophe considered by humans as a natural disaster plays the role of a stimulator of development for the system N. For instance, typhoons bringing heavy destruction to land, generate zones of upwelling in the ocean and create conditions for high productivity of phytoplankton due to the lifting of biogenic elements from depths to surface waters. As a result, the trophic pyramid of the zone of the typhoon's impact becomes complex, leading to the growth of biocomplexity of the ecosystem of this zone. According to Szathmáry and Griesemer (2003), the ratio between living and non-living substances determines the vitality of the natural system and answers the question of whether it is alive. The natural system accomplishes the transition between its extreme states due to its changing complexity, whose indicator can serve as a forerunner to a critical state. In particular, such transitions can be caused by climate change (Greenland *et al.*, 2003). The successful search for such indicators depends on our knowledge of the laws of the living world and its evolution. The basic concepts of ecology developed by Beeby and Brennan (2003) suggest the conclusion that living systems of any level respond to an approaching natural catastrophe.

The problem of interaction of various elements and processes in the global Nature–Society System (NSS) has recently attracted the attention of many scientists. Attempts to assess and forecast the dynamics of this interaction have been made by many different sciences. One of these attempts is the "Biocomplexity" program announced by the US National Science Foundation, which plans during 2001–2005 to study and understand the interaction between the dynamics of the complexity of biological, physical, and social systems and the tendencies of present environmental changes. Within this program, the complexity of the system interacting with the environment is connected with the phenomena appearing in cases of contact of a living system with the environment on a global scale.

Biocomplexity is a derivative of biological, physical, chemical, social, and behavioral interactions of the environmental subsystems, including living organisms and the global population. As a matter of fact, the notion of biocomplexity is closely connected with the laws of biospheric functioning as a unity of the forming ecosystems and natural–economic systems of various scales, from local to global. Therefore, to define biocomplexity and assess it, a combined formalized description is needed of biological, geochemical, geophysical, and anthropogenic factors, and processes taking place at a given level of the spatial–temporal hierarchy of scales.

Manifestation of biocomplexity is a characteristic indicator of all systems of the environment connected with life. The elements of this manifestation are studied within the theory of stability and survivability of ecosystems. It should be noted that the formation of biocomplexity includes indicators of the degree of mutual modification of the interacting systems, and it means that biocomplexity should be studied with due regard to both spatial and biological levels of organization. The difficulty of this problem is determined by complexity of behavior of an object under study, especially if the human factor is taken into account due to which the number of stress situations in the environment is constantly increasing.

Humankind has accumulated knowledge about environmental systems. Use of this knowledge to study the biocomplexity is possible within the synthesis of a global model which reflects the laws of interaction of the environmental elements and permits an evaluation of the efficiency of realization of scenarios of human society development based on the data of surface and satellite measurements. It is this problem that forms the basis of all questions posited by the "biocomplexity" program.

Studies of the processes of human–nature interaction are directed, as a rule, toward understanding and evaluation of the consequences of this interaction. Reliability and accuracy of such assessments depend on criteria taken as the basis in conclusions, examination, and recommendations. At present, there is no agreed upon technique to choose such criteria for lack of a single scientifically substantiated approach to the ecological criteria of economic impacts on the environment. The accuracy of the ecological examination of the existing and planned production systems, as well as the representativeness of the data of global geoinformation monitoring, depend on the choice of such criteria.

Processes taking place in the environment can be represented as a totality of interactions between its subsystems. Since a human is one of its elements, it is impossible, for example, to simply divide the environment into interacting biosphere and society: everything on Earth is correlated. The problem is to find mechanisms to describe relevant correlations and interdependences which would reliably reflect the environmental dynamics and answer the questions formulated in the "biocomplexity" program:

(1) How does the complexity of biological, physical, and social systems in the environment occur and change?

(2) What are the mechanisms for spontaneous development of many events in the environment?
(3) How do environmental systems with living components, including those created by humans, respond and adapt to stress situations?
(4) How does information, energy, and matter move within the environmental systems and through their levels of organization?
(5) Is it possible to forecast the adaptability of the system and to give prognostic estimates of its changes?
(6) How does humankind affect and respond to biocomplexity in natural systems?

Many other, also important and significant questions can be added to this enumeration. For instance, of what level of complexity should the space-borne observational systems be so that their information is complete enough to reliably assess the state of the environment, at least at the moment of receiving information? Also important is the question of an optimal location of the geoinformation monitoring means at different levels of its established organization. It is also important to understand which forerunners of natural disasters can be measured using existing space-borne systems.

Environmental biocomplexity is, to some extent, an indicator of the interaction between its systems. In this connection, the scale Ξ of biocomplexity can be introduced, which varies from conditions when all interactions in the environment cease (get broken), to a level when they correspond to the natural process of evolution. Thus, we obtain an integral indicator of the environmental state, on the whole, with bioavailability, biodiversity, and survivability taken into account. This indicator characterizes all the kinds of interactions between environmental components. So, for instance, in biological interactions connected with relations of the type "predator–prey" or "competition for energy resource", there is some minimal level of available food when it becomes practically inaccessible, and the consumer–producer interaction stops. The chemical and physical processes of interaction of environmental elements also depend on the sets of certain critical parameters.

All this emphasizes the fact that biocomplexity refers to categories difficult to measure and express quantitatively. However, we shall try here to give formalized quantitative evaluations. Proceeding to the quantitative numerical scale Ξ, we postulate that there are relations of the type $\Xi_1 < \Xi_2$, $\Xi_1 > \Xi_2$, or $\Xi_1 \equiv \Xi_2$. In other words, there is always a value of this scale ρ which determines the level of biocomplexity $\Xi \rightarrow \rho = f(\Xi)$, where f is some transformation of the notion of biocomplexity into a number.

3.2 BIOCOMPLEXITY INDICATORS

3.2.1 Determination of biocomplexity indicator

Let us try to find a satisfactory model which will transform the verbal definition of biocomplexity into notions and features that can be formalized and converted. With this aim in view, we select in the NSS m elements – subsystems of the lowest level, whose interaction is determined by binary matrix function $A = \|a_{ij}\|$, where: $a_{ij} = 0$, if elements i and j do not interact; $a_{ij} = 1$, if elements i and j interact. In a general case, the indicator a_{ij} can be interpreted as a level of interaction of elements i and j. Then any point $\xi \in \Xi$ is determined as the sum:

$$\xi = \sum_{i=1}^{m} \sum_{j>i}^{m} a_{ij}$$

This formula can be modified with the introduction of weight coefficients for all NSS elements. The character of these coefficients depend on the nature of the elements. Therefore, we select in the NSS two basic types of elements: living (plants included) and non-living elements. Living elements are characterized by density expressed in number of species per unit area (volume) or by biomass concentration. Vegetation is characterized by type and share of occupied area. Non-living elements are divided according to the level of their concentrations related to the area or volume of space they occupy. In a general case, to each element i some coefficient k_i is ascribed which corresponds to its significance. As a result, we obtain a specification for a calculated formula in transition from the notion of biocomplexity to the scale Ξ of its indicator:

$$\xi = \sum_{i=1}^{m} \sum_{j>i}^{m} k_j a_{ij} \tag{3.1}$$

It is clear that $\xi = \xi(\varphi, \lambda, t)$ where φ and λ are geographic latitude and longitude, respectively, and t is the current time. For some territory Ω the indicator of biocomplexity is determined as an average value:

$$\xi_\Omega(t) = (1/\sigma) \int_{(\varphi,\lambda)\in\Omega} \xi(\varphi, \lambda, t) d\varphi d\lambda$$

where σ is the area of the territory Ω.

Thus, the indicator $\xi_\Omega(t)$ plays a role of the integral index of NSS complexity, reflecting the individuality of its structure and behavior at each time moment t in a space Ω. In accordance with the laws of natural evolution, a decrease (increase) of the parameter ξ_Ω will follow an increase (reduction) of biodiversity and the ability of natural–anthropogenic systems to survive. Since a decrease of biodiversity breaks the closedness of biogeochemical cycles and leads to an increase of load on irreversible resources, the binary structure of the matrix A favors resource-depleting technologies, and the vector of energy exchange between the NSS subsystems shifts to a state when the level of its survivability drops.

The NSS consists of elements–subsystems B_i ($i = 1, \dots, m$), whose interaction forms in time depending on many factors. The NSS biocomplexity is the sum of the structural and dynamic complexities of its elements. In other words, the NSS biocomplexity is formed during the process of interaction of its parts $\{B_i\}$. In due course, the subsystems B_i can change their state and, hence, the topology of their bonds will change, also. The evolutionary mechanism for adapting of subsystems B_i

to the environment suggests the hypothesis that each subsystem B_i, independent of its structure, possesses the structure $B_{i,S}$, behavior $B_{i,B}$, and goal $B_{i,G}$. Thus, $B_i = \{B_{i,S}, B_{i,B}, B_{i,G}\}$. The goal $B_{i,G}$ of the subsystem B_i is to achieve a certain preferable state. The expedience of the structure $B_{i,S}$ and the purposefulness of behavior $B_{i,B}$ of the subsystem B_i is estimated by its efficiency at reaching the goal $B_{i,G}$.

This can be exemplified by the process of migration of nekton elements. Fish migrate toward a maximum gradient of nutritional ration with due regard to the possible limitations of water medium parameters (temperature, salinity, oxygen concentration, pollution, etc.). Hence, the goal $B_{i,G}$ of the elements of nekton is to enlarge their ration, and their behavior $B_{i,B}$ relates to the calculation of the trajectory of shift in the process of shoal formation, which for each kind of nekton element can be presented in terms of $B_{i,S}$.

Since the interaction of subsystems $\{B_i\}$ is connected with chemical and energy cycles, it is natural to suppose that each subsystem B_i organizes the geochemical and geophysical transformations of matter and energy in order to preserve its stability. The formalized approach to this process rests on the supposition that in the NSS structure an exchange takes place, between subsystems B_i, of some amount V of spent resources for some amount W of consumed resources ((V, W)-exchange). Natural disasters prevent a group of subsystems B_i of a certain region from accomplishing a profitable (V, W)-exchange during some time. In a general case, $W = W(V, B_i, \{B_k, k \in K\})$, where K is a multitude of the numbers of subsystems making contact with subsystem B_i. Denote $B_K = \{B_k, k \in K\}$. Then the interaction of the subsystem B_i with its environment B_K will result in the following (V, W)-exchanges:

$$W_{i,0} = \max_{B_i} \min_{B_K} W_i(V_i, B_i, B_K) = W_i(V_i, B_{i,opt}, B_{K,opt})$$

$$W_{K,0} = \max_{B_K} \min_{B_i} W_K(V_K, B_i, B_K) = W_K(V_K, B_{i,opt}, B_{K,opt})$$

Hence, when determining the levels V_i and V_K, the goal of the subsystem B_i becomes slightly vague. Since there are limiting factors in nature, in this case it is natural to assume the presence of some threshold $V_{i,\min}$, the reaching of which would mean that the energy resource of the subsystem would stop being used in order to get an external resource (i.e., at $V_i \leq V_{i,\min}$ subsystem B_i moves on to the regime of regeneration of the internal resource). In other words, at $V_i \leq V_{\mathrm{i,min}}$ the biocomplexity indicator $\xi_\Omega(t)$ decreases due to broken connections between subsystem B_i and other subsystems. In a general case, $V_{\min}$ is the structural step function (i.e., the transition a_{ij} from the state $a_{ij} = 1$ to the state $a_{ij} = 0$ does not take place for all j simultaneously). In reality, in any trophic pyramid the relationships "predator–prey" stop when the concentration of prey decreases below some critical level. In other cases the interaction of subsystems $\{B_i\}$ can stop depending on various combinations of their parameters. The formalized description of the possible situations of interaction of the subsystems $\{B_i\}$ can be carried out using a simulation model of the NSS functioning.

3.2.2 Modeling results

The current world is administratively divided into 270 countries with a population of 6.4 billion people (with a population annual increment of 1.14%). The real growth of the Gross Domestic Product (GDP) averages 3.8% with the following distribution over the sectors of the global economy: agriculture – 4%, industry – 32%, and human services – 64%. Some quantitative characteristics of individual regions are given in Tables 3.1 and 3.2. It is seen that the state of the environment, practically at any point on the planet, changes considerably, with the consequences reflecting themselves both in people's living standards and in an enhancement of the general degradation of nature. Information about the state of the environment in all global regions can be found at: *http://www.cia.gov/cia/publications/factbook/index.html.*

Determine the coefficients k_i in (3.1) with the following relationship:

$$k_i = \frac{W_{i,0} + \psi_i}{W_{K,0} + \sum_{s \in K} \psi_s}$$

We use the introduced indicator ξ_Ω to describe the global dynamics within a set of some anthropogenic scenarios which are thought possible in the near future. Figures 3.1 and 3.2 demonstrate the results of the computation. It is seen that the indicator of biocomplexity reflects some trends in the development of global natural disasters, such as a reduction of wild forests areas, overexploitation of marine resources, increasing the rate of urbanization, and degradation of ploughed lands. These and other negative processes in the NSS affect the processes of biogeochemical interaction between biospheric and geospheric elements, changing the regional levels of (V, W)-exchanges.

Figure 3.1 shows the role of forest vegetation in the change of global biocomplexity. This correlation is governed by a dependence of biogeochemical cycles of greenhouse gases on the state of the forests. At present, the areal extent of forest is estimated at $\sigma_{L0} \approx 40.3$–41.8 million km^2 (Watson *et al.*, 2000) with 1% attributable to national parks and protected forest. The agricultural area is $\sigma_{X0} \approx 19.5$ million km^2. As seen from Figure 3.1, the rate of deforestation increase raises the CO_2 concentration in the atmosphere by 31% and destabilizes the biogeochemical cycles of greenhouse gases (GHGs) in general. This causes a destabilization of the global structure of (V, W)-exchanges. Even with a 10% reduction of forested territories by the year 2050, the concentration of atmospheric CO_2 could increase by 44% by the end of the 21st century, and a 10% increase in the area of forest could reduce CO_2 by 15%. With a 30% growth of forested areas by 2050, atmospheric CO_2 will be reduced by 53% by 2100 with respect to its value associated with preserving the rates of impact on forest ecosystems that have occurred in the late 20th century. Hence, oscillations of forested areas within $\pm 10\%$ can markedly change the dynamics of many global components. The biocomplexity indicator reflects the integral effects of such changes. In fact, the global model of the nature–society system (GMNSS) makes it possible to evaluate these dependences by the space of the distribution of

Table 3.1. Key indicators of (V, W)-exchanges for individual regions.

Significant indicator of regional development	Region						
	Globe	Russia	USA	China	Japan	Mozambique	Brazil
Rate of increase of GDP (%)	3.8	7.3	3.1	9.1	2.7	7.0	−0.2
Rate of increase of industrial production (%)	3.0	7.0	0.3	30.4	3.3	3.4	0.4
Investments (% of GDP)	–	18.2	15.2	43.4	23.9	47.8	18.0
Export/import (US$ billions)	6,421/ 6,531	134.4/ 74.8	714.5/ 1,260	436.1/ 397.4	447.1/ 346.6	0.795/ 1.142	73.28/ 48.25
Electric power (billions of kWh per year):							
production	14,930	915	3,719	1,420	1,037	7.19	321.2
consumption	13,940	773	3,602	1,312	964.2	1.39	335.9
export		21.16	18.17	10.30	0	5.8	0
import		7.00	38.48	1.80	0	0.5	37.2
Production/consumption of oil (millions of barrels per day)	75.57/ 75.57	7.3/ 2.6	8.1/ 19.7	3.3/ 4.6	0.02/ 5.3	0/ 0.09	1.6/ 2.2
Production/consumption of natural gas (billions of m^3 per day)	2,578/ 2,555	581/ 408	548/ 641	30.3/ 27.4	2.52/ 80.42	0.06/ 0.06	5.95/ 9.59
Ploughed land (millions km^2)	15.981	1.246	1.753	1.436	0.046	0.040	0.589

elements and processes characteristic of the NSS dynamics. A more detailed consideration of these appearing problems is given in Kondratyev *et al.* (2003b, 2004b).

The curves in Figure 3.2 reflect an integral complexity of processes in the NSS, when different interactions of its elements of biological, physical, chemical, geophysical, economic, and social character are manifested through levels of (V, W)-exchanges. We see that real data are grouped near the solid curve. Variations in regional levels of GDP do not change the character of this dependence. Table 3.3 gives a comparative analysis of the distribution of biocomplexity and its dynamics for different regions. It is seen that the main causes of negative development of the regions are deforestation, pollution of water basins and soils, as well as the development of cities. During the next 45 years, only Australia, Canada, Central Africa, China, Japan, and some regions of South East Asia will preserve favorable conditions for sustainable development. Other regions are characterized by a decreasing

Table 3.2. Characteristics of some regions used in GMNSS for calculations of biodiversity indicators (*http://www.cia.gov/cia/publications/factbook/index.html*).

Region	Possible natural disasters	Some indicators of the environmental state
Australia	Cyclones along the coast, severe droughts, and forest fires.	Limited freshwater supplies, soil erosion caused by overgrazing, urbanization, desertification, destruction of the world's largest coral reefs, reduction of many unique species of animals and plants.
England	Floods, winter hurricanes.	In 1998/1999 and 1999/2000, 8.8% and 10.3% of domestic waste was recycled, respectively; by 2015, 33% will be reached. By 2010, CO_2 emissions will be reduced by 12.5% against the year 1990.
Egypt	Periodic droughts, frequent earthquakes, floods, landslides, dust and sand storms.	Loss of agricultural lands due to urbanization and sand intrusion, desertification, increasing soil salinity, pollution of coral reefs with oil products, limited freshwater supplies.
India	Droughts, short-term floods, hurricanes, and earthquakes.	Deforestation, soil erosion, desertification, overgrazing, pollution of water resources with domestic and agricultural sewage, atmospheric pollution by transport and industry.
Spain	Periodic droughts.	Pollution of the Mediterranean Sea with sewage and oil, deforestation, desertification, atmospheric pollution.
Kazakhstan	Mudflows in the region of Alma-Ata, earthquakes.	Excessive water diversion for irrigation; dust, salt, and sand storms, pollution of the Caspian Sea, soil salination, atmospheric pollution, danger for human health from sedimentation of radioactive elements and toxic chemicals.
China	Typhoons, floods, tsunamis, earthquakes, droughts.	Atmospheric pollution with GHGs and sulfur oxides, acid rain, desertification, soil erosion, deforestation, pollution of water bodies with untreated sewage.
Russia	Severe frosts, floods, earthquakes, volcanic activity on Kamchatka, forest fires.	Deforestation, soil erosion, danger of radioactive pollution of the environment, pollution of the atmosphere and water bodies with industrial and domestic sewage.
USA	Tsunamis, volcanic eruptions, earthquakes, hurricanes, tornados, landslides.	Deforestation, acid rain, CO_2 emissions from fuel burning, atmospheric pollution, water resource pollution.
Japan	Forest fires, volcanic activity, typhoons, tsunamis.	Acid rain and water body acidation, degradation of water ecosystems.

biocomplexity of the environment and lowering level of survivability. This conclusion follows from reduction of intensities of (V, W)-exchanges (Kondratyev *et al.*, 2002b). Figure 3.3 explains how regional (V, W)-exchanges could vary with varying GDP. It is seen that the values of V and W converge with increasing GDP and differ

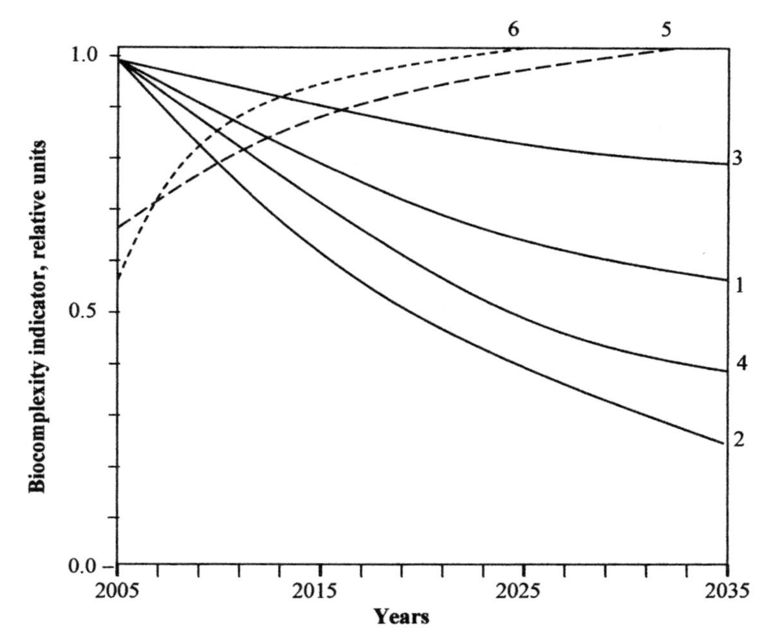

Figure 3.1. Dependence of the biocomplexity indicator on the anthropogenic strategy of the impact on forest ecosystems: 1 – the rate of change of the area of forest remaining at the level existing in 1970–2000. 2 – by the year 2050 all forests disappear. 3 – by the year 2050 the area of forests decreases by 10%; 4 – by 50%. 5 – by the year 2050 the area of forest increases by 10%; 6 – by 30%.

from a stable state when the rate of GDP growth decreases. It means that economic parameters correlate with the complexity indicator, and this dependence is a function of a global nature-protection strategy. For complete analysis of all the appearing versions of possible dependences, a consolidation of global databases is needed.

Finally, the results in Figure 3.3 show a non-linear dependence of biocomplexity of the regional part of the NSS on the correlation between the area of ploughed lands and investments. This dependence shows itself weakly on large territories of developed regions where investments do not exceed 20% of the GDP, and it shows itself acutely in other cases.

3.2.3 Remarks

Thus, biocomplexity is an important informative integral characteristic of the NSS's development, which provides a simple algorithm for the evaluation of the interaction between its living and non-living elements, and gives a quantitative scale for the complex presentation of key mechanisms for the socio-economic development of the regions. It can be expected that further improvement of the biocomplexity indicator will be due to the consideration of partial indicators of the type NDVI

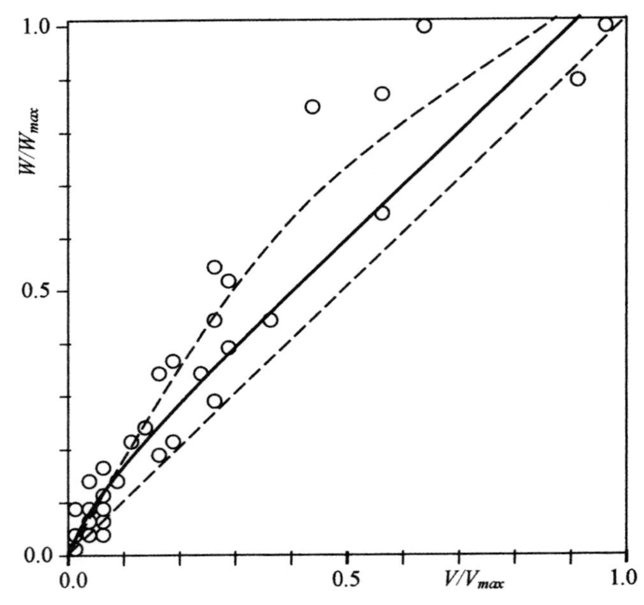

Figure 3.2. The export–import correlation depending on nature use strategy. Circles denote real data (a fragment of which is given in Table 3.1). The solid curve approximates the existing (V, W)-exchanges. The dashed curves correspond to an area of variation of (V, W)-exchanges with GDP by $\pm 5\%$.

(Normalized Difference Vegetation Index) or LAI (Leaf Area Index), measured in the satellite-monitoring regime. Together with considered structural and functional interactions of the NSS subsystems, parameterized at a representative spatial level of discretization of the biosphere and socio-economic structure of the world, the index ideology of the estimation of the state of the NSS has a great perspective. The basis for such a development has been given in Kondratyev *et al.* (2002, 2003c, 2004a), where the GMNSS version has been described oriented toward taking global data from different sources.

As shown by preliminary calculations of the dependence of the biocomplexity indicator on NSS characteristics, a great contribution to its formation is made by the indicators of forestation of territories. Unfortunately, there are differences in definitions for a forest. For instance, in European countries a forest is a territory for which the plant density, with trees which are not less than 7 m in height, exceeds 20%. In other countries this threshold decreases to 10%. Also, there is no established classification of forested territories. Most prevalent are three categories: wild forest, artificial forest (plantation), and bush (the height of plants is from 0.5 to 7 m). Table 3.4 characterizes the situation in the countries of the Mediterranean basin, where despite the development of anthropogenic processes, starting in 1961, the area of forest grows. By 1994 it reached 84 million hectares.

Table 3.3. Comparative analysis of biocomplexity indicators for different regions within the existing regional anthropogenic strategies.

Region	ξ_ω^*				Commentaries concerning the key reason for biocomplexity change
	2005	2010	2020	2050	
Australia	0.56	0.57	0.58	0.59	Urbanization and desertification limit the environmental diversity.
Belgium	0.41	0.39	0.38	0.34	The repercussions of neighboring countries reduces biodiversity.
Brazil	0.58	0.63	0.59	0.53	Deforestation in the Amazon Basin, land and wetland degradation destroys the habitat.
Bulgaria	0.24	0.22	0.21	0.18	Acid rain and soil contamination from heavy metals intensifies deforestation.
Canada	0.57	0.59	0.61	0.60	Acid rain affects lakes and lowers forest productivity.
Central Africa	0.58	0.58	0.59	0.61	Anthropogenic deforestation and flash floods prevent regional development.
Central Asia	0.39	0.38	0.35	0.33	Desertification processes predominate.
China	0.78	0.81	0.84	0.87	Population growth causes negative consequences for natural systems.
France	0.29	0.28	0.26	0.23	Agricultural runoff and acid rain limit the country's progress.
Germany	0.31	0.30	0.28	0.27	Flora and fauna are damaged by sulfur emissions.
Japan	0.22	0.23	0.24	0.24	Threat for aquatic life and degradation of water quality creates obstacles to progress.
Mexico	0.49	0.47	0.43	0.39	Deforestation, widespread erosion, and deteriorating agricultural lands mean the country's development is delayed.
North Africa	0.19	0.17	0.16	0.14	Soil degradation processes lead to the simplification of natural ecosystems.
Russia	0.95	0.91	0.87	0.81	Uncontroled deforestation and urbanization and air pollution accelerate the environmental degradation process.
South Africa	0.39	0.37	0.34	0.31	Soil erosion and river pollution by agricultural runoff lead to losses of biodiversity.
South East Asia	0.88	0.89	0.93	0.96	Deforestation, soil erosion, and overgrazing reduce the rate of progress.
Spain	0.43	0.41	0.39	0.36	Effluents from the offshore production of oil and gas and deforestation are a cause of biocomplexity decrease.
Ukraine	0.51	0.48	0.45	0.41	Inadequate use of arable land brings large deviations from optimal (V, W)-exchange.
USA	0.63	0.62	0.60	0.59	Growth in fossil fuel consumption and non-perfect management of natural resources causes degradation to the living medium.

Note: $\xi_\omega^* = \xi_\omega/\xi_{\omega,\max}, \xi_{\omega,\max} = \max_{\omega\in\Omega} \xi_\omega$.

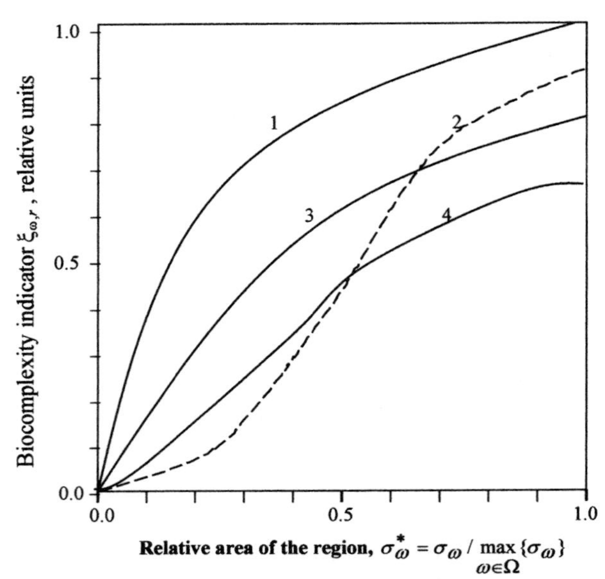

Figure 3.3. The dependence of the biocomplexity indicator on the state of the region. Numbers on the curves correspond to scenarios: 1 – developed regions with areas of cultivated lands <20%; 2 – regions with levels of investments <20% of the GDP; 3 – regions of high agricultural production (the area of cultivated lands >20%); and 4 – regions with levels of investments >20% of the GDP.

Table 3.4. Estimates of the forest cover of some territories of the Mediterranean basin (1,000 ha).

Country	1961	1980	1994	Forest cover of territories (%)
Albania	1,266	–	1,046	38.2
Algeria	2,800	4,384	3,950	1.7
Bosnia-Herzegovina	–	–	2,710	53.2
Greece	2,474	2,619	2,620	20.3
Egypt	31	31	34	0.03
Israel	89	116	126	6.1
Spain	12,900	15,598	16,137	32.3
Italy	5,847	6,355	6,809	23.2
Cyprus	123	123	123	13.3
Lebanon	92	85	80	7.8
Libya	485	600	840	0.5
Morocco	7,505	7,790	8,970	20.1
Syria	402	466	484	2.6
Slovenia	–	–	1,077	53.5
Tunisia	405	540	676	4.4
Turkey	20,170	20,199	20,199	26.2
France	11,614	14,614	15,015	27.3
Yugoslavia	–	–	1,789	17.3

3.3 BIOCOMPLEXITY OF WATER ECOSYSTEMS AND CONDITIONS FOR THEIR DESTRUCTION

3.3.1 The World Ocean as a complex hierarchical aquageosystem

The study of the structure and functioning of the ocean's ecosystems becomes one of the most important and rapidly developing directions for marine biology. Its various aspects are being developed in many countries within the International Biological Programme. One of the problems of this study is to obtain the possibility to forecast the system's behavior as a result of changing some of its parameters. However, due to uniqueness and great spatial extent of the World Ocean's ecosystems, it is difficult to quantitatively evaluate all the elements of the system at different moments of its development and in different regions of the ocean, and the more so, to assess the impact of their change on the functioning of the system as a whole. Therefore, the use of a model approach is one way to solve these problems.

The World Ocean, covering 71% of the Earth's surface, represents 1% of the total food consumption of humankind, the remaining 99% being obtained from cultivated lands. From rough estimates, the total biomass of nekton constitutes 5.3 billion tonnes. The industrial fishing from the World Ocean is estimated at 70 million tonnes per year, which constitutes 20% of the protein consumed by humankind. Industrial fishing is close to a threshold ($\approx$ 90–100 million tonnes per year). However, this is not a threshold for the industrial capabilities of the ocean ecosystems in general, since reserves of krill and other biological objects are poorly utilized by humankind.

This inconsistency between the role of land and ocean ecosystems in food production is explained, first of all, by the fact that on land the cultural economy is intensively developing, something which is poorly developed in the seas and oceans. Therefore, many possible ways to raise the ocean's bioproductivity have not been used. First, humankind uses mainly the top trophic levels of natural communities of the World Ocean – fish and whales. Each subsequent trophic level receives only a $\sim$0.1 share of the energy accumulated by the preceding level. On land, two levels of organisms are used (plants, herbivorous animals), and in the ocean and seas – up to five levels. Direct use of non-fish industrial objects will make it possible to strongly increase the amount of protein products obtained from the ocean.

Second, the problem arises of the transition from free fishing to a cultural method of economic activity in the World Ocean (i.e., the problem of an artificial increase of productivity of biological communities of the ocean). For this purpose, first of all, it is necessary to study the methods of control of the output of the final product in the biological systems of the World Ocean. To determine the rational impacts on the ocean communities, it is necessary to study their structure and functioning, to have a clear idea about their production processes, and transformation of substances and energy fluxes at different trophic levels of the ocean's ecosystems. It is necessary to develop a theory of control in the biological systems of coastal waters and the open ocean, which differ both in natural hydrophysical and biogeochemical parameters and also the degree to which they are anthropogenically forced.

Marine communities are complex biological systems consisting of populations of individual species whose interaction results in the dynamic development of the community. Their spatial structure is strongly determined by the composition of numerous biotic and abiotic factors which depend on a totality of the oceanic fields. The latter are determined by the laws of the oceanic general circulation which is the sum of rising tides and falling tides, the zones of convergence and divergence, wind and thermohaline fluxes, etc.

In the late 20th century, an urgent problem was addressed: to forecast the ocean systems' dynamics under conditions of growing anthropogenic forcing (chemical poisoning, mechanical destruction of living organisms, environmental change) and to evaluate their role in the dynamics of the whole biosphere. Recent studies of the climatic impact of GHGs have shown that the role of the World Ocean in this process is underestimated. In particular, Kondratyev and Johannessen (1993) gave data on the role of the Arctic basins of the World Ocean in the formation of the global CO_2 cycle. This suggested that this role had been assessed incorrectly. This is connected with the fact that a consideration of the biological and gravitational processes, together forming a system which is pumping over CO_2 from the atmosphere to the deep layers of the ocean, did not correspond to reality in the earlier models of the global biogeochemical carbon cycle. Therefore, a specification of the models of the regime of this pump's functioning, with due regard to climatic zones, can strongly affect the prognostic estimates of the greenhouse effect (Kondratyev *et al.*, 2004b).

The ocean ecosystems affect the intensity of biogeochemical cycles through the atmosphere–water boundary and it is usually parameterized based on observational data. However, in this impact the vertical structure of processes taking place in the ocean plays a substantial role. The character of these processes depends strongly on external phenomena, including tsunamis, typhoons, thunderstorms, earthquakes, and tornados. According to Legendre and Legendre (1998), in the Arctic zones of the World Ocean the patchy structure of the springtime blossoming of phytoplankton is determined by winter conditions of ice formation and the subsequent process of thawing. In other zones such external circumstances are factors of pollution of the atmosphere and ocean surface, changing the conditions for phytoplankton's existence and also the carbonate system's functioning.

Phytoplankton is at one of the primary stages of the trophic hierarchy of the ocean's ecosystem. As observations have shown, the World Ocean has a patchy structure formed by a combination of heterogeneous spatial distributions of illumination, temperature, salinity, nutrient element concentrations, hydrodynamic characteristics, etc. The vertical structure of phytoplankton distribution is less diverse and has more universal properties. These properties manifest themselves through the presence of 1–4 maxima of phytoplankton biomass in a depth of ocean water.

Variability of topology of the patchy character and vertical structure is connected with the seasonal cycles and has been experimentally well studied in many climatic zones of the World Ocean. Model qualitative and quantitative indicators of this variability have been found. Combined distributions of abiotic, hydrological, and biotic components of the ocean ecosystems have been studied.

The complexity and interrelationships of all processes through the oceanic depth considerably complicate a search for the laws of formation of plankton spots and establishment of correlations between various factors regulating the intensity of trophic relationships in the ecosystems of the oceans. For instance, in many studies a close relationship has been established between primary production and amount of phytoplankton. At the same time, this relationship breaks down depending on a combination of synoptic situations and illumination conditions. It turns out that the degree of breakdown depends considerably on the combination of groups of phytoplankton (Legendre and Legendre 1998).

An analysis of the accumulated observational data for the assessment of production of seas and oceans, and an attempt of many experts to reveal the laws of production formation inherent to various water basins, have led to numerous specific laws of local relationships between productivity and environmental parameters.

An efficient method for studying the vertical structure of the ocean's ecosystems is numerical modeling. Creation of the model requires knowledge of the structure of the trophic relationships in the ecosystem, features of hydrological conditions, and information about other characteristics of the environment. The modeling experience has shown the possibility of efficient forecasting of the dynamics of the World Ocean communities. These models can be exemplified by a 3-D model of the ecosystem of the Peru upwelling (Krapivin, 1996) and the Okhotsk Sea (Kondratyev *et al.*, 2002b), etc. In all these models the unit of parameterization of the ecosystem's vertical structure is the central one.

3.3.2 General principles for ocean ecosystem model synthesis

Let us describe each element of the ocean ecosystem A by some number of parameters, and transfer the connection between elements to the connection between respective parameters. Then, on the whole, ecosystem A can be characterized by a totality N of parameters $x^*(t) = \{x_j(t), j = 1, \ldots, N\}$, depending on time. The structure $|A(t)$ and behavior $\overline{A}(t)$ of ecosystem A, which we can observe more or less in detail, are functions of these parameters. Therefore, the ecosystem $A(t) = \{|A(t)|, \overline{A}(t)\}$, itself as a totality of structure and behavior, is the function of these parameters $A(t) = F(x^*(t))$. Hence, according to this dependence, as time t changes, ecosystem A will be characterized by the trajectory in the $(N+1)$-dimensional Euclidean space. The model ecosystem $A(t)$ will be an abstract formation $A_M(t) = F_M(x^{*,M})$, which depends on $M \leq N$ components of the vector $x^*(t)(\{x^{*,M}(t)\} \in \{x^*(t)\})$ and takes into account not all connections between them.

The closer M is to N and the more adequately the bond between the components of the vector $x^{*,M}(t)$ are considered, the less is the disagreement between trajectories of ecosystem $A(t)$ and its model $A_M(t)$. This disagreement can be measures by any natural measure, for instance, by a maximum of absolute difference of respective coordinates for the final time interval. In other words, we introduce some goal functional $V = Q(\{x_i(t)\})$ along the trajectory of the ecosystem $A(t)$. The form of the function Q is determined by the character of the requirements made by system A

to the environment. It is assumed that the natural evolutionary process leads to an optimal system, and therefore the system $A_{\text{opt},M}(t)$, which provides an extremum of the functional V we shall call an optimal model of the ecosystem A.

The degree of disagreement between trajectories of ecosystem A and an optimal model $A_{\text{opt},M}(t)$ is affected by the degree of correspondence of the chosen goal functional V of the actual goal $\underline{A}$ of ecosystem A. The multitude $\underline{G} = \{\underline{A}_1, \ldots, \underline{A}_r, \ldots, \underline{A}_m\}$ of possible plausible goals $\{\underline{A}_r\}$ of ecosystem A can be constructed on the basis of experience gained by oceanologists. Then, introducing:

$$A_{\text{opt},M,r} = g_r(\underline{A}_r), \underline{A}_r \in \underline{G}$$

we obtain a limited set of possible optimal systems $A_{\text{opt},M,r}(r = 1, \ldots, m)$, whose trajectories together with the trajectory of ecosystem A are in the space of possible trajectories. Determining $A_{\text{opt},M,r0}$, whose trajectory has a minimal disagreement with $A(t)$, we can find the most plausible goal of the ecosystem $\underline{A}_{r0} = (g_r)^{-1}(A_{\text{opt},M,r0}(t))$.

The construction of the numerical model of the ocean ecosystem A requires, according to the principles mentioned above, either a detailed description of the whole set of its states, or creation of a sufficiently complete set of numerical models of the totality of processes of energy exchange between trophic levels taking place in A and interactions of biotic, abiotic, and hydrophysical factors. Of course, certain assumptions on the character of balance relationships in the ecosystem A are supposed.

As the basic assumption, it is supposed that in ecosystem A, a single original source of energy and substance for all forms of life is the energy of solar radiation (E). According to numerous theoretical and observational studies, a penetration of solar light into deep layers of the ocean follows an exponential law:

$$E(\varphi, \lambda, z, t) = uE_0 \exp\left[-\int_0^z \{\delta p(\varphi, \lambda, x, t) + \beta d(\varphi, \lambda, x, t) + \nu Z(\varphi, \lambda, x, t)\}dx - \alpha z\right] + (1 - u)E_0 \exp(-\zeta z) \quad (3.1)$$

where $E_0 = E(\varphi, \lambda, 0, t)$ is the ocean surface illumination; α is the coefficient of absorption of light by filtered seawater; δ, β, and ν are coefficients of light extinction due to shadowing by phytoplankton (p), detritus (d), and zooplankton (Z), respectively; and u and ζ are parameters selected in a concrete situation for the best approximation of $E(\varphi, \lambda, z, t)$ to a real pattern of changes of illumination with depth. Note that here the effect of the biomass of other trophic levels on water transparency is negligibly small.

The level of illumination affects the rate of photosynthesis R_p. It is known that R_p as a function of E has a maximum at some optimal value $E_{\max}$, decreasing when illumination increases or decreases from this critical value. Maximum R_p at different latitudes φ is located at depths varying as a function of season (i.e.,, sun elevation). In the tropical zones, this depth-dependent variability of the location of the photosynthesis maximum is most pronounced. The average location of photosynthesis maximum is at depths of 10–30 m, and in open water basins it can be observed at depths below 30 m. Here $E_{max} = 65$–$85\,\text{cal cm}^{-2}\,\text{day}^{-1}$. Beginning from depths where $E = 20$–$25\,\text{cal cm}^{-2}\,\text{day}^{-1}$, photosynthesis decreases in proportion to E. An apparent suppression of phytoplankton with light is observed at $E > 100\,\text{cal cm}^{-2}\,\text{day}^{-1}$. These estimates are different at northern latitudes, where photosynthesis is at a maximum on the surface, as a rule.

The rate of photosynthesis at a depth z, depends on water temperature T_w, concentration of nutrient elements n, and phytoplankton biomass p, as well as on other factors, which are not considered here. To express this dependence, various equations are used which in some form reflect the limiting role of the elements E, n, and p. Bearing in mind that $\partial p/p\partial z \to 0$ at $n \to 0$ and $\partial p/p\partial z \to$ const. with growing n, we take the following function as the basis to describe photosynthesis intensity at depth z:

$$R_p(\varphi, \lambda, z, t) = k_0(T_W)K_T f_2(p) f_3(n) \tag{3.2}$$

where

$$K_T = Af_1(E), A = kA_{max}/E_{max}, f_1(E) = E \times \exp[m(1 - E/E_{max})]$$

$$f_2(p) = [1 - \exp\{-\gamma_1 p\}], f_3(n) = [1 - \exp\{-\gamma_2 n\}]^{\theta}$$

where k is the coefficient of proportionality; $k_0(T_W)$ is the function characterizing the dependence of the rate of photosynthesis on water temperature T_W; A_{max} is an assimilation number in the region of maximum photosynthesis (an increment per unit weight of phytoplankton organisms); $\gamma_1, \gamma_2, \theta$, and m are constants whose choice can determine the species characteristics of the phytoplankton elements. For A_{max} the following estimation is valid:

$$A_{max} = \begin{cases} 5.9E_{max} & \text{in the region of maximum photosynthesis} \\ 2.7E_{max} & \text{for other regions} \end{cases}$$

According to this estimate, an assimilation number of trophic phytoplankton in the region of maximum photosynthesis averages about 11–$12\,\text{mg C hr}^{-1}$. For the Peru upwelling $A_{max} = 6.25\,\text{mg C hr}^{-1}$. The light saturation of photosynthesis in the equatorial regions is reached at $9\,\text{cal cm}^{-2}\,\text{day}^{-1}$.

As for the dependence $k_0(T_W)$, it is known that the specific intensity of phytoplankton photosynthesis with temperature changing from low to higher values, first grows, reaching a maximum in some range of temperatures optimal for p, and then, as temperature grows further, starts lowering. In the vicinity of a maximum, the following approximation is often used: $k_0(T_W) = \exp\{(T_W - T_{W,\text{opt}})\ln(\theta_0)\}$, $0 < \theta_0 \leq 2$. The dependence of the rate of photosynthesis on the concentration of nutrient elements $n(\varphi, \lambda, z, t)$ (phosphorus, silicon, nitrogen, and other salts), expressed above with an exponential term is, of course, more complicated. Nutrient elements are one of the most important parts of the ecosystem, since it is they that regulate the energy flux in it. The supplies of nutrient elements are spent in the process of photosynthesis at a rate R_n, usually approximated with the expression $R_n = \delta R_p$, where δ is the coefficient of proportionality. The supply of nutrient elements is replenished when they rise from deep waters where they result from chemical processes of decomposition of dead organic matter. The process of decomposition of dead organic matter is controled by several abiotic conditions characteristic of different climatic zones of the World Ocean. The vertical flux of nutrient elements is determined by conditions of water mixing. In the tropical zones, where the vertical structure of water has a clearly expressed 3-layer configuration with a layer where the temperature jumps suddenly (the thermocline), the vertical motion of nutrient elements is confined to this layer. In the water basins where the thermocline propagates to depths of 40–100 m, the upper layer is usually poor in nutrient elements, and their input into this layer takes place only in the zones of upwellings. In this case, the average rate of the vertical penetration of water beneath the thermocline varies within the range 10^{-3} to 10^{-2} cm s^{-1}, and in the zones of upwelling (breaking of thermocline) it can reach 0.1 cm s^{-1}.

3.3.3 Equations of the World Ocean ecosystem dynamics

The entire thickness of the ocean's water is considered a single biocenosis in which the basic connecting factor is the flux of organic matter produced in surface layers and then penetrating to maximum depths of the ocean. It is assumed that all parameters of the model can change depending on place and time, and their parametric description is made proceeding from average characteristics (i.e., with deterministic models). Assume that nutritional bonds between trophic levels are described by Ivlev's model (Nitu *et al.*, 2000a) (i.e., the consumption of various kinds of nutrient by the ith trophic level is proportional to their efficient biomasses). With due regard to the diagram of nutritional bonds and the structure of the trophic pyramid of the standard ocean ecosystem shown in Figure 3.4, we consider each trophic level in more detail.

Bacterioplankton b, plays an important role in the trophic chains of the ocean. From available estimates, not less than 30% of bacterioplankton mass is in natural aggregates of the size >3–5 μm due to which this biomass becomes available to meet the nutritional requirements of filtrators. This fact should be taken into account when developing a model of the ecosystem, since in many regions of the World Ocean the production of bacteria is comparable with production of phytoplankton. Bacteria occupying a special place in the trophic pyramid, exhibit a varying exchange, strongly decreasing with the nutrient deficit, which is followed by a respective decrease of the rate of their growth. Detritus d, and dissolved organic matter (DOM) g, emitted by phytoplankton, are the main nutrients for bacteria. As a result, the ration for bacteria can be described with Ivlev's formula:

$$R_b = k_b b[1 - \exp(-k_{1,d} d - k_{1,g} g)]$$

where $k_b, k_{1,d}$, and $k_{1,g}$ are experimentally measured coefficients.

The equation describing the dynamics of bacterioplankton biomass is written as:

$$\frac{\partial b}{\partial t} + V_\varphi \frac{\partial b}{\partial \varphi} + V_\lambda \frac{\partial b}{\partial \lambda} + V_z \frac{\partial b}{\partial z} = R_b - T_b - M_b - \sum_{s \in \Gamma_b} C_{bs} R_s + k_{2,\varphi} \frac{\partial^2 b}{\partial \varphi^2} + k_{2,\lambda} \frac{\partial^2 b}{\partial \lambda^2} + k_{2,z} \frac{\partial^2 b}{\partial z^2}$$

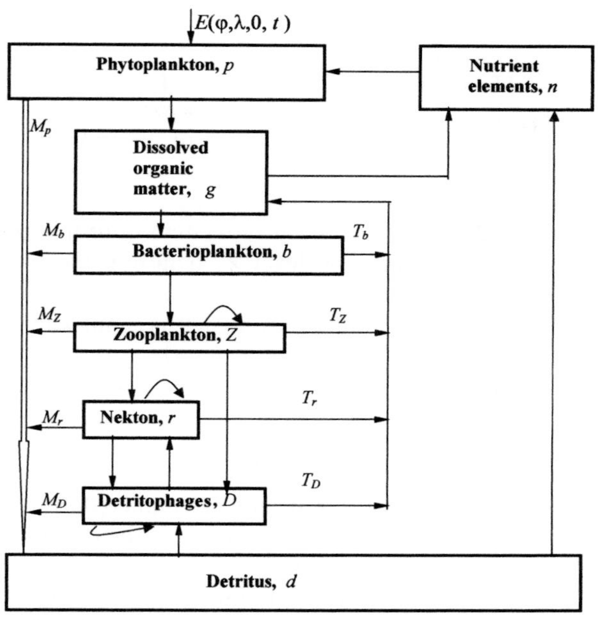

Figure 3.4. The standard block-scheme of energy connections in the trophic pyramid of the ocean ecosystem. The arrows show directions of energy fluxes.

where T_b and M_b are losses of bacterioplankton biomass due to energy exchange with the environment and dying-off, respectively; Γ_b is a multitude of trophic subordination of bacterioplankton (in a standard case Γ_b consists of one element Z); $C_{b,s}$ is the share of bacterioplankton in the nutritional ration of the sth element of the ecosystem. The parameters T_b and M_b we describe with relationships:

$$T_b = t_b b \qquad M_b = \max\{0, \mu_b(b - \underline{B}_b)^{\xi}\}$$

where t_b is the specific expenditure on the exchange with the medium; μ_b is the rate of bacteria dying-off; $\underline{B}_b$ and ξ are constants determining the dependence of the intensity of bacteria dying-off on their concentration. The coefficient $k_2 = (k_{2,\varphi}, k_{2,\lambda}, k_{2,z})$ determines the process of turbulent mixing of the ocean medium.

The dynamic equation for phytoplankton biomass is:

$$\frac{\partial p}{\partial t} + V_\varphi \frac{\partial p}{\partial \varphi} + V_\lambda \frac{\partial p}{\partial \lambda} + V_z \frac{\partial p}{\partial z} = R_p - T_p - M_p - \sum_{s \in \Gamma_p} C_{ps} R_s + k_{2,\varphi} \frac{\partial^2 p}{\partial \varphi^2} + k_{2,\lambda} \frac{\partial^2 p}{\partial \lambda^2} + k_{2,z} \frac{\partial^2 p}{\partial z^2}$$

where Γ_p is a multitude of trophic subordination of phytoplankton; C_{ps} is the share of phytoplankton biomass in the nutritional ration of the elements of the sth trophic level of the ecosystem; T_p are expenditures on the energy exchange with the environment; and M_p is the dying-off of the phytoplankton cells. The latter parameters are determined with relationships:

$$M_p = \max\{0, \mu_p(p - \underline{p})^\theta\} \qquad T_p = t_p p$$

where t_p is the specific expenditure on respiration of phytoplankton cells; μ_p is the coefficient of phytoplankton dying-off; and $\underline{p}$ and θ are coefficients characterizing the dependence of the rate of phytoplankton cells dying-off on their concentration.

An important trophic part of the ocean ecosystem is zooplankton. The scheme in Figure 3.4 demonstrates the integral level Z, denoting the presence of a great number of sub-levels with intersecting trophic connections. Zooplankton consumes phyto- and bacterioplankton. But zooplankton itself is a nutrient for many animals denoted in Figure 3.4 as nekton and detritophages D.

The zooplankton ration may be described with Ivlev's formula (Nitu *et al.*, 2000b):

$$R_Z = k_Z(1 - \exp[-\nu\overline{B}]) \tag{3.3}$$

where $\overline{B}$ is the biomass of available nutrient ($\overline{B} = \max\{0, B - \overline{B}_{\min}\}$); k_Z is the maximum value of ration with excess nutrient; and ν is the coefficient characterizing the level of starvation. Maximum ration is assumed to be equal to nutritional requirements which, in their turn, are determined by exchange intensity T_1 and maximum possible increment P_1. Both latter parameters are related by the coefficient $q_2 = P_1/(P1 + T_1)$, so that we obtain: $k_Z = T_1 u(1 - q_{2,\max})$, where $1/u$ is the nutrient assimilation $q_{2,\max} = \max q_2$.

The formula to calculate R_Z means that at a small amount of nutrient the ration of zooplankton grows in proportion to the amount of nutrient, then as the ration approaches a maximum of k_Z, its dependence on B weakens. Since in fact, one trophic level is not completely eaten out by another level, the parameter R_Z has a limit: $R_Z = 0$ at $B \leq B_{\min}$, where $B_{\min}$ is the minimum unconsumed nutrient biomass. In the formula to calculate M_p the parameter $\underline{p}$ plays the same role, but in the process of phytoplankton cells dying-off.

Thus, a change of zooplankton biomass follows the law described by the following differential equation:

$$\frac{\partial Z}{\partial t} + V_\varphi \frac{\partial Z}{\partial \varphi} + V_\lambda \frac{\partial Z}{\partial \lambda} + V_z \frac{\partial Z}{\partial z} = R_Z - T_Z - M_Z - H_Z - \sum_{s \in \Gamma_Z} C_{Zs} R_s$$
$$+ k_{2,\varphi} \frac{\partial^2 Z}{\partial \varphi^2} + k_{2,\lambda} \frac{\partial^2 Z}{\partial \lambda^2} + k_{2,z} \frac{\partial^2 Z}{\partial z^2}$$

where Γ_Z is the multitude of trophic subordination of zooplankton; C_{Zs} is the share of zooplankton biomass in the nutrient ration of the sth trophic level; and H_Z, T_Z, and M_Z are parameters of zooplankton biomass losses due to unassimilated nutrient, expenditures on respiration, and dying-off, respectively. The latter three parameters are described with relationships:

$$H_Z = h_Z R_Z \qquad T_Z = t_Z Z \qquad M_Z = (\mu_Z + \mu_{Z,1} Z) Z$$

where the coefficients h_Z, t_Z, μ_Z, and $\mu_{Z,1}$ are determined empirically for a concrete species of zooplankton.

As seen from these equations, zooplankton is considered as a passive element of the ecosystem, subject to only physical processes of motion with moving water masses. However, zooplankton is known to migrate mainly vertically. In this model, a simple mechanism is used to simulate the process of the vertical migration of zooplankton. Divide the water thickness into two layers: $0 \le z \le z_0$ and $z_0 < z \le H$. Assume that zooplankton migration between these layers is of nutritional character (i.e., part of zooplankton from the layer $[z_0, H]$ can meet its nutritional requirements in the layer $[0, z_0]$). So that with $B_{\min}$ taken into account, the whole vertical profile $B(\varphi, \lambda, z, t)$ is considered.

The coefficients $C_{as}(a = p, Z)$ in equations for phytoplankton and zooplankton are determined on supposition that the consumption of various kinds of nutrient by the sth trophic level is proportional to their efficient biomasses:

$$C_{as} = k_{sa}\overline{B}_a \left[\sum_{a \in S_s} k_{sa}\overline{B}_a \right]^{-1}$$

where $\overline{B}_a$ is the efficient biomass of the ath nutrient; S_s is the multitude of trophic subordination of the sth component; and k_{sa} is the coefficient of proportionality which determines the significance of the ath element.

According to the scheme in Figure 3.4, equations to describe the dynamics of the biomass of nekton, detritophages, detritus, DOM, and nutrient elements will be:

$$\frac{\partial r}{\partial t} = R_r - H_r - T_r - M_r - \sum_{s \in \Gamma_r} C_{rs}R_s$$

$$\frac{\partial D}{\partial t} = R_D - H_D - T_D - M_D - \sum_{s \in \Gamma_D} C_{Ds}R_s$$

$$\frac{\partial d}{\partial t} + V_\varphi \frac{\partial d}{\partial \varphi} + V_\lambda \frac{\partial d}{\partial \lambda} + V_z \frac{\partial d}{\partial z} = M_b + M_D + M_r + M_p + M_Z + H_Z + H_r + H_D - \mu_d d - C_{dD}R_D + k_{2,\varphi}\frac{\partial^2 d}{\partial \varphi^2} + k_{2,\lambda}\frac{\partial^2 d}{\partial \lambda^2} + k_{2,z}\frac{\partial^2 d}{\partial z^2}$$

$$\frac{\partial n}{\partial t} + V_\varphi \frac{\partial n}{\partial \varphi} + V_\lambda \frac{\partial n}{\partial \lambda} + V_z \frac{\partial n}{\partial z} = \mu_d d - \delta R_p + k_{2,\varphi}\frac{\partial^2 n}{\partial \varphi^2} + k_{2,\lambda}\frac{\partial^2 n}{\partial \lambda^2} + k_{2,z}\frac{\partial^2 n}{\partial z^2}$$

$$\frac{\partial g}{\partial t} + V_\varphi \frac{\partial g}{\partial \varphi} + V_\lambda \frac{\partial g}{\partial \lambda} + V_z \frac{\partial g}{\partial z} = T_p + T_b + T_r + T_D + T_Z - C_{gb}R_b + k_{2,\varphi}\frac{\partial^2 g}{\partial \varphi^2} + k_{2,\lambda}\frac{\partial^2 g}{\partial \lambda^2} + k_{2,z}\frac{\partial^2 g}{\partial z^2}$$

Here $H_a = (1 - h_a)R_a$ is the unassimilated nutrient of the ath element $(a = r, D)$; $T_a = t_a\, a$ are expenditures on energy exchange; $M_a = (\mu_a + \mu_{a,1}a)a$ is the dying-off; ρ_g is the indicator of the rate of replenishing the supplies of nutrient elements due to disintegration of DOM; and δ is the coefficient of consumption of nutrient elements in the process of photosynthesis.

It has been assumed that elements of nekton and detritophages do not move with the moving water masses. These elements are supposed to migrate in space, independent of hydrophysical conditions in the medium. Consider two possible versions of modeling the process of migration. The first version is connected with an addition into right-hand parts of some equations of the terms describing the turbulent mixing with coefficients $k_2^* > k_2$. In other words, the process of migration is identified with the process of intensive turbulent mixing (i.e., it is random). However, the process of fish migration is governed by an expediency of the choice of direction of their motion. According to the biological principle of adaptation, migration of fish follows the principle of complex maximization of an efficient nutritional ration with preserved parameters of the domain within their living conditions. Hence, the first migration at characteristic rates ensures their presence in places with the most favorable nutritional and other (temperature, salinity, dissolved oxygen, chemicals concentration) abiotic conditions. This means that fish migrate in the direction of maximum gradient of efficient nutrient with limited medium parameters.

3.3.4 Estimation of the Okhotsk Sea's ecosystem biocomplexity

The Okhotsk Sea as a unique ecosystem has attracted attention in many studies. This is connected with the desire to understand the climatic and ecological processes taking place here and to establish their connection with global processes, especially in the Arctic Basin. For Russia and Japan, it is also important for economic reasons. Therefore, the synthesis of the Okhotsk Sea ecosystem model (OSEM), which would make it possible to give a prognostic estimate of the state of the ecosystem depending on global and regional changes of the environment is, of course, an urgent problem. Using this model, it is possible to establish some laws of the dynamics of the trophic pyramid of the sea and to understand mechanisms of regulation of the marine community due to external forcing.

The OSEM structure and the scheme of energy fluxes in the Okhotsk Sea ecosystem are shown in Figures 3.5 and 3.6, respectively. The equations of the trophic graph are written like the equations given in Section 3.4.3. A change of the oxygen regime is approximated by solving the equation:

$$\frac{\partial O}{\partial t} + V_\varphi \frac{\partial O}{\partial \varphi} + V_\lambda \frac{\partial O}{\partial \lambda} + V_z \frac{\partial O}{\partial z} = \xi_p R_p - \xi_Z T_Z - \xi_r T_r - \xi_D T_D - \xi_d R_d + O_A$$

where O_A is the function describing the oxygen exchange at the atmosphere–sea boundary; ξ_p is the indicator of oxygen production in photosynthesis; $\xi_i (i = Z, r, D)$ is the coefficient of oxygen expenditure on respiration of hydrobiont i; and ξ_d is the indicator of oxygen expenditure at detritus decomposition.

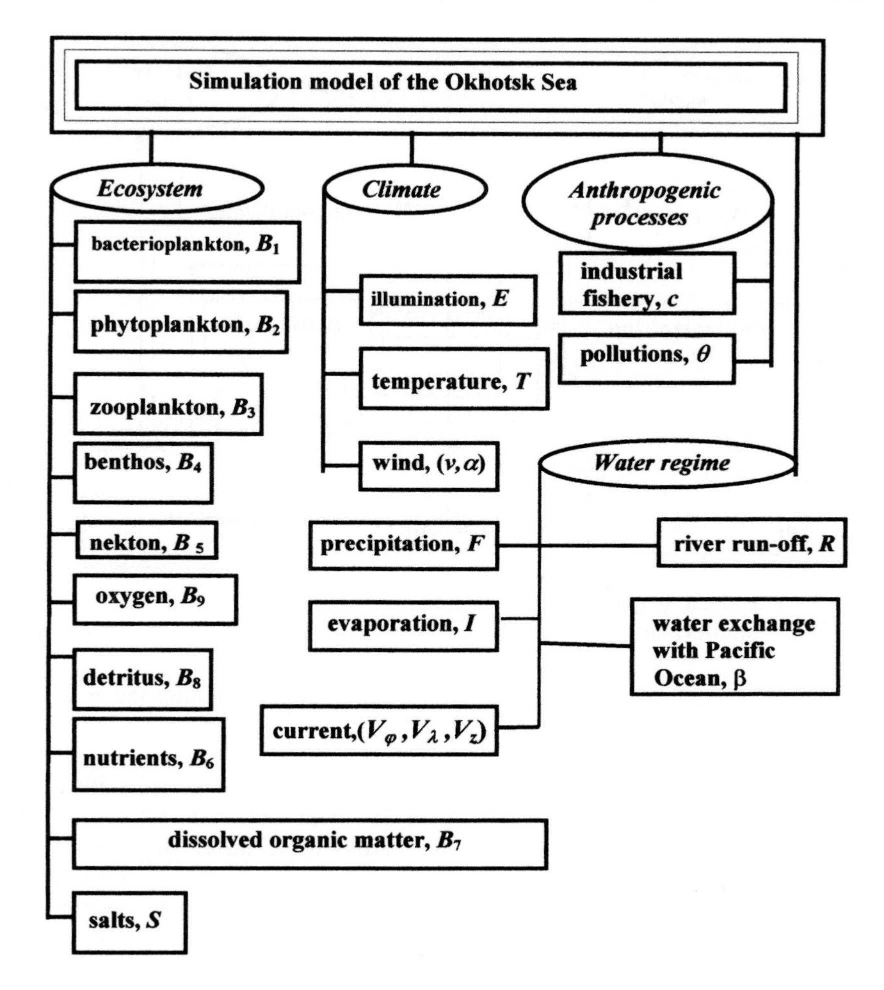

Figure 3.5. The OSEM structure.

The spatial and seasonal dynamics of salinity is described with balance relationship:

$$V_i \frac{dS_i}{dt} = \sum_{\substack{j=1 \\ j \neq i}}^{N} [\alpha_{ij} W_j S_j + \mu_{ij}(S_j - S_i)] + (1 - \beta_i) W_i S_i + f_i \quad (i = 1, K, N)$$

where N is the number of the volume segments $V_i = \Delta\varphi \times \Delta\lambda \times \Delta z_i$ appearing with division of the water body of the Okhotsk Sea in the process of discretization of latitude φ, longitude λ, and depth z; μ_{ij} is the volume coefficient of diffusion equal to $\delta_{ij}\sigma_{ij}/L_i$ (L_i is the average extent of adjacent boundary of the ith segment; σ_{ij} is the area of the contact of segments i and j; δ_{ij} is the coefficient of turbulent diffusion);

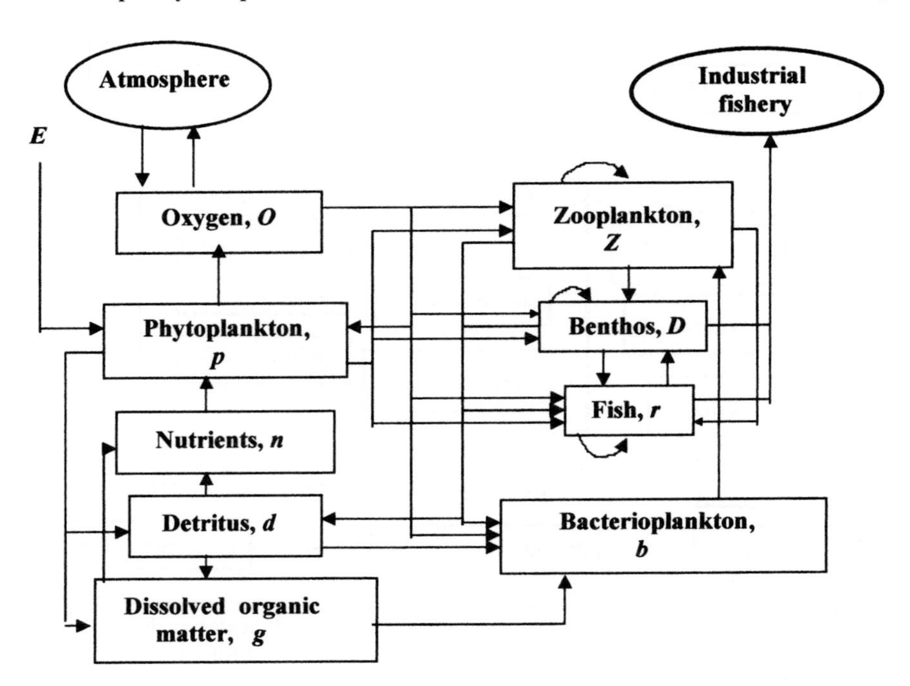

Figure 3.6. The trophic pyramid of the Okhotsk Sea ecosystem forming the basis for the OSEM.

f_i is the function of the source describing an external input of salts into the ith segment (with precipitation F, river run-off R, from the Pacific Ocean β, from bottom sediments with lifting waters); W_j is the total water flux from the jth segment; α_{ij} is its share directed to the ith segment ($\alpha_{ij} = 0$ for non-contacting segments); and β_i is the share of the flux W_i directed to the outside environment.

The temperature regime of the Okhotsk Sea waters can be described with the thermal balance equation:

$$\rho c V_i \frac{dT_i}{dt} = \sum_j W_{ij,A} + \sum_j W_{ij,D} - W_{i,S} + \Omega_{i,1} + \Omega_{i,2}$$

where T_i is the average temperature of water in the ith segment (°C); $W_{ij,A}$ is the input of heat with water flux from the jth segment into the ith one; $W_{ij,D}$ is the heat exchange between the ith and jth segments due to the turbulent motion of water masses; $W_{i,S}$ is the total output of heat from the jth segment; $\Omega_{i,1}$ is the input of heat with water fluxes across the external boundary; $\Omega_{i,2}$ is the heat exchange across the external boundaries not connected with water fluxes; ρ is the seawater density (g cm^{-3}); and c is the heat capacity ($\text{cal g}^{-1} \text{deg}^{-1}$).

The assumed scheme of calculations of the Okhotsk Sea fields is, in fact, a simulation. This semi-empirical approach is acceptable for the climatic zone of the Okhotsk Sea and does not require complicated calculations. Exchange processes

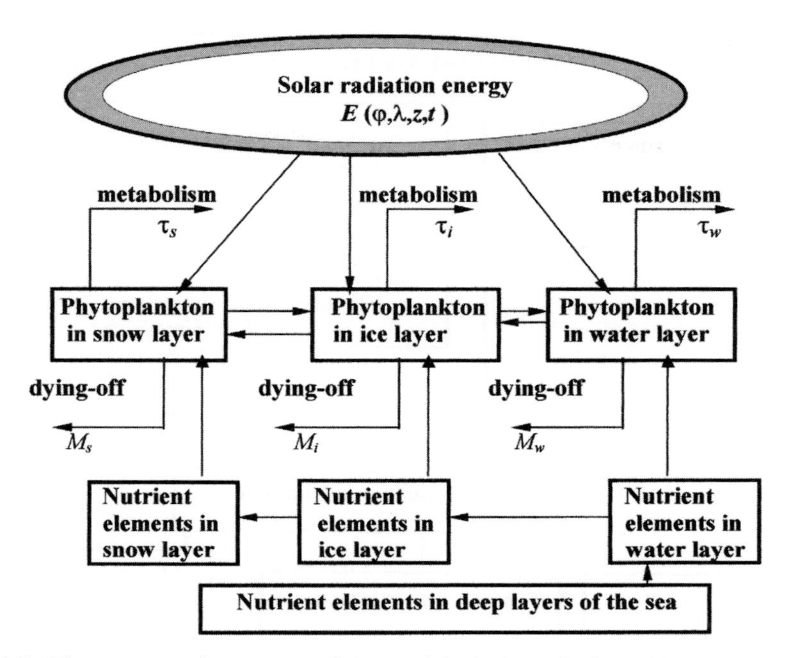

Figure 3.7. The conceptual structure of the model of phytoplankton biomass kinetics under climatic conditions of the Okhotsk Sea region. The source of external energy is solar radiation $E(\varphi, \lambda, z, t)$, whose intensity changes in latitude φ and longitude λ, with the sea depth z, and in time t.

with water masses between segments $V_i(i = 1, \dots, N)$ are simulated with due regard to data on directions and velocities of currents.

The functioning of the Okhotsk Sea ecosystem is considerably determined by severe climate conditions. Most of the sea is covered with ice 80–100 cm thick during 6–7 months in a year. Therefore, in the OSEM, the vertical structure of the Okhotsk Sea is presented as a section shown in Figure 3.7. With due regard to this scheme, in the zones of sea freezing, the energy fluxes shown in Figure 3.2 become complicated due to a division of the phytoplankton component into three possible media: snow, ice, and water.

The OSEM makes it possible, following the scheme in Figure 3.2, to synthesize the system of monitoring of the Okhotsk Sea basin and to obtain prognostic estimates of its various parameters. The principal structure of this system is shown in Figure 3.8. The regime of the step-by-step monitoring will provide economic use for the fisheries. The OSEM use is exemplified in Figures 3.9 and 3.10. The accuracy of such forecasts depends on the information base adequacy. The Japan meteorological service pays serious attention to filling in this database. With this aim in view, in the city of Mombetsu, on the northern coast of the Hokkaido Island, a system has been created to monitor the ice condition of the Okhotsk Sea, which includes a marine stationary station and a surface center of space data processing. On this basis, GIMS technology can be efficiently utilized.

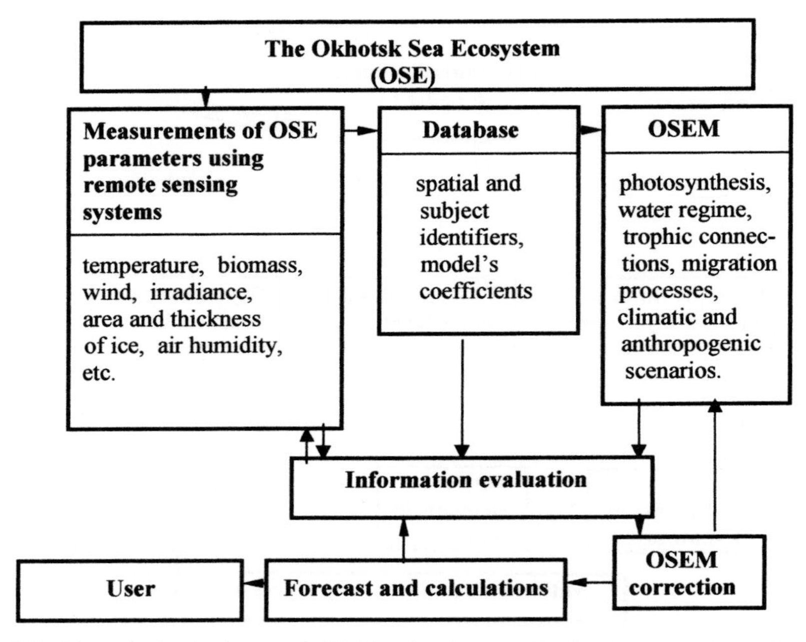

Figure 3.8. The principal scheme of GIMS-echnology realization in the system of assessment of the Okhotsk Sea ecosystem's survival.

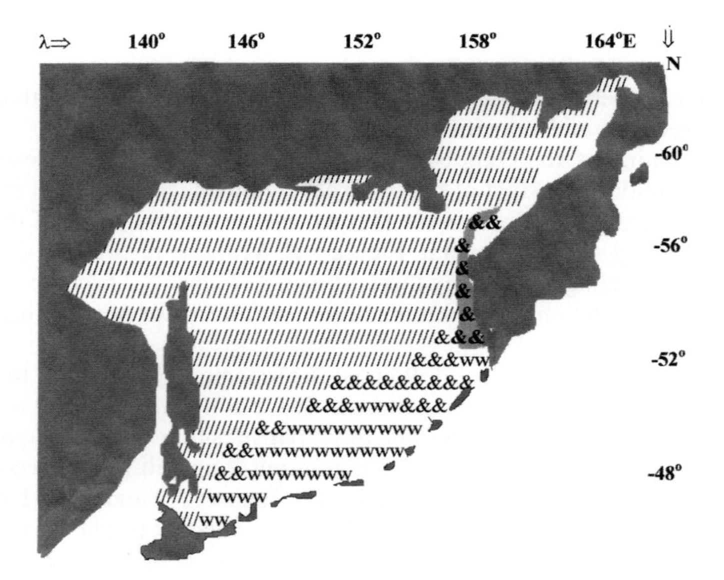

Figure 3.9. Forecast of the ice situation in the Okhotsk Sea for February 2004 from initial data of the Japan meteorological system for November 2003. The difference between forecast and data of satellite monitoring is marked with solid symbols. Notes: / = stable ice; & = unstable ice (ice cover with polynyas); *w* = open water.

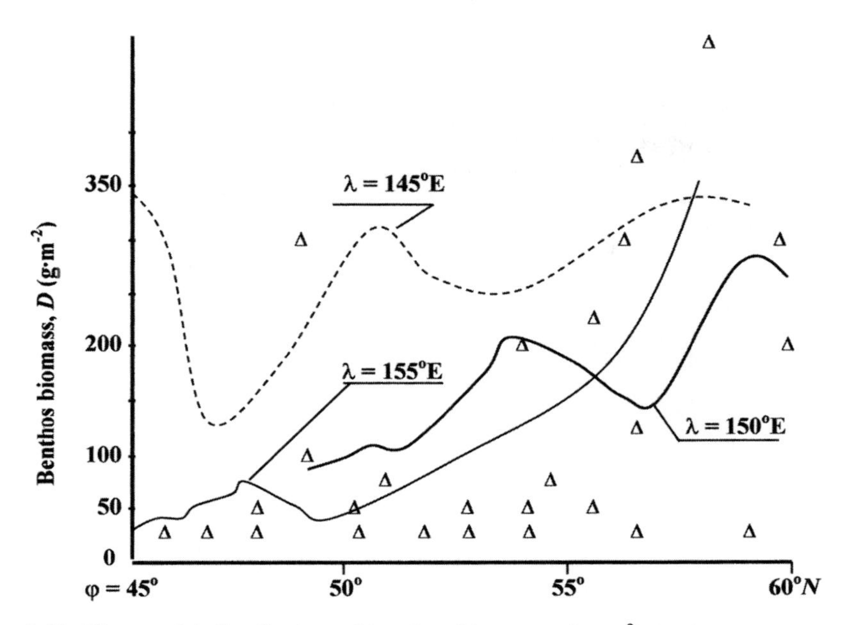

Figure 3.10. The spatial distribution of benthos biomass $D(\mathrm{g\,m^{-2}})$ in the Okhotsk Sea. The symbol Δ marks experimental or theoretical estimates published elsewhere over the last 5 years.

Useful information for industrial fishing and specialists in the Okhotsk Sea fish resources protection can be obtained from the biocomplexity indicator. Its calculations are exemplified in Figures 3.11 and Table 3.5. This method of calculation of the indicator introduced into GIMS makes it possible, with minimal expenditures, to estimate in detail the animal populations, and forecast consequences of anthropogenic forcing on the ecosystem.

3.3.5 Biocomplexity of the upwelling ecosystem

The World Ocean zones of deepwater uprising called upwelling are characterized by a high organic productivity. Upwelling appears as a result of wind-induced recession of surface waters from the shore, due to divergent fluxes, or at outflow of water from land for other reasons. The speed of water uprising and the upwelling stability are determined by a set of synoptic parameters. The most characteristic value of the vertical speed of water uprising in the zone of upwelling is $0.77 \times 10^{-3}\,\mathrm{cm\,s^{-1}}$. The depths, from which water uprising begins, vary widely, but depths about 200 m prevail. Concrete conditions of the origin, development, and functioning of the upwelling ecosystem are determined by its geographic location and, of course, climate.

Consider the ocean ecosystem in the zone of upwelling. Consider its impact proceeding from the concept of successive development of the community from the moment of its formation in the region of deepwater uprising to the climax

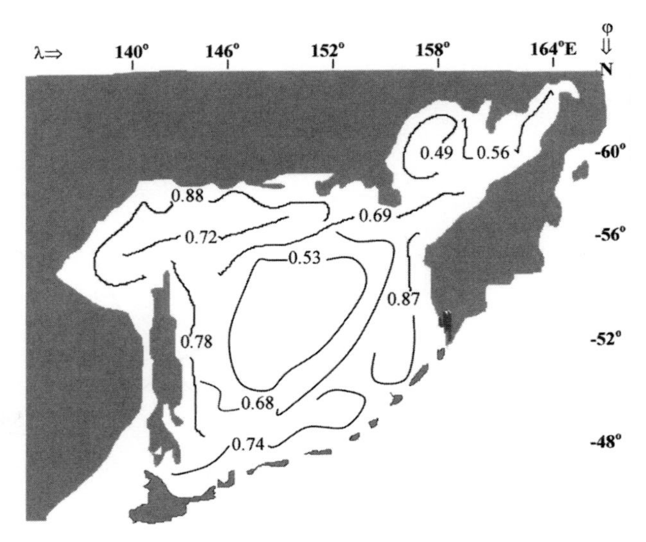

Figure 3.11. The spatial distribution of the biocomplexity indicator $\xi^* = \xi/\xi_{max}$ of the Okhotsk Sea aquageosystem (Table 3.1) in the spring–summer period calculated using the method described in Section 3.2.

Table 3.5. The trophic pyramid of the Okhotsk Sea ecosystem considered in calculations of biocomplexity indicator (Figure 3.11).

Consumers of energy and matter	Sources of energy and matter																		
	B_1	B_2	B_3	B_4	B_5	B_6	B_7	B_8	B_9	B_{10}	B_{11}	B_{12}	B_{13}	B_{14}	B_{15}	B_{16}	B_{17}	B_{18}	B_{19}
Phytoplankton, B_1	0	0	0	0	0	0	0	0	0	0	0	0	0	0	0	0	0	1	0
Bacterioplankton, B_2	0	0	0	0	0	0	0	0	0	0	0	0	0	0	0	0	0	0	1
Microzoa, B_3	1	1	0	0	0	0	0	0	0	0	0	0	0	0	0	0	0	0	1
Microzoa, B_4	1	1	0	0	0	0	0	0	0	0	0	0	0	0	0	0	0	0	0
Carnivores, B_5	0	1	1	1	1	0	0	0	0	0	0	0	0	0	0	0	0	0	1
Zoobenthic animals, B_6	1	1	1	1	1	0	0	0	0	0	0	0	0	0	0	0	0	0	1
Flat-fish, B_7	0	0	0	0	0	1	1	1	1	1	1	1	1	1	1	1	1	0	0
Coffidae, B_8	0	0	0	0	1	1	1	1	1	1	1	1	1	1	1	1	1	0	0
Ammodytes hexapterus, B_9	0	0	0	0	0	1	1	0	0	1	1	1	0	0	0	1	0	0	0
Mallotus, B_{10}	0	0	0	0	1	1	1	0	0	0	1	0	0	0	0	0	0	0	0
Theragra chalcogramma, B_{11}	0	0	0	0	1	0	0	0	0	1	1	1	0	0	0	0	0	0	0
Salmonidae, B_{12}	0	0	0	0	0	1	1	0	0	1	1	1	1	0	0	1	0	0	0
Coryphaenoides, B_{13}	0	0	0	0	1	1	1	1	1	1	1	1	1	1	1	1	1	0	0
B_{14}	0	0	0	0	1	1	1	1	1	1	1	1	1	1	1	1	1	0	0
B_{15}	0	0	0	0	1	1	0	0	0	0	0	0	0	0	1	0	0	0	0
Crabs, B_{16}	0	0	0	0	0	1	0	1	1	1	0	1	0	0	0	1	1	0	0
Laemonema longipes, B_{17}	0	0	0	0	1	1	0	0	0	0	0	0	0	0	0	0	1	0	0
Biogenic salts, B_{18}	0	0	0	0	0	0	0	0	0	0	0	0	0	0	0	0	0	0	1
Detritus, B_{19}	1	1	1	1	1	1	1	1	1	1	1	1	1	1	1	1	1	0	0
People, B_{20}	1	1	1	1	1	1	1	1	1	1	1	1	1	1	1	1	1	1	1

Notes: B_{14} = *Reinchardti ushippoglossoi des matsuurae*; B_{15} = *Clupeapallasi pallasi* Val.

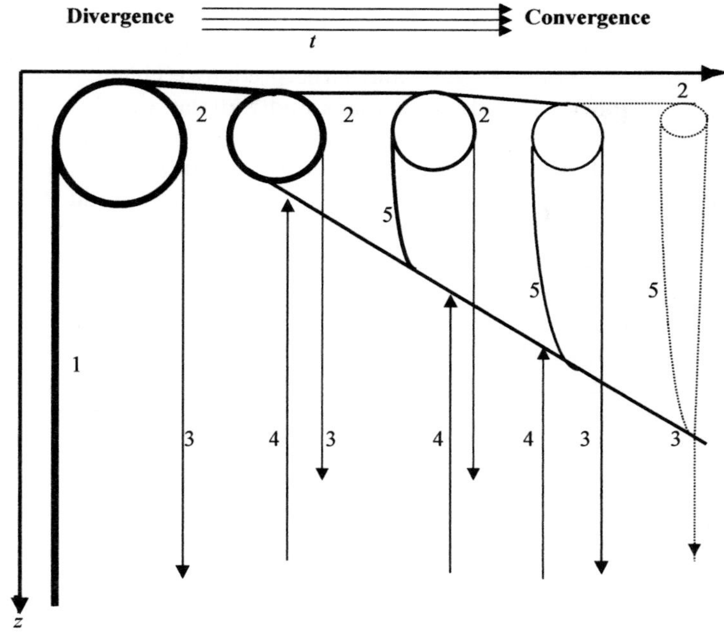

Figure 3.12. A scheme of the cycle of nutrient elements and organic matter in a succession of an ecosystem in the zone of upwelling after Vinogradov (1975). Notations for the characteristic zones of development for the upwelling ecosystem are: 1 – lifting of nutrients elements and DOM with water from deep layers in the zone of upwelling; 2 – their multiple use in production/destruction cycles of the surface layer communities; 3 – loss of nutrient elements with lowering organic remains and migrating organisms; 4 – turbulent lifting of nutrients and DOM from deep layers and their delay in the zone of lower maximum; 5 – inclusion of nutrients lifted from the layer of lower maximum with migrating organisms into the production cycles of the surface community.

state in the oligotrophic region of convergence, and synthesize using an upwelling ecosystem model (UEM). Between these moments the system develops in time and moves, respectively, together with the water masses. In this case, both the energy supply in the community and the structure (spatial, trophic, specific) of the community change. The main moments of these changes have been studied in general in numerous expeditions. Therefore, the correspondence of model calculations to these ideas can serve as one of the criteria of the model's adequacy.

Assume that the ecosystem from the zone of upwelling develops uniformly horizontally, so that changes are observed both in depth z, with a step $\Delta z = 10$ m, and in time, with a step $\Delta t = 1$ day. It is assumed that the horizontal velocity of water flow from the zone of upwelling is constant and equals to $V = V_\varphi = V_\lambda$, so that the removal of a unit volume of water from the zone of upwelling is $\Delta r = (\Delta\varphi^2 + \Delta\lambda^2) = V \times \Delta t$. Schematically, this situation is shown in Figure 3.12.

The state of the ecosystem at each horizon, $z =$ const., is determined by illumination $E(z, t)$, concentration of nutrient elements $n(z, t)$ and detritus $d(z, t)$, biomass of phytoplankton $p(z,t)$, bacteria $b(z,t)$, protozoa $Z_1(z,t)$, microzooplankton $Z_2(z,t)$, small filtrators $Z_3(z,t)$, large filtrators $Z_4(z,t)$, small predatory cyclopoids $Z_5(z,t)$, predatory calanoids $Z_6(z,t)$, and large predatory chaetogratha and polychaetes $Z_7(z,t)$. Protozoa include infusorians and radiolarians. Microzooplankton includes nauplii of all copepods. Studies of nutrients of tropical plankton animals have shown that small filtrators (up to 1 mm in size) include such plankton animals as *Oikopleura*, copepods *Clausocalanus, Paracalanus, Acartia,* and *Lucicutia*, and small ostracods *Conchoecia*. Large filtrators include copepods larger than 1 mm, such as *Undinula, Pleuromamma, Centropages, Temora, Scolecythrix*, etc., as well as baby euphausiids, pteropods, etc. In the trophic pyramid the impact of the elements of nekton is negligibly small. Nutritional bonds between these elements are shown schematically in Figures 3.13 and 3.14.

It is assumed that 30% of the bacteria biomass are in natural aggregates >3–5 μm in size, which can be eaten and assimilated by filtrators (Z_3, Z_4). Nauplii (Z_2),

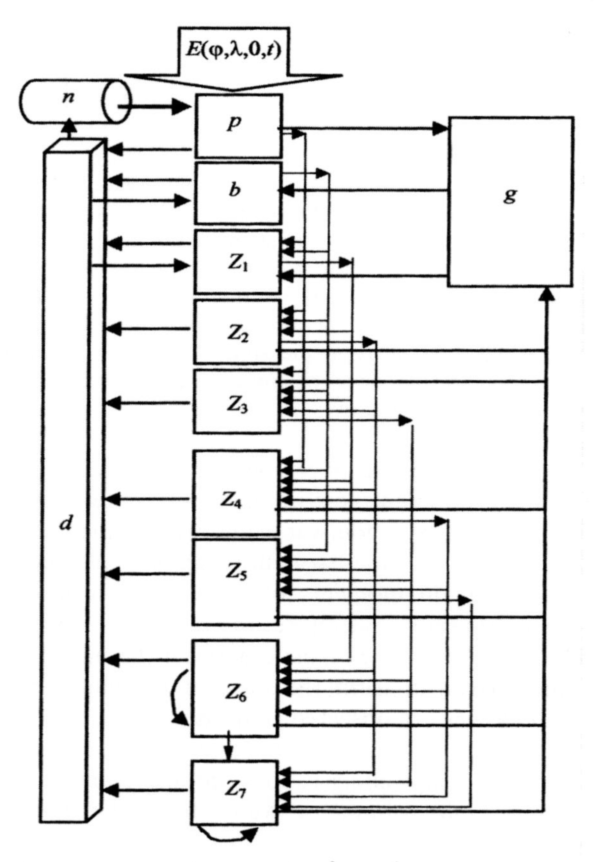

Figure 3.13. The scheme of energy fluxes (cal m^{-3} day^{-1}) through the community of pelagic organisms residing in the 200-m layer in the oligotrophic tropical region in the World Ocean.

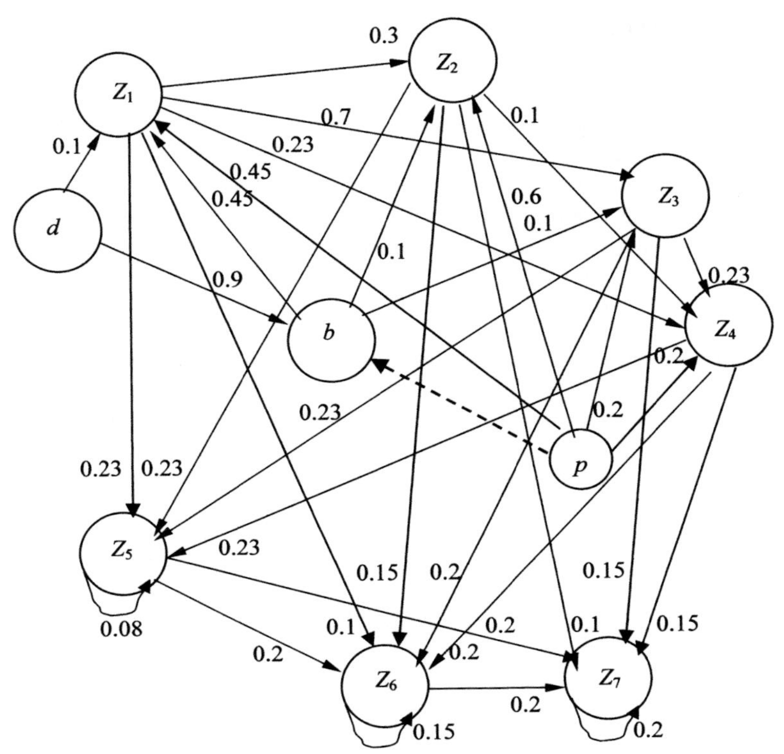

Figure 3.14. The scheme of trophic loads between elements of the upwelling ecosystem. The letter identifiers correspond to Figure 3.10. The arrows indicate the values of the coefficients C_{ij}.

protozoa (Z_1), and small filtrators (Z_3) can also feed on non-aggregated bacterioplankton. Trophic bonds between elements are described on the basis of the energy principle. Biomass (the rate of producing and exchange) rations are expressed in energy units (cal m^{-3} or cal m^{-2}).

Also, as in a general model of the ocean ecosystem, the source of energy and matter in a community is the primary production of phytoplankton (R_p). The energy of solar radiation (E) and nutrient elements (n) are obtained from outside. The vertical structure of the water medium is described by the 3-layer model: the layer above the thermocline, the layer of the most drastic gradients, and the layer of lower gradients beneath the thermocline. Locations of the upper (z_b) and lower (z_l) boundaries of the layer of the highest gradients are assumed to be the following:

$$z_b = \begin{cases} 10 + 2.2t \text{ at } 0 \leq t \leq 50 \text{ days} \\ 120 + 0.6(t - 50) \text{ at } t > 50 \text{ days} \end{cases} \qquad z_l = \begin{cases} 30 + 2.4t \text{ at } 0 \leq t \leq 50 \text{ days} \\ 150 + 1.4(t - 50) \text{ at t>50 days} \end{cases}$$

Thus, the thermocline at an initial time moment $t = 0$ is located in the 10–30-m layer lowering gradually down to 120–150 m on the 50th day and down to 150–190 m on the 100th day.

Replenishing the supply of biogenic elements in the 0–200-m layer takes place due to detritus decomposition and emission of unassimilated nutrient by the living organisms of the community and due to a receipt from deep layers ($z > 200$ m) as a result of turbulent mixing (coefficient k_2) and vertical advection (velocity V_z). The content of nutrients in organic matter is constant and constitutes 10%.

With due regard to assumed suppositions, the equation for photosynthesis is written as:

$$Rp = k_T[1 - 10^{-0.25p\gamma(t)}][1 - 10^{-0.1n}]^{0,6}$$

where the function $\gamma(t)$ characterizes the temporal dependence of the production/biomass ratio (P/B is the coefficient) for phytoplankton. The maximum P/B value for phytoplankton at the point of upwelling ($t = 0$) is assumed to be 5, then the P/B coefficient decreases, and on the 15th day it reaches unity, and then does not change. Hence, the equation of phytoplankton biomass dynamics is written as:

$$\frac{\partial p}{\partial t} = R_p - t_p p - \mu_p p - \sum_{j \in \omega_0 \setminus p} C_{pj} R_j + k_2 \frac{\partial^2 p}{\partial z^2} + (V_z - w_p) \frac{\partial p}{\partial z}$$

where the coefficient w_p (cm day^{-1}) describes the process of gravitational sedimentation and is evaluated by :

$$w_p = \begin{cases} 50 \text{ at } 0 \le z \le z_b, \\ 10 \text{ at } z_b < z < z_l, \\ 30 \text{ at } z \ge z_l. \end{cases}$$

Nutrient for bacteria is detritus and DOM emitted by phytoplankton (allochthonous DOM is not taken into account). In the equation for DOM, the following approximation is considered:

$$T_p + T_b + T_Z + T_r + T_D = 0.3Rp$$

It is also assumed that:

$$\mu_b = 0.01 \qquad t_b = 0.75[1 - 10^{-0.2d - 0.3R_p}] \qquad R_b = 3b(1 - 10^{-0.2d - 0.3R_p})$$

A limitation is established that during one day, bacterioplankton can consume not more that 10% of all detritus in the same water layer. Bearing in mind the abovesaid, the equation for nutrient elements is written in the form:

$$\frac{\partial n}{\partial t} = -0.1R_p + 0.1d + 0.05 \sum_{i \in \omega_0} t_i B_i + k_2 \frac{\partial^2 n}{\partial z^2} + V_z \frac{\partial n}{\partial z}$$

where $\omega_0 = \{p, b, Z_j(j = 1 \div 7)\}$ and B_i is the biomass of the ith element.

Coefficients k_2 (cm^2 day^{-1}) and V_z (cm day^{-1}) are prescribed as follows:

$$V_z = \begin{cases} 5 & \text{at} \quad 0 \le z \le z_b \\ 0.1 & \text{at} \quad z_b < z < z_l \\ 1 & \text{at} \quad z \ge z_l \end{cases} \qquad k_2 = \begin{cases} 200 & \text{at} \quad 0 \le z \le z_b \\ 50 & \text{at} \quad z_b < z < z_l \\ 150 & \text{at} \quad z \ge z_l \end{cases}$$

The formula to calculate the solar radiation energy for the case considered is simplified:

$$E(z_n, t) = E_0 \times 10^{-\alpha z_n}$$

where z_n is the depth of the nth layer (m);

$$\alpha = 0.01 + 0.001 \sum_{k=0}^{n} [p(z_k, t) + d(z_k, t)]/z_n$$

According to Vinogradov *et al.* (1975), $E_{\max} = 70\,\text{cal}\,\text{cm}^{-2}\,\text{day}^{-1}$. A marked suppression of phytoplankton is observed at $E > 100\,\text{cal} \times \text{cm}^{-2}\,\text{day}^{-1}$. Hence, (3.2) when describing the photosynthesis intensity at depth z will change with the following dependence taken into account:

$$K_T = 0.041E \times 10^{0.25[1-E/E_{\max}]}$$

For this case, the formula parameterizing the bacterioplankton biomass dynamics is transformed:

$$\frac{\partial b}{\partial t} = R_b - \mu_b b - t_b b - \sum_{j \in \omega_0 \setminus (p,b)} C_{bj} R_j + k_2 \frac{\partial^2 b}{\partial z^2} + (V_z - w_p) \frac{\partial b}{\partial z}$$

The rate of change of the biomass of each zooplankton element $\partial Z_i/\partial t (i = 1\text{–}7)$ is determined by the level of consumption (ration) $R_{Z,i}$, nutrient assimilation $1/_{uZ,i}$, expenditure on energy exchange $T_{Z,i} = t_{Z,i} Z_i$, the rate of dying-off $\mu_{Z,i} = 0.01 Z_i$, consumption of the ith element by the jth one with the coefficient C_{ij}, and age transition of nauplii into copepod of filtrators and predators. Thus,

$$\frac{\partial Z_i}{\partial t} = \Omega_{m,i} + (1 - h_{Z,i}) R_{Z,i} - (t_{Z,i} + \mu_{Z,i}) Z_i - \sum_{j \geq i}^{7} C_{ij} R_{Z,j}, (i = \overline{1,7})$$

where $h_{Z,i} = 1 - 1/u_{Z,i}$ and $\Omega_{m,i}$ characterizes the age transition of nauplii into another category of zooplankton. The transition of nauplii Z_2 into copepod of filtrators (Z_3, Z_4) takes place with the following intensity:

$$\Omega_{m,3} = Z_2 Z_3 \{15(Z_3 + Z_4 + Z_5 + Z_6)\}^{-1}$$

and

$$\Omega_{m,4} = Z_2 Z_4 \{20(Z_3 + Z_4 + Z_5 + Z_6)\}^{-1}$$

and for predatory copepods (Z_5, Z_6) with the intensity:

$$\Omega_{m,4} = Z_2 Z_5 \{15(Z_3 + Z_4 + Z_5 + Z_6)\}^{-1}$$

$$\Omega_{m,6} = Z_2 Z_6 \{20(Z_3 + Z_4 + Z_5 + Z_6)\}^{-1}$$

To calculate the rations of zooplankton elements with (3.3), the coefficient ν is determined from the condition $\nu = 0.01/B_{\min}$. In calculation of coefficients C_{ij} it is assumed that nutritional requirements refer to different nutritional objects in proportion to biomasses of the latter with nutrition selectivity taken into account. Coefficients C_{ij} for all elements of the ecosystem are shown in Figure 3.14.

The daily migrations of zooplankton are simulated by adding, to the daily ration of the elements Z_j ($j = 4$–7) in the 0–50-m layer, a certain share (k_Z) of the total nutritional requirements for the same elements of the community but located in the layer 50–200 m. The coefficient k_Z is assumed to be dependent on time:

$$k_z = \begin{cases} 0.02 + 0.0016t & \text{at} \quad t \leq 50\,\text{days} \\ 0.1 & \text{at} \quad t > 50\,\text{days} \end{cases}$$

Here various scenarios of migration and nutrition of zooplankton elements are possible. The only reliably known fact is that some share of zooplankton elements from deep layers meet part of their nutritional requirements in the upper layers of the ocean.

In this case the equation for detritus biomass is written as:

$$\frac{\partial d}{\partial t} = \sum_{i \in \omega_0} (H_{Z,i} + M_{Z,i}) - \sum_{j \in \omega_0 \backslash p} C_{dj} R_j + k_2 \frac{\partial^2 d}{\partial z^2} - \mu_d d + (V_z - w_d) \frac{\partial d}{\partial z}$$

here the coefficient of the rate of gravitational sedimentation of detritus (cm day^{-1}) is written in the form:

$$w_d = \begin{cases} 25 & \text{at} \quad z_b \leq z \leq z_l \\ 50 & \text{at} \quad z < z_b, z > z_l \end{cases}$$

For calculations we assume that all elements of the ecosystem at the point of upwelling are vertically uniformly distributed: $n(z,0) = 250\,\text{mg m}^{-3}$, $d(z,0) = 0$, $b(z,0) = 1$, $p(z,0) = 0.5$, $Z_i(z,0) = 0.01$–$0.5 (i = 2$–$7)$, $Z_1(z,0) = 0.0001\,\text{cal m}^{-3}$. Note that variations of these values, within their increasing or decreasing by a factor of 50, practically do not affect the character of the dynamics of the whole system, and in 50 days the system always approaches the same level of values of the biomasses of all components (i.e., as if it "forgets" about the initial fluctuations). At the upper ($z = 0$) and lower ($z = 200\,\text{m}$) boundaries of the zone of modeling, zero gradients were chosen for all elements, except n and d. For them the boundary conditions are as follows:

$$\left.\frac{\partial n}{\partial z}\right|_{z=0} = 0; \left.\frac{\partial d}{\partial z}\right|_{z=0} = 0; \left.\frac{\partial d}{\partial z}\right|_{z=200} = -0.5; n(200,t) = 250\,\text{mg m}^{-3}$$

In evaluation of the change of the system in time t, it is assumed that it takes more than 60 days for water masses to pass the length from the place of upwelling to the oligotrophic zone (the climax state of the system). Figure 3.15 shows a temporal change of concentration (biomass) of the system's elements and, hence, with moving away from the zone of upwelling. It is seen that biomasses of phytoplankton and bacteria grow most rapidly. Phytoplankton reaches maximum development on the 5–10th day, and bacteria – on the 10–15th day. From an absolute value, maximum phytoplankton (4,500 cal m^{-2}) is characteristic of intensive tropical upwellings near the equator in the eastern sector of the Pacific Ocean or near the western coast of America.

Small filtrators are somewhat backward in their development, and large filtra-

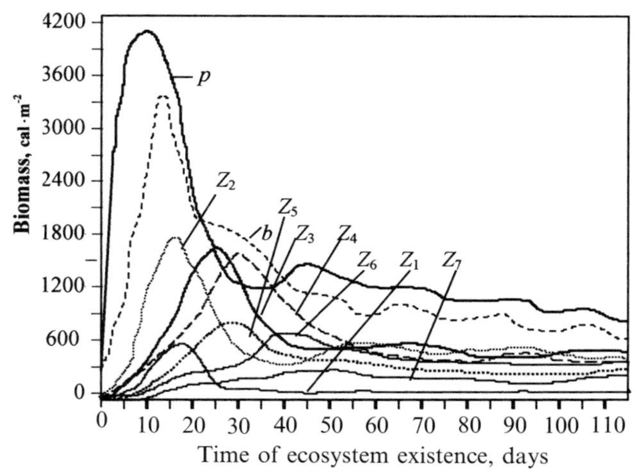

Figure 3.15. The temporal dependence of the total biomass of living elements of an ecosystem in the 0–200-m layer.

tors develop still slower. The biomass of the latter reaches a maximum only on the 30th day of the system's existence. Nevertheless, their combined impact on phyto- and bacterioplankton, along with the retarded growth of the latter due to an exhausted supply of biogenic elements, leads to a drastic reduction of the biomass of phytoplankton and bacterioplankton (i.e., feedbacks to restore the supplies of free mineral forms of nutrients contribute little to R_p and R_b). After 40 days of development, phytoplankton exists mainly at the expense of nutrient elements which rise to the euphotic zone through the layer of discontinuity as a result of turbulent mixing, and further its biomass decreases rather slowly. Here, mechanisms start functioning, stabilizing the community due to external energy fluxes.

Predators are still more inert than filtrators, and the biomass of their different groups reaches a maximum only on the 35–50th day. By this time, that is, on the 50–60th day, the system becomes quasi-stationary, with a low concentration of all living elements and a balance ratio between photosynthesis intensity and the inflow of nutrient elements from the lower water layers across the pycnocline. In due course, against a background of oscillations of the biomass of all living elements, its further gradual degradation takes place. Such a quasi-stationary, mature state of the community, characterized by low variability in time and hence in space, is inherent to oligotrophic areas of the tropical regions of the World Ocean, and first of all, chalistatic zones of planetary convergences.

Note should be taken of some relationships between the elements of the model. First of all, the model developed here clearly demonstrates a change with the moving away from the zone of upwelling and separation of maxima of the biomass of phytoplankton, filtrators, and predators which has been emphasized by many experts. One fact that attracts attention is that an overwhelming majority of filtrators are observed in the regions close to upwelling, whereas in the oligotrophic regions far from upwelling, predators constitute about half the total mass of zooplankton

Table 3.6. Comparison of estimates of the biomass of the upwelling ecosystem's elements (cal m^2) modeled and measured during the 50th voyage of the ship *Vitiaz* (*www.vitiaz.ru*).

Element of the ecosystem	Community at a middle stage of maturity (30–40 days)			Ripe community (60–80 days)	
	Model data		Field measurements	Model data	Field measurements
	$\Delta t = 30$ days	$\Delta t = 40$ days	($\Delta t = 35$ days)	($\Delta t = 70$ days)	($\Delta t = 70$ days)
Phytoplankton, *p*	1,319	1,092	2,000	827	900
Bacteria, *b*	1,673	864	4,100	564	2,180
Nauplii, Z_2	394	303	321	300	–
Small filtrators, Z_3	1,338	612	525	290	74
Large filtrators, Z_4	1,416	726	420	252	164
$Z_3 + Z_4$	2,754	1,338	945	542	238
Predatory *Cyclopoida*, Z_5	624	491	495	203	236
Predatory *Calanoida*, Z_6	288	600	610	191	175
Predatory *Chaetognatha* and *Polychaetha*, Z_7	184	183	15	102	51
$Z_5 + Z_6 + Z_7$	796	1,274	1,110	496	462

(Figure 3.15). The same relationships are observed in the tropical regions of the ocean.

It is of interest to compare not only a general qualitative change of the process of the upwelling ecosystem's development but also some absolute values obtained from the model and oceanic studies. This has been done in Table 3.6. Of note is the fact that the actual amount of bacteria turned out to be much greater than model calculations. Apparently, this results from the fact that the model leaves out of account the growth of bacteria due to a consumption of allochthonous organic matter. However, this fact requires a thorough experimental study. In other respects, Table 3.6 characterizes suppositions and descriptions assumed in the model as acceptable for practical application. This is also confirmed by the curves characterizing the dynamics of the ecosystem's vertical structure. Calculations showed that in the period close to the formation of the system ($t = 5$ days), when the total amount of phytoplankton approaches a maximum, its biomass is uniformly high in the 0–50-m layer. All the other living elements of the system have a more or less clearly expressed and rather narrow (in depth) maximum, connected with the layer of the thermocline. However, on the 10th day, the supply of nutrients in the upper layer becomes almost completely exhausted, however, at a depth of 10–20 m the phytoplankton biomass remains at a maximum. Still deeper, near the upper boundary of the layer of discontinuity, due to biogenic elements coming across the thermocline, a second, "lower" maximum starts forming, which is still poorly expressed at this time. Such a two-maximum structure also forms in the vertical distribution of other elements of the system. It is especially clearly expressed on the 20–30th day, when both maxima become clear. Due to a lowering of the layer of discontinuity to 55–75 m, both clearly differ in depth.

With the deteriorating biogenic background of the surface layer, the vertical transport of nutrient elements from beneath the thermocline plays a growing role, and as a result, the lower maximum becomes greater in absolute value than the upper maximum. Then, with further lowering of the thermocline to 80–100 m and deeper, the illumination at its upper boundary becomes insufficient for intensive development of phytoplankton. A situation occurs when the lower maximum "breaks away" from the thermocline, and its location is determined by the limit of illumination from above and by the flux of nutrient elements from below. Such a location of the lower maximum of phytoplankton above the thermocline is characteristic of oligotrophic regions (Vinogradov *et al.*, 1970) and, hence, the proposed model describes adequately the qualitative pattern of the spatial structure of the upwelling ecosystem in its dynamics.

In a more mature state of the system ($t > 50$–60 days) the oligotrophic and ultra-oligotrophic regions are characterized by an almost complete disappearance of the upper maximum. In the model, all living elements of the system retain only the lower maximum lying well above the thermocline, this time due to the increased water transparency – deeper than in the meso-oligotrophic region. In fact, this is characteristic, primarily, of phytoplankton. Zooplankton, predatory zooplankton in particular, even in oligotrophic regions often retains not only the lower but also the upper maximum. This is connected with the process of vertical migration, which should be more accurately described using the respective model unit.

Hence, we have a problem of evaluation of the model results reliability. This problem can be solved by changing the model's parameters. Though the results demonstrated in Figures 3.15 and 3.16 and in Table 3.6 do not contradict available ideas about the behavior and structure of the upwelling ecosystem, nevertheless, it is important to know which changes in certain parameters most affects the

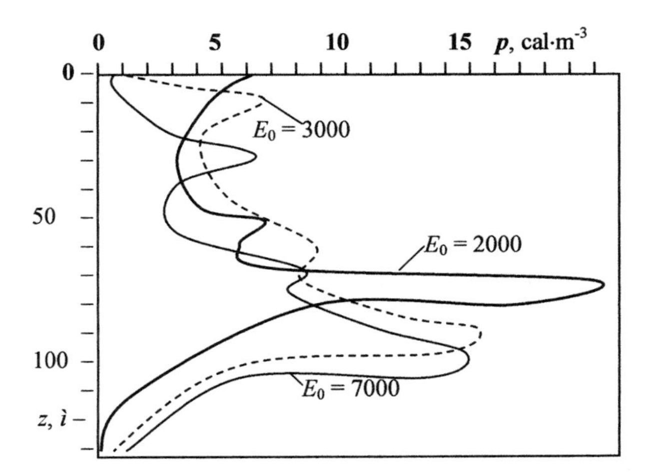

Figure 3.16. Dependence of the ecosystem's vertical structure on the illumination E_0 (kcal m^{-2} day^{-1}) at the 30th day after the upwelling.

developed systems, as well as to determine situations in which the system looses stability.

The system is consisted stable in the time interval $(0, T)$ if the phase coordinated determining it (in this case the biomass of elements B_i $(i = 1, \dots, n)$) are within the "psychologically permissible limits":

$$B_{i,\min} \leq B_i \leq B_{i,\max}(i = 1, \dots, n) \tag{3.4}$$

With this assumption applied to a biologically correctly constructed model of the community determined by the graph of trophic bonds of its elements, one can show that with convergence of the biomass of any element of a community to infinity, $B_i \to \infty$, there will be only one element for which the biomass $B_j \to 0$. It follows from this that the right-hand part of (3.4) becomes unnecessary (i.e., it is enough to satisfy only the left-hand part of the inequality). The same conditions hold for meeting the total requirements of stability:

$$B_{\min} \leq \sum_{i=1}^{n} B_i$$

where $B_{\min}$ is some minimum total biomass. Thus, the community model is considered stable if at a fixed trophic structure in the time interval $(0, T)$ the above inequality is satisfied. Calculation of $B_{\min}$ is connected with summing-up minimal unconsumed biomasses for all trophic levels.

First of all, based on the constructed model, we made a series of calculations varying the coefficients of the model in order to analyze their impact on the ecosystem's behavior. Note that there is no need to evaluate the model parameters with similar accuracy. Some of them, without detriment to final results, can be determined approximately, whereas the determination of others needs special attention. Some estimates of the consequences of changes in various model parameters are given in Tables 3.7–3.10. As seen, a deviation of nutrient assimilation (u^{-1}) by $\pm 20\%$ from the value prescribed in the model leads to an increase of the r.m.s. deviation of the model's trajectory from real values by not more than 4% for $Z_3 + Z_4$ and by 70% for Z_2. Also, of less importance are deviations of the coefficient characterizing an expenditure on exchange. Similar deviations in T and H cause comparable deviations in the pattern of the vertical distribution of zooplankton elements. Microzooplankton and filtrators are more sensitive to deviations in T and H than predators. Hence, T and H should be more thoroughly determined for the lowest trophic levels of the ecosystem compared with the higher ones.

It is seen from Table 3.9 how weakly a change of initial conditions reflects on the concentration of the remaining elements of the ecosystem, especially in the case of its relatively long lifetime. In general, similar model calculations show that the impact of a drastic violation of the initial relationships between trophic levels (in the case of Table 3.10, a multiple increase of the biomass of predatory elements) tells on the development of a community only for a short time – in the initial period ($t = 20$ days).

After examination of Table 3.10 it might seem unnatural that an increase in the

Table 3.7. The variability of the distribution of phytophages and nauplii with variations of some model parameters.

Δ	Model	$u^{-1} - 20\%$	$u^{-1} + 20\%$	$t_Z - 2\sigma$	$t_Z + 2\sigma$	$\alpha + 50\%$	$K_2 = 0.3$	$K_2 = 0.2$
Δ_1	0.29	0.35	0.43	0.76	3.36	7.9	1.44	2.6
Δ_2	0.75	0.91	1.27	1.02	1.71	1.78	3.5	6.0

Notes: Z_2, Z_3, Z_4 = observed values after Vinogradov *et al.* (1970); K_2 = coefficient of the use of assimilated nutrients for growth; $\Delta_1 = [(1/200)\int_0^{200}(Z_3(z,40) + Z_4(z,40) - Z_3 - Z_4)dz]^{1/2}$, $\Delta_2 = [(1/200)\int_0^{200}(Z_2(z,40) - Z_2)dz]^{1/2}$.

Table 3.8. Mean square deviations Δ of the trophic level's biomass with varying expenditures of energy exchange (T) and non-assimilated nutrients (H).

Δ_i	$T - 20\%$	$T - 10\%$	$T + 10\%$	$T + 20\%$	$H - 20\%$	$H - 10\%$	$H + 10\%$	$H + 20\%$
Δ_2	1.7	1.2	0.6	1.9	2.4	0.7	1.9	2.6
Δ_3	0.3	0.2	0.3	1.0	0.5	0.5	0.9	1.3
Δ_4	0.5	0.4	0.6	1.3	0.9	0.4	0.6	1.5
$\Delta\sum_{i=2}^{4} Z_i$	0.6	0.5	0.6	1.2	0.9	0.6	1.0	1.8
Δ_5	1.4	0.9	0.7	1.2	0.9	0.6	1.2	1.1
Δ_6	0.1	0.1	0.1	0.1	0.1	0.1	0.1	0.2
Δ_7	0.8	0.4	0.1	0.2	0.5	0.2	0.1	0.2
$\Delta\sum_{i=5}^{7} Z_i$	1.8	1.2	1.2	1.3	1.2	0.6	1.1	1.3

Notes: Z_i and $Z_{i,\text{exp}}$ = model value without and with a change of the parameter, respectively;

$$\Delta_i = \left\{\frac{1}{200}\int_0^{200} [Z_i(z,40) - Z_{i,\text{exp}}(z,40)]dz\right\}^{1/2}$$

number of predators leads, at chosen moments in time, to an increase of phytophagan biomass, the latter causing growth of phytoplankton on the 5th day. This effect is connected with the phase shift in the dynamic curves of the changes of biomasses of all elements. The initial prevalence of predators leads to a sharp reduction in the number of phytophagans during the first several days of the ecosystem's existence, which leads to a maximum level of phytoplankton being reached rapidly and causes growth in the amount of phytophagans. It is important here that the biomass of elements changes little compared with the sharp initial change of predator biomass. To assess the reliability of the results of modeling and to plan for further studies, it is of course necessary to carry out a full-scale study of the most crucial (sensitive) parameters of the model. Figure 3.16 shows the effect of oscillations of solar irradiance on the vertical structure of the ecosystem on the 30th day of its existence. It is seen that an increase in the ocean surface irradiation leads to an insignificant lowering of the lower maximum, to a faster expenditure of nutrients above the layer of discontinuity, and to a stronger suppression of phytoplankton near the surface. With diurnal irradiance of 2,000 kcal m^{-2}, the maximum rate of

Table 3.9. Mean square deviations Δ of the vertical distribution of various elements of the ecosystem on the 30th day of its existence, with changing initial concentration of nutritional salts. Notes are given in Table 3.7.

Δ_i	Changes in $n(z, t_0)$		
	−50%	−20%	+20%
Δ_2	5.0	0.6	0.9
Δ_3	0.4	0.4	0.3
Δ_4	1.3	0.1	0.1
$\Delta\sum_{i=2}^{4} Z_i$	1.5	0.4	0.4
Δ_5	1.1	0.5	0.4
Δ_6	1.1	0.04	0.03
Δ_7	0.1	0.03	0.02
$\Delta\sum_{i=5}^{7} Z_i$	1.2	0.5	0.4

Table 3.10. The dynamics of the ecosystem's structure with changing initial concentration of predators (cal m^{-2}). The model uses the initial concentration $Z_5(z, t_0) = Z_6(z, t_0) = Z_7(z, t_0) = 0.1\ \text{cal m}^{-3}$.

Element	Day 5			Day 30		
	Model	Initial biomass of predatory zooplankton $(Z_5 + Z_6 + Z_7)$ is increased n times		Model	Initial biomass of predatory zooplankton $(Z_5 + Z_6 + Z_7)$ is increased n times	
		$n = 100$	$n = 1000$		$n = 100$	$n = 1000$
Bacteria, b	1,789	2,263	2,263	2,269	2,159	1,995
Phytoplankton, p	3,264	3,462	3,794	3,971	2,086	2,614
Protozoa, Z_1	31	12	3	18	3	2
Microzooplankton, Z_2	178	106	101	109	265	465
Small phytophagans, Z_3	70	85	80	80	201	415
Large phytophagans, Z_4	67	14	74	145	156	191
Predatory *Cyclopoida*, Z_5	18	14	9	114	119	53
Predatory *Calanoida*, Z_6	1	25	41	14	25	22
Predatory *Chaetognatha*, Z_7	36	234	183	213	437	821

Table 3.11. Phytoplankton biomass with varying initial values of nutrient concentrations ($cal\,m^{-3}$).

Layer (m)	$n(z, t_0) = 100\,mg\,m^{-3}$			$n(z, t_0) = 100\,mg\,m^{-3}$		
	Lifetime of the system (days)					
	10	50	100	10	50	100
0–10	0.3	0.1	0	0.3	0.1	0
10–20	31.7	8.7	10.9	37.4	16.1	15.4
20–30	39.6	9.4	10.7	289.7	7.8	10.4
30–40	39.4	9.2	10.6	207.7	7.8	10.4
40–50	36.6	9.7	10.9	295.1	7.9	10.7
50–60	36.7	9.5	10.9	153.7	7.1	10.5
60–70	41.7	11.4	10.9	62.1	18.9	8.8
70–80	13.4	15.7	11.2	15.6	10.3	17.9
80–90	1.9	31.4	17.0	2.1	39.0	17.9
90–100	0.6	3.6	12.8	0.6	4.2	1.4
100–150	3.1	2.1	2.1	3.1	2.2	2.4
150–200	0.9	0.4	0.5	0.8	0.7	0.7

photosynthesis is observed near the surface. Irradiance oscillations from 2,000 to 7,000 $kcal\,m^{-2}\,day^{-1}$ with a sufficient amount of nutrients only affects the vertical shifting of the distributions of phyto- and zooplankton, practically without changing their form. It means that the proposed model can be used to study the upwelling zones in a wide range of geographic latitudes. Note that a change of the indicator of irradiance reduction (α) leads to the same effect as a change of irradiance.

The impact of changes in the concentration of nutrient elements at an initial moment on the dynamics of phytoplankton biomass is shown in Table 3.11. It follows from this table that oscillations of $n(z, 0)$ from $100\,mg\,m^{-3}$ or greater do not practically affect the state of the system at all times, but they affect only the initial period of its development. An important parameter of the ecosystem is the P/B-coefficients for phytoplankton and other elements. For phytoplankton this indicator depends on the function $\gamma(t)$. The above prescribed temporal dependence of the P/B-coefficient is not always confirmed practically. Numerous oceanologists suppose that the P/B-coefficient for phytoplankton in oligotrophic waters can be higher than in eutrophic waters at the initial moments of the system's existence. At any rate, with the P/B-coefficient assumed to decrease not to unity but to 0.5 on the 15th day, the system practically dies on the 35th day, since such elements as large predators (Z_6 and Z_7) have no time to develop as a result of a lack of nutrients. An increase of the P/B-coefficient for phytoplankton leads to a faster impoverishment of the nutritional background in the upper layer and, hence, to an accelerated disappearance of the upper maximum. Figure 3.17 gives an example when $R_p/p \equiv 3$. It turns out that both in this case and in other situations, when the P/B-coefficient for

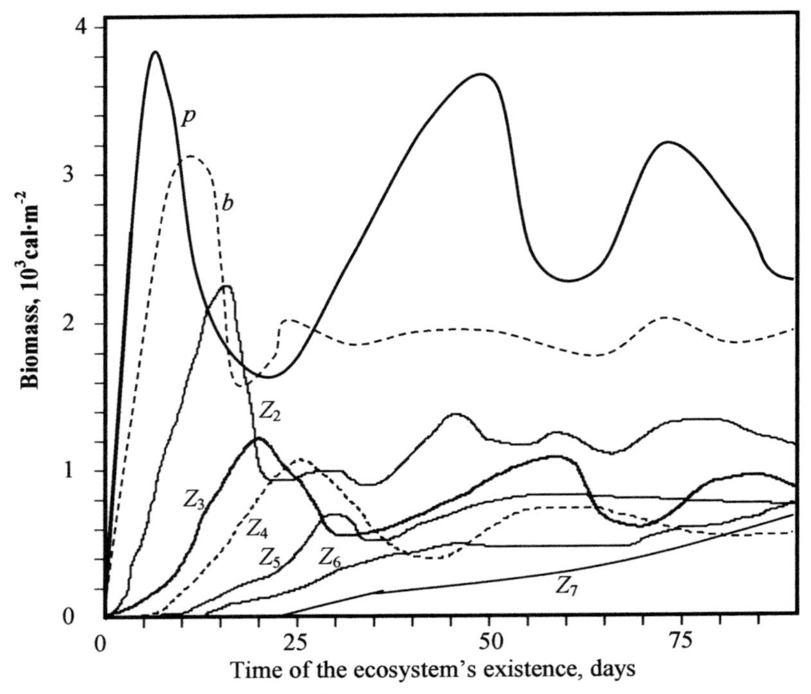

Figure 3.17. The temporal change of biomass of the ecosystem's components when the P/B-coefficient of phytoplankton always equals 3.

phytoplankton exceeds unity, the phytoplankton biomass oscillates considerably, reaching values comparable in magnitude with the concentration of plankton at initial moments of the system's development – and the biomass of predators steadily grows. Situations occur that are not observed in reality, which raises doubts with respect to the existence of high P/B-coefficients after 15 days of the ecosystem's development.

Finally, to assess the adequacy of the model of the upwelling ecosystem, considered separately from other ecosystems because of the absence in the model of nekton predators, it is necessary to determine the role of the high trophic levels. Also, the problem arises of the limits for aggregating the elements of the ecosystem taken into account. Unfortunately, the available information on trophic bond rigidities does not enable one to answer these questions. It is only possible to consider situations based on hypotheses and suppositions. One such situation is shown in Figure 3.18. It is seen that as a result of the absence of predatory elements in the system, the amount of phytophagans increases without any control, and the system looses stability on the 50th day. The conclusion can be drawn that "cannibalism" of predators and their consideration in the model is a stabilizing factor. It becomes clear that an introduction of nekton in the model, whose biomass is small in the open waters of the World Ocean, should not affect the results of modeling shown above.

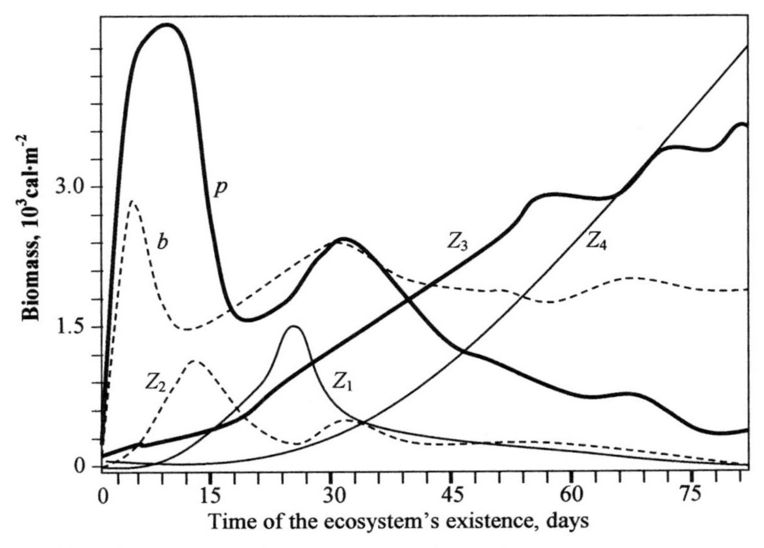

Figure 3.18. Hypothetic temporal changes of biomass of the ecosystem's components in the absence of predators.

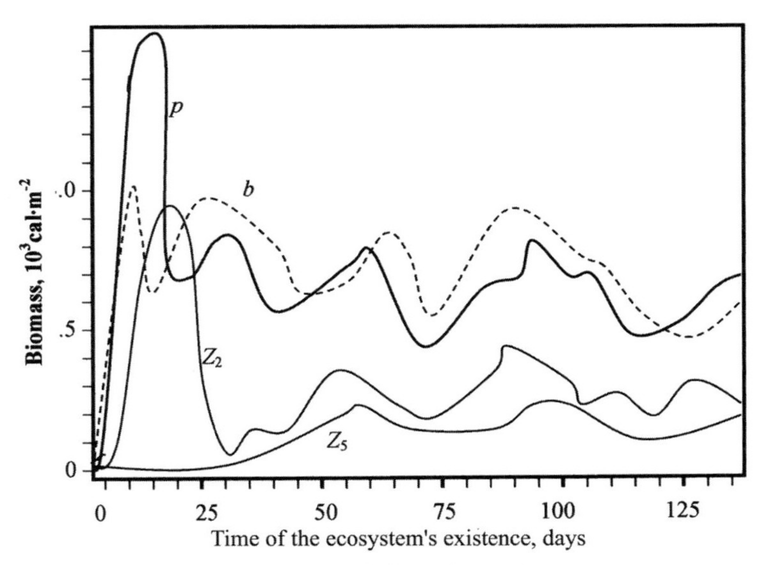

Figure 3.19. Dependence of the biomass of the ecisystem's components on the time of its existence with regular (every 5 days) intrusions of nutrients into the upper mixing layer.

During a study of the Peru upwelling ecosystem shown in Figure 3.19 the role has been assessed of the vertical rate of deepwater uprising. The value $V_z = 10^{-3}\,\text{cm s}^{-1}$ is most widely used. The V_z parameter is difficult to measure directly, and for most of the ocean regions it is either unknown or estimated incor-

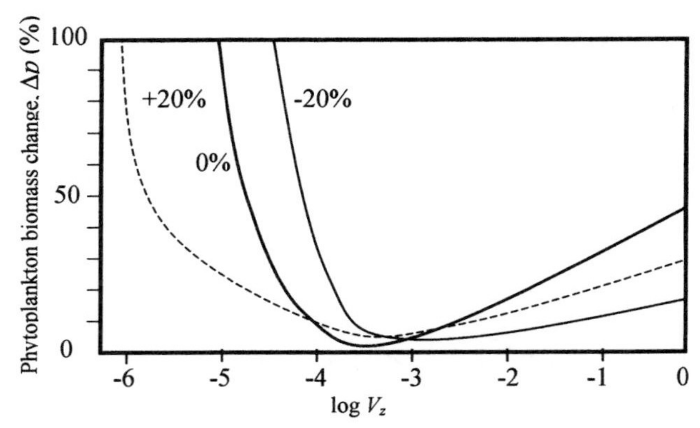

Figure 3.20. Effect of the oscillations in the rate of vertical water lifting on the behavior of the Peruvian Current ecosystem (PCE). The curves indicate the deviations of V_z (cm s^{-1}) values in the formula for the calculation of the biomass of nutrients.

rectly. The existing estimates are distributed within the range 10^{-4}–0.1 cm s^{-1}. In the case considered, the effect of errors in evaluation of V_z is demonstrated by the curves in Figure 3.20. As for the case of the Peru upwelling, here the pattern of deviation from the initial vertical distribution of the elements of the modeled community does not change significantly within the range $10^{-4} \leq V_z \leq 10^{-2}$ and even 0.1 cm s^{-1}. It is however violated with more intensive and mainly with a slower (10^{-5}–10^{-6} cm s^{-1}) water uprising.

4

Natural disasters and humankind

4.1 THE NATURE–SOCIETY SYSTEM AND THE DYNAMICS OF ITS DEVELOPMENT

4.1.1 Civilization development trends

A globalization of humankind's impact on the environment with the continuing growth of the population and a great uncertainty of the prospects for civilization development have raised numerous principal problems of economic, political, and social character (Lomborg, 2001; Malinetskiy *et al.*, 2005). One of them is an assessment and prediction of natural disasters emphasizing the role of anthropogenic processes in their origin. Clearly, this problem cannot be selected and discussed separately from other problems of society–nature interaction. Nevertheless, to obtain a set of relevant problems, it is necessary from the broad spectrum of natural–anthropogenic processes to select the key aspects, an understanding of which will enable one to answer the principal questions of sustainable development of the global Nature–Society System (NSS). Though the notion of "sustainable development" has not been clearly defined, the term "sustainable development" is widely used in scientific literature. Various discussions on the origin and meaning of this term have long taken place among specialists in various fields. They include widely known decisions of international conferences in Stockholm (1972), in Rio de Janeiro (1992), and in Johannesburg (2002), as well as an accomplishment of several international programs, such as the International Geosphere–Biosphere Programme (IGBP), International Human Dimensions Programme (IHDP), the World Climate Research Programme (WCRP), the International Biodiversity Programme DIVERSITAS, the Global Carbon Project (GCP), etc. (Braswell *et al.*, 1996).

The principal conclusion of the July 2001 Amsterdam Conference "Challenges to Changing Earth" is that "the Earth System behaves as a single, self-regulating system of physical, chemical, biological and human components. The interactions and feedbacks between the component parts are complex and exhibit multi-scale temporal and spatial variability". In reality, diverse important impacts of human activities on a global scale can be traced both on land and in the World Ocean. In some spheres, the anthropogenic changes in the environment have become comparable or even exceeded the limits of natural environmental changes. "Human-driven changes cause multiple effects that cascade through the Earth System in complex ways. These effects interact with each other and with local- and regional-scale changes in multidimensional patterns that are difficult to understand and even more difficult to predict". Therefore, human activities have the potential to switch the Earth System to irreversible modes of operation that may prove unfavorable and even impossible for development of living matter.

> *In terms of some key environmental parameters, the Earth System has moved well outside the range of natural variability exhibited over the last half million years at least. The nature of changes now occurring simultaneously in the Earth System, their magnitudes and rates of change are unprecedented. The Earth is currently operating in a no-analog state.*

These extracts from the decisions of the Amsterdam Conference suggest that "an ethical framework for global stewardship and strategies for Earth System management are urgently needed. The accelerating human transformation of the Earth's environment is not sustainable". This problem can be solved by developing new scientific directions in global ecology, (e.g., ecoinformatics as a possible solution) (Arsky *et al.*, 1992; Krapivin and Kondratyev, 2002; Krapivin and Potapov, 2002). No doubt, this scientific direction should comprise a wide spectrum of natural and social sciences, creating an efficient mechanism for using available methods and technologies and ensuring their goal-directed development. A certain basis for this unification has been already created (Gorshkov, 1990; Kondratyev, 1998, 1990; Gorshkov *et al.*, 2000; Yanovsky, 1999; Gerstengarbe, 2002; Berner and Hollerbach, 2001).

Following the work of Steffen and Tyson (2001), we shall analyze some formulations aimed at defining the Earth System as a complex of interacting physical, chemical, biological, and human activity components, which determine the processes of transportation and transformation of substances and energy, and thus keeps the planet habitable.

> *The Earth is a system that life itself helps to control. Biological processes interact strongly with physical and chemical processes to create the planetary environment, but biology plays a much stronger role than previously thought in keeping Earth's environment within habitable limits.*

It should be mentioned that this thesis has undoubtedly a fundamental meaning, but in this context it would be rather appropriate to refer to the conception of the biotically regulated environment extensively elaborated in Gorshkov (1990) and Gorshkov *et al.* (2000).

> *Global change is much more than climate change. It is real, it is happening now and*

it is accelerating. Human activities are significantly influencing the functioning of the Earth System in many ways; anthropogenic changes are clearly identifiable beyond natural variability and are equal to some of the great forces of nature in their extent and impact.

We have to emphasize that in the contemporary world, when the scientific discussion on the extent of anthropogenic pressure on climate becomes transformed into a political discussion reaching the level of state leaders, it should be pointed out that global climate change deserves much keener attention in the context of the Kyoto Protocol, which has been done in a number of publications by Russian scientists (Kondratyev and Demirchian, 2001; Kondratyev *et al.*, 2001, 2003a; Kondratyev, 2002). However, we agree with Steffen and Tyson (2001) in a sense that climate is not the chief item in global change problems. In this connection, the following facts should be mentioned:

- Over a few generations humankind has been exhausting fossil fuel reserves that were generated over several hundred million years.
- Nearly 50% of the land surface has been transformed with significant negative consequences for biodiversity, nutrient cycling, soil structure and biology, and climate.
- More nitrogen is now fixed synthetically (anthropogenically) and applied as fertilizers in agriculture than is fixed naturally in all terrestrial ecosystems.
- More than half of all accessible freshwater is used by humankind, and underground water resources are being depleted especially rapidly.
- The concentrations of greenhouse gases, including CO_2 and CH_4, have substantially increased in the atmosphere.
- Coastal and marine habitats are being dramatically altered; the areas of wetlands have almost halved.
- About 22% of marine fisheries are overexploited, and an additional 44% are at the limit of exploitation.
- Both on land and in the World Ocean intensive processes of biodiversity reduction take place.
- "The human activity causes numerous and intensive effects on the Earth System in complex ways. Therefore, global changes cannot be understood in terms of a simple cause–effect paradigm. Effects of human activities propagate in the Earth System interacting with each other and in the processes of various scales."
- "The Earth system's dynamics are characterized by critical thresholds and abrupt changes. Human activity can unintentionally intensify changes with catastrophic consequences for the Earth System. Apparently, such changes could be avoided in the case of depletion of the stratospheric ozone layer. The Earth System has operated in different quasi-stable states, with abrupt changes occurring between them over the last half million years. Human activities clearly have the potential to switch the Earth System to alternative modes of operation that may prove irreversible."

Though the last two theses are very important, we must keep in mind that until now there are only preliminary quantitative assessments and only fragmentary data exist – there is not enough information for concrete judgement about the potential changes. This is true, for instance, in the case of the impact on the global ozone content in the stratosphere (Kondratyev and Varotsos, 2000). In this context, the problem of the anthropogenic impact on the global biogeochemical cycles of carbon, sulfur, nitrogen, and other chemical compounds deserves serious attention (Kondratyev, 2000a,b; Krapivin and Kondratyev, 2002; Kondratyev *et al.*, 2002a; Zavarzin and Kolotilova, 2001; Kashapov, 2002; Alverson, 2000; Irion, 2001).

> *The Earth is currently operating in a no-analog state. In terms of key environmental parameters, the Earth System has recently moved well outside the range of the natural variability exhibited over at least the last half million years. The nature of changes now occurring simultaneously in the Earth System, their magnitudes and rates of change are unprecedented.*

Such general assessments are important for developing an efficient technology to assess global environmental changes (Zerchaninova and Potapov, 2001). In the context of perspectives of contemporary civilization development, a key factor is undoubtedly the continuing global population increase: if during the second half of the 20th century the global population doubled, at the same time the grain crop production tripled, the energy consumption increased four-fold, and economic activity increased five-fold. Gorshkov (1990) has clearly demonstrated that *homo sapiens* have long left their ecological niche, which has brought forth the first signs of a developing ecological catastrophe.

Taking into account these conceptual circumstances, Steffen and Tyson (2001) have formulated two important conclusions regarding how society should respond to an emerging ecological threat and work out respective scientific developments aimed at an adequate understanding of the processes that govern the variability of the Earth System:

- Ethics of global stewardship and strategies for Earth System management are urgently needed. The inadvertent anthropogenic transformation of the planetary environment is, in effect, already a form of management, or rather dangerous mismanagement. This transformation is not sustainable. Therefore, the business-as-usual way of dealing with the Earth has to be replaced as soon as possible by planned strategies of an adequate management.
- A new system of global environmental science is gradually emerging. The largely independent efforts within the framework of various international programs form the basis for Earth System science being capable of solving these problems. This new science will employ innovative integration methodologies of its organization into a global system with transnational infrastructures, and a continuing dialogue with stakeholders around the world.

Analysis of published suggestions, opinions, and decisions of international governmental and non-governmental organizations concerning the strategy of studies and development of environmental sciences in conditions of increasing anthropogenic forcings suggests the conclusion that there is no complete and objective list of priorities for the problems of global change. This means that humankind has no scientifically grounded strategy for interaction with the environment.

4.1.2 Contemporary global ecodynamics

The accumulated data on global environmental changes enable one to formulate a general pattern of these changes and characterize possible trends for decades to come (Brown *et al.*, 2001; Losev, 2001).

- The area of natural land ecosystems shrinks at a rate of 0.5–1.0% yr^{-1}. By the beginning of the 1990s about 37% of these ecosystems remained. This trend will lead to a total abolition.
- From the estimates for 1985 the rate of human consumption of pure primary biological production constituted 40% on land and 25% in the whole biosphere. In the future these estimates will reach 80–85% and 50–60%, respectively.
- Between 1972 and 1995 the greenhouse gas (GHG) concentrations grew at a rate of tenths of a percent to several percent annually. Concentrations of CO_2 and CH_4 are expected to grow at the expense of an increasing rate of biota destruction.
- An annual 1–2% depletion of the ozone layer, the growth of the ozone hole in the Antarctic, and the growth of areas of other ozone holes. This trend will remain even with ceasing emissions of chlorofluorocarbons.
- Shrinking areas of forests, especially tropical forests. During the period 1990–2004 the forest areas shrunk at a rate of 13 mln ha yr^{-1}, and the ratio of "afore-station/deforestation" was 1:10. By 2030 the areal extent of the mid-latitude forests will decrease, and that of the tropical forests will halve.
- An increase of the arid areas including 40% of land, the growth of technogenic desertification caused by the broken moisture cycle, deforestation and soil contamination. This will lead to a reduction in agricultural lands per capita, fertility reduction, and a growing acidity and salinity of soil.
- During the last decades of the 20th century the level of the World Ocean was rising by 1–2 mm yr^{-1}. This rate may increase to 7 mm yr^{-1}.
- Natural disasters and technogenic accidents in the late 20th century were characterized by an annual increase of 5–7% with increasing damage of 5–10%, and an increasing number of victims by 6–12%. Preservation and intensification of these trends is a natural result of global processes.
- Biological species in the second half of the 20th century disappeared 100–1,000 times faster than anytime in the past. The trend may increase as the biosphere gets destroyed.
- There was a qualitative depletion of land waters caused by increasing volumes of sewage, growing numbers of point and areal sources of pollution, a broadening spectrum of pollutants, and an increase in pollutant concentrations. With this trend preserved, by 2030 more than 60% of the population will suffer water shortage.

- There is a persistent growth in the mass and number of pollutants accumulated in the environment and organisms. A migration of pollutants, especially radionuclides in food chains, intensifies.
- A worsening quality of life, a growing number of diseases connected with the spoiling of humankind's ecological niche, environmental pollution (genetics included), and the appearance of new diseases. Poverty, food shortage, high infant mortality, high levels of morbidity, inadequate provision of the developing countries with pure water, a high accident rate, a growing consumption of medicine, an increase of allergic diseases in the developed countries, a decrease of immune status. In the 21st century these trends will remain with an increase in the territory claimed by infectious diseases and the appearance of new ones (HIV/AIDS is of special concern).
- The global distribution, including through humans, of supertoxicants through the trophic chains. Damage to the human endocrine system will worsen the effectiveness of the human reproductive system, the brain, and other vital organs. All this will result in a growing number of people unable to conceive children.
- Artificial introduction and accidental invasion of alien species into ecosystems. The transport of the pests and diseases of plants, animals, and humans, and a reduction of biodiversity. These trends will continue in the 21st century.
- Changes observed in the World Ocean: destruction of reefs, reduction of mangrove ecosystems, depletion of fish supplies as a result of intensive fishing, reduction of whale populations, pollution of inland and coastal waters, and "red tides".

These trends suggest two principal conclusions: (1) in the second half of the 20th century there was a continuing worsening of environmental characteristics; (2) relevant expected trends will not only remain negative in the future but may even worsen. Thus, the trends of unsustainable development will intensify. To confirm this, we shall give some quantitative estimates.

Table 4.1 illustrates the dynamics of the global population, the principal features of which are the continuing growth of the absolute size of the population, with its anticipated rate of growth decreasing (Lutz *et al.*, 2004). Of course, the mean tendencies mask the strongest regional differences. Practically in all industrial countries the size of the population is either slowly increasing or decreasing. The USA is an exception, where the increment of population is 1% yr^{-1} partly due to immigration, which constitutes one-third. A total of 95% of the global increase in population in 2000 was due to developing countries including Asia (57% with an absolute increment of 45 million), Africa (23%), Latin America (9%), and the Middle East (5%). About half the global increase in 2000 resulted in India, China, Pakistan, Nigeria, Bangladesh, and Indonesia. A number of factors favor a reduction of birth rate and the rate of population growth, these include: improvement of the economic situation and the public health system in many countries; progress in women's education alongside their increasing status; easy access to contraceptives; etc. So, for instance, in Iran the rate of population growth decreased from 3.2% in

Table 4.1. The global population for 1950–2005 with a forecast for 2020.

Year	G	ΔG	Year	G	ΔG
1970	3.9642	58.1	1988	5.1478	74.0
1971	4.0218	57.6	1989	5.2226	74.8
1972	4.0805	58.7	1990	5.2986	76.0
1973	4.1398	59.3	1991	5.3757	77.1
1974	4.2050	65.2	1992	5.4540	78.3
1975	4.2661	61.1	1993	5.5333	79.3
1976	4.3284	62.3	1994	5.6138	80.5
1977	4.3913	62.9	1995	5.6955	81.7
1978	4.4552	63.9	1996	5.7785	83.0
1979	4.5100	54.8	1997	5.8626	84.1
1980	4.5859	75.9	1998	5.9478	85.2
1981	4.6526	66.7	1999	6.0343	86.5
1982	4.7202	67.6	2000	6.1223	87.0
1983	4.7889	68.7	2001	6.2113	89.0
1984	4.8587	69.8	2002	6.3017	90.4
1985	4.9294	70.7	2003	6.3933	91.6
1986	5.0011	71.7	2004	6.4865	93.2
1987	5.0738	72.7	2005	6.4886	87.4
			2020	7.4437	84.5

Notes: G = population size (billions of people), ΔG = population growth (millions of people per year).

1986 to 0.8% in 2000. The decrease of birth rate has been caused by a negative factor, such as the spread of AIDS (3.1 million people died in 2004 due to AIDS, and the number of people living with HIV in 2004 was 39.4 million).

Naturally, during the second half of the 20th century the global Gross Domestic Product (GDP) continued growing (though its rate slowed) (Table 4.2). Relative growth of the scales of the global economy in 2000 constituted 4.7% and exceeded the 1999 level (3.4%). The Gross World Product (GWP) reaching US$43 billion has provided an average GWP per capita of US$7102. This progress has been achieved due to a successful development of economies in the USA and Western Europe, a recovery of the economy in Asia after the 1997 financial crisis, and the economy in Latin America after the 1998 crisis, as well as a marked improvement of the situation in the countries with transitional economies. Note that the growth of the economy in China in 2000 constituted 7.5% and was the most significant in Asia. Figures for India, Pakistan, and Bangladesh are equal, respectively, to 6.7%, 5.6%, and 5%, which is especially important for the region with its population exceeding 1.2 billion, where the per-capita income is about US$1 per day or even less.

Fossil fuel consumption is a significant indicator of the development of both the economy and technologies (Table 4.3). As seen from this table, the year 2000 was the second consecutive year showing a decrease in the global level of total fossil fuel consumption after a long period of growth (which exceeded a factor of 4 starting in 1950). The global production of coal, oil, and natural gas in 2001

Table 4.2. GWP (1950–2001).

Year	U	U_G	Year	U	U_G
1950	6.7	2,641	1984	28.1	5,890
1955	8.7	3,112	1985	29.1	5,993
1960	10.7	3,516	1986	30.1	6,101
1965	13.6	4,071	1987	31.2	6,216
1970	17.5	4,708	1988	32.5	6,375
1971	18.2	4,805	1989	33.6	6,470
1972	19.1	4,933	1990	34.2	6,492
1973	20.3	5,157	1991	34.7	6,468
1974	20.8	5,174	1992	35.4	6,499
1975	21.1	5,154	1993	36.1	6,538
1976	22.1	5,312	1994	37.3	6,663
1977	23.0	5,432	1995	38.6	6,791
1978	24.0	5,573	1996	40.1	6,964
1979	24.8	5,672	1997	41.7	7,139
1980	25.3	5,688	1998	42.7	7,202
1981	25.8	5,698	1999	44.0	7,337
1982	26.1	5,664	2000	46.0	7,566
1983	26.9	5,728	2001	46.9	7,617

Notes: U = net GWP (US\$$10^9$, 1999); U_G – per-capita GWP (US\$, 1999).

Table 4.3. Global fossil fuel consumption (1950–2002) (10^6 tons of oil equivalent).

Year	C	O	G	Year	C	O	G
1950	1,074	470	171	1985	2,107	2,801	1,493
1955	1,270	694	266	1986	2,143	2,893	1,504
1960	1,544	951	416	1987	2,211	2,949	1,583
1965	1,486	1,530	632	1988	2,261	3,039	1,663
1970	1,553	2,254	924	1989	2,293	3,088	1,738
1971	1,538	2,377	988	1990	2,270	3,136	1,774
1972	1,540	2,556	1,032	1991	2,218	3,138	1,806
1973	1,579	2,754	1,059	1992	2,204	3,170	1,810
1974	1,592	2,710	1,082	1993	2,200	3,141	1,849
1975	1,613	2,678	1,075	1994	2,219	3,200	1,858
1976	1,681	2,852	1,138	1995	2,255	3,247	1,914
1977	1,726	2,944	1,169	1996	2,336	3,323	2,004
1978	1,744	3,055	1,216	1997	2,324	3,396	1,992
1979	1,834	3,103	1,295	1998	2,280	3,410	2,017
1980	1,814	2,972	1,304	1999	2,163	3,481	2,069
1981	1,826	2,868	1,318	2000	2,217	3,519	2,158
1982	1,863	2,776	1,322	2001	2,555	3,511	2,164
1983	1,914	2,761	1,340	2002	2,298	3,529	2,207
1984	2,011	2,809	1,451				

Notes: C = coal, O = oil, G = gas.

reached 7,956 mln t of oil equivalent at a growth rate of 1.3%. Now the contribution of fossil fuel consumption in the commercial production of energy is about 90%, the share of coal being 25%. This mean-global share is gradually decreasing, but in the USA the consumption of coal (~25% of global use) increased in 2000 by 1.6% due to the development of electric power stations functioning on coal. Contrary to this, the coal consumption in China (which is also ~25% of global use) decreased by 3.5% (from 1996 the decrease reached 27%). In India, the third largest coal consumer (7%), the level of consumption in 2000 increased by 5.4%.

Global oil consumption (Table 4.3) (its contribution to energy production is 41%) has increased by 1.1%, and in the USA, the leading petroleum user (26%), it has increased only by 0.1%, whereas in the Asian countries of the Pacific region (27% of global oil consumption) the use of oil has increased by 2.6%. The respective figures for Western Europe are 22% and 0.2%. As for natural gas (24% of commercial energy production), its level of consumption has increased by 2.1%. In the USA (where 27% of global natural gas is consumed), the scale of consumption has increased by 2.4%. The consumption of natural gas in the Baltic countries has also strongly increased: Lithuania – 29%, Estonia – 30%, and Latvia – 45%. Leaders of a 16% growth of natural gas consumption in Asia and Western Europe turned out to be South Korea and Spain. The last years are characterized by high prices of oil, with a maximum (US$44.5 per barrel) in 2004, which has drawn attention to analysis of perspectives of oil and gas extraction on the Arctic shelf of Russia and Alaska, though the Alaskan resources are comparatively small, and huge investments are needed for Russia.

According to the International Energy Agency, the total fossil fuel consumption can increase by 57% (2% yr^{-1}) by the year 2020, preserving a 90% share of consumption in energy production. The supposed global trend of coal consumption increase will constitute 1.7% yr^{-1}, with two-thirds of this increase falling on India and China. Petroleum will remain the main source of energy (increasing by 1.9% yr^{-1}) and its share in primary energy production will reach 40%, but the highest rate of consumption increase (2.7% yr^{-1}) is characteristic of natural gas consumption (mainly due to the gas-operating electric power stations).

There is the prospect for change in energy sources, though the available trends in this direction are characterized by low rates of generation. Nevertheless, the rate of increasing generation of energy by nuclear power stations in 2001 constituted 0.4%, and on the whole, this increase reached 7% during the last decade of the 20th century. The energy generation due to winds and use of photogalvanic solar arrays grows more rapidly.

Data of Table 4.4 reflect the dynamics of CO_2 emissions into the atmosphere and its increasing concentration in the atmosphere. Starting from 1950, the atmosphere has gained an amount of CO_2 equivalent to 217 bln t C with an annual four-fold increase of emissions. An important fact is that the fossil fuel use per unit production, calculated with CO_2 emissions taken into account, decreases: in 2000 this decrease constituted 3.6% (this figure is equivalent to 148 t C per US$1 million GWP), and during 50 years it reached 41%. Contrary to the Kyoto Protocol, CO_2 emissions in the western industrial countries increased during the period from 1990

Table 4.4. Global CO_2 emissions due to fossil fuels (1960–2002), its concentration in the atmosphere, and air temperature.

Year	T_a	H_1	C_A
1960	14.01	2,535	316.7
1965	13.90	3,087	319.9
1970	14.02	3,997	325.5
1975	13.94	4,518	331.0
1976	13.86	4,776	332.0
1977	14.11	4,910	333.7
1978	14.02	4,961	335.3
1979	14.09	5,249	336.7
1980	14.16	5,177	338.5
1981	14.22	5,004	339.8
1982	14.06	4,961	341.0
1983	14.25	4,944	342.6
1984	14.07	5,116	344.2
1985	14.03	5,277	345.7
1986	14.12	5,439	347.0
1987	14.27	5,561	348.7
1988	14.29	5,774	351.3
1989	14.19	5,882	352.7
1990	14.37	5,953	354.0
1991	14.32	6,023	355.5
1992	14.14	5,907	356.4
1993	14.14	5,904	357.0
1994	14.25	6,053	358.9
1995	14.37	6,187	360.9
1996	14.23	6,326	362.6
1997	14.40	6,422	363.8
1998	14.56	6,407	366.6
1999	14.32	6,239	368.3
2000	14.31	6,315	369.4
2001	14.46	6,378	370.9
2002	14.52	6,443	372.9

Notes: $H_1 = CO_2$ emissions to the atmosphere (10^6 t C), $C_A = CO_2$ concentration in the atmosphere (ppm), T_a = temperature (°C) (Vital, 2003).

by 9.2%, and in the USA – by 13% on average. The respective contribution of the developing countries has reached 22.8%. According to the Intergovernmental Panel on Climate Change (IPCC), by 2020 the annual global CO_2 emissions can reach 9–12.1 bn t C and by 2050 they will be within 11.2–23.1 bn t C. Therefore, the problem of global climate change due to CO_2 emissions is discussed in detail and studied by many authors (Kondratyev *et al.*, 2002b, 2003c, 2004b).

The most important component of economic development is grain production (Table 4.5), which has recently started decreasing from 1,869 mln t in 1999 to 1,840 mln t in 2000. The corn crop in 2000 turned out to be 2% below the

Table 4.5. Global grain production (1950–2000).

Year	P	P_G	Year	P	P_G
1950	631	247	1984	1,632	342
1955	759	273	1985	1,647	339
1960	824	271	1986	1,665	337
1965	905	270	1987	1,598	318
1970	1,079	291	1988	1,549	303
1971	1,177	311	1989	1,671	322
1972	1,141	295	1990	1,769	335
1973	1,253	318	1991	1,708	318
1974	1,204	300	1992	1,790	328
1975	1,237	303	1993	1,713	310
1976	1,342	323	1994	1,760	314
1977	1,319	312	1995	1,713	301
1978	1,445	336	1996	1,872	324
1979	1,411	322	1997	1,881	322
1980	1,430	321	1998	1,872	316
1981	1,482	327	1999	1,869	311
1982	1,533	332	2000	1,840	303
1983	1,469	313			

Notes: P = net grain production (10^6 t), P_G = per-capita consumption (kg).

maximum grain production reached in 1997 (1,881 mln t). The main reason of this decrease was a reduced corn crop in China from 391 mln t in 1998 to 353 mln t in 2000 (i.e., by about 10%). This has resulted from low grain prices that discouraged the farmers from raising the grain production, as well as drought and water deficit in the northern region of the country. Grain production in the USA, second highest producer to China, increased during the same period from 332 to 343 mln t, mainly due to the increased yield of maize.

The mean global grain yield decreased slightly in 2000 (2.75 t ha^{-1}) compared with 1999 (2.77 t ha^{-1}), when it reached its multiyear maximum. During preceding years there were only slight fluctuations about an average of 2.75 t ha^{-1}, but in 2000 the consumption per capita decreased to 303 kg, which was 13% below maximum level of grain consumption per capita in 1984. This decrease was mainly due to the Western European nations, the former USSR, and Africa. In 2000 and during the previous two years the maize production (588 mln t) for the first time exceeded the level of wheat production (580 mln t), whereas the production of rice constituted 401 mln t. About 43% of the global maize yield has been spent by the USA on stock-breeding. Using wheat production as an index, leadership in that area is the following: China, India, and USA. China also dominates in rice production. The USA export more than 75% of their maize to other countries. The main exporter of wheat is the USA, then France, Canada, and Australia. The leading role in the export of rice is with China, Thailand, USA, and Vietnam. Until recently, Japan

Table 4.6. Foreign debt of the countries of the Former Soviet Union and Eastern Europe (1970–1999).

Year	V	Year	V
1970	0.26	1985	1.47
1971	0.29	1986	2.57
1972	0.32	1987	2.73
1973	0.37	1988	1.68
1974	0.42	1989	1.70
1975	0.51	1990	1.77
1976	0.59	1991	1.80
1977	0.74	1992	1.85
1978	0.86	1993	1.97
1979	0.98	1994	2.15
1980	1.07	1995	2.30
1981	1.18	1996	2.35
1982	1.23	1997	2.40
1983	1.33	1998	2.61
1984	1.35	1999	2.57

Note: V = sum of debt (US$$10^9$, 1999).

had been the main importer of wheat, but during the last few years Brazil, Iran, and Egypt have overtaken Japan.

One of the key indicators of the level of socio-economic development is amount of foreign debt. Table 4.6 totals the data for the developing and Eastern European countries, as well as for the former USSR republics. As seen, a continuously growing debt, at a maximum in 1998, has stabilized, somewhat lowering in 1999 (US$2.57 bln) due to inflation. Serious changes in the financial trends in Brazil, Indonesia, Russia, and South Korea after 1996 are largely explained by the financial crises in 1997 and 1998.

To understand and predict the global changes in the environment, of great importance is the data on human health (Brown *et al.*, 2001). In Table 4.7 a fragment of the most demonstrative trend of AIDS propagation is given. This disease alone is able to reduce the size of the able-bodied population, especially concerning the nations of Africa and the Caribbean Basin. Expenses on health services reaching about US$3 bln vividly demonstrate the urgency of the problem. However, these expenses have been distributed rather unevenly. The nations with a low and average income and with 84% of the global population only share 11% of global expenses spent on health services, despite the fact that 93% of the damage caused by diseases falls within these countries. If the expenses on health services throughout the world average 5% GWP, in the USA they reach 13.7%, in Somali they only reach 1.5%. Respectively, the annual expenses per capita range from US$50 in poor countries to US$4,100 in the USA. Therefore, there is a significant probability of the development of catastrophic consequences due to such imbalances in protection against diseases.

Table 4.7. Regional HIV and AIDS statistics and features (end of 2004).

Region	Adults and children living with HIV (1,000 people)	Number of women living with HIV (1,000 people)	Adults and children newly infected with HIV (1,000 people)	Adult prevalence (%)	Adults' and childrens' deaths due to AIDS (1,000 people)
North America, Western and Central Europe	1,600	420	64	0.4	23
Middle East and North Africa	540	250	92	0.3	28
Oceania	35	7.1	5	0.2	0.7
Latin America	1,700	610	240	0.6	95
Eastern Europe and Central Asia	1,400	490	210	0.8	60
Asia	8,200	2,300	1,200	0.4	540
Caribbean	440	210	53	2.3	36
Sub-Saharan Africa	25,400	13,300	3,100	7.4	2,300
Total	*39,315*	*17,587.1*	*4,964*	*1.5*	*3,082.7*

One of the key indicators of socio-economic development is the level of militarization. Table 4.8 characterizes the dynamics of nuclear armaments as an indicator of technical progress. Here the cause–effect feedbacks of the world's "nature–society" system are directly regulated by political aspects. Militarization is one of the poorly controled processes whose consequences can exhibit global impacts on the environment. For instance, large-scale oil fires in Iraq are one of the explanations for a sharp increase of temperature on the European continent in the winter of 2004–2005.

To summarize the results of the global civilization development during the second half of the 20th century, it is necessary first of all to underline that this was a period of unprecedented rapid changes in the global size of the population, in the biosphere, economy, and society as a whole. The world has become more rich economically but poorer from an ecological point of view. The following most important trends have been observed during this period:

- The global population has exceeded 6 billion, increasing by 3.5 million during the last half century (i.e., more than doubling). Most of the growth has taken place in developing countries, which are already overpopulated. The growth was especially rapid in the urban areas where the population increased almost four-fold. A remarkable feature of the demographic dynamics in the industrial countries is an increasing share of older people.
- The growth of world economy has exceeded (approximately seven-fold during 50 years) the rate of population growth, which provides (on average) a

Table 4.8. World nuclear arsenal (1945–2000).

Year	*Y*	Year	*Y*
1945	2	1983	66,979
1950	303	1984	67,585
1955	2,490	1985	68,633
1960	20,368	1986	69,478
1965	39,047	1987	68,835
1970	39,691	1988	67,041
1971	41,365	1989	63,645
1972	44,020	1990	60,236
1973	47,741	1991	55,772
1974	50,840	1992	52,972
1975	52,323	1993	50,008
1976	53,252	1994	46,542
1977	54,978	1995	43,200
1978	56,805	1996	40,100
1979	59,120	1997	37,535
1980	61,480	1998	34,535
1981	63,054	1999	31,960
1982	64,769	2000	31,535

Note: Y = number of nuclear warheads.

significant increase of living standards but at the same time allows 1.2 billion people still to live in severe poverty with 1.1 billion not having access to pure drinking water.

- The global grain yield has nearly tripled since 1950, allowing people to enrich their diet, but leading to dangerous ecological consequences: lowering the level of sub-soil water and intensifying natural water pollution due to a massive use of fertilizers and pesticides.
- Only a small part of primary forest (boreal and tropical) has been preserved on the planet, more than a half of the wetlands and over a quarter of coral reefs have been lost. This was accompanied by considerable damage to biodiversity and, more seriously, the mechanisms for the biotic regulation of the environment have been broken (Gorshkov *et al.*, 2000).
- The anthropogenic impacts on the global climate and ozone layer are of great concern, though there are serious uncertainties on established trends.

What was especially important in the second half of the 20th century was that, as a rule, observed trends in socio-economic and ecological dynamics were unpredictable. As the world becomes more and more complicated and inhomogeneous, environmental prediction becomes more difficult but necessary, since an adequate planning aimed at minimizing the risk and maximizing favorable conditions for life depends on its accuracy. In particular, to resolve some problems arising here, the Climate

Change Science Program (CCSP) poses the following key questions:

(1) What are the most important feedbacks (and their quantitative relationships) between the dynamics of ecosystems and global change (first of all, climate)?
(2) What are the possible consequences of the impact of global change on ecosystems?
(3) What are the possibilities of providing sustainable development of ecosystems and the respective needs of society in an ecosystem's production, with due regard to supposed global change?

In answer to these questions one can only mention some preliminary results. As has been mentioned above, according to the results of satellite remote sensing, starting from 1980, the global primary production has decreased by more than 6%, about 70% of this decrease ocurred at high latitudes. During this period, the sea surface temperature (SST) in the Atlantic and Pacific Oceans of the northern hemisphere increased, respectively, by 0.7°C and 0.4°C. However, a decrease of primary production in the Antarctic Ocean was not connected with climate change. This decrease of primary production can reflect a decrease in the level of the carbon sink due to photosynthesis attenuation in high-latitude oceans. It is still unclear whether these changes are part of a long-term trend, or if they are a response to climate variations on a scale of decades.

4.1.3 Sustainable development

The paradox of the contemporary situation is that despite the unprecedented scale of discussions on sustainable development, the world is continuing to follow the path of unsustainable development. The definition of the term "sustainable development" proposed by the International Commission on Environment and Sustainable Development ("Bruntland Commission") as "a development that provides the needs of the present generations of mankind, and at the same time, gives the opportunity to the coming generations to satisfy their needs as well" is too general and non-constructive. This term should, of course, reflect the key role of the biospheric dynamics, which has been mentioned in the works by Timofeef-Resovsky (1961), Losev (2001), Kondratyev (1998, 1999), Gorshkov (1990), Gorshkov *et al.* (2000). In these publications it has been shown that the communities of organisms within the biocenosis and the whole biosphere regulate the chemistry of the most important media – atmosphere, hydrosphere, and soil. A necessity has also been emphasized to determine the limits of permissible disturbances in these media. Losev (2001) points to the following four problems to be solved to move on to sustainable development.

The first problem is a clash of our civilization with nature. The modern civilization is the result of 10,000 years of spontaneous human development, when people could build their history without any restrictions defined by the biosphere. However, by the beginning of the 20th century as a result of continuing expansion, humanity had gone beyond the limits, with subsequent global environmental changes. As a result, the 20th century has become the century of severe ecological crisis, which has not been realized yet by many people and politicians, though its negation or underestimation is incomparable with the global nature-protection infrastructure created during the last 25 years, environmental protection expenses, and application of resource-preserving technologies.

The ecological crisis is a signal that humanity cannot continue at random with its development but must coordinate its history with the laws of the biosphere from which man is inseparable. First of all, development should be coordinated with the law of energy flow distribution in the biosphere that determines the limits of the corridor of ecologically safe civilization development.

Signals from the biosphere destroyed by man either do not yet sufficiently affect the majority of people or they do not connect these signals with the rapidly developing environmental crisis. At the same time, as long as the virgin ecosystems exist on our planet, there is hope that irreversible changes have not yet started, and this process can be stopped and reversed. But these signals should be adequately perceived as a guide for action. None of the national programs or strategies on sustainable development contain an adequate response to such signals because there is no theoretical scientific basis. Some ideas on the development of such a basis and technology to perceive signals from the biosphere have been discussed in Kondratyev *et al.* (2000).

The second problem is economical. It is the development of the economy that has led to the destruction of natural mechanisms of environmental stabilization and to the development of the ecological crisis. The ecological costs of civilization have begun to convert into economical, social, and demographic ones. They have reduced the world economies' efficiency, investing activity, slowed down the rate of growth, stimulated poverty, broken genetic programs, undermined human health, etc. In its interaction with nature, humankind still remains an appropriator, not only using natural resources free of charge but also thoughtlessly destroying the biosphere – the basis of life. In the 21st century, humankind will have to radically change its attitude toward the organization of production cycles (Manwell *et al.*, 2002; Stevens and Verhe, 2004).

The third problem of sustainable development is the social sphere. The existing socio-economic system not only has led civilization to clash with nature but also has not solved the social problems. The gap between poor and rich people is widening, the difference between the living standards in poor and rich countries is growing. It means that the global socio-economic system does not correspond to the requirements for the sustainable development of humankind.

The fourth problem is the demographic crisis, which leads to a decrease in the areas of arable land and the volume of food production per capita. It intensifies the ecological crisis, which, in turn, deteriorates an important social indicator – human health. The social inequality of people destabilizes the situation in developing countries; causes conflicts and migration of populations; gives rise to extremism and terrorism; and finally can evoke global-scale conflicts.

Thus, modern civilization is in a state of system crisis, which has been repeatedly discussed at a number of international conferences. The way out of this crisis is in the transition to sustainable development, not a mythical solution fixed in numerous international and national documents, but a real change of the vector of civilization development. This crisis is somehow connected with the problem of the growing atmospheric CO_2 concentration, which worries both economists and politicians. It is important to predict the consequences of global climate changes to work out the strategies of regional and global economic and political initiatives in the future.

In this regard, scientists of the USA are accomplishing a complex program to study the global carbon cycle. The main goals of this program are as follows:

- To develop techniques to measure, monitor, and model components of the carbon cycle in nature in order to synthesize an integrated control system for these components.
- To work out a scientific basis to assess the ability of land and oceanic ecosystems to extract and accumulate CO_2, with the existing anthropogenic activity taken into account.
- To develop techniques to identify and assess the sources of CO_2 and other GHGs on regional and global scales.
- To search for an algorithm to describe the functions of these sources in the future and to provide information on the prognostic estimates of the climate trends.
- To assess the potential strategies of carbon extraction from the environment.
- To develop an efficient technology for an objective planning of environmental studies in the future and a realization of the decisions when managing land ecosystems.
- To create a unified information base for the available and planned environment research program.

4.2 CORRELATION BETWEEN NATURAL AND ANTHROPOGENIC COMPONENTS UNDER CONDITIONS OF NATURAL DISASTERS

4.2.1 Anthropogenic factors in global ecodynamics

The human (H)–nature (N) interaction is a function of a broad complex of factors functioning both in human society and the environment (Kondratyev, 1999; Demirchian and Kondratyev, 1998; Kondratyev *et al.*, 1997; Marchuk and Kondratyev, 1992; Gorshkov *et al.*, 2002; Jenkins *et al.*, 2005). The principal feature of this interaction is the globalization of the human effect on the environment, with a rapid development of megalopolises concentrating most of the sources of anthropogenic impact on the environment within their territories. Almost half the population of the Earth live in cities occupying only about 3% of the land area. This trend of concentration in the cities and the transformation of the latter into megalopolises will intensify, judging from numerous expert estimates. The number of large cities (with populations of 3–8 million) and megalopolises (>8 million people) in the 21st century will increase in the developing countries, with limited resources available for a search and realization of optimal urban infrastructures. And therefore the ecological conditions of the territories with an 80% global population increment will cause worry and demand additional investments (Kondratyev, 1996).

Many cities are contaminated by waste, huge territories have been transformed into waste heaps, and the discharge of contaminated sewage to the hydrosphere and the emission of polluting matter and gases to the atmosphere have intensified. Processes of transformation and even vegetation cover destruction have intensified. Of course, human economic activity has always taken place during historical development. Since the pre-historical epochs until now people have gradually increased their influence on nature, intensified the use of mineral resources, contaminated the environment, and broken the connections between various natural events established in the process of evolution, which has led, in particular, to a biodiversity reduction, deforestation of huge territories, soil salination, and depletion of biological resources of the seas and oceans. In the period from Palaeolithic epoch to the epoch of feudalism these forcings had been first local and then regional. The technical progress in the 18th and 19th centuries had contributed to a broadening of the scale of human impacts on nature, causing an extermination of some species of animals, wholesale destruction of forests, and rapacious exploitation of other natural resources. In the 20th century these forcings have become of global scale and dangerous for the biosphere. The problem arose of the search for the strategy of an optimal interaction of the systems H and N. In this connection, due to the accomplishment of numerous international and national programs of environmental studies, databases have been accumulated that make it possible to assess the level and direction of anthropogenic processes (Tables 4.9 and 4.10).

The principal regularity noted by many experts (Kondratyev, 1998; Watson *et al.*, 2000) is that the rate of the impact of human civilization on the biosphere depends directly on the population growth. Such categories as produced energy, amounts of consumed mineral resources, and investments depend directly on the size of the population (Demirchian and Kondratyev, 1998). The result is that the global-scale forcing of society on natural processes has reached a level where the survival of humans is in doubt. To survive, the human race should learn to co-evolve with nature, and to do that, it is necessary to thoroughly study it (Kondratyev, 1992, 1993; Kondratyev *et al.*, 1992).

According to the general opinion of experts in various sciences, the basic trends in human economic activity are to regulate the global processes in the biosphere, such as production of energy, industrial materials, and food. On average, an efficiency of these processes calculated per capita tends to increase. For instance, in the USA the per-capita amount of energy increases annually by ~2.5%. The mean global energy increment exceeds that of the population by 2%. So, for the territory of the FSU the average population increment constituted 1.4% yr^{-1}, and the annual growth of energy production is estimated at 7%. On the whole, the energy potential has been, so far, one of the characteristics of the level of the human societies' development. Other parameters determining the state of the production capabilites and economies of regions, territories, and countries, depend directly on energy potential.

The atmosphere, hydrosphere, and soil are becoming more and more contami-

Table 4.9. Trends in the impact on natural resources and the environment.

Indicators	Trends
Fossil fuels and the atmosphere	In 2002, the global level of the use of coal, oil, and natural gas was 4.7 times higher than in 1950. The level of CO_2 concentration in the atmosphere in 2002 was 18% higher than in 1960, and, apparently, 31% higher than before the industrial revolution (1750).
Degradation of ecosystems	During the last decades, more than a half of the Earth's wetlands (from coastal marshes to intra-continental lowlands exposed to floods) has been lost due to various measures for economic mastering. About half initial forests have been destroyed, 30% of the remaining forests degraded. In 1999, the scale of the use of wood as a fuel and in industry had more than doubled compared with 1950.
Sea level	In the 20th century the World Ocean level rose by 10–20 cm (at an average rate of increase of 1–2 mm per year) due to melting of continental glaciers and thermal expansion of water masses (under conditions of climate warming).
Soil/land surface	About 10–20% of the agricultural lands have degraded in many respects (loss of fertility), which has brought a decrease (during the last 50 years) in yield of 18% on cultivated lands and 4% on pastures.
Fishery	In 1999, fish catches increased 4.8 times compared with 1950. Creation of the modern fishing fleet has led to a 90% catch of tunny, cod, marlin, swordfish, shark, halibut, flounder, and cramp fish.
Water	Over-exploitation of groundwater has led to a decrease in levels in many agricultural regions of Asia, North Africa, the Middle East, and the USA. The impact of sewage or fertilizers, pesticides, oil bi-products, heavy metals, stable phosphor-organic compounds, and radioactive substances has led to a substantial decrease in groundwater quality.

nated. Practically, every principal geospheric component within which there is life, suffers from growing anthropogenic forcing. For instance, during the last years more than 4 mln t of oil has been discharged into the World Ocean with the concentration of mercury in the oceans has almost doubled.

Since all the domains of the geosphere are interconnected, a destructive forcing on one of them threatens all the others. Therefore, for example, burning of the final product of sewage purification cannot be considered a solution to the problem of pollutant control, since this technology only transfers the load from the hydrosphere to the atmosphere. In this sense, also inefficient is the cleansing of food organic discharge from soil by pumping into sub-soil water, because as a result, the rivers become oversaturated with bacteria and oxygen depleted. In short, transporting the pollutants into any domain of the geosphere damages all spheres of life. Though, of course, all spheres react differently to contamination, and a re-distribution of anthropogenic forcings among various geospheres can be an element governing the environment. Therefore, when analyzing the processes of pollution generation, it is necessary to determine the complex characteristics of the anthropogenic

Table 4.10. Global production of electric power.

Year	Nuclear power stations (GW)	Wind farms (MW)	
	Capacity	Total	Growth
1980	135	10	5
1981	155	25	15
1982	170	90	65
1983	189	210	120
1984	219	600	390
1985	250	1,020	420
1986	276	1,270	250
1987	297	1,450	180
1988	310	1,580	130
1989	320	1,730	150
1990	328	1,930	2,200
1991	325	2,170	240
1992	327	2,510	340
1993	336	2,990	500
1994	338	3,490	730
1995	340	4,780	1,290
1996	343	6,070	1,290
1997	343	7,640	1,570
1998	343	10,150	2,510
1999	346	13,930	3,780
2000	349	18,450	4,520
2001	352	24,930	6,480
2002	357	31,650	6,720

processes. For this purpose we shall consider each geosphere and formally describe the processes of their pollution.

One of the aspects of human impact on the Earth's climate is atmospheric pollution. Along with the GHG biogeochemical cycles that break the natural heat balance of the Earth, a relevant global model should take into account other pollutants that change the optical properties of the atmosphere, such as industrial dust, combustion gases, radioactive substances, smoke, molecular admixtures, and vapors of water and other liquids. For instance, molecular admixtures entering the atmosphere can markedly change its temperature (by up to 5°C), the most dangerous admixtures being those whose absorption spectrum is in the atmospheric "transparency windows". Atmospheric molecular pollutants are, for instance, methane, water vapor, sulfur dioxide, hydrogen sulphide, ozone, carbon monoxide, nitrogen oxides, etc. These components are divided into stable and short-lived (Table 4.11). The short-lived pollutants, such as SO_2 or mercury vapor in the atmosphere are unstable and cannot propagate for long distances, being unable to transform local anomalies into global ones. On the contrary, long-lived pollutants like CO_2 tend to

Table 4.11. Lifetime characteristics of some atmospheric components.

Component	Residence time
Carbon dioxide	5 years
Carbon oxide	0.1–3 years
Water vapor	10 days
Sulphur dioxide	3 days
Ozone	10 days
Hydrogen chloride	3–5 days
Nitric oxide	5 days
Nitrogen dioxide	5 days
Nitrous oxide	120 years
Ammonia	2–5 days
Methane	3 years
Freons	50–70 years

accumulate in the atmosphere and propagate over large territories. Therefore, their concentration during long time periods becomes uniform in the global atmosphere.

Dust and aerosols get to the atmosphere from many sources. Small solid particles are emitted to the atmosphere together with gases from internal-combustion engines and industrial smoke from metallurgical works and power stations. About half the solid particles ejected with combustion gases contain lead that tends to propagate for long distances. For example, anthropogenic lead was discovered in Greenland's mainland ice. A dangerous component of atmospheric pollution is mercury whose dispersed liquid particles are harmful to all living beings. From the available estimates the total amount of mercury emitted annually into the atmosphere reaches 1,100 t. Mercury sulphate with concentrations of 10 $ml\,l^{-1}$ kills living organisms within several days, and even in lower concentrations it substantially violates the life cycles. For example, tests on mollusks (*Monodonta articulata*) with a concentration of mercury salts (sulphate, acetate or chloride) of 0.01 $mg\,l^{-1}$ was found to reduce O_2 consumption. A discharge of mercury and copper into the sea in California killed gigantic focus algae which were only exposed to negligible concentrations of these metals over four days (0.05 g of mercury and 0.1 g of copper per ton of water). Mercury, in concentrations dangerous to humans, is found in fish and mollusks in more and more water bodies of the World Ocean (the coastal waters of the USA, Sweden, Norway, Canada, and the Mediterranean) (McIntyre, 1999).

Also dangerous are radioactive substances – some of them can remain in the atmosphere for a long time. Of special danger are accidents at atomic reactors and atomic bombs explosions – these being the most intensive sources of radioactive elements.

Both technogenic and natural processes affect changes in the chemical and optical properties of the atmosphere. Table 4.12 lists some sources of atmospheric

Table 4.12. Sources of atmospheric pollution (Brenninkmeijer and Rockmann, 1999; Demirchian *et al.*, 2002; Grigoryev and Kondratyev, 2001).

Sources of pollution	Pollutant
Natural	
Volcanoes, fumaroles, solfataras	Gases, volcanic dust, mercury vapor
Natural emissions of natural gas and oil	Hydrocarbons
Deposits	
Mercury	Mercury vapor
Sulphides	Sulfur dioxide
Radioactive ores	Radon
Blowing away from the sea and ocean surfaces	Chlorides, oil, sulphates
Underground coal fires	CO_2, CO, SO_2, hydrocarbons
Natural forest and steppe fires	Smoke
Plant transpiration	Water vapor, aromatic and other volatiles
Anthropogenic	
Burning of solid and liquid organic matter	CO_2, CO, SO_2, lead, hydrocarbons, mercury vapor, cadmium, nitric oxides
Metallurgy of black, non-ferrous, and rare metals	Dust, SO_2, mercury vapor, metals
Atomic industry	Radioactive substances
Nuclear explosions	Radioactive isotopes
Cement industry	Dust
Construction explosions	Dust
Human-caused forest and steppe fires	Smoke
Oil- and gas-fields	Hydrocarbons
Automotive transport	CO, soot, nitric oxides

pollution. Each source is specific. This list is constantly renewed due to appearing new spheres of human activity.

Atmospheric pollution changes human and animal habitats and causes climatic variations. The spectrum of human activity broadens toward higher comfort of living, and this requires an increase of energy consumption and productivity of natural and man-made ecosystems (Table 4.13). Feedbacks appear that require changes in the technologies of natural resource exploitation. On the other hand, the increase of atmospheric pollution grows linearly with the growing size of the population.

The next important sphere of anthropogenic activity manifestation is the pollution of the World Ocean with oil products, organic and inorganic compounds, and various chemicals. From the available estimates, the total amount of oil products anually entering the oceans varies within 32–60 mln t (Klubov *et al.*, 2000). Oil pollution is not uniform in the oceans, in many water bodies it exceeds the admissible norm by thousands of times ($0.5\,mg\,l^{-1}$). From

Table 4.13. Some characteristics of the energy parameters of the biosphere.

Characteristic	Estimation
Energy of the Earth's mass (Joule)	10^{46}
Global consumption of all kinds of energy (kJoule yr^{-1})	$(2.1–3) \times 10^7$
Energy utilized by the biosphere (Joule yr^{-1})	10^{20}
Energy used by humans (Joule yr^{-1})	10^{21}
Energy of tides and ebbs (kW-hr yr^{-1})	6.4×10^{14}
Energy of rivers (kW-hr yr^{-1})	23×10^{12}
Energy of wind (kW-hr yr^{-1})	15×10^{23}
Solar energy consumption (%):	
falling on the Earth's surface	100
accumulated in photosynthesis	0.1
used as food	0.001
Energy needs satisfied due to burning of the fossil products of photosynthesis (%)	95
Consumption of photosynthesis products as food (%)	1
Coefficient of performance:	
stock-breeding	0.2
agriculture:	
extensive	20
intensive	2
greenhouse	0.02
Energy spent on nitric fertilizer production by the chemical fixation of atmospheric nitrogen (%)	30
Annual rate of the energy consumption increase (%)	4
Volume of burnt organic fuel (10^9 t per year)	6–7
Annual expenditure of natural gas (10^{12} m^3)	1.4
Geological supplies of coal and brown coal (10^9 t)	$(14–15) \times 10^3$
Energy content in these supplies (10^9 Joule)	201×10^{12}
Natural gas supplies (10^{12} m^3)	145–340
Explored oil deposits (10^9 t)	65
Extracted oil (10^9 t)	147
Global oil output (10^9 t yr^{-1})	7.4
Share of the sea's oil-products in the global oil output (%)	19–20
Extraction of mineral fuel from the ocean bottom:	
oil (10^6 t yr^{-1})	450
gas (10^9 m^3 yr^{-1})	180
Known supplies of combustibles in units of conditional fuel (10^9 t)	27×10^3
Volume of extracted conditional fuel (10^9 t yr^{-1})	6
Total energy of fossil fuels (kJoule)	10^{20}
components (%):	
oil	10
gas	10
coal	80
The ratio of the amount of explored deposits of fossil combustibles to their annual extraction:	
oil	37
gas	41
coal	720
Energy emitted by burning timber (kJoule yr^{-1})	2×10^6
Power of all energy constructions over the globe (kW)	10^{10}
Share of the total anthropogenic energy emissions over the Earth compared with its total radiation budget	0.02
Means of energy expenditure (%):	
heat power stations and central heating	30–35
industry	30–35
transport	25–30
household needs	5–10

estimates of many authors, the concentrations of oil hydrocarbons in the surface microlayer of the World Ocean exceeds maximum permissible norms by a factor of 19.8 in the Atlantic Ocean, 3.6 in the Pacific Ocean, 4.4 in the Indian Ocean, 29.5 in the North Sea, 25 in the Norwegian Sea, and 4.8 in the Caribbean Sea. The property of oil is such that after a leakage it spreads over the water surface forming a film 1.02×10^{-3} mm thick. The areal extent of an oil spot originating from 1 tonne of reaches 12 km^2. At present about 30% of the World Ocean surface is covered with oil film. This film hinders the atmosphere–ocean gas exchange as well as decreases the coefficient of water transparency. Deposited oil is accumulated in bottom sediments, which violates the vital functions of benthos animals and bacteria. Important trophic and energetic feedbacks in the structure of the marine ecosystem become broken. Damage by oil pollution is transported through the trophic chains of marine ecosystems (Stephens and Keeling, 2000).

Along with the oil pollution of the World Ocean in the late 20th century, dangerous toxic pollutants of the marine domain have become appreciable, such as lead, mercury, arsenic, and other heavy metals, which damage the marine organisms accumulating in them. The latter makes these organisms unfit for consumption. Quantitatively, the background natural concentrations of heavy metals in numerous water bodies of the oceans exceed five–ten-fold the admissible levels, and in some places even reach dangerous levels. So, with the total input of mercury to the oceans being 5,000 t yr^{-1} due to its local use in chemical and pulp-and-paper industry the highest concentrations of mercury are observed in the coastal waters of Japan, Scandinavia, the Netherlands, and Canada.

Approximate estimates suggest an annual input of lead to the World Ocean of 40,000 t, half of which is deposited from the atmosphere. This half contains a substance especially dangerous for marine organisms – tetraethyl lead, usually added to petroleum. Lead, like mercury, can accumulate in marine organisms.

The World Ocean pollution spectrum is rather broad. Solid waste and sewage play a marked role in the total pollution of the oceans. From rivers and ships a huge amount of rubbish enters the oceans. All this leads to the lumbering of the oceanic domain and can worsen living conditions for marine animals.

Along with pollutants directly harmful to marine domains, huge amounts of biogenic elements are introduced into the oceans together with everyday sewage and fertilizers washed out from the fields (e.g., various compounds of nitrogen, phosphorus, and silicon). In some cases such inflows are harmless or even useful, but they can often lead to negative consequences causing eutrophication, or secondary biological pollution. In this case an excessive reproduction and development of algae takes place that consumes all the oxygen dissolved in water, after which masses of living organisms die. The eutrophication processes can take place without anthropogenic interference (e.g., in the region of the Peru upwelling). These natural processes occur in a balanced regime developed in the process of evolution and, on the whole, are not dangerous for the ocean ecosystems dynamics. Bearing in mind that the annual global input of sewage has exceeded 450 km^3 and in the future it may increase by 10–15 times due to population growth, the eutrophication processes will become all-embracing with so far unpredictable consequences.

Therefore, one of the goals of ecoinformatics is to assess these processes and their consequences. Clearly, the technogenic pollution processes and natural self-cleaning ability of the oceanic waters are in a constant conflict. A complex of physical, chemical, and biological processes of pollutant decomposition, and their transformation with resulting non-toxic substances, provide a powerful barrier to anthropogenic forcing in the oceans. However, the rivers, for example, have already lost this ability since the water medium does not have time to clean itself completely, and pollutants are accumulating. As a result, many rivers have been transformed into ditches.

The World Ocean is capable of self-cleaning such stable pollutants as oil. Due to the atmosphere–water surface interaction, oil gradually evaporates: during the first 24 hours by 15%, during the next ten days by ~20%. Storm winds forming foam crests on the water surface favor an input of sea spray into the atmosphere also carrying oil particles. Solar UV radiation, chemical oxidation, and bacterial decomposition also disperse oil products. Micro-organisms use oil hydrocarbons as a source of carbon and energy, and transform these hydrocarbons into bacterial cells, carbonic acid, and water. Besides this, the micro-organisms and many hydrobionts promote the transport of oil to the ground.

Anthropogenic forcings on the environment should be assessed together with natural processes, in order to work out a technology to reliably predict the consequences of human activity. For this purpose it is necessary to analyze the sphere of anthropogenic activity – utilization of natural resources. On the whole, there are the following groups of resources: fossil (geological and mineral), water, soil, vegetation, fauna, atomic, climatic, planetary, and cosmic, including chemical compounds of the Earth's crustal elements, deposits containing oil, gas, coal, and various salts, forests, etc. Natural resources are divided into renewable and non-renewable, exhaustible and inexhaustible. The renewable resources are soil, vegetation, and fauna. The non-renewable resources are fossil fuels. Solar radiation and wind energy are inexhaustible resources. Water resources and atomic energy are practically globally inexhaustible but can become scarce in some regions. Depending on time and the scientific–technical progress, the anthropogenic processes can change both their scale and direction.

Assessments of fossil fuel supplies make it possible to predict, to some extent, the levels of their anthropogenic utilization. Numerous valuable elements of industrial value are contained in the Earth's crust in relatively small amounts. For instance, the share of aluminum and iron is, respectively, 1/24 and 1/30. Almost half of all chemicals contain oxygen, land, so far, being the basic source of minerals and raw materials for humankind. Chemical elements from the World Ocean are used in small amounts.

At the dawn of their industrial activity humans had used only 19 chemical elements and their compounds, in the 18th century – 28, in the early 20th century – more than 100. With the development of science and technology the use of minerals and their extraction has intensified. So, during the last 100 years the annual consumption (and hence extraction) of coal, iron, manganese, and nickel has increased 50–60-fold, that of tungsten, aluminum, molybdenum, and potassium by

200–300-fold. Annually, about 100 bln t of ore and mineral fuel are extracted, as well as more that 300 mln t of mineral fertilizers.

The per-capita amount of extracted fuel is the principal indicator of energy supply. In 1972 the per-capita amount of conditional fuel constituted about 2 bln tonnes, but in the early 1970s the annual output of conditional fuel constituted almost 6 bln tonnes, spent as follows: transport – 25/30%, heat power stations – 30/35%, industry – 30%, everyday needs – 5/10%.

The basic types of minerals are fuel-energy resources, ores of black and alloy metals, non-ferrous and noble metals, and non-metal raw materials. According to an assumed classification, the trends of extracting these resources are thus. Fuel energy raw materials are oil, gas, and coal. In the 20th century the output of oil and gas increased from 10 to 1,800 mln t and from 5 to 1,300 bln m^3, respectively. At the same time, the output of coal increased only 4 to 5 times. The development trends of the output of other minerals such as iron and manganese ores, chromites, copper, lead, tin, molybdenum, etc., are growing. According to the available estimates, this pace of extraction will completely exhaust acceptable supplies of aluminum in 570 years, iron – in 250 years, zinc – in 23 years, lead – in 19 years, and by the mid-21st century most of the metals will disappear completely. The danger of a possible realization of this prediction makes one intensify the extraction and use of fossil non-metals, such as asbestos, black lead, mica, melted spar, etc.

Noble metals (platinum, gold, silver) are special among minerals. Their production is small by volume but they are important not only in the development of industry and scientific instrument making but also in the formation of the socio-economic relationships in the human society.

Consumption of natural resources also includes the use of freshwater and forest. The global water balance has been discussed by Krapivin and Kondratyev (2002). It includes a component that affects the natural water cycle. In particular, water used for household needs returns to the cycle in quite a different way than the natural exchange between water and individual biospheric elements. Water consumption for household needs is followed by irreversible losses, which, for instance, in the USA and Russia constitute 10–20%. Water supply in agriculture is characterized by 20–40% of irretrievable losses. On the whole, the volume of irretrievable water consumption by the population exceeds 20 $km^3 yr^{-1}$ or 17%. Industrial water consumption constitutes 500 $km^3 yr^{-1}$ on a global scale.

Especially large fluctuations in the water cycle are caused by irrigation. The total amount of water used for irrigation is determined by physico-geographical conditions, agricultural crops, technical equipment of irrigation systems and their application technology. The general characteristic of water consumption dynamics is given in Tables 4.14 and 4.15.

Along with the quantitative change of the water balance in the 20th century there has developed a process of exhaustion of water resources connected with the increasing pollution of natural waters. Most sewage containing harmful chemicals is discharged into the hydrophysical network without preliminary purification, causing a change in the water resource quality and making it unfit for further use without purification. This circumstance has led to a preferential use of underground, less

Table 4.14. Dynamics of the global-scale water consumption ($km^3\ yr^{-1}$).

Years	Water consumption W_G and % of irretrievable losses p_G in different spheres						Total water consumption and irretrievable losses	
	Municipal economy		Industry		Agriculture			
	W_G	p_G	W_G	p_G	W_G	p_G	W_G	p_G
1900	20	25.0	30	6.7	350	74.3	430	67.5
1940	40	20.0	120	5.0	660	72.7	870	61.0
1950	60	18.3	190	4.7	860	73.3	1,190	59.1
1960	80	17.5	310	4.8	1,510	76.7	1,990	63.2
1970	120	16.7	510	3.9	1,930	78.9	2,630	61.5
1975	150	16.7	630	4.0	2,100	76.2	3,080	60.0
1985	250	15.2	1,100	4.1	2,400	79.2	3,970	56.4
1995	320	15.1	1,560	3.9	2,760	78.4	4,750	54.9
2000	440	14.8	1,900	3.7	3,400	76.5	6,000	50.0

Table 4.15. Characteristic of the regional water consumption in the 20th century.

Region	Population		Industry		Agriculture		Σ	b
	a	b	a	b	a	b		
Europe	29	11	160	5	125	67	314	24
Asia	40	15	60	6	1,400	79	1,500	74
Africa	4	18	3	7	110	82	117	78
North America	41	21	270	30	210	61	521	28
South America	4	15	8	10	55	82	67	74
Australia	1	10	8	8	13	77	22	49

Notes: a = amount of consumed water (km^3); b = percentage of irreversible losses; Σ = total amount of consumed water (km^3).

polluted waters. As a result, conditions have been created for a sharp violation of the natural water cycle in the biosphere.

Forest is one of the catastrophically exhausted natural resources. Forests are in a process of being exterminated globally. For example, in the USA only 18 mln ha of the original 365 mlm ha of forested areas remain (Watson *et al.*, 2000). The area of tropical forests has decreased by two-thirds. In many countries, forested areas occupy a small part of their territory: in India – 18%, Greece – 15%, Spain – 12.5%, China – 9%, and Cuba – 8%. On Madagascar the forests have been almost completely exterminated.

Soils, as an independent natural body, are one of the natural resources of great importance to the biosphere. Mother rock, living organisms (plants, animals, and micro-organisms), climate, and relief are all factors of soil formation. The anthropogenic forcing on soil is manifested through its impoverishment, making it subject to erosion. Deep cast ground and open mining is a factor promoting soil destruction. Fertile soils become barren and deprived of vegetative areas, transforming them into the so-called "industrial deserts".

Thus, anthropogenic factors in present conditions are versatile in their functions and affect practically all the natural processes. Clearly, a complete account of the whole spectrum of anthropogenic forcings on the environment is impossible because of a lack of required data. In modeling the global carbon cycle a unit is needed in the models responsible for the parameterization of anthropogenic processes and providing their consideration in simulation experiments.

4.2.2 Modeling of demographic processes

The demographic situation determines the dynamics of anthropogenic processes. Therefore, attempts to develop models of demographic processes are very important. The existing predictions of changes in the size of the population and variations in its spatial distribution enable one to synthesize the scenarios to be used in a global model as well as to try to solve the problem of its verification. An adequate model of the global demographic process requires an extensive database covering the characteristics of changes in the standards of demographic behavior, detailed information on intensities of demographic processes in various regions of the planet, assessments and criteria of demographic policy, and, especially important, the structural indicators of human society. Table 4.16 characterizes a fragment of the existing data on the global structure of the population of 270 regions. Table 4.17 shows the level of basic economic parameters of many countries.

Many experts have considered the problem of parameterization of demographic processes. The demographic processes are considered as a part of planetary biospheric processes in which *Homo sapiens* play the role of a user of biocenoses products and a regulator of energy and matter fluxes among them.

4.2.2.1 Matrix model of the population size dynamics

In accordance with the possibilities and needs of the global model, the unit of population dynamics includes the impact of the following factors (Logofet, 2002):

- Per-capita food provision F (calculated as the sum of several shares of vegetation and animal population of the region as well as fish catch).
- The share A of animal protein in the human diet (determined from the contribution to F from animals and fish).
- The level of the public medical service M (in this model version it is in proportion to per-capita financing).
- Genetic load for the human population G (grows slowly with population development and depends on the level of the environmental pollution).

The sexual structure of the population and the processes of population migration

Table 4.16. Some characteristics of the populations for individual regions.

Region	Area (10^3 km^2)	Size of population (millions)	Middle age	Rate of population growth (%)	Age structure of the population (%)		
					0–14	15–64	65 or over
Australia	7,686.85	19.91	36.3	0.9	20.1	67.2	12.7
England	244.82	60.27	38.7	0.29	18.0	66.3	15.7
Bangladesh	144.0	141.34	21.5	2.08	33.5	63.1	3.4
Brazil	8,511.97	184.1	27.4	1.11	26.6	67.6	5.8
Vietnam	329.56	82.69	24.9	1.3	29.4	65.0	5.6
Egypt	1,001.45	76.12	23.4	1.83	33.4	62.3	4.3
India	3,287.59	1,065.07	24.4	1.44	31.7	63.5	4.8
Indonesia	1,919.44	238.45	26.1	1.49	29.4	65.5	5.1
Iran	1,648.0	69.02	23.5	1.07	28.0	67.2	4.8
Spain	504.75	40.28	39.1	0.16	14.4	68.0	17.6
Italy	301.23	58.06	41.4	0.09	14.0	66.9	19.1
Canada	9,984.67	32.51	38.2	0.92	18.2	68.7	13.1
China	9,596.96	1,298.85	31.8	0.57	22.3	70.3	7.4
Columbia	1,138.91	42.31	25.8	1.53	31.0	63.9	5.1
Nigeria	923.77	137.25	18.1	2.45	43.4	53.7	2.9
Mexico	1,972.55	104.96	24.6	1.18	31.6	62.9	5.5
Pakistan	803.94	159.2	19.4	1.98	40.2	55.7	4.1
Russia	17,075.2	143.78	37.9	−0.45	15.0	71.3	13.7
Sudan	2,505.81	39.15	17.9	2.64	43.6	54.1	2.3
USA	9,631.42	293.03	36.0	0.92	20.7	66.9	12.4
Peru	1,285.22	27.54	24.6	1.39	32.1	62.8	5.1
Tanzania	945.09	36.59	17.6	1.95	44.2	53.2	2.6
Turkey	780.58	68.89	27.3	1.13	26.6	66.8	6.6
Ukraine	603.7	47.73	38.1	−0.66	15.9	68.7	15.4
France	547.03	60.42	38.6	0.39	18.5	65.1	16.4
Ethiopia	1,127.13	67.85	17.4	1.89	44.7	52.5	2.8
South Africa	1,219.91	42.72	24.7	−0.25	29.5	65.3	5.2
Japan	377.84	127.33	42.3	0.08	14.3	66.7	19.0

between regions are ignored. The age structure contains three groups (0–14 years, 15–64 years, 65 years and older). There is also a fourth group of the population (invalids and disabled persons sitting outside of these three age groups). For terminology convenience the fourth group is considered as not affecting the age structure.

Let S_t be the age structure at a time moment t, then the population size dynamics can be described by the following matrix equation:

$$S_{t+1} = D \times S_t$$

where D is the demographic 4×4 matrix including the effect of the factors F, A, M,

Table 4.17. Characteristic of regional ratios of the export/import type (US$$10^9$).

Region	Export (W)	Import (V)	Region	Export (W)	Import (V)
Russia	134.4	74.8	USA	714.5	1,260.0
China	436.1	397.4	Japan	447.1	346.6
Mozambique	0.8	1.1	Brazil	73.3	48.3
Byelorussia	9.4	11.1	Austria	83.5	81.6
South Korea	201.3	175.6	Estonia	4.1	5.5
North Korea	1,044.0	2,042.0	Ukraine	23.6	23.6
Kuwait	22.3	9.6	New Zealand	15.9	16.1
England	304.5	363.6	Oman	11.7	5.7
Vietnam	19.9	22.5	Belgium	182.9	173.0
Finland	54.3	37.4	United Arab Emirates	56.7	37.2
Peru	9.0	8.2	Guatemala	2.8	5.7
Norway	67.3	40.2	Indonesia	63.9	40.2
Malaysia	98.4	74.4	Turkey	49.1	62.4
South Africa	36.8	33.9	Hong Kong	225.9	230.3
Spain	159.4	197.1	Venezuela	25.9	10.7
Germany	696.9	585.0	Bahrain	6.5	5.1
Sweden	102.8	83.3	Greece	5.9	33.3
Turkmenistan	3.4	2.5	Chile	20.4	17.4
The Netherlands	253.2	217.7	Canada	279.3	240.4
Luxemburg	8.6	11.6	Italy	278.1	271.1
France	346.5	339.9	Nigeria	21.8	14.5
Denmark	64.2	54.5	Madagascar	700.0	920.0
Australia	68.7	82.9	Laos	332.0	492.0
Morocco	8.5	12.8	Thailand	76.0	65.3
Uruguay	2.2	2.0	India	57.2	74.2
Cambodia	1.6	2.1	Poland	57.6	63.7
Pakistan	11.7	12.5	Mexico	164.8	168.9
Kazakhstan	12.7	8.6	Mongolia	524.0	691.0
Panama	5.2	6.6			

and G:

$$D = \begin{Vmatrix} d_{11} & d_{12} & 0 & 0 \\ d_{21} & d_{22} & 0 & 0 \\ 0 & d_{32} & d_{33} & 0 \\ d_{41} & d_{42} & d_{43} & d_{44} \end{Vmatrix}$$

Matrix D differs from Leslie's traditional matrix for the model of a population by age structure, first, by the inequality to zero of its diagonal elements, explained by the overlapping of the next generations, and, second, by the last line that reflects the non-age character of the fourth group. The diagonal elements of matrix D are determined from apparent balance relationships:

$$d_{ii} = 1 - \mu_i - \sum_{j>i}^{4} d_{ji} \qquad (i = 1\text{–}4) \tag{4.1}$$

which include the coefficients of mortality μ_i of the ith group, descending functions of the per-capita food provision, and the medical services level, $\mu_i = \mu_i(F, M)$ with:

$$\lim_{F,M \to \infty} \mu_i(F, M) = \mu_{i,\min} > 0$$

where $\mu_{i,\min}$ characterizes a minimum of physiological mortality with optimal food provision and medical services.

It is supposed that the reproductive potential of the human population is entirely concentrated in the second age group. The d_{12} coefficient value is considered to be a regional constant. Note that the birth-rate coefficient is a complex and poorly studied function of many variables, ethnic traditions being one of the important variables. The birth-rate is affected by religion, for instance, the Moslems consider children to be their wealth. And in South Asia the religious norms order each family to have at least one son. Catholicism also affects markedly the birth-rate index. Therefore, if the function μ_i is well parameterized by the statistical data, the coefficient d_{12} is still an unidentified parameter. Nevertheless, the d_{12} value for different global regions can be estimated from the birth-rate data.

The coefficients of transition to the following age groups d_{21} and d_{32} are determined from the duration of the respective age group and the hypothesis of a uniform distribution of ages within the group, namely: $d_{21} = 1/15$; $d_{32} = 1/50$.

Genetically stipulated diseases and a deficit of protien in childrens' food are the main causes of transitions from a younger age group to an invalid group. Thus, $d_{41} = d_{41}(G, A)$ is the age function of its arguments. Coefficients d_{42} and d_{43} are small compared with d_{41} and determined by the sum of the vital parameters of the environment, the level of medical services, and other indicators of anthropogenic medium (investments). In a first approximation, suppose $d_{4k} = \Delta_k \mu_k (k = 23)$, where $\Delta k \ll 1$.

Demographic matrix D has a certain set of properties which make it possible to reveal the typical trends of demographic process. A maximum eigenvalue is:

$$\lambda_{\max}(D) = \max\{\lambda_1; d_{33}; d_{44}\} > 0$$

where

$$\lambda_1 = 0.5(d_{11} + d_{22}) + 0.25[(d_{11} + d_{22})^2 + d_{12}d_{22} - d_{11}d_{22}]^{1/2}.$$

It follows from (4.1) that if $\lambda_{\max}(D) \geq 1$, then $\lambda_{\max}(D) = \lambda_1$ (i.e., the rate of *Homo sapiens*' population growth in the neighborhood of a stationary age distribution is determined by the λ_1 value. The stationary age structure is calculated as an eigenvector corresponding to $\lambda_{\max}(D)$. The coefficients and modifying dependences of matrix D calculated, as a rule, from the data of Stempell (1985), give the dominating eigenvector $P_D = (0.369; 0.576; 0.05; 0.005)$. The first three components of vector P_D coincide, to within second sign, with the global data on the age complement of the global population.

The matrix version of the global model demographic unit has a time step of 1 year. An inclusion of this unit in the chain of other global model units with an arbitrary time step $\Delta t < 1$ year is needed as an adjusting procedure. This is possible, for instance, if the demographic unit is included only at time moments of 1-year multiples. In this version the spasmodic changes of the population size will cause some imbalance in continued trajectories of the global system. Another procedure free of this drawback consists of the use of a prognostic equation:

$$S_{t+\Delta t} = S_t + \Delta t(D - I)S_t \tag{4.2}$$

where I is an identity matrix. One can demonstrate that for the age structure vectors coinciding in direction with the eigenvector of matrix D, the main term of a relative error accumulated during 1 year with a time step $\Delta t = 1/n$ constitutes $(1 - 1/n)(\eta - 1)^2/2$. Let the right-hand part of (1.1) be denoted by $f(S, \Delta t)$. Then if e is the eigenvector of matrix D with an eigenvalue η, then with the time step $\Delta t = 1/n$:

$$f(e, 1/n) = e + (1/n)(D - I)e = (1 + [\eta - 1]/n)e.$$

It is apparent that $S_{t+n\Delta t} = f^{(n)}(S_t, 1/n)$. And with $S_t = ef^{(n)}(e, 1/n) = (1 + [\eta - 1]/n)^n e$. But if $\Delta t = 1$, then $S_{t+1} = f(e, 1) = \eta e$, so that a relative error resulting from a subdivision of the 1-year time interval is:

$$(1 + [\eta - 1]/n)^n - \eta = n(n - 1)(\eta - 1)^2/(2n^2) + 0((\eta - 1)^3) \quad \text{Q.E.D.}$$

In other words, the accuracy of an approximation of (4.2) rises with Δt approaching to unity.

4.2.2.2 Differential model of population dynamics

The matrix model considered above enables one to use the demographic statistics by age groups but requires knowledge of many parameters, which leads to uncertainties in the global model. Therefore, a second version of the demographic unit is suggested which simulates the dynamics of only the total size of the population. This version assumes that an impact of numerous environmental factors and social aspects on the population size dynamics G_i in the ith region is manifested through birth-rate R_{Gi} and mortality M_{Gi}:

$$dG_i/dt = (R_{Gi} - M_{Gi})G_i \quad (I = 1, \ldots, m) \tag{4.3}$$

Birth-rate and mortality depend on food provision and its quality, as well as environmental pollution, living standards, energy supply, population density, religion, and other factors. Within the global model all these factors will be taken into account following the principle:

$$R_G = (1 - h_G)k_G G H_{GV} H_{GO} H_{GG} H_{GMB} H_{GC} H_{GZ}$$

$$M_G = \mu_G H_{\mu MB} H_{\mu G} H_{\mu FR} H_{\mu Z} H_{\mu C} H_{\mu O} G + \tau_{GO} \tau_{GC}^{\omega(G)}$$

where for simplicity purposes we shall omit the i index attributing the relationship to the ith region; h_G is the coefficient of food quality ($h_G = 0$ when the quality of food is perfect); coefficients k_G and μ_G point to levels of birth-rate and mortality, respectively; indices τ_{GO} and τ_{GC} characterize the dependence of the regional population mortality on O and C – indicators of the environmental state (within the global model it is the content of O_2 and CO_2 in the atmosphere) manifesting through human physiological functions; coefficient $\omega(G)$ characterizes an extent of the influence of population density on mortality (in the present conditions $\omega(G) \approx 0.6$); and functions $H_{GV}(H_{\mu FR})$, $H_{GO}(H_{\mu O})$, $H_{GC}(H_{\mu C})$, $H_{GMB}(H_{\mu MB})$, $H_{GG}(H_{\mu G})$, and $H_{GZ}(H_{\mu Z})$ describe, respectively, an impact on birth-rate (and mortality) of the environmental factors, such as food provision, atmospheric O_2 and CO_2 concentrations, living standards, population density, and environmental pollution. Functions $H_{\mu O}$ and $H_{\mu C}$ approximate the medico-biological dependences of mortality on the atmospheric gas composition. Let us consider all these functions in more detail. For this purpose we shall formulate a number of hypotheses concerning the forms of dependences of mortality and birth-rate on various factors.

The results of numerous studies with national specific features taken into account enable one to assume the following dependence as an approximation of the function H_{GV}: $H_{GV} = 1 - \exp(-V_G)$, where V_G is an efficient amount of food determined as a weighted sum of the components of the *Homo sapiens* food spectrum:

$$V_{Gi} = k_{G\Phi i}\Phi + k_{GFi}\left(F_i + \sum_{j\neq i} a_{Fji}F_i\right) + k_{Gri}I_i(1 - \theta_{Fri} - \theta_{uri})$$
$$+ k_{GLi}L_i + k_{GXi}\left[(1 - \theta_{FXi})X_i + (1 - \nu_{FXi})\sum_{j\neq i} a_{Xji}X_j\right]$$

Here Φ is the volume of vegetable food obtained from the ocean; F – protein food; L – food supplied from forests; X – vegetable food produced by agriculture; I – fishery products; coefficients $k_{G\Phi}$, k_{GF}, k_G, k_{GL}, and k_{GX} are determined using the technique described in Chapter 6; α_{Fji} and α_{Xji} are, respectively, shares of the protein and vegetable food in the ith region available for use by the population of the ith region; θ_{FXi} and ν_{FXi} are shares of vegetable food produced and imported by the ith region, respectively, to produce protein food; and θ_{Fri} and θ_{uri} are shares of fishery spent in the ith region on protein food production and fertilizers, respectively.

With an increasing food provision the population mortality drops to some level determined by the constant $\rho_{1,\mu G}$ at a rate $\rho_{2,\mu G}$, so that $H_{\mu FR} = \rho_{1,\mu G} + \rho_{2,\mu G}/F_{RG}$, where the normalized food provision F_{RG} is described by the relationship $F_{RG} = F_{RG}(t) = V_G/G(F_{RG}(t_0) = F_{RGO} = V_G(t_O)/G(t_0))$. Similarly, we assume that the birth-rate depending on the living standard M_{BG} of the population is described by the function with saturation, so that a maximum of birth-rate is observed at low values of M_{BG}, and at $M_{BG} \to \infty$ the birth-rate drops to some level determined by the value of a_{*GMB}. The rate of transition from maximum to minimum birth-rate with changing M_{BG} is set by the constants $a_{1,GMB}$ and $a_{2,GMB}$:

$$H_{GMB} = a_{*GMB} + a_{1,GMB}\exp(-a_{2,GMB}M_{BG})$$

where

$$M_{BG} = (V/G)\{[1 - B - U_{MG} - U_{ZG}]/[1 - B(t_0) - U_{MG}(t_0) - U_{ZG}(t_0)]\}$$
$$\times [E_{RG}(t)/E_{RG}(t_0)]$$

$$E_{RG}(t) = 1 - \exp[-k_{EG}M(t)/M(t_0)]$$

The dependence of mortality on the living standard is described by the decreasing function:

$$H_{\mu MB} = b_{1,\mu G} + b_{2,\mu G} \exp(-b_{*\mu G} M_{BG}).$$

This function shows that population mortality with an increasing per-capita share of capital drops with the rate coefficient $b_{*\mu G}$ to the level $b_{1,\mu G}$.

The birth-rate and mortality in certain limits are, respectively, decreasing and increasing functions of population density:

$$H_{GG} = G_{1,G} + G_{*G} \exp(-G_{2,G} Z_{GG}) \qquad H_{\mu G} = \theta_{1,\mu G} + \theta_{2,\mu G} Z_{GG}^{\omega \mu G}$$

where $Z_{GG} = G(t)/G(t_0)$.

Finally, an important aspect of the *Homo sapiens* ecology is the environmental state. In this connection the problems of anthropobiocenology have been widely discussed in scientific literature and many authors have tried to determine the required regularities. Without going into the details of these studies, most of which cannot be used in the global model, we shall confine ourselves to the following dependences:

$$H_{GZ} = l_{1,G} \exp(-l_{*G} Z_{RG}), H_{\mu Z} = n_{1,\mu G} + n_{2,\mu G} Z_{RG}$$

$$\tau_{GC} = \begin{cases} \tau_{1,GC} + \tau_{2,GC}(C_a - C_{1,G}) & \text{for} \quad C_a > C_{1,G} \\ \tau_{1,GC} & \text{for} \quad 0 \le C_a \le C_{1,G} \end{cases}$$

$$H_{\mu O} = f_{1,\mu G} + f_{2,\mu G}/O(t), H_{\mu C} = \exp(k_{\mu G} C_a)$$

$$H_{GO} = 1 - \exp(-k_{GO} O), H_{GC} = \exp(-k_{GC} C_a), Z_{RG} = Z(t)/Z(t_0)$$

$$\tau_{GO} = \begin{cases} \tau_{1,GO} & \text{for} \quad O > O_{1,G} \\ \tau_{2,GO} - (\tau_{2,GO} - \tau_{1,GO}) O/O_{1,G} & \text{for} \quad 0 \le O \le O_{1,G} \end{cases}$$

where C_a is the atmospheric CO_2 concentration, $C_{1,G}$ and $O_{1,G}$ are safe-for-human levels of the atmospheric CO_2 and O_2.

4.2.3 Parameterization of anthropogenic processes

The biosphere Ω as a complicated unique system is functioning following the laws of co-evolution of its subsystems, the human society H and nature N being the basic subsystems. The impact of human activity on nature, being of comparatively small scale, can be apparently assessed only with newly developed technology. Clearly, for this purpose a system approach is needed to formalize the ecological, technological, economic, and geopolitical interactions of the H and N subsystems. In general, the system H has at its disposal technologies, science, economic structure, size of population, etc. The system N has a set of mutually dependent processes, such as climatic, biogeocenotic, biogeochemical, geophysical, etc.

From the viewpoint of the theory of systems, H and N are open systems. Their division is a conditional procedure aimed at selecting controled and non-controled components of the environment. Without going into philosophical and methodical aspects of this procedure, we assume that both systems are symmetrical from the viewpoint of their simulation (i.e., we assume that each system has a goal, structure, and behavior of its own). Let $H = \{H_G, H_S, |H|\}$ and $N = \{N_G, N_S, |N|\}$, where H_G and N_G are the goals of the systems, H_S and N_S are strategies of the systems' behavior, and $|H|$ and $|N|$ are the systems' structure, respectively. Then, the H–N interaction can be described by the process of (V, W)-exchange in that each of the systems, to reach its goal, spends resources V and, in exchange, obtains a new resource of an amount W. Each system is aimed at an optimization of the (V, W)-exchange with another system (i.e., to maximize W and minimize V).

Now write the equations for (V, W)-exchange:

$$W_H(H^*, N^*) = \max_{\{H_s, |H|\}} \min_{\{N_s, |N|\}} W(H, N) = \min_{\{N_s, |N|\}} \max_{\{H_s, |H|\}} W(H, N) \qquad (4.4)$$

$$W_H(H^*, N^*) = \min_{\{H_s, |H|\}} \max_{\{N_s, |N|\}} W(H, N) = \max_{\{N_s, |N|\}} \min_{\{H_s, |H|\}} W(H, N) \qquad (4.5)$$

where H^* and N^* are optimal systems. Here, in contrast to traditional game-theory models, there exists a power spectrum of the $H-N$ interaction covering the final intervals of changes in the pay-offs W_H and W_N depending on the aggressiveness of each of them. A concrete definition of the pay-off function requires a certain systematization of the mechanisms of the human and natural co-evolution. One of the widely used models of the balanced development of the world community and nature subjected to criterion (4.4) and (4.5) consists in the identification of the system H with a totality of large cities with adjacent industrial and recreation zones. There are numerous considerations and formal descriptions of such structures. In particular, there exists a well-known method of logic-information modeling of the processes of rational nature use, and a simulation method of controling the ecological–economic systems (Ougolnitsky, 1999). According to these methods, to solve a concrete problem, it is necessary to conceptualize the information base of the model and to select most general relationships between the elements of the interacting systems. This procedure is completed with enumeration of all functional elements of the systems and determination of the capacity loads on their elements. The whole procedure is concluded with a synthesis of the simulation model, which, within assumed assumptions, is an instrument of investigation. In the case considered we assume that the structure of the system H includes population G, pollutants Z, and natural resources M, (i.e., $|H| = \{G, Z, M\}$). Similarly, the structure of the system N consists of elements such as the climate parameters – temperature T, environmental quality Q, areas of forests σ_L, and agricultural lands σ_X, (i.e., $|N| = \{T, Q, \sigma_L, \sigma_X\}$). The behavioral strategy of the system H is formed from the distribution of investments into the retrieval of resources U_{MG}, the struggle with pollution U_Z, and agricultural investments U_{BG} (i.e., $H_S = \{U_{MG}, U_{ZG}, U_{BG}\}$). The behavioral strategy of the system N is identified with the rate of investment ageing T_V, population mortality μ_G, agricultural productivity H_X, the cost of resources retrieval G_{MG}, and time constant of biospheric self-cleaning of pollutants T_B, i.e.,:

$$N_S = \{T_V, \mu_G, H_X, G_{MG}, T_B\}$$

Equations (4.4) and (4.5) are basic equations for the model of survival. In general, this model is formulated in terms of the theory of the evolutionary technology of modeling. If all possible states of the biosphere with acceptable conditions for human life constitutes the multitude $\Gamma = \{\Gamma_i\}$, then as a result of the effect of C_k on the biosphere, two outcomes are possible: $C_k(\Gamma_i) \rightarrow \Gamma_i \in \Gamma$ and $C_{Jl}(\Gamma_{jl} \rightarrow \Gamma_{bl} \notin \Gamma)$. When the sequence of biospheric states $\{C_k(\Gamma_i)\} \in \Gamma$, then we can speak about the persistent co-evolution of the system $H \cup N$. No doubt, there is a problem of adequacy between real processes and their simplified presentation as a model. Nevertheless, despite the philosophical doubt in expedience of mathematical modeling for the perspective assessment of the kinetics of biospheric parameters, the model approach has proved to be profitable. The use of the biospheric model instead of the biosphere itself is convenient, first, because there is more information about the model than about the biosphere, second, because the model is easier to handle, and third, because direct experiments with the biosphere are dangerous. All these aspects are the subject of studies for global ecoinformatics aimed at achieving a sufficient similarity between the observed behavior of the system $H \cup N$ and the model. This is possible with the constant renewal of databases and the broadening of knowledge in accordance with the technological formation of a multitude of biospheric parameters suggested by Krapivin and Kondratyev (2002).

Trends in human activity are determined by a great number of factors. All of them are reduced to economy and energetics, whose interconnections can be described by the linear regression $V = k_{Ve}e + b_{Ve}$, where V is the regional budget, and e is the energy produced in this region. The constant coefficients k_{Ve} and b_{Ve} reflect the specific features of economic activity and its efficiency. The general scheme of a possible model of socio-economic processes includes V_i as the size of the funds in the ith region, which this region can use when planning its anthropogenic activity:

$$dV/dt = G_{VG}G_{VMG} - V/T_{VG}$$

where T_{VG} is the time constant of basic fund deterioration, and coefficients G_{VG} and V_M determine the rate of the fund generation. The function V_M defines the dependence of the rate of fund generation on the living standard M_B of the population of the region. This dependence is described with a logarithmic function:

$$V_{MG} = k_{MGV} \ln(1 + k_{1,MGV}M_{BG})$$

where the coefficient k_{MGV} is chosen from the condition:

$$k_{MGV} \ln(1 + k_{1,MGV}M_{BG}(t_0) = 1$$

In this case the coefficient G_{VG} represents the volume of the funds generated per-capita at the moment t_0 in the region. The time constant of the basic fund deterioration is a function of the scientific–technical progress, and is considered in the model as a controling parameter.

The basic fund determines the intensity and direction of anthropogenic activity. In particular, the generation of pollutants and their utilization are substantially regulated by this distribution.

Let Z be the concentration of an arbitrary pollutant and Z_{VG} the rate of pollutant assimilation due to a realization of the technologies of environmental purification. Then:

$$dZ/dt = Z_{VG} - Z_{TV}$$

where $Z_{TV} = Z_T + Z_V$, Z_T is the natural rate of the pollutant decomposition, and Z_V is the rate of pollutant assimilation due to environmental purification.

Assume that the rate of pollution generation in the region is in proportion to population density with coefficient k_Z and depends on the per-capita volume of the fund $V_{RG} - V/G$:

$$Z_{VG} = k_Z G Z_{*VG}$$

where $Z_{*VG} = Z_{VG,\max}[1 - \exp(-G_{1,ZG} V_{RG}]$.

The rate of an artificial assimilation of pollutants is determined by the capital share U_{ZG} to be used to intensify the cleaning processes in the environment: $Z_V = U_{ZG}V/G_{ZV}$ where G_{ZV} is the cost of the environmental cleaning of unit pollution.

The rate of the natural removal of harmful wastes is in direct proportion to the pollutant concentrations and in indirect proportion to the decomposition time T_Z: $Z_T = Z/T_Z$. Naturally, the processes of generation and utilization of pollutants are more complicated. Therefore, the respective unit of the global model needs further development, considering a new knowledge in the field of pollution utilization. The generated pollutant is scattered in the environment getting partially into the atmosphere, soil, and the hydrosphere. Possible ways of pollutant propagation in these media are described in other chapters. However, for a complex analysis of a pollutant an equation for the global model is required which would reflect the general state of the atmosphere. Let us introduce the atmospheric turbidity index B measured by the weight of admixtures in the atmosphere per unit area of the biospheric surface. A change in this index is determined by the share N_B with regards to general pollution Z generated in the region, by the amount of waste of energy-generating enterprises (coefficient N_A), and by the water vapor content in the atmosphere W_A and the rate of its natural cleaning $1/TB$, assumed to be equal to the inverse value of the time TB of sedimentation of dust and smoke particles and water droplets. The effect of water vapor on atmospheric transparency is manifested through the derivative $dW_A/dt = \rho_B dT/dt$ where T is the atmospheric temperature and ρ_B is the coefficient. As a result, the dynamic equations for the system variable B will be:

$$dB/dt = N_B Z + N_A b_{GC} G + \rho_B \, dT/dt - B/T_B + B_n$$

where B_n is the rate of atmospheric pollution resulting from natural processes (rock weathering, erosion, volcanism, etc., $\sim$0.78 t km^{-2} yr^{-1}).

The horizontal transport of pollutants and the respective change in B are simulated with an account of wind speed and the spatial grid $\Delta\varphi \times \Delta\lambda$. The parameter T_B is calculated with the formula $T_B = h_e/v_s$ where h_e is the efficient height of the pollutant scattering in the atmosphere and v_s is the rate of particle sedimentation calculated with the Stokes formula: $v_s = 2g\rho_k r^2/(9\varsigma)$. Here g is the acceleration of gravity, ρ_k is the particle density, r is the particle radius, and ς is

Table 4.18. Estimation of the annual volume of particles with radius <20 μm emitted to the atmosphere.

Type of particles	Flux of particles ($10^6\,t\,yr^{-1}$)
Natural particles, soil and stone particles	100–500
Particles from forest fires and forestry waste burning	3–150
Sea spray	300
Volcanic dust	25–150
Particles resulting from gas production:	
sulphates of H_2S	130–200
ammonium salt of HN_3	80–270
nitrates of NO_x	60–430
hydrocarbons of vegetable compounds	75–200
Artificial particles	10–90

the coefficient of molecular viscosity. For instance, for particles with $r \in [10^{-5}, 5 \times 10^{-3}]$ cm and $\rho_k \in [1, 4]\,g\,cm^{-3}$ we obtain $v_s = 10^2\,cm\,s^{-1}$, which for calm atmospheric conditions gives $T_B = 7$ hours. Some characteristics of atmospheric aerosols are given in Tables 4.11 and 4.18.

Most complicated is the scattering process of pollutants in water and soil. These processes are simulated in the respective chapters in studies of the hydrospheric components and associated processes. The relationships between natural and anthropogenic formation of air or water quality, deduced and applied in simulation experiments, contain the parameters determined by the pace of expenditure and renewal of natural resources. The pace dynamics are determined by the rate of population growth and civilization development. On average, the annual global acceleration of natural resources consumption constitutes about 4% with a 2% mean annual population increment. Hence, the value of per-capita natural resource consumption R_{MG} is an important indicator of current civilization development, since the annual consumption of accessible resources constitutes $\Delta M = R_{MG}G$. The consumption of natural resources increases with growing living standards, but the extent of this growth decreases with increasing investments into science. Clearly, this relationship is many sided, and has been poorly studied. Therefore, to consider this problem, assume the following parameterization:

$$R_{MG} = m_G \ln(1 + M_{BG})$$

where the parameter m_G is the time function set in a simulation experiment as a scenario.

The protection and restoration of natural resources requires certain material expenses. Assume that the effect of these measures is in proportion to the share U_{MG} of the respective investments and inversely proportional to the cost G_{MG} of the retrieved conditional unit of resources:

$$dM/dt = \Delta M + U_{MG}V/G_{MG}$$

The investments spectrum also includes the sphere of agricultural production, including expenses on scientific investigations into selection, plant physiology, agrotechnology and reclamation, as well as solving technical and social problems. Let qV be the share of capital invested in agriculture. We obtain:

$$dq_V/dt = U_{BG}B_{FG}B_{qG} - q_V/T_B$$

where the controling parameter $U_{BG} = U_{BG}(t)$ in simulation experiments is described by the scenario, and the remaining components are determined by the following dependences:

$$B_{FG} = \exp(-b_{BG}F_{RG}) \qquad B_{qG} = b_{1,BG} + b_{2,BG}(q_{MG}/q_{FG})^{\alpha SG}$$

$$q_{MG} = b_{1,qG} + b_{2,1G}(M_{BG})^{\alpha qG} \qquad q_{FG} = a_{qG}(F_{RG})^{\beta qG}$$

4.2.4 Megalopolises

The present demographic situation is characterized by a new global phenomenon which transforms itself into a global process, the rate and scale of which are increasing. Namely, the process of urbanization and increase of the number of megalopolises. The size of the urban population approaches a one-half of the global population. Such a localized anthropogenic load will, of course, bring forth problems of stability to the natural and natural–technogenic systems on global scales.

A megalopolis is an area on the Earth's surface area with high urbanization, developed industries, and other human activity attributes concentrated over a limited territory (Henderson and Thisse, 2004). The number of such territories in the world is constantly growing and their total area increases with the growing size of the global population. Characteristic examples of large megalopolises are Moscow, Tokyo, New York, Ho Chi Minh, etc. Some typical estimates of the environmental parameters in megalopolises are given in Table 4.19. For instance, the Moscow

Table 4.19. Comparative characteristics of the developed and developing megalopolises.

Parameter	Value of parameter	
	Moscow	Ho Chi Minh
Area of megalopolis (km^2)	100	31
Population size in megalopolis (10^3 people)	8,894	8,563
Area of green plantations (km^2)	14	16
Number of enterprises in megalopolis (10^3)	25	14
Number of pollution sources (10^3)	100	70
Emissions of harmful matter to the atmosphere ($10^3\ t\ yr^{-1}$)	1,153	248
Open water bodies extent on the megalopolis territory (km)	398	101
Volume of polluted sewage ($10^6\ m^3\ yr^{-1}$)	3,306	226

megalopolis is characterized by high concentrations of sources of anthropogenic pollution over a limited territory (energy enterprises, chemical industry, and automobile transport). Their share in the total emissions constitutes 90.4%. The polluted air plume from Moscow can be observed at a distance of about 100 km from the city. The state of the water bodies within the megalopolis is determined by the input of sewage, surface run-off, and run-off from industrial enterprises to the Moskva and Yausa Rivers, as well as into 70 rivers and springs in the territory of the megalopolis. The concentration of chemicals in the Moskva River both in the city and downstream varies widely depending on the season. For instance, the copper ion content varies during the year between 0.004 to 0.013 $mg\,l^{-1}$ (4–13 permissible concentration level – PCL) with a maximum in the spring. The content of oil products varies within 0.25 to 0.6 $mg\,l^{-1}$ (5–12 PCL). A similar situation exists in the small Yausa River where the concentration of copper varies from 0.007 to 0.12 $mg\,l^{-1}$ and that of oil products between 0.38 to 0.7 $mg\,l^{-1}$. The ecological service of the megalopolis monitors local concentrations of pollutants forming a respective time series of data on the state of the environment. There is a global network of megalopolises, a consideration of which in the global model results in an increase in its accuracy. Many megalopolises are in the stage of formation. Ho Chi Minh is an example of a young megalopolis as too are the adjacent provinces of South Vietnam – Dong Nai, Bin Zyong, and Baria-Vung Tau. This megalopolis occupies an important place in the economy of Vietnam covering 75% of the industrial production of South Vietnam and 50% of the whole country (Fung and Le, 1997). The development of the megalopolis infrastructure includes services of ecological and sanitary control, which makes possible some perspective planning of the environment.

An analysis of the data on the structure of the environment of megalopolises of the world suggests the conclusion that for its complex assessment it is possible to develop a sample system of the models simulating the transport and propagation of pollutants in the atmosphere and in water bodies. The input information for this model can be both from monitoring systems and the global model. An important indicator of the state of megalopolisis is housing density. Thuy *et al.* (2004) used satellite lidar measurements to form databases on housing density of some regions of Tokyo and showed that such measurements, and the use of geographic information system (GIS)-technology, enable one to map housing with an indicated size and height of houses, which is important for operational decision making in case of earthquakes, floods, or other emergency events. Combinations of these measurements with mappings of the state of buildings, using the method developed in Yano *et al.* (2004), enables rescue services to operatively distribute efforts to reduce victims and economic damage in the case of an emergency.

Characteristic linear dimensions of a megalopolis constitute tens of kilometers. This means that to simulate the processes of atmospheric transport of pollutants a model of the Gaussian type can be used. To assess the quality of the atmosphere in the megalopolis it is sufficient to form a composition of the Gaussian streams and at the points of their intersections to summarize the concentrations of the respective types of pollutants.

Let the source of pollutant of type s be located at a point (0, 0) in the coordinate system (x, y, z) where the 0_x axis is along the direction of the wind and the 0_z axis is normal to the Earth's surface and is at a height h. Then the concentration C_s of the pollutant of type s at any point in (x, y, z) can be calculated with the formula:

$$C_s(x, y, z) = Q(2\pi U\sigma_y\sigma_z)^{-1}\exp\left\{-y^2/2(\sigma_y)^2 - (z+H)^2/2(\sigma_z)^2 - (z-H)^2/2(\sigma_z)^2\right\} \tag{4.6}$$

where $H = h + \Delta h$; $\sigma_y = (U/[2k_y(x)])^{1/2}$; $\sigma_z = (U/[2k_z(x)])^{1/2}$; k_y and k_z are the coefficients of eddy diffusion corresponding to axes y and z; U is the wind speed; and the parameters of Δh are calculated from the Briggs formula: $\Delta h = \Delta_1$ for unstable and neutral conditions of the lower atmospheric layer and $\Delta h = \Delta_2$ for stable conditions of the lower atmospheric layer:

$$\Delta_1 = \begin{cases} 1.6F^{1/3}x^{2/3} & \text{for} \quad x \le x_p \\ 1.6F^{1/3}(3.5x_p)^{2/3} & \text{for} \quad x > 3.5x_p \end{cases}$$

$$\Delta_2 = \begin{cases} \min\{2.6(FU^{-1}/S_n)^{1/3}, 1.6F^{1/3}(3.5x_p)^{2/3}\} & \text{for} \quad U > 1.4\,\mathrm{m\,s^{-1}} \\ 5.3F^{1/4}S_n^{-3/8} - D/2 & \text{for} \quad U \le 1.4\,\mathrm{m\,s^{-1}} \end{cases}$$

$F = g(T_s - T_a)V/(\pi T_s)$ is the buoyancy coefficient; $S_n = (g/T_a) \times (\partial T_a/\partial z + 0.01)$ is the coefficient of stability; $\partial T_a/z$ is the potential temperature gradient at the point of pollutant emission; T_s is the temperature of the emitted gas (°K); T_a is the ambient air temperature (°K); D is the diameter of the pollution source (tubes, etc., m); V is the gas expenditure per unit time ($\mathrm{m^3\,s^{-1}}$); g is the gravity acceleration; and x_p is the distance along the wind direction where eddy diffusion dominates over blowing away of the pollutant:

$$x_p = \begin{cases} 14F^{5/8} & \text{for} \quad F \le 55 \\ 34F^{2/5} & \text{for} \quad F > 55 \end{cases}$$

To calculate the pollutant concentration at an arbitrary point in the megalopolis, it is necessary to take into account the impacts of all the sources. For this purpose we choose an arbitrary system of coordinates $x0y$. Assume that the wind direction vector is at an angle α to the $0x$ axis. Denote the coordinates of the pollution sources Q_i $(i = 1, \ldots, n)$ by (a_i, b_i). Then the net concentration of the pollutant of type z at an arbitrary point (u, v) at height z will be:

$$C_s(u, v, z) = \sum_{i=1}^{n} C_s(u, v, z, a_i, b_i)$$

where $C_s(u, v, z; a_i, b_i)$ is the concentration of the pollutant of type s at the point (u, v) at a height z due to a source Q_i. Knowledge of the angle α makes it possible to reach an agreement between the coordinates of the systems $u0v$ and $x0y$.

The hydrological cycle parameters for a megalopolis are calculated using a general model described by Krapivin and Kondratyev (2002). However, simplifications are possible here due to the absence of some elements of the standard scheme or changes in their weighting. The water flow by canals (pipelines) is most characteristic of a megalopolis. In this case we have the following equation:

$$\sigma^{-1}\partial Q/\partial t - Q\sigma^{-2}\partial\sigma/\partial t + Q\sigma^{-1}\left[\sigma^{-1}\partial Q/\partial x + Q\sigma^{-2}(\partial\sigma/\partial x + \partial\sigma/\partial y \times \partial y/\partial x)\right] + g(\partial z/\partial x + \partial y/\partial x + p) + E_{IL} = 0 \quad (4.7)$$

where z is the level of the canal; y is the water depth in the canal measured from the level z; p is the canal bottom friction; E_{IL} is the impact of inflows and outflows; σ is the effective area of the canal's cross section at the point with the x coordinate; x is the water flow direction; and Q is the water flow at the point x.

In addition to (4.7), an equation is needed for the water medium continuity: $\partial\sigma/\partial t + \partial Q/\partial x = q$, where q is the intermediate inflow per unit time per unit length of the canal's section.

A change of temperature T_w of an element of the water volume V is described by the relationship:

$$V\partial T_w/\partial t = (lc^{-1}\rho^{-1})\partial(\sigma_x\Delta_x\partial T_w/\partial x)/\partial x - l\partial(Q^* T_w)/\partial x + [SH_{al} + \theta]c^{-1}\rho^{-1} - T_w\partial V/\partial t$$

where l is the length of the canal's section; S is the section surface area bordering with the atmosphere; ρ is the water density; c is the specific heat capacity of water; σ_x is the area of the cross section of the canal at point x; Q^* is an advective heat flux through the section's boundaries; θ is the outer heat sources; Δ_x is the coefficient of eddy diffusion; and H_{al} is the heat flux velocity through unit surface area.

The heat flux through the surface on the border with the atmosphere is the sum of the following components: $H_{al} = E_1 + E_2 - E_3 - E_4 - E_5$, where E_1 and E_2 are fluxes of the incoming short-wave and long-wave radiation, respectively; E_3 is the outgoing long-wave radiation flux; E_4 is the evaporation-driven outgoing heat flux; and E_5 is the convective heat exchange. The E_1 parameter is a complex function of cloudiness, dust-loading of the atmosphere, sun location, and other geophysical parameters. For the E_2 flux we use the dependence $E_2 = 1.23 \cdot 10^{-16}[1 + 0.17\chi^2](T_a + 273)^6$, where χ is the amount of cloud. The E_3 flux is estimated from the expression $E_3 = \varepsilon\sigma_1(T_a + 273)^4$, where ε is the water emissivity ($\sim$0.07) and σ_1 is the Stephan–Boltzman constant ($= 1.357 \times 10^{-8}$ cal m^{-2} s^{-1} °C^{-4}). For $T_a \in [0°, 30°]C$ an approximation of $E_3 = 7.36 \times 10^{-2} + 1.17 \times 10^{-3} T_a$ is valid.

Heat losses through evaporation are presented as the dependence $E_4 = \rho L e_a$, where ρ is the water density (kg m^{-3}), L is the latent heat of evaporation (kcal kg^{-1}), and E_a is the rate of evaporation (m s^{-1}). The latter parameter is approximated by the relationship $E_a = (a + bW)(p_s - p_a)$, where W is the above-surface wind speed, p_s is the pressure of the saturated vapor over the surface (hPa), p_a is the vapor pressure in the atmosphere (hPa), and a and b are empirical coefficients. The p_s parameter is well approximated by the linear function $p_s = \alpha_p + \beta_p T_a$ whose coefficients are given in Table 4.20.

Finally, the convective heat transport $E_5 = R_4 E_4$, where $R_4 = 6.1 \times 10^{-4}$

Table 4.20. Coefficients of the temperature dependence of water vapor in the atmosphere.

T_a (°C)	α_p	β_p
0–5	6.05	0.522
5–10	5.10	0.710
10–15	2.65	0.954
15–20	−2.04	1.265
20–25	−9.94	1.659
25–30	−22.29	2.151
30–35	−40.63	2.761
35–40	−66.90	3.511

$p_a(T_a - T_d)$ $(p_s - p_a)$ is the atmospheric pressure (hPa), and T_d is the dry air temperature (°C). We obtain:

$$H_{al} = \mu_{al} - \lambda_{al} T_a$$

where $\mu_{al} = E_1 + E_2 - 7.36 \times 10^{-2} - \rho L(a + bW)(\alpha_p - p_a - 6.1 \times 10^{-4} p_a T_d)$ and $\lambda_{al} = 1.17 \times 10^{-3} + \rho L(a + bW)(\beta_p + 6.1 \times 10^{-4} p_a)$.

The atmosphere–water gas exchange is described with the equation $V\partial C/\partial t = SK_C(C^* - C)$, where $K_C = K_2\theta^{(T_a-20)}$, C is the concentration of the water-solved gas, C^* is the concentration of saturated dissolved gas at a local temperature, K_2 is the coefficient of gas re-airing at 20°C, and θ is the temperature coefficient. The K_2 coefficient at known wind speed W ($\mathrm{m\,s^{-1}}$) and coefficient of molecular diffusion D_g is calculated with the formula:

$$K_2 = D_g \times 10^6[200 - 60W^{1/2}]^{-1}$$

where the coefficient D_g is equal to 2.04×10^{-9} for O_2 and 1.59×10^{-11} for CO_2.

The system of geoinformation monitoring in the megalopolis, based on the data of measurements of the environmental parameters and on (4.4) and (4.5), reveals, analyzes, and assesses the most vulnerable places of the megalopolis's structure in the case of natural catastrophes as reactions of the system N to the behavior of the system H. As a result, an interaction takes place between the external shell of the global model, with details of the description of the environment by the scale $\Delta\varphi \times \Delta\lambda$, and the model of the megalopolis where the spatial resolution is more detailed. The resulting diverse schemes of agreement between the global model and the megalopolis model are determined both by its specific features and by the formal system circumstances. Their study and development are the subject of future investigations on the realization of constructive ideas discussed in the monograph by Kondratyev (1999).

Finally, discussing the phenomenon of the megalopolis, it is necessary to emphasize its role as a concentrator of waste, the problems of processing and utilization of which in many countries has become urgent in order to preserve the human habitat. Now, one can speak about the impending waste catastrophe. For instance, in Russia alone, dust-heaps have accumulated more than 80 bln t of solid waste which, after decomposing and/or burning, introduce a multitude of chemicals harmful for human health to the atmosphere. Washing-out of chemical compounds from dust-heaps leads to their penetration to the hydrosphere, which violates the quality of drinking water and poisons the habitats of aquatic animals.

4.2.5 Scenarios of anthropogenic processes

As experience shows, the development of a strategy of global change monitoring utilizing the present-day developments of science is impossible using any models that simulate the functioning of the nature–society system. Hopefully, the developed methods of modeling the type of evolutionary technology (Bukatova et al., 1991) give hope to creating a global model which, based on the accumulated knowledge, will make it possible to obtain reliable enough long-term predictions of global change and to substantiate thereby conditions for the sustainable development of the environment. Obviously, success in this direction depends on the complex use of modeling technologies and environmental observations. So far, the global models require units, which would realize a parameterization of some processes in the nature–society system in the form of scenarios. The construction of the latter is the result of the expert knowledge and a certain level of imagination on the part of an investigator. This problem has been discussed in detail by Alcamo *et al.* (2001), Gyalistras (2002), and in a special IPCC report (Watson *et al.*, 2000). To some extent, the Kyoto Protocol can serve as one of the scenarios. Some countries of the European Economic Community (EEC), who signed this protocol, agreed to reduce the GHGs (carbon dioxide, methane, nitrogen oxides, hydrofluorocarbons, perfluorocarbons, sulfur hexafluoride) emissions by 2008–2012 by 12.5%, with CO_2 emissions decreasing (due to the assumption) by 2010 by 20% compared with 1990 levels. In each country the diverse scenarios of this type bring forth a high level of global uncertainty, the solution of which is impossible without certain generalizations. Therefore, we shall dwell upon some generalizing scenarios.

The use of World Ocean products in the 20th century was characterized by an increasing trade of fish and other elements of higher trophic levels with the trade intensity doubling every 10–15 years. A maximum catch was in 1970 at a level of 61 mln t. The per-capita consumption of sea products varied between 7.2 to 11.8 kg. Numerous forecasts of the possible per-capita use of World Ocean products give about $12\,\mathrm{kg\,yr^{-1}}$. Therefore, assume a hypothetical dependence of the intensive fishing of nekton in each ith region:

$$\lambda_{GRi} = \begin{cases} \lambda_{G0i} - (\lambda_{G0i} - \lambda_{G\infty i})(t - t_0)/(t_{\lambda Gi} - t_0) & t_0 \leq t \leq t_{\lambda Gi}, \\ \lambda_{G\infty i}, & t > t_{\lambda Gi} \end{cases}$$

where λ_{G0i} is the fishing intensity at the moment t_0 (e.g., in 1970 $\lambda_{G0i} \approx 0.0286$; in 1999 $\lambda_{G0i} \approx 0.0312$) and $\lambda_{G\infty i}$ is the maximum possible trade intensity reached at the moment $t_{\lambda Gi} \geq t_0$. The λ_{Gri} coefficient characterizes the fishing trade in the ith region and is considered in the demographic unit of the global model. A human being is considered one of the higher trophic levels with respect to nekton. It is assumed that in the multitude Γ there is the element s_H, for which the relationship is valid:

$$C_{rs_H} R_{s_H} = r \sum_{i=1}^{m} \lambda_{GRi}$$

The contribution of the fishing industry I_i in the diet of *Homo sapiens* V_G is determined by the ratio of the volume of products fished from the World Ocean to agricultural products.

An intensity of food production on land depends on the area covered with agricultural crops and on their productivity. Clearly, within the global model, at the level considering details of all processes and elements, it is impossible to predict all the directions of agricultural development. Therefore, all the processes of its increasing productivity have been generalized, which reflects the general trends. Let us introduce for agricultural formation the identifier k and consider that in any region, part of the territory can be occupied with cultivated lands:

$$\sigma_{ki} = \begin{cases} \sigma_{koi} + (\sigma_{k^*i} - \sigma_{koi})(t - t_0)/(t_{kiS} - t_0) & t_0 \le t \le t_{kiS} \\ \sigma_{k^*i} & t > t_{kiS} \end{cases}$$

$$H_{kiS} = \begin{cases} 1 + (H_{\mathrm{k^*i}} - 1)(t - t_0)/(t_{kBiS} - t_0) & t_0 \le t \le t_{kBiS} \\ H_{k^*i} & t > t_{kBiS} \end{cases}$$

where $\sigma_{ki} \le \sigma_i$ is the area of land cultivated in the ith region and H_{kiS} is the indicator of changes in agricultural productivity with respect to the time moment t_0 $(H_{kiS} = R_{ki}(\varphi, \lambda, t)/R_{ki}(\varphi, \lambda, t_0))$.

According to our suppositions, in the ith region the area σ_{ki} under agricultural crops, starting from the moment t_0, changes following the linear law from σ_{k0i} to σ_{k^*i} in the time interval to t_{kiS}. The time for reaching the level σ_{k^*i} depends on the amount of investments into agriculture with an inversely proportional coefficient t_{k^*i} and on other factors affecting the time constant t_{ki}: $t_{kiS} = t_0 + t_{ki} + t_{k^*i}/(q_{Vi}V_i)$. Possible changes in the agricultural productivity H_{kiS} are also approximated by a linear law where the value H_{k^*i} shows how many times productivity in the ith region can change from t_0 to t_{kBiS}. The reserves for increasing H_{ki} are rather large. Even highly productive plants such as sugar cane, during the process of photosynthesis, consume annually only about 2% of solar energy reaching the Earth's surface; crops consume 1%, and other plants even less. Plant physiology and agro-technology have broad possibilities for a multiple increase of agricultural productivity and, hence, food amounts for *Homo sapiens*. Assume that to reach the level H_{k^*i}, the population of the ith region has time $t_{kBiS} - t_0$:

$$t_{kBiS} = t_0 + \bar{t}_{kBi} + \hat{t}_{kBi}/(q_N V_i)$$

where the constituent $\bar{t}_{kBi}$ is independent of the agricultural investments and $\hat{t}_{kBi}$ characterizes the efficiency of these investments.

According to the V_{Gi} food spectrum for *Homo sapiens*, a certain role is played here by forest ecosystems, the general trend in changes of their areas being characterized by a negative derivative. On a global scale, forest resources are constantly depleting, so that on the border of the 20th and 21st centuries the total area of forest constitutes $\sigma_L = 4,184$ mln ha, with a 31% density (about 28.3% of the total land area).

Two principal processes are observed in forestry: deforestation and aforestation. Describe the totality of these processes using the scenario of forest area change:

$$\sigma_{Li} = \begin{cases} \sigma_{L0i} + (\sigma_{L^*i} - \sigma_{L0i})(t - t_0)/(t_{Li} - t_0) & t_0 \le t \le t_{Li} \\ \sigma_{L^*i} & t > t_{Li} \end{cases}$$

This dependence foresees that the area under forest in the ith region, until moment t_{Li}, varies linearly from σ_{L0i} to σ_{L^*i} and then remains constant. Values of the input parameters constitute the freedom of choice in simulation experiments.

To complete the formulation of the global scenario describing the level of food provision for *Homo sapiens*, consider the law of changing stock-breeding productivity. Suppose that with the constant constituent k_{FSi}, it increases in proportion to agricultural investments q_{Vi} with coefficient k_{F^*i} and varies in time with an exponential law of change:

$$k_{Fi} = k_{F3i} + k_{F^*i} q_{Vi} + (k_{F1i} - k_{F3i})[1 - \exp\{-k_{F2i}(t - t_0)]$$

The direction of anthropogenic activity is determined by the intensity of pollution generation, time of pollution utilization, cost of environmental cleaning, amount of investments into the renewal of resources and into the prevention of the environmental pollution, investments into industry and agriculture, rate of the natural resource expenditure, and search for new sources, etc. The hypothetical trends of these processes should be set in order to realize the predictions with the use of the global model.

Let the intensity of pollution vary between k_{Z0i} at time moment t_0, and k_{Z1i} at moment t_{Z^*i}. Similarly vary the time of waste utilization T_{Zi} and the cost of cleaning of a unit pollution G_{ZVi}, respectively, between the values $T_{Z0i} = T_{Zi}(t_0)$, $T_{Z1i} = T_{Zi}(t_{Zi})$, and $G_{2ZVi} = G_{ZVi}(t_0)$, $G_{1ZVi} = G_{ZVi}(t_{ZVi})$. Since the estimates of these indicators can vary for many reasons and over different times, the simplest scenario for the experiment will be the following functional presentations:

$$k_{Zi} = \begin{cases} k_{Z0i} - (k_{Z1i} - k_{Z0i})(t - t_0)/(t_{Z^*i} - t_0) & t_0 \le t \le t_{Z^*i} \\ k_{Z1i} & t > t_{Z^*i} \end{cases}$$

$$T_{Zi} = \begin{cases} T_{Z0i} - (T_{Z1i} - T_{Z0i})(t - t_0)/(t_{Zi} - t_0) & t_0 \le t \le t_{Zi} \\ T_{Z1i} & t > t_{Zi} \end{cases}$$

$$G_{ZVi} = \begin{cases} G_{2ZVi} - (G_{2ZVi} - G_{1ZVi})(t - t_0)/(t_{ZGi} - t_0) & t_0 \le t \le t_{ZGi} \\ G_{1ZVi} & t > t_{ZGi} \end{cases}$$

Since it is rather difficult to specify the spheres of human activity, the spectrum of investments is confined here to the functions U_{ZG}, q_v, and U_{MG}. The strategy of the investments distribution for each region is determined by two-step functions of time:

$$U_{MGi} = \begin{cases} U_{MG1i} & t_0 \le t \le t_{MGi} \\ U_{MG2i} & t > t_{MGi} \end{cases}$$

$$U_{ZGi} = \begin{cases} U_{ZG1i} & t_0 \le t \le t_{ZVi} \\ U_{ZG2i} & t > t_{ZVi} \end{cases}$$

where t_{MGi} and t_{ZVi} are the moments of the economic policy change in the ith region in the field of investments into the renewal of natural resources and prevention of pollution, respectively. Here the term "renewal of resources" denotes the processes favoring an increase of M. In particular, this is a change of mineral resources for others, with their significance recalculated for the levels of significance of the previous ones.

Agricultural investments are one of the most important components of investment. To obtain a high productivity of cultural crops, it is necessary to completely satisfy their needs for water and mineral matter. This means that high yields require great amounts of energy. The situation is the same with protein food production, where an efficiency of the transformation of the vegetable food energy into the energy of meat and fat is about 10% or, in other words, 10 calories of energy are spent on the production of 1 calorie of protein food. This energy production requires some share of investment and consumption of the fossil products of photosynthesis. Hence, here the investments and energy expenditures are mutually dependent. The global model unit that describes the agricultural investment is constructed by setting the dependences of agricultural productivity on energy expenditures. The investment parameter of the control U_{BGi} is described by the two-step function:

$$U_{BGi} = \begin{cases} U_{BG1i} & t_0 \le t \le t_{BGi} \\ U_{BG2i} & t > t_{BGi} \end{cases}$$

where t_{BGi} is the moment of change of the investment policy in agriculture in the ith region. The T_{Bi} parameter characterizing the time of assimilation of agricultural investments in the ith region can also change at moment t_{Bi}:

$$T_{Bi} = \begin{cases} T_{B1i} & t_0 \le t \le t_{Bi} \\ T_{B2i} & t > t_{Bi} \end{cases}$$

At present the need for energy by humankind is 90% satisfied due to burning of the fossil products of photosynthesis (coal, oil, gas) and only 10% due to hydro- and electric power stations. In the future these relationships should change drastically due to a mastering of new technologies of energy production and, in particular, a raising of the efficiency of solar energy use. Here, the processes of accumulation and transformation of solar energy through improving the technology for capturing World Ocean energy (waves, tides, ebbs, currents, etc.) can play an important role. Of course, this moment will be different for various regions of economic development. In the global model this is reflected by a substitution of the initial value of the component M_i for a new M_{0GMi} value. The t_{GMi} parameter is the function of the multiplier m_{Gi} determined by the ratio of the natural resource supplies to their annual expenditure at a given time moment. It can be regulated in each region by different means, with numerous aspects of human activity taken into account. Without dwelling upon details, take the following scenario of a possible change of the multiplier m_{Gi}:

$$m_{Gi}(t) = \begin{cases} m_{G0i} - (m_{G0i} - m_{G\infty i})(t - t_0)/(t_{Mi} - t_0) & t_0 \le t \le t_{Mi} \\ m_{G\infty i} & t > t_{Mi} \end{cases}$$

Within the time interval $[t_0, t_{MGi}]$ in each region the cost G_{MGi} of the natural resources renewal can vary linearly from the value G_{MG0i} to the value G_{MG1i}:

$$G_{MGi}(t) = \begin{cases} G_{MG0i} - (G_{MG0i} - G_{MG1i})(t - t_0)/(t_{MGi} - t_0) & t_0 \le t \le t_{MGi} \\ G_{MG1i} & t > t_{MGi} \end{cases}$$

The basic capital $V_i (i = 1, \ldots, m)$ invested into the development of industry, science, agriculture, construction, and other spheres of human activity become used, leading to a decreased efficiency. To parameterize these processes, let us introduce "the ageing time" term T_{VG} ($\sim$40 years). An assessment of this parameter is a characteristic of each region. As before, assume that T_{VG} varies linearly from the value T_{VG0i} to the value T_{VG1i} within the time interval $t \in [t_0, t_{Vi}]$ and then remains constant:

$$T_{VGi}(t) = \begin{cases} T_{VG0i} - (T_{VG0i} - T_{VG1i})(t - t_0)/(t_{Vi} - t_0) & t_0 \le t \le t_{Vi} \\ T_{VG1i} & t > t_{Vi} \end{cases}$$

4.3 TECHNOGENIC SOURCES OF NATURAL DISASTERS

The observed increase of the number of natural disasters is connected with the growing role of anthropogenic factors that determine the environmental conditions for the origin of critical situations. The anthropogenically induced natural disasters include forest fires, desertification, deforestation, dust storms, floods, snow avalanches and snowslides, reduced biodiversity, etc.

The interaction between the processes in the environment is the main mechanism for the appearance of emergency situations through the fault of humankind. One of the causes of a natural balance violation is environmental pollution of the atmosphere, in particular. For this reason, laws in most developed countries try to regulate the volumes of emitted pollutants. The international community understands the seriousness of the problem and tries to find solutions. Respective attempts have been made by a number of international conferences with participation of scientists and politicians (Kondratyev, 1993, 1996, 1999).

Air pollution is a serious problem of the current ecodynamics. In particular, the products of burning are emitted to the atmosphere in the form of various oxides of sulfur and nitrogen, and as a result, sulfuric and nitric acids form. Acid rain negatively affects both vegetation and soil. The biotic cycle of land ecosystems are broken, which can lead to unpredicted changes in the biogeochemical cycle of greenhouse gases. In this aspect, anthropogenic impacts on forest ecosystems are especially dangerous. Irrational human economic activity has led to an irreversible disappearance of virgin jungles. A similar danger threatens the Siberian taiga, which plays a decisive role in the gas balance of the atmosphere. Along with direct forest cutting, humankind also affects forests through the degradation of the atmosphere with sulfur dioxide and nitrogen oxides thus reducing tree vitality. Sulfur dioxide and nitrogen oxides are emitted to the atmosphere by electric power stations, boiler-works, metallurgic and chemical plants, and motor transport. Both these gases, combining in the atmosphere with water vapor, give sulfuric and nitric acids, which, falling onto the ground, infiltrate into the soil and corrode the root systems of trees, and over time, leach calcium, potassium, and other nutrients. Falling onto the canopy of trees, these acids break the process of respiration, circulation, and assimilation of nutrients, which eventually leads to death for the trees.

Contamination of the land surface is a very dangerous anthropogenic impact for the environment. This process is characterized by a retarded response of soil ecosystems, since pollutants gradually accumulated in soils begin to act when some critical threshold of their concentration is exceeded. In such cases, the response of land ecosystems manifests itself through poisoning living organisms, mass diseases, and epidemics. The scale of soils contamination in each region of the planet is determined by the volumes of waste emitted to the environment and the state of industry for their processing. A special kind of soil contamination is radioactive precipitation. For instance, the most characteristic soil contamination resulted from the Chernobyl accident in 1986. Its area is estimated at 4 mln ha (Grigoryev and Kondratyev, 2001).

Contamination of soil with industrial waste is a serious factor affecting demographic processes. Harmful chemical compounds through plants and water enter the food chain which leads, every year, to 5.2 million deaths, especially among children. Most urgent is the problem of contamination of soils and water resources in developing countries, where less than 10% of the urban and industrial waste are refined. Particularly dangerous to human health are pesticides. But their role is dual. On the one hand, they exterminate pests and promote an increase of the harvest. On the other hand, pesticides accumulate in soils and enter humans and animals causing degradation and diseases.

Anthropogenically induced natural disasters include a large number of events which take place as accidents at industrial enterprises (i.e., disasters regarding transport systems, oil tankers, dams and pipelines). Especially large-scale pollution is caused by accidents at oil pipelines.

The continuing growth of the number of anthropogenic disasters brings forth the complicated economic problem of optimizing measures of their prevention and reducing their consequences. Many countries solve this problem using life and property insurance against natural disasters. The geographic distribution of insurance payments correlates with the location of the most dangerous zones and the level of economic development of the regions. So, in 2001, one of the worst years for insurance companies, insurance payments to countries most subject to catastrophes constituted: USA – US$5.05 bln, Germany – US$500 mln, and

Taiwan and Japan – US$600 mln. The complexity of the problem of insurance against natural disasters consists in that most catastrophes are connected with weather processes which do not follow a statistical balance and, hence, traditional statistic technologies of forecasts turn out to be ineffective and unreliable.

By the beginning of the 21st century, the number of large-scale natural disasters had increased by a factor of 2.9 compared with the 19th century. During the same period the economic and insurance losses increased by factors of 7.7 and 14.3, respectively (Berz, 1999; Munich Re, 1999). Table 4.21 illustrates the dynamics of the relationship between economic and human losses from natural disasters. It is seen that this relationship is unstable and does not follow the laws of formal statistics.

The present ecodynamics and its laws have become less predictable due to the growing role of the anthropogenic factor in the origin of natural disasters. This trend is observed when attempts are made to apply methods of system analysis to studies of various processes in the NSS. In particular, to overcome the appearing difficulties in coordination of the results of interdisciplinary studies, during the last years synergetics has been successfully used, which makes it possible to identify universal regularities in complicated systems with inherent catastrophic behavior. For instance, it was found that in specially formed phase spaces, various types of natural disasters have similar ranging characteristics described by exponential laws. In other words, synergetics gives a promising approach to formulating and solving the main problem of searching for a "trigger" (Vorobyev *et al.*, 2003) for the origin of a natural disaster.

Lisienko *et al.* (2002) have analyzed the interaction between economy and ecology over a confined territory and proposed some constructive approaches to a balanced reduction of the ecologo-economic damage. Solution of this problem is reduced to a system optimization of the territorial infrastructure. As a result, a scenario of development of the territory is developed and first-priority goals of production modernization are determined which take into account the hierarchy of economic and ecological priorities and, primarily, human health.

4.4 DEMOGRAPHIC PREREQUISITES OF NATURAL DISASTERS

As has been mentioned above, the scale of damage from a natural disaster depends on economic and social factors, as well as on ethnopsychological features of the perception of a dangerous phenomenon, relevant information, timely protective measures, and efficiency in overcoming the consequences of an unfavorable situation. On the whole, with a growing amount of emergency situations in the environment, the respective risk for the planetary population grows. As Grigoryev and Kondratyev (2001) have justly noted, the main causes of enhancing risks and scales of catastrophes are the growth of the population size and the development of various infrastructures. Population density, especially in megalopolises, increases the risk. The high population density has played a decisive role in enhancing the consequences of many natural disasters, especially earthquakes and floods. For instance,

Table 4.21. Comparison of economic losses and human victims caused by natural and technogenic disasters for the period 1970–2000.

Year	Natural disasters		Technogenic disasters	
	Number of victims	Insurance losses (US$$10^9$)	Number of victims	Insurance losses (US$$10^9$)
1970	365,063	2.643	3,195	1.763
1971	15,308	0.500	3,010	1.316
1972	16,971	2.222	5,159	1.760
1973	5,609	2.089	4,473	2.303
1974	13,092	4.749	4,767	2.308
1975	3,838	1.444	3,273	2.267
1976	297,138	1.940	3,993	2.555
1977	15,021	0.994	3,904	3.288
1978	46,730	1.824	3,835	2.444
1979	4,952	3.557	18,523	4.929
1980	12,145	1.801	5,194	4.063
1981	11,982	1.015	6,435	1.548
1982	10,538	3.833	3,590	2.989
1983	6,867	5.209	3,954	2.214
1984	5,151	2.886	6,749	1.874
1985	51,012	4.795	5,092	2.766
1986	6,594	1.939	5,092	2.801
1987	13,020	8.707	10,422	3.745
1988	41,410	3.346	7,234	6.271
1989	7,932	12.234	7,416	6.889
1990	60,984	18.611	7,031	4.247
1991	156,600	14.646	6,570	4.256
1992	13,427	29.230	7,222	5.189
1993	20,719	8.235	9,340	5.225
1994	9,783	20.960	8,430	4.805
1995	20,703	16.335	7,622	2.490
1996	13,923	9.006	8,231	4.785
1997	14,371	4.485	7,641	3.057
1998	34,932	16.002	9,747	4.226
1999	98,152	28.131	7,182	4.309
2000	7,767	7.548	9,694	3.049

there are 746 cities in Russia subject to floods and 500 cities located in zones of origins of hurricanes and waterspouts (Vladimirov *et al.*, 2000).

The risk of natural disasters is increased by the drastic enlargement of developed territories and settlements in regions dangerous to life. At present, about half the global population live on coastal regions subject to ecological disasters. For instance, 25% of the Former Soviet Union extends over regions especially dangerous from the seismic point of view. There is no region in Russia where the probability of a natural disaster such as earthquake, flood, forest or peatbog fires, showdrift, avalanche, landslide, and hurricane would be negligibly small.

Among the social causes of enhanced risk of catastrophes is human poverty and an economic obsolescence of many countries, as well as a poorly developed political structure. For example, in such countries as Guatemala, China, and Russia (of the "perestroika" period), natural disasters have recently caused numerous human victims for the reasons mentioned above. A rather representative example is Russia where during the last 15 years, de-industrialization and serious changes in the social sphere have led to a 5-year decrease in human lifespans and a drop in per-capita GDP, with the volume of industrial production decreasing by more than 50%. As a result, the risks of large losses have increased, and the losses themselves from both natural disasters and anthropogenic accidents have drastically increased.

To parameterize the process of fear of expecting a natural disaster, Vladimirov *et al.* (2000), based on a synergetic approach, proposed a dynamic model which reflects the dependence of fear on the probability of a natural disaster and its expectation. This model, together with the model of ecosystem functioning in emergency situations can simplify the search for a behavioral strategy of the NSS's anthropogenic component. One of the possible ways of realizing such a model is to synthesize the monitoring system for megalopolises. The monitoring of their sustainable development will make it possible to forecast the risks and losses from natural and natural–anthropogenic catastrophes. In these zones, the correlation between natural and anthropogenic processes is great, as too are the possibilities of taking measures to prevent catastrophies by changing some structures within a megalopolis.

On the whole, the relationship between a confined space and growing population density is one of the important characteristics of increasing risk of large losses due to natural and anthropogenic disasters, and at the same time, serves as a pre-condition for the postponement of disasters, such as an extension of territories useless for life, a transition of the climate system to the state where the frequency of occurrence of known types of disasters changes the ecosystem's productivity. It is known that in many large cities on the planet the content of organic and inorganic compounds in the atmosphere exceeds maximum permissible concentrations. A heavy industrial load on the environment grows in proportion to the growth of the population size. Viewing numerous indicators, it can be seen that humankind has approached a critical threshold. For instance, further reduction of forest areas may lead to irreversible consequences. A change of forest areas estimated at 40.3–41.84 mln km^2 could substantially affect the gas exchange in the atmosphere. Even if by 2050 forest area is reduced only by 10%, the atmospheric CO_2 content by the end of the 21st century could increase by 44%. On the contrary, a 10% increase in forested area will reduce the concentration of atmospheric CO_2 by 15%. This means that oscillations of forested areas in the biosphere even within $\pm 10\%$ can substantially change the dynamics of many components of the global ecosystem. Moreover, the global model of the nature–society system (GMNSS) makes it possible to assess the consequences of the replacement of forest ecosystems with other biomes.

Table 4.22 exemplifies the calculation of CO_2 sinks to vegetation cover within

Table 4.22. Dynamics of CO_2 assimilation by plants within Russia. Carbon emissions in 1990 over this territory are assumed to be $1.6\,\mathrm{Gt\,C\,yr^{-1}}$ with an annual change following the Keeling scenario (Krapivin and Vilkova, 1990).

Soil–plant formation (Figure 4.1)	Rate of CO_2 assimilation ($10^6\,\mathrm{t\,C\,yr^{-1}}$)				
	Years				
	1990	2000	2050	2100	2150
A	2.6	2.8	6.7	7.1	6.9
C	3.7	4.6	10.9	12.0	12.1
M	4.0	5.1	12.4	14.5	13.8
L	3.2	3.9	9.2	10.3	10.4
F	11.2	14.8	43.6	47.2	44.2
D	31.6	39.9	110.6	121.9	109.3
G	23.3	29.2	72.2	73.4	70.5
R	5.2	6.2	13.1	13.8	10.7
W	4.7	5.1	8.2	8.8	7.9
V	0.7	0.7	0.9	1.1	0.8
@	2.4	2.6	3.7	3.9	2.9
S	0.6	0.7	1.2	1.4	1.0
Q	1.5	1.6	2.2	2.3	1.8
Total	*94.7*	*117.20*	*294.9*	*317.7*	*292.3*

Russia. Such calculations with the use of GMNSS enable one to see the dynamics of the mosaic of CO_2 fluxes in the atmosphere–plant–soil system. Knowledge of this mosaic makes it possible to assess the role of concrete types of soil–plant formations (Figure 4.1) in the regional carbon balance and on this basis of calculating the global fluxes of CO_2 at the atmosphere–land boundary. Similar calculations are also possible for the atmosphere–ocean system.

Table 4.23 demonstrates the consequences of changing the global structure of the soil–plant formations to the dynamics of CO_2 absorption by vegetation. It is seen that an artificial change of vegetation cover changes substantially the balance of the components in the global carbon cycle. Clearly, such experiments require a thorough preparation of data for the possibility of transformation of vegetation cover, with climatic zones and biocenologic compatibility taken into account. Nevertheless, such hypothetical experiments are useful for a general evaluation of the possible ranges of anthropogenic forcing on the global carbon cycle. For instance, natural and anthropogenic cataclysms connected with forest fires introduce considerable changes into this cycle each year since they change many fluxes and supplies of carbon over large territories. Tables 4.23 – 4.26 give estimates of the deviations of the content of carbon in the basic biospheric reservoirs due to forest fires in different zones. It is seen that large-scale impacts on land biota become damped during 60–100 years. The biosphere is more stable with respect to forcings on forests of southern latitudes and more sensitive to violations of forests at moderate latitudes. The conclusion can be

```
  W    150   120    90    60    30    0     30    60    90   120   150    E
N ....................****. **********........................................ N
  ....................AAA. . **********. . . . . **. . . . . . . . . . . . . . . A. . . . . . . . . . . . . . .
  ............AAA. AAAA. . . . . *********. . . . . . . . . . . . . C. . . . . AAAAA. . . . . A. . . . . . . .
  . . . . CCCC. . . . . CCA. A. AAAC. . . ******. . . . . . . . . LLL. . . . . C. CCCCCMMMMCAAAAAAAA. . . . .
  MC. FFFMFMFFFFCCCCCAA. . MM . ***. . . L. . . . . . DDDFF. FFFFCLLLFFFMFFFFFMAAAMMMMMMM
  . . . . LLFFFFCCCCCCCC. . . C. . . . . L. . . . . . . . . . RG. DDDDDDDDDFFFFDDDDDDDDDMMFFMCCM .
  . . . . . L. . . FFFFGGGFF. . . CFC. . . . . . . . . . . +. . . R. RGGGGGGGDGGGGGGGGDDDDDDM . . L. . . .
  . . . L. . . . . . +RGGGGGDDF. FFFF. . . . . . . . . . ++. ++RRRRXXXXGXXXXGGGGGGDDGDDD. . . L. . . .
50. . . . . . . . . . . RRXWXXGGDDDDD. D. . . . . . . . . . . +++++XXWWVVVVVVVGGGGGGGGGDDD. . . . . . . . 50
  . . . . . . . . . . . . RXWWX++. +++G. . . . . . . . . . . . +++++XXWWV. S. @@QQSGVVVVGXRR. . . . . . . . . .
  . . . . . . . . . . . . BBBXXX++++. . . . . . . . . . . . . . ++. . P++. . WQ. KS&QQQSSSSSSXX++. . +. . . . . . .
  . . . . . . . . . . . . RBUWXP+PP. . . . . . . . . . . . . PP. . . . P. WWWW&K&&QQSSSSSQXX +. . P. . . . . . . .
  . . . . . . . . . . . . . @UUUPPPP. . . . . . . . . . . . . . &&&&. . . . . &&WUUUUUQBBQQ+++. . . P. . . . . . . . .
30. . . . . . . . . . . . . @@@UP. . P. . . . . . . . . . . . . HHHHHHHHHHHHHHTHHHJPQQQQPPPP. . . . . . . . . . . . . 30
  . . . . . . . . . . . . . . @@@. . . . . . . . . . . . . . . . . HHHHHHHHH. HH. TTTHNNYZZPPPP. . . . . . . . . . . . .
  . . . . . . . . . . . . . . . U@. . . J. . . . . . . . . . . . TTTTTTTTTTTTTTT. . . NNI. YYYI. . . . . . . . . . . . . .
  . . . . . . . . . . . . . . . . YYZ. . . J. . . . . . . . . . TTTTTTTTTTT. TT. . . . JN. . YYY. . . . . . . . . . . . . . .
  . . . . . . . . . . . . . . . . . . YZ. . . . . . . . . . . . . JJJJJTJJJNTT. . . . . NN. . . YY. . Y. . . . . . . . . . . .
10. . . . . . . . . . . . . . . . . . . . Z. JJ. . . . . . . . . . JJJJJJJINNTT. . . . . N. . . . Y. . . . . . . . . . . . . . . . 10
  . . . . . . . . . . . . . . . . . . . . . ZJJZ. . . . . . . . . . YJZYJJJJJJT. . . . . . . . . . Z. . . . . . . . . . . . . . . .
  . . . . . . . . . . . . . . . . . . . . . ZZZZJ. . . . . . . . . . . . . ZYYYYJNN. . . . . . . . . . Z. . Z. . . . . . . . . . . . .
  . . . . . . . . . . . . . . . . . . . . QZZZZJZZ. . . . . . . . . . YYZYJJJ. . . . . . . . . . . . Z. Z. . . . . . . . . . . . .
  . . . . . . . . . . . . . . . . . . . . WZZZZJJJJJ. . . . . . . . . JYYNNJ. . . . . . . . . . . . . . . . . . ZZ. . . . . . .
10. . . . . . . . . . . . . . . . . . . . ZZZZJJJJ. . . . . . . . . . JNNNN. . . . . . . . . . . . . . . . . . . . . . . Z. . . . . . 10
  . . . . . . . . . . . . . . . . . . . . . WWZZZJJY. . . . . . . . . . JNNNN. . Z. . . . . . . . . . . . . . . NN. . . . . . . . .
  . . . . . . . . . . . . . . . . . . . . . . WWNZYJY. . . . . . . . . . JJJNJ. J. . . . . . . . . . . . . . . . NNNNN. . . . . . .
  . . . . . . . . . . . . . . . . . . . . . . TNNYYY. . . . . . . . . . . TJJJJ. N. . . . . . . . . . . . . . JJTTJJJ. . . . . . .
  . . . . . . . . . . . . . . . . . . . . . . TNNYP. . . . . . . . . . . . HHJE. . . . . . . . . . . . . . . . JJTTTTNN. . . . . .
30. . . . . . . . . . . . . . . . . . . . . . UUEY. . . . . . . . . . . . . . HEE. . . . . . . . . . . . . . . . . HHTTTJJ. . . . . . 30
  . . . . . . . . . . . . . . . . . . . . . . UUEI. . . . . . . . . . . . . . E. . . . . . . . . . . . . . . . . . U. . . UUP. . . . . .
  . . . . . . . . . . . . . . . . . . . . . . UEE. . . . . . . . . . . . . . . . . . . . . . . . . . . . . . . . . . . . . . . . . P.
  . . . . . . . . . . . . . . . . . . . . . +VV. . . . . . . . . . . . . . . . . . . . . . . . . . . . . . . . . . . . P. . . . +. .
  . . . . . . . . . . . . . . . . . . . . . +V. . . . . . . . . . . . . . . . . . . . . . . . . . . . . . . . . . . . . . . . . . .
50. . . . . . . . . . . . . . . . . . . . . +E. . . . . . . . . . . . . . . . . . . . . . . . . . . . . . . . . . . . . . . . . . 50
  . . . . . . . . . . . . . . . . . . . . . . +. . . . . . . . . . . . . . . . . . . . . . . . . . . . . . . . . . . . . . . . . . .
S . . . . . . . . . . . . . . . . . . . . . . . . . . . . . . . . . . . . . . . . . . . . . . . . . . . . . . . . . . . . . . . . . . . S
  W    150   120    90    60    30    0     30    60    90   120   150    E
```

Figure 4.1. The spatial distribution of soil–plant formation over the geographical grid 4° × 5°. Notations is given in Table 4.22.

drawn that the northern hemisphere forests up to 42°N play an important role in the stabilization of biospheric processes.

The scenario of deforestation, as seen from the studies of many experts, attracts maximum interest in global studies of the carbon cycle and associated climate change. A diversity of possible real situations of land cover transformation is so broad that it is impossible to evaluate all the consequences. Note only that, for instance, the destruction of all northern taiga and middle taiga forests (types *F*, *D*) in 50 years will lead to a 53% increase in atmospheric CO_2 with subsequent

Table 4.23. Dynamics of the ratio of integral rates of atmospheric (H_6^C) CO_2 assimilation by vegetation cover with the natural distribution of soil–plant formations (Figure 4.1) and with its transformation according to the scenario in the second column.

Scenario		H_6^C (changed Figure 4.1)/H_6^C (Figure 4.1)		
		Years		
Figure 4.1	Changed Figure 4.1	2003	2020	2050
A	*L*	2.79	2.14	1.97
C	*L*	0.96	0.94	0.95
M	*L*	1.67	1.15	1.01
F	*D*	1.68	1.57	1.11
G	*D*	2.11	1.67	1.45
R	*D*	3.98	3.70	2.87
P	*Z*	3.17	2.68	2.43
U	*Z*	22.52	20.73	17.95
W	*Z*	23.12	19.44	16.32
E	*Z*	100.14	77.75	68.54
H	*Z*	194.56	155.50	138.39
Q	*Z*	799.14	777.50	751.26
Y	*Z*	1.43	1.39	1.23
N	*Z*	69.98	62.20	56.59
J	*Z*	5.89	5.09	4.67
T	*Z*	26.54	25.92	23.58
I	*Z*	17.88	16.37	14.91
#	*Z*	0.91	1.12	0.97

Table 4.24. Model estimates of the carbon content deviation under conditions of all coniferous forests of the northern hemisphere being burned (to 42°N).

Post-impact years	Deviations in the carbon content (Gt)			
	Atmosphere	Soil	Upper ocean	Deep ocean
0	140.9	−5.4	15.5	0.1
10	104.8	−33.1	29.8	3.1
20	83.1	−44.1	21.5	7.4
30	63.4	−43.5	19.0	8.5
40	47.2	−39.7	14.6	10.4
50	34.2	−34.6	11.6	11.7
60	24.1	−29.4	8.3	12.7
70	16.3	−24.7	6.2	13.5
80	10.2	−20.5	4.5	14.0
90	5.6	−17.1	3.3	14.3
100	2.1	−14.2	2.3	14.5
200	−9.0	−3.4	−0.8	13.5

Table 4.25. Model estimates of the carbon content deviation under conditions of all forests of the northern hemisphere being burned (to 42°N).

Post-impact years	Deviations in the carbon content (Gt)			
	Atmosphere	Soil	Upper ocean	Deep ocean
0	230.8	−7.9	24.9	0.1
10	173.2	−30.9	47.9	4.9
20	138.9	−67.6	39.2	10.0
30	107.9	−89.3	31.1	13.8
40	82.0	−64.3	24.1	16.8
50	60.9	−56.9	18.4	19.1
60	44.2	−49.0	13.9	20.7
70	31.1	−41.6	10.3	21.9
80	20.9	−35.0	7.5	22.8
90	12.9	−29.4	5.4	23.3
100	6.9	−24.7	3.7	23.6
200	−12.7	−5.9	−1.7	21.7

Table 4.26. Model estimates of the carbon content deviation under conditions where all tropical forests are burned.

Post-impact years	Deviations in the carbon content (Gt)			
	Atmosphere	Soil	Upper ocean	Deep ocean
0	396.0	−20.0	42.2	0.2
10	261.6	−93.7	73.9	8.0
20	162.2	−84.8	48.0	14.9
30	90.6	−38.6	28.2	19.1
40	45.3	−36.5	15.0	21.3
50	18.3	−21.6	7.5	22.4
60	2.9	−12.4	3.0	22.8
70	−5.8	−7.0	0.5	22.8
80	−10.6	−4.2	−0.9	22.6
90	−13.2	−2.6	−1.7	22.5
100	−14.5	−1.9	−2.1	21.8
200	−13.2	−2.3	−1.9	17.7

negative consequences for the H_6^C flux. Similar consequences result from the destruction of all wet evergreen and broad-leaved tropical forests (types Z, Y). Only in this instance, the indicated increase of atmospheric CO_2 will be reached in 20 years.

Land cover structure changes not only because of mankind's intervention. In some regions of the globe, hurricanes cause considerable change in the carbon balance of forest ecosystems. So, in the USA every three years two hurricanes occur which accelerate the transition of the living biomass of trees into dead organic matter. In the USA forests loose annually 20 Tg C, with 10–15% of this due to single hurricanes (McNulty, 2002). Hence, hurricanes accelerate the return of carbon to the atmosphere, and their consideration over the whole planet is needed for more accurate estimates of numerous fluxes of the global carbon cycle.

4.5 BALANCE BETWEEN NATURE AND HUMAN SOCIETY

4.5.1 Global changes: real and possible in the future

4.5.1.1 Global models of Club of Rome and Forrester

The problem of global environmental change is the subject of discussion by many scientists (Krapivin and Kondratyev, 2002; Krapivin, 1993; Kondratyev *et al.*, 2002b; Lomborg, 2001; Nitu *et al.*, 2000a,b, 2004; Pielke, 2001a,b; Svirezhev, 2002). Among them are pessimistic and optimistic estimates of the future joint development of humankind and the environment. Natural, social, and economic trends show that humankind strives for sustainable development, trying to find compromises between indicators of its wellbeing and the environment. Though the notion of global change can be considered established enough, despite the preserved numerous terminological differences (it refers especially to the definition of "sustainable development"), note that we speak here mainly about the interaction between society (socio-economic development) and nature. Most substantial features of global change are multi-component in character, interactive, and non-linear. These features complicate the prognostic estimates to such an extent that the notion of prediction has recently been ousted by a more vague concept of "scenarios" or "projections". The uncertainty of scenarios increases due to the fact that, as a rule, there are no probabilistic estimates for various scenarios (this completely refers, for instance, to the problems of global climate changes).

One possible mechanism to overcome these uncertainties is a new scientific direction intensively developing over the last years – global ecoinformatics, within the framework of which information technologies have been created, which ensure a combined use of various data on the past and present state of the NSS. The creation of a model of the NSS functioning, which is based on knowledge and available data and blends with the adaptive–evolutionary concept of the geoinformation monitoring, which enables one to realize an intercorrection of the NSS model and the regime of the global data collection, can be considered an important step in global ecoinformatics. As a result, there is a possibility to set forth the problem of optimization of planning the organizational–behavioral structure of the NSS, which enables one to hope that it is possible to ensure the purposeful global changes for the benefit of humans and without damage to nature, and what is most important, to create international mechanisms for the coordinated nature-use behavior of the entire population on Earth.

One of the important scientific directions of ecoinformatics is the development of models of various processes taking place in the NSS. Modeling refers to spheres of knowledge. It is connected with the fact that the model has ensured an understanding of the correlations between the NSS's fragments and therefore made it possible to see a single whole in the mosaic of the processes which appear isolated at first sight. An especially important property of the model is its ability to reflect the present duality of the anthropogenic component. On the one hand, a human is an element of nature and its behavior is determined by nature, but on the other hand, he can project his behavior and through this he can transform the natural environment into an artificial one.

Three decades ago (in 1972) the first report of the Club of Rome (CR) (Meadows *et al.*, 1972) was published which caused worldwide interest. In 1972 an International Conference on the Environment was held in Stockholm, which also caused wide international resonance. Later on, developments in the problems of global change were mainly concentrated within the framework of the IGBP and the WCRP, which then were supplemented with the International Human Dimensions Programme (IHDP). These and attendant international programs of environmental studies have enabled scientists to create extensive global databases on various NSS components and, in particular, to identify their trends during the last decades.

Let us briefly analyze the results of developments in the problems of global observations for the last three decades and (importantly) answer why, despite huge effort and enormous expense (reaching many billions of US dollars), the global ecological situation not only has not improved but has even worsened. But first of all, we shall briefly discuss the models developed by Forrester and the CR.

The book titles *Limits to the Growth* (Meadows *et al.*, 1972) and *The World Dynamics* reflected an appearance of a new fundamental concept: development of human society (population size and growing scales of human activity) cannot be boundless and has already approached certain limits – first of all, from the viewpoint of the levels of use of irreversible natural resources. The main goals of the CR developments consisted in analysis of global demographic dynamics and estimates of natural resources, as well as in substantiation of the model of global ecodynamics and possible scenarios of the ecodynamics of the future. Meadows *et al.* (1972) formulated the principal goals:

> *The goal of the project is to study a complex of problems of concern for all nations: poverty among abundance; environmental degradation; loss of confidence in public institutions; uncontroled expansion of cities; unreliable employment; estranged young people; neglect of traditional values; inflation and other economically destructive phenomena.*

Three decades ago the CR participants justly emphasized the complex and interactive character of the problems which include technical, social, economic, and political aspects.

Instead of an introduction, the authors of *Limits to the Growth* quoted the address by U. Thant, the UN Secretary-General:

> *I do not want to overdramatize the events, but I only want, based on information I have got as the UN Secretary-General, to make the conclusion that the UN*

participants have possibly ten years at their disposal to regulate their old disagreements and to start a global cooperation in order to curb the arms, to improve the environment, to restrain the explosive growth of population, and to stimulate efforts in the sphere of the socio-economic development. If the global cooperation does not become a first-priority goal, I am very much afraid that all these problems will reach stunning levels, beyond our capabilities to control them.

Now, three decades later, the conclusion might be drawn that U. Thant (like D. Meadows) adhered to the concepts of catastrophism: at first sight, it seems that with all the problems of the development of present civilization, the world survived and did not reach a critical point of irreversible negative trends. Analysis of realities of the present global ecodynamics does not permit such an optimistic conclusion. As for the initial CR developments and U. Thant's speech, one can only assess their foresightedness highly. But there are further grounds for such an assessment because further development of international scientific events has demonstrated the evolution from such low-success (though large-scale) events as the Second UN Conference on Environment and Development (Rio de Janeiro, 1992) and the UN Special Session (New York, 1997) to the fiasco of the World Conference on Sustainable Development (Johannesburg, 2002) against a background of a continuously worsening global ecological situation. Below we shall briefly discuss the basic content of the CR report, whose authors begin with analysis of the global demographic dynamics.

From 1650 onwards, when the global population size was about 0.5 billion people, the population increment constituted 0.3% annually, which corresponded to a population doubling over 250 years. By 1970, the size of population had reached 3.6 billion, and the rate of increment increased to 2.1% yr^{-1} (respectively, the period of doubling decreased to 36 years): the increase in population became "superexponential". An important demographic indicator is lifetime, which in 1650 constituted about 30 years. By 1970 it reached 53 years and continued increasing. Meadows *et al.* (1972) pointed out the lack of prospects for stabilizing the global population size by the year 2000.

The second important parameter is the rate of industrial development: between 1965 and 1968 the rate of growth averaged 7% yr^{-1} (the respective per-capita indicator was 5%). It is important to note that this growth was mainly concentrated in industrial countries where the rate of population growth was, however, relatively low (in this way, a tremendous contradiction formed between developed and developing countries, characteristic of the present-day). The data of Table 4.27 illustrates the rate of growth of GDP in different countries. The CR report justly emphasizes the low probability that this dynamic growth of both population and GDP will remain even until the end of the 20th century, since many factors of this dynamic growth will suffer changes. It is of interest, however, that the example of Nigeria, where one should expect the end of inter-ethnic conflicts and, respectively, growth in the economy, proved lame, reflecting a continuation and deepening of the socio-economic conflicts on the African continent (convincingly demonstrated in the growth of foreign debt). Of interest are estimates (Meadows *et al.*, 1972) of possible levels of per-capita GDP in 2000 (Table 4.27). Though the reliability of

Table 4.27. Rates of population size and economy growth in different contries.

Country	G	ΔG	Ψ (1968)	$\Delta\Psi$ (1961–1968)	Ψ (2000)
China	730	1.5	90	0.3	100
India	524	2.5	100	1.0	140
USSR	238	1.3	1,100	5.8	6,330
USA	201	1.4	3,980	3.4	11,000
Pakistan	123	2.6	100	3.1	250
Indonesia	113	2.4	100	0.8	130
Japan	101	1.0	1,190	9.9	23,200
Brazil	88	3.0	250	1.6	440
Nigeria	63	2.4	70	−0.3	60
Germany	60	1.0	1,970	3.4	5,850

Notes: G = population in 1968 (10^6); ΔG = rate of population increase between 1961–1968 (%); Ψ = per-capita GDP (US$); $\Delta\Psi$ = per-capita annual rate of GDP increase (%).

such extrapolative estimates cannot be high, they are in shocking disagreement with reality, which testifies to the extreme complexity of political and socio-economic predictions. One of the key aspects of socio-economic dynamics has turned out to be quite predictable: the process of economic growth was followed by an increased contrast between rich and poor countries.

Answering the question "what is necessary to provide a persistent growth of the economy and population size up to the year 2000 and further?", Meadows *et al.* (1972) analyze separately the aspects of both physical (material) and social needs. Satisfaction of material needs requires, first of all, a solution of the problems of sufficient food, drinking water, and sustainable natural resources.

Although reliable data on feeding conditions are absent, it is supposed in the CR report that about 50–60% of the population of the developing countries (about one-third of the global population) are underfed. The area of arable soils constitutes about 3.2 bln ha, of which about 50% (which represents the best) are cultivated. According to Food and Agriculture Organization (FAO) data (the international organization facing the problems of food and health), the cultivation of the remaining soils is economically inexpedient. It follows from available estimates that even if all arable soils are used, with a preserved rate of population growth, an acute shortage of agricultural soils will occur by 2000 (below we will see that this prediction has proved incorrect). In this connection the CR report emphasizes that humankind has limited capabilities for preventing catastrophic consequences of exponential population growth in conditions of limited resources.

Meadows *et al.* (1972) obtained various prognostic estimates using the global model, the structural description of which causes no objections, but as follows from the subsequent publications (Krapivin and Kondratyev, 2002; Kondratyev *et al.*, 2002b), the model does not take into account numerous undisputable feedbacks (both direct and indirect) between society and nature, and what is of principal importance, their spatial heterogeneity. Nevertheless, the predictions made have brought forth many problems for specialists in the field of global modeling and has prompted them to develop efficient technologies for environmental control. Moreover, it is apparent that the problems discussed in Meadows *et al.* (1972) cannot be solved without using the systems of the global monitoring of the environment.

4.5.1.2 The current state of the nature-society system

It should be mentioned that during the last 30 years, due to the efforts of many scientists, the key priorities of global ecodynamics have been formulated and perspective trends in solutions of numerous new problems have been outlined. Clearly, to work out a global strategy of sustainable development, a constructive formalized approach to the description of the NSS is needed, which would take into account its multi-dimensional and multi-component nature, as well as non-linearity and interactivity of the processes taking place in it. Many feedbacks in the NSS have strengthened and prevail over other feedbacks. So, from available estimates, about 1.2 billion people live now on less than one dollar per day; about 3 million people die every year from AIDS; 100–150 million people suffer from asthma; 2.4 billion people need better sanitary living conditions; 150–300 million hectares of cultivated soils (these are 10–20% of all agricultural soils) have become degraded; and more than 2 billion people live in conditions with deficits of drinking water, food, and normal dwellings. Finally, there is a trend toward increasing population mortality due to intensified terrorism and various technogenic catastrophes. All this changes the concept of the global model and requires a search for new information technologies for the control of trends in the NSS (Table 4.28). Moreover, the problem of insurance against natural anomalies has become extremely important. As seen from Table 4.29, with unpredictable growth of frequency of occurrence of natural disasters, the problem of insurance becomes more complicated. Its solution requires a clear determination of the level of danger as an indicator of economic development of a given region and its socio-cultural level. For instance, such data as the ratio of losses and the amount of people exposed to danger in natural disasters during the last years show an absence of any regularity of their formation both in space and in time. So, on a global scale, in 1980 and 1990 the number of deaths due to all natural disasters constituted 86,328 and 75,252, with 147 and 211 million people exposed to danger, respectively. Apparently, here it is important to take into account the simultaneous relationship between many indicators of the readiness of a region to help itself to resist the catastrophe with characteristics of the phenomenon itself. The non-uniform development of regions manifests itself through the different socio-economic costs of natural disasters. For instance, in 1999 insurance payments for damages from natural disasters in Austria, Germany, and Switzerland constituted 42.5%, and in Venezuela – only 4%.

The CR model reflects only a narrow spectrum of feedbacks (levels) in the NSS (population, finance, pollutions, food production, mineral resources). Besides this, the model neither reflects a direct role of the biospheric feedbacks nor considers the spatial heterogeneity of these feedbacks. Therefore, it could not be objectively pre-

Table 4.28. The current state of the basic components of the global NSS.

Global NSS component	Estimate of the component as of the end of the 20th century
Grain production:	
total (10^6 t yr^{-1})	1,836
per-capita (kg/person/yr)	302
Meat production:	
total (10^6 t yr^{-1})	232
per-capita (kg/person/yr)	38.2
Area of irrigated lands:	
total (10^6 ha)	274
area per 1,000 people (ha)	45.7
Fossil fuel expenditure (10^6 t of oil equivalent):	
coal	2,186
oil	3,504
gas	2,164
Power generation by:	
nuclear power stations (GW yr^{-1})	348
wind systems (MW yr^{-1})	18,100
Global mean temperature (°C)	14.3
Carbon emissions due to fossil fuel burning (10^6 t C yr^{-1})	6,480
Partial CO_2 emission to the atmosphere (ppm)	370.9
Metal production (10^6 t yr^{-1})	902
Round timber production (10^6 m^3 yr^{-1})	3,336
Oil spillage due to anthropogenic activity (10^3 t yr^{-1})	48.6
Gross production:	
total (US\$$10^{12}$ yr^{-1})	44.9
per-capita (US$/person/yr)	7,392
Foreign debt of developing countries and countries of the:	
former Eastern block (US\$$10^{12}$)	2.53
Global population:	
total (10^9)	6.08
annual natality (10^6)	77

dictive even within the framework of successfully formulated scenarios, which, nevertheless, introduced into the model numerous uncertainties with a broad spectrum of possible issues. A comparison of Meadows' model with that of the global dynamics by Forrester reveals their conceptual identity both in their diagrams of feedbacks and ideology. Depending on variations of the initial suppositions (scenarios) on limited or unlimited irreversible resources, as well as on stabilization of the population size, the results of prognostic estimates of the state of the NSS components coincide qualitatively for both models, but differ substantially from the reality of the end of the 20th century. The basic difference consists in the

Table 4.29. Correlation between real economic losses and insurance payments in the USA, estimated by the number of most significant natural disasters.

Decade	Number of events	Economic damage (US$10^9)	Insurance payments (US$10^9)
1950–1959	12	30	3.7
1960–1969	15	41	7.4
1970–1979	17	45	7.8
1980–1989	13	50	18.0
1990–1999	35	219	82.0
1994–2003	39	161	66.0

estimation of population size whose rate of growth in the 1990s stabilized at about 80 million people per year, passing a maximum of 87 million people per year in the late 1980s and returning to 1970 levels in the beginning of the 21st century. This is explained by the fact that the time correlations considered in the CR model (and in Forrester's model) have suffered unpredictable changes in connection with stirring up the earlier weekly manifested feedbacks in the NSS. Understanding of the structure and importance of heterogeneous and complex feedbacks in the present world has changed rapidly, together with their rather difficult to predict dynamics. So, due to an extension of trade relations, transport, and information networks during the last 30 years, the spatial correlation between ecological, demographic, political, and economic events has grown.

One of the substantial differences between the RC model and the present world is connected with the concepts of the use of mineral resources and food production. It is clear now that such components of the global NSS's functioning as alternative energy sources, energy-saving systems and technologies, birthrate and mortality, production structure of people and their migration, response of nature to anthropogenic forcing, and many other key indicators need parameterization and reflection in the model scheme. For instance, in the sphere of food production the role of aquaculture during 1984–1999 grew distinctly. During this period, production of global aquaculture grew by almost 400%, from 6.9 mln t in 1984 to 33.3 mln t in 1999. This growth was heterogeneous both in space and in production components. For instance, the share of fish products increased from 19% in 1990 to 31% in 2001, 68% of which were produced by aquaculture firms in China. The global model component responsible for fossil fuels should be rather complex, too. There are about ten well-studied alternative mechanisms of economical use and substitution. This aspect is present in the CR model in the form of several primitive scenarios.

A general idea about the present state of the NSS is demonstrated in Table 4.28. It shows that the decrease or stabilization of the per-capita food production, as of the early 21st century, predicted by the CR model has proved to be inaccurate. This is natural because food production is characterized by more complicated cause-and-effect feedbacks than those considered in the models by the CR and Forrester. It has to be pointed out that although production of grain, meat, and other components of human food, suffered increases and decreases, it gives no reasons for pessimistic predictions for the decades to come. As has been mentioned above, production of fish products sharply increased during the last years. Clearly, the global model should consider the bioproductive processes in the World Ocean and inland water basins in order to have a chance of estimating the limits of their capabilities to produce food. As for grain, at the turn of the century its production decreased and its consumption increased. So, for instance, the per-capita grain production in 2001 constituted 299 kg, which was below the indicator reached in 1984 by 14%. At the same time, if we consider the long-range trend of grain production, between 1950 and 1984 it increased by 38%. The observed prevalence of grain consumption over its production and the respective decrease of its global-scale supplies are only a short-term fluctuation of the process of food production. In fact, other constituents of human food are characterized by positive gradients in their production, though distributed non-uniformly over countries and continents. For instance, meat production (beef, pork, and poultry) in 2001 grew at a rate of 2%. Beginning in 1950, the per-capita meat production doubled from 17.2 $kg\,yr^{-1}$ to 38.2 $kg\,yr^{-1}$ in the year 2000. In 2001, a maximum of pork and poultry production was observed.

Finally, as for one of the key indicators of the NSS's state, namely, non-renewable natural resources, here the CR's concept of their continuous reduction and their limiting role in the development of other NSS levels differs from real trends at the turn of the century. The real global-scale extraction of coal, oil, and natural gas grows at a rate of 1–2% yr^{-1} (in 2001 it was 1.3%). The volumes of fossil fuel consumption are also growing, though non-uniformly over countries and differing in types of fuel consumed. On the whole, the global increase of consumption of oil, coal, and gas constitutes 0.2%, 3.2%, and 1.2%, respectively. Besides this, the use of nuclear, solar, and wind energy is growing, and energy-saving technologies are rapidly developing. At the end of the 20th century there was a rapid growth of industrial production of photoelectric cells transforming sunlight into electricity. In this way, the world gets more than 1,140 MW, and there is an increasing trend toward solar energy. Therefore, the dependence of global dynamics on energy resources should be parameterized, not on the basis of simplified models, but taking into account the whole spectrum of the available information on the nature of multiple feedbacks in the NSS and especially the trends of scientific–technical progress.

4.5.1.3. Perspectives of the global model development

Over 30 years, after the appearance of the CR predictions and the respective global model, a serious step forward has been made in the field of global modeling. The new approach proposed in Krapivin and Kondratyev (2002) was based on the idea of the NSS as a self-organizing and self-structuring system, the correlation of whose elements in time and space is ensured by the process of natural evolution. The anthropogenic constituent in this process is aimed at breaking this integrity. Attempts to formally parameterize the process of co-evolution of nature and humans as elements of the biosphere are connected with a search for a single description of all the processes in the NSS, which would unite the efforts of various branches of knowledge about the environment. Such synergism serves as the basis of many studies on global modeling.

Let us cover the Earth's surface Ω with a geographical grid $\{\varphi_i, \lambda_j\}$ with digitization steps $\Delta\varphi_i$ and $\Delta\lambda_j$ by latitude and longitude, respectively, so that within a cell of land surface $\Omega_{ij} = \{(\varphi, \lambda)\colon \varphi_i \leq \varphi \leq \varphi_i + \Delta\varphi_i; \lambda_j \leq \lambda \leq \lambda_j + \Delta\lambda_j\}$ all the processes and elements are considered as homogeneous and are parameterized by point models. In the case of a water surface in the territory of a cell Ω_{ij} the water masses are stratified into layers Δz_k thick (i.e., 3-D volumes are selected $\Omega_{ijk} = \{(\varphi, \lambda, z)\colon (\varphi, \lambda) \in \Omega_{ij}, z_k \leq z \leq z_k + \Delta z_k\}$, inside of which all elements of the ecosystem are distributed uniformly). Finally, the atmosphere over the site Ω_{ij} by height h is digitized either as levels of atmospheric pressure or as layers Δh_s thick.

Interactions in the NSS are considered as interactions between natural and anthropogenic components within these spatial structures and between them. The complex model of the NSS realizes the spatial hierarchy of hydrodynamic, atmospheric, ecological, and socio-economic processes with the division of the whole volume of the environment into structures Ω_{ij} and Ω_{ijk}. The cells of this division are the supporting grid in numerical schemes, for solutions based on dynamic equations, or in synthesis of data series in the evolutionary type learning procedures.

The cells Ω_{ij} and Ω_{ijk} are heterogeneous in parameters and functional characteristics. Through this heterogeneity the global model is referenced to databases. Moreover, to avoid an excess of structure of the global model, it is supposed a priori that all elements taken into account in the model, and the NSS processes, have a characteristic spatial digitization. Ambiguity of spatial digitizations in various units of the global model is removed at an algorithmic level of agreement of data fluxes from the monitoring system. As a result, the model's structure is independent of the structure of the database and, hence, does not change with a changing of the latter. A similar independence between the model's units is also provided. This is realized by data exchange between them only through inputs and outputs under the control of the basic dataway. In the case of turning off one or several units, their inputs are identified with the corresponding inputs to the database. Then the model operating in the regime of a simulation experiment can be schematically represented by the process, whereby the user's choice of a spatial image is formed of the modeled medium and of the regime of the simulation experiment control. Of course, in this case the user should have a certain knowledge base and understand the method of its structuring. For instance, a list can be used of the key problems of the global ecology or lists of the NSS elements recommended for study.

The character of the spatial structure of the global model is determined by the database. The simplest version of the point model is realized with the initial information in the form of averaging over the land surface and the whole World Ocean basin. The spatial heterogeneity is considered through various forms of space digitization. The base form of the spatial division of land and oceans is a heterogeneous grid $\Delta\varphi \times \Delta\lambda$. A realization of a real version of the use of the model is provided by an integration of the cells Ω_{ij}, so that various forms of the spatial structure of the considered elements and biospheric processes can be present in each unit. Such a flexible setting of the spatial structure of the biosphere makes it possible to easily adapt the model to heterogeneities in databases and to perform simulation experiments with an actualization of individual regions.

Depending on the special features of the considered natural process, the structure of the regional divisions can be identified with climatic zones, continents, latitudinal belts, socio-administrative structure, and natural zones. For climatic processes, many scientists are oriented toward regions with dimensions $\Delta\varphi = 4^\circ$ and $\Delta\lambda = 5^\circ$, the biogeocoenotic processes are studied at $\Delta\varphi = \Delta\lambda = 0.5^\circ$, the socio-economic structure is represented by nine regions, the atmospheric processes in the biogeochemical cycles of long-lived elements are approximated by point models, and the functioning of the ocean ecosystems are described by heterogeneous digitization of the shelf zone into cells Ω_{ij} with the selection of four parts of the World Ocean pelagians.

The structure of the division of the Earth's surface into regions Ω_{ij} covers all the enumerated versions. It means that the general scheme of digitization of the processes in the NSS foresees a hierarchy of levels including the global, continental, regional, landscape, local levels, etc. The scheme of independent inclusion of units at all these levels, with their combination through parametric interfaces, does not prevent an increase in the amounts of the model's units due to the introduction of new components, which specify the models of the processes under consideration. The model of the upper level can serve as an information base for the model of the lower level and vice versa. The results of modeling at the lower level can be used to form the information base for models of higher levels. This mechanism of the information exchange between the models of various levels reduces the requirements of the global database and broadens the capabilities of the NSS model.

The structure of the global model includes some auxiliary units, which provide an interaction between the user and the model. In particular, these are units which realize the algorithms of the spatial–temporal interpolation or coordinate the user's actions with a bank of scenarios. Note that some scenarios can be transformed by the user into a rank of model units. Such a duality (excessiveness) is characteristic of scenarios of climate, demography, anthropogenic activity, scientific–technical progress, and agriculture. The user's interface makes it possible to select the structure (Ω_{ij}) in the default mode or have the required spatial structure formed from the base elements by averaging and interpolation.

Thus, the synthesis of the global model version requires a preliminary analysis of the present situation with the global databases and knowledge bases. Here the specialists face principal difficulties and, first of all, the absence of adequate knowledge about climatic and biospheric processes as well as an uncoordinated database on the global processes on land, in the atmosphere, and in the oceans.

Another principal difficulty is connected with the inability of present-day science to formulate the requirements for global databases required to reliably assess the state of the environment and to give a reliable forecast of its development for a sufficiently long time period. Moreover, there is no technology to form databases aimed at creating the global model.

Many scientists have made attempts to answer these questions. One of the efficient ways to solve these problems is a single planetary adaptive geoinformation monitoring system (GIMS), which has a hierarchic structure of data collection and forms a multi-level global database. The adaptive character of this system is provided by correcting the regime of data collection and by changing the parameters and structure of the global model.

The global GIMS can be created taking into account the existing structure of databases whose formation continues within the framework of the IGBP and numerous national ecological and nature-protection programs. The developed system of world data centers favor a rapid use of accumulated information about the global processes, and simplifies the GIMS synthesis. However, a significant progress in this direction, connected with large economic expenses, cannot lead to a successful solution of the problem of global environmental control. Though this phase cannot be avoided.

According to Kondratyev (1998), to control the global geobiosystem of the Earth, regular observations of specified key variables are needed. With an increasing probability of drastic global change, the spectrum of these variables will vary, and the global predicting system should be constantly modernized. The methodological substantiation of an adequate information content of variables for the monitoring system can be objective only in the case of GIMS functioning. Many of the enumerated variables can be calculated using the respective models, and there is no need to measure them. However, so far, measurements are planned in parallel with model development, and there are no results in the field of planning global experiments that would raise hopes. As follows from some studies (Kondratyev and Galindo, 2001; Kondratyev *et al.*, 2002a), the bases of global knowledge and data make it possible to synthesize and develop GIMS series (Kondratyev *et al.*, 2000).

An inclusion of the global model into the GIMS structure enables one to consider it as an expert system. It means that there is a possibility of the complex analysis of numerous elements of the NSS in conditions of realization of hypothetic situations, which can appear for natural or anthropogenic reasons. The scheme in Figure 4.2 reflects the basic elements taken into account in the NSS global model (GMNSS). A concrete realization of each unit of the GMNSS is determined by the level of knowledge of the processes reflected in that unit. The units responsible for modeling the biogeochemical and biogeocoenotic processes are described with balance equations.

Let $\psi_s(t)$ be the information content of an element ψ in a medium S at a moment t. Then, following the law of preservation of matter and energy, we write the following balance equation:

$$\frac{d\psi_s}{dt} = \sum_j H_{js} - \sum_i H_{si},$$

where fluxes H_{js} and H_{si} are, respectively, incoming and outgoing fluxes with respect to the medium S. Summing up is made by external media i and j interacting with S. In fact, the medium S implies elements of digitization of the environment by latitude φ, longitude λ, depth z, and height h. A variety of functional parameterizations of fluxes H_{pq} is determined by the level of knowledge of physical, chemical, and biological features of the element ψ.

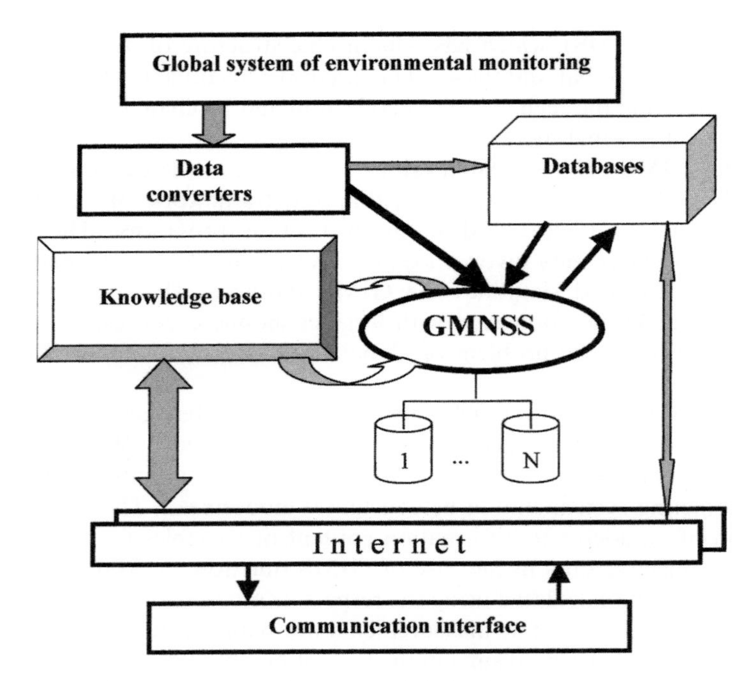

Figure 4.2. The principal scheme of the use of GIMS-technology to synthesize the global system of environmental control using standard means of telecommunications and GMNSS.

A parameterization of the processes of photosynthesis, dying-off, and respiration of plants in land ecosystems is based on a knowledge of phytocoenology, which includes information about external and internal system connections within the vegetation community. These are temperature dependences of photosynthesis and evapotranspiration of plants, gas exchange processes between plants and the atmosphere, impacts of the solar radiation energy on the processes of growth and exchange, relationships between plants and processes in the soil, and the interaction of vegetation covers with the hydrological cycle.

The GMNSS units responsible for the parameterization of climatic and anthropogenic processes are of complex character (i.e., partially described by equations of motion and balance, and partially an evolutionary model is constructed for them, based only on the observational data).

4.5.1.4 Preliminary conclusions

Having considered approaches to an assessment of the NSS dynamics, from Forrester and Meadows to recent publications (Krapivin *et al.*, 1982; Krapivin and Kondratyev, 2002; Kondratyev *et al.*, 2002a,b, 2003a,b,c,d,e, 2004a,b,c,d), one can reach the conclusion that an appreciable progress in a search of ways to reach global sustainable development can be made with a systems approach to the multi-functional monitoring of the NSS. Thanks to the CR authors, more than 30 years ago the contradiction was first emphasized between the growth of population size and limited natural resources, and for the first time after the works by Vernadsky (1944a,b) an attempt has been made to use numerical modeling to study the NSS evolution. Of course, the CR model oversimplifies real internal connections in the NSS, describing interactions of its elements by averaged indirect relationships, without direct account of economic, ecological, social, and political laws. The possibility of such an account appeared later in connection with the studies in the field of simulation and evolutionary modeling and theory of optimization of interaction for complex systems resulting in the creation of the methods and algorithms of prognostic estimation of dynamic processes in conditions of a priori uncertainty. However, the problem of creation of a global model, adequately representative of the real world, still cannot be solved even in the present conditions. On the one hand, a complete consideration of all the NSS parameters leads to insurmountable multivariance and information uncertainty with unremovable problems. Besides this, in the spheres such as the physics of the ocean, geophysics, ecology, medicine, sociology, etc., an adequate parameterization of real processes will be always problematic because it is impossible to have a complete database. Nevertheless, a search for new efficient ways to synthesize the global system of NSS control, based on adaptive principles of the use of the global model and renewed databases, seems to be perspective and raising hopes for reliable forecasts of the NSS dynamics. Preliminary calculations with the use of the GMNSS have shown that the role of biotic regulation in the NSS has been underestimated, and the forecasts, for instance, of the levels of the greenhouse effect have been overestimated. Therefore, in this book an attempt has been made to synthesize the GMNSS taking into account the earlier experience and accumulated databases, as well as knowledge about the environment and human society.

Further improvement of the GMNSS will be connected with the balanced development of studies both in the sphere of parameterizations in the NSS and in modernization of the Earth observation systems, covering the whole thematic space of the NSS:

- Solar–land interactions (physical mechanisms of the transport of mass, momentum, and energy in the geosphere).
- Atmospheric dynamics (atmospheric chemistry, atmospheric physics, meteorology, hydrology, etc.).
- Dynamics of the World Ocean and coastal zones (winds, circulation, sea surface roughness, color, photosynthesis, trophic pyramids, pollution, fisheries).
- Lithosphere (geodynamics, fossil fuel and other natural resources, topography, soil moisture, glaciers).
- Biosphere (biomass, soil–plant formations, snow cover, agriculture, interactions at interfaces, river run-off, sediments, erosion, biodiversity, biocomplexity).
- Climatic system (climate parameters, climate-forming processes, radiation balance, global energy balance, greenhouse effect, long-range climate forcings, delay of climate effects).

- Socio-political system (demography, geopolicy, culture, education, population migration, military doctrines, religion, etc.).

4.6 STRATEGIC ASPECTS OF PREVENTION OF NATURAL DISASTERS

Catastrophes in general and natural disasters in particular, threaten human life. Therefore, their prediction, in order to attempt to prevent them, is a necessary element of the current science of processes taking place in the NSS. Assessment of the risk of a natural disaster origin, and resulting damage, requires the use of interdisciplinary analysis of a large volume of information about different aspects of the NSS functioning. First of all, the ecological safety considered in the loosely interpreted section of the NSS is closely connected with the socio-economic level of development of a society on a given territory. According to Vladimirov *et al.* (2000), for instance, for conditions in Russia, numerous indicators of societies' development are in supercritical states. For instance, Russia allocates funds for ecological safety that are 50 times less than those allocated in Germany. Moreover, most of the indicators of the NSS development in Russia indicate a negative trend in the direction of slackening the readiness of Russian society to prevent natural disasters and to overcome their consequences. Whereas in many developed countries most of the natural cataclysms refer to a category of "natures whims", in many regions of Russia these cause considerable damage. For instance, the 13 October 2004 40-cm snowfall in Cheliabinsk paralyzed the city for a whole day.

Undiscovered mysteries about many natural phenomena require their formalized description in order that it is possible to predict them. Since all processes taking place in the NSS are somehow connected, the desire to construct the GMNSS with a broad set of functions is one of the possible ways to solve the global problem of the prediction of emergency situations in the environment. The mean statistical dependences created to describe correlations between the frequency of occurrence of natural disasters and their consequences are in many cases an efficient means to make strategic decisions on the prevention of natural disasters. In this sense, the law of Richter–Gutenberg is well known and realized in the USA in the following manner:

$$\log N = \begin{cases} 1.65 - 1.35F & \text{for floods} \\ 1.03 - 1.39F & \text{for tornados} \\ 0.45 - 0.58F & \text{for hurricanes} \\ -0.55 - 0.41F & \text{for earthquakes} \end{cases}$$

where F is the logarithm of the average number of annual human victims for the last 100 years; N is the number of events.

Knowledge of the laws of natural disaster origins, the ability to predict catastrophic events, and the availability of mechanisms for early warning do not ensure the complete protection of humans and their infrastructures from losses and destruction. It is necessary that these constituents be claimed by people who understand the risk and realize danger. Vladimirov *et al.* (2000) introduce the notion of "safety culture" which reflects the behavioral code, moral code, and emotional response to cataclysms.

Risk, as a measure of danger evaluated with consideration of a multitude of factors, can serve as a reference point for solving the problems of controling the totality of potential factors capable of violating the human habitat and changing the conditions of the functioning of the society. Risk is a more capacious notion than the probability of natural disasters. It includes the probability of an undesired event and the volume of losses from its realization. In other words, risk reflects a measure of danger of a natural phenomenon, which includes the estimates of the levels of undesiredness in different aspects. At the same time, the risk has also a subjective component, which is assessed using the formal method of decision making, taking into account intuitive estimates of the situation and psychological norms of perception of the environment. Numerous problems are solved at the level of adoption of legislative acts which reflect the general principles of danger assessment.

4.7 THE LEVEL OF ECONOMIC DEVELOPMENT AND THE SOCIAL INFRASTRUCTURE OF THE REGION AS INDICATORS OF THE EFFICIENCY OF NATURAL DISASTER RISK CONTROL

An optimization of the risk of insurance against natural disasters becomes more and more urgent with every year, since economic losses increase and become poorly predicted. For instance, the insurance payments alone for destruction caused by Hurricane Hugo in 1989 constituted US$5 bln, exceeding by more than 50% the greatest losses from any preceding natural disasters. However, three years later the insurance losses from Hurricane Andrew exceeded the losses from Hurricane Hugo by almost 4 times.

The problem of risk control as a quantitative measure of the NSS properties, to ensure the safety of a habitat over a given territory, is a complicated mathematical problem of optimization, which is successfully solved using the methods of logico–probabilistic (LP) theory (Solozhentsev, 2004) based on construction, solution, and study of the LP-function of risk and safety of the systems with complex structures. The mathematical basis for these studies is Boolean algebra and its use together with methods of mathematical modeling of processes taking place in the NSS. The informative characteristics of such an approach to estimation of the risk of natural disaster origins are determined by the accuracy of the forecast of the development of natural–anthropogenic processes on a given territory and the reliability of proposed scenarios which substitute for individual elements of this forecast.

According to Solozhentsev (2004), to assess the efficiency of risk control one should use discrete multitudes of parameters $\{Z_j\}$ affecting the efficiency, and the scale Y of the efficiency indicator. In this case, an introduction of the goal function $F = N_{1c} + \cdots + N_{kc}$, where N_{jc} is the number of correctly identified events of the scale Y in class j, and maximization F enable one to determine the weights of factors affecting the efficiency. The complexity of this problem is explained by the absence of reliable data on statistical characteristics of these factors. The obvious difficulties can be overcome using the methods of estimation of the level of self-organization and self-regulation of natural systems (Ivanov-Rostovtsev *et al.*, 2001) and synergetics (Chernavsky, 2004). In particular, the use of Double Self-Organization (D-SELF) theory gives the possibility to introduce a generalized characteristic of the NSS state in the form of a non-linear function $\xi_0 = \sqrt{\xi_i \xi_i^*}$, where ξ_0, ξ_i, and ξ_i^* are the NSS parameters meeting three axioms:

- discreteness of elements within and outside the NSS in their interaction;
- hierarchy of the elements within and outside the NSS; and
- contingency between discrete elements of the hierarchy of the structure within the system (ξ_i) and outside (ξ_i^*) in the form of dependence $\xi_0 = \sqrt{\xi_i \xi_i^*}$.

Synthesis of the model D-Self as applied to the NSS, is an independent problem and it is not considered here. Note only that according to Ivanov-Rostovtsev *et al.* (2001), the method of modeling the evolutionary dynamics of natural systems based on D-Self technology has features in common with the technology of evolutionary modeling (Bukatova *et al.*, 1991; Bukatova and Makrusev, 2004). In both cases the evolution of a natural system is considered as a discrete process of a change of its states in some parametric space. In the case considered, the economic indicators and the social infrastructure of the territory are selected into a separate level of the parametric hierarchy and considered as indicators of the efficiency of risk control. Other characteristics of the NSS functioning are considered external parameters.

The frequency and intensity of natural disasters accelerate, which is noticeable from the data of Table 1.4. Respectively, economic damage from natural disasters grows (Tables 4.29, 4.30; Figure 4.3). Therefore, detailed analyses of the relationships between natural disaster parameters and such characteristics of the regional NSS as infrastructure, level of economic development, state of the social system, population education, and religious aspect are an important stage of development of a regional strategy to overcome the consequences of natural disasters. Losses from natural disasters during the period 1987–1997 reached US$700 bln, and this means that for poor countries overcoming the consequences of natural disasters in the years to come will become an unsolvable problem. For instance, in 1998, total economic losses from natural disasters reached US$65.6 bln, with 66% of the sum falling on developing countries. It is clear that economic mechanisms for support of the regions that suffer from natural disasters should be optimized considering the real state of the NSS. From World Bank estimates, poverty is characterized by the following

Table 4.30. Insurance losses from heavy natural disasters (US$$10^9$).

Year	Economic damage from natural disasters
1992	24.4
1993	10.0
1994	15.1
1995	14.0
1996	9.3
1997	4.5
1998	15.0
1999	22.0
2000	7.5
2001	11.6
2002	10.0
2003	12.75
2004	43.815

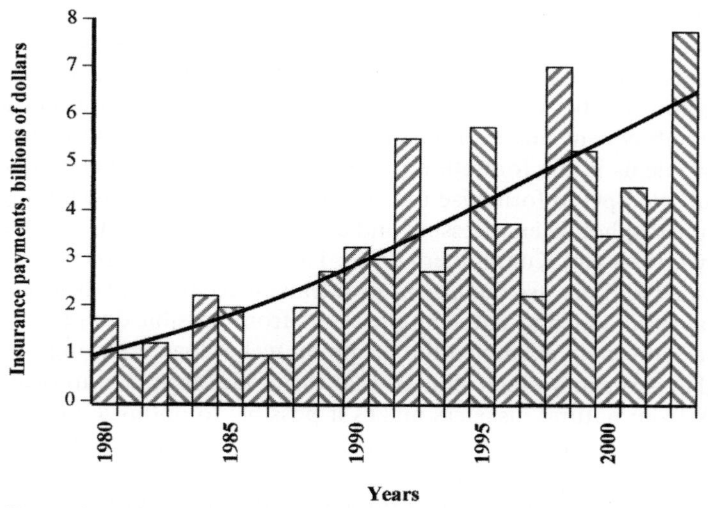

Figure 4.3. Dynamics of the insurance payments in the USA for thunderstorm damage (Phelan, 2004).

indicators:

- one-third of the population of developing countries are the poor, with 18% of them beggars;
- about 50% of the poor and 50% of the beggars live in southern Asia;
- poverty prevails in rural areas; and
- the means of subsistence for the poor is mainly obtained from agriculture.

Infrastructure development plays an important role in poverty reduction. But it is the infrastructure itself that is subject to heavy destruction by natural disasters. So, in Asia, almost 70% of natural disasters are floods, each flood causing damage estimates of about US$15 bln, the main contribution to which is made by losses in agriculture. Therefore, a problem arises of risk reduction by transformation of the infrastructure. In developing countries, the solution of this problem is connected with an optimization of the use of soil resources and a development of strategies of ecological character. It is necessary to take into account the indicator of the growth of population density in developing countries. The quantitative and qualitative dynamism in urban development leads to a super-proportional increase of vulnerability of such territories. This tendency in developing countries is provoked by economic interests or socio-political circumstances of a given region – within a small area a large amount of people are concentrated – which drastically raises the risk of loss of life under the conditions of a natural disaster. On the other hand, with this situation it is easier to build protective structures. Therefore, in time, the problem of optimization of a regional infrastructure becomes more urgent (Kasyanova, 2003; Gurjar and Lilieveld, 2005). To some extent, the plan of accomplishing decisions at the World Summit on Sustainable Development (Johannesburg, South Africa) adopted at the 17th Plenary Meeting on 4 September 2002 has promoted the solution to this problem. This plan foresees the mobilization of technical and financial support to developing countries to reach a balanced economic, social, and ecological development for all global regions.

The relationship between economic losses from natural disasters and a real guarantee of subsequent financial investment, to liquidate the consequences, is one of the key problems in the formation of strategies for insurance companies. One of the successful attempts to formalize the appearing processes is a technology of using models of catastrophes when assessing and controling the risks of emergency events, developed by Grossi and Kunreuther (2005). The authors lay emphasis on the risk of natural disasters and discuss the urgent problems of controling risk from terrorism. The study has been aimed at reducing the losses from possible dangers in the future. Pennsylvania University examined the proposed technology, carrying out a number of numerical experiments which showed that the model approach to planning of the financial risk from natural disasters makes it possible to optimize insurance against emergency events. The necessity of this is seen from the data of Table 4.31 which demonstrates a great heterogeneity in the regional distribution of the ratio between insurance payments and total losses from natural disasters. With the distribution of losses by types of events (Table 4.32) taken into account, it becomes clear that the model optimization can substantially increase the effect of insuring against these events.

Table 4.31. Characteristics of natural disasters recorded in 2003.

Region	Number of events	Number of victims	Economic losses		
			Total	Insurance	Ratio of insurance payments to total losses
Africa	57	2,778	5,158	0	0
America	206	946	21,969	13,274	0.6
Asia	245	53,921	18,230	600	0.03
Australia/Oceania	65	47	628	246	0.39
Europe	126	20,194	18,619	1,690	0.09
Global	699	77,886	64,604	15,810	0.24

Table 4.32. Distribution of natural disasters and resulting losses in 2003.

Type of event	Share of a given type of event (%)	Victims (%)	Total economic damage (%)	Insurance losses (%)
Earthquakes, volcanic eruptions	12	61	10	1
Hurricanes, tornados, storms	43	2	39	76
Floods	28	5	21	8
Other natural disasters	17	32	30	15

5

The monitoring of natural disasters

5.1 THE PRESENT TECHNICAL MEANS OF ENVIRONMENTAL CONTROL

At the turn of the millennium the problem of the human society–environment relationship has become urgent, attracting attention of not only ecologists but even politicians of most of the developed countries. Suffice it to enumerate the international conferences on the environment and sustainable development, and it becomes clear that prospects of global ecodynamics cause a concern to the world community. According to Grigoryev and Kondratyev (2001), during the last decades the risk of origins of large-scale ecological catastrophes both caused by man and resulting from nature's protective response has increased.

Natural and anthropogenic ecological disasters have a historical aspect. Volcanic eruptions, earthquakes, floods, climate changes, droughts, locust plagues, and forest fires have taken place over the entire history of civilization, but until the mid-20th century they had been mainly of spontaneous character. With the development of present-day civilization they have been supplemented by destructive forcings on nature, such as deforestation, landscape change, contamination of inland water basins and the World Ocean, change of air composition, decrease of biodiversity, etc. As a result, new types of disasters have occurred, which include desertification, degradation of soil resources, dust storms, reduction of bio-productivity, etc.

The beginning of the 21st century put forward an urgent problem of the estimation of the risk of ecological catastrophes, prediction, and preventative measures. In other words, the problem of the control of ecological disasters has become of primary importance. This has become feasible with the availability of the required information about the past, present and future state of environmental objects, including natural, natural–technogenic, and anthropogenic systems. Solution of this problem depends on many factors, among which are the development of global Earth-observing systems and an accumulation of ecological databases, as well as the study of the dynamics of various processes in the nature–society system (NSS) with a systematization of the knowledge that is gained. Of course, this problem refers to the most complicated and versatile problems of today. It should become an object requiring the attention of all the sciences, as well as the humanities.

Methods of local diagnostics of the environment can provide a complex assessment of the state of a natural object or process, especially when this environmental element covers a vast area. Any technical means of environmental data accumulation makes it possible to obtain only fragmentary information in both space and time. In particular, the remote sensing systems widely used in flying laboratories and Earth resources satellites provide data series which are referenced geographically to flight routes. Retrieval of information in the space between measurement routes is only possible using methods of spatial–temporal interpolation, the development of which results from many scientists applying methods and algorithms of simulation modeling.

From the viewpoint of the complex problem of environmental diagnostics, it is important to synthesize a system which would combine such functions as data collection, using remote and *in situ* methods, and their analysis and accumulation with subsequent thematic processing. This system can provide systematic observations and assessment of the state of the environment, carry out prognostic diagnostics of changes to environmental components due to economic activity and, if necessary, analyze the development of anthropogenic processes taking place in the environment giving subsequent warning about undesirable changes in the characteristics of natural subsystems. Such functions of environmental monitoring can be realized with the use of methods of simulation modeling which ensure the synthesis of the model of the natural system under study.

Development of the models of biogeochemical, biocenotic, hydrophysical, climatic, and socio-economic processes in the environment, which would ensure a synthesis of the images of its subsystems, always requires the formation of systems of automated processing of the monitored data and creation of respective databases. As numerous relevant studies have shown, there are balanced criteria for information selection which take into account the hierarchy of cause-and-effect connections in the biosphere, including an agreement of tolerances and depth of spatial digitization when describing the atmosphere, land ecosystems, and hydrosphere, as well as the degree of their elemental detailing.

Application of mathematical modeling for the systems of satellite monitoring, as shown by numerous studies, can give practical effect only with the developing of a single network of data, conjugated with the NSS model. The conceptual scheme of the adaptive regime of monitoring organized in this way dictates an acceptance of the architecture of the monitoring system, which would combine knowledge of different sciences into one system and give the possibility of flexible control of this knowledge. This is possible by combining geographic information system (GIS)-technology, methodologies of expert systems, and simulation modeling.

The GIS provides the processing of geographic data, connection with databases, and a symbolic presentation of the topology of territories under study. The broadening of GIS to geoinformation monitoring system (GIMS) by the GIMS = GIS + Model scheme changes some functions of the user's interface of computer cartographic systems, including the prognostic estimates based on the functioning of the environmental subsystems. Available measurements of the subsystem parameters can be used both to evaluate the model's coefficients and to directly obtain prognostic estimates using the evolutionary technology method.

Development and application of the ideas of GIMS-technology which foresees a combination of methods and algorithms of mathematical modeling with *in situ* and remote measurements of environmental characteristics, as experience shows, is possible on the basis of synthesis of *in situ* observations, and flying and mobile laboratories. In the future, such complexes will solve the following important problems (Von Oosteron *et al.*, 2005):

- Prediction of the time of onset and of the degree of danger of natural disasters, accidents, and technogenic catastrophes.
- Control of the dynamics of accidents and catastrophes, including heavy meteorological conditions, and the issuing of information to decision makers.
- Evaluation of the consequences of accidents and catastrophes for cities, agricultural and forest/marsh areas, and marine and coastal flora and fauna.
- Generation of instructions to rescue services at search and rescue operations.

GIMS-technology will make it possible to solve problems of monitoring territories of large industrial centers. Problems such as:

- The study of seasonal parameters of the elements of urban and suburban landscapes, geophysical fields and local anomalies of different origins, a revealing of the laws of the interaction of their phenomenological and topological characteristics, and a presentation of the results of studies in the form of thematic maps of standard scale.
- The development of a methodology to assess the ecological and sanitary state of residential areas, industrial areas, forest/park and suburban zones, water basins and rivers, heating mains and product pipelines, transport and power line networks.
- The study of the seasonal and diurnal dynamics of the characteristics of garbage tips and industrial waste sites, and sources of contamination of the Earth's land cover, air, and water basins.
- The solution of inverse problems and the development of statistical criteria of similarity, as applied to local anthropogenic and geophysical features of urban and suburban territories, the near-surface atmosphere, cloudiness and the ozone layer, and the dynamics of pollution within their elements.

The monitoring systems based on GIMS-technology should be realized based on the data of the following five supplementing stages with a provision for technical, methodical, and metrological requirements:

(1) Surveying from space vehicles:
 –estimation of the global ecological characteristics of the region under study (with an interest in the underlying levels of ecological monitoring taken into account);
 –optical transparency, and the state of minor gas and aerosol components of the atmosphere;
 –integral characteristics of technogenic pollution of air and water basins, and the state of the ozone layer; and
 –the state of the Earth's cover and urban environmental objects.
(2) Studies from multi-purpose and specialized flying laboratories:
 –ecological characteristics of the regional environment (with an interest in the third and fourth levels of ecological monitoring taken into account);
 –energetic and polarization characteristics of microwave fields of radiation and reflection of elements of the urban and other landscapes;
 –thermal and radar images of the territory of a city, megalopolis, forest fire, and other territories;
 –images and parameters of the thermal field of the landscape in the near- and far-IR regions; and
 –optical transparency, moisture content, and other characteristics of the atmosphere when sounded in nadir and zenith.
(3) Helicopter and light-motor flying laboratories. They provide data for low-altitude (80–500 m) ecological monitoring to fix the following parameters:
 –local ecological characteristics of the region (with the interest of mobile and stationary surface or flowing observation platforms taken into account);
 –pollution of the near-surface atmosphere with gas and solid aerosols;
 –background and anomalies of environmental radioactivity, thermal fields (IR and microwave regions) of individual sites of a region; and
 –antigen pollution of the air in the zone of human respiration.
(4) Motor-car, ship, and stationary points of observation. These points determine and specify, within some routes and sectors of the region, the location and structure of local ecological anomalies.
(5) The fifth level of ecological monitoring is provided with the use of GIMS-technology:
 –collection, annotation, and storage of data from all the four levels of ecological monitoring in a single spatial–temporal system of coordinates;
 –express analysis, processing and referencing to characteristic points of location and mapping of measured and calculated estimates of the state of the environmental components with a selection of their special features;
 –identification of the sources of pollution and other violations of the environment, assessment of their dynamics and prediction of the consequences of their impact on ecological and sanitary situations, revealing their means of migration and concentrations; and
 –preparation of information for the user in standard forms, ensuring a

Table 5.1. Instrumental equipment of the space observatory Aqua (Parkinson, 2003).

Measuring system	Characteristics of the measuring system
The Atmospheric Infrared Sounder (AIRS)	It has 2,382 high-resolution channels, 2,378 channels measuring IR radiation in the range 3.7 to 15.4 μm, other channels cover visible and near-IR regions (0.4–0.94 μm).
The Advanced Microwave Sensing Unit (AMSU)	It is a 15-channel gauge to measure the upper atmospheric temperature; radiation in the range 50 to 60 GHz and at frequencies 23.8, 34.4, and 89 GHz; water vapor, and precipitation. Spatial resolution of 40–45 km.
The Humidity Sounder for Brazil (HSB)	HSB is a 4-channel microwave device measuring humidity (183.31 GHz) and radiation (150 GHz). Horizontal resolution of 13.5 km.
The Clouds and the Earth's Radiant Energy System (CERES)	CERES is a broadband 3-channel scanning radiometer. One channel measures reflected solar radiation in the range 0.3–5.0 μm. Two other channels (0.3–100 μm and 8–12 μm) measure reflected and emitted radiant energy at the top of the atmosphere.
A Moderate-Resolution Imaging Spectroradiometer (MODIS)	MODIS is a 36-channel scanning radiometer in the visible and IR ranges (0.4–14.5 μm), aimed at obtaining biological and physical information about the atmosphere–land system.
An Advanced Microwave Scanning Radiometer for EOS (AMSR-E)	AMSR-E is a 12-channel scanning passive radiometer to record land cover radiation at frequencies 6.9, 10.7, 18.7, 23.8, 36.5, and 89.0 GHz with regard for horizontal and vertical polarization of the signal. The antenna's diameter is 1.6 m, scanning period is 1.5 s.

reconstruction of an objective pattern of ecological and sanitary situations, as well as substantiating rational solutions to current and perspective nature–protection problems.

One of the current systems of ecological monitoring is the space observatory Aqua developed at NASA, within the EOS (Earth Observing System) program, launched on 4 May 2002 and orbiting around the Earth every 98.8 minutes (Parkinson, 2003). The Aqua equipment is demonstrated in Table 5.1. In the main, the system measures the water components in the surface atmospheric layer and on the Earth's surface, dividing them into liquid, vaporous, and solid phases of water, as well as giving information on the climate system. Six basic measuring subsystems enable one to obtain data, the use of which, for instance, in GMNSS, can provide an evaluation and forecast of the state of the global NSS.

Now we briefly mention the characteristics and capabilities of the space observatory Aqua. The subsystem AIRS/AMSU/HSB produces air temperature and humidity profiles up to 40 km. The horizontal resolution varies from 1.7 to 13.5 km depending on the channel used. Besides this, this subsystem ensures the diagnostics of solar radiation fluxes in a cloud layer and measures the content of liquid, vaporous, and solid phases of water in clouds, and evaluates the rain rate. Also, the subsystem gives information on atmospheric minor gas components, land surface and water temperatures, the heights of the tropopause, the stratospheric structure and composition, as well as some characteristics of cloud layers and long-wave radiation (LWR) and short-wave radiation (SWR) fluxes. Of course, the subsystem's data should be processed using respective algorithms. Creation of such algorithms can broaden considerably the significance of information obtained from the subsystem.

The CERES subsystem has been tested on satellites TRMM (the Tropical Rainfall Measuring Mission) and Terra launched by NASA in 1997 and 1999, respectively. Scanning is carried out in fixed and rotating azimuth planes. This makes it possible to easily evaluate, for instance, the greenhouse effect of water vapor in the upper troposphere as well as to calculate many important characteristics of clouds and aerosols.

Many problems of determination of the causes of global climate change cannot be solved for lack of data on energy exchange processes in the systems "atmosphere–ocean" and "atmosphere–land". The functions of MODIS provide information on various characteristics of these systems. In particular, for the oceans, measurements are made of primary production, photosynthetically active radiation, concentration of organic matter, surface layer temperature, the state of sea ice and its albedo, and chlorophyll concentration. For land, pure primary production, type of Earth cover, change of vegetation index, temperature of the surface and its radiance, as well as state of snow cover and its albedo are determined. In the atmosphere, measurements are made of optical thickness and microphysical properties of the cloud layer, optical characteristics and size distribution of aerosols, the ozone content, total supply of precipitable water, as well as profiles of temperature and water vapor. Its spatial resolution constitutes between 250 m to 1 km depending on the problem to be solved.

The MODIS subsystem has been first installed on the satellite Terra. The satellite Aqua carries a modified version of MODIS–Terra. In particular, it includes two lines 10.78–11.28 μm and 11.77–12.27 μm, which broaden the possibilities of measurements of environmental temperatures.

The AMSR-E subsystem was developed by the Japan National Aeronautics Agency (NASDA). It is a successor to the SMMR (Scanning Multichannel Microwave Radiometer), SSM/I (Special Sensor Microwave/Imager), MSR (Microwave Scanning Radiometer), and TMI (TRMM Microwave Imager). Such a powerful information subsystem on Aqua will make it possible to evaluate the rain rate, to calculate the total water content in air column over a given territory, to measure the sea surface temperature (SST) and wind speed over the sea, to map the sea ice concentration (with an indication of its temperature) and soil moisture, and to calculate snow cover depths whilst estimating the water content in it. The spatial resolution here varies between 5 and 56 km.

The information flux from the satellite Aqua goes to, and is recorded at, two ground stations: Alaska Ground Station and Svalbard Ground Station. Then the data are transmitted to the Goddard Center for Space Data Processing. In parallel, practically in the real time, the AMSR-E data are transmitted to NASDA and Japan's Information Center for Fisheries, where they are used for numerical weather forecasts and evaluation of fish supplies in the industrial zones of the World Ocean.

The space-borne, Earth-observing systems developed during the last decades cover practically all significant spheres of the NSS functioning. It should be noted here that, unfortunately, these means have been created without interaction with global modeling results. Nevertheless, at present, the space-borne systems provide data on the following parameters (Klyuev, 2000; Krapivin *et al.*, 2004; Savinykh and Tsvetkov, 2001; Kramer, 1995; Summer *et al.*, 2003):

- interactions in the Sun–Earth system (physical mechanisms to control the motion of matter between biospheric and geospheric formations);
- atmospheric dynamics (physical and chemical processes, meteorological parameters, hydrological fluxes);
- dynamics of the oceans and coastal regions (circulation, wind-driven roughness, water temperature and color, surface layer productivity, photosynthesis, transformation of inorganic matter to organic substances, phytoplankton, chlorophyll, marine nutritional chains, fish congestions, structure of ocean populations);
- processes in the lithosphere (parameters of the Earth's rotation, tectonic activity, dynamics of continents and glaciers, soils moisture, topographic structures, resources);
- functioning of the biosphere (characteristics of atmosphere–biosphere interaction, global structure of soil–plant formations, biomass distribution, agricultural structures, forest areas, snow-covered areas, structures of rivers and other water systems, precipitation, food supplies, pollutions); and
- the climate system's dynamics (Earth's radiation budget, atmospheric minor gas components, greenhouse gases, temperature, cloudiness, long-range climate trends).

The remote-sensing data are the main source of operational information for the systems of control of global ecological, biogeochemical, hydrophysical, epidemiological, geophysical, and even demographic situations on Earth. By now, the remote geoinformation monitoring of the environmental objects and processes provides large data volumes to solve many problems of the NSS control. However, unfortunately, the problem of coordinated requirements as to global databases and the structure of satellite monitoring has not yet been solved. Development of methods to process remote-sensing data lags behind the progress in technical equipment of the satellite-monitoring systems. It is expected that the use of GIMS-technology will make it possible to overcome this lag and to balance the algorithmic and technical aspects of remote sensing.

The Earth-observing systems enumerated in Table 5.2 are only a small part of a large spectrum of space-borne systems with diverse characteristics directed at accumulation of data on the state of the elements of land and the World Ocean,

Table 5.2. Some systems for environmental observation and their equipment.

System	Characteristic of the system
AARGOS	Global atmosphere observing system.
ADEOS	Improved satellite Earth observing system equipped with an advanced radiometer of visible and near-IR range (AVNIR), ocean color and temperature scanner (OCTS).
ERS-1,2	European satellites of remote sensing of the environment used within the ESA programme.
EOS	Earth observing system equipped with clouds and ERB sensors, altimeter, acoustic atmospheric lidar, laser wind meter.
UARS	NASA satellite launched in 1999 to study the upper atmosphere.
MOS	Sea-observing satellites of NASDA equipped with VNIRs and microwave scanning radiometer.
ENVISAT	ESA satellite to observe the environment.
ERBS	NASA satellite to study the Earth's radiation budget.
GMS	Geostationary meteorological satellite.
GOES	Geostationary satellite to observe the environment.
GOMS	Russian geostationary satellite to observe the environment.
JERS	Japan–NASA satellite to study the Earth's resources.
Aqua (NASA–EOS)	Equipped with six sensors characterized in Table 5.1.
Landsat 4,5	Equipped with thematic mapper in the visible and near-IR region, launched in 1982 and 1984. Spatial resolution 30 m.
Landsat 7	Equipped with advanced thematic mapper, launched by NASA in 1999. Spatial resolution 15 m.
SPOT 1-3	Satellites of the French Space Agency equipped with multi-spectral and pan-chromatic sensors with resolutions 20 m and 10 m, respectively.
IKONOS	Partially functioning satellite launched in 1999, equipped with multi-spectral and pan-chromatic sensors with resolutions 4 m and 1 m, respectively.
Terra	Launched by NASA in 1999, carries ASTER radiometer with resolution 15 m.

as well as on atmospheric parameters. Such systems include the satellite observatories Aqua and Terra as elements of the EOS. These systems are aimed at control of the Earth's health as a planet. They measure a great number of the characteristics of the ground system, ocean, and atmosphere. The Sumatra–Andaman earthquake of 26 December 2004, with its consequent tsunami, was the first great earthquake to be observed with modern instruments.

Table 5.2 contains information about some space-borne observation systems. These systems have been developed in the process of developing technologies of remote sensing and data processing. Of course, the technology of sounding in the optical region has turned out to be most developed. However, during recent years, technologies of remote sensing in the microwave region have been intensively

Table 5.3. Briefly described characteristics of some remote-sensing devices (Kramer, 1995).

Device	Frequency range	Resolution
• Advanced Airborne Hyperspectral Imaging Spectrometer (AAHIS)	440–880 nm	3 nm
• Airborne Imaging Spectrometer (AIS)	900–2400 nm	9.3 nm
• Compact Airborne Spectrographic Imager (CASI)	418–926 nm	2.9 nm
• Compact High-Resolution Imaging Spectrograph Sensor (CHRISS)	430–860 nm	11 nm
• Airborne Ocean Color Imager Spectrometer (AOCI)	443, 520, 550, 670, 750 11,500 nm	20 nm 100 nm 2,000 nm
• Wide-Angle High-Resolution Line-Imager (WHIRL)	595 nm	20 nm
• Airborne Imaging Microwave Radiometer (AIMR)	37 and 90 GHz	3 dB
• Airborne Multichannel Microwave Radiometer (AMMR)	10, 18.7, 21, 37, and 92 GHz	0.5 K
• Advanced Microwave Precipitation Radiometer (AMPR)	10.7, 19.35, 37.1, and 85.5 GHz	0.2–0.5°C
• Airborne Water Substance Radiometer (AWSR)	23.87, 31.65 GHz	0.1 K
• CRL Radar/Radiometer	9.86, 34.21 GHz	0.5 K
• Dornier SAR (DO-SAR)	5.3, 9.6, 35 GHz	1 m
• Electromagnetic Institute Radiometer (EMIRAD)	5, 17, 34 GHz	1 K
• Electromagnetic Institute SAR (EMISAR)	5.3 GHz	2×2 m
• Electronically Steered Thinned Array Radiometer (ESTAR)	1.4 GHz	±(3–4 m)
• Microwave Radiometer RADIUS	0.7, 1.425, 5.475, and 15.2 GHz	0.5–1.5 K

developed (Table 5.3). Passive sensing of the environment with the use of microwave radiometers gives efficient results in assessing the thermal characteristics and structure of land cover. Active radiometry gives information about the physical structure and electric properties of environmental objects. The combined sounding in the optical and microwave regions will enable one to raise the efficiency of remote-sensing systems at the expense of a mutual overcoming of shortcomings inherent to each range. The advantage and availability of the use of polarization effects in measuring systems is also apparent.

Synthesis of the systems of satellite monitoring of the environment is not only a complicated scientific–technical problem but also rather expensive. Therefore, the search for efficient solutions for this problem requires the use of the system approach to analysis of information fluxes needed for reliable prognostic estimation of the trends of global ecodynamics. On the one hand, this requires schematic solutions for the planning of measurements, and on the other hand, development of adaptive technologies of monitoring data processing. One of the perspective approaches to the problem of measurement planning is the radio-occultation method developed by Yakovlev (1998, 2001). The idea of this approach consists in the simultaneous use of two satellites, one carrying a signal-emitting system, the other carrying the receiver. Propagation of the signal along the satellite-to-satellite path leads to a dynamic change of its characteristics, which makes it possible to retrieve the atmospheric parameters along the path of the signal propagation. Since this path constantly changes its position with the satellite's motion, a situation is created when data are constantly obtained on the impacts of the atmosphere at different altitudes on the distribution of radio waves of a chosen interval, and on the basis of these data, by solving inverse problems, density, pressure, temperature, water content, and other atmospheric parameters are determined. The system is most informative when radio waves propagate along the path of a geostationary satellite–orbital vehicle. For this configuration of the monitoring system, efficient theoretical models have been developed of the dependence of refraction attenuation, angle of refraction, frequency, and absorption of the radio beam by the atmosphere. The transition from a binary structure of the system of atmospheric radio sensing to a multiple satellite system will make it possible to solve the problem of optimization of the global system of environmental monitoring.

There are a lot of examples of successful environmental monitoring, solving the problems of the diagnostics of situations of the origin of emergency phenomena. In particular, in some of the recent publications by Stevens *et al.* (2001a,b, 2003, 2004), methods have been described for satellite control of the zone of volcanic activity, based on recording the shifting of soil and changing of local topography. Systematical monitoring of volcanic eruptions began with observations of the Mount St. Helens eruption, May 1980.

The retrospective topographic mapping of volcanos enables one to identify slopes which may be dangerous in the case of eruptions, as well as to assess the state of lava deposits and to create information bases on geometric configurations of volcanic flows. One of the most efficient instruments to solve these problems is the scanning Interferometric Synthetic Aperture Radar (InSAR) which provides Earth's surface sensing in the presence of clouds and volcanic ash.

Digital models of the altitude profiles based on the data of ERS–InSAR measurements provide mapping with a horizontal resolution of $\sim$30 m, and with measurements using several radars this resolution can be reduced to 1–3 m. This makes it possible to contour the areas of risk in the zones of active volcanos, and the use of computer models makes it possible to create digital images of relief and simulate the paths of lava flows. The Japanese Advanced Space-borne Thermal Emission and Reflection Radiometer (ASTER), carried by the Terra/NASA satellite, with a resolution of about 15 m in the near-IR interval, provides stereo-images in the 60-km band, which together with the ERS–InSAR data gives spatial images of the zone of impact of a volcano.

The surfaces of lava flows are complicated and heterogeneous in spatial measurement. Knowledge of lava characteristics at different sections is important in solving the problems of the prediction of volcanic activity. Methods of interferometry and data of measurements with the synthetic aperture radar make it possible to retrieve the topography and deformation characteristics, to an accuracy of centimeters, due to differential measurements of the phase constituent of the reflected signal. Stevens *et al.* (2001b) have shown that SAR interferometry provides a reliable source of information about the behavioral dynamics of erupted lava flows. In this case such characteristics are retrieved as levels of thermal compression of the lava flow and its deformation in time. Using such measurements in the zone of the Mt. Etna Volcano (Sicily) it has been shown that the 1983 lava flow (55 m thick) compressed every year by 30 mm over 10–14 years. The older lava flows which erupted in 1981 and 1971 also went on compressing but at a slower rate. Knowledge of this dynamic makes it possible to predict the paths of lava flows for the next volcanic eruption. As Stevens and Wadge (2003) note that, technologies using interferometry to control volcanic zones have not been widely used in space-borne systems. This is partially connected with the presence of several unsolved technological and theoretical problems of overcoming the interference of clouds, snowfalls, and other interfering factors (Stevens and Scott, 2002).

Tupper *et al.* (2004) discussed the capabilities of the existing space-borne systems to detect and trace clouds of dust and gases that have erupted from volcanos. The processed GMS-5/VISSR, MODIS, and AVHRR data, as well as aircraft and ground measurements, were used to identify volcanic clouds. It was shown that the geostationary satellite GMS-5 and the satellite Terra made it possible to detect and identify ice crystals, CO_2, and aerosols in the erupted cloud.

Satellite systems are successfully used to control and prevent forest fires. For instance, automated detection of the location of forest fires is accomplished with the AVHRR radiometer included in the measuring complex carried by NOAA satellites. The algorithm of detection is based on the recording of radiation temperature in the channel 3.7 μm and the difference in radiation temperatures in the channels 3.7 and 11.0 μm. When these parameters exceed some threshold, a conclusion is drawn about some temperature anomaly on the land surface. Sobrino *et al.* (2004) have shown that the 6-channel thematic cartographer carried by the Landsat-5 satellite makes it possible to detect temperature gradients on the land surface of up to 1K.

The zones of tropical cyclone formation are well systematized, and therefore their forecast becomes possible due to the regular monitoring of these zones with the use of specialized patrol aircraft and satellite systems. Information obtained from aircraft flying hundreds of kilometers from the coast, enable one to detect the initial moments of a cyclone's origin and its direction. Also important are radars located on coastal sites that are most subject to hurricane impacts – enabling one to fix and trace a tropical hurricane at distances of up to 400 km.

5.2 METHODOLOGICAL AND INFORMATIVE CAPABILITIES OF NATURAL DISASTER MONITORING SYSTEMS

It is well known that climate is a totality of environmental phenomena connected with the formation of the human habitat and expressed in characteristics such as temperature, precipitation, humidity, composition and motion of the atmosphere, oceans' dynamics, etc. The climate system has three dimensions: space, time, and human perception. Each of them is historically connected with a certain totality of processes in the NSS, and therefore the creation of a global system consisting of models and technical means is urgent for obtaining reliable estimates of the NSS state and especially for forecasting its future trends. One of the very complicated and contradictory elements of this system is the global climate system, since its synthesis requires solution of the problem of parameterization of the interaction between the NSS components. The complexity of this problem is determined by the multi-dimensional character of the phase space of the climate system. The developed models divide this space into atmosphere, land, and ocean. Depending on the spatial–temporal division and assumed simplifications and approximations, models of different types have appeared. This is connected both with the presence of developed conceptual and theoretical approaches and with available empirical information. Nevertheless, the following four characteristics of the most important components of the climate system can be selected, which are present in the models: radiation, dynamics, processes at the interface between the two media, and spatial–temporal resolution. Depending on orientation, the models are divided into energy-balance, radiative–convective, statistical dynamical, and 3D-models of atmospheric general circulation.

In connection with the global character of environmental problems, their solution requires joint efforts of the international community, which has been recently manifested through international and national programs to study global change and the development of space-borne systems of geoinformation monitoring (Malinnikov *et al.*, 2000). Scientists from the USA pay special attention to these problems. So, from 1995, the US National Scientific Fund (NSF) financed the Methods and Models for Integrated Assessment (MMIA) program within the framework of which the role of large-scale technogenic and socio-political processes in the dynamics of physical, biological, and human systems was studied. The NSF supports more than 20 research programs, in some way connected with global environmental change. Among them are such programs as GGD (Greenhouse Gas Dynamics), US GLOBEC (US Global Ocean Ecosystems Dynamics), EROC (Ecological Rates of Change), and others. Special emphasis is laid on studies of the high-latitude environment, and Arctic systems, in particular (with special emphasis on the International Polar Year Programme).

An important aspect of studies of global environmental change is the technical equipment of the data collection systems, organization of their functioning (Kondratyev, 1999; Maksudova *et al.*, 2000), and generation of data in more compatible formats. One such format is the WEFAX-format (Kramer, 1995) which is used by the system of geostationary satellites (Meteosat-5 and 7, INSAT, FY-2B Wefax, GMS, GOES–West, GOES–East) ensuring the daily flux of 400 images of the Earth's surface with the spatial resolution from 2.5 to 10 km. Among other widely used formats for images transmission are APT, HRPT, and PDUS. The polar-orbit satellites APT give, on average, 12 images of the Earth's surface per day at frequencies of 136.5 MHz (NOAA-12 and 15) and 137.62 MHz (NOAA-14 and 17). The HRPT formats provide the 1.1-km resolution in five spectral intervals (two visible and three IR) at a rate of ~12 images per day. The PDUS is used in the Meteosat series mainly to map the European territory with a resolution of 2.5 km. Here a widespread problem arises of the agreement of the monitoring data series in the case of irregular observations. To solve this problem, Timoshevsky *et al.* (2003) suggest a method of model self-organization based on the problem-oriented search for an optimally complicated model which describes these series or a phenomenon represented by them.

The fragments mentioned above indicate that the capabilities of a remote sensing method are sufficiently developed to provide the GMNSS with real-time mode data for an operational assessment of the environmental state and a prediction of its key characteristics. However, the role of the available flow of remote-sensing data in studies of the global environmental change is confined mainly to the detection and recording of particular phenomena without an analysis of their global role (Chen *et al.*, 2003; Del Frate and Ferrazzoli, 2003; Dong *et al.*, 2003; Levesque and King, 2003; Timoshevsky *et al.*, 2003).

Thus, the observed accelerated data accumulation, representing a gigantic scientific–technical progress, and the global human forcing on nature have put forward the problem of comprehensive assessment of the state of the environment and analysis of the possibilities of long-range forecasts of environmental change. Scientific studies in the field of anthropogenic forcing on the environment have led to the conclusion that the problem of objective control of dynamics of natural phenomena and the human impact on these phenomena can be solved only by creating a single system of geoecoinformation monitoring based on the GMNSS. Within the international and national programs of the NSS studies, voluminous databases have been accumulated of environmental parameters, as well as information on the dynamics of natural and anthropogenic processes of various scales, and models have been developed of biogeochemical, biogeocenotic, climatic, and demographic processes. The technical basis for global geoinformation monitoring includes an efficient means of collection, recording, accumulation, and processing of data obtained from spaceborne, airborne, ground, and sea platforms.

However, despite considerable progress in many spheres of nature monitoring, the main problem remains unsolved, consisting of a substantiation of an optimal combination of all technical means, development of an efficient and economic structure of monitoring, and preparation of reliable methods of prognostic estimation of the environmental dynamics under conditions of anthropogenic forcing. Experience gained from recent studies testifies to the possibility of developing a global model which can be used (in the adaptive regime of its application) to substantiate recommendations on the structure of monitoring and formation of requirements to databases (Kondratyev *et al.*, 2002b).

The monitoring system should be based on a regional segment including the bank of models of biogeochemical, biogeocenotic, and hydrodynamic processes in the biosphere with due regard to their spatial heterogeneity and directed at a combination of the available regional databases as well as the use of existing means of environmental observation. On the basis of cooperation with specialists who have developed climatic and socio-economic models, it is necessary to develop a model of the NSS. Using this model, one can evaluate the volume and quality of available databases and give recommendations on the development of their structure.

As a subsequent improvement of the model, it is necessary to study the NSS's interaction with processes taking place in the adjacent space, and to take an active role in the development of a global model taking into account the processes in the magnetosphere and the impact of space.

The final goal is to create a system able to forecast the development of natural processes and to evaluate the long-term consequences of environmental forcing of different scales. This system should cover various problems of protection of the environment on regional and global scales with a realization of the functions of ecological examination of the projects changing the Earth's cover, hydrological regimes, and atmospheric composition.

The functional structure of the monitoring system is based on the idea of distributed network architecture. Practical realization of the idea of the NSS model requires an accomplishment of problem-oriented complex studies, among which the following problems are important:

- Systematization of data on regional and global change and formation of complex ideas of biospheric processes and structures of biospheric levels. Development of the conceptual model of the biosphere as a component of the global geoecoinformation monitoring system.
- Inventory and analysis of the existing banks of ecological data and the choice of the structure of the bank of regional and global data.
- Classification of spatial–temporal characteristics and cause-and-effect connections in the biosphere in order to develop the scope of coordination of spatial and temporal scales of ecological processes.
- Accumulation of the banks of models of sample ecological systems, biogeochemical, biogeocenotic, hydrological, and climatic processes.
- Study of the processes of biosphere–climate interaction. Search for the laws of the impact of solar activity on biospheric systems of different spatial scales.
- Systematization of information on local ecosystems. Description of geophysical and trophic structures (regionally) taking into account an agreement with spatial–temporal scales.
- Construction of models of regional and global biotechnological fields and development on the basis of algorithms for the synthesis of the industrial unit of the global model.
- Formation of a bank of scenarios of co-evolutionary development of the biosphere and human society. Development of demographic models. Parameterization of the factors of scientific–technical progress in the context of the use of biospheric resources. Modeling the perspectives of the urban building and the use of the Earth's resources.
- Search for new information technologies of global modeling which provide a reduction in the requirements of databases and bases of knowledge. Development of an architecture and algorithmic principles of functioning of

computing systems of neuro-like elements, accomplishing an evolutionary processing of information with high speed and efficiency.

- Further development of the concept of ecological monitoring and construction of a theoretical base of ecoinformatics. Development of methods and the criteria of assessment of stability of regional and global natural processes. Analysis of stability of the biospheric and climatic structures.
- Synthesis of the NSS model and development of a computer means to carry out calculations to asses the consequences of realization of various scenarios of anthropogenic activity.

Analysis of the results of recent studies of the problems mentioned above has shown that for successful global modeling it is necessary to develop new methods of system analysis of complicated natural processes as well as methods of data processing directed at the synthesis of the balanced criteria of information sampling and consideration of the hierarchy of the cause-and-effect connections in the NSS. The global model should be based on a totality of regional-scale models in which the region is considered a natural subsystem interacting with the global NSS through biospheric, climatic, and socio-economic feedbacks. The respective model describes this interaction and functioning of all the processes in the subsystem at various levels of the spatial–temporal hierarchy. The model covers natural and anthropogenic processes characteristic of a sample region and is initially based on the existing information. The model's structure is used in the adaptive regime.

The combination of the system of information collection about the environment, the model of regional geoecosystem functioning, the system of computer cartography, and means of artificial intelligence should result in development of the Geoinformation Monitoring System of a sample region (GIMS-region) which would be able to solve the following problems:

- To assess the impact of global change on the regional environment.
- To assess the role of environmental change in the region on changes to the climate and the biosphere and in the adjacent regions.
- To assess the ecological state of the atmosphere, hydrosphere, and soil–plant formations in the region.
- To form and renew databases on the ecological, climatic, demographic, and economic parameters of the region.
- To operatively map the landscape.
- To forecast the ecological consequences of anthropogenic forcings on the environment of the region and its adjacent territories.
- To typify the Earth's cover, natural disasters, inhabited areas, surface pollutions of landscapes, hydrological systems, and forest tracts.
- To assess the level of safety for populations in the region.

Substantiation of the GIMS-region is connected with the selection of components of the biosphere, climate, and social medium characteristic of a given level of spatial hierarchy. Successive actions in the development of the GIMS-region project and its realization proceed from experience of creation of GISs and construction of mathematical models of the functioning of large-scale natural–economic systems. This experience determines an expediency of the following GIMS-region structure:

- Subsystem of data collection and express analysis.
- Subsystem of primary processing and accumulation of data.
- Subsystem of computer mapping.
- Subsystem to assess the state of the atmosphere.
- Subsystem to assess the state of soil–vegetation cover.
- Subsystem to assess the state of water medium.
- Subsystem to assess the ecological safety and risk to human health in the region.
- Subsystem to identify the causes of violation of ecological and sanitary situations.
- Subsystem of intelligent support.

Based on the simulation–evolution technology of modeling, a sample information-computing complex is formed aimed at introducing methods, algorithms, and procedures of evolutionary structural synthesis of discrete–continuous descriptions of objects characterized by information uncertainty, temporal dynamics, stochastic character, inadequate, and contradictory observational data. Further improvement to this technology is connected with the introduction of neuro-technology as an alternative method of synthesis of future realizations of the NSS models in the monitoring systems, and its transition to the level of intellectual computer system – of higher rank than existing expert systems.

Analysis of the state of databases and their replenishment shows that the existing monitoring systems lack a closed network of the means of environmental monitoring used presently. Based on the experience of aerospace monitoring centers applied in many countries, it is suggested to use a sample system of autonomous multi-purpose and specialized ground and aircraft centers for the assessment of the state of elements of natural and anthropogenic landscapes, forecasts of ecological and sanitary states of inhabited areas, environmental diagnostics, anomalous phenomena, natural disasters, and technogenic catastrophes. A network of such centers consistent with the NSS model can solve the following problems:

- Thematic cartography of the results of ground and aerospace measurements and surveys carried out with *in situ* and remote methods in the optical, IR, and microwave regions.
- Providing current estimates and forecasts of the ecological and sanitary state of inhabited, industrial, and field zones; sources of drinking water and rivers; agricultural, forest, and hunting areas; garbage tips and industrial waste tips; accumulation of mineralized and polluted water as potential sources of emergencies.
- Producing a terrestrial–seasonal inventory and typifying quasi-homogeneous and anomalous areas and areas of floristic background and soils, water basins and the atmosphere, geophysical fields of different natures, map referencing, identification of the sources of pollution of surface cover, air, and water basins and their impact on inhabited areas and environmental elements.

- Prompt examination and high-accurate fixation of the coordinates of small-sized areas and objects with non-ordinary reflecting/emitting characteristics, detection of anomalous phenomena, natural disasters and technogenic catastrophes, dynamics of the consequences of such events in inhabited areas, as well as environmental parameters.
- Instruction to accomplish search and rescue operations on land and in the sea, collection of expensive metal remains of rockets and river and sea ships, assessment of fire risks of taiga and peatbogs, oil–gas extraction, mining and other enterprises, and identify potential sources of accidents at hydrocarbon product pipelines.
- Combining of the results of aircraft observations with space images and ground measurements to reconstruct an objective picture of ecological and sanitary situations and to substantiate the many ways of rational use of the Earth's resources; agrarian, forest and industrial potential, water basins and rivers, and transport and irrigation systems.

Autonomous aircraft centers are organized on the basis of small, medium, and unmanned flying vehicles. They are equipped with radars, radiometers, TVs, laser and photographic means of observations, as well as systems of referencing observed results to characteristic points with local coordinates.

At present, it is possible to carry out the project of a single system of regional models of the most important natural processes, including the processes taking place in the magnetosphere. Knowledge and accumulated data make it possible to realize both the local and the regional levels of the spatial hierarchy. Such an exploitation of the system will enable one to formulate the requirements of databases which will ensure the transition to a more detailed spatial division of natural systems and processes, even with the role of individual elements of the ecosystems taken into account.

The ecological monitoring system is known to include three main directions:

(1) *Ecosociology* – the study of the impact of human society on the environment and interactions in the NSS.
(2) *Ecoinformatics* – the recording, storage, transmission, analysis, synthesis, modeling, and delivery of information on the state of the environment.
(3) *Ecotechnology* – the development of technologies and technical means with minimum sufficient forcings on nature and human organism.

Ecosociology includes:

- Analysis of social structures affecting the environment, their motives, goals, methods, and technical means.
- Analysis of the necessity and sufficiency of anthropogenic impacts of social structures on the environment.
- Analysis of an optimal allocation of industrial and agrarian infrastructures and inhabited areas in the environment.
- Development of eco-protective measures taking into account the socio-regional, socio-national, and socio-global interests.

Ecoinformatics includes obtaining diagnostic, prognostic, and practical information on the state of the environment and its change with anthropogenic and natural forcings. The obtained information should contain the following:

- Quantitative estimates of the state of the environment and natural resources.
- Quantitative estimates of the anthropogenic impact and detection of harmful and excessive production.
- Current state and maintenance of regional (in some cases district) simulation models of interactions in the systems "production–medium". "agriculture–medium" , and "habitation–medium".
- Current state and maintenance of the regional ecological model.
- Current state used by the ecological model with national and international cooperation.
- Current state that makes it possible, based on the simulation ecological model, to construct prognostic schemes of the impact of harmful and excess production on the regional environment.
- Current state used in prognostic modeling of the environmental change in the development of detected resources.
- Current state used, on the basis of simulation and prognostic models, in working out the recommendations for development or reduction of production and inhabited areas in the region.

The ecoinformation-monitoring system is constructed as a 3-level scheme including space-borne and ground observations as well as aircraft observations. A functional scheme of the system consists of the following main components:

- Complex system of observations (data collection) including the satellite, aircraft, and ground segments which provide regular *in situ* and remote measurements of the environmental parameters.
- Reception/transmission subsystems based on the satellite and ground means with the use of satellite, radio-relay, wire, and optical-fiber communication channels.
- Subsystems of primary data processing from different observation systems.
- Subsystems of data (data banks) accumulation and systematization.
- Groups of subsystems to solve the applied (thematic) problems of monitoring.

By their purposefulness and information–technical characteristics, the space-borne observation systems refer to the following classes:

(a) Systems of non-operational detailed observations from space vehicles with the use of data of multi-zonal and topographic surveys.
(b) Systems of operational observations divided into sub-classes:
 –Systems of planned surveys for global operational control of the state of the atmosphere and the Earth's surface with a low spatial resolution of continuous (quasi-continuous) observations, with the use of geostationary space vehicles and regular observations, with the use of space vehicles at high-elliptic and circular, near-polar orbits.

–Systems of detailed observations ensuring the obtaining of data for the chosen objects or regions on the Earth's surface with high or middle spatial resolutions of continuous (at high-elliptic or circular orbits) and regular (at mid-latitude circular orbits) operation.

At present, for most of the projects for medium-orbital space vehicles both sub-goals have been reached with the use of one satellite. In the system of aircraft observations, to organize a precise sounding of the regions with complicated ecological situations, the aircraft-scanner is required to carry adequate equipment and have sufficient functioning both in manned and unmanned regimes. Operational support of the regional ecological model requires the development of a high-productive, information-computing network. It should be based on a distributed network of current computer means with respective software.

The regional system of ecological monitoring should become the core of ecological examination in the territorial control. The system of ecological monitoring will provide, for each citizen of the region, an access to information on the state of the environment (generally available sites or pages in any global network, for instance, Internet). Based on this knowledge, one can make a competent decision on many vitally important problems both in social and in personal spheres.

The *ecotechnological* direction includes:

- Assessment of real and potential dangers of the functioning productions connected with emissions of substances harmful to both the environment and humans.
- Assessment of technological processes and detection of the origins of these processes that are emitting substances.
- Improving technological processes with preserved parameters of the final product whilst reducing the harmful effect on the environment and humans.
- Development of instrumentation realizing the improved technological process.
- Instruction of staff operating under new conditions and with new machinery.
- Constant control regarding the observation of technological discipline.
- Qualified examination of technological processes and respective equipment.

Development of production with minimum effect on the environment and humans is connected with the use of high-intellectual computer technologies which make it possible to control all production processes and technological equipment.

All enumerated directions of ecological activity can be concentrated in a single organizational structure – a regional center of ecological monitoring which, according to the scheme in Figure 5.1, is an element of the global GIMS. The regional center of ecological monitoring should consist of three structural units corresponding to the main directions of ecological activity: ecosociology, ecoinformatics, and ecotechnology. Each of them, in correspondence with its goals, can be divided into several sections:

(1) Ecosociology.
  –Analysis and forecasts in the NSS.

–Nature-protection measures.
–Decision making and interaction with governing structures.

(2) Ecoinformatics.
–Reception and analysis of space-derived information.
–Aero-survey with a mobile group of aircraft.
–Ground observations with a network of sensors and ground stations.
–Control of human health.
–Modeling and maintenance of the regional ecological model.
–Information-computing center.

(3) Ecotechnologies.
–Eco-analysis of industrial enterprises.
–Eco-analysis of extractive industries.
–Keeping technologies under control.
–Examination of technologies and technical decisions.

One of the examples of practical realization of the environmental control technology described above, to detect critical situations and take preventive measures, is the system of decision making as applied to the assessment of the risk of forest fires in the Mediterranean basin (Iliadis, 2005). This system is based on modeling the data series on the state of the environment using the cluster analysis and learning algorithms. The system has been realized in Greece, but its subject orientation can be adapted to global scales.

5.3 EXPERT LEVEL OF DECISION-MAKING SYSTEMS IN THE CASE OF NATURAL DISASTERS

5.3.1 Statistical decision-making in monitoring systems

The environmental monitoring regime can foresee situations of decision-making in a real mode of time, based on information accumulated before the moment of decision-making, or as a result of analysis of the database fragments without referencing to the current time. The statistical analysis of several events monitored by the system can be carried out by numerous methods, the use of which is determined in each case by a totality of probabilistic parameters characterizing the phenomenon under study. However, the instability and parametric uncertainty in situations, when each measurement requires much effort and expenditure, requires a search for new methods of decision-making based on observational data, fragmentary both in time and in space.

With development of alternative methods of making statistical decisions, the problem of an objective estimation of parameters of the processes taking place in the environment has obtained a new substantiation. It is possible to consider and compare two approaches to this problem: a classical approach based on the a priori limited number of observations, and a sequential analysis based on step-by-step decision-making. Development of computer technologies makes it possible to realize both approaches in the form of a single system of making statistical decisions.

Making decisions on detection of some effect in the process of continuous monitoring of the environment depends on the scheme of measurements organization. The classical approach directs the observation system to a collection of a fixed volume of data, with their subsequent processing revealing certain effects or properties in the space under study. The method of sequential analysis does not divide these stages but alternates them. In other words, the monitoring data are processed after each measurement. Hence, the algorithmic load in the successive procedure changes dynamically, whereas in the classical case, data are processed at a final stage. From the viewpoint of the formation of the structure of an automated decision system, these approaches should be realized in the form of individual units chosen in the regime of a dialogue with the system's operator. The operator's decision can change in the process of monitoring, but to do this, the operator should be an expert in this sphere. Therefore, in the proposed structure of the decision system the operator's functions are reduced to the control of the parameters of the procedure of the choice between alternative hypotheses.

The decision theory makes us consider the above-formulated problem of using sequential analysis in the procedure of the choice between competing hypotheses in a more general form and with due regard to all aspects appearing in such problems. The list of problems arising before a decision-maker covers the following themes:

- Selection of an efficient criterion of parameter evaluation.
- Revealing the character of probabilistic characteristics of the process under study.
- A priori estimation of possible losses in the precision of decisions.
- Forecast of the dynamic stability of the results of the decision made.

One of the features of the automated statistical-decision system is its broad spectrum of functions, which should, if possible, provide solutions of most of the intermediate and final assessments of the situation in the monitoring regime. Therefore, the system should include the following units:

- Visualization of measurement data in the form of a direct image, histogram, sum and frequencies over time intervals.
- Calculation of statistical characteristics of observation results (mean-square value, dispersion, variance, standard deviation, moments of the third and higher order, excess, coefficient of asymmetry, entropy, etc.).
- Construction of empirical probability density and distribution function.
- Construction of a theoretical distribution of probabilities with its estimation by one of the statistical criteria.
- Calculation of characteristics for the procedure of acceptance of a hypothesis by classic Newman–Pearson method (Rasch *et al.*, 2004) with the resulting estimate of the procedure by the current sampling volume.
- Calculation of characteristics of the procedure of sequential analysis and visualization of its state.

- Realization of the functions of the operator's access to the system's units at any stage of the system's functioning with the possibilities of making decisions to change the parameters of the procedure or ending measurements.
- Visualization of the decision made.

This set of units provides the formation of the model of the measurement procedure depending on available a priori information about parameters and the character of the process under study. At the same time, the operator can check the correctness of the input information and promptly change the monitoring strategy. On the whole, the set of indicated units constitutes an automated statistical-decision system.

The successive procedure of making statistical decisions is based on the Wald distribution. In studying the sum x_n of independent similarly distributed variables, two dual problems of evaluation of the distribution function $P(x_n < x) = F_1(x)$ both at fixed n, and at its random value. In both cases the value of x_n is compared with some threshold C. However, in the second case this problem is transformed into the problem of studying the distribution $P(\nu < n) = F_2(n)$ of a random number ν of items leading to an initial exceeding by x_ν of threshold $C = C(\alpha, \beta)$: $x_i < C(i = 1, 2, \ldots, \nu - 1)$; $x_\nu \geq C$, where α and β are errors of the first and second order, respectively.

In accordance with the central limit theory at $n \to \infty$ in the first (classic) case $F(x)$ is described by the normal distribution. In the second case the Wald distribution is realized, whose density is determined on half-line $[0, \infty)$ and has one maximum at the point $x = m_c$:

$$w_c(y) = c^{1/2} y^{-3/2} (2\pi)^{-1/2} \exp[-0.5c(y + y^{-1} - 2)]$$

Let us calculate the derivative of the Wald distribution function and equate it to zero: $dw_c(y)/dy = 0$. This equation is solved as follows:

$$y = m_c = [(9 + 4c^2)^{1/2} - 3]/(2c)$$

The location of the $w_c(y)$ maximum shifts depending only on the value of one parameter c, so that always $m_c < 1$. At $c \to 0$, $m_c \to 0$, and at $c \to \infty, m_c \to 1$:

$$w_c(m_c) = (3/(2\pi)^{1/2})(1/[m_c(1 - m_c^2)^{1/2}]) \exp\{-3(1 - m_c)/[2(1 + m_c)]\}$$

Hence, at $c \to 0$ or $c \to \infty$, the parameter $w_c(y)$ degenerates into delta-function $\delta(y)$ or $\delta(y - 1)$, respectively.

5.3.2 Functions of environmental monitoring systems

The monitoring system is constructed via a standard man–machine dialogue scheme with the use of a hierarchical menu. The principal scheme of the system has the form of some transformation F, whose structure and content are determined by a set of units performing individual operations on analysis and transformation of measurement data. At the system's input are measurement data from n of sources (sensors). Decisions on the reliability of the hypothesis (H_0 or H_1) can be made with respect to either source or some totality of them, $m \leq n$. Various regimes of the operator–

system interaction are foreseen. The end of the decision-making procedure is followed by information on the length of the used data sequence for each turned-on information channel. In the case of turning-on of two or more channels, the final acceptance of one of the hypotheses is made based on the weighted majority of decisions made for individual channels. With an appearance of obstructions to decision-making, the operator is warned about situations of uncertainty, and can change the weights of channels and introduce corrections into other parameters.

To realize the procedures of acceptance of complicated hypotheses, when the number of possible decisions is beyond the binary situation, a recurrent procedure is carried out for realization of a multi-criterion problem of decision making in the form of a composition of simple binary situations. The base procedure realizes the function decision $(\gamma, \xi, \alpha, \beta G, Y, \theta_0, \theta_1, N, R)$, where γ is the controling parameter; ξ is the input flux of measured values of the controled parameter; G is the set of parameters of f_a distribution; Y is the type of f_a distribution; N is the threshold level of the number of observations for an automated change of the procedure of sequential analysis for the classic procedure; and R is the indicator of the decision made ($R = 0$ – the H_0 hypothesis holds; $R = 1$ – the H_1 hypothesis holds; $R = 2$ – no decision). Introduction of N makes it possible to avoid a situation when the decision can be absent no matter for how long. When the number of measurements exceeds the threshold N, the system proceeds to decision-making by the classic procedure. The γ parameter controls the regimes of the use of the decision function. Various regimes of realization of the decision-making procedure are possible: $\gamma = 0$ – the operator's interference is not planned; $\gamma = 1$ – corrections of the procedure are foreseen. Depending on the γ, the system is adjusted to the respective regime of control of the units. With $\gamma = 0$, many intermediate operations are neglected, and the rate of the system's operation grows. At $\gamma = 1$, the system activates all its functions and practically at each step the operator has a priority to change the procedure parameters. The system remembers all actions of the operator and in case of contradictions, informs him of them. The γ parameter also governs the regime of the input of data from information channels of different profiles. There is a possibility to choose the structure of information channels, for which a decision is made. The decision can be made either for individual channels or for their group. Also, the regime of delivery of the results of realization of the procedure of hypotheses acceptance is under control. The operator is informed about the state of final conclusions of individual units. By channel number, the user can inquire about statistical characteristics, the type of distribution, and other parameters. The system blocks an absurd inquiry, such as one about statistical characteristics or probabilistic distribution with small volumes of samples. In the case of the uncertainty situation, when making a decision the operator is recommended to change the structure of the identifier of the considered channels, change their weights, or perform other operations according to his decision.

5.3.3 Processing of multi-channel information

Methods of local diagnostics of the environment do not permit a complex assessment of the state of a natural object or process, especially if this element of the environment covers a vast area. In the problems of geoinformation monitoring, situations occur when it is necessary to make a decision in real time, with limited capabilities of the applied means of data processing (speed, memory, rate of data transmission, etc.). In this connection, below we shall consider the model of such a situation and calculate its characteristics. The general scheme of sounding of the environmental elements supposes that the data flux $\{T_B\}$ from the jth sensor can be analyzed both independently and together with the data from other channels. There is a digital processor of the type found in a personal computer with the speed V operations per second. The problem is set to recognize an unknown vector parameter $T_B = \{T_{B1}, \ldots, T_{Bn}\}$, which characterizes the state of the controled object (e.g., in the problems of microwave monitoring it is brightness temperatures). To solve this problem, an algorithm is proposed of the parallel-in-time analysis of the components of the vector T_B, which makes it possible to cut down the data-processing time, but in this case the data fluxes should be coordinated. The procedure consists of a continuous formation of the vector T_B structure by gradual adding of components whose estimates by local criteria are attributed to T_B: $(T_{B1}), (T_{B1}, T_{B2}), \ldots, (T_{B1}, T_{B2}, \ldots, T_{Bi}), \ldots, (T_{B1}, \ldots, T_{Bk})$.

Two situations are possible: the system has time to analyze the whole chain of vectors without a delay in individual channels, and also channels with some time delays. In a general case, two types of delay are possible: in time and in volume. Delays constant in time we denote as $\tau_1, \ldots, \tau_k$, and delays in volume – as $m_1, m_2, \ldots, m_k$. In the first case, the process of the choice of the vector T_B will consists in that between channels i and $i+1$, during the time τ_i, the estimates of T_{Bj} can accumulate, which in time passes to the chain $\{T_{Bj}\}$. The amount of delayed estimates of T_{Bj} in each channel is accidental and therefore the problem arises of reservation of buffer memory for each channel. In the second case, the variate is the delay time $\tau_i (i = 1, \ldots, k)$, and volumes of the buffers m_i are fixed. In both cases it is necessary to assess the probability of not missing a real value of the vector T_B and to find optimal values of τ_i and m_i corresponding to a maximum level of this probability.

Let in each channel the volume of observations be estimated by the value $n_i (i = 1, \ldots, k)$ and the probability of the first-order error in decision making be α_i. Then the probability of appearance during the time $t_i = r_i n_i S_i$ of estimates of T_{Bi}, which pretend to be the constituents of the final decision will be:

$$P\{\mu_i = S_i\} = C_{r_i}^{S_i} \alpha_i^{S_i} (1 - \alpha_i)^{r_i - S_i} = \nu_i(S_i)$$

At $\mu_i \leq r_i \alpha_i$ the estimates arrive rarely and have time to move from the ith to the $(i+1)$th decision unit without delay. But if $\mu_i > r_i \alpha_i$, then the estimates from the ith unit arrive often, and between their inputs to the $(i+1)$th unit the latter has no time to examine all versions of T_{Bi+1}. Therefore, the estimates are delayed. The probability that the number of estimates will not exceed the average value of $r_i \alpha_i$ by more than ε_i, is:

$$P\{\mu_i \leq r_i \alpha_i + \varepsilon_i\} = \sum_{s=0}^{r_i \alpha_i + \varepsilon_i} \nu_i(s)$$

Denote as M_i the memory volume for the delay for each channel. Then the following condition should be met:

$$r_i\alpha_i + \varepsilon_i \leq M_i$$

In this case the probability of not overfilling the memory M_i will be:

$$P\{\mu_i \leq M_i\} = \Phi(u_i)$$

where:

$$u_i = \frac{\varepsilon_i}{\sqrt{r_i\alpha_i(1-\alpha_i)}} = \frac{M_i - r_i\alpha_i}{\sqrt{r_i\alpha_i(1-\alpha_i)}}$$

$$\Phi(u_i) = \frac{1}{\sqrt{2\pi}}\int_{\infty}^{u_i} \exp\left[-\frac{t^2}{2}\right] dt$$

Using the Boole formula (Borwein *et al.*, 2004), one can calculate the probability of not overfilling the memory M_i at the ith channel with delay τ_i in the process of sorting all values of the ith component and continuous output of values in regular time intervals in the $(i+1)$th decision unit. Denote as $P_{1,\ldots,N_i}$ the probability of not overfilling the memory. Then:

$$P_{1,2,\ldots,N_i} \geq 1 - N_i[1 - \Phi(u_i)]$$

This probability should differ from unity by not more than δ_i. Then we obtain the equation to evaluate the delay τ_i:

$$\Phi(u_i) = 1 - \frac{\delta_i}{N_i}$$

or

$$r_i^2 - 2\frac{r_i}{\alpha_i}\left[M_i + (1-\alpha_i)\ln\frac{N_i}{\delta_i}\right] + \left(\frac{M_i}{\alpha_i}\right)^2 = 0$$

In the multi-channel system of monitoring, along with these peculiarities of decision making, the problem arises of evaluation of the noise-loaded parameter in the communication channel. This problem is especially important in remote transmission of the monitoring data. In this connection, we study the situation when the communication channel is characterized by the Gauss additive and Rayleigh multiplicative noise.

5.3.3.1 *Reception of noise-loaded signals*

One of the important problems of the systems of remote radiometric sensing of the environment is the transmission of useful signals from noise-loaded channels and its reconstruction at the receiver. This problem is studied by many scientists with different assumptions made of the character of the information channel. The most general approach to this problem has been developed by Kotelnikov (1956). Let us have a communication noise-loaded channel containing only an additive component ν which follows the Gauss law, with $M\nu = 0$ and $D\nu = \sigma^2$. At the input of the channel is the signal $X = f_\lambda(t)$, the known (usually continuous) function of time t, and the transmitted parameter λ (the way of coding the parameter). Then the output signal will be: $Y(t) = f_\lambda(t) + \nu(t)$. With respect to the function f_λ we assume the existence of a first λ derivative.

An ideal reception of the transmitted parameter can be realized by optimal discrete statistics of a volume n of heterogeneous sampled values of $Y(t)$, namely, $y_1, \ldots, y_n$ are measured. The sampled values are processed by the method of maximum likelihood, which gives the best estimates of the λ parameter (the Fisher method). For this purpose, we form the likelihood equation, first finding the likelihood function. According to our assumptions, the probability density of the deviate $Y(t)$ will be:

$$p_\lambda(y) = \left(\sqrt{2\pi\sigma}\right)^{-1} \exp[-0.5\sigma^{-2}(y - f_\lambda(t))^2] \tag{5.1}$$

The likelihood function is written as:

$$g(y_1, \Lambda, y_n; \lambda) = \left(\sqrt{2\pi\sigma}\right)^{-n} \exp\left\{-0.5\sigma^{-2}\sum_{j=1}^{n}[y_j - f_\lambda(t_j)]^2\right\} \tag{5.2}$$

The likelihood equation is written as:

$$L' = \sigma^{-2}\sum_{j=1}^{n}[y_j - f_\lambda(t_j)]f'_\lambda(t) = 0 \tag{5.3}$$

In a general case, (5.3) is transcendent with respect to the evaluated parameter $\lambda = \lambda_*$. Therefore, it is expedient to apply an efficient method of sequential approximations. As a first approximation, we choose some value of the parameter $\lambda = \lambda_1$ and find the following approximation from the formula: $\lambda_2 = \lambda_1 + h_1$, where h_1 is the correction determined by the form of the function (5.2) and the volume of information contained in a heterogeneous sample $y_1, \ldots, y_n$. We obtain:

$$h_1 = \frac{L'(y_1, \Lambda, y_n; \lambda_1)}{I(\lambda_1)} \tag{5.4}$$

where

$$I(\lambda_1) = \sum_{k=1}^{n} M_\lambda \left(\frac{p'_k}{p_k}\right)^2$$

p_k is the probability density of the parameter y_k. We have:

$$I(\lambda_1) = \int_{-\infty}^{\infty}\sum_{k=1}^{n}\left(\frac{p'_k}{p_k}\right)^2 p_k dy = \frac{1}{\sigma^2}\sum_{k=1}^{n}[f'_{\lambda_1}(t_k)]^2$$

The final relationship for evaluation of the next approximation of the parameter λ will be:

$$\lambda_2 = \lambda_1 + \frac{\sum_{j=1}^{n}[y_j - f_\lambda(t_j)]f_{\lambda_1}(t_j)}{\sum_{k=1}^{n}[f'_{\lambda_1}(t_k)]^2}$$

In a similar way we move to a third approximation, etc. In general, if the mth approximation of λ_m has been found, then the $(m+1)$th approximation can be obtained from the formula:

$$\lambda_{m+1} = \lambda_m + \frac{\sum_{j=1}^{n}[y_j - f_{\lambda_m}(t_j)]f'_{\lambda_m}(t_j)}{\sum_{k=1}^{n}[f'_{\lambda_m}(t_k)]^2} \tag{5.5}$$

These sequential approximations are reduced to some threshold $\lambda_{**} \approx \lambda_*$ which is assumed as an estimate for λ_*. Of course, the rate of convergence depends on a good choice of λ_1.

Evaluation of λ_* by a method of maximum likelihood is distributed asymptotically normally with mathematical expectation $M\lambda_* \approx \lambda_*$ and dispersion $D\lambda_* = 1/I(\lambda)$, where $I(\lambda)$ is the Fisher information indicated above. Then the confidence interval λ_* will be:

$$P = P\left\{|\lambda - \lambda_*| < u\sqrt{D\lambda_*}\right\} = \frac{1}{\sqrt{2\pi}}\int_{-u}^{u} \exp\left(-\frac{z^2}{2}\right) dz$$

Further generalizing the problem, we consider the case where at the input of the channel is the signal $X = f(\lambda_1, \ldots, \lambda_k; t)$, which is the known function of time t and the group of transmitted parameters $\lambda_1, \ldots, \lambda_k$. With respect to the function f we assume the existence and continuity of partial derivatives $\partial f/\partial\lambda_j (j = 1, \ldots, k)$. The output signal will be: $Y(t) = f(\lambda_1, \ldots, \lambda_k; t) + \nu(t)$, with $M\nu = 0$, $D\nu = \sigma^2$. Below we take $\lambda = \{\lambda_1, \ldots, \lambda_k\}$. Then (5.1) and (5.2) will not change, and the likelihood equation (5.3) is written as a system of equations:

$$\sum_{j=1}^{n}[y_j - f(\lambda; t_j)]\frac{\partial f(\lambda; t_j)}{\partial\lambda_i} = 0 \quad (i = 1, \Lambda, k) \tag{5.6}$$

In some cases the system (5.6) can be solved by elementary means. However, in a general case, the (5.6) is solved approximately. In this case the (5.5) is written as:

$$\lambda_i^{(m+1)} = \lambda_i^{(m)} + \sum_{j=1}^{n}\frac{\partial f(\lambda^{(m)}; t_j)}{\partial\lambda_i}[y_j - f(\lambda^{(m)}; t_j)] \Big/ \sum_{j=1}^{n}\left[\frac{\partial f(\lambda^{(m)}; t_j)}{\partial\lambda_i}\right]$$

where some values of the parameters λ are chosen as a first approximation $(m = 0)$. At $m \to \infty$, $\lambda_i^{(m)} \to \tilde{\lambda}_i$ $(i = 1, \ldots, n)$.

The above-considered version of additive noise is an approximation of the real situation of the radiometric systems' functioning. Therefore, we also study the version of multiplicative noise. Let us have a 1-beam channel with multiplicative noise $\alpha(t)$, whose probability density follows the law of Rayleigh (with respective normalization of the amplitude):

$$p_\alpha(x) = (x/\gamma^2)\exp[-0.5x^2/\gamma^2]$$

and with negligibly low Gauss additive noise ($\sigma^2 \approx 0$). Then the output signal will be $Y(t) = f_\lambda(t)\alpha(t)$.

Let $f_\lambda(t) > 0$. The probability density of the deviate $Y(t)$ will be:

$$p_Y(y) = \frac{y}{\gamma^2 f_\lambda^2(t)} \exp\left\{-\frac{1}{2}\left[\frac{y}{f_\lambda(t)\gamma}\right]^2\right\}$$

Then we form the likelihood function:

$$g(y_1, \ldots, y_n; \lambda) = \frac{1}{\gamma^{2n}} \prod_{i=1}^{n} \frac{y_i}{f_\lambda^2(t_i)} \exp\left\{\sum_{j=1}^{n}\left[-\frac{y_j^2}{2\gamma^2 f_\lambda^2(t_j)}\right]\right\}$$

The likelihood equation is written as:

$$L'(\lambda) = \sum_{j=1}^{n}\left\{\left[-\frac{y_j^2}{2\gamma^2 f_\lambda^2(t_j)} + \ln\frac{y_j}{\gamma^2 f_\lambda^2(t_j)}\right] \times \left(\frac{y_j^2}{\gamma^2 f_\lambda^2(t_j)} - 2\right)\frac{f_\lambda'(t_j)}{f_\lambda(t_j)}\right\} = 0$$

The denominator in (5.4) is estimated from the formula:

$$I(\lambda) = 4\sum_{k=1}^{n}\left[\frac{f_\lambda'(t_k)}{f_\lambda(t_k)}\right]^2$$

Hence, the formula for sequential approximations of the estimates of the parameter λ will be written as:

$$\lambda_{m+1} = \lambda_m + L'(\lambda_m)/I(\lambda_m)$$

5.3.3.2 The Kotelnikov integral

In the procedure of making statistical decisions an important stage is signal decoding (Borodin *et al.*, 1998). Consider the ensemble M of equi-probable signals represented in the time interval T with the time functions of similar output S and suppose that these signals are transmitted by channel with continuous time and additive Gauss white noise, with spectral density N_0 and channel capacity $c = S/N_0 \log_2 \mathrm{e}$ bits per second. Then the output signal can be decoded with the error probability described with the Kotelnikov integral (1956):

$$P_e = 1 - \int_{-\infty}^{\infty} \Phi'\left(x - \sqrt{2Q}\right)\Phi^{M-1}(x)\,dx \tag{5.7}$$

where:

$$\Phi(x) = (2\pi)^{-1/2}\int_{-\infty}^{\infty} \exp(-t^2/2)dt \qquad Q = cT$$

For the integral (5.7) we obtain an approximate formula which expresses itself through the normal distribution. With this goal in view, we obtain the lower and upper values of this integral. To obtain the lower value, we use the inequality:

$$\int_{-\infty}^{a} \Phi^{M-1}(x)\Phi'\left(x - \sqrt{2Q}\right)dx < \Phi^{M-1}(a)\Phi\left(a - \sqrt{2Q}\right) \tag{5.8}$$

which is valid for any value of a.

As seen from (5.7), the function $\Phi^{M-1}(x)$ aims for unity at $x \to \infty$ and aims for zero at $x \to -\infty$. With $M \gg 1$ in some proximity to the point $x = x_0$, the function $\Phi^{M-1}(x)$ suddenly passes from small values to values close to unity. Therefore, using the asymptotic formulas for the probability integral, we can write the following inequality:

$$1 - \frac{M-1}{\sqrt{2\pi}x}\exp\left[-\frac{x^2}{2}\right] < \Phi^{M-1}(x) < 1 - \frac{M-1}{\sqrt{2\pi}x}\left(1 - \frac{1}{x^2}\right)\exp\left[-\frac{x^2}{2}\right]$$

Introducing these relationships into (5.7) and (5.8), we obtain:

$$\begin{aligned} P_e > P_{e1} &= \Phi\left(x_0 - \sqrt{2Q}\right)[1 - \Phi^{M-1}(x_0)] + \frac{M-1}{\sqrt{2}}\exp\left(-\frac{Q}{2}\right) \\ &\times \left\{\left[1 - \Phi\left(x_0\sqrt{2} - \frac{\sqrt{2Q}-1}{\sqrt{2}}\right)\right]\exp\left[-\frac{1}{2}\left(\sqrt{2Q} - \frac{1}{2}\right)\right]\right. \\ &\left. - \frac{1}{x_0^3}\left[1 - \Phi\left(x_0\sqrt{2} - \sqrt{Q}\right)\right]\right\} \end{aligned}$$

To obtain the upper values, we use the following apparent inequality:

$$\Phi^{M-1}(x) > \begin{cases} 1 - (M-1)/(\sqrt{2\pi}x)\exp(-x^2/2) & x \geq x_0 \\ 2^{1-M} & 0 \leq \times < x_0 \\ 0 & x < 0 \end{cases}$$

Introducing this inequality into (5.7), we obtain:

$$\begin{aligned} P_e < P_{e2} &= (1 - 2^{1-M})\Phi\left(x_0 - \sqrt{2Q}\right) + \frac{M-1}{2\sqrt{\pi}x_0}\exp(-Q/2) \\ &\times \left[1 - \Phi(x_0\sqrt{2} - \sqrt{Q})\right] \end{aligned} \tag{5.9}$$

The formula (5.9) can be specified, if we assume:

$$\Phi^{M-1}(x) \approx 2^{2-M}\left(0.5 + (M-1)x/\sqrt{2\pi}\right) \quad 0 \leq x < x_0$$

Then:

$$\begin{aligned} P_e \approx P_{e3} &= \left(1 - 2^{1-M} - \frac{M-1}{2^{M-2}}\sqrt{\frac{Q}{\pi}}\right)\Phi\left(x_0 - \sqrt{2Q}\right) + \frac{M-1}{\pi 2^{M-1}} \\ &\times \left[\sqrt{\pi Q}\Phi\left(-\sqrt{2Q}\right) - e^{-Q} + \exp\left\{-0.5\left(x_0 - \sqrt{2Q}\right)^2\right\}\right] \\ &+ \frac{M-1}{2\sqrt{\pi}x_0}\exp(-Q/2)\left[1 - \Phi\left(x_0\sqrt{2} - \sqrt{Q}\right)\right] \end{aligned}$$

At $M \gg 1$ from (5.9) we have:

$$P_e < P_{e4} = \Phi\left(x_0 - \sqrt{2Q}\right) + \left(2\sqrt{\pi}x_0\right)^{-1}\exp(\ln M - Q/2) \times \left[1 - \Phi\left(x_0\sqrt{2} - \sqrt{Q}\right)\right]$$

In the formulas obtained above for approximate calculations of the integral (5.7), the parameter x_0 remained uncertain. Let the estimate (5.9) be at a minimum. Then for x_0 we obtain the following equation:

$$(1 - 2^{1-M})\Phi'\left(x_0 - \sqrt{2Q}\right) - \frac{M-1}{\sqrt{2\pi}x_0}\exp(-Q/2)\Phi'\left(x_0\sqrt{2} - \sqrt{Q}\right) - \frac{M-1}{2\sqrt{\pi}x_0^2}\exp(-Q/2)\left[1 - \Phi\left(x_0\sqrt{2} - \sqrt{Q}\right)\right] = 0 \quad (5.10)$$

For practical application, it is expedient to have more simplified formulas. For this purpose we use in (5.9) the following estimates:

$$\Phi\left(x_0 - \sqrt{2Q}\right) < \left[\sqrt{2\pi}\left(\sqrt{2Q} - x_0\right)\right]^{-1}\exp\left[-0.5\left(\sqrt{2Q} - x_0\right)^2\right] \quad Q > x_0^2/2$$

$$1 - \Phi\left(x_0\sqrt{2} - \sqrt{Q}\right) < \begin{cases} 1 & x_0^2 < Q/2 \\ \dfrac{\exp\left[-0.5\left(x_0\sqrt{2} - \sqrt{Q}\right)^2\right]}{\sqrt{2\pi}\left(x_0\sqrt{2} - \sqrt{Q}\right)} & x_0^2 \geq Q/2 \end{cases}$$

Introducing these expressions into (5.9) and putting forward the condition that the exponents in the resulting expression be equivalent, we obtain the following estimate for x_0:

$$x_0 \approx \tilde{x}_0 = \begin{cases} \sqrt{2\ln(M-1)} & 1/4 < u < 1 \\ \sqrt{2Q}\left[1 - \sqrt{1 - 2u}/\sqrt{2}\right] & 0 \leq u \leq 1/4 \end{cases}$$

where $u = \ln(M-1)/Q$.

Similarly, we evaluate the parameter P_e:

$$P_e < \tilde{P}_e = \begin{cases} \dfrac{1}{\sqrt{2\pi Q}}\exp\left[-Q\left(\dfrac{1}{2} - \dfrac{\ln M}{Q}\right)\right] & 0 \leq \dfrac{\ln M}{Q} \leq \dfrac{1}{4} \\ \dfrac{1}{2\sqrt{\pi Q}}\exp\left[-Q\left(1 - \sqrt{\dfrac{\ln M}{Q}}\right)^2\right] & \dfrac{1}{4} < \dfrac{\ln M}{Q} < 1 \end{cases}$$

Table 5.4 gives estimates of x_0 obtained from solutions of (5.10) and estimates of $\tilde{x}_0$. It is seen that $P_{ei}(x_0) < P_{ei}(\tilde{x}_0)$ and with an increase of M the difference between $P_{ei}(x_0)$ and $P_{ei}(\tilde{x}_0)$ decreases. As seen from Table 5.5, the values of P_{ei} at $M \gg 1$ and $Q > \ln M$ well approximate P_e.

Table 5.4. Exemplified calculation of the x_0 parameter by solving (5.10) and its comparison with the estimate from (5.11).

Q	$M = 32$		$M = 256$		$M = 512$	
	x_0	$\tilde{x}_0$	x_0	$\tilde{x}_0$	x_0	$\tilde{x}_0$
2	1.96	2.63	2.71	3.33	2.92	3.53
6	1.98	2.63	2.71	3.33	2.93	3.53
10	2.01	2.63	2.72	3.33	2.93	3.53
14	2.06	2.63	2.73	3.33	2.94	3.53
20	2.15	2.70	2.76	3.33	2.96	3.53

Table 5.5. Comparison of different estimates of probability of signal decoding.

Q	$M = 16$			$M = 256$		
	P_e	P_{ej}	$\bar{P}_e$	P_e	P_{ej}	$\bar{P}_e$
10	0.9×10^{-2}	0.2×10^{-1}	0.2×10^{-1}	0.6×10^{-1}	0.1	0.1
20	0.5×10^{-4}	0.1×10^{-3}	0.1×10^{-3}	0.7×10^{-3}	0.2×10^{-2}	0.1×10^{-2}
30	0.3×10^{-6}	0.8×10^{-6}	0.6×10^{-6}	0.5×10^{-5}	0.1×10^{-4}	0.9×10^{-5}

Table 5.6. Estimates of Kotelnikov's integral.

Q	$M = 16$	$M = 32$	$M = 256$
10	0.925×10^{-2}	0.164×10^{-1}	0.635×10^{-1}
20	0.557×10^{-4}	0.111×10^{-3}	0.716×10^{-3}
30	0.301×10^{-6}	0.616×10^{-6}	0.471×10^{-5}
40	0.208×10^{-8}	0.416×10^{-8}	0.333×10^{-7}
50	0.125×10^{-10}	0.251×10^{-10}	0.201×10^{-9}
60	0.771×10^{-13}	0.154×10^{-12}	0.123×10^{-12}
70	0.481×10^{-15}	0.962×10^{-15}	0.770×10^{-14}
80	0.303×10^{-17}	0.606×10^{-17}	0.485×10^{-16}
90	0.193×10^{-19}	0.385×10^{-19}	0.308×10^{-18}

It turns out that the function:

$$\bar{P}_e = \frac{1}{4}\sum_{i=1}^{4} P_{ei}$$

uniformly approximates P_e at any values of M and Q. Table 5.6 lists the values of the integral (5.7).

In conclusion, consider the case of a signal with unknown phase. In this case the error probability will be:

$$P_e = 1 - \int_0^\infty W_q'(x) W_0^{M-1}(x) dx$$

where:

$$W_q(x) = \int_0^x z \exp(-0.5[z^2 + q^2]) I_0(qz) dz \qquad q = \sqrt{2Q}$$

From this we obtain:

$$\begin{aligned} P_e &= 1 - \sum_{k=0}^{M-1} (-1)^k C_{M-1}^k \exp(-q^2/2) \int_0^\infty x I_0(qx) \exp[-x^2(k+1)/2] dx \\ &= 1 - \exp(-q^2/2) \sum_{k=0}^{M-1} (-1)^k C_{M-1}^k \int_0^\infty x \exp\{-x^2(k+1)/2\} \sum_{i=0}^\infty \frac{(qx/2)^{2i}}{(i!)^2} dx \\ &= 1 - \exp(-q^2/2) \sum_{k=0}^{M-1} \sum_{i=0}^\infty (-1)^k \frac{C_{M-1}^k}{i!(k+1)} \left[\frac{q^2}{2(k+1)}\right]^i \end{aligned}$$

Taking into account the apparent inequality:

$$\exp\left[\frac{q^2}{2(k+1)}\right] = \sum_{i=0}^\infty \frac{1}{i!} \left[\frac{q^2}{2(k+1)}\right]^i$$

we obtain finally:

$$P_e = \sum_{k=1}^{M-1} (-1)^{k+1} C_{M-1}^k \frac{1}{k+1} \exp\left[-\frac{q^2}{2} \frac{k}{k+1}\right]$$

5.4 TECHNOLOGY SEARCHING FOR ANOMALIES IN THE ENVIRONMENT

An approaching of the moment of origin of a natural disaster is characterized by placing of the vector $\{x_i\}$ into some cluster of multi-dimensional phase space X_c. In other words, proceeding to the quantitative determination of this process, we introduce a generalized characteristic $I(t)$ of a natural disaster and identify it with the graduated scale Ξ, for which we postulate the presence of relationships of the type $\Xi_1 < \Xi_2$, $\Xi_1 > \Xi_2$ or $\Xi_1 \equiv \Xi_2$. It means that $I(t) = \rho$ has always a value which determines the level of the proximity of the origin of a natural disaster of a given type: $\Xi \to \rho = f(\Xi)$, where f is some transformation of the notion of "natural disaster" into a number. As a result, the value $\theta = |I(t) - \rho|$ determines the expected time interval before the catastrophe.

Let us try to find a satisfactory model to turn the verbal portrait of "natural disaster" into notions and indicators subject to a formalized description and transformation. With this goal in view, we select m elements–subsystems of the lowest

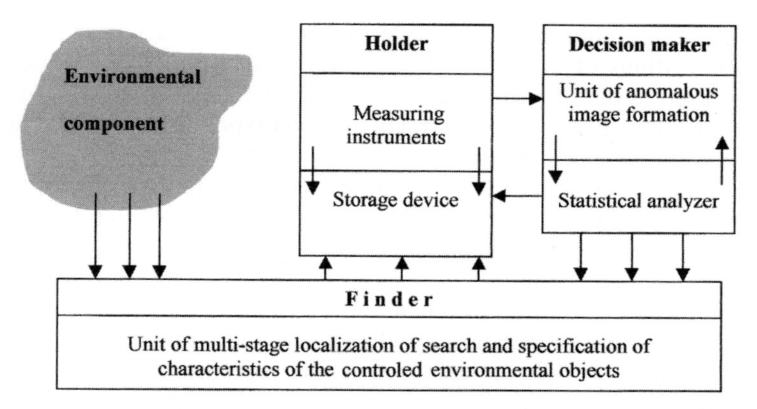

Figure 5.1. The structural scheme of the monitoring system for search and detection of natural anomalies (Krapivin and Mkrtchyan, 2002; Bondur *et al.*, 2005; Kondratyev and Krapivin, 2004).

level in the system $N \cup H$, the interaction between which we determine with the matrix function $A = \|a_{ij}\|$, where a_{ij} is the indicator of the level of dependence of relationships between subsystems i and j. Then the characteristic $I(t)$ can be determined as the sum:

$$I(t) = \sum_{i=1}^{m} \sum_{j>i}^{m} a_{ij}$$

It is clear that in a general case $I = I(\varphi, \lambda, t)$. For a limited territory Ω with the area σ the indicator I is determined as an average:

$$I_\Omega(t) = (1/\sigma) \int_{(\varphi,\lambda)\in\Omega} I(\varphi, \lambda, t) d\varphi d\lambda$$

Introducing the characteristic I_Ω enables one to propose the following scheme of monitoring and prediction of natural disasters. Figure 5.1 demonstrates a possible structure of the monitoring system with functions of search, forecast, and tracing of natural disasters. The system has three levels: fixing, decision making, and searching. The respective units have the following functions:

(1) Regular control of environmental elements aimed at collection of data on their state in the regime allowed by the applied technical means.
(2) Fixation of suspicious environmental elements, the value of their indicator $I_\Omega(t)$ corresponds to the interval of the danger of a given type of natural anomaly.
(3) Formation of dynamic series $\{I_\Omega(t)\}$ for a suspicious element to make a statistical decision on its noise and signal character, and in the latter case, checking of the suspicious element by criteria of the next level of accuracy (including of the vector $\{x_i\}$ into the cluster, etc.).
(4) Making the final decision on the approach of the moment of origin of natural disasters with delivery of information to the respective services of the environmental control.

The efficiency of this monitoring system depends on the parameters of the measuring means and algorithms of the observational data processing. Here the model of the environment used in parallel with the formation and statistical analysis of the series $I_\Omega(t)$ and adapted to the monitoring regime in correspondence with the scheme in Figure 5.2 plays an important role.

From the above-introduced criterion of an approaching natural catastrophe it is seen that the form and behavior of $I_\Omega(t)$ are characteristic of each type of process taking place in the environment. One of the complicated problems consists in determination of these forms and their respective classification. For instance, often occurring dangerous natural phenomena such as landslides and mudflows have characteristic indicators of preliminary changes of relief and landscape which are recorded from satellites in the optical interval and, together with the data of aircraft survey and ground measurements of relief slopes, expositions of slopes and state of the hydro-network, make it possible to predict them several days before their realization. However, limited possibilities of the optical region under cloud and vegetation cover conditions should be broadened by introducing the systems of remote sensing in the microwave part of the electromagnetic spectrum. Then, in addition to indicators of landslides and mudflows, one can add the informative parameters such as soil moisture and biomass. An increase of soil moisture is known to lead to landslides and an increase of biomass indicates the strengthening of the restraining role of vegetation cover against the moving of rocks. It is especially important in the control of snow–rock or simply snow avalanches. An accumulation of the catalogue of such indicators for all possible natural disasters, and introducing it into the knowledge base of the monitoring system, is a necessary stage for raising its efficiency.

Knowledge of the totality of informative indicators $\{x_i^j\}$ of a natural disaster of the jth type and a priori determination of its cluster X^j in the space of these indicators makes it possible in the process of satellite monitoring to calculate the velocity v_j of the approaching of the point $\{x_i^j\}$ to the center X^j, and thus to calculate the time of catastrophe occurrence. Other algorithms of prediction of natural disasters are possible, too. For instance, a forest fire can be predicted using the dependence of the thermal emission of forests at different wavelengths on the moisture content of forest combustible materials arranged, as a rule, in layers. Knowledge of this dependence gives a real possibility for evaluating the risk of forest fire with due regard to the moisture content of the vegetation cover and upper soil layer (Kondratyev and Grigoryev, 2004).

Studies of numerous scientists have shown that there is a real possibility to evaluate the fire risk of water-logged forests taking into account the moisture content of the vegetation cover and upper soil layer, using microwave sensing in the range 0.8–30 cm. The multi-channel sensing makes it possible, using algorithms of cluster analysis, to solve the problem of forest classification by fire risk category. The efficiency of these methods depends on a detailed description in the model of the forest structure which reflects the state of the canopy and density of trees. The most dangerous and difficult-to-detect fires are ground fires. In this case the 3-layer model of the soil–trunk–canopy system coordinated with the fire risk indicator

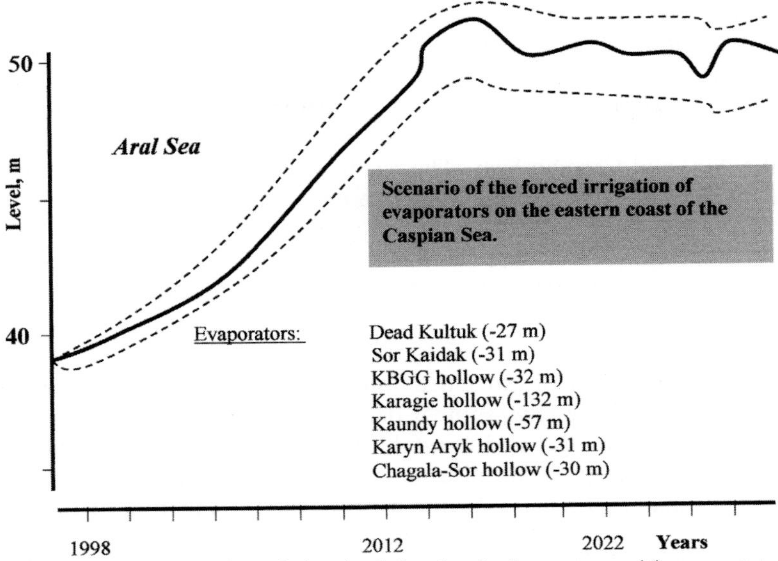

Figure 5.2. Possible dynamics of the Aral Sea levels (in meters with respect to the World Ocean level) as a result of the impact of the influence on the hydrological regime of the territory of the Aral–Caspian ecosystem starting in 1998 (Kondratyev *et al.*, 2002).

$I(\lambda_1, \lambda_2) = [T_b(\lambda_1) - T_b(\lambda_2)]/[T_b(\lambda_1) + T_b(\lambda_2)]$ is efficient. For instance, with $\lambda_1 = 0.8$ cm and $\lambda_2 = 3.2$ cm the indicator I changes approximately from -0.25 in the zone without danger of catching fire to 0.54 in the zone of fire. In the region of the appearance of the first indicators of litter catching fire, $I \approx 0.23$. The parameter I depends weakly on the distribution of the layers of the forest combustible materials, such as lichen, moss, grass rags, dead pine-needles and leaves.

Realization of the indicated 3-level regime of decision making on the approaching natural disaster depends on the coordination of spatial and temporal scales of the monitoring system with characteristics of a natural phenomenon. Most efficient for decision making are delayed-action natural disasters which can happen after decades have passed. Such expected catastrophes include the ozone holes, global warming, desertification, biodiversity reduction, the Earth's overpopulation, etc. The base problem of reliable prognostic estimates of the occurrence of such catastrophes, or regional-scale undesirable natural phenomena initiated by their approach, can be solved using the GMNSS whose input data are information from renewed data bases and the current satellite and ground measurements.

Use of the GMNSS in some studies has shown that this technology makes it possible not only to forecast the delayed-action catastrophes but also to produce scenarios for their prevention. An illustration is the scenario of reconstruction of the water regime of the Aral–Caspiy system (Krapivin and Phillips, 2001). Figure 5.2 shows the final result of the use of the GMNSS to solve this problem. It is seen that an easily realized procedure of irrigation of some lowlands on the eastern coast of the Caspian Sea, without subsequent anthropogenic forcing, can abruptly change the hydrology of the territory between the Aral and Caspian Seas. Of course, this result is only a demonstration of the ability of the GMNSS to evaluate the consequences of the realization of scenarios of the impact on the environment. In this connection, numerous problems regarding the organization of studies arise which can be solved within the complex scientific–technical program of monitoring of the zone of the impact of the Aral and Caspian Seas. The results of analysis of natural disasters in the zone of the Aral Sea are discussed in more detail in Chapter 7.

5.5 TAXONOMIC CAPABILITIES OF NATURAL DISASTER MONITORING SYSTEMS

The efficiency of the system of natural monitoring with detection and tracing of catastrophic processes in the environment depends on the ability of these systems to determine the level of danger and to establish the character of a natural disaster. The most important category of disasters is connected with tropical cyclones (Table 5.7) and tsunamis. Figure 5.3 demonstrates the propagation of the tsunami waves reconstructed on the basis of historical data and space-borne observations. A lot of experience has been gained in the use of satellite images to monitor different catastrophes. At the 25th International Asian Conference on Remote Sensing held in November 2004 in Thailand, a special time slot was organized during which the results of satellite monitoring of natural disasters was discussed.

Ahmad *et al.* (2004) have shown an efficiency of microwave measurements of environmental parameters which enable one to map the development of floods with some regions of Malaysia as an example. Floods in this country are regular, and their occurrence is connected with the seasons of monsoon winds. The eastern regions of Malaysia are subject to floods in the season of the north-western monsoons (November–January), and floods on the western coast happen in March–May and September–October. The temporal regularity and territorial referencing of floods in Malaysia make it possible to plan satellite monitoring with the application of measurements made with SAR carried by the satellite Radarsat (1995). Microwave sensors to monitor floods in the tropical latitudes are more advantageous than the optical systems because they can penetrate dense clouds.

The Indian satellite monitoring service has achieved progress in flood monitoring (Roy, 2004). One of the problems in India is regular floods and landslides in the basin of the Ganges and Brahmaputra Rivers. The watershed territory of these rivers covers about 110 mln km^2 (43% of all river basins and 60% of all water resources in India). The launching of five Indian satellites IRS (Indian Remote Sensing Satellite) has made it possible to create a national system of control of dangerous natural phenomena – the Disaster Management System (DMS) – which provides regular information about floods, landslides, soil erosion, and forest degradation. Of course, more complicated problems regarding warning about these natural phenomena arise here, for which efficient methods of space-borne measurement processing are required. Vansarochana (2004), taking as an example the landslide zone in Thailand located between the mountains of Doi Inthahon (2,590 m) and Doi Pui

Table 5.7. List and characteristics of tropical cyclones in the Atlantic sector following the chronology of their origin in 2003 (Phelan, 2004).

Cyclone	Lifetime of the cyclone	Maximum wind speed (km h^{-1})
1. Tropical Hurricane Ana	21–24 April	65
2. Tropical depression	11–12 June	55
3. Tropical Hurricane Bill	25 June–1 July	90
4. Hurricane Claudette	8–16 July	145
5. Hurricane Danny	16–20 July	120
6. Tropical depression	19–21 July	55
7. Tropical depression	25–27 July	55
8. Hurricane Erika	14–17 August	120
9. Tropical depression	21–22 August	55
10. Hurricane Fabian	27 August–8 September	235
11. Tropical Hurricane Grace	30–31 August	65
12. Tropical Hurricane Henri	3–8 September	70
13. Hurricane Isabel	6–19 September	260
14. Tropical depression	8–10 September	55
15. Hurricane Juan	25–29 September	170
16. Hurricane Kate	25 September–7 October	200
17. Tropical Hurricane Larry	2–6 October	90
18. Tropical Hurricane Mindy	10–13 October	70
19. Tropical Hurricane Nicholas	13–23 October	115
20. Tropical Hurricane Odette	4–7 December	95
21. Tropical Hurricane Peter	9–11 December	115

Luang (1,496 m), has demonstrated the possibility to warn about a landslide occurrence using the spatial relief modeling based on calculations of three characteristics: (1) the rate of separation of two control points on the land surface; (2) the change of the slope of this surface, and (3) the spatial correlation reflecting changes in the structure of ground or its oscillations.

Yano and Yamazaki (2004) proposed a discrete scale with five gradations for identification of the levels of building destruction resulting from an earthquake. The first level corresponds to insignificant damage without structural destruction of the building (cracks in some walls, falling of small pieces of plaster and structures from the upper parts of the walls). The second level adds moderate destruction of the buildings (numerous cracks in the walls, falling of plaster over large areas and partial destruction of chimneys). The third level includes large and numerous cracks in the walls, roof damage, fractures in the foundations, and destruction of non-structural annexes and other outhouse elements. Serious structural violations to the building and other large-scale destruction is identified as the fourth level on the scale. Finally the fifth level represents the situation of the total destruction of buildings. Applying this classification to the case of the earthquake that happened on 26 December 2003 in the south-east of Iran (magnitude 6.5) and led to serious losses (26,271 people died, 30,000 people wounded, and 75,600 people lost their homes). Images of the

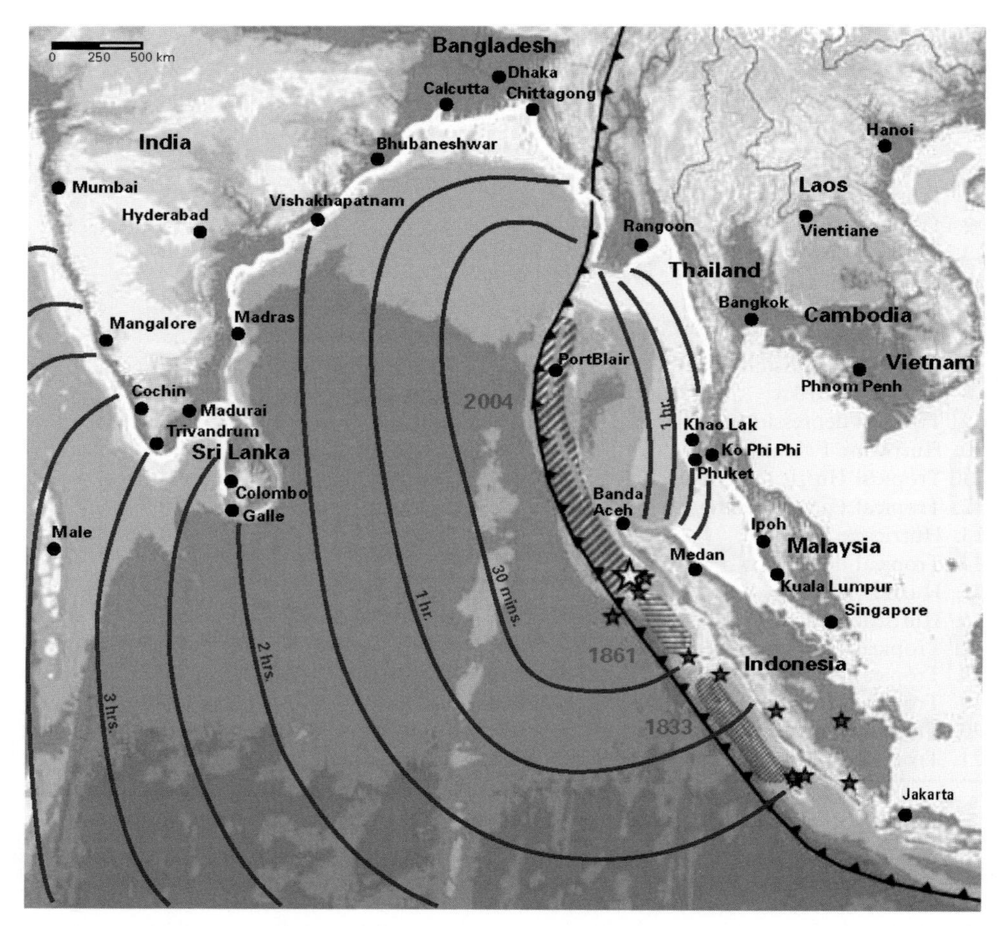

Figure 5.3. Trajectories of the tsunami wave motion due to the largest earthquakes (1833, 1861, and 2004). The epicenter of the quake on 26 December 2004 is marked by the large star. Small stars represent quakes that have occurred since 1973 with magnitudes of 7.0 or above. The ribbed line corresponds to the plate boundary. Shaded areas are rupture surfaces (Munich Re, 2005a).

outskirts of the most damaged city of Bam obtained from satellites Quick Bird (30 September 2003 and 3 January 2004) and IKONOS (27 December 2003) have demonstrated an efficiency of the introduced scale for the operational estimation of the earthquake consequences.

Desertification and droughts are of serious concern for many global regions. The latter bring large economic losses, and therefore their monitoring with the possibility of a warning is urgent. A drought is known to form under the influence of factors such as temperature, rain rate and rain duration, surface wind strength, and soil properties. Especially important is the combination of temperature and rain rate. Based on the control of this combination, Zhu Xiaoxiang *et al.* (2004) proposed a model of drought based on the calculations of two indices from temperature and the Normalized Differential Vegetation Index (NDVI) measured with AVHRR NOAA/NASA. The integral indicator of an arid territory is calculated from the formula $DI = r_1 D_1 + r_2 D_2$, where r_1 and r_2 are weighting coefficients reflecting the impact of vegetation and temperature on the development of the process of drying-up of the territory; $D_1 = 100 \times (\mathrm{NDVI} - \mathrm{NDVI}_{\min})/\ (\mathrm{NDVI}_{\max} - \mathrm{NDVI}_{\min})$; $D_2 = 100 \times (T_{b,\max} - T_b)/(T_{b,\max} - T_{b,\min})$; T_b is brightness temperature. It is shown that the dryness index DI identifies well the states of the territory: $DI \geq 40$ – no drought, $20 \leq DI < 40$ – weak drought, $5 \leq DI < 20$ – moderate drought, and at $DI < 5$ – severe drought. Mapping of the territory of China with the use of this scale has shown that the space-derived maps provided the possibility of operatively detecting dangerous regions and planning measures for the reduction of damage from drought.

The study of Lim *et al.* (2004) is dedicated to the development of the method to control zones with anomalous temperatures. Based on measurements with the use of the spectrometer MODIS carried by the satellites Terra and Aqua they calculated the fire-risk indicator. The space-recorded radiation is the sum of atmospheric radiation and emission of the Earth's surface. Individual components of the recorded signal contain re-reflected solar radiation. Following the traditional method of division of all components of the MODIS signal, it is easy to calculate the level of temperature in each pixel of the Earth's surface and to map the temperature. Such maps are drawn every week for the whole USA territory with indication of five levels of dryness and the degree of the impact of increased temperature on the state of water systems, crop harvests, pastures, and hayfields. Figures 5.4–5.6 demonstrate images from the satellite SPOT, from which one can determine the contours of water systems and, with regular receipt of such images, trace the dynamics of the development of various events connected with the behavior of water systems.

One of the principal annual concerns for nature-protection services in many countries is the problem of prediction of forest fires and assessment of conditions of their propagation. Clearly, to solve relevant problems and to reduce the forest fire risk, it is necessary to develop technologies for preventive detection of the regions with an increased fire risk. The scale of forest fire depends on several factors such as water content in plants, topography, and wind direction. The moisture supply in plants is the most substantial factor of the rate of forest fire generation. Maki *et al.* (2004) studied the relationship of various indices and established, under laboratory conditions, that Vegetation Dryness Index (VDI), Water Deficit Index (WDI), and Normalized Difference Water Index (NDWI) can be successfully used in solving the problems regarding warning about possible forest fire.

To assess the water content in plants, indicators are used such as fuel moisture content (FMC) and equivalent water thickness (EWT):

$$\mathrm{FMC}/100\% = (\mathrm{FW} - \mathrm{DW})/\mathrm{FW}, \mathrm{EWT}(\mathrm{g\,cm^{-2}}) = (\mathrm{FW} - \mathrm{DW})/\sigma$$

where σ is the leaf area in the canopy (cm^{-2}), FW is the wet weight of plants (g), and DW is the dry weight of plants (g).

NDWI, WDI, and VDI are calculated from the following formulas:

Figure 5.4. Photo of the delta of the Rhine River (France) taken on 4 April from the satellite SPOT-4 with a resolution of 6 × 6 km.

$$\mathrm{NDWI} = (T_{b,NIR} - T_{b,SNIR})/(T_{b,NIR} + T_{b,SNIR})$$

$$\mathrm{WDI} = 1 - \Delta_1/\Delta_2$$

$$\mathrm{VDI} = 1 - \Delta_3/\Delta_4$$

where $T_{b,NIR}$ is brightness temperature in the near-IR region, $T_{b,SNIR}$ is brightness temperature in the short-wave IR region, Δ_1 is the distance on the phase plane (NDWI, NDVI) between the states of deficit (B) and excess (C) of water under conditions of thin vegetation, Δ_2 is the distance between the states of thick (A) and thin (B) vegetation in case of deficit of water, Δ_3 and Δ_4 are distances between measured levels of NDVI and NDWI and between the sides of the

Figure 5.5. Photo of the dam on the Paraná River in Argentina taken from the satellite SPOT on 4 December 1993.

Figure 5.6. Change of the Jamura tributary bed of the Brahmaputra River in Bangladesh recorded by the satellite SPOT during May 1987 to March 1989.

Table 5.8. Estimates of coefficients in relationships (5.11) (Maki *et al.*, 2004).

Type of plants	a_{EWT}	b_{EWT}	R_{EWT}	a_{FMC}	b_{FMC}	R_{FMC}
Nerium oleander var.indicum	−0.106 97	0.622 68	0.82	155.89	−34.087	0.06
Betula platyphylla var.japonica	−0.031 36	0.333 9	0.88	135.59	212.72	0.26
Liriodendron tulipifera	−0.009 507 3	0.175 21	0.87	68.312	539.65	0.56
Other types	−0.024 424	0.279 63	0.79	119.42	204.69	0.28

parallelogram with vertexes A, B, C, and D (thick vegetation and excess of water) of the plane (NDWI, NDVI) at NDVI = const., respectively. The NDWI is calculated based on measurements from the satellite SPOT in four channels: 430–470 nm, 610–680 nm, 780–890 nm, and 1580–1750 nm. For NDVI evaluation, the readings of sensors carried by satellites of the Landsat type measuring in the red and near-IR regions are usually used (Fong and Liang, 2003).

Maki *et al.* (2004) proposed a linear approximation:

$$EWT = a_{EWT} + b_{EWT}NDWI, FMC = a_{FMC} + b_{FMC}NDWI \quad (5.11)$$

with respective coefficients for some types of plants listed in Table 5.8.

6

Prediction of natural disasters

6.1 FORMALIZED DESCRIPTION AND PARAMETERIZATION OF NATURAL DISASTERS

A natural disaster is one of the manifestations of the environmental "strategy" which provides the global ecodynamics formation (Syvorotkin, 2002; Kondratyev *et al.*, 2004b). Therefore, to characterize and study natural disasters, the nature–society system (NSS) should be considered a single process of atmospheric and geospheric–biospheric phenomena. To evaluate atmospheric phenomena, constructive approaches have been developed in meteorology based on the combined use of the systems of data collection and analysis. In particular, a method has been developed based on the synthesis of synoptic maps for different altitudes. A first map was drawn for 1686 (Woo, 1999). These maps include data on cloudiness (density, altitude, and type), atmospheric pressure, wind speed and direction, temperature, amount of liquid and solid precipitation, air humidity and transparency, as well as many other atmospheric phenomena (thunderstorms, fog, haze, etc.). Such maps, for instance, for the USA, are drawn every hour. In other countries they are drawn not so often. At any rate, most countries use the World Meteorological Organization (WMO) standards, which make it possible to use synoptic maps in any regions of the globe. With the coordinates of the meteorological stations known, the scale and boundaries of air masses (atmospheric fronts) can be easily estimated, and then, comparing near-ground and altitudinal synoptic maps, one can forecast an appearance of emergency situations. Many problems arise at this stage, however, which are the subject of synoptic meteorology. These problems are connected with the fact that the development of atmospheric processes is chaotic, with special spatial–temporal characteristics not always well parameterized.

One of the sections of global meteorology is weather forecasting made from data on the global structure of the atmosphere. These data, reflecting the horizontal and vertical development and motions of synoptic objects, are used as input information for mathematical models directed at the description of the laws of atmospheric motion. The range of forecasts and their accuracy depend on the quality of observations, applied mathematical apparatus, computer capability, and the scale of predicted meteorological phenomenon. The forecast accuracy is also determined by special features of the atmosphere which cannot be described only with the use of thorough observations and more accurate models. It is a question of the non-stationary character of many atmospheric processes, to parameterize which, new technologies of modeling should be developed (Krapivin and Potapov, 2002).

The forecast of different atmospheric phenomena on different spatial–temporal scales requires an estimation of the level of their chaotic nature, and depending on that, a selection of the respective model. For instance, the diurnal forecast of atmospheric pressure is sufficiently accurate, but unfavorable meteorological phenomena, such as squalls, thunderstorms, and tornados are difficult to predict even several hours before their occurrence. The existing satellite technologies make it possible to fix them with a high accuracy and in good time, which enables one to extrapolate their motion during one to two hours.

The turbulent nature of the atmosphere strongly limits the ability of geoinformation systems. In making long-range forecasts the scientists face the necessity to use global databases, the formation of which have recently become a task of first priority (Barnett, 2003). One of the important elements of the global database is sea surface temperature (SST) whose variability is characterized by time intervals from several weeks to months, which provides the possibility of a stable detection of regions with anomalous temperatures and rain rate. An important task of global modeling is simulation of a totality of correlated units of feedbacks in the system of biogeochemical cycles (Kondratyev and Krapivin, 2004a,b).

One of the forms of parameterization of natural disasters is proposed by the chaos theory, which makes it possible to formalize the complexity of interaction of natural and natural–technogenic systems. Numerous models are aimed at consideration of a limited collection of processes with the use of simplified schemes of interaction of the environmental elements. Wright and Erickson (2003) analyzed the efficiency of such models as the Integrated Assessment (IA), the Carbon Emissions Trajectory Assessment (CETA), and the Model for Evaluating Trajectory of Regional and Global Effects (MERGE). These models are directed at searching the optimal levels of carbon emissions balanced with economic indicators of the region. The shortcoming of such models consists in that the ecological component in them is presented at the level of scenario, which makes them primitive, and doubts can be raised about the respective conclusions. In particular, an optimization of carbon emission taxes and relevant Intergovernmental Panel on Climate Change (IPCC) reports contain rather contradictory and ambiguous recommendations, which has been discussed in detail by Kondratyev (2002a,c, 2004a). Nevertheless, theoretical models of chaos can help from a methodological viewpoint in the synthesis and analysis of more complicated models of natural disasters (Hide *et al.*, 2004).

Finally, note an important and practically non-formalized phenomenon directly connected with an occurrence of many types of natural disasters – abyssal degassing of the Earth and its impact on the biogeochemical cycles of greenhouse gases.

Syvorotkin (2002) proposed a degassing model which describes, on the methodological level, mechanisms of the impact of processes of abyssal degassing on the environment, causing natural disasters both indirectly through a change in the ozone layer and directly through volcanic activity and earthquakes as well as through direct impact of emissions of abyssal gases on the habitats of humans and animals. An example of direct impact is the case of the spontaneous emissions of gases (with a prevailing content of CO_2) from Nios Lake located in the crater of Cameroon Volcano on 21 August 1986, which led to the deaths of 1,700 people and 1,000 cattle in adjacent areas.

Fluid fluxes from the Earth's core form methane and carbon monoxide in the mantle which, entering into the water media due to a sharp decrease of temperature, take part in the reactions: $2CO = C + CO_2$ and $4CO + CH_4 = 4C + CO_2 + H_2O$. Carbon deposits and CO_2 enter the atmosphere or partially dissolves in water (Marakushev, 1995). Along with this, due to the bottom hydrothermal load or direct release of abyssal gases, helium and hydrogen enter the atmosphere. The hydrogen mechanism of ozone destruction by the reaction formula $H_2 + O_3 = H_2O + O_2$ activates different feedbacks leading to drastic changes in the habitat. Under conditions of mass degassing this mechanism can transfer all the supplies of ozone into water. Of course, a rapid realization of this scenario is impossible because of the huge supplies of ozone and oxygen. According to Syvorotkin (2002), under certain conditions, the World Ocean level could rise by 20 m over 1 million years and by 100–200 m over 5–6 million years.

The process of abyssal degassing is regulated by gravitational impacts between the Sun–Earth–Moon system. This is confirmed by a correlation of earthquakes with the period of lunar–solar tides. The synchronous nature of the processes in the lithosphere and hydrosphere with the magnetic field fluctuations generated in the liquid core governs the fluid fluxes which promote a heating and decompressing of the mantle manifested through tectomagnetic processes that control life within the system "planet Earth" (SPE). An increase or decrease of the rate of abyssal degassing depends on many cosmic processes and, hence, it is outer space that governs the life of the NSS as an element of the SPE.

The final goal of natural disaster forecasts is to reduce their damage. Therefore, the interest of specialists in different scientific spheres, in the problem of predictability of anomalous natural events, is increasing. Huge economic and human losses occur with unexpected natural disaster phenomenon. For instance, in the early morning of 15 January 1995 the strongest earthquake recorded in Japan occurred at Port Kobe. More than 5,500 people perished under the ruins. Though the earthquake lasted for only 14 seconds, the economic losses exceeded US$100 bln. If the earthquake had been predicted, this damage could have been far less. In contrast to earthquakes, floods and hurricanes can be predicted several days before their occurrence. But practically all types of natural disasters can be attributed to an accidental event with an unknown law of probability distribution, and the forecast of such events by traditional deterministic or statistical methods cannot provide acceptable reliability. Among the perspective approaches to analysis of available data series on emergency events is the method of evolutionary modeling (Bukatova *et al.*, 1991).

Zimmerli (2003) compares the natural disaster forecast with the game of dice, the model of which can form a strategy of insurance against natural disasters. This model makes it possible to describe the risks of a natural disaster. The model input parameters contain four groups of data analyzed by the respective model unit:

- *The danger level*. How often and with what intensity do emergency events occur?
- *Vulnerability*. What is the damage at a given intensity of the event?
- *Cost distribution*. Where are various types of insurance objects and what is their cost?
- *Insurance conditions*. What share of the damage is insured?

6.2 GLOBAL MODEL OF THE NSS AS AN INSTRUMENT FOR NATURAL DISASTER PREDICTION

In connection with the different aspects of environmental change taking place during the last decades, many experts put forward numerous conceptions of the NSS global description, and models of various complexity have been developed to parameterize the dynamics of the characteristics of the biosphere and climate. The availability of a large database for these characteristics enables one to consider and estimate the consequences of a possible realization of different scenarios of the development of the subsystem H. Traditional approaches to the synthesis of global models are based on consideration of a totality of balance equations, which include parameters $\{x_i\}$ in the form of functions, arguments, coefficients, and conditions of transition between parametric descriptions of processes taking place in the environment. Also, other approaches are applied, which use evolutionary and neuronet algorithms. Organization of the global model of the $N \cup H$ system functioning can be presented as a conceptual scheme in Figure 6.1. This scheme is realized by introducing the geographical grid $\{\varphi_i, \lambda_j\}$ with steps of sampling of land surface and the World Ocean of $\Delta\varphi_i$ and $\Delta\lambda_j$ in latitude and longitude, respectively, so that within a pixel $\Omega_{ij} = \{(\varphi, \lambda): \varphi_i \leq \varphi \leq \varphi_i + \Delta\varphi_i, \lambda_j \leq \lambda \leq \lambda_j + \Delta\lambda_j\}$ all processes and elements of the NSS are considered as homogeneous and parameterized by point models. The choice of the size of pixels is determined by several conditions governed by the spatial resolution of satellite measurements and the available global database. In the case of water surface, in the pixel Ω_{ij} the water mass is divided into layers by depth z (i.e., 3-D volumes $\Omega_{ijk} = \{(\varphi, \lambda, z): (\varphi, \lambda) \in \Omega_{ij}, z_k \leq z \leq z_k + \Delta z_\mathrm{k}$ are selected within which all elements are distributed uniformly). Finally, the atmosphere over the pixel Ω_{ij} is digitized by altitude h either by atmospheric pressure levels or by characteristics of the layers of altitude Δh_s.

It is clear that the development of a global model is only possible using knowledge and data on the international level. Among numerous models, most adequate is the model described in Kondratyev *et al.* (2004a). The block-scheme of this model is shown in Figure 6.2. The synthesis of a model of such scale requires consideration of the existing models of various partial processes from climatology, ecology, hydrology, geomorphology, etc. (Wainwright and Mulligan 2003).

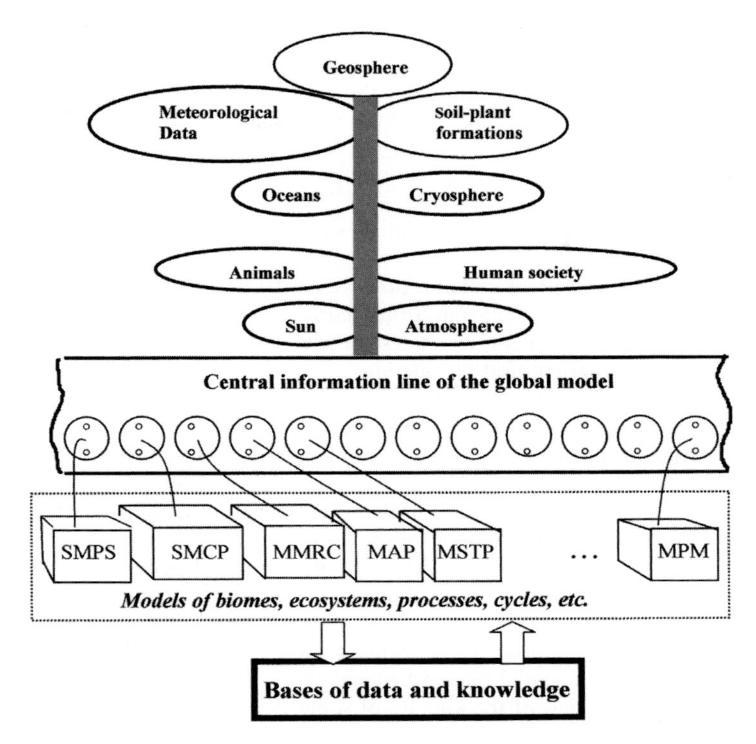

Figure 6.1. The information–functional structure of the global NSS model. Abbreviated terms are detailed in Table 6.1.

An adaptive procedure of introducing the global model into the system of geoinformation monitoring has been proposed in Kondratyev *et al.* (2003b). This procedure is schematically shown in Figure 6.2.

Use of the global model to assess the consequences of climate warming over Europe is exemplified by the regional high-resolution model developed at the Denmark Meteorological Institute to predict the level of floods. Based on the consideration of scenarios of greenhouse gas (GHG) emissions, calculations of precipitation were made for the periods 1961–1990 and 2071–2100. It was shown that though the summer precipitation will decrease over most of Europe, on the whole, their future general level will exceed the present estimates by 95–99% (Christensen and Christensen, 2004). To solve a similar problem, Semmler and Jacob(2004) used the regional climate model REMO 5.1. To describe the boundaries of the area studied with a spatial resolution of 0.5°, the climate model HadAM3H was used. The SST and sea-ice distribution was described with the use of the HadAM3H1 observational data. The modeling results have shown that compared with 1960–1990 the 2070–2100 total precipitation will increase by 50%, and in the region of the Baltic Sea – by 100%.

Development of a universal comprehensive global model, which could combine the key processes in the environment into a single description, requires the

Table 6.1. Characteristic of the global model of the nature–society system (GMNSS) units.

Identifier for the units in Figures 6.1 and 6.2	Characteristic of the unit's functions
SMPS	A set of models of the population size dynamics with regard for the age structure.
SMCP	A set of models of climatic processes with differently detailed consideration given to the parameters and their correlations.
MMRC	Model of the mineral resources control.
MAP	Model of agricultural production.
MSTP	Model of scientific–technical progress.
CGMU	Control of the global model units and database interface.
AGM	Adjustment of the global model to simulation experiment conditions and its control.
PSR	Preparation of simulation results to visualization or other forms of account.
MBWB	Model of the biospheric water balance.
MGBC	Model of the global biogeochemical cycle of carbon dioxide.
MGBS	Model of the global biogeochemical cycle of sulfur compounds.
MGBO	Model of the global biogeochemical cycle of oxygen and ozone.
MGBN	Model of the global biogeochemical cycle of nitrogen.
MGBP	Model of the global biogeochemical cycle of phosphorus.
SMKP	A set of models of the kinetics of some types of pollutants in different media.
SMWE	A set of models of water ecosystems in different climatic zones.
MHP	Model of hydrodynamic processes.
SMSF	A set of models of soil–plant formations.
MPM	Model of processes in the magnetosphere.

development of new technologies of global modeling and monitoring. As Syvorotkin (2002) justly noted, the joint efforts of specialists in different sciences are needed. Syvorotkin (2002) was the first among geologists to propose the project of the cybernetic model, formulating a fundamental scheme of the interdisciplinary co-operation in the Earth system study. An important step forward is the broadening of the notion of "environment". The cybernetic model is considered a highly-organized open system which continuously interacts with the circumsolar space. In the terms considered here, the cybernetic model performs the (V, W)-exchange with space, reaching a simulation of states by taking the required energy from space. As one of the strategies, the cybernetic model uses the structuring of the planet by division into planetary fractures and tectonic zones and changes the range of altitudinal separations (at present 20 km). From this point of view, stable atmospheric and oceanic planetary fluxes generated by solar energy are mechanisms for the distribution of transformed external energy, which determines the character of the physico-geographic and sedimentation processes in the cybernetic model. One of the cybernetic model strategy components is also the scientific–technical progress which has

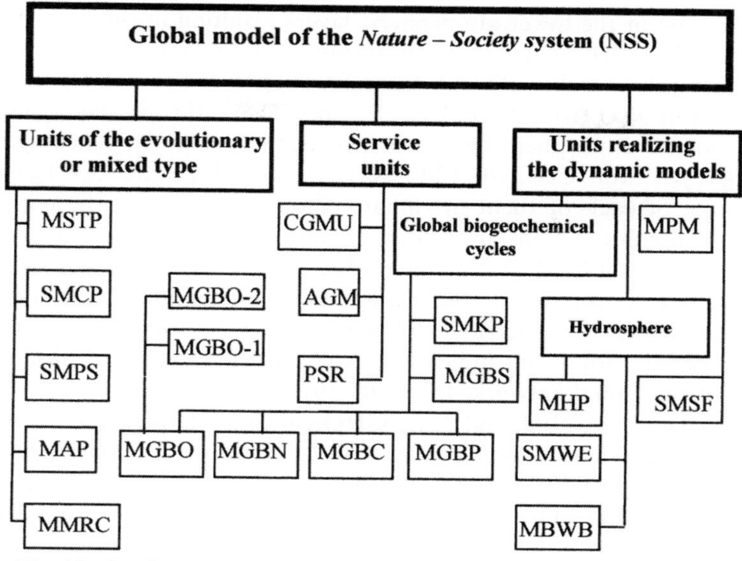

Figure 6.2. The block-scheme of the GMNSS. Abbreviated terms are detailed in Table 6.1.

led humankind to mastery of the energy source, in principle, similar to the external source – the thermonuclear energy of the Sun. Syvorotkin (2002) calls it an epic step of the cybernetic model "to achieve an unachievable goal – independence from the environment".

Note that the NSS, being a part of the Earth system, can be considered a closed object of (V, W)-exchange with space as well as part of the Earth system which includes the core and the mantle as sources of planetary energy formed due to the process of gravitational differentiation and radioactive decay. The GMNSS should be improved on this methodological basis, which has some mechanisms for the (V, W)-exchange regulation via the consideration of correlations of the biogeochemical cycles of carbon, methane, ozone, water, oxygen, nitrogen, sulfur, and phosphorus. The broadening of this series, and establishing the functional connections between all spheres of the Earth's system, is possible by developing interdisciplinary studies. Here an attempt has been made to synthesize the GMNSS elements responsible for the simulation of the complex of biogeochemical cycles of some important elements.

6.3 DEVELOPMENT, PROPAGATION, AND CONSEQUENCES OF CATASTROPHIC WAVE PROCESSES IN THE BIOGEOCHEMICAL CYCLES OF GREENHOUSE GASES

6.3.1 Introduction

Atmospheric gases H_2O, CO_2, O_3, N_2O, CH_4, CFC-11 (CCl_3F), and CFC-12 (CCl_2F_2), being absorbers of IR radiation of the Earth's surface and atmosphere, favor the heating of the lower atmospheric layers resulting in the so-called "greenhouse effect". The GHG concentration during the industrial period, starting from 1850, has changed substantially: CO_2 from 280 ppm to 364 ppm, CH_4 from 700 ppb to 1721 ppb, and N_2O from 275 ppb to 312 ppb. The anthropogenic constituent of their input to the atmosphere has been increasing constantly, reaching values of $8.1 \pm 1.0\,\mathrm{Pg\,C\,yr^{-1}}$ for CO_2, 70–$120\,\mathrm{Tg\,C\,yr^{-1}}$ for CH_4, and 3–$8\,\mathrm{Tg\,N\,yr^{-1}}$ for N_2O. The role of each of them in climate change depends on the concentration and estimated lifetime, for instance, for CH_4 at 10 years, N_2O at 100 years, CFC-1 at 50 years, and CFC-12 at 102 years. The CO_2 lifetime changes from tens to thousands of years depending on the medium of its interaction (Ledley *et al.*, 1999).

The quantitative characteristic of the present trends of GHG content in the atmosphere has the following indicators. Analysis of the data of observations at 50 stations of the global network, from late 1970, revealed a continuous growth in atmospheric CO_2, CH_4, N_2O, and other GHGs, with a relative increase of CO_2 and N_2O almost equal, whereas the increase of CH_4 concentration markedly decreased and almost stopped. As for the contributions to the formation of radiative forcing (RF), in the case of CO_2 it exceeds 60%, and of CH_4 it is less than 20%. The rate of growth of CO_2 concentration averaged over the last several decades constitutes about $1.5\,\mathrm{ppm\,yr^{-1}}$ with a strong inter-annual variability. A consideration of the results of observations of the CO_2 exchange between land and atmospheric ecosystems in the USA (within the program AmeriFlux) has shown that in this case, carbon is assimilated by ecosystems, which function as atmospheric CO_2 sinks. From the data for the time intervals of 3–10 years, the annual assimilation of carbon varied within 2–$4\,\mathrm{t\,ha^{-1}}$ for forests and about $1\,\mathrm{t\,ha^{-1}}$ or less in the case of crops and grass cover. An accomplishment of the program AmeriFlux carried out from 1996 has also made it possible to obtain regular data of micrometeorological and biological observations important for understanding the processes of formation of the carbon cycle and its determining biological factors on synoptic, seasonal, and inter-annual scales. In phase with these data, estimates of the levels of water–energy exchange have been obtained.

A discussion of the problem of the greenhouse effect cannot be constructive without a complex consideration of CO_2 cycle feedbacks with biogeochemical processes with the participation of other elements, such as nitrogen, sulfur, phosphorus, methane, ozone, etc. The processes of CO_2 absorption from the atmosphere are affected by diverse natural and anthropogenic factors which can manifest themselves through a long chain of cause-and-effect bonds. For instance, acid rain affects the state of vegetation cover, and the latter affects the CO_2 exchange at the atmosphere–land boundary. On the whole, precipitation acidity is controled by the totality of chemical reactions in which nitrogen, sulfur, and carbon dioxide take place.

$$H_2O + CO_2 \rightarrow H_2CO_3; SO_2 + H_2O + 1/2O_2 \rightarrow H_2SO_4$$

Dissolved NH_4 through nitrification raises the surface water acidity, and NO_x oxidation gives HNO_3.

The nitrogen and phosphorus fertilization in agriculture changes the role of cultural plants in atmospheric CO_2 assimilation and affects the rate of the soil's organic matter disintegration. In totality, many chemicals, especially GHGs, getting to the environment from anthropogenic sources, become an object of not only biogeochemical analysis but also economic consideration. Such a many sided analysis was made in connection with CH_4 at the Second International Conference in Novosibirsk in 2000 (Bazhin, 2000; Byakola, 2000). Such bonds should be thoroughly systematized and parameterized. Without it, the estimates of the role of the biosphere in CO_2 assimilation from the atmosphere cannot be considered reliable. Complex studies in this direction are carried out, for instance, in some laboratories in the USA. Measurements of spatial and temporal distributions of gases involved in the global CO_2 cycle are made using flying laboratories and specialized stationary platforms. An accumulation of relevant information will make it possible to reveal dependences needed for the global model.

This confirms the fact that fragmentary studies of the global carbon cycle, not based on the constantly complicated model of the type described in Krapivin and Kondratyev (2002), will raise doubts. For global-scale conclusions of the type of the Kyoto Protocol Recommendations, this doubtfulness becomes dangerous, with unpredictable, global-scale consequences. Nevertheless, such conclusions and estimates are necessary. Unfortunately, most of the international programs on the subject discussed are not aimed at development of global modeling technology and do not concentrate efforts of specialists on the synthesis of a single NSS simulation model (Figure 6.2).

The existing global models are still simple and poorly provided with databases. There are three directions in the global modeling discussed in Kondratyev *et al.* (2002c, 2003a, 2004b), Sellers *et al.* (1996a,b), Boysen (2000), Krupchatnikov (1998), and Degermendji and Bartsev (2003). Each of these studies lacks one or several components, but on the whole, from the conceptual point of view, they complement each other. This gives prospects for their combination and, hence, synthesis of the global model, which takes into account the most important processes in the NSS.

One of the poorly studied (practically neglected) factors in global models is the photochemical system of the atmosphere. Knowledge of the laws of changes of the incoming solar radiation intensity in connection with its absorption by gases and aerosols is important. Initially, of importance here is the role of molecular nitrogen, ozone, water vapor, nitrogen oxide, sulfur dioxide, nitrogen dioxide, carbon dioxide, and other gases. Also, the developed models practically ignore the role of marshes which are a powerful sink for atmospheric CO_2 ($\sim$400.67 g m^{-2} yr^{-1}). Therefore, below some models are proposed which enable one to parameterize part of the elements indicated above.

Ozone is quite special in a series of biogeochemical cycles. Several studies (Syvorotkin, 2002; Gorny *et al.*, 1988; Nerushev, 1995) reveal a clear correlation between the state of the ozone layer and natural disasters. It has been established that at the fronts of cyclones the concentration of ozone sharply decreases. For instance, Nerushev (1995) discovered a feedback between the monthly mean number of tropic typhoons that occurred in Japan over the period of 30 years, and the ozone content over Japan. Also, the stratospheric ozone concentration decreased during El Niño years. On the whole, the concentration of stratospheric ozone is affected by atmospheric, geologic, biospheric, anthropogenic, and cosmic processes, so that, for instance, due to the ozone layer destruction a stratospheric cooling takes place, which is transmitted to the lower layer of the atmosphere, which leads to anomalous phenomena of the type of hurricane winds and sudden changes of temperature.

6.3.2 Special features of global biogeochemical cycles

An interaction of abiotic factors of the environment and living organisms of the biosphere is accompanied by continuous cycles of matter in nature. Various species of living organisms assimilate substances needed for their growth and life support, emitting into the environment the products of metabolism and other complex mineral and organic compounds of chemical elements in the form of unassimilated food or dead biomass. The biospheric evolution has resulted in a stable chain of global biogeochemical cycles, whose breaking in the second half of the 20th century has posed many principal problems, such as unpredictable climate changes due to the greenhouse effect enhancement, a decrease of biodiversity, progressing desertification, to name but a few. The questions "what happens with the Earth's climate?" and "what are the prospects for ozone layer depletion?" remain unanswered despite huge economic expenditure. Now it is clear that these and other nature-protection questions cannot be answered without developing an efficient global system of monitoring based on the global model of the NSS, one of the basic units of which simulates the biogeochemical cycles of the main chemicals in the biosphere. This approach will enable one to assess the anthropogenic fluxes of pollutants and to estimate the permissible emissions of hydrogen, chlorine, fluorine, methane, and other chemicals into the environment.

The global biogeochemical cycle of CO_2, described in this book, has been the center of attention for scientists during the last decades. Specialists from many countries try to answer the following questions:

(1) What are the concentrations of CO_2 that can be expected in the future with the existing or predicted rates of organic fuel burning?
(2) What climatic changes can result from the increasing concentrations of CO_2?
(3) What are the consequences of such climatic changes in the biosphere?
(4) What can mankind undertake in order to either reduce the negative consequences of climate change or to prevent them?

As Barenbaum (2002) notes, the development of a reliable technology for the calculation of the dynamic characteristics of the global carbon cycle is impossible without a consideration of the geochemical processes in the geosphere and if the structure of geophysical processes taking place on Earth have not been taken into account. In the process of global cycle formation, moving carbon repeatedly crosses the Earth's surface. Above the Earth's surface, which plays the role of "geochemical barrier", carbon circulates mainly as CO_2, and beneath the surface as CH_4. Moreover, this circulation is governed by the water cycle. Apparently, the structure of seismic zones should be taken into account too (De Boer, 2004, 2005).

Clearly, nowadays, according to rough model estimates, industrial civilization should be searching for new sources of energy, which would reduce the rates of organic fuel burning and, hence, reduce the external impacts on natural biogeochemical cycles.

During the last decade the notion of "greenhouse effect" has been used in numerous publications to represent the problems of global climate change on the Earth. This term implies a totality of descriptions of the effects appearing in the climatic system and connected with a number of natural and anthropogenic processes. This notion of "greenhouse effect" refers to an explanation of changes in the thermal regime of the atmosphere caused by the effect of some gases on the process of the solar radiation absorption as well as long-wave radiation transfer. Many gases are characterized by a high stability and long residence time in the atmosphere. Carbon dioxide is one of them – its time of residence in the atmosphere is estimated at 2–3 years.

Numerous long-term observations in various latitudinal belts show a high level of correlation between temperature and CO_2 content. The atmosphere–ocean interaction contributes most into this dependence. Though the atmosphere and the ocean are in equilibrium with respect to CO_2 exchange, this equilibrium still gets regularly broken. The most serious causes of which include: (1) SST variations; (2) changes in the ocean volume; and (3) changes in the regime of the vertical circulation of the ocean. In general, an efficiency of these causes can be characterized by the following ratio of the forcing on CO_2 concentrations in the atmosphere. The first cause contributes about 65% into the change of CO_2 partial pressure in the atmosphere (p_a). The remaining 35% is brought about by the second and third causes. Quantitatively, this relationship is characterized by a 6% increase of the atmospheric CO_2 partial pressure per 1°C increase in temperature of the upper layer of the ocean. Also, a 1% decrease of the ocean volume increases p_a by 3%.

An assessment of the greenhouse effect requires a complex consideration of the interaction of all processes of energy transformation on Earth. However, within the diversity of the processes (from astronomical to biological) that affect the climatic system over various timescales, there exists a hierarchy of significance. But this hierarchy cannot be constant, since the role of some processes can vary over a wide range regarding their significance to climatic variations. In fact, the impact of the greenhouse effect is determined by an exceeding of surface temperature T_L over the efficient temperature T_e. The Earth's surface temperature T_L is a function of surface emissivity κ. The effective temperature T_e is a function of the emissivity α of the atmosphere–land–ocean system. In a general case, the parameters κ and α depend on many factors, in particular, on the CO_2 concentration in the atmosphere. There are a lot of simple and complicated numerical models where attempts have been made to parameterize these dependences. Unfortunately, there is not a single model that would meet the requirements to adequately and reliably describe the prehistory of the climatic trends on Earth. Nevertheless, one can state that the greenhouse effect depends non-linearly on the difference $T_L - T_e$ (i.e., on atmospheric transmission, especially in the long-wave region). The more CO_2 in the atmosphere, the weaker the atmospheric transmission. The strongest effect of CO_2 on the atmospheric transmission is in the long-wave region of 12–18 μm. This effect is weaker in the wavelength intervals 7–8, 9–10, 2.0, 2.7, and 4.3 μm. It is clear that with the increasing partial pressure of CO_2 in the atmosphere the role of various bands of CO_2 will grow, and this means that with intensified CO_2 absorption bands the upward long-wave radiation flux will decrease. At the same time, the downward long-wave radiation flux on the Earth's surface will increase. From the available estimates, a reduction of the global mean upward flux, and increase of the downward flux are estimated at 2.5 and 1.3 $W\,m^{-2}$, respectively.

Thus, to estimate the level of the greenhouse effect due to CO_2 and other gases, it is necessary to know how to predict their concentration in the atmosphere with account of all the feedbacks in their global biogeochemical cycle (Watson *et al.*, 2000; Grant, 2004; Grogan *et al.*, 2001; Guaduong and Masao, 2001; Grossman, 2001). This problem touches upon several spheres of science – biogeochemistry, geochemistry, soil science, ecology, agrochemistry, geology, oceanology, physiology, and radiochemistry. The present methods of global ecoinformatics enable one to combine knowledge accumulated in these fields (Kondratyev e*t al.*, 2002a, 2003c, 2004a).

Of course, global cycles of chemicals should be studied not only to be able to assess the climatic consequences of anthropogenic activity, but also to understand the prospects of environmental dynamics from the viewpoint of its quality and potential to support life. Since the cycles of chemicals in nature are closely connected with living organism activity, one can single out the geological, biogenic, and biological cycles. The biogenic cycle includes sub-cycles, such as biogeochemical, biogeocenotic, and geochemical (Haan *et al.*, 2001; Heans, 2001). Geochemical processes due to abyssal degassing affect the content of stratospheric ozone, whose depletion affects vegetation cover dynamics negatively. The state of the ozone layer depends on the combination of more than 20 processes, but one of the key processes is the planet's degassing. The main channels of degassing are known. The connection between the accumulation of hydrocarbons and abyssal fractures of the rift type, makes it possible to close the global biogeochemical cycles. Of course, to construct the respective global model of the nature–society system (GMNSS) units, it is necessary to establish the cause-and-effect connections between base chemical compounds. In particular, Diadin and Gushchin (1998) drew attention to gas hydrates which, with a certain combination of thermobaric conditions, start to rapidly disintegrate and, via degassing channels, enter the atmosphere. It is known that supplies of methane gas-hydrates exceed the content of methane in the atmosphere by a factor of 3,000. This means that with an intensive anti-freezing of gas-hydrate deposits, huge inputs of CH_4 to the atmosphere are possible. Representing a special danger are huge supplies of gas-hydrates preserved in multiyear permafrost in the region of the Siberian magnetic anomaly. At present, the Siberian ozone maximum is fixed here, which somehow stabilizes the state of the frozen ground.

6.3.3 Carbon cycle

The problem of the global carbon cycle (GCC) has been in the focus during the last decades in connection with numerous, often speculative, explanations of the role of CO_2 in future climate change. Unfortunately, an objective assessment of this role is still absent. Recently published studies (Kondratyev, 2000, 2004a,b; Kondratyev and Krapivin, 2004b; Krapivin and Chukhlantsev, 2004) have summed up the first results of development of the formalized technology for the assessment of the greenhouse effect due to CO_2, with due regard to the role of the land and ocean ecosystems. An interactive connection has been demonstrated between the global carbon cycle in the form of CO_2 and climate change. The formalization of this connection is based on the synthesis of the global model of the NSS functioning, with the spatial distribution of the elements of this system taken into account, making it possible to reduce into a single correlated scheme the cause-and-effect connections of carbon fluxes between the different biospheric and geospheric reservoirs.

An objective formalization of the biospheric sources and sinks of CO_2 as functions of the environmental parameters, and a consideration of the actual role of anthropogenic processes, becomes possible due to the recent studies of many experts who have developed models with different degrees of detailed descriptions of spatially distributed carbon fluxes and their interaction with NSS components (Kondratyev *et al.*, 2004a).

In the recently published first report on the international project on the global carbon cycle, the GCP (Global Carbon Project) (Canadel *et al.*, 2003), the strategy has been initially formulated with interdisciplinary cooperation within a broad spectrum of environmental problems considered in the context of the global system of nature–society interaction, with special emphasis on the necessity to develop methods and information technologies to analyze the carbon–climate–society system (CCSS). The central goal is a consideration of the following five aspects of the general problem of the global carbon cycle, including:

- A study of the GCC which should be based on the integration of natural and anthropogenic components by interactive analysis of interactions between energy systems based on fossil fuels, biogeochemical carbon cycles, and the physical climate system.
- The development of new methods of analysis and numerical modeling of the integrated carbon cycle.
- Global studies of the carbon cycle carried out with due regard to the results of national and regional research program on the studies of the carbon cycle and its reservoirs.
- The strategic problem of searching for ways to stablize regional development to achieve a stabilized concentration of CO_2 in the atmosphere.
- Classification of all countries as developed and developing allows the more detailed identification of industrial, economic, and energy sectors of the NSS taking into account their characteristics as sources of antrhopogenic emissions.

In addition to this, we enumerate the key directions of developments within the CCSP program (*Our Changing Planet*, 2004):

- What are the specific features of the spatial–temporal variability of the sources and sinks of carbon on the territory of North America on timescales from seasonal to centennial, and what processes have a prevailing effect on the carbon cycle dynamics?
- What are the respective features of variability and their determining factors in the case of ocean components (sources and sinks) of the carbon cycle?
- How do local, regional, and global processes on the land surface (including land use) affect the formation of the sources and sinks of carbon (in the past, present, and future)?
- How do the sources and sinks of carbon vary on land, in the ocean, and in the atmosphere on timescales from seasonal to centennial, and how can the respective information be used to get a better understanding of the laws of the GCC formation?
- What will be the future changes in concentrations in the atmosphere of CO_2, methane, and other carbon-containing GHGs, and how will sources and sinks of carbon on land and in the ocean change?
- How will the Earth's system, and its components, respond to the various versions of the choice of strategy of regulation of the carbon content in the environment, and what information is needed to answer this question?

The carbon cycle is closely connected with the climate, water, and nutrient cycles, and also photosynthesis production on land and in the ocean. Therefore, all studies of the GCC that neglect such connections, are inevitably doomed to failure and, hence, cannot give even approximately adequate estimates of the consequences of anthropogenic emissions of carbon to the environment. For this reason, many international projects on the study of the greenhouse effect, and its impact on climate, have failed, like the Kyoto Protocol, which was intended to regulate CO_2 emissions. Therefore, the GCP instills hope to making progress by planning the interdisciplinary studies of the GCC. These studies can be divided into three:

- Formation of a strategy for GCC studies and evaluation of its variability.
- Analysis of connections between the causes and consequences in studies of mechanisms of interaction of natural and anthropogenic sources and sinks of CO_2 with the environment.
- Identification and quantitative estimation of the evolutionary processes in the CCSS.

The first GCP report (Canadel *et al.*, 2003) formulates the goal of the coming decadal period of GCC study, which ideologically combines the earlier isolated programs of the IGBP (the International Geosphere–Biosphere Programme), IHDP (the International Human Dimensions Programme), and WCRP (the World Climate Research Programme). The authors of this report substantiated the detailed scheme of the cause-and-effect connections between the climate–biosphere system's components, and pointed to a necessity of their joint consideration in order to raise

Table 6.2. Potentials of relative global warming due to various greenhouse gases (EPA, 2001).

Gas	Potential	Gas	Potential
CO_2	1	HFC-227es	2,900
CH_4	21	HFC-236fa	6,300
N_2O	310	HFC-4310mee	1,300
HFC-23	11,700	CF_4	6,500
HFC-125	2,800	C_2F_6	9,200
HFC-134a	1,300	C_4F_{10}	7,000
HFC-143a	3,800	C_6F_{14}	7,400
HFC-152a	140	SF_c	23,900

the level of reliability of the estimates and forecasts of the climatic impact of CO_2. All these problems had been discussed before in numerous publications (Kondratyev, 2004a; Kondratyev and Krapivin, 2003a; Kondratyev *et al.*, 2003a). Unfortunately, in the first GCP report the role of other GHGs, whose contribution in the nearest future can exceed the role of CO_2, is still underestimated. The list of GHGs, such as methane, nitrogen monoxide, hydrofluorocarbons, perfluorocarbons, and hydrofluoroethers increases with time. Moreover, these gases are more efficient in the formation of the greenhouse effect. Their total equivalent emissions in 1990 constituted 3.6 Gt of CO_2, and by the year 2010 the level of 4.0 Gt CO_2 will be exceeded (Bacastow, 1981). At the same time, the anthropogenic CO_2 emissions are estimated at 6 GtC yr^{-1}. Table 6.2 demonstrates the contributions of different GHGs to climate change. According to EPA (2001), the levels of contributions of anthropogenic emissions of some GHGs to the enhancement of the greenhouse effect from the beginning of the industrial revolution constituted: CO_2 – 55%, CH_4 – 17%, O_3 – 14%, N_2O – 5%, and others – 9%.

All this testifies to the fact that continuing, sufficiently primitive (in most cases) descriptions of the GCC and a practical absence of even such a parameterization for other GHGs cannot lead to reliable estimates of possible future climate changes due to anthropogenic activity within the NSS. The idea of identification of locations and power of the sources and sinks of CO_2 on land and in the World Ocean, declared in the GCP report (Canadel *et al.*, 2003), has not been supported by the development of new information technologies for complex analysis of the Earth's radiation budget.

Nevertheless, note should be taken that the basic postulate of the GCP promotes a better understanding of the GCC, directing the base conception of its studies at combining natural and anthropogenic components with the use of the developed methods, algorithms, and models. The main structure of the carbon cycle is determined by its fluxes between the basic reservoirs including carbon in the atmosphere (mainly in the form of CO_2), oceans (with division into surface, intermediate, and deep layers and bottom deposits), land ecosystems (vegetation, litter, and soil), rivers and estuaries, and fossil fuels. All these reservoirs should be studied with due regard to their spatial heterogeneity and dynamics under the influence of natural and anthropogenic factors based on accumulated knowledge, according to which:

- Anthropogenic carbon emissions grow constantly from the beginning of industrial development, reaching 5.2 GtC in 1980 and 6.3 GtC in 2002.
- The content of the main GHGs – CO_2, CH_4, and N_2O in the atmosphere has increased from the year 1750 by 31%, 150%, and 16%, respectively. About 50% of the CO_2 emitted to the atmosphere due to fuel burning has been assimilated by vegetation and the oceans.
- The observed distribution of atmospheric CO_2 and oxygen/nitrogen relationship show that the land sink of carbon prevails in northern and middle latitudes over the oceanic sink. In tropical latitudes, emissions of CO_2 to the atmosphere are substantial due to the use of the Earth's resources.
- The inter-annual oscillations of CO_2 concentration in the atmosphere follow changes in the use of fossil fuels. The intra-annual variability of atmospheric CO_2 concentration correlates more with the dynamics of land ecosystems than with that of the ocean ecosystems.
- The regional flow of carbon due to the production and trade of crops, timber, and paper, in 2000 constituted 0.72 GtC yr^{-1}. The global pure carbon flux at the atmosphere–ocean boundary observed in 1995 was estimated at 2.2 GtC ($-19\% = \pm 22\%$) with the intra-annual variation about 0.5 GtC. Maximum amplitudes of CO_2 flux oscillations in the atmosphere–ocean system are observed in the equatorial zone of the Pacific Ocean.
- An approximate picture of distribution of the ocean sources and sinks of atmospheric CO_2 is known: tropical basins of the oceans are sources of CO_2, and high-latitude water basins are CO_2 sinks. The role of rivers is reduced mainly to the transport of carbon to the coastal zones of the World Ocean ($\sim$1 GtC $year^{-1}$).

The most important section of the GCP is global environmental monitoring with an accumulation of detailed information on land biomes production, CO_2 fluxes at the atmosphere–ocean boundary, and volumes of anthropogenic emissions. The space-borne sounding of CO_2 with the use of the Atmospheric Infrared Sounder (AIRS) at the space-borne laboratory EOS–Aqua, launched by NASA on 4 March 2002 to the altitude 705 km, and the Infrared Atmospheric Sounder Interferometer (IASI) carried by the satellite METOP (METeorological Operational Polar) plays a special role (Nishida *et al.*, 2003). Other functioning or planned space vehicles will be used to evaluate CO_2 fluxes from the data of indirect measurements of environmental characteristics. In particular, these are aims of the satellite TIROS-N (Television Infrared Observational Satellite N) and instrumentation SCIMASACHY (Scanning Imaging Absorption Spectrometer for Atmospheric Cartograghy). The latter spectrometer launched in 2002 provides a high-spectral resolution within absorption bands of GHGs such as CO_2, CH_4, H_2O (accuracy 1%) and N_2O, CO (accuracy 10%), with a surface resolution of 30–240 km depending on latitude.

Traditionally, ground measurements will be continued, their particular goal being to substantiate national strategies of the use of the Earth's resources, including the development of forestry and agriculture, stock breeding and crop growing.

The GCP program foresees an extensive study of physical, biological, biogeochemical, and ecophysiological mechanisms of the formation of carbon fluxes in the environment. A deeper understanding of these mechanisms and their parameterization will make it possible to specify the GCC models and related climate changes. Broadening the respective base of knowledge will make it possible to specify the following information about these mechanisms:

- The atmosphere–ocean carbon exchange is mainly controled by physical processes including the mixing between the surface and deep layers of the ocean through the thermocline. Biological processes favor the transport of carbon from the ocean surface layers to deep layers and further to bottom deposits. The biological "pump" functions due to phytoplankton photosynthesis.
- A complex of feedbacks controls the interactive exchange of energy, water, and carbon between the atmosphere and land surface, causing a response of these fluxes to such disturbances as transformation of land cover or oil pollution of the World Ocean. Physiological responses of plant communities to changes in temperature and humidity of the atmosphere and soil.
- The carbon sink in the northern hemisphere depends on the growth of forest, climate change, soil erosion, fertilization, and accumulation of carbon in freshwater systems. Unfortunately, the processes taking place here have been poorly studied, and factual information is practically absent. The power of the land sink of carbon can increase under certain conditions of climate change. When the atmospheric CO_2 concentration exceeds the 550-ppm level, many processes on land ecosystems become short of nutrients and water, and therefore the photosynthetical accumulation of carbon by land plants becomes physiologically saturated.
- The number of key factors that determine directions and amplitudes of CO_2 fluxes between the atmosphere and the land ecosystems is limited by several factors:
  - –extreme climatic phenomena such as droughts, serious drifts of seasonal temperatures, solar radiation changes due to large-scale inputs of aerosol to the atmosphere (e.g., volcanic eruptions or large-scale fires similar to those which took place in Iraq during recent military operations);
  - –forest and other fires which introduce large-scale and long-term changes in carbon cycle characteristics (for a global pure primary production of about $57\,\mathrm{GtC\,yr^{-1}}$, about 5–10% is emitted to the atmosphere as a result of wood burning);
  - –land use leading to a change of the boundaries of biomes and a change in their types (change from evergreen forest to coniferous, forests to pastures, meadows to urban areas);

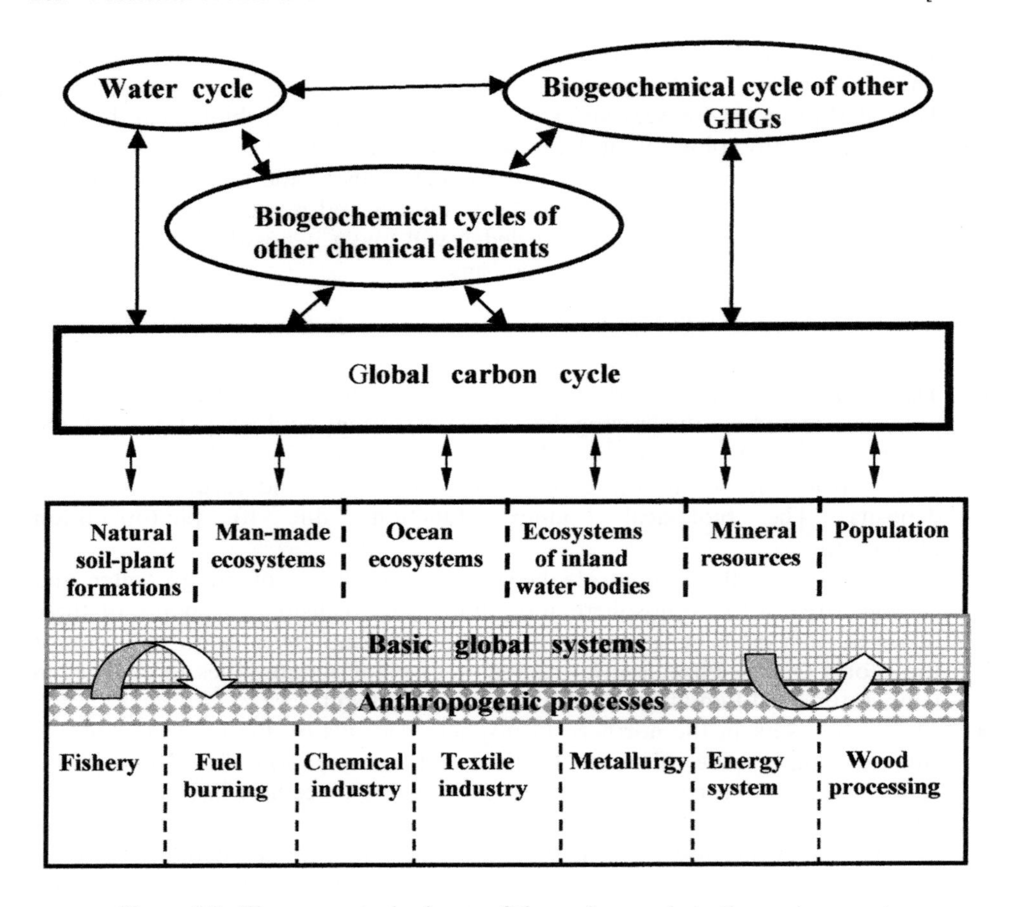

Figure 6.3. The conceptual scheme of the carbon cycle in the environment.

–reduction of biodiversity and a change in the structures of communities, changing the character of their impact on the nutrient, carbon, and water cycles.

- Phenomena like El Niño or the thermohaline circulation in the North Atlantic lead to a global instability in the process of energy–matter exchange, which should be reflected in parameterization of non-linear feedbacks.

The future dynamics of the carbon exchange in the NSS will be determined by the strategy of interaction of natural and anthropogenic factors. For the first time, a broader approach to this problem has been developed (Kondratyev *et al.*, 2002b, 2003c, 2004b) which, unfortunately, has been neglected by the GCP authors.

In a number of recent studies (Bartsev *et al.*, 2003; Vilkova, 2003; Kondratyev *et al.*, 2002, 2003a, 2004a) the GCP is proposed to be considered in the context of its interaction with other processes in the NSS. As shown in Figure 6.3, the carbon cycle correlates with a multitude of natural and anthropogenic factors whose interaction

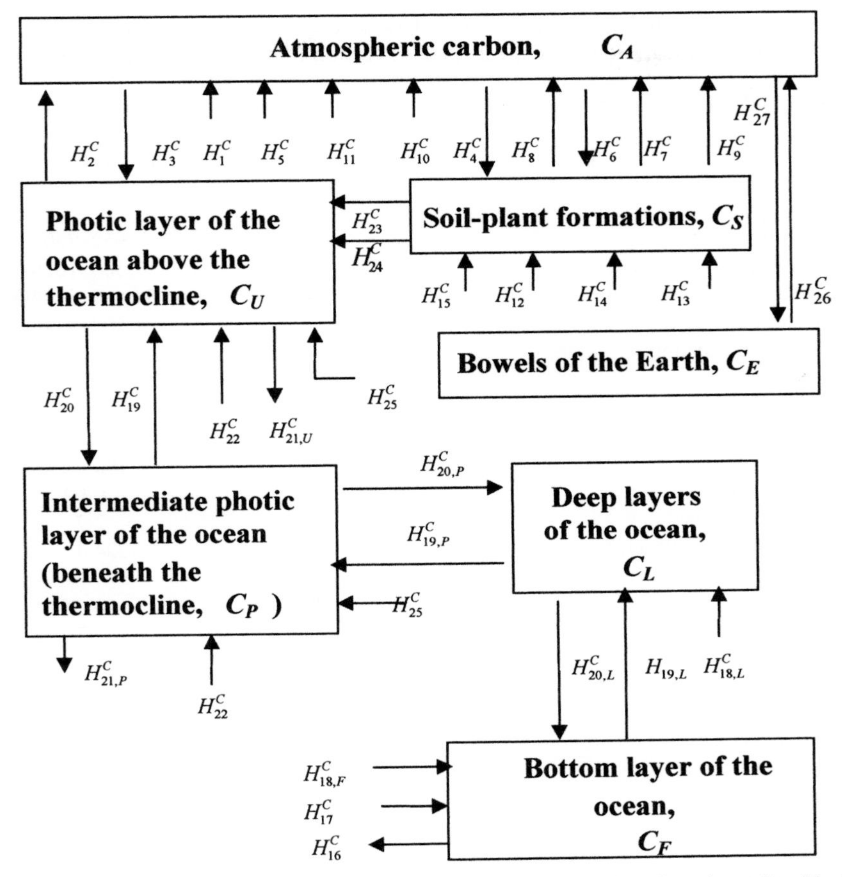

Figure 6.4. The block-scheme of the global biogeochemical cycle of carbon dioxide in the "atmosphere–land–ocean" system. The CO_2 reservoirs and fluxes are described in Table 6.3.

forms the dynamics of the key processes in the NSS. For CO_2, such processes are exchanges at the boundaries of the atmosphere with land surfaces and sea and ocean basins. It is clear that the CO_2 dynamics in the biosphere can be analyzed with available data on the spatial distribution of its sinks and sources. The present level of knowledge makes it possible to set and solve the problem of specification of the impact of the greenhouse effect on climate and to decrease thereby the interval of uncertainty in estimates of future climate change. However, the applied GCC model should reflect not only the spatial mosaic of its reservoirs, sinks, and sources, but should also provide a dynamic calculation of their abilities. The earlier calculations with the use of GCC models have not adequately taken into account information on the state and classification of land cover and the variability of the World Ocean basin. Therefore, the scheme in Figure 6.4 and Table 6.3 is aimed at overcoming the shortcomings of other models. The system of balance

Table 6.3. Reservoirs and fluxes of carbon as CO_2 in the biosphere considered in the simulation model of the global biogeochemical cycle of carbon dioxide shown in Figure 6.4.

Reservoirs and fluxes of carbon dioxide	Identifier	Estimate of reservoir (10^9 t) and flux ($10^9\,t\,yr^{-1}$)
Carbon:		
atmosphere	C_A	650–750
photic layer of the ocean	C_U	580–1020
deep layers of the ocean	C_L	34,500–37,890
soil humus	C_S	1,500–3,000
Emissions from burning:		
vegetation	H_8^C	6.9
fossil fuel	H_1^C	3.6
Desorption	H_2^C	97.08
Sorption	H_3^C	100
Rock weathering	H_4^C	0.04
Volcanic emanations	H_5^C	2.7
Assimilation by land vegetation	H_6^C	224.4
Respiration:		
plant	H_7^C	50–59.3
human	H_{10}^C	0.7
animal	H_{11}^C	4.1
Emission:		
soil humus decomposition	H_9^C	139.5
plant roots	H_{15}^C	56.1
Vital activity:		
population	H_{12}^C	0.3
animals	H_{13}^C	3.1
Plants dying	H_{14}^C	31.5–50
Bottom deposits	H_{16}^C	0.1–0.2
Solution of marine deposits	H_{17}^C	0.1
Detritus decomposition:		
photic layer	H_{22}^C	35
deep layers of the ocean	H_{18}^C	5
Upwelling with deep waters	H_{19}^C	4
Downwelling with surface waters and due to gravitational sedimentation	H_{20}^C	40
Photosynthesis	H_{21}^C	69
Groundwater flow	H_{23}^C	0.5
Surface runoff	H_{24}^C	0.5–0.6
Respiration of living organisms in the ocean	H_{25}^C	25
Degassing processes	H_{26}^C	21.16
Tectonic subduction	H_{27}^C	1.3

equations for such a scheme is written as:

$$\frac{\partial \alpha_S^i(\varphi,\lambda,z,t)}{\partial t} + V_\varphi \frac{\partial \alpha_S^i(\varphi,\lambda,z,t)}{\partial \varphi} + V_\lambda \frac{\partial \alpha_S^i(\varphi,\lambda,z,t)}{\partial \lambda} + V_z \frac{\partial \alpha_S^i(\varphi,\lambda,z,t)}{\partial z}$$

$$= \sum_{j\in\Omega_S} H_{jS} - \sum_{m\in\Omega_S} H_{Sm}, (i = 1,\dots,N); \quad (6.1)$$

where S is the carbon reservoir in the ith cell (pixel) of spatial digitization, φ is the latitude, λ is the longitude, z is the depth, t is the time, H_{jS} is the carbon sink from the jth reservoir to reservoir S, H_{Sm} is the carbon sink from reservoir S to the mth reservoir, Ω_S is the multitude of carbon reservoirs bordering reservoir S, N is the number of carbon reservoirs, and $V(V_\varphi, V_\lambda, V_z)$ is the rate of exchange between reservoirs.

In (6.1) the rate V and fluxes H are non-linear functions of the environmental characteristics. These functions have been described in detail in Krapivin and Kondratyev (2002). We shall only specify them. First of all, the elements of the biogeocenotic unit of the global model shown in Figures 6.3 and 6.4 should be specified. With this end in view, let us cover the whole land surface Σ with a homogeneous grid of geographic pixels $\Sigma_{ij} = \{(\varphi,\lambda): \varphi_{i-1} \le \varphi \le \varphi_i; \lambda_{j-1} \le \lambda \le \lambda_j\}$ with boundaries in latitude $(\varphi_{i-1},\varphi_i)$, longitude $(\lambda_{j-1},\lambda_j)$, and area σ. The number of pixels is determined by the available database (i.e., by the choice of grid sizes $(\Delta\varphi,\Delta\lambda)$: $i = 1,\dots,n$; $n = [180/\Delta\varphi]$; $j = 1,\dots,k$; $k = [180/\Delta\lambda]$). Each pixel can contain N types of surfaces, including the types of soil–plant formations, water basins, and other objects. The dynamics of the vegetation cover of the sth type follows the law:

$$\frac{dB_s}{dt} = R_s - M_s - T_s \quad (6.2)$$

where R_s is photosynthesis, M_s and T_s are losses of biomass B_s due to dying-off and evapotranspiration, respectively.

The components of the right-hand part of (6.2) are functions of the environmental characteristics – illumination, temperature, air and soil humidity, and atmospheric CO_2 concentration. There are several ways and forms of parameterization of these functions. One of them is the model of Collatz *et al.* (1990, 1991, 1992), which served the basis for the global biospheric model SiB2 (Sellers *et al.*, 1996b). Temperature, humidity, and rate of moisture evaporation in the vegetation cover and soil depend on the biospheric parameters and energy fluxes in the atmosphere–plant–soil system. By analogy, with electrostatics, the notion of "resistance" is introduced, and fluxes are calculated from a simple formula: flux = potential difference/resistance. The SiB2 model takes into account the fluxes of sensible and latent heat through evaporation of water vapor in plants and soil, and CO_2 fluxes are divided into classes C_3 and C_4, which substantially raises the accuracy of parameterization of the functions in the right-hand part of (6.2). According to Collatz *et al.* (1991), three factors regulate the function R_s: efficiency of the photosynthetical enzymatic system, amount of photosynthetically active radiation (PAR) absorbed by cellulose chlorophyll, and ability of plant species to assimilate and transmit to the outside medium the products of photosynthesis. An application of the Libich principle (Kondratyev *et al.*, 2002a; Nitu *et al.*, 2000a) and consideration of the data on the distribution of the types of vegetation cover by pixels $\{\Sigma_{ij}\}$, on partial pressures of CO_2 and O_2, the temperature and density of the atmosphere, and the level of illumination, makes it possible to calculate fluxes H in (6.1) for all pixels of land.

The model of the carbon cycle in the atmosphere–ocean system has been described in detail by Pervaniuk and Tarko (2001). It is based on the same grid of geographic pixels but combined by the zonal principle according to the Tarko classification (Tarko, 2001, 2005). The ocean thickness is considered a single biogeocenosis in which the main binding factor is the flux of organic matter produced in the surface layers which then penetrates down to the deepest layers of the ocean. In this medium the carbonate system, the parametric description of which has been given in Kondratyev *et al.* (2004b), is a regulator of carbon fluxes.

One of the principal questions concerning atmosphere–ocean CO_2 exchange consists in the fact that the role of hurricanes has not been studied in sufficient details. Perrie *et al.* (2004) studied the hurricane influence on the local rates of air–sea CO_2 exchange. Hurricanes affect the thermal and physical structure of the upper ocean. Air–sea gas transfer includes processes such as upper ocean temperature changes and upwelling of carbon-rich deep water. Observations have shown that SST and CO_2 partial pressure can decrease by 4°C and 20 μatm due to the influence of a hurricane, respectively. Perrie *et al.* (2004) proposed a model to parameterize the CO_2 flux H_3 with the following formula:

$$H_3 = k_L \alpha \Delta[CO_2]$$

where α is the solubility of CO_2 and $\Delta[CO_2]$ is the difference between its partial pressure in the atmosphere and sea upper layer. Parameter k_L (cm h^{-1}) is determined with one of the following correlations depending on the wind speed:

$$k_L = \begin{cases} 0.31 U_{10}^2 (S_c/660)^{-0.5} & \text{for hurricanes 1–3 category} \\ 0.0283 U_{10}^3 (S_c/660)^{-0.5} & \text{for hurricanes 4–5 category} \end{cases}$$

where S_c is the Schmidt number (Hasegawa and Kasagi, 2001, 2005) and U_{10} is the wind speed at the altitude 10 m (m s^{-1}).

Introducing the wave spectrum's peak frequency ω_p, the air-side friction velocity $u*$, and kinematic viscosity ν, the parameter k_L can be calculated by the formula:

$$k_L = 0.13 \left[\frac{u_*^2}{\nu \omega_p} \right]^{0.63}$$

In actual fact, parameter k_L is formed from two components: $k_L = k_{L1} + k_{L2}$, where k_{L1} and k_{L2} are the wave-breaking and the interfacial terms, respectively. The terms

k_{L1} and k_{L2} are calculated with the use of the following formulas:

$$k_{L1} = u_*^{-1}\left[\sqrt{\rho_w/\rho_a}\left(h_w S_{cw}^{0.5} + \kappa^{-1}\ln\{z_w/\delta_w\}\right) + \alpha\left(h_a S_{ca}^{0.5} + c_d^{-0.5} - 5 + 0.5\kappa^{-1}\ln S_{ca}\right)\right]$$

$$k_{L2} = fV\alpha^{-1}\left[1 + \left(e\alpha S_c^{-0.5}\right)^{-1/n}\right]^{-n}$$

where $f = 3.8 \times 10^{-6} U_{10}^3$; α is the gas solubility; subscript "a" ("w") denotes air-(water-) side; ρ is the density; z is the measurement depth; δ is the turbulent surface layer thickness; κ is the von Kármán constant; c_d is the drag coefficient; $h \equiv \Lambda\varphi^{-1}R_r^{0.25}$; Λ is an adjustable constant; R_r is the roughness Reynolds number; φ is an empirical function that accounts for buoyancy effects on turbulent transfer in the ocean; and V, e, and n are empirical constants equal to 14, 1.2 and 4,900 cm h^{-1} in the GasEx-1998 field experiment (Perrie *et al.*, 2004), and may need readjustment for other data sets.

The principal significance has fact that hurricanes form upwelling zones where air–water gas exchange acquires other characters. Hales *et al.* (2005) studied atmospheric CO_2 uptake by a coastal upwelling system located in the US Pacific coast off Oregon using high-resolution measurements of partial pressures of CO_2 and nutrient concentrations in May through to August 2001. This showed that the dominance of the low-CO_2 waters over the shelf area makes the region a net sink during the upwelling season due to:

- upwelled water that carries abundant preformed nutrients;
- complete photosynthetic uptake of these excess nutrients and a stoichiometric proportion of CO_2; and
- moderate warming of upwelled waters.

Received numerical estimations are:

- The North Pacific's eastern boundary area represents a sink of atmospheric CO_2 representing 5% of the annual North Pacific CO_2 uptake.
- By mid-August, the partial pressure of CO_2 in subsurface waters increased 20–60%, corresponding to a 1.0–2.3% total dissolved CO_2 increase, due to respiration of settling biogenic debris.

Many parameters of the global carbon cycle model are measured in the satellite monitoring regime, which makes it possible to apply an adaptive scheme of calculation of the greenhouse effect's characteristics (Figure 6.5). This scheme makes it possible to "teach" the model by the continuous regime of correction of its structure and parameters. Satellite measurements in the visible and near-IR regions provide operational estimates of PAR and vegetation characteristics such as canopy greenness, area of living photosynthetically active elements, soil humidity and water content in the elements of vegetation cover, CO_2 concentration on the surface of leaves, etc. The regime of prediction of the vegetation cover biomass in each pixel Σ_{ij} and comparisons with satellite measurements enables one to correct some fragments of the model, for instance, using the doubling of its units or their parametric adjustment to minimize discrepancies between prediction and

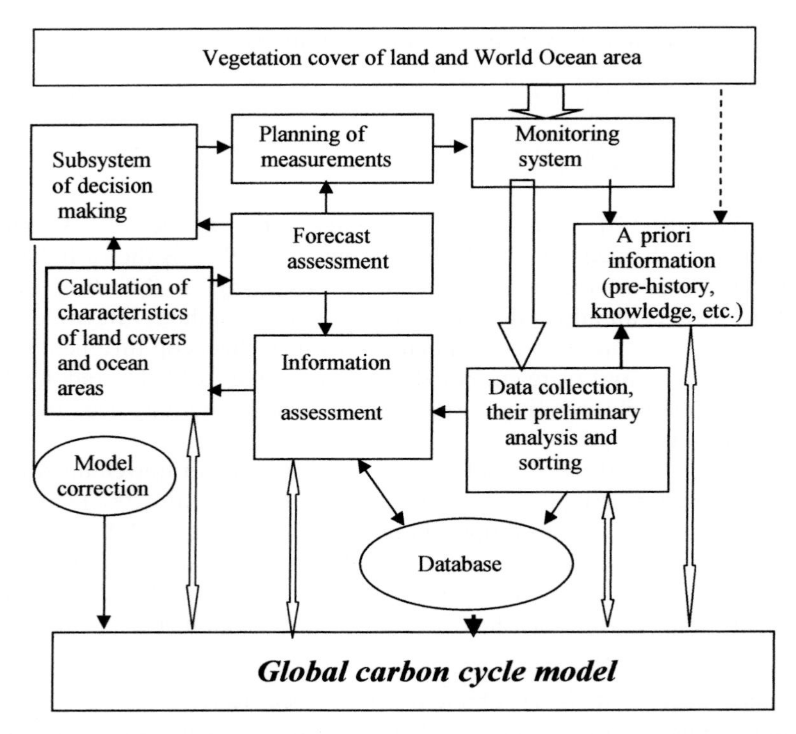

Figure 6.5. An adaptive regime of greenhouse effect monitoring with assessment of the role of vegetation cover of land and World Ocean areas.

measurements (Figure 6.5). In particular, to calculate the primary production, there are some semi-empirical models, which can be used by sample criterion in different pixels. There is a certain freedom of choice in the estimation of evaporation from vegetation cover (Wange and Archer, 2003).

The key component of the global CO_2 cycle is its anthropogenic emissions to the environment. The main problem studied in this connection by most scientists consists of the assessment of the ability of the biosphere to neutralize an excess amount of CO_2. Table 6.4 and Figures 6.6 and 6.7 illustrate the modeling results. It is seen that of the 6.3 GtC emitted to the atmosphere by industry, 41.3% remains in the atmosphere, and ocean and land vegetation absorbs 20.2 and 38.5%, respectively. Taking as a basis the dependence of air temperature changes on CO_2 variation (Mintzer, 1987):

$$\Delta T_{CO_2} = -0.677 + 3.019 \ln[C_a(t)/338.5],$$

for the realistic scenario in Figure 6.6, we obtain $\Delta T_{CO_2} \leq 2.4°C$. This substantially specifies the estimate $\Delta T_{CO_2} \leq 4.2°C$ published by many authors and assumed in the Kyoto Protocol.

As seen from Figure 6.6, the discrepancy between the forms of distribution of

Table 6.4. Model estimates of excessive CO_2 assimilation over the Russian territory. A more detailed classification of soil–plant formations is given in Table 6.5.

Soil–plant formation	Flux of assimilated carbon as CO_2 ($10^6\,t\,C\,yr^{-1}$)
Arctic deserts and tundras, sub-Arctic meadows and marshes	2.2
Tundras	3.3
Mountain tundra	3.6
Forest tundra	2.8
North taiga forests	10.8
Mid-latitude taiga forests	31.2
South taiga forests	22.9
Broad-leaved–coniferous forests	4.8
Steppes	3.6
Alpine and sub-alpine meadows	1.1
Deserts	2.2

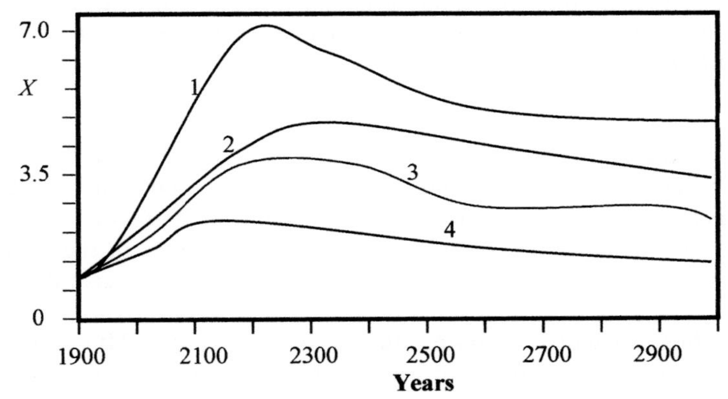

Figure 6.6. Forecast of CO_2 concentration in the atmosphere with different scenarios of mineral resources expenditure: 1 – pessimistic scenario (Bacastow, 1981), 2 – optimistic scenario (Bjorkstrom, 1979), 3 – scenario of Intergovernmental Panel on Climate Change (IPCC) (Dore *et al.*, 2003), 4 – realistic scenario (Demirchian and Kondratyev, 2004). $X = C_a(t)/C_a$ (1900).

the CO_2 absorption curve and vegetation index suggests an idea that in the southern hemisphere the structure of model pixels and their fitting with observational data should be specified. Nevertheless, an introduction to the GCC model of the pixel mosaic has made it possible to evaluate the role of some types of ecosystems and territories of Russia in the greenhouse effect regulation. Table 6.4 demonstrates the role of taiga in the territory of Russia in this regulation. On the whole, the model enables one to consider various scenarios of land cover changes and study the dependence of CO_2 partial pressure in the atmosphere on their structure. For

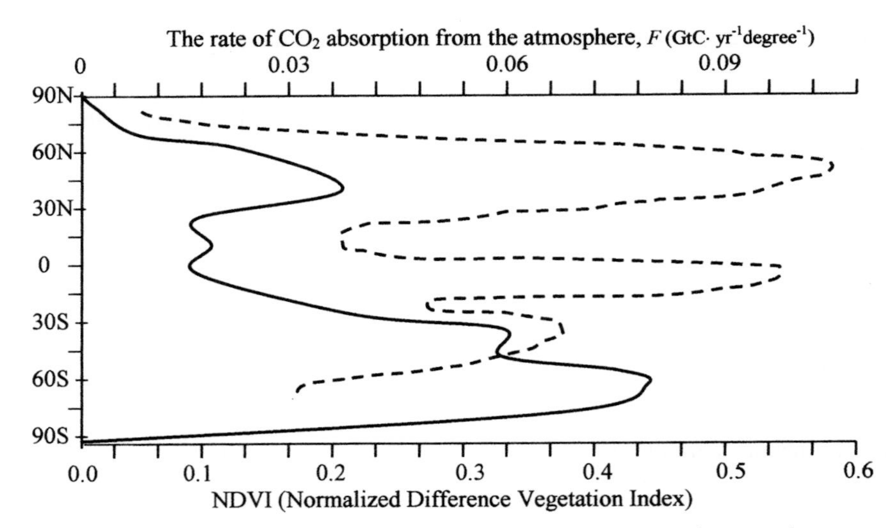

Figure 6.7. The latitudinal distribution of the rate F (GtC yr^{-1} degree^{-1}) of carbon absorption (solid curve) from the atmosphere and vegetation index (dashed curve). Types and spatial distribution of the soil–plant formations are determined in Table 6.5. Industrial emissions of CO_2 are assumed to equal 6.3 GtC yr^{-1}.

instance, if by the year 2050 the forest areas are reduced only by 10% with respect to 1970 ($\sim$42 mln km^2), then by the end of the 21st century the content of atmospheric CO_2 can increase by 46.7%, with stable anthropogenic emissions of carbon about 6 GtC yr^{-1}. On the contrary, broadening of the forest areas in the northern hemisphere by 10% will reduce the anthropogenic impact on the greenhouse effect by 14.8%.

The problem of global warming due to the growth of GHG concentrations is synonymous with the problem of sustainable development of civilization. The approach proposed here enables one to synthesize the accumulated data and knowledge of the GCC and other GHGs into a single monitoring system. Unfortunately, the initiated international program on the GCC study (Canadel *et al.*, 2003), like other similar programs of global character, has not been aimed at developing a constructive information technology able to substantially raise the reliability of prognostic estimates of future climate change. Nevertheless, ideas and approaches of the Russian specialists published recently (Bartsev *et al.*, 2003; Krapivin and Kondratyev, 2002; Tarko, 2005; Kondratyev *et al.*, 2002c, 2003a, 2004b) as well as models by American scientists (Collatz *et al.*, 1990, 1991, 1992; Sellers *et al.*, 1996a,b) will make it possible, though not within the GCP, to overcome the existing isolation of GCC studies and create a global model able, in the operational regime of satellite monitoring, to give reliable estimates of the role of regions in greenhouse effect dynamics. Such a model will be a tool to work out an efficient strategy of land use, and it will lead, of course, to making substantiated (in contrast to the Kyoto Protocol) international decisions.

Table 6.5. Identifier of the types of soil–plant formations following the classification after Bazilevich and Rodin (1967).

Type of the soil–plant formation	Symbol
Arctic deserts and tundras	*A*
Alpine deserts	*B*
Tundras	*C*
Mid-latitude taiga forests	*D*
Pampas and grass savannahs	*E*
North taiga forests	*F*
South taiga forests	*F*
Sub-tropical deserts	*G*
Sub-tropical and tropical grass–tree thickets of the tugai type	*I*
Tropical savannahs	*J*
Saline lands	*K*
Forest tundra	*L*
Mountain tundra	*M*
Tropical xerophytic open woodlands	*N*
Aspen–birch sub-taiga forests	*O*
Sub-tropical broad-leaved and coniferous forests	*P*
Alpine and sub-alpine meadows	*Q*
Broad-leaved coniferous forests	*R*
Sub-boreal and saltwort deserts	*S*
Tropical deserts	*T*
Xerophytic open woodlands and shrubs	*U*
Dry steppes	*V*
Moderately arid and arid (mountain including) steppes	*W*
Forest steppes (meadow steppes)	*X*
Variably humid deciduous tropical forests	*Y*
Humid evergreen tropical forests	*Z*
Broad-leaved forests	+
Sub-tropical semi-deserts	&
Sub-boreal and wormwood deserts	@
Mangrove forests	#
Lack of vegetation	*

Revealing the key factors of global change by modeling the global ecodynamics faces some problems connected with the choice of the form and methods of modeling. The appearing problems have been discussed in studies by Bartsev *et al.* (2003), Kondratyev *et al.* (2003c), Sorokhtin (2001), and others. The main problem here is a combination of parameters of the developed models with available data and knowledge bases as well as the choice of a compromise between complicated structures of the models and their semi-empirical realizations. It is clear that the processes of choice of technology of modeling and interpretation of the results obtained are of mutual character. In this process one can indicate the important parameters of the NSS which affect the global ecodynamics (Kondratyev *et al.*, 2004b). However, as follows from the publications of Barenbaum (2002, 2004) and Yasanov (2003), the description of the GCC lacks an important fact connected with the carbon buried in geological structures and its income from space. Therefore, in perspective, when synthesizing the global model of the carbon cycle, it is necessary to consider a more detailed combined description of the biospheric and lithospheric parts of this cycle. Clearly, the lifetime of carbon in each sub-cycle should be estimated more accurately. The lithospheric part of the GCC includes its transformation during the processes of long interactions and conversions to methane, oil, coalified deposits, etc. Depending on temperature and pressure, hydrocarbons can be oxidized and become the main component of underground fluids and magmas. In this way the carbon cycle correlates with the cycles of methane and water.

6.3.4 Sulfur cycle

The global model of the biosphere can be improved by increasing the number of biogeochemical cycles taken into account. A necessity to include a unit describing the sulfur fluxes in natural systems is dictated by the dependence of biotic processes on the content of sulfur in the compartments of the biosphere. The available data on supplies and fluxes of sulfur compounds in the atmosphere, soils, vegetation cover, and hydrosphere enable one to formulate numerical relationships to simulate the global sulfur cycle.

The global sulfur cycle consists of the mosaic structure of local fluxes of its compounds with other elements formed due to water migration and atmospheric processes. The conceptual schemes of the global and regional cycles of sulfur have been described in detail by many authors (Nitu *et al.*, 2004; Xu and Carmichael, 1999; Stein and Lamb, 2000). However, the existing models have been developed for autonomous functioning and usage, which makes it difficult to include them in the global model without substantial changes to their parametric space. The model of the global sulfur cycle proposed here has been derived in the form of a unit with inputs and outputs, which enables one to combine it with other units of the global model via their inputs and outputs.

In contrast to hydrogen, sulfur compounds cannot be attributed to long-lived elements of the biosphere. Therefore, in the unit of sulfur the spatial digitization of its natural and anthropogenic reservoirs should be planned to reflect the local distributions of sulfur in the vicinity of its sources, and should enable one to estimate the intensities of the inter-regional fluxes of sulfur compounds. The version of the sulfur unit proposed here, in contrast to the known hydrodynamic models of long-range transport, takes into account the fluxes of sulfur compounds between the hydrosphere, atmosphere, soil, and biota. The model does not consider the vertical stratification of the atmosphere. The characteristics of sulfur fluxes over

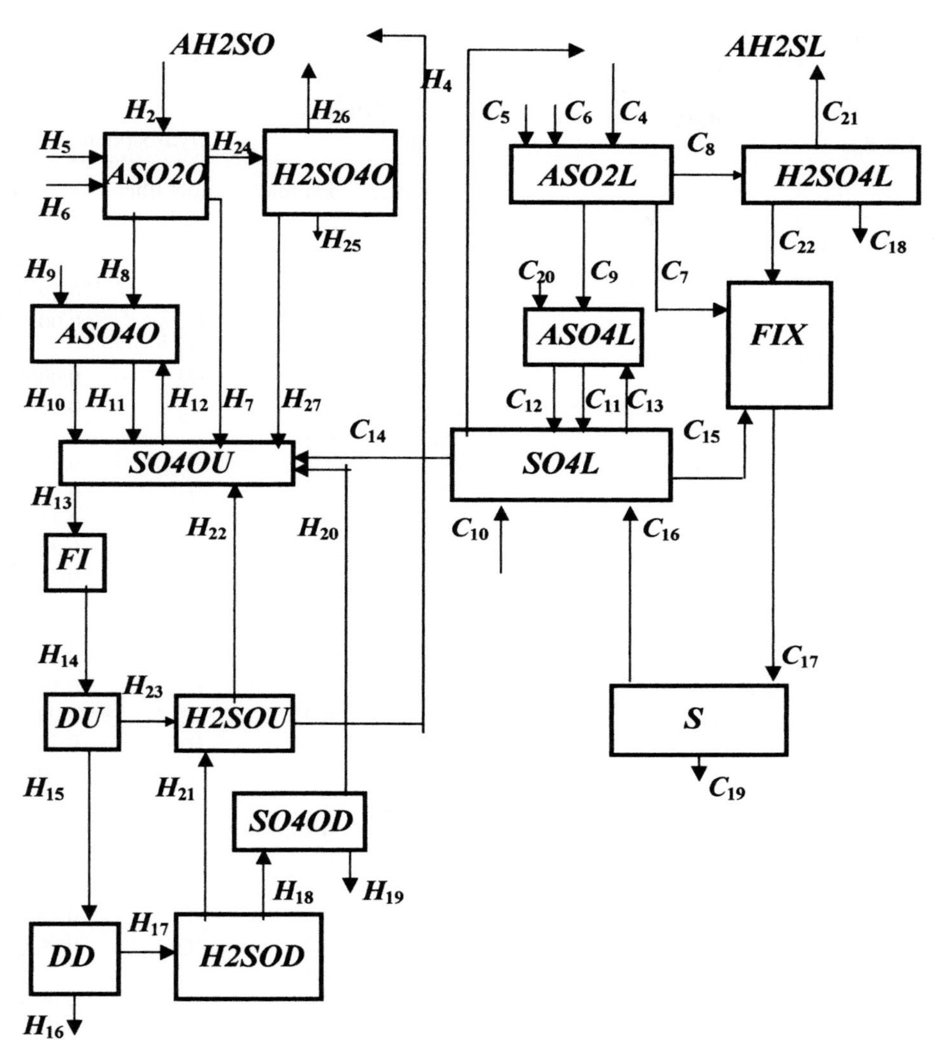

Figure 6.8. The scheme of sulfur fluxes in the environment. Notation given in Tables 6.6 and 6.7.

land and oceans averaged vertically are calculated. The spatial digitization of the biosphere and the World Ocean corresponds to the criterion inherent to the global model. The block-scheme of the model of the biogeochemical cycle of sulfur is shown in Figure 6.8, a description of the fluxes of sulfur compounds is given in Table 6.6. This scheme is realized in every cell Ω_{ij} of the Earth's surface and in every compartment Ω_{ijk} of the World Ocean. The interaction between the cells and the compartments is organized through the climate unit of the global model.

Table 6.6. Characteristics of land and hydrospheric fluxes of sulfur in the structure of Figure 6.8. Numerical estimates of fluxes ($mg\,m^{-2}\,day^{-1}$) are obtained by averaging over the respective territories (Krapivin and Nazarian, 1997).

Sulfur flux	Land		Hydrosphere	
	Identifier	Estimate	Identifier	Estimate
Volcanic invasions:				
H_2S	C_1	0.018	H_3	0.0068
SO_2	C_5	0.036	H_5	0.0073
SO_4^{2-}	C_{20}	0.035	H_9	0.0074
Anthropogenic emissions:				
H_2S	C_2	0.072	H_1	0.00076
SO_2	C_6	0.92	H_6	0.038
SO_4^{2-}	C_{10}	0.47		
Oxidation of H_2S to SO_2	C_4	1.13	H_2	0.3
Oxidation of SO_2 to SO_4^{2-}	C_9	1.35	H_8	0.16
Dry sedimentation of SO_4^{2-}	C_{12}	0.37	H_{11}	0.11
Fall-out of SO_4^{2-} with rain	C_{11}	1.26	H_{10}	0.38
Biological decomposition and emission of H_2S to the atmosphere	C_3	1.03	H_4	0.31
Assimilation of SO_4^{2-} by biota	C_{15}	0.41	H_{13}	1.09
Biological decomposition and formation of SO_4^{2-}	C_{16}	1.13	H_{17}	0.43
			H_{23}	0.12
Sedimentation and deposits	C_{18}	0.22	H_{15}	0.98
	C_{19}	0.11	H_{16}	0.55
			H_{19}	0.0076
			H_{25}	0.036
Wind-driven return to the atmosphere	C_{13}	0.25	H_{12}	0.33
Replenishing sulfur supplies due to dead biomass	C_{17}	0.86	H_{14}	1.1
Assimilation of atmospheric SO_2	C_7	0.46	H_7	0.18
Washing-out of SO_2 from the atmosphere	C_8	0.27	H_{24}	0.061
River run-off of SO_4^{2-} to the ocean	C_{14}	1.17		
Transition of gas-phase H_2SO_4 to H_2S	C_{21}	0.018	H_{26}	0.0076
Assimilation of the washed-out part of atmospheric SO_2 by biota	C_{22}	0.036	H_{27}	0.015
Oxidation of H_2S to SO_2 in water medium			H_{18}	0.045
			H_{22}	0.19
Advection of SO_2			H_{20}	0.38
Advection of H_2S			H_{21}	0.37

Therefore, the equations of the sulfur unit lack the terms reflecting the dynamic pattern of the spatial transformation of the sulfur reservoirs. With account of notations assumed in Figure 6.8 and in Tables 6.6 and 6.7, the equations describing the balance relationships between the reservoirs of sulfur compounds will be written in the form:

$$\frac{dAH2SL}{dt} = C_1 + C_2 + C_3 - C_4 + C_{21}$$

$$\frac{dASO2L}{dt} = C_4 + C_5 + C_6 - C_7 - C_8 - C_9$$

$$\frac{dASO4L}{dt} = C_9 + C_3 + C_{20} - C_{11} - C_{12}$$

$$\frac{dS}{dt} = C_{17} - C_{16} - C_{19}$$

$$\frac{dSO4L}{dt} = C_{10} + C_{11} + C_{12} + C_{16} - C_3 - C_{13} - C_{14}$$

$$\frac{dFIX}{dt} = C_7 + C_{15} - C_{17} + C_{22}$$

$$\frac{dH2SO4L}{dt} = C_8 - C_{18} - C_{21} - C_{22}$$

$$\frac{dAH2SO}{dt} = H_1 + H_3 + H_4 + H_{26} - H_2$$

$$\frac{dASO2O}{dt} = H_2 + H_5 + H_6 - H_7 - H_8 - H_{24}$$

$$\frac{dASO4O}{dt} = H_8 + H_9 + H_{12} - H_{10} - H_{11}$$

$$\frac{\partial SO4OU}{\partial t} + v_z \frac{\partial SO4OU}{\partial z} + k_z \frac{\partial^2 SO4OU}{\partial z^2} = H_7 + H_{10} + H_{11} + H_{20} + H_{22} + H_{27} + C_{14} - H_{12} - H_{13}$$

$$\frac{\partial H2SOU}{\partial t} + v_z \frac{\partial H2SOU}{\partial z} + k_z \frac{\partial^2 H2SOU}{\partial z^2} = H_{21} + H_{23} - H_4 - H_{22}$$

$$\frac{\partial H2SOD}{\partial t} + v_z \frac{\partial H2SOD}{\partial z} + k_z \frac{\partial^2 H2SOD}{\partial z^2} = H_{17} - H_{18} - H_{21}$$

$$\frac{\partial SO4OD}{\partial t} + v_z \frac{\partial SO4OD}{\partial z} + k_z \frac{\partial^2 SO4OD}{\partial z^2} = H_{18} - H_{19} - H_{20}$$

$$\frac{\partial DU}{\partial t} + v_z \frac{\partial DU}{\partial z} + k_z \frac{\partial^2 DU}{\partial z^2} = H_{14} - H_{15} - H_{23}$$

$$\frac{\partial DD}{\partial t} + v_z \frac{\partial DD}{\partial z} + k_z \frac{\partial^2 DD}{\partial z^2} = H_{15} - H_{16} - H_{17}$$

$$\frac{\partial FI}{\partial t} + v_z \frac{\partial FI}{\partial z} + k_z \frac{\partial^2 FI}{\partial z^2} = H_{13} - H_{14}$$

$$\frac{\partial BOT}{\partial t} = H_{16} + H_{19}$$

where v_z is the velocity of the vertical water motion in the ocean (m day^{-1}) and k_z is the coefficient of the turbulent mixing (m^2 day^{-1}).

The above equations in each cell of the spatial division of the ocean surface are supplemented with initial conditions (Table 6.7). The boundary conditions for the equations are zero. The calculation procedure to estimate the sulfur concentration

Table 6.7. Some estimates of sulfur reservoirs which can be used as initial data.

Reservoir	Identifier in equations	Quantitative estimate of sulfur reservoir ($mg\,m^{-2}$)
Atmosphere over the oceans:		
H_2S	AH2SO	10
SO_2	ASO2O	5.3
SO_4^{2-}	ASO4O	2
Atmosphere over land:		
H_2S	AH2SL	36.9
SO_2	ASO2L	17.9
SO_4^{2-}	ASO4L	12.9
Land:		
SO_4^{2-}	SO4L	11.2
biomass	FIX	600
soil	S	5,000
Photic layer of the World Ocean:		
H_2S	H2SOU	1.9
SO_4^{2-}	SO4OU	19×10^7
biomass	FI	66.5
MOB	DU	730
Deep layers of the World Ocean:		
H_2S	H2SOD	2×10^6
SO_4^{2-}	SO4OD	3.4×10^9
MOB	DD	13,120

consists of two stages. At each time moment t_i, initially for all cells Ω_{ij}, these equations are solved by the method of quasi-linearization, and all reservoirs of sulfur are estimated for $t_{i+1} = t_i + \Delta t$, where a time step Δt is chosen from the condition of convergence of the calculation procedure. Then at the moment t_{i+1} with the use of the climate unit of the global model these estimates are specified with account of the atmospheric transport and ocean currents over the time Δt.

The sulfur supplies in the reservoirs are measured in $mg\,S\,m^{-3}$, the sulfur fluxes have the dimensionality $mg\,S\,m^{-3}day^{-1}$. The sulfur supplies in the water medium are calculated with the volumes of the compartments Ω_{ijk} taken into account. To estimate the sulfur supplies to the atmosphere, it is assumed that an effective thickness of the atmosphere h is an input parameter either introduced into the model by the user or prescribed as the constants from Table 6.7, or received from the climate unit of the global model. The quantitative estimates of the fluxes in the right-hand parts of the equations are obtained in different units of the GMNSS. The anthropogenic fluxes of sulfur H_1, H_6, C_2, C_6, and C_{10} are simulated in the unit of scenarios. The fluxes H_3, H_5, H_9, C_1, C_5, and C_{20} are prescribed either by the climate unit or formed in the unit of scenarios. The accuracy of different functional presentations of the fluxes in the equations corresponds to the accuracy of similar fluxes of the biogeochemical cycles of hydrogen, phosphorus, and nitrogen.

The rate of emission of H_2S into the atmosphere at humus decomposition is described by the linear function $C_3 = \mu_1(pH) \cdot SO_4L \cdot T_L$, where μ_1 is the proportion coefficient depending on soil acidity ($day^{-1} \cdot K^{-1}$) and T_L is the soil temperature (K). The initial value of SO_4L is estimated from the humus supply considering that the content of sulfur in humus is prescribed by the parameter a_g (%). According to the available observations of the input of H_2S to the atmosphere from the ocean, the flux H_4 varies widely from low values to high values at the transition between the stagnant waters to zones of upwellings. The flux H_4 is assumed to be the function of the ratio of the rates of H_2S oxidation in the photic layer to the rate of the vertical uprising of water. Therefore, to describe the H_4 flux, we use the parameter t_{H2SU}, which reflects the lifetime of H_2S in the water medium: $H_4 = H_2SU/t_{H2SU}$. Determine the value of t_{H2SU} as a function of the rate of the vertical advection v_z and concentration of oxygen O_2 in the upper layer Z_{H2S} thick: $t_{H2SU} = \mathrm{H_2SOU} \cdot v_z(\theta_2 + \mathrm{O_2})/[\mathrm{O_2}(\theta_1 + v_z)]$, where the constants θ_1 and θ_2 are determined empirically, while the value of O_2 is either calculated by the oxygen unit of the GMNSS or prescribed from the global database.

The reaction of oxidation of H_2S to SO_2 in the atmosphere, on land, and over the water surface is characterized by the rapid process of the reaction of hydrogen sulfide with atomic and molecular oxygen. At the same time, the reaction of H_2S with O_3 in the gas phase is slow. It is impossible to describe within the global model the diversity of the situations appearing here, however, an inclusion of the fluxes H_2 and C_4 into the unit of sulfur enables one to take into account the correlation between the cycles of sulfur and oxygen. These fluxes are parameterized with the use of the indicator t_{H2SA} of the lifetime of H_2S in the atmosphere: $C_4 = A\mathrm{H_2}/t_{H2SA}$, $H_2 = A\mathrm{H_2}/t_{H2SA}$. The mechanism to remove SO_2 from the atmosphere is described by the fluxes H_7, H_8, H_{27}, C_7, C_8, and C_9. Schematically, this mechanism consists of a set of interconnected reactions of SO_2 with atomic oxygen under the influence of various catalysts. A study of the succession of reactions enables one to estimate the lifetime of SO_2 for oxidation over land t_{SO2L} and water surfaces t_{SO2A1}. This makes it possible to assume the following parameterizations of the fluxes H_8 and C_9: $H_8 = ASO_2O/t_{SO2A1}$, $C_9 = ASO_2L/t_{SO2L}$.

Sulfur dioxide is assimilated from the atmosphere by rocks, vegetation, and other Earth covers. Over the water surface this assimilation is connected with the intensity of turbulent gas fluxes and surface roughness. We describe a dry deposition of SO_2 over the vegetation by the model $C_7 = q_2RX$, where $q_2 = q_2' \cdot ASO_2L/(r_{tl} + r_s)$, in which r_{tl} is the atmospheric resistance to the SO_2 transport over the vegetation of the l type ($\mathrm{day\,m^{-1}}$); r_s is the surface resistance of the s type to the SO_2 transport ($\mathrm{day\,m^{-1}}$); RX is the produce of vegetation of the X type ($\mathrm{mg\,m^{-2}\,day^{-1}}$) (calculated by the biogeocenotic unit of the global model); and q_2' is the proportionality coefficient. The parameters r_{tl} and r_s are functions of the types of soil–vegetation formations and estimated, respectively, at 0.05 and 4.5 for the forests, 0.9 and 3 for grass cover, 0.5 and for bushes, 0.8 and 1 for bare soil, 1.9 and 0 for water surfaces, and 2 and 10 for snow cover.

The process of washing-out of SO_2 from the atmosphere with a changing to H_2SO_4 and a subsequent neutralization on the surface of l type is described by the function: $C_8 = q_{1l}W \cdot ASO_2L$ with the Langmuire coefficient q_{1l} and precipitation intensity $W(\varphi, \lambda, t)$. An interaction of acid rain with the Earth's surface elements is reflected in the scheme in Figure 6.8 by the fluxes C_{18}, C_{21}, and C_{22} for land and H_{25}, H_{26}, and H_{27} for water surfaces. To parameterize these fluxes, assume the hypothesis that the reservoirs of H_2SO_4L and H_2SO_4O are spent in proportion to the outfluxes, and the coefficients of this proportion are the controling parameters of the numerical experiments: $C_{18} = h_1 \cdot H_2SO_4L$, $C_{22} = h_2 \cdot RX \cdot H_2SO_4L$, $C_{21} = h_3T_a \cdot H_2SO_4L$, $H_{25} = h_6 \cdot H_2SO_4O$, $H_{26} = h_4T_a \cdot H_2SO_4O$, $H_{27} = h_5 \cdot RFI \cdot H_2SO_4O$, $h_1 + h_2 \cdot RX + h_3T_a = 1$, $h_4T_a + h_5 \cdot RFI + h_6 = 1$, where $T_a(\varphi, \lambda, t)$ is the air surface temperature.

Parameterize the fluxes H_7 and H_{24} by the relationships: $H_7 = ASO_2O/t_{SO2A2}$, $H_24 = q_{1l}W \cdot ASO_2O$ where t_{SO2A2} is the lifetime of SO_2 over the water surface (day^{-1}).

Sulfates interacting with the ecosystems and establishing an interaction between the sulfur cycle and other biogeochemical processes are one of the most important elements in the global cycle of sulfur. Numerous complicated transformations of sulfates in the environment are described by a set of fluxes H_7, H_8, H_{10}, H_{11}, H_{12}, C_9, C_{11}, C_{12}, and C_{13} for the atmospheric reservoir and the fluxes H_{13}, H_{18}, H_{19}, H_{20}, H_{22}, C_3, C_{14}, C_{15}, and C_{16} for land and the World Ocean.

Physical mechanisms for the transport of sulfates from the atmosphere to the soil and water medium are connected with dry and wet deposition. An efficient model of the wet removal of particles and gases from the atmosphere was proposed by Langmann (2000): a substitution of the mechanism of the aerosols and gases by a simplified binary model enables one to combine it with other units of the global model: $H_{10} = \mu W \cdot ASO_4O$, $H_{11} = \rho v_O \cdot ASO_4O$, $C_{11} = b_3W \cdot ASO_4L$, $C_{12} = d_1v_a \cdot ASO_4L$, where v_O and v_a are the rates of the aerosol dry deposition over the water surface and land, respectively; μ, b_3, ρ, and d_1 are the constants.

The return of sulfates from the soil and water medium to the atmosphere is connected with rock weathering and sea spray above the rough water surface: $C_{13} = d_2 \cdot \text{RATE} \cdot SO_4L$, $H_{12} = \theta \cdot \text{RATE} \cdot SO_4U$, where $\text{RATE}(\varphi, \lambda, t)$ is the wind speed over the surface (m s^{-1}) and d_2 and θ are the empirical coefficients.

The flux C_{14} relates the surface and water reservoirs of sulfur. Let σ be the share of the river system area on land and d_3 – the proportion coefficient, then $C_{14} = d_3W \cdot SO_4L + (C_{11} + C_{12})\sigma$.

The surface part of the sulfur cycle is connected with the functioning of the atmosphere–vegetation–soil system. Plants adsorb sulfur from the atmosphere in the form of SO_2 (fluxes C_7 and C_{22}) and assimilate sulfur from the soil in the form of SO_4^{2-} (flux C_{15}). In the hierarchy of the soil processes two levels can be selected defining the sulfur reservoirs as "dead organics" and "SO_4^{2-} in soil". The transitions between them are described by the flux $C_{16} = b_2ST_L$, where the coefficient $b_2 = b_{2,1}b_{2,2}$ reflects the rate $b_{2,1}$ of transition of sulfur contained in dead organics into the form assimilated by vegetation The coefficient $b_{2,2}$ indicates the content of sulfur in dead plants.

The fluxes of sulfur in the water medium according to studies by Bodenbender *et al.* (1999), depend on the biological processes in the water bodies and constitute an isolated part of the global cycle of sulfur that contains only the fluxes that connect it with the atmospheric and surface cycles. Rough estimates show that the rates of the sulfur cycle in the water of the seas and oceans do not play a substantial role for the remaining parts of its global cycle. And though this specific feature does exist, for the purity of the numerical experiment, in the proposed model the internal hydrospheric fluxes of sulfur compounds are separated in space and parameterized with the same details as other fluxes of sulfur in the atmosphere and on land. This excessiveness is important for other units of the global model as well. In particular, it is important for the parameterization of photosynthesis whose rate *RFI* affects the closure of other biogeochemical cycles. Finally, assume: $H_{13} = \gamma \cdot RFI$, $H_{14} = b \cdot MFI$, $H_{15} = f \cdot DU$, $H_{16} = p \cdot DD$, $H_{17} = q \cdot DD$, $H_{18} = H_2SOD/tH_2SOD$, $H_{19} = u \cdot \mathrm{SO_4}D$, $H_{20} = a_1 v_D \cdot \mathrm{SO_4}D$, $H_{21} = b_1 v_D \cdot H_2SOD$, $H_{22} = H_2SOU/t_{H2SOU}$, $H_{23} = g \cdot DU$, where *MFI* is the mass of dead phytoplankton, t_{H2SOU} and t_{H2SOD} is the time of complete oxidation of H_2S in the seawater in the photic and deep layers, respectively; γ, b, f, p, q, u, a_1, b_1, and g are the constants.

The anthropogenic input to the sulfur unit is realized through the prescribed fluxes C_2, C_6, C_{10}, H_1, and H_6 in the form of the functions of the spatial coordinates and time.

6.3.5 Nitrogen cycle

The MGNC unit simulating the fluxes of nitrogen in the environment is necessary in the global biospheric model for several indisputable reasons: nitrogen compounds can affect environmental conditions, change food quality, affect climate, and transform hydrospheric parameters. An abundant use of nitrates leads to water pollution and deteriorates the quality of food products. It is well known that an intensive exploitation of soils, without taking into account the consequences of the misuse of nitrogen fertilizers, breaks the stability of the agri-ecosystems and human health. Besides this, nitrogen protoxide (N_2O), nitrogen dioxide (NO_2), and nitrogen oxide (NO), being minor gas components (MGCs) of the atmosphere, substantially affect the formation of the processes of absorption of optical radiation in the atmosphere. Small deviations in their concentrations can cause significant climatic variations near the Earth's surface.

The nitrogen cycle is closely connected with the fluxes of hydrogen, sulfur, and other chemicals. The global cycle of nitrogen as one of the nutrient elements is a mosaic structure of the local processes of its compounds formed due to water migration and atmospheric processes. The present-day nitrogen cycle is strongly subject to anthropogenic forcings manifested through direct and indirect interference. Therefore, the construction of an adequate model of the nitrogen cycle's nature should be based on the description of the whole complex of natural processes and also those initiated by humans.

The natural sources of nitrogen oxides are connected with the vital functions of bacteria, volcanic eruptions, as well as several atmospheric phenomena (e.g., lightening discharges). The biogeochemical cycle of nitrogen includes processes such as fixation, mineralization, nitrification, assimilation, and dissimilation. The structural schemes of these processes have been described in detail by many authors (Ehhalt, 1981; Ronner, 1983). Their level of complexity is determined by the goal of the study, availability of data on the rates of transformation of nitrogen-containing compounds and their supplies, and by the level of detailing, etc.

Nitrogen transport in the biosphere is driven by a complicated meandering structure of fluxes including a hierarchy of cycles at various levels of life organization. From the atmosphere, nitrogen enters the cells of micro-organisms, from which it moves into soil and then to higher plants, animals, and humans. Survivability of living organisms results in the return of nitrogen to the soil, from which it either re-enters plants and living organisms or is emitted to the atmosphere. A similar scheme of nitrogen oxide cycling is inherent in the hydrosphere. The characteristic feature of these cycles is their openness connected with the available processes of removal of nitrogen from the biospheric balance into rocks, from where it returns much more slowly. Taking into account the nature of the nitrogen cycle in the biosphere and its reservoir structure, enables one to formulate a global scheme of nitrogen fluxes.

A comparative analysis of the model schemes of the flux diagram of nitrogen compounds in nature proposed by various experts makes it possible to construct the block-scheme presented in Figure 6.9. Here the atmosphere, soil, lithosphere, and hydrosphere are considered as nitrogen reservoirs. The first three reservoirs are described by 2-D models, and the hydrosphere is described by a 3-D multi-layer model. The characteristics of nitrogen fluxes between these reservoirs are given in Table 6.8. The equations of the model are written as:

$$\frac{\partial N_A}{\partial t} + V_\varphi \frac{\partial N_A}{\partial \varphi} + V_\lambda \frac{\partial N_A}{\partial \lambda} = H_1^N + \begin{cases} H_{20}^N - H_{16}^N & (\varphi, \lambda) \in \Omega_O \\ H_7^N + H_{19}^N - H_8^N - H_9^N & \\ \quad + H_{22}^N - H_2^N - H_{10}^N & (\varphi, \lambda) \in \Omega/\Omega_O \end{cases}$$

$$\frac{\partial N_{S1}}{\partial t} = H_8^N + H_6^N - H_3^H$$

$$\frac{\partial N_{S2}}{\partial t} = H_2^N + H_3^N + H_5^N + H_9^N - H_6^N - H_7^N - H_{11}^N - H_{21}^N$$

$$\frac{\partial N_U}{\partial t} + v_\varphi \frac{\partial N_U}{\partial \varphi} + v_\lambda \frac{\partial N_U}{\partial \lambda} = H_{16}^N + H_{4,U}^N + H_{18,U}^N + H_{11}^N - H_{17,U}^N$$
$$- H_{20}^N - H_{14,UP}^N - H_{15,UP}^N$$

$$\frac{\partial N_P}{\partial t} + v_\varphi \frac{\partial N_P}{\partial \varphi} + v_\lambda \frac{\partial N_P}{\partial \lambda} = H_{18,P}^N + H_{4,P}^N + H_{14,UP}^N + H_{15,PL}^N$$
$$- H_{17,P}^N - H_{14,PL}^N - H_{15,UP}^N$$

$$\frac{\partial N_L}{\partial t} = Q_L + H_{12,L}^N + H_{14,PL}^N + H_{15,LF}^N - H_{14,LF}^N - H_{15,PL}^N$$

$$\frac{\partial N_F}{\partial t} = Q_F + H_{12,F}^N + H_{23}^N + H_{14,LF}^N - H_{13}^N - H_{15,LF}^N$$

where $V(V_\varphi, V_\lambda)$ is the wind speed, $v(v_\varphi, v_\lambda)$ is the current velocity in the ocean, and Q_L and Q_F are functions describing the mixing of the deep waters of the ocean.

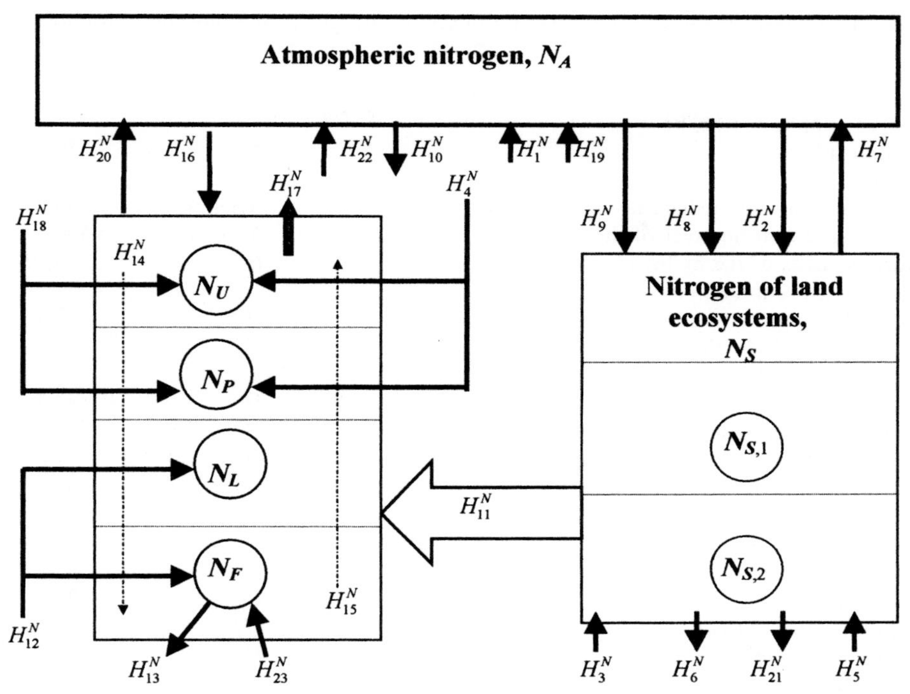

Figure 6.9. The scheme of nitrogen fluxes in the GMNSS. Notation given in Table 6.8.

To simplify the calculation scheme presented in Figure 6.9, advection processes in the equations can be described by a superposition of the fluxes H_{14}^N and H_{15}^N. The computer realization of the equations of the nitrogen unit introduces into its equations some corrections for the agreement between the dimensionalities of the variables in conformity with the spatial digitization of Ω. Therefore, the estimates of the fluxes H_i^N given below should be corrected following this criterion.

6.3.5.1 The atmospheric part of the nitrogen cycle

The atmospheric part of the nitrogen cycle is an example of a complicated mechanism of transformation of gas substances characterized by an intricate set of fluxes on the borders between the basic reservoirs of nitrogen. Nevertheless, the results obtained by many experts make it possible to formulate clear ideas about the flux diagram of nitrogen compounds in the atmosphere.

Nitrogen resides in the atmosphere both in a free state (N_2) and in the form of various compounds – ammonia (NH_3), nitrogen protoxide, nitrogen oxide, nitrogen dioxide, and other nitrogen oxides (NO_3, N_2O_3, N_2O_4, N_2O_5), which play an intermediate role in chemical reactions. From the available estimates, the active part of atmospheric nitrogen constitutes 3.92×10^{12} t (i.e., $N_A = 0.77 \times 10^4\,\mathrm{t\,km^{-2}}$). Detailing the atmospheric reactions of nitrogen is incomplete because of

Table 6.8. Characteristics of reservoirs and fluxes of nitrogen in the biosphere (Figure 6.9).

Reservoirs (Gt) and fluxes (10^6 t yr^{-1})	Identifier	Estimate
Nitrogen supplies:		
atmosphere	N_A	39×10^5
soil	N_S	280
photic and intermediate layer of the ocean	$N_U + N_P$	2,800
deep and bottom layer of the ocean	$N_L + N_F$	36,400
Natural sources of the hydrosphere	H_1^N	0.392
Technogenic accumulation:		
fuel burning	H_2^N	22.8
fertilizers production	H_9^N	41.8
Input due to dead organisms:		
on land	H_3^N	42.2
in upper layers of the World Ocean	H_{18}^N	5
in deep layers of the World Ocean	H_{12}^N	7.8
Input due to organisms functioning:		
on land	H_5^N	0.1
in the World Ocean	H_4^N	0.3
Biological fixation:		
on land	H_6^N	20.3
in the World Ocean	H_{17}^N	10
in the atmosphere	H_{10}^N	40
Denitrification:		
on land	H_7^N	52
in the World Ocean	H_{20}^N	49.8
Atmospheric fixation:		
over land	H_8^N	4
over the World Ocean	H_{16}^N	3.6
Runoff from land into the World Ocean	H_{11}^N	38.6
Precipitation	H_{13}^N	0.5
Vertical exchange processes in the oceans:		
descending	H_{14}^N	0.2
ascending	H_{15}^N	7.5
Anthropogenic emissions to the atmosphere	H_{19}^N	15
Removal of nitrogen from the nitrogen cycle due to sedimentation	H_{21}^N	0.2
Input of nitrogen to the atmosphere during rock weathering	H_{22}^N	0.217
Input of nitrogen to the water medium with dissolving sediments	H_{23}^N	0.091

insufficiently studied sources and behavior of various forms of ammonia. The most important reactions in the atmosphere are the following:

$$NO_2 \rightarrow NO + O, \varphi K_a = 0\text{–}25\,\text{hr}^{-1}$$

$$O_3 + NO \rightarrow NO_2 + O_2, K_1 = 1,320\,\text{ppm}\,\text{hr}^{-1}$$

The photochemical equilibrium is described by the relationship:

$$(NO)(O_3)/(NO_2) = \varphi K_a/K_1$$

The time of relaxation in this case constitutes 16 s and therefore the equilibrium between NO, NO_2, and O_3 in the atmosphere can be considered stable. However, the equilibrium $N_2 + O_2 \leftrightarrow 2NO$ under anthropogenic conditions is connected with the transformation of NO into NO_2 during several hours. Therefore, here too, from the viewpoint of global modeling, a separate consideration of the components NO and NO_2 is inexpedient. In other words, we shall consider nitrogen of the atmosphere as a generalized component of the global model.

Atmospheric fixation. As a result of various physico-chemical processes, taking place in the atmosphere, free nitrogen can move from the atmosphere to soil and water bodies. Fixation of atmospheric nitrogen due to electrical charges and photochemical processes constitutes annually not more than $0.035\,t\,km^{-2}$ (from more accurate estimates, it may constitute $0.027\,t\,km^{-2}$ for land and $0.01\,t\,km^{-2}$ for the oceans). Since the nitrogen fluxes due to atmospheric fixation are mainly determined by meteorological conditions, it is quite natural to consider them independently for each region of land and each water body of the World Ocean in the form of the functions of temperature and precipitation.

The flux H_{16}^N of nitrogen fixed in the atmosphere over any ocean basin is described by the relationship:

$$H_{16}^N = [\lambda_1(\theta_1)^{\Delta T} + \lambda_2 R_W]N_A$$

where ΔT is the atmospheric temperature variation, θ_1 is the indicator of the temperature dependence of the rate of atmospheric fixation of nitrogen, R_W is precipitation, and λ_1 and λ_2 are the coefficients.

The equation of atmospheric fixation over a land site Ω_{ij} is written by analogy to H_{16}^N:

$$H_{8,ij}^N = [\lambda_3(\theta_1)^{\Delta T} + \lambda_4 R_{W,ij}]N_A$$

where λ_3 and λ_4 are coefficients.

To estimate the coefficients $\lambda_i (i = 1, \dots, 4)$, as a first approximation one can use average data on nitrogen fluxes and precipitation. If we assume $H_{16}^N = 9.96 \times 10^{-3}\,t\,km^{-2}\,yr^{-1}$, $H_8^N = 0.027\,t\,km^{-2}\,yr^{-1}$, estimate local precipitation over the ocean and land at $1.01\,m\,yr^{-1}$ and $0.24\,m\,yr^{-1}$, respectively, and the convective precipitation over the ocean and land at $0.19\,m\,yr^{-1}$ and $0.116\,m\,yr^{-1}$, respectively, we obtain $\lambda_1 = 0.004\,98$, $\lambda_2 = 0.004\,58$, $\lambda_3 = 0.0135$, and $\lambda_4 = 0.0285$. These estimates are easily specified with account of local data at a fixed time moment for smaller regions and water bodies.

Input of nitrogen into the atmosphere from geospheric sources. The flux of nitrogen H_1^N is determined by the geothermal activity of the Earth. Its estimates testify to the necessity to consider this constituent in the global model. In particular, for instance, in the nitrogen fumaroles of Vesuvius the content of nitrogen by weight constitutes 98%, in gases of the lavas of the Hawaiian volcanoes there is only 5.7% of nitrogen, and over the globe an input of juvenile nitrogen averages

$0.4 \times 10^6\,\text{t yr}^{-1}$. Let H_1^N be a function of time approximating a statistical series of observations. A more strict account of this flux of nitrogen in the model can be realized by using the algorithms of parameterization of random processes, for instance, with the use of evolutionary modeling. However, within the global model, orientated toward describing the processes with a time step of decades, it is enough to use a consecutive approximation with an average annual step-by-step discretization.

The flux H_1^N can be, to some extent, interpreted as a compensation for the fluxes H_{13}^N and H_{21}^N.

6.3.5.2 The surface part of the nitrogen cycle

The nitrogen supplies on land consist of the assimilable nitrogen in soil $N_{S2} \approx 0.19 \times 10^4\,\text{t km}^{-2}$, in plants ($12 \times 10^9$ t), and living organisms (0.2×10^9 t). A diversity of nitrogen fluxes is formed here of the processes of nitrification, denitrification, ammonification, fixation, and river run-off. The intensities of these fluxes depend on climatic conditions, temperature regime, moistening, as well as the chemical and physical properties of soil. Many qualitative and quantitative characteristics of these dependences have been described in the literature. Let us consider some of them.

Nitrification. The return of nitrogen to the cycle due to living micro-organisms is one of the stabilizing natural processes. Simplifying the whole process of the transformation of ammonia salts into nitrates, let us present the activity of heterotrophic micro-organisms and saprophags as a generalized process of organic matter decomposition. The rate of organic matter decomposition and nitrification increases with increasing temperature, reaching its optimal value at $T_a = 34.5°\text{C}$. Therefore, for the flux H_3^N an approximation $H_3^N = \lambda_{N\kappa} M_\kappa$ can be assumed, where M_κ is the rate of dying-off of the component κ, and $\lambda_{N\kappa}$ is the content of nitrogen in the component κ.

Denitrification. The processes of denitrification of H_7^N on land are important channels for the input of nitrogen to the atmosphere. The intensity of these processes depends on temperature, humidity, pollution of soil with poisonous chemicals, and pH. The quantitative and functional characteristics of these dependences have been well studied. Within the global model it is possible to take into account only the factors of temperature and humidity:

$$H_7^N = \lambda_6 \theta_2^{\Delta T} W_S \frac{N_S}{k_1 + N_S}$$

where W_S is soil moisture, θ_2 is the temperature coefficient, and λ_6 and κ_1 are empirical parameters. If we assume $H_7^N = 0.318\,\text{t km}^{-2}\,\text{yr}^{-1}$, then $\lambda_6 = 0.496$, $k_1 = 0.556$.

Biological fixation. In the biological cycle of nitrogen, of importance are the processes of its fixation by micro-organisms and plants whose intensity is estimated at $148 \times 10^6\,\text{t yr}^{-1}$. The rate of fixation, depending on the character of the medium, can vary reaching $3 \times 10^9\,\text{t yr}^{-1}$ in highly productive regions. The nitrogen flux H_{10}^N depends on the distribution of vegetation cover and can be described by the equation $H_{10}^N = \sigma_\kappa \lambda_\kappa R_\kappa / \sigma_{ij}$, where σ_κ is the area under vegetation of κ type on the territory Ω_{ij} of the area σ_{ij}, R_κ is the productivity of plants of κ type, and λ_κ is the coefficient.

Fixation of nitrogen by plants directly from the soil via the root systems (flux H_6^N) occupies a principal place in the cycle of nitrogen, especially in territories with cultural vegetation. For instance, an increase of the share of legumes in agriculture can raise H_6^N to 35 t km^{-2} yr^{-1}. Therefore, a consideration of this flux in the model is necessary and can be realized in the following form:

$$H_6^N = \sigma_\kappa R_\kappa \mu_\kappa / \sigma_{ij}$$

where μ_κ is a constant.

The rate of assimilation of nitrogen by the roots of plants is known to depend much on the soil temperature regime, decreasing slightly with temperatures of 8–10°C and dropping rapidly at temperatures below 5–6°C. The motion of nitrogen from the roots to the upper parts of the plants slows down, too.

On land, plants assimilate annually about 30×10^6 t N from the atmosphere and more than 5.3×10^6 t N directly from the soil. Approximate estimates of the productivity of various types of vegetation average $R_\kappa = 710$–$3,243$ t km^{-2} yr^{-1}. Hence, from the above equation we have $\mu_\kappa = 0.134 \times 10^{-5}$–$0.506 \times 10^{-4}$.

The loss of nitrogen through washing-out from soils. On global scales, the means of nitrogen migration include the transport of its compounds between land and oceans due to water run-off. The annual input of nitrogen from land to the World Ocean is estimated at 38.6×10^6 t. If the total sink to the ocean from land is described by the function W_{SO}, then the nitrogen flux H_{11}^N can be approximated by the expression:

$$H_{11}^N = \lambda_N N_{S2}[1 - \exp(-k_N W_{SO})]$$

where λ_N and k_N are the coefficients. The functional form foresees that the nitrogen flux from land to the ocean is equal to zero in the absence of the run-off and stabilizes at a level λ_N with the run-off volume considerably increasing. To estimate the parameters λ_N and k_N, it is necessary to take into account the spatial heterogeneity of the types of soil–vegetation formations, relief, and other geophysical parameters. In particular, the content of nitrogen compounds in water differs as a function of the run-off territory. The river waters in the forest regions with temperate climates contain 0.4 mg l^{-1} of nitrates, for the arid climate this value is 1.45 mg l^{-1}. The concentration of nitrates increases sharply in drainage waters of the irrigation systems (5.5 mg l^{-1}), in the river waters of thickly populated regions (25 mg l^{-1}), and reaches a maximum in the soil solutions of the salted irrigated soils (200 mg l^{-1}). Ground waters contain from 10 to 100 mg l^{-1} of nitrates. With the total run-off of water into the World Ocean reaching 50×10^3 km^3, 30% of which constitutes the underground run-off, the total flux of nitrogen per unit area of the ocean will be 0.107 t yr^{-1}. Assuming $W_{SO} = 0.337$ m and that a 95% level of the sink saturation is reached at a five-fold increase of W_{SO}, we obtain $k_N = 1.367$, $\lambda_N = 0.708$.

The surface part of the nitrogen cycle contains a constant process of removal of nitrogen from the biosphere into deposits, in particular, as a result of accumulation of saltpeter on the Earth's surface through erosion and alkalifying. From the available estimates, $H_{21}^N \approx 3.9 \times 10^{-4}\,\mathrm{t\,km^{-2}\,yr^{-1}}$, with $H_{21}^N < H_{22}^N$, but $H_{21}^N + H_{13}^N \cong H_1^N + H_{22}^N + H_{23}^N$. This relationship follows from the fact that during the geological period the loss of nitrogen had been balanced by its input. Of course, in the present biosphere, with the changing intensity of most of the fluxes enumerated in Table 6.8, this balance has started breaking due to increasing fluxes H_9^N and H_{19}^N.

6.3.5.3 The hydrospheric part of the nitrogen cycle

In seawater, nitrogen is present in the form of dissolved gas, ions of ammonium NH_4^+, nitrite NO_2^-, nitrate NO_3^-, and in the form of various organic compounds. Inorganic nitrogen compounds are assimilated by algae and phytoplankton and thus transfer into organic forms that serve as food for living organisms. An expenditure of inorganic nitrogen supplies is compensated by atmospheric precipitation, river run-off, and mineralization of organic remains in the processes of living organism functioning and dying-off. The nitrogen cycle in the seawater is presented schematically in Figure 6.10. Of course, not all nitrogen fluxes existing in nature have been taken into account here. A diversity of the ways of nitrogen transformation in water has been studied inadequately, though the available information is sufficient for the global model. The processes, such as replenishing the nitrogen supplies in water due to lysis of detritus and functioning of living organisms, the nitrogen exchange between photic and deep layers of the ocean, as well as nitrogen fixation at photosynthesis and denitrification, have been thoroughly studied and described in

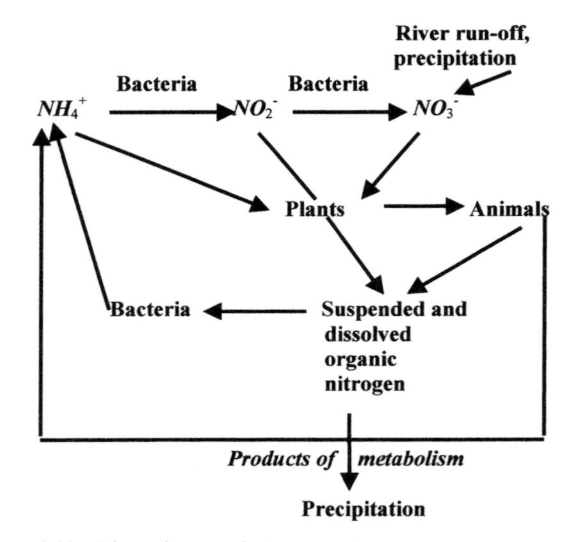

Figure 6.10. The scheme of nitrogen fluxes in the marine medium.

the literature (Valiela and Bowen, 2002; Roda *et al.*, 2002; Pauer and Auer, 2000). Also, there are rough estimates of nitrogen supplies to the ocean, according to which one can assume, on average, that $N_U = N_P = 0.77 \times 10^4\,\mathrm{t\,km^{-2}}$ and $N_L = N_F = 10^5\,\mathrm{t\,km^{-2}}$. More detailed spatial distributions of nitrogen supplies in the hydrosphere can be calculated from the data on biomass, dissolved organic matter, and concentration of dissolved oxygen. The volume relationships of dissolved nitrogen are related to the volume of oxygen as $\mathrm{ml}\,N_2\,\mathrm{l^{-1}} = 1.06 + 1.63\,\mathrm{ml\,O_2\,l^{-1}}$.

The nitrogen supplies in water bodies are replenished due to bacterial decomposition of organic sediments and dissolved organic matter. Consider the component D denoting the content of dead organic matter in the water medium. On its basis, one can write the following relationship: $H_{18}^N = \lambda_D D(\varphi, \lambda, z, t)$, where λ_D is the indicator of the nitrogen content and the rate of detritus lysis. The free nitrogen supplies in the water medium are also replenished in the process of functioning of various organisms. With account of phytoplankton Φ and nekton r, we have:

$$H_4^N = H_r^N + H_\Phi^N;\ H_r^N = \lambda_r T_r;\ H_\Phi^N = \lambda_\Phi T_\Phi$$

where T_r and T_Φ are the characteristics of metabolic processes, respectively, in nekton and phytoplankton, and λ_r and λ_Φ are the coefficients. To determine the averages of these coefficients, assume $T_r = 0.194\,\mathrm{t\,km^{-2}\,yr^{-1}}$, $T_\Phi = 0.125\,\mathrm{t\,km^{-2}\,yr^{-1}}$, and $H_r^N = H_\Phi^N = 0.83 \times 10^{-3}\,\mathrm{t\,km^{-2}\,yr^{-1}}$. Then $\lambda_r = 0.004\,28$, $\lambda_\Phi = 0.006\,64$.

The process of denitrification in the water medium delivers considerable amounts of nitrogen to the atmosphere: $H_{20}^N = \lambda_5(\theta_2)^{\Delta T} N_U(\varphi, \lambda, z, t)$, where λ_5 and θ_2 are constants.

The biological fixation of nitrogen in the water medium constitutes about $10 \times 10^6\,\mathrm{t\,yr^{-1}}$, reaching $20.7 \times 10^6\,\mathrm{t\,yr^{-1}}$ in the photic layer of the ocean, and $(36\text{–}1,800) \times 10^4\,\mathrm{t\,km^{-3}\,yr^{-1}}$ in small lakes. For the World Ocean, $H_{17}^N = 0.0277\,\mathrm{t\,km^{-2}\,yr^{-1}}$, on average. Assuming $H_{17}^N = \lambda_R R_\Phi$, where R_Φ is the phytoplankton produce averaging $168.8\,\mathrm{t\,km^{-2}\,yr^{-1}}$, we obtain $\lambda_R = 0.164 \times 10^{-3}$.

The characteristic feature of the nitrogen cycle in the water medium is its transport due to gravitational sedimentation, vertical convection, turbulent diffusion, and convergence. The processes of nitrogen transport by migrating animals are almost negligible and can be neglected in the global model. The simplest form of description of the vertical fluxes of nitrogen is reduced to the model $H_{14}^N = \lambda_\kappa \Delta N_\kappa$, $H_{15}^N = \lambda_\rho \Delta N_\rho$, where $\kappa = (U, P, L)$, $\rho = (P, L, F)$.

6.3.5.4 *Anthropogenic factors of the nitrogen cycle*

The present contribution of human activity to the general biospheric cycle of nitrogen has reached a level when the consequences of the introduced changes become unpredictable and probably rather catastrophic (Mosier *et al.*, 2004). The epidemiological studies testify to the growth of respiration diseases in territories with high concentrations of nitrogen and sulfur oxides as well as photochemical oxidizers. The harmful effect of nitrogen oxides on living organisms starts manifesting itself when the level 940 μkg m^{-3} is exceeded. In general, the consequences of the nitrogen pollution of the biosphere are more complicated. For instance, on the one hand, the technogenic accumulation of nitrogen from the atmosphere due to industrial fertilizer production plays a positive role by raising the productivity of land and water ecosystems, and on the other hand, it causes an undesirable eutrophication of water basins. Removal of nitrogen from the atmosphere for industrial and agricultural needs is compensated for by a technogenic input of nitrogen into the atmosphere with the burning of solid and liquid fuels. A considerable share is contributed here by the automobile and other transport emitting nitrogen oxides reaching, for instance, in the USA, 11.7×10^6 t yr^{-1}. However, even an observance of this physical equilibrium cannot substitute for the chemico-biological balance. Therefore, in this multifunctional hierarchical set of global fluxes of nitrogen, the most vulnerable aspects should be revealed, which is only possible within a well-planned numerical experiment.

The quantitative estimate of the main stages of the nitrogen cycle with account of anthropogenic factors enables one to see an integral effect of breaking the global balance of nitrogen. But from the available data on the global distribution of the violations of the nitrogen cycles it is impossible to reliably assess the contribution of the industrial synthesis of nitrogen compounds and their scattering over the globe into its biogeochemical cycle. During the last years of the 20th century the industry increased the total amount of nitrogen circulating in the biosphere by 50%. As a result, the natural equilibrium between the processes of nitrification and denitrification has turned out to be violated with an excess of nitrates reaching 9×10^6 t.

Preliminary estimates of an increasing anthropogenic pressure on the nitrogen fluxes between the biospheric elements suggest a hypothesis about a strong correlation existing between the fertilizers production H_9^N and population density G, technogenic accumulation of nitrogen at fuel burning H_2^N and mineral resources expenditure R_{MG}, anthropogenic input of nitrogen into the atmosphere H_{19}^N, and intensity of emissions of general pollution Z_{VG}. The quantitative characteristics of these dependences can be obtained from known trends. From some estimates, the amount of nitrogen oxides emitted into the atmosphere is proportional to the weight of the used fuel with a 4% annual increasing trend. The scales of industrial fixation of nitrogen for the last 40 years increased by a factor of 5, reaching a value that could have been fixed by every ecosystem on the Earth before the use of the present agricultural technology. In 1968 the global industry introduced about 30×10^6 t of fixed nitrogen, and in 2000 this value reached 1 billion.

Let us formalize these correlations using the following models:

$$H_9^N = \min\{U(K)\bar{G}(K,t); N_A\sigma_K/\sigma\}$$

$$H_2^N = \lambda_{AG}R_{MG}, H_{19}^N = \lambda_{GA}Z_{VG}$$

where K is the number of an economic region, $\bar{G}$ is the average population density of the K region, and σ_K is the area of the K region. The coefficients U, λ_{AG}, and λ_{GA} are determined from analysis of available information. If we assume that $H_2^N = 0.154$ t km^{-2} yr^{-1}, $H_{19}^N = 0.102$ t km^{-2} yr^{-1}, and $H_9^N = 0.283$ t km^{-2} yr^{-1},

Table 6.9. Basic reactions of the global biogeochemical cycle of nitrogen and their characteristics.

Reaction	Reaction formula	Energy output (kcal)
Ammonification	$CH_2NH_2COOH + 3/2\ O_2 \rightarrow 2CO_2 + H_2O + NH_3$	176
Nitrification	$NH_3 + 1/2O_2 \rightarrow HNO_2 + H_2O$	66
	$KNO_2 + 1/2\ O_2 \rightarrow KNO_3$	17.5
Fixation	$N_2 \rightarrow 2N$	−160
	$2N + 3\ H_2 \rightarrow 2\ NH_3$	12.8
Respiration	$C_6H_{12}O_6 + 6O_2 \rightarrow 6CO_2 + 6H_2O$	686
Denitrification	$C_6H_{12}O_6 + 6\ KNO_3 \rightarrow 6CO_2 + 3H_2O + 6KOH + 3N_2O$	545
	$5\ C_6H_{12}O_6 + 24\ KNO_3 \rightarrow 3OCO_2 + 18\ H_2O + 24KOH + 12\ N_2$	570
	$5\ S + 6\ KNO_3 + 2\ CaCO_3 \rightarrow 3K_2SO_4 + 2\ CaSO_4 + 2CO_2 + 3\ N_2$	132

then at $\bar{G} = 24.4$ people km^{-2}, $R_{MG} = 30.5$ conv. units km^{-2} yr^{-1}, and $Z^{VG} =$ 3.39 t km^{-2} yr^{-1}, we obtain $U = 0.283$, $\lambda_{AG} = 0.504 \times 10^{-2}$, and $\lambda_{GA} = 0.03$.

The anthropogenic interference into the nitrogen cycle can also have medico-biological consequences expressed, for instance, through increased mortality with growing amounts of NO_2 at 190–320 μkg m^{-3}, if a living organism is exposed for one hour per month. From the data of the World Health Organization the natural background concentration of NO_2 over the continents constitutes 0.4–9.4 μkg m^{-3}.

Finally, the human impact on the nitrogen cycle can affect the structure and intensity of the energy exchange in the biosphere. As seen from Table 6.9, a lot can be done in this direction depending on the intensification of either reaction.

6.3.6 Phosphorus cycle

One of the GMNSS components is a unit that parameterizes the biospheric phosphorus cycle, which, together with the cycles of other chemicals, plays an important role in the formation of global productivity. In contrast to nitrogen, the main reservoir of phosphorus in the biosphere is not the atmosphere but the rocks and other deposits formed in the past geological epochs, which, being subject to erosion, emit phosphates. Besides this, there are other mechanisms of the return of phosphorus to the biospheric cycle, but, as a rule, they are not that efficient. One of these mechanisms is fishing, returning to land from the hydrosphere about 60×10^3 t P yr^{-1}, as well as an extraction of phosphorus-containing rocks estimated at 1–2 $\times 10^6$ t P yr^{-1}.

The present cycles of phosphorus are closed by its fluxes to the bottom deposits in the World Ocean via sewage, as well as with the coastal and river run-off. The flux diagram of the model of the global phosphorus cycle is presented in Figure 6.11,

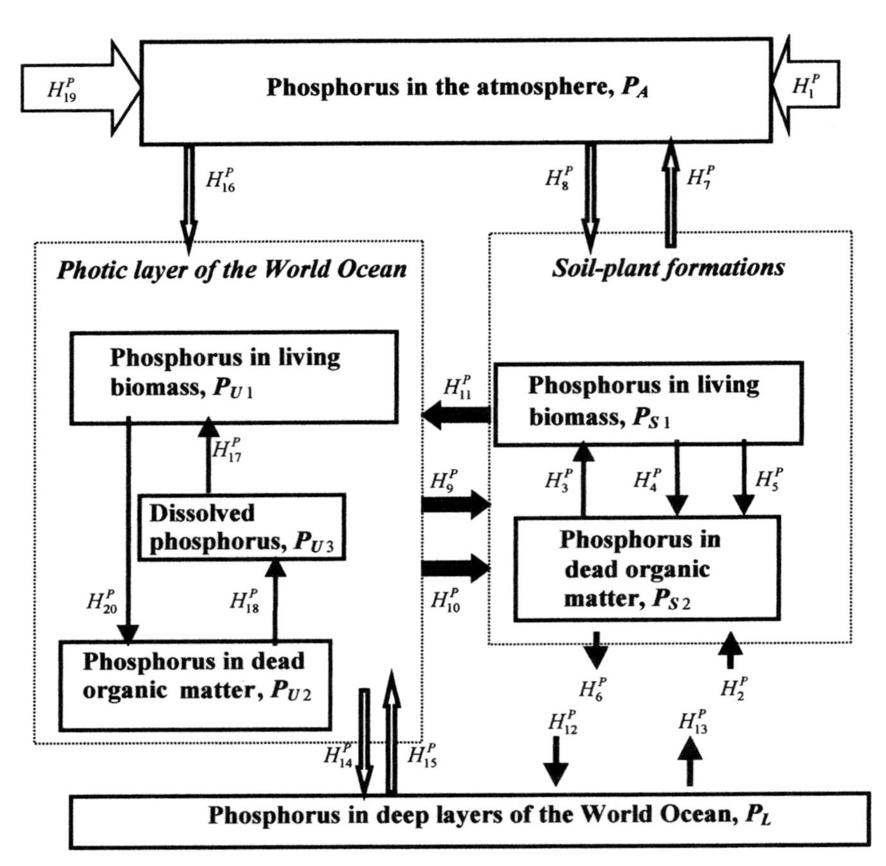

Figure 6.11. The block-scheme of the model of the global biogeochemical cycle for phosphorus in the biosphere. Notation is given in Table 6.10.

according to which the balance system of equations will be:

$$\frac{\partial P_A}{\partial t} + V_\varphi \frac{\partial P_A}{\partial \varphi} + V_\lambda \frac{\partial P_A}{\partial \lambda} = H_1^P + H_{19}^P + \begin{cases} -H_{16}^P, & (\varphi, \lambda) \in \Omega_O \\ H_7^P - H_8^P, & (\varphi, \lambda) \in \Omega/\Omega_O \end{cases}$$

$$\frac{\partial P_U}{\partial t} + v_\varphi \frac{\partial P_U}{\partial \varphi} + v_\lambda \frac{\partial P_U}{\partial \lambda} + v_z \frac{\partial P_U}{\partial z} = H_{11}^P + H_{15}^P + H_{16}^P - H_9^P - H_{10}^P$$

$$\frac{\partial P_L}{\partial t} + v_\varphi \frac{\partial P_L}{\partial \varphi} + v_\lambda \frac{\partial P_L}{\partial \lambda} + v_z \frac{\partial P_L}{\partial z} = H_{12}^P + H_{14}^P - H_{13}^P - H_{15}^P$$

$$\frac{\partial P_S}{\partial t} = H_2^P + H_8^P + H_9^P + H_{10}^P - H_6^P - H_7^P - H_{11}^P$$

where $P_U = P_{U1} + P_{U2} + P_{U3}, P_S = P_{S1} + P_{S2}$.

Table 6.10. Characteristics of fluxes ($10^6\,t\,yr^{-1}$) and reservoirs (10^6 t) of phosphorus in the biosphere (Krapivin, 2000b).

Reservoirs and fluxes of phosphorus	Identifier of MGBP unit	Estimate
Phosphorus supplies:		
in the atmosphere	P_A	3
on land	P_S	1,546
in the photic layer of the World Ocean	P_U	2×10^4
in deep layers of the World Ocean	P_L	12×10^4
Volcanic eruptions	H_1^P	0–2
Fertilization	H_2^P	19
Assimilation by plants	H_3^P	45.34
Input with dead vegetation	H_4^P	39.34
Input due to organisms functioning:		
on land	H_5^P	5
in the World Ocean	H_{20}^P	81.5
Transition into an unassimilatable form	H_6^P	2.9
Weathering	H_7^P	5
Precipitation:		
onto land	H_8^P	1.8
into the oceans	H_{16}^P	2
Removal with fishing	H_9^P	0.06
Removal by birds	H_{10}^P	0.04
Washing-out and run-off to the World Ocean	H_{11}^P	4–14
Input due to detritus lysis:		
in the photic layer of the World Ocean	H_{18}^P	550
in the deep layers of the World Ocean	H_{12}^P	159
Exchange between photic and deep layers of the World Ocean:		
upwelling	H_{15}^P	96.1
downwelling	H_{14}^P	22
Sedimentation	H_{13}^P	13–83.9
Rock weathering	H_{19}^P	1
Photosynthesis	H_{17}^P	630–1,300

With this detailing we have:

$$\frac{\partial P_{U1}}{\partial t} + v_\varphi \frac{\partial P_{U1}}{\partial \varphi} + v_\lambda \frac{\partial P_{U1}}{\partial \lambda} + v_z \frac{\partial P_{U1}}{\partial z} = H_{17}^P - H_9^P - H_{10}^P - H_{20}^P$$

$$\frac{\partial P_{U2}}{\partial t} + v_\varphi \frac{\partial P_{U2}}{\partial \varphi} + v_\lambda \frac{\partial P_{U2}}{\partial \lambda} + v_z \frac{\partial P_{U2}}{\partial z} = H_{20}^P - H_{18}^P$$

$$\frac{\partial P_{U3}}{\partial t} + v_\varphi \frac{\partial P_{U3}}{\partial \varphi} + v_\lambda \frac{\partial P_{U3}}{\partial \lambda} + v_z \frac{\partial P_{U3}}{\partial z} = H_{11}^P + H_{18}^P - H_{17}^P$$

$$\frac{\partial P_{S1}}{\partial t} = H_3^P - H_4^P - H_5^P$$

$$\frac{\partial P_{S2}}{\partial t} = H_2^P + H_4^P + H_5^P + H_8^P - H_3^P - H_6^P - H_7^P$$

Determine now the functional and dynamic characteristics of the fluxes of phosphorus (Table 6.10) based on analysis of the existing ideas about their nature. The atmospheric cycle is governed by rock weathering, volcanic eruptions, and by washing-out of phosphorus by precipitation. From the available estimates (Carlsson *et al.*, 1997; Jun and Shin, 1997), the phosphorus content in the lithosphere constitutes 0.093%, and the processes of weathering deliver annually into the atmosphere between 0.67 and 5.06 mg P cm^{-3} yr^{-1}. Every year, volcanic eruptions introduce to the atmosphere about 0.2 mln t P. Since these processes are of a complicated stochastic nature and their models are absent, in the first approximation the fluxes H_1^P and H_{19}^P will be considered constant.

The continental cycle of phosphorus is determined by ten fluxes (Figure 6.11) closed by a single component P_S indicating the phosphorus supplies on land in the soil–vegetation formations and in animals. The supplies of phosphorus in soils are replenished due to the fluxes H_l^P $(l = 2, 4, 5, 8, 9, 10)$. The loss of phosphorus from the soil is determined by the fluxes H_j^P $(j = 3, 6, 7, 11)$. With a more complicated detailing of the surface reservoirs of phosphorus and consideration of more ingenious effects in the interaction of these reservoirs, the classification of the surface fluxes of phosphorus becomes more complicated. Within the assumed parameterization of the surface reservoirs of phosphorus the following functional presentations of the fluxes $\{H_i^P\}$ can be considered:

$$H_2^P = p_1 GMG_0^{-1}M_0^{-1} \quad H_3^P = p_V R_V \quad H_4^P = p_4 M_L \quad H_5^P = p_5 H_F$$

$$H_6^P = p_6 P_{S2}/P_{S2,0} \quad H_9^P = p_3 I \quad H_7^P = p_7 \theta^{\Delta T} P_{S2}/P_{S2,0}$$

$$H_{11}^P = p_2 P_{S2}[1 - \exp(-k_{su} W_{so})]/P_{S2,0}$$

where p_V is the content of phosphorus in the living biomass of plants, p_4 and p_5 is the content of phosphorus in organic matter of vegetable and animal origin, respectively, I is the amount of produce extracted from the ocean, G is the population, M is the mineral resource, R_V is the produce of vegetation photosynthesis, H_F is the biomass of unassimilated food of animals, $M_L = \mu_V L$, μ_V is the rate of vegetation dying-off, L is the vegetation biomass, θ is the temperature coefficient of the rate of the dead organic matter decomposition on land, ΔT is the surface atmospheric temperature (SAT) variation with respect to the control value, W_{so} is the volume of the river run-off into the oceans, and $p_i (i = 1, \ldots, 7)$ are the constants. The index "0" in G_0, M_0, and $P_{S2,0}$ attributes these parameters to some control time moment t_0, when all parameters of $MGPC$ are known.

Describe the hydrochemical cycle of phosphorus by a totality of fluxes $H_k^P (k = 9\text{–}18, 20)$: $H_{12}^P = p_{14} R_{DL}$, $H_{13}^P = p_8 P_L/P_{L,0}$, $H_{14}^P = p_9 P_U/P_{U,0}$, $H_{15}^P = p_{10} P_L/P_{L,0}$, $H_{16}^P = p_{12} R_{WO} P_A/P_{A,0}$, $H_{17}^P = p_{13} R_\Phi$, $H_{18}^P = p_{15} R_{DU}$, and $H_{20}^P = p_{16} M_\Phi$, where R_D is the rate of dead organic matter decomposition, R_{WO} is precipitation over the ocean, R_Φ is the produce of phytoplankton and other living organisms in the ocean, M_Φ is the rate of living biomass dying-off, and $p_i (i = 8\text{–}16)$ are the proportion constants.

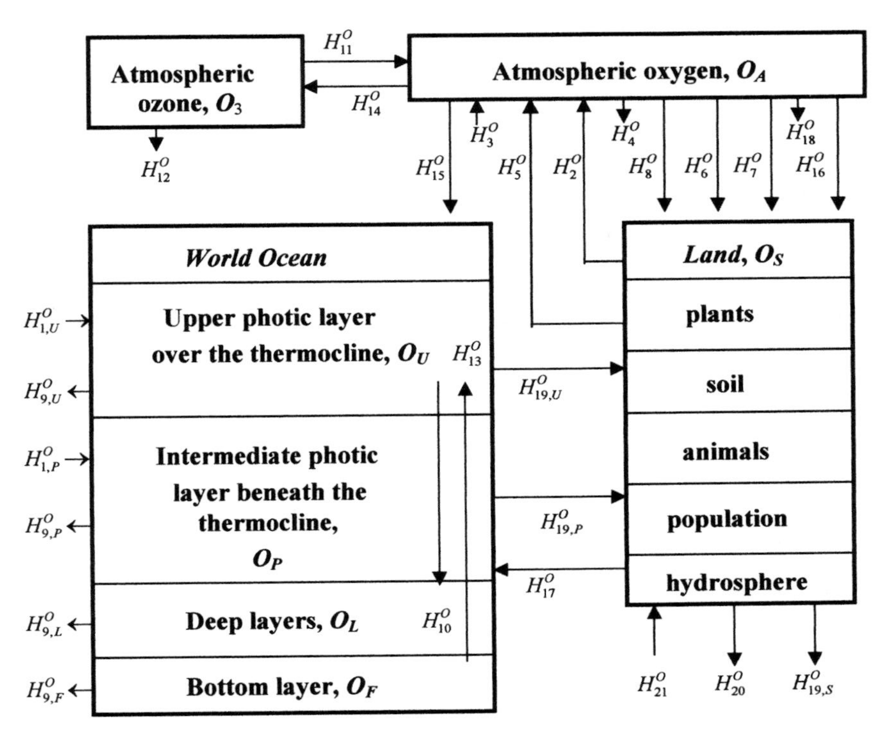

Figure 6.12. Oxygen fluxes in the biosphere. Estimates and notations of fluxes are given in Tables 6.11 and 6.12.

The presented model can be used in any global model of the biosphere. In order to do that, it is necessary to evaluate its parameters as functions of spatial coordinates and time steps. Estimates shown in Table 6.10 are too rough. Therefore, a more detailed classification is needed of all functions and coefficients introduced into the model equations. This approach is most efficient in determination of key moments in the global ecodynamics. Introduction of the proposed model as a GMNSS unit, and subsequent calculations in the regime of adaptive sorting of alternative descriptions of partial dependences of elements indicated in Table 6.10, can solve many problems in the assessment of the environmental stability.

6.3.7 Biospheric balance of oxygen and ozone

The oxygen cycle in nature is composed of characteristic biogeochemical transitions between the reservoirs of basic constituents circulating in the biosphere. Therefore the block-scheme of the oxygen exchange is similar to those of sulfur, nitrogen, carbon, and phosphorus (Figure 6.12). However, oxygen refers to the constituents spread over the globe most widely, which makes it one of the substantial components

Table 6.11. Estimates of characteristic parameters of the biogeochemical cycles of oxygen and ozone in nature (Krapivin, 2000a; Kondratyev and Varotsos, 2000).

Parameter	Estimate
Total mass in the atmosphere (10^9 t):	
oxygen	1.184×10^6
ozone	3.29
Oxygen supply in the Earth's crust (t)	$13 \cdot 10^{18}$
Concentration of oxygen in the present atmosphere by volume (%)	20.946 ± 0.02
Oxygen supply in the ocean waters (t):	
in a bound state	1.28×10^{18}
in dissolved compounds	2.6×10^{15}
in a free state	7.48×10^{12}
Partial density of ozone at normal pressure and temperature ($\mu g\, m^{-3}$)	2.14
Income processes in oxygen balance (10^9 $O_2\, yr^{-1}$):	
photosynthesis in the oceans	56
photosynthesis on land	174
natural growth	152
agricultural crops	21
algae	1
water photolysis	0.013
smelting of metals from oxides	0.3
Expenditure processes in oxygen balance (10^9 $O_2\, yr^{-1}$):	
organics oxidation in the World Ocean	59.2
organics oxidation on land	
by sub-aerial natural systems	151.2
in reservoirs, lakes, rivers	1.07
in oil, coal, and gas burning	16
in the use of raw materials, fodder, food, and biofuel	29.5
in anthropogenic decrease of humus	1.5
on dumps and burial places	2.6
other processes	0.4
Fossilization in the Earth's crust with the continents growth	0.15
The ratio of the rate of deposition of suboxidized organic carbon to vegetation productivity	2×10^{-4}
Photodissociation of water vapor in the atmosphere, t/(km^2 yr)	0.008
Distribution of organic carbon deposits (%):	
continents	69.6
shelf and continental slope	29.0
the World Ocean floor	1.4
Relative water vapor concentration in the stratosphere	5×10^{-6}
Ozone lifetime, days:	
northern hemisphere	39
southern hemisphere	85
Minimum permissible concentrations of oxygen in the atmosphere (% of the present level):	
humans	31
mammals	14–31
birds	22–31
reptiles	10–20
amphibians	3.1–11
fish	2.5–11
bacteria	1

Table 6.12. Estimates of reservoirs and fluxes of oxygen and ozone in the GMNSS oxygen unit used in its verification (Kondratyev and Varotsos, 2000).

Reservoirs (t km^{-2}), fluxes (t km^{-2} yr^{-1})	Identifier	Evaluation
Oxygen in the upper photic layer of the World Ocean	O_U	0.8×10^8
Oxygen in the transition layer of the World Ocean	O_P	0.7×10^9
Oxygen in the deep World Ocean	O (?)	3×10^4
Oxygen in the bottom layer of the World Ocean	O_F	9×10^3
Oxygen in the atmosphere	O_A	0.24×10^7
Oxygen in the surface part of the hydrosphere	O_S	0.6×10^8
Ozone	O_3	0.23
Photosynthesis in the World Ocean	H_1^0	108–388
Photosynthesis on land	H_2^0	70–100
Photodecomposition of water in the atmosphere	H_3^0	0.008
Oxidation processes in the atmosphere	H_4^0	0.009
Respiration of plants	H_5^0	0.07–0.1
Respiration of animals	H_6^0	50–60
Respiration of humans	H_7^0	70–80
Oxidation–restoration processes in soil	H_8^0	1
Oxidation processes in the World Ocean	H_9^0	164
Descending of the oxygen-saturated waters	H_{10}^0	190
Decomposition and destruction of O_2 in the atmosphere	H_{11}^0	
Formation of O_3 from NO_2	H_{12}^0	0.23–22.2
Lifting of dissolved oxygen in the upwelling zones	H_{13}^0	36
Decomposition and destruction of ozone in the atmosphere	H_{14}^0	1.48–1.66
Exchange on the atmosphere–ocean border	H_{15}^0	18–140
Exchange on the atmosphere–inland water body border	H_{16}^0	18–140
Transport of oxygen into the ocean by river runoff	H_{17}^0	50
Anthropogenic consumption of oxygen	H_{18}^0	60–90
Expenditures of O_2 on metabolism of aquatic animals	H_{19}^0	0.2
Oxidation processes in the continental water bodies	H_{20}^0	90–200
Photosynthesis in the continental water bodies	H_{21}^0	100–400

of the biogeochemical cycles. Its amount in the Earth's crust, including the hydrosphere, reaches 49% by mass. The lithosphere (without the ocean and the atmosphere) contains 47.2% of oxygen, water – 88.89%. In the ocean water, oxygen constitutes 85.82%, with living substances – 65% by mass. These estimates testify to a significance of oxygen for the biosphere. About 39×10^{14} t O_2 circulate in the biosphere.

Oxygen is present in the biosphere in the form of molecular oxygen (O_2), ozone (O_3), atomic oxygen (O), and as a constituent of various oxides. On the one hand, oxygen maintains life on Earth due to the process of respiration and the formation of the ozone layer, and on the other hand, it is itself the product of organism functioning. This fact hinders a simulation of its cycle, since it requires a synthesis of parameterizations of various processes.

6.3.7.1 Oxygen sources

Presently and in the geological past there have been two sources of oxygen – endogenic and photosynthetical. Without dwelling upon the respective scientific discussions and all existing concepts, we shall try to describe the sources of oxygen in the present-day biosphere, following numerous studies in this field (Bgatov, 1988; Biutner, 1986; Phothero, 1979).

The basic source of atomic oxygen is the photosynthesis of plants, whose equation has the form:

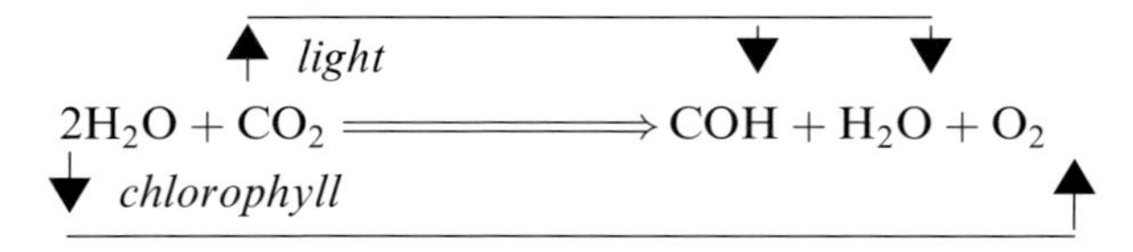

Photosynthesis produces annually over 50×10^9 t of oxygen (i.e., of an order of 3.3×10^{-4}% of its supply in the atmosphere). Hence, we see that only due to photosynthesis can the oxygen supplies in the atmosphere be totally renewed during a time period of 300,000 years. About 80% of the total amount of oxygen produced by photosynthesis results from phytoplankton functioning, and land vegetation communities produce only 20%. If we denote by $R_\Phi(\varphi, \lambda, t)$ and $R_\kappa(\varphi, \lambda, t)$ the produce of phytoplankton Φ and land surface of the κ type at the Earth's surface point (φ, λ) at a time moment t, then the oxygen fluxes to the hydrosphere, and from land to the atmosphere, can be described by relationships:

$$H_1^O = a_\Phi R_\Phi \qquad H_2^O = a_\kappa R_\kappa \qquad H_{21}^O = a_S R_\Phi$$

where the coefficients a_Φ, a_κ, and a_S depend on phytoplankton species and the type of vegetation. For their averaging we use the data on the fluxes: $H_1^O = 140\,\mathrm{t\,O_2\,km^{-2}\,yr^{-1}}$, $H_2^O = 70\,\mathrm{t\,O_2\,km^{-2}\,yr^{-1}}$, $H_{21}^O = 600\,\mathrm{t\,O_2\,km^{-2}\,yr^{-1}}$, $R_\Phi = 401.3\,\mathrm{t\,km^{-2}\,yr^{-1}}$, $R_\kappa = 102.4\,\mathrm{t\,km^{-2}\,yr^{-1}}$. Then $a_\Phi = 0.35$, $a_\kappa = 0.68$, $a_S = 1.49$. Of course, these estimates have a considerable spatial and temporal scatter. In particular, using the data on the productivity of some oceans given in Biutner (1986), for instance, we obtain values of the coefficient a_Φ for the oceans.

Apart from photosynthesis, photolysis can be a source of oxygen in the atmosphere (i.e., the decomposition of water vapor under the influence of UV radiation in the upper layers of the atmosphere). However, the intensity of this source in present conditions is negligible. Nevertheless, denote this flux by $H_3^O = a_H W_A$, where W_A is the water vapor content in the atmosphere and a_H is the empirical coefficient. If we assume that in the upper layers of the atmosphere a constant share of W_A can reside, then at $H_3^O = 0.0039\,\mathrm{t\,O_2\,km^{-2}\,yr^{-1}}$ and $W_A = 0.025\,\mathrm{m}$, we have $a_H = 1.56 \times 10^{-7}\,\mathrm{yr^{-1}}$.

6.3.7.2 Processes of oxygen assimilation

The oxidation processes both on land and in the water medium are the basic consumers of oxygen on Earth. The ability of oxygen to react with many elements in the Earth's crust forms the fluxes of oxygen leaving the biospheric reservoirs. The balance between the income and expenditure fluxes of oxygen has been reached in the course of biospheric evolution.

Oxygen is spent on respiration of plants, animals, humans, and on the dead organic matter decomposition both in the hydrosphere and on land. To parameterize the income parts of the oxygen balance, we use the following models: $H_5^O = a_1 T_\kappa$, $H_6^O = a_2 T_F$, $H_7^O = a_3 T_G$, $H_{19}^O = a_6 T_R$, $H_8^O = a_Q R_D$, $H_{20}^O = a_5 R_S$, where T_m is the energy expenditure on respiration $(t = \kappa, F, G, R)$ and R_ζ is the rate of the dead organic matter decomposition $(\zeta = Q, D, S)$.

6.3.7.3 Ozone

Atmospheric ozone constitutes 0.64×10^{-6} of the atmospheric mass and belongs to a group of optically active gases. It absorbs UV solar radiation in the range 200–300 nm, strongly affecting thereby the thermal regime of the stratosphere. Besides this, ozone has a number of vibration–rotation bands of absorption in the IR spectral region (9.57 μm) and partially absorbs visible radiation in the Chappuis band (0.6 μm). The formation and destruction of ozone has been described in detail (Kondratyev and Varotsos, 2000; Wotawa *et al.*, 2000; Woodbury *et al.*, 2002).

Ozone forms in the upper stratosphere from molecular oxygen under the influence of UV solar radiation. In the lower stratosphere and troposphere, the source of ozone is the decomposition of nitrogen dioxide under the influence of UV and visible radiation. The formation of the vertical profile of ozone concentration is connected with its meridional and vertical transports. The general characteristic of this profile is the total amount of ozone measured by the thickness of its layer in Dobson units (1DU = 0.001 cm).

Ozone destruction is due to a complex set of photochemical reactions with participation of the compounds of hydrogen, nitrogen, and chlorine. From the available estimates, 50–70% of ozone is destroyed by nitrogen compounds, 20–30% – by oxygen O, 10–20% – by water-containing compounds of HO_x, and less than 1% – by chlorine compounds. The prevailing role of nitrogen compounds in ozone destruction has been confirmed (Wauben *et al.*, 1997) for all latitudes. The equation of photochemical equilibrium between concentrations of ozone and nitrogen oxides is $[NO][O_3]/[NO_2] = \mu$, where the equilibrium constant μ depends on the solar radiation intensity and can range from 0 to 0.02.

There are various approaches to a parameterization of the process of formation and destruction of the ozone layer. The complexity of derivation of the dynamic models of the ozone cycle in the atmosphere is connected with its participation in more than 75 chemical reactions, a qualitative and quantitative description of which is impossible without deriving detailed models of numerous minor gas components of the atmosphere. Nevertheless, there are empirical models of the ozone layer, which make it possible, in the present climatic situation, to obtain adequate spatial distributions of ozone. For instance, Bekoriukov and Fedorov (1987) derived a simple empirical model of the total ozone content confirmed by observational data for the southern hemisphere:

$$O_3(\varphi, \lambda) = \sum_n \sum_{n \leq m} P_n^m(\varphi)\left[a_{n,m} \cos(m\lambda) + a_{n,-m} \sin(m\lambda)\right]$$

where P_n^m are non-normalized spherical functions of the n degree of the m order, and $a_{n,m}$ and $a_{n,-m}$ are the empirical coefficients whose values are given in Bekoriukov and Fedorov (1987).

There are also static models to describe the vertical profile of the ozone density distribution. One such model is the Kruger formula (Bgatov, 1988):

$$O_3(h) = 51.4 \exp[-(h - 40)/4.2](\mu\text{kg m}^{-3})$$

The ozone models also include the models of Hessvedt-Henriksen, Grovs-Matingli, and others. Detailed discussions of such models are given in Biutner (1986).

The simplest dynamic model of the ozone layer can be written in the form of a balance equation that reflects its income–expenditure components. The ozone supplies are replenished due to the impact of UV radiation on oxygen ($H_{11}^O = e_3 O_A$) and nitrogen dioxide ($H_{12}^O = e_2 N_A$). The ozone layer is destroyed at a rate $H_{14}^O = O_3/T_3$, where T_3 is the lifetime of ozone molecules depending on atmospheric pollution: $T_3 = T_3^O - e_1 B$. The lifetime T_3^O of the ozone molecules in a pure atmosphere averages 50–60 days. Nitrogen oxides participating in the cycle running opposite to the H_{12}^O cycle of ozone destruction contribute much to the magnitude of B.

The stratospheric ozone depletion also takes place due to the input of hydrogen from the Earth's interior in the processes of volcanic, geothermal, and seismic activity. The hydrogen hypothesis of the Earth's ozone layer depletion proposed by Syvorotkin (2002) is rather convincing, though it has not been adequately studied. The experimental and theoretical estimates of the possibilities of hydrogen penetration from the Earth to the stratosphere as part of the products of gas emissions testify to the reality of this hypothesis. The spatial coincidence of the zones of reduced concentration of stratospheric ozone and hydrogen emissions favor this hypothesis. For instance, Golubov and Kruchenitsky (1999) detected the ozone layer destruction by deep emissions of hydrogen during industrial nuclear explosions in Yakutia.

6.3.7.4 The anthropogenic impact on the biospheric oxygen cycle

Studies of the history of biospheric evolution reveal a close correlation between the oxygen production intensity and the development of life on Earth. And though the expected relative oscillations of the oxygen concentration in the near future do not exceed 10%, the assessed impacts on the biosphere do not cover all potential anthropogenic trends, and therefore cannot be considered reliable. Therefore, let us analyze the constituents of the possible mechanisms for the violation of the natural balance of oxygen. Naturally, our concern is not only for an increase but also a decrease of the oxygen content in the atmosphere.

The oxygen cycle is complicated by its ability to take part in a lot of chemical reactions giving a multitude of epicycles. This fact makes the oxygen cycle sufficiently stable but hinders an assessment of this stability.

Anthropogenic forcing on numerous epicycles of oxygen manifests itself both directly through its involvement in other cycles of substances at fuel burning and production of various materials and indirectly, through environmental pollution and biospheric destruction. Therefore, a parameterization of the anthropogenic impact on the oxygen balance is accomplished within other units of the global model. The flux H_{18}^{O} taken into account in the scheme in Figure 6.12 completely covers the direct consumption of oxygen both in industry and in agriculture. Assume $H_{18}^{O} = y_1 R_{MG}$, where R_{MG} is the rate of the natural resources expenditure and y_1 is the coefficient ($\approx$0.084).

The fluxes H_{15}^{O} and H_{16}^{O} are strongly affected by anthropogenic forcings. Their variations are caused by discharges of high-temperature industrial sewage containing considerable amounts of oxidizers, as well as by the oil polluted water bodies. The quantitative characteristics of the change of oxygen dissolved in water as a function of temperature have been well studied (Wang *et al.*, 1978). The empirical formula to calculate the concentration of the seawater-dissolved oxygen has the form: [O_2 dissolved] $= 80/(0.2T_O - 7.1)$, where [O_2] is expressed in mg l^{-1}, T_O – in °C. The estimates of oxygen solubility in water are well known.

The fluxes H_9^{O} and H_{19}^{O}, balancing in natural conditions the oxygen fluxes into the water medium, at an anthropogenic forcing increase, as a rule, due to more active aerobic bacteria and increasing metabolic needs of animals. For instance, a 10°C increase of water temperature increases the oxygen expenditure on respiration of marine animals by a factor of 2.2.

One of the negative manifestations of the anthropogenic impact on the oxygen cycle is a depletion of the ozone layer, especially marked in the polar regions. There are various hypotheses on the causes of sharply changing concentrations of ozone, as well as discussions on the so-called "ozone hole" over the Antarctic. The main cause of all violations is connected with the progressing human activity accompanied by the growing volumes of long-lived components emitted to the atmosphere (e.g., freons). The consequences of these violations are very serious, and the real scales of danger threatening life on Earth can be estimated only with the use of the global model of the climate–biosphere–society system.

A diversity of the anthropogenic impacts on the global biogeochemical cycle of oxygen is determined by the direct and indirect causes of breaking the natural balance of oxygen. According to the equation of photosynthesis, the gram-molecular amounts of assimilated CO_2 and emitted O_2 are equal. Also equal are the gram-molecular amounts of assimilated O_2 and emitted CO_2 for dead organic matter decomposition and fuel burning. Hence, for time periods of tens and hundreds of years, a change of the CO_2 amount in the atmosphere is accompanied by the same change of O_2, but in the opposite direction. For instance, a CO_2 doubling in the atmosphere leads to a decrease in the amount of O_2. But, since the volume concentration of CO_2 in the atmosphere is now estimated at 0.031%, and that of O_2 at 20.946%, in this case a decrease of O_2 will constitute only 0.15% of the total O_2 content in the atmosphere.

Imagine the following situation. Let the total biomass of the biosphere ($\sim 9.6 \times 10^{11}$ t C), all organic matter in the soil ($\sim 14 \times 10^{11}$ t C), and all fossil chemical fuel, the explored supplies of which constitute 128×10^{11} t of conditional fuel (64×10^{11} t C), be burnt. Then the amount of CO_2 in the atmosphere increases by a factor of 12.5, and that of O_2, respectively, decreases but only by 1.75%. Hence, the amount of oxygen over hundreds of years remains practically constant.

However, it should be borne in mind that the region of excess anthropogenic emissions of CO_2 and, hence, O_2 assimilation is concentrated in small areas of cities and forest fires. Since the concentrations in the atmosphere do not equalize instantly, a gradient of O_2 concentrations can appear around these areas, when the oxygen provision will be insufficient for animals and humans. Therefore, the model of the global oxygen cycle reflecting the spatial heterogeneities in the distributions of O_2 concentrations enables one to reveal such dangerous territories.

An interaction between the cycles of oxygen, nitrogen, sulfur, phosphorus, and carbon manifests itself through the processes of oxidation and decomposition. The available level of detailing the global model units does not permit one to reflect all the diversity of these processes. Therefore, in the simplest case, when only averaged characteristics of the oxygen cycle components are taken into account, the scheme in Figure 6.12 of the global O_2 fluxes can be presented schematically (Figure 6.13). The indicated stability of the O_2 concentration in the atmosphere makes it possible to simplify a description of the oxygen unit, using a single balance equation:

$$\frac{\partial O}{\partial t} + V_\varphi \frac{\partial O}{\partial \varphi} + V_\lambda \frac{\partial O}{\partial \lambda} = k_0 R_F + k_L R_L - \nu_L T_L - b_G G - \nu_F T_F - \nu_G T_G - \mu_Q R_Q$$

where k_0 and k_L are the indicators of the rate of O_2 emission due to photosynthesis in the ocean and on land, respectively; ν_s is the indicator of the role of respiration of

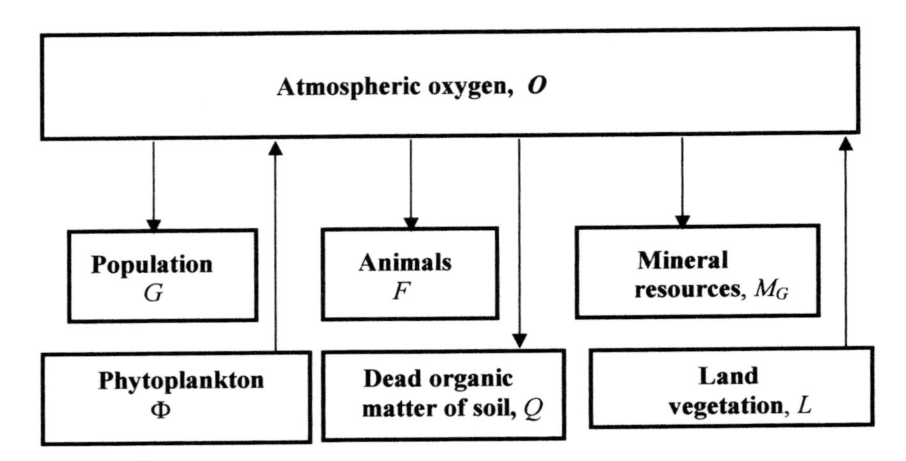

Figure 6.13. A simplified scheme of the biogeochemical cycle of oxygen in the biosphere.

land vegetation ($s = L$), animals ($s = F$), and humans ($s = G$) in the removal of oxygen from the atmosphere; and μ_Q is the rate of O_2 consumption at the decomposition of the soil dead organic matter.

6.3.7.5 *The model block-scheme of the oxygen balance in the biosphere*

Assuming the scheme of the oxygen fluxes in nature shown in Figure 6.12 to be balanced, let us write the equations of the model in the following traditional form of the balance relationships:

$$\frac{\partial O_A}{\partial t} + V_\varphi \frac{\partial O_A}{\partial \varphi} + V_\lambda \frac{\partial O_A}{\partial \lambda} = H_2^O + H_3^O + H_{14}^O + H_{15}^O + H_{16}^O - \sum_{i=4}^{8} H_i^O - H_{11}^O - H_{18}^O$$

$$\frac{\partial O_3}{\partial t} + V_\varphi \frac{\partial O_3}{\partial \varphi} + V_\lambda \frac{\partial O_3}{\partial \lambda} = H_{11}^O + H_{12}^O - H_{14}^O$$

$$\frac{\partial O_S}{\partial t} = H_{21}^O - H_{16}^O - H_{17}^O - H_{19,S}^O - H_{20}^O$$

$$\frac{\partial O_U}{\partial t} + V_\varphi \frac{\partial O_U}{\partial \varphi} + V_\lambda \frac{\partial O_U}{\partial \lambda} = H_{1,U}^O + H_{13,PU}^O + H_{17}^O - H_{9,U}^O - H_{10,PU}^O - H_{15}^O - H_{19,U}^O$$

$$\frac{\partial O_P}{\partial t} + V_\varphi \frac{\partial O_P}{\partial \varphi} + V_\lambda \frac{\partial O_P}{\partial \lambda} = H_{1,P}^O + H_{10,PU}^O + H_{13,PL}^O - H_{9,P}^O - H_{10,PL}^O - H_{13,PU}^O - H_{19,P}^O$$

$$\frac{\partial O_L}{\partial t} = Q_L + H_{10,PL}^O + H_{13,LF}^O - H_{9,L}^O - H_{10,LF}^O - H_{13,PL}^O$$

$$\frac{\partial O_F}{\partial t} = Q_F + H_{10,LF}^O - H_{9,F}^O - H_{13,LF}^O$$

Here Q_L and Q_F denote the oxygen fluxes resulting from mixing of the deep and bottom layers of the ocean. The oxygen exchange between the hydrosphere and the atmosphere (fluxes H_{15}^O and H_{16}^O) depend on its partial pressures above the water–air border. The directions of the fluxes H_{15}^O and H_{16}^O depend on the relationship between temperatures T_a, T_U, and T_S. Due to high concentrations of oxygen in the atmosphere, the partial pressure varies negligibly, and therefore the fluxes H_{15}^O and H_{16}^O can be considered to depend only on oscillations in the concentrations of O_U and O_S: $H_{15}^O = k_{OU}(T_U - T_a)O_U$; $H_{16}^O = k_{OS}(T_S - T_a)O_S$. If we assume $O_U = 5.5\,\text{ml}\,\text{l}^{-1}$ and $O_S = 2.1\,\text{ml}\,\text{l}^{-1}$, then at $T_U - T_a = T_S - T_a = 2°\text{C}$ and $H_{15}^O = 18\,\text{t}\,\text{km}^{-2}\,\text{yr}^{-1}$, $H_{16}^O = 140\,\text{t}\,\text{km}^{-2}\,\text{yr}^{-1}$ we obtain $k_{OU} = 0.5 \times 10^{-4}\,\text{km}°\text{C}^{-1}\,\text{yr}^{-1}$ and $k_{OS} = 0.1 \times 10^{-2}\,\text{km}°\text{C}^{-1}\,\text{yr}^{-1}$. A more accurate model of the ocean–atmosphere gas exchange has been considered by Biutner (1986).

The ocean layers exchange oxygen by circulation processes, and as a result, depending on latitude, longitude, and season, the intensity of the fluxes H_{10}^O and H_{13}^O can sharply change. In any case, this intensity mainly depends on the velocities v_A of the vertical water upwelling and v_H of its downwelling. In the zones of upwellings the flux H_{13}^{O} prevails and, on the contrary, the flux H_{10}^{O} prevails in the zones of convergence. The velocities v_A and v_H range from 0 to 0.1 m s^{-1}. The most characteristic values of these velocities range from 10^{-2} to 10^{-4} m s^{-1}. For instance, near California, $v_A \approx 0.77 \times 10^{-5}$ m s^{-1}, and in the region of the Bengal upwelling $v_A = 0.25$ m s^{-1}. Thus, for the fluxes H_{10}^{O} and H_{13}^{O} we use the following approximations: $H_{10}^{O} = \lambda_{\kappa\gamma}(O_\kappa - O_\gamma)$ and $H_{13}^{O} = \mu_{\gamma\kappa}(O_\gamma - O_\kappa)$, where $\lambda_{\kappa\gamma}$ and $\mu_{\gamma\kappa}$ are the coefficients of local mixing ($\kappa = U, P, L; \gamma = P, L, F$).

In the shelf zone of the ocean the oxygen balance formation is affected much by the run-off from the continents. Despite the complexity of this process, the following simple parameterization can be accepted for this case:

$$H_{17}^{O} = n_{SU}(O_S - O_U)$$

Some estimates of the parameters and fluxes considered in the oxygen model are given in Tables 6.11 and 6.12.

6.3.7.6 The role of aviation in the changes of the ozonosphere

The problem of monitoring and predicting the ozone layer dynamics in the atmosphere is as equally important as the problem of the atmospheric greenhouse effect (Varotsos and Kondratyev, 1998; Popovicheva *et al.*, 2000). In both cases there are contradictory estimates of the causes of the ecological danger of observed levels of ozone concentration, ozone destruction, or greenhouse gases. Despite such contradictory estimates, often not without a political nuance, these problems attract the attention of the experts in the fields of natural sciences, who try to create some information technologies to ensure a high level of objectivity and reliability when estimating the consequences of the anthropogenic interference into the global biochemical cycles of ozone, carbon dioxide, methane, water vapor, and other minor MGCs. Here we consider a narrow but important problem of the ozone layer changes over a limited territory caused by transport aviation emissions over this territory. This problem has recently attracted growing attention. The impact of flights of subsonic (altitudes 9–13 km) and supersonic (16–20 km) aircraft on the ozonosphere has become substantial, at least, on a regional scale. More so as the volumes of global air transportation increase by almost 5% annually, and the amount of emitted nitrogen oxides, sulfur compounds, and other MGC increases by about 4% annually. According to average global estimates, the NO_x emissions ($NO + NO_2$) constitute now about 500 kt N yr^{-1} with a predicted increase up to 1,100 kt N yr^{-1} by 2015.

A totality of substances ejected by aircraft engines to the atmosphere include components, such as H_2SO_4, HNO_3, HNO_2, HNO, Cl, NO_3, ClO_2, CO, CO_2, CH_4, N_2O, H_2O, SO_2, SO_3, N_2O_5, CH_3Cl, Cl_2, CH_3NO_2, CH_3NO_3, $BrONO_2$, HNO4, and $ClONO_2$, many of which cause the formation of polar stratospheric clouds, affect markedly the aerosol composition of the atmosphere, and intensify the greenhouse effect.

Analysis of the distribution of these components in the atmosphere requires an understanding of many factors in the field of photochemistry and atmospheric dynamics. Unfortunately, the available information about the rates of reactions with participation of the enumerated substances as well as about the coefficients of micro-macro-turbidity, and local synoptic characteristics, are limited by data averaged in time and space. In this connection, many experts apply various simplifications, which make it possible to overcome information uncertainties.

Atmospheric ozone chemistry has been well studied (Kondratyev and Varotsos, 2000). Nevertheless, this knowledge is insufficient to derive a model of the biogeochemical cycle of ozone whose adequacy would not raise doubts. The main cause is a time-dependent nature of the environmental processes. Unfortunately, the use of both simple and complicated climate models that take into detailed account the compounds of atmospheric chemistry of ozone, do not give acceptable results. Therefore, one of the possible approaches to raise the reliability of the estimates of the state of the ozone layer over a given territory is an application of technology which combines measurements and modeling and takes into account the expert estimates. In this case, to assess the vertical ozone profile, all available information can be used on its formation and destruction, and the additional background information about anthropogenic and natural processes can be obtained from established correlations or scenarios.

One of the difficulties in the synthesis of the model of ozone dynamics and observational data is a necessity to adequately describe the location of the tropopause. An uncertainty about the aircraft flight routes of subsonic aircraft relative to the tropopause appears because of inaccurate determination of the height of the tropopause. This is very important, since depending on whether the flight route is below the tropopause or in the stratosphere, the photochemical reactions with the participation of ozone differ. With supersonic aviation, this problem is solved uniquely: all routes lie in the stratosphere. Therefore, to exclude from the model the elements of instability, a hypothesis is assumed on the seasonal change of the tropopause altitude, for instance, following the binary law: in spring and summer Z_1, and in autumn and winter Z_2.

This approach excludes the instability of the obtained estimates. Though, of course, there are many successful results of modeling the ozone photochemistry. Among them, a number of models of the Lagrange type are rather efficient, which take into account up to 75 chemical elements and compounds. Also efficient is the 3-D model MOZART (Model for Ozone and Related chemical Tracers) (Kondratyev and Varotsos, 2000).

The problem of estimation of the vertical profile of the ozone concentration in the atmosphere over the limited territory is considered in a narrow aspect, taking into account only one anthropogenic source for the effect on the ozonosphere – aviation. Since the set of substances ejected by aircraft engines is rather large, the consequences of aircraft flights over this part of the Earth's surface include numerous changes of both the gas and aerosol composition of the atmosphere. Of course, the scale of these changes is determined by the intensity of the aircraft flights and density of the flight corridors. Here a method of integration of these causes is proposed and an information system of expert level is developed, which can be used in the systems of regional and national monitoring. Besides this, the presentation of

Table 6.13. Characteristics of the SSROC units.

Unit	Functions of the unit
AADB	Algorithmic adaptation of the database to the structure of the controled territory. A matrix structure is developed with elements reliably attaching the environmental elements to geographical coordinates, configuration of the territory of the region, location on it of objects such as airports, as well as division of the territory into land and water surfaces.
IRDB	Information renewal of the database. A possibility is provided of an operational change of configuration and appearance of the territory of the region described in the AADB.
CIF	Control of information fluxes between the SSROC units. Dimensions of the model parameters are coordinated; dimensions of input data are coordinated with the scales assumed in the SSROC. For instance, the formula 1 ppmv = 10^{-3}[M/[$M_i\rho$]mkg m^{-3}, where M_i is the molecular weight of the ith chemical element. The type of formulas 1 mkg O_3 m^{-3} $\Rightarrow 0,467 \times 10^{-7}$ atm–cm are also recalculated.
CSBF	Control of the system block functions. Depending on the availability of the needed information in the database about correlations between various processes, a version is selected from the alternative versions, that does not contradict the database.
PCMD	Parametric coordination of the models and database. Signals from the user's interface are analyzed for an efficient removal from the database of the coefficients of models, or in the case of disagreements the model is substituted for the scenario.
MPR	Modeling the photochemical reactions in the flight corridor with selection of three stages: (1) in the nearest zone after an ejection of the products of fuel combustion from the nozzle of the aircraft engine; (2) scattering the jet of combustion products; (3) complete mixing with the surrounding atmosphere.
MOST	Model of ozone spreading and transformation in the interaction of the flight passageway (vortex trail) with the surrounding atmosphere.
MOFD	Model of ozone formation and destruction with account of all flight corridors over the controled territory.
CCAB	Calculation of corrections for the atmospheric balance of ozone due to an account of the effects of land cover and sea surface.
FBOL	Formation of the background ozone level either from the data of regional and global monitoring systems or with the use of the respective model.
SF	Formation of scenarios of location of the flight corridors and of the timetables of their load.
CIUI	Control of the information user's interface. Provision of computer experiments.

the input information and the spectrum of chemical reactions have been limited. An operator can specify a list of substances contained in exhaust gases.

Units of the simulation system for the regional ozonosphere control (SSROC) are enumerated in Table 6.13. The controled territory Ω is covered with a geographical grid with steps $\Delta\varphi$ and $\Delta\lambda$ by latitude and longitude, respectively. Over each cell

$\Omega_{ij} = \{(\varphi, \lambda) : \varphi_i \le \varphi \le \varphi_{i+1}; \lambda_j \le \lambda \le \lambda_{j+1}; i = 1, \ldots, N; j = 1, \ldots, M\}$ a possible location (altitude and time) of the flight corridor is determined. For this purpose, in the database for SSROC an indicator is formed of the flight load on Ω (type of engine, fuel, velocity, and flight altitude). The background concentration of ozone and the meteorological situation are assumed to be taken from the data of regional, national, and global systems of environmental monitoring.

The ozone concentration as a function of the spatial coordinates and time is calculated by the formula:

$$\frac{\partial O_3(\varphi, \lambda, z, t)}{\partial t} = Q + S + U - P - R \tag{6.3}$$

where z is the altitude, and the functionals in the right-hand part of the equation describe the following processes of ozone formation: Q is a change of the ozone concentration due to atmospheric motion and gravitational sedimentation, P and U are photochemical destruction and formation of ozone outside the passageway, respectively, and R and S are photochemical destruction and formation of ozone within the flight corridor.

Equation (6.3), with initial data for a time moment $t = t_0$, is solved with account of the mosaic of the flight corridors. The functionals R and S are calculated for $\Omega_{i,j,k} = \{(\varphi, \lambda, z) : (\varphi, \lambda) \in \Omega_{ij}; z_k \le z \le z_{k+1}\}$ at a time moment t only in the presence of aircraft. Three zones are considered for the interaction of the products of fuel combustion in the trail of the working aircraft engine: (1) immediately behind all engines (time duration Δt_1), (2) at a stage of mixing the exhaust gases with the atmosphere (Δt_2), and (3) the means of penetration of the mixture to large-scale reservoirs (Δt_3). Hence, after the aircraft transit the passageway exists during a time period $\Delta t = \Delta t_1 + \Delta t_2 + \Delta t_3$, after which $R(\varphi_i, \lambda_j, z, k, t) = S(\varphi_i, \lambda_j, z, k, t) \equiv 0$ and in (6.3) the functional Q, U, and P start working.

During a time period Δt a lot of the various processes of transformation of substances ejected by the engines occur within the flight corridor. There is a term "index of transformations in a jet", which is an integral estimate of the concentrations of these substances as a function of time. Let an exhaust take place at moment t_0 (the moment of aircraft transit over a given point of the Earth's surface). Then the index of transformation of chemicals in the trail after the flight can be presented by a 3-step function:

$$J_N(t) = \begin{cases} J_{N1} & \text{for} \quad t_0 \le t < t_0 + \Delta t_1 \\ J_{N2} & \text{for} \quad t_0 + \Delta t_1 \le t < t_0 + \Delta t_1 + \Delta t_2 \\ J_{N3} & \text{for} \quad t_0 + \Delta t_1 + \Delta t_2 \le t < t_0 + \Delta t_1 + \Delta t_2 + \Delta t_3 \end{cases}$$

The values $J_{Ni} (i = 1, 2, 3)$ depend on the time of day, season, and on many other parameters (temperature, altitude, geographical coordinates, etc.). The empirical estimates of J_{Ni} are introduced to the database of the system and used in calculations of $J_N(t)$. With further improvement of the SSROC it will be expedient to include a unit, which would give theoretical estimates of J_{Ni}.

Since the aircraft flies at a velocity V_a along the route $x(\varphi, \lambda, z)$, at a time moment t_0 it is at some point x_0, and all its engines eject $V_M(t_0)$ of a substance of M type. Taking into account the above expression, we obtain:

$$V_M(t) = \begin{cases} L_1 & \text{for} \quad t_0 \leq t < t_0 + \Delta t_1 \\ L_2 & \text{for} \quad t_0 + \Delta t_1 \leq t < t_0 + \Delta t_1 + \Delta t_2 \\ L_3 & \text{for} \quad t_0 + \Delta t_1 + \Delta t_2 \leq t < t_0 + \Delta t \end{cases}$$

where

$$L_1 = V_M(t_0) - \frac{t - t_0}{\Delta t_1}[V_M(t_0) - J_{N1} V_M(t_0)]$$

$$L_2 = J_2 V_M(t_0) + \frac{t_0 + \Delta t_1 + \Delta t_2 - t}{\Delta t_2} V_M(t_0) \cdot (J_{N1} - J_{N2})$$

$$L_3 = J_3 V_M(t_0) + \frac{t_0 + \Delta t_1 + \Delta t_2 - t}{\Delta t_3} V_M(t_0) \cdot (J_{N2} - J_{N3})$$

After some time Δt the apparent effect of the aircraft flight is considered to cease, and all processes of transformation and destruction of ozone within the flight corridor once again become natural. The zone of the vortex trail behind the flying aircraft has a circular section of diameter r, and during the time period $\tau = \Delta t_1 + \Delta t_2$ its interaction with the surrounding atmosphere can be considered negligibly small. At the third stage, this interaction begins with the slight contact of the two media. At any rate, the scenario of interaction of the flight corridor with the surrounding atmosphere needs to be specified and developed, in particular, by forming a set of scenarios.

NO_x is the most important component of exhaust gases. During the lifetime of the aircraft trail NO_x becomes oxidized by hydroxyl OH, present in the vortex trail, giving HNO_3 and HO_2NO_2. As laboratory studies have shown, the processes of formation and destruction of ozone are also affected markedly by heterogeneous mechanisms of the impact on atmospheric chemistry. This impact manifests itself both within the flight corridor and in a free atmosphere. In particular, the reaction $N_2O_5 + H_2O = 2HNO_3$ with sulfate aerosols mainly resulting from the aircraft flight, reduces the rate of ozone destruction due to the NO_x cycle, but raises the role of Cl_x and HO_x in O_3 destruction.

The second important component of the exhaust gases is SO_2, whose ejection by the engines doubles the area occupied by sulfate particles in the atmosphere of the flight corridor, which leads to an increase of O_3 losses. In a number of field experiments (Kraabol and Stordal, 2000) on the F-16 fighter and in laboratory experiments with the F-100 engine using several types of aviation fuel with a high (~1,150 ppm S), moderate (~170–300 ppm S), and low (~10 ppm S) content of sulfur, the SO_2 emission changed from $2.49\ \mathrm{g\,SO_2\,kg^{-1}}$ for the fuel with a high sulfur content to $0.01\ \mathrm{g\,SO_2\,kg^{-1}}$ for the fuel with a low sulfur content. For these experiments the following relationship was derived:

$$0.02 \leq \frac{[SO_3]}{[SO_2] + [SO_3]} \leq 0.14$$

The results of studies carried out by Weisenstein *et al.* (1998) show that the mechanisms of evolution of the composition of the engines exhaust gases at their interaction with the atmosphere have been poorly studied, and therefore it is important to further develop the kinetic models describing the role of aircraft flights in changing the atmosphere.

The functional Q in (6.3) is written following the traditional scheme:

$$Q = -v_\varphi \frac{\partial O_3}{\partial \varphi} - v_\lambda \frac{\partial O_3}{\partial \lambda} - v_z \frac{\partial O_3}{\partial z} + \frac{\partial}{\partial \varphi}\left(D_\varphi \frac{\partial O_3}{\partial \varphi}\right) + \frac{\partial}{\partial \lambda}\left(D_\lambda \frac{\partial O_3}{\partial \lambda}\right) + \frac{\partial}{\partial z}\left(D_z \frac{\partial O_3}{\partial z}\right)$$

where $V(V_\varphi, V_\lambda, V_z)$ is the wind speed and $D(D_\varphi, D_\lambda, D_z)$ is the coefficient of eddy diffusion.

The units CCAB, MOFD, and MOST divide the functional Q with account of the output information from the unit AADB. As a result, the air mass mixing is realized at two stages: (1) mixing of the atmospheric compartment of the zone of aircraft flight with the environment; and (2) mixing of the cells $\{\Omega_{i,j,k}\}$ selected by the unit AADB. At the first stage the volumes and location of the zone of the effect of the aircraft are calculated:

$$\omega = \{(\varphi, \lambda, z) \colon \varphi_0 \le \varphi \le \varphi_1; \lambda_0 \le \lambda \le \lambda_1; z_0 \le z \le z_1\}$$

where $\varphi_1 = \varphi_0 + V_{a\varphi}\Delta t/k_\varphi; \lambda_1 = \lambda_0 + V_{a\lambda}\Delta t/k_\lambda$; $\Delta z = z_1 - z_0$ is the diameter of the zone of the impact of aircraft ($\Delta z = r$), φ_0 and λ_0 are latitude and longitude of the aircraft location at a time moment t_0, κ_φ and κ_λ is the number of kilometers within 1° of latitude and longitude, respectively, and z_1 and z_0 are the lower and upper boundaries of the flight corridor.

The obtained space ω agrees with the adjacent multitude of atmospheric units $\{\Omega_{ijk}\}$. Then the ozone content is averaged over ω and the adjacent compartments $\{\Omega_{ijk}\}$, with their volumes taken into account.

The second stage realizes a 2-step procedure of re-calculation of the ozone concentration over the whole space $\Xi = \{(\varphi, \lambda, z) \colon (\varphi, \lambda) \in \Omega; 0 \le z \le z_H\}$, where z_H is the altitude of the atmospheric boundary layer ($z_H \sim 70$ km), consideration of which is important in estimating the state of the regional ozonosphere. These two steps correspond to the vertical and horizontal constituents of atmospheric motion. This division is made for convenience, in order that the user of the expert system could choose a synoptic scenario. According to the available estimates (Karol, 2000; Kraabol *et al.*, 2000; Meijer and Velthoven, 1997), the processes of vertical mixing prevail in the dynamics of the ozone concentration. It is here that due to uncertain estimates of D_z there are serious errors in model calculations. Therefore, the units CCAB, MOFD, and MOST provide the user with the principal possibility to choose various approximations of the vertical profile of the eddy diffusion coefficient (D_z). The database of SSROC contains the versions of estimates of D_z ($m^2 s^{-1}$) used already by many experts in the models of the ozone dynamics ($\approx 3\, m^2 s^{-1}$;

$0.5(V_\Phi + V_\lambda)$; $\approx 10\,\text{m}^2/\text{s}$ for $0 \le z \le 10\,\text{km}$ $\approx 10 + 2.57(10\text{-}z)$ for $10 < z \le 13\,\text{km}$; $\approx 70\text{–}1.22(70\text{-}z)$ for $13 < z \le 70\,\text{km}$).

Since the model retrieval of the background situation over a given territory has numerous versions that require information on regional and global processes, the SSROC foresees the possibility to apply different models of ozone dynamics. However, the scenario is considered basic, formed by the operator from the data of the regional or global ozonometric network. The operator can prescribe a discrete function $O_{3,i,j,k,s} = Q_3(\varphi_i, \lambda_j, z_k, t_s)$, a priori considering the value of $O_{3,i,j,k,s}$, an average estimate of the function $Q_3(\varphi_i, \lambda_j, z_k, t_s)$ in space Ω_{ijk}, for a time period $t_{s-1} \le t \le t_s$.

Possibilities are foreseen to choose various versions of the approximation of the O_3 function over the whole territory by a number of latitudinal and meridional distributions. In this case only the ozone–air mixing ratio can be prescribed. From this ratio the O_3 function is reconstructed, provided the SSROC database contains information on the vertical profile of air density at any point $(\varphi, \lambda) \in \Omega$. The respective computer dialog procedure makes this choice automatic.

One of the important elements of the SSROC functioning is the replenishing of the database over the period of its adaptation to conditions within a given region. The database includes information about the flight timetable over the region and other characteristics of the engines, cruising altitudes, air speeds, location of airports, air routes, etc.

All background information is concentrated in the form of matrix structures, such as $F = \|f_{ij}\|, D = \|d_{ij}\|, C = \|c_{ij}\|, B = \|b_{ij}\|$, where f_{ij} is a vector whose components contain all the needed information for the ith arrival flight (f_{i1} – time of landing, f_{i2} – arrival direction, f_{i3} – type of engine, f_{i4} – air speed, f_{i5} – cruising altitude, f_{i6} and f_{i7} – latitude and longitude of the airport of entry and other possible characteristics); the d_{ij} vector contains similar information about departing aircraft, the c_{ij} vector describes information about air routes and transit aircraft, and finally, the b_{ij} vector decodes the f_{i3} component, giving the volume of burnt fuel, its type, and the composition of exhaust gases. The transit flights, with intermediate stopovers at airports in the region, are taken into account by the identifiers F and D separately (before their landing and after take-off).

Model and experimental estimates of the impact of transport aircraft on the atmosphere and climate, which can be obtained with the use of SSROC will make it possible, with available synoptic information, to solve the problem of optimization of the transport flight corridors and flight timetables. Considering the various scenarios of the aircraft load on the regional ozonosphere using the SSROC, it is possible to determine the location of the flight corridors, which, in other similar conditions, will reduce the consequences of this load. A possibility appears to specify the compounds of other biogeochemical processes with participation of greenhouse gases.

6.3.8 Methane cycle

Methane like carbon dioxide belongs to the family of greenhouse gases. The spectrum of its natural and anthropogenic sources is wide, and its greenhouse effect efficiency exceeds 20 times that of CO_2, though its concentration (~1.6 ppm) in the atmosphere is about 200 times less than that of CO_2 (Dementjeva *et al.*, 2000). Before human interference, the natural cycle of methane had been balanced with respect to climate. With extraction of natural combustible gases consisting of 90–95% methane, humankind has contributed a great deal to this cycle's instability and uncertainty. Most of the experts estimate the level of the global emission of methane into the atmosphere at 535×10^6 t CH_4 yr^{-1}, of which 375×10^6 t CH_4 yr^{-1} is of anthropogenic origin (50×10^6 t CH_4 yr^{-1} being from rice fields). The anthropogenic input of methane is expected to increase within the next 20–30 years, though in some developed regions, measures have been taken to reduce the anthropogenic emissions of methane into the atmosphere. Nevertheless, the concentration of methane in the present atmosphere increases seven times faster than the growth of CO_2 concentration, its amount increasing annually by 2% (i.e., by 2020 the amount of methane in the atmosphere could double compared with 2000, which, from numerous estimates, will lead to a global warming by 0.2–0.4°C). As in the case of CO_2, these estimates will remain rather doubtful and contradictory until the global model mentioned above is synthesized. However, with present-day knowledge, only the first steps are possible in modeling all these features of the global cycle of CH_4.

The sources of methane are oil, sedimentary and ejected rocks, bottom sediments of lakes, seas, oceans, and other objects of the hydrosphere, as well as soil, peatbogs, and rice plantations. Numerous processes of methane transformation (70–80%) are of biogenic origin, mostly affected by humans. Of course, the significance of these processes varies depending on many natural and anthropogenic parameters. The relationships of individual elements of correlation between the cycles of CO_2, CH_4, and other chemicals vary, too. At any rate, it is clear that depending on the strategy of the NSS, in due course the composition of the terrestrial atmosphere could change substantially. As an example, the burning of 1 m^3 of methane extracts from the atmosphere 2 m^3 of O_2. From open dust-heaps and municipal and industrial sewage, the atmosphere annually gains about 2% anthropogenic methane (~$270–460 \times 10^6$ t C). A certain contribution to the studies of spatial and temporal variability of methane has been made by the Second International Conference on the problems of methane held in Novosibirsk in 2000. The proceedings of this conference contain concrete data on the sources of methane in many regions of the globe. For instance, according to Byakola (2000), within the framework of the United Nations Environment Programme (UNEP)/Gross Domestic Product (GDP), an inventory of the sources and sinks of CO_2 and CH_4 has been made for the territory of Uganda (236×10^3 km^2). In Uganda, the main anthropogenic sources of methane are agriculture, municipal sewage, and biomass burning. In 1990, stock breeding and rice fields in Uganda contributed 205.45×10^3 t CH_4 and 23.45×10^3 t CH_4, to the atmosphere respectively. Agricultural waste burning added 3.55×10^3 t CH_4.

Naturally, Uganda can be recommended to reduce greenhouse gas emissions, but it is not clear what the threshold is for these emissions. Of course, in the future, stock breeding and rice production in Uganda will develop, increasing thereby the volumes of CH_4 emitted to the atmosphere. Hence, a balanced correlation should be sought between the economy of the country and the state of the environment. This problem can be solved by new technologies in nature use (Krapivin and Kondratyev, 2002). In particular, one of the ways to reduce CH_4 emissions is secondary use of organic wastes, for instance, in paper production. In Uganda, up to 16% of urban waste is used in paper production.

The gas transport systems are one of the powerful anthropogenic sources of CH_4. The study by Coconea *et al.* (2000) contains information about methane emissions from pipelines in Romania, the country that signed the Lisbon Protocol in 1994 and now supports the Kyoto Protocol. Romania was the first country in Europe where, in 1917, a 50-km pipeline was laid to transport natural gas. At present, natural gas constitutes 37% of the energy resources of the country, the share of oil and coal constituting 32.6% and 15.2%, respectively. Therefore, the problem of anthropogenic input of CH_4 from the territory of Romania into the atmosphere is rather urgent. Here, like in Uganda, technology plays an important role, reducing by 38.9% the leakage of methane from pipelines (over the last 20 years this constituted 55.35%, in 1994, with respect to the leakage in 1987). On the whole, both extraction and distribution of coal, oil, and gas in the territory of Romania contribute 56% to the total amount of CH_4 emitted from this territory. Agriculture takes second place (29%).

One of the significant sources of CH_4 is Russia, which contributes to the atmosphere about 47×10^6 t CH_4 yr^{-1}, this flux is expected to reach 78×10^6 t CH_4 yr^{-1} by 2025. This increase will be caused by the developing infrastructure of the gas, oil, and coal industry. On global scales, these trends will be practically observed in all countries. In Table 6.14 the contribution of the coal industry to CH_4 production is estimated for various global regions. These estimates are determined by technologies used in the coal industry. On the average, the contributions of various sources to the coal industry itself constitute: 70% – underground ventilation in the coal mines; 20% – underground drainage; 5% – surface loading and unloading operations; 4% – opencast mining of deposits; 1% – derelict mines.

The global cycle of methane has been studied inadequately, and therefore its modeling faces a lot of unsolved problems. Methane fluxes from waterlogged territories have been studied best. These fluxes constitute about 20% of the total input of methane to the atmosphere from all sources. Note that almost 80% of the sources of methane are of biological nature, so that the anthropogenic interference into its natural cycle is also possible through violation of various biospheric processes. In particular, on waterlogged territories, methane only forms due to biological processes. The hydrospheric sources of methane can be presented by a multi-layer model (Figure 6.14). This scheme describes the vertical structure of most of the water bodies. Methane forms in the layer of bottom deposits due to bacteria functioning, and in the zone with oxygen, methane is partially oxidized giving carbon dioxide $CH_4 + 2O_2 \rightarrow CO_2 + 2H_2O + E$. Bacteria taking part in methane oxidation uses the released energy E for organic matter synthesis. The remaining methane reaches the atmosphere and, in contrast to CO_2, does not return to the water medium. This is somehow connected with the fact that CH_4 solubility in water is almost 40 times lower than that of CO_2. The lifetime τ_H of methane in the atmosphere is estimated at

Table 6.14. Emissions of methane by the coal industry in different countries (Gale and Freund, 2000).

Country	Coal production (10^5 t yr^{-1})	CH_4 emissions (10^6 t yr^{-1})	Specific rate of CH_4 emission (kg of CH_4 per ton of coal)
Australia	229	0.8	3.5
England	68	0.5	7.4
Germany	280	1.0	3.6
India	263	0.4	1.5
China	1,141	7.7	6.7
Poland	199	0.6	3.0
Russia	539	4.5	8.3
USA	859	4.3	5.0
Czechoslovakia	88	0.3	3.4
South Africa	182	1.0	0.5
Total	*3,848*	*21.1*	*42.9*

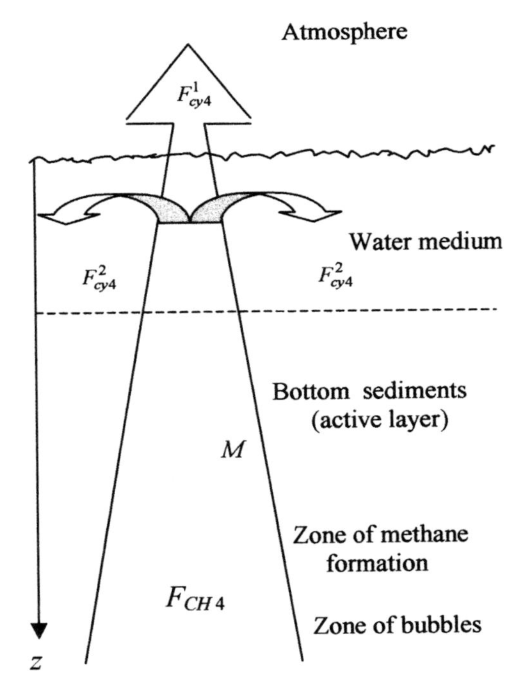

Figure 6.14. The block-scheme of the formation and transport of methane in waterlogged areas. Notation: F^1_{cy4} – methane flux on the atmosphere–water border; F^2_{cy4} – oxidation of methane in the aerobic zone; F_{CH4} – methane power source; M – methane concentration.

about 5 years. Its extraction from the atmosphere takes place due to the participation of methane in photochemical reactions, resulting in methane oxidation first to CO, and then to CO_2. The cycle $CO/OH/CH_4$ plays an important role in the cycle of methane:

$$OH + CH_4 \rightarrow CH_3 + H_2O$$

$$OH + CO \rightarrow CO_2 + H$$

The participating OH-radicals form in the atmosphere during water vapor photolysis. As a result, the simplest diagram of methane oxidation in the atmosphere is the following:

$$CH_4 \xrightarrow{\swarrow OH} CO \rightarrow CO_2$$

Human interference into the processes described by this diagram breaks the natural stability of the $CH_4/CO/CO_2$ balance. In particular, the reclaiming of marshes is one such destabilizing factor. For instance, the drainage of 20% of marshes leads to a natural reduction of CH_4 emissions by 20%, and on the whole, the amount of methane is reduced by 4%, which does not practically influence climate, but does cause changes in the biogeochemical cycles of ozone and carbon dioxide – with unpredictable consequences. These estimates are important in reaching a final conclusion about the level of the integral greenhouse effect. However, the solution of this problem is connected with many factors, a neglecting of which may lead to serious errors. For instance, the CH_4 flux on the atmosphere–marsh border depends on the vertical profile of the temperature in the marsh body. In a simplest case, if we denote $T_W(z,t)$ as the temperature at a time moment t at a depth z and write the equation of heat conductivity:

$$\frac{\partial T_W(z,t)}{\partial t} = a^2 \frac{\partial^2 T_W(z,t)}{\partial z^2}$$

where $a^2 = Kc^{-1}\rho^{-1}$, K is the coefficient of heat conductivity, c is specific heat capacity, ρ is the medium density. Then an estimation of the F^1_{CH4} flux as a time function becomes dependent of a multitude of the poorly assessed characteristics of the environment.

Let the marsh surface temperature vary cyclically with frequency ω and amplitude A, decreasing with depth: $T_W(0,t) = A(z) \cdot \cos(\omega t)$, where:

$$A(z) = A(0) \times \exp\left(-\sqrt{\frac{\omega}{2a}}z\right)$$

Solution of the heat conductivity equation enables one to trace the temperature variations $T_W(z,t)$ and suggests the conclusion that in this case these variations weakly depend on $T_W(0,t)$. Even if $T_W(0,t)$ increases by 2°C, then the amplitude of temperature changes with depth will rapidly decrease to 0.97°C, 0.33°C, and 0.01°C at depths of 40 cm, 2 m, and 3 m, respectively. Hence, with a 2°C increase in the average global atmospheric temperature the flux F^1_{CH4} will increase by not more than 1.4%.

Comparing the global significance of the CO_2 and CH_4 cycles in the atmosphere–marshes system, note that the CO_2 cycle promotes a climatic stabilization, whereas the CH_4 cycle intensifies climate changes. With a climate warming the marshes assimilate part of the CO_2 from the atmosphere and reduce thereby the greenhouse effect. On the contrary, when the climate warms due to increasing F_{CH4}^{1}, the greenhouse effect intensifies.

The west Siberian region of Russia is characterized by numerous intensive natural and anthropogenic sources of methane formation. These are marshes, tundra, permafrost, and oil and gas deposits. In this region the F_{CH4}^{1} flux varies widely both during the period of a year and shorter time periods. From measurements carried out by Jagovkina *et al.* (2000a,b) at the coastline of Yamal in June 1996, the CH_4 concentration in the atmosphere at a height of 2 m varied from 1.83 ppmv on 18 June to 1.98 ppmv on 23 June, with average daily variations of 0.032 ppmv.

The peatbogs of Siberia occupy quite a special place in the global cycle of methane. They play a unique role in the biogeochemical cycles of methane and carbon dioxide. On the one hand, they are a non-anthropogenic source of CH_4 and CO_2, but on the other hand, they are intensive assimilators of carbon from the atmosphere and extract it from the natural cycle for a long time. The marshes of west Siberia, for instance, contain 20–30% of the global carbon supplies. The intensity of CH_4 emissions from the marshes is, on average, almost 2,000 times weaker than that of CO_2. Some 35–50% of all methane emitted from the territory of Russia emanates from the marshes. The west Siberian marshes emit to the atmosphere not more than 1.7×10^6 t CH_4 yr^{-1}, which does not exceed 1% of the global CH_4 flux. The spatial heterogeneity of the F_{CH4}^{1} flux is high, which is determined by the diverse characteristics of the marsh ecosystems. In particular, the upper oligotrophic coniferous–shrubby sphagnous swamps emit 0.9–10 mg C m^{-2} h^{-1} (Dementjeva, 2000). This estimate is rather approximate, since the scattering of such estimates by various authors constitutes hundreds of percent. For instance, a drained sphagnous swamp of the transitional type can emit 142–204 g C m^{-2} hr^{-1}, and the rush–sphagnous bogs – 83.5–309 mg C m^{-2} hr^{-1}.

The main mechanism for the formation of methane in a marsh is connected with the functioning of special groups of micro-organisms. Part of the methane due to diffusion is emitted to the atmosphere, but most of the methane remains in the peat layer and is gradually emitted to the atmosphere.

Humankind is interfering with the natural biogeochemical balance of greenhouse gases practically all over the world. One aspect of influence is a reduction in the areas of marshes and their transformation into agricultural fields. Diverse human agricultural activity adds 20% of all the anthropogenic fluxes of greenhouse gases to the atmosphere. For instance, in the USA it is 30%. Stock breeding contributes considerably to this flux. In California and Wisconsin each hectare of pasture emits annually 502 kg CH_4 (or 10,511 kg CO_2) and 134 kg CH_4 (or 2,814 kg CO_2), respectively. In New Zealand such emissions of CH_4 are estimated at 291 kg CH_4 (or 6,110 kg CO_2) (Johnson and Ulyatt, 2000).

Among the Kyoto Protocol signatories, England takes ninth place by volume of reduced emissions of greenhouse gases. The decreasing trend of methane emissions is part of the general reduction of emissions of six greenhouse gases (CO_2, CH_4, N_2O, hydrofluorocarbons, perfluorocarbons, sulfur hexafluoride) from 1990. In 2000 greenhouse emissions decreased by 15% compared with 1990. By 2010, the CH_4 emissions will constitute 20,134 t yr^{-1}. This reduction will be reached mainly due to new technologies in processing waste and in the coal industry. On the whole, in England, according to the developed scenario, emissions of methane by 2010 will decrease by 14% in agriculture, by 82% in the coal industry, by 29% in the oil and gas industry, and by 73% in the waste-processing industry. The possibility of realization of this scenario is confirmed by the decreasing CH_4 trend in 1998 compared with 1990. For instance, during this period, emissions of methane in the coal industry have decreased by 64%, and in waste processing by 29%. In 1990, the share of waste processing in England constituted 32% of all CH_4 emissions, only 3% of these emissions being connected with sewage processing.

In agriculture, emissions of CH_4 in England constituted $1{,}037 \times 10^3$ t in 1990 and 998×10^3 t in 1998. The scenario of a reduction in the F^1_{CH4} flux in England due to improved technologies in agriculture foresees emissions of $902\text{–}983 \times 10^3$ t CH_4 in 2010. The CH_4 emission from burning agricultural waste is completely excluded, and in the stock breeding, emissions of methane are reduced by 8% compared with 1990.

In the coal industry in England, emissions of methane in 1990 constituted 819×10^3 t with the main contribution to this flux made by underground operations. This constituted 24% of the whole flux of methane from England. In 1998 the F^1_{CH4} flux decreased to 264×10^3 t and by 2010 it should decrease to 218×10^3 t. A similar decreasing trend in methane emissions from England remains in the oil and gas industry, too. According to the scenario, the contribution of these sectors of energy production into the F^1_{CH4} flux will decrease from 540×10^3 t in 1990 to $349\text{–}464 \times 10^3$ t in 2010 (Meadows, 2000).

According to Bazhin (2000), the F^1_{CH4} flux in every water basin with a vertical stratiform structure forms in an active layer beneath the water layer. Practically all aquageosystems have such a structure. The layer, where methane forms, has two areas. In the bottom area located at a depth h, methane has the form of bubbles. Above this layer, due to diffusion, the concentration of methane decreases, and the bubbles disappear. Denote as $D_{CH4}(z)$ the coefficient of methane diffusion at depth z, then the stationary behavior of the whole system shown in Figure 6.14 is described by the equation:

$$\frac{d}{dz}\left[D_{CH4}(z)\frac{d}{dz}M(z)\right] - F_{CH4}(z) + F^1_{CH4} + F^2_{CH4} = 0$$

Model calculations and field measurements performed by Bazhin (2000), show, for instance, that on the rice fields $h_b = 1.3\,\text{m}$, $F_{CH4} = (1.3 - 1.7) \times 10^{-12}\,\text{mol m}^{-3}\,\text{s}^{-1}$. According to Khalil *et al.* (2000), the rice fields play a significant role in the gas balance of the atmosphere due to emissions of CH_4, CO, N_2O, H_2, and $CHCl_3$. For instance, in China the rice fields deliver these gases to the atmosphere at the following rates (mg m^{-2} h^{-1}): CH_4 – 900–50,000; CO – 80–100; H_2 – 5–30; N_2O –

50–1,000; CHCl – 1–8. A wide scatter of these estimates is explained by highly unstable fluxes of these gases due to rice-growing technology. For instance, the use of sulfates on the rice fields increases emissions of methane by 12.0–58.9% depending on other characteristics of these fields (Liping *et al.*, 2000).

Thus, an estimation of the F^1_{CH4} flux as a function of a given territory with account of natural and anthropogenic processes taking place there requires first of all a detailed inventory of methane sources as well as natural and technogenic systems functioning on this territory. Examples of such an inventory given above serves as the basis for development of studies in this direction.

The dynamics of the CH_4 content H_A in the atmosphere can be parameterized by a simple balance relationship:

$$\frac{\partial H_A}{\partial t} + V_\varphi \frac{\partial H_A}{\partial \varphi} + V_\lambda \frac{\partial H_A}{\partial \lambda} = F^1_{CH4}(t, \varphi, \lambda, \Xi) - \frac{H_A(t, \varphi, \lambda)}{\tau_H}$$

where Ξ is the identifier of the type of the natural or technogenic system.

6.4 INTERACTIVITY OF NATURAL DISASTERS AND ELEMENTS OF THE GLOBAL WATER BALANCE

Among all water reservoirs on Earth the World Ocean is the largest, with a present volume exceeding 50 times the water volume in glaciers, which are the second largest water reservoirs. This comparison is important for understanding the relationship between the hierarchical stages of water basins and for the determination of their structure in the model. Within the a priori scenarios of anthropogenic activity and possible changes in the biosphere, the relationship between these stages is really important. For instance, the Antarctic has accumulated 1.6% of the supplies of all water on Earth. Comparison of these supplies with the volume of the Arctic Ocean which has 20% less ice cover than the Antarctic, suggests the conclusion about an inadequacy of the global model of the hydrological cycle, which neglects the role of the Antarctic (Keeling and Visbeck, 2001). These relationships become especially important if considering climate scenarios.

The water cycle problems are of critical importance not only in the context of climate change studies but (even to a greater degree) as a factor of life support on Earth. Due to respective feedbacks, the water cycle functions as an integrator of various processes in the NSS. The related questions include, in particular, the following:

- What are the mechanisms and processes that are responsible for the formation and variability of water cycles and how much are they subject to anthropogenic forcings?
- How are interactions between the global water cycle and other cycles (carbon, energy, etc.) controled through the processes of feedbacks and how do these processes change in time?

- What are the uncertainties of forecasts of the annual change and inter-annual variability as well as long-term "projections" of various parameters (components) of the water cycle and how is it possible to reduce the levels of such uncertainties?
- What are the probable consequences of water cycle variability on different spatial–temporal scales for the human activity and ecosystems, and how can this variability affect the Earth's system influencing the transport of deposits and nutrients as well as the biogeochemical cycles?
- What are possibilities for using information on the global carbon cycle for decision making in the field of ecological policy with respect to water resources?

The successful launch of the satellite GRACE in March 2002 to monitor the Earth's gravity field has made it possible for the first time to obtain global information on the spatial–temporal variability of the gravity field determined by the non-uniform distribution of masses within the Earth's crust. These data make it possible, in particular, to study the gravity field variability determined by such factors as variations of the surface and deep-water currents, river run-off, and subsoil water, as well as the water exchange between ice sheets and oceans. Analysis of the GRACE data will make it possible to trace changes in the extent and volume of water masses contained in continental water basins (in large water reservoirs, lakes, and underground water) as well as such changes as shifting of the zones of warm water masses in the Pacific Ocean (El Niño) and the dynamics of tectonic plates. The GRACE data were successfully reproduced by numerical modeling of temporal changes of the water balance in the basin of the Mississippi River. The capabilities of the remote sensing of water supplies opens up prospects for monitoring the dynamics of the regional water balance and, respectively, obtaining (on national and international scales) the estimates needed to regulate water resources.

In January 2003 the satellite ICESat was launched to obtain information on the ice cover, clouds, and land topography. The data from this satellite, covering both polar regions, provides information on land topography and vegetation cover, ice sheet dynamics as well as water cycles, aerosols, and clouds characteristics. Together with the data from the Terra and Aqua satellites, variable information will be obtained on the Greenland ice sheet and on inter-annual variations of the sea ice cover in the Arctic. In the case of the Aqua data, of importance is a higher spatial resolution and more versatile multi-spectral information than those achieved earlier. The satellite data will supplement the results of aircraft observations.

The hydrology and sea currents of the Southern Ocean with account of the effect of glacier cover have been described in numerous monographs, and circulation models of different complexities and degrees of detailing have been derived to simulate them. Such models for the World Ocean, on the whole, are based on accounts of configurations of the non-penetrating boundaries and topology of the straits. Numerous numerical experiments with such models have made it possible to reveal the principal structure of the global oceanic circulation consisting of a hierarchy of the closed ring circulations with centers of uprising and descending waters, and including the geometry of water basins with straits between them. To

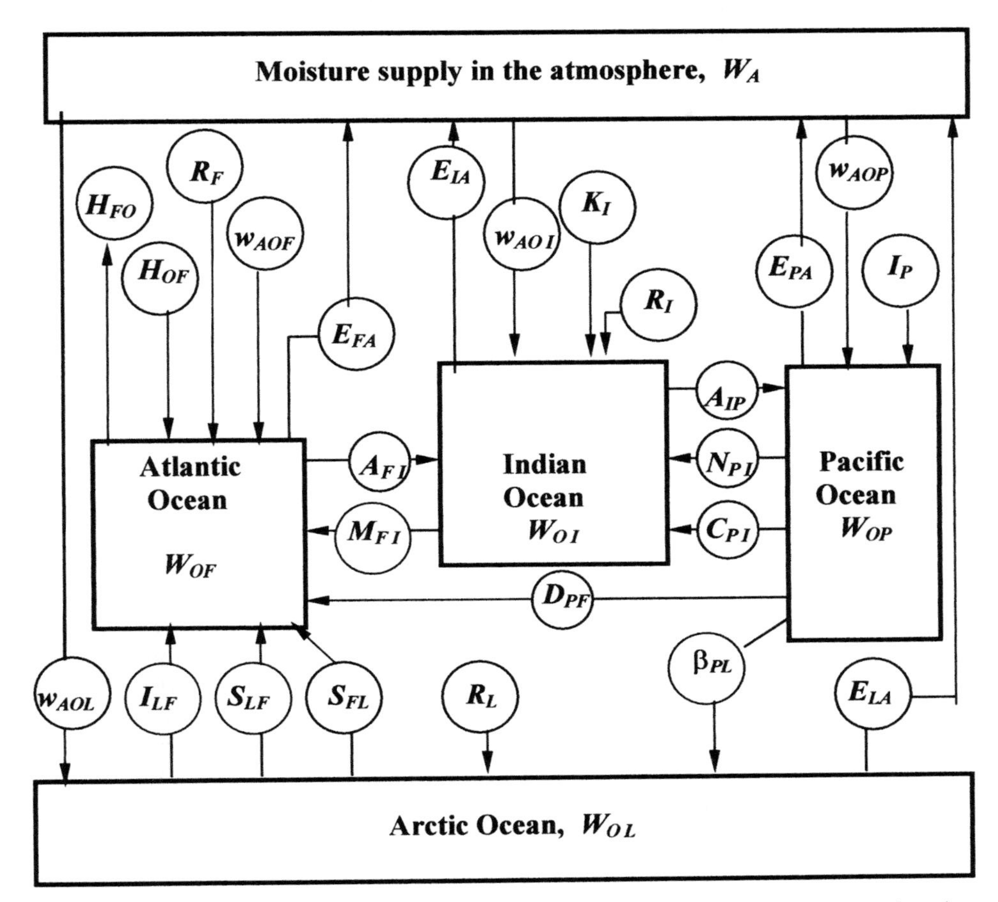

Figure 6.15. Elements of the global water balance with the role of the ocean taken into account. Notation: w_{AOL}, w_{AOF}, w_{AOI}, w_{AOP} – precipitation; H_{FO}, H_{OF} – Strait of Gibraltar; R_F, R_I, R_L – rivers; E_{FA}, E_{IA}, E_{PA}, E_{LA} – evaporation; A_{FI}, A_{IP} – Antarctic Current; M_{IF} – the Cape Igolny Current; C_{PI} – the East Australian Current; β_{PL} – Bering Strait; I_{LF} – Arctic ice; D_{PF} – Drake's Passage; I_P – Antarctic ice; N_{PI} – Indonesian Seas; S_{LF}, S_{FL} – straits.

describe the water circulation in the southern basin, it is necessary that Drake's Passage be taken into account. A scheme for the hydrological field circulation in the World Ocean, acceptable for simulation, has been proposed by Seidov (1987) and Chahine (1992). The model is a system of equations and boundary conditions taking into account the outline of the shores, the bottom relief, as well as ice formation and melting. However, on a global scale, to simulate the oceanic circulation, a simplified scheme is necessary, reflecting mainly the role of the straits. Such a scheme is shown in Figure 6.15. The quantitative characteristics of the constituents of this scheme are given in Table 6.15. The final unit responsible for the modeling of the World Ocean

Table 6.15. Quantitative estimates of water flows in Figure 6.15 (10^3 km^3 yr^{-1}).

Flow	Estimate	Flow	Estimate
w_{AOL}	3.6	I_{LF}	0.57
S_{LF}	436	S_{FL}	400
R_L	5.14	A_F	0.3
H_{OF}	23.97	R_F	19.33
w_{AOF}	72.5	E_{FA}	96.6
A_{FI}	6,780.24	M_{IF}	952
D_{PF}	5,771.09	C_{PI}	437
N_{PI}	66.86	O_I	0.765
A_{IP}	6,338.74	E_{IA}	115.4
w_{AOI}	84	R_I	5.386
K_I	0.005	R_P	13.12
E_{PA}	200.4	w_{AOP}	206.7
I_P	0.975	E_{LA}	1.7
β	80.5		

circulation has the following form:

$$\sigma_{OF}\frac{dW_{OF}}{dt} = H_{OF} + R_F + I_{LF} + S_{LF} - S_{FL} - H_{FO} - A_{FI} + M_{IF} + D_{PF} + (w_{AOF} - E_{FA})\sigma_{OF} + A_F$$

$$\sigma_{OI}\frac{dW_{OI}}{dt} = A_{FI} + C_{PI} + N_{PI} + K_I + R_I + (w_{AOI} - E_{IA})\sigma_{OI} - A_{IP} - M_{IF}$$

$$\sigma_{OP}\frac{dW_{OP}}{dt} = A_{IP} + R_P + (w_{AOP} - E_{PA})\sigma_{OP} + I_P - B_{PL} - D_{PF} - C_{PI} - N_{PI}$$

$$\sigma_{OL}\frac{dW_{OL}}{dt} = R_L + B_{PL} + (w_{AOL} - E_{LA})\sigma_{OL} + S_{FL} - I_{LF} - S_{LF}$$

$$\sigma\frac{dW_A}{dt} = (E_{PA} - w_{AOP})\sigma_{OP} + (E_{FA} - w_{AOF})\sigma_{OF} + (E_{IA} - w_{AOI})\sigma_{OI} + (E_{LA} - w_{AOL})\sigma_{OL}$$

Within this large-scale approach to the formation of the Biosphere Water Balance Model (BWBM) ocean unit (Krapivin and Kondratyev, 2002), the dependences of the fluxes of water in its different phases on the environmental parameters remain uncertain. Apparently, the mass exchange between the reservoirs s and l can be described by the simplest linear scheme: $w_{sl} = |W_{OS}\sigma_{OS} - W_{OL}\sigma_{OL}|/T_{sl}$, where T_{sl} is the time for equalizing the levels W_{OS} and W_{OL}, and σ_{OS} and σ_{OL} are the areas of water basins s and l. For the scheme in Figure 6.15 we have:

$$A_{FI} = \max\{(V_{OF} - V_{OI})/T_{FI}, 0\} \qquad M_{FI} = \max\{(V_{OI} - V_{OF})/T_{IF}, 0\}$$

$$A_{IP} = \max\{(V_{OI} - V_{OP})/T_{IP}, 0\} \qquad N_{PI} = \max\{(V_{OP} - V_{OI})/T_{PI}, 0\}$$

$$C_{PI} = \max\{(V_{OP} - V_{OI})/T^*_{PI}, 0\} \qquad D_{PF} = \max\{(V_{OP} - V_{OF})/T_{PF}, 0\}$$

$$S_{LF} = \max\{0, (V_{OL} - V_{OF})/T_{LF}\} \qquad S_{FL} = \max\{0, (V_{OF} - V_{OL})/T_{FL}\}$$

$$\beta_{PL} = \max\{0, (V_{OP} - V_{OL})/T_{PL}\}$$

where $V_{OS} = W_{OS}\sigma_{OS}, (S = F, I, P, L)$.

To estimate the K_I flux, we took into account the information on the moisture balance in the region of the Red Sea. According to available estimates, the input of water to the Red Sea via the Suez Canal and by precipitation can be neglected. Not a single river flows into the Red Sea. The main component of the K_I flux through Bab el Mandeb is rather persistent. Hence, we can assume $K_I = \max\{0, w_{AK}\sigma_{KMP} - E_{KMA}\sigma_{KM}\}$, where w_{AK} and σ_{KMP} are the level and the area of the mainland run-off to the Red Sea, respectively, and E_{KMA} is the evaporation from the area σ_{KM} of the Red Sea. The water expenditure through the Strait of Gibraltar H_{FO} (H_{OF}) is determined by the relationship of the levels of W_{OF} and the Mediterranean Sea. In order not to complicate the structure of the model, the level of water in the Mediterranean Sea is determined by its watershed and the difference between precipitation and evaporation. Since the intra-annual distribution of the water inflow into the Atlantic Ocean varies within 20%, we can reliably assume: $W_{FO} = -W_{OF} = \text{const}$.

6.4.1 Regional model of the water balance

Emergency hydrological phenomena occur on a given territory in the case of considerable deviations of the regional water balance components from their average levels. These deviations can be predicted using the respective model of the regional water balance introduced into the GMNSS. Consider the scheme of Figure 6.16 as the basis for modeling the hydrological regime of a limited territory Ω_L, occupied by the aquaecosystem under study. Each territory has the river network, water bodies, and land. According to the landscape–hydrological principle, to derive a simulation model in the zone of the hydrological systems functioning, it is necessary to select the facies, which are connected with typifying the floristic background (the appearance of which is determined by the micro-relief), type and properties of the soil, surface moisture, depth of ground water, and other factors. In general, the territory Ω_L is characterized by the presence of m facies, and the water network has n heterogeneous sites. Bearing this in mind, according to the scheme in Figure 6.16, the closed system of balance equations has the form:

$$\sigma_{ij} dW_{A,ij}/dt = E_{ij} - R_{ij} + \sum_{k=1}^{n}(V_k - B_k S_k) + D_{ij} + \sum_{l=1}^{m}(L_l + T_l - W_l \sigma_l) \tag{6.4}$$

$$S_k dG_k/dt = Y_k - V_k + B_k S_k - H_k + J_k + \sum_{l=1}^{m}(K_{lk} - F_{kl} - V_{kl} - M_{kl})$$
$$- \Gamma_k + S_k(C_{k-1} V_{k-1}/\Delta_{k-1} - C_k V_k/\Delta_k) \tag{6.5}$$

$$\sigma_l d\Phi_l/dt = \sum_{k=1}^{n} Z(F_{kl} + V_{kl} + M_{kl})$$
$$+ \sum_{k=1}^{m} \psi_l^k \theta_l - L_l - T_l - P_l - \theta_l + N_l + W_l \sigma_l \tag{6.6}$$

$$\sigma_{ij} dG_{ij}/dt = I_{ij} - Z_{ij} - D_{ij} + \sum_{k=1}^{n}(H_k - J_k) + \sum_{l=1}^{m}(P_l - N_l) \tag{6.7}$$

In (6.4)–(6.7) the following notations are assumed: σ_{ij}, σ_l, and S_k are the areas of the territory Ω_{ij}, of the lth facies and the kth compartment of the river network in km^2, respectively; Δ_k is the linear size of the kth compartment of the river network, km; $W_{A,ij}$, G_k, and Φ_l are, respectively, the levels of water in the atmosphere, in the kth compartment of the river network, and the lth facies on the territory Ω_{ij}; θ_{ij} is the level of ground water, m; ψ_l^k is the share of the run-off from the kth facies, getting to the territory of the lth facies; with the remainder of the notations given in the scheme in Figure 6.16.

An application of the model to other regions is realized via the variables E, R, Y_i, Γ_i, I, Z. Besides this, in analysis of the concrete situation, the configuration of the waterway and the level of the water body are taken into account. The required equations are written similarly to those above, proceeding from the condition of the water volume balance. Functionally, all fluxes in the scheme in Figure 6.16 can be described based on the laws of hydrodynamics and with account of available observational information. The inflow E_{ij} and outflow R_{ij} of moisture can be determined from remote-sensing data. Between measurements, information is used regarding wind speed V_{ij}, and the functions E_{ij} and R_{ij} are calculated using the formulas: $E_{ij} = E_{H,ij}$; $R_{ij} = W_{A,ij} l^*/(l^* + k_1 V)$, where $l_* = 2\sqrt{\sigma}/H$, E_H is the atmospheric moisture on the windward border of the region, and k_1 is the constant coefficient reflecting the contribution of wind to the circulation of precipitation.

Information on precipitation and run-off is introduced into the information catalogs of the hydrometeorological services. Based on these data, the respective model units can be derived. Assuming that the distribution of precipitation is proportional to relevant areas, we obtain:

$$B_k = W_{A,ij}\sigma_k/\sigma_{ij};\ W_l = W_{A,ij}\sigma_l/\sigma_{ij}$$

The model of river runoff formation should take into account the watershed

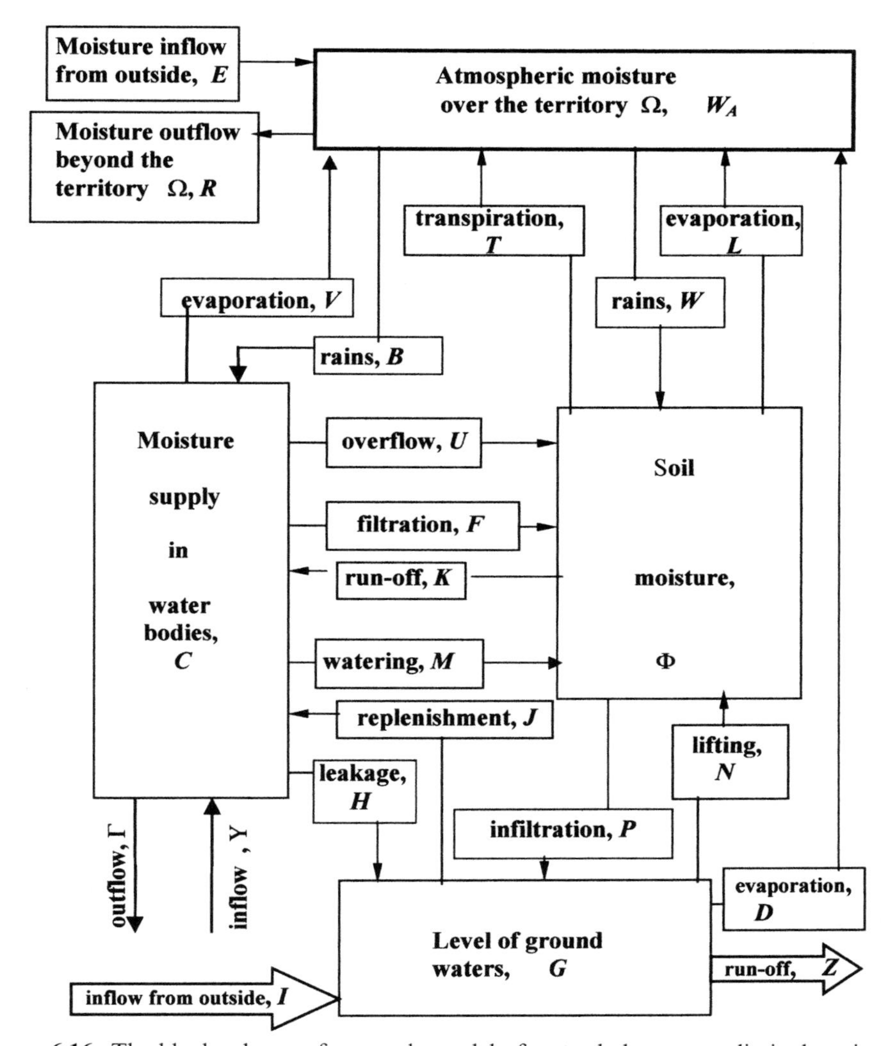

Figure 6.16. The block-scheme of a sample model of water balance on a limited territory.

topography and the spatial distribution of its soil characteristics, as well as special features of vegetation cover. Let $\theta_l = (g_l + K_l \exp[-a_l X_l - C_l A_l])\sigma_l$, where X_l and A_l are, respectively, the vegetation density (m km^{-2}) and the soil layer thickness (m) over the area σ_l; g_l is the coefficient of the relief specific run-off in the lth facies; k_l is the coefficient of water penetration through vegetation and soil cover over the area σ_l; and a_l and C_l are the coefficients of precipitation retained by plants or soil in the lth facies, respectively. The parameters of this dependence can be determined from field measurements that establish, for a given type of soil and plants, a connection between precipitation intensity, the rate of uptake of water by the soil, and soil water resistance. So, for takyrs, for instance, the run-off is equal to precipitation. This rather rough approximation can be specified, since the radiometric methods make it possible to classify the soil moisture, at least, into three types: firmly bound, loosely bound, and free water. The bound water is a film of moisture adsorbed by the surface of ground particles with the film thickness being 6–8 molecular layers. The content of bound water is estimated at 2–3% in sand, and 30–40% in clay and loess. The bound water cannot be assimilated by plants, it does not dissolve salts. In the models considered these specific features are taken into account when determining the respective coefficients of evaporation and transpiration.

The run-off θ_l is distributed between facies, and in the form of return water K_{lk}, flows into the river. In a general form this is reflected in (6.6) through the coefficients of the run-off distribution $\psi_s^l\left(\sum_{s=1}^{m+2}\psi_s^l = 1\right)$, where ψ_{m+1}^l is the share of the run-off from the lth facies beyond the region and ψ_{m+2}^l is the share of the run-off from the lth facies into the river. The coefficients $\omega_{lk}\left(\sum_{k=1}^{n}\omega_{lk} = 1\right)$ characterize the run-off distribution from the lth facies by the river compartments and are determined by the landscape relief and the spatial location of the facies and the waterway compartments. Thus, $K_{lk} = w_{lk}\psi_{m+2}^l\theta_l$.

Evaporation from the soil surface can be described by the formulas of Hitchcock, Horton, Weissman, and others (Bras, 1990). For instance, the formula by Priestly and Taylor for latent heat of evaporation q_E is: $q_E = \alpha S(q^* - q_i)/(S + \gamma)$, where q_i is the soil heat flux (W m^{-2}); q^* is the net radiation flux (W m^{-2}); $\gamma = 0.066 \times 10^3$ Pa K^{-1} is the psychometric constant; S is the slope of the curve of the temperature dependence of the saturated moisture pressure (Pa K^{-1}):

$$\alpha = \begin{cases} 1.06 & \text{for a wet soil} \\ 1.04 & \text{for a dry soil} \\ >1.26 & \text{at a warm air advection over a wet surface} \end{cases}$$

The Horton formula gives:

$$V = 0.36[(2 - \exp\{-0.44\theta\})l_V - l_a] \quad (\text{mm day}^{-1})$$

where θ is the wind speed (m s^{-1}), l_V is the vapor pressure near the water surface, and l_a is the water vapor elasticity.

The Rower formula is written as:

$$V = 0.771(1.465 - 0.007\rho)(0.44 + 0.26\theta)(l_V - l_a) \quad (\text{mm day}^{-1})$$

where ρ is atmospheric pressure (mm Hg).

A diversity of the forms for parameterization of the dependence of the rate of evaporation from the soil surface on the environmental parameters enables one to easily adapt the model of the water balance to the information base.

The flux T in Figure 6.16 reflects the impact of vegetation cover on the hydrological regime of the territory. One of the simple models of transpiration is the following dependence:

$$T = y(24\alpha^* + \beta^*) \quad (\text{cm day}^{-1})$$

where y is the specific water return of the soil, α^* is the rate of the ground water lifting (cm h^{-1}), and β^* is the daily change in the level of ground water (cm).

Determine the constituents of the block-scheme in Figure 6.16 characterizing the processes of leakage and filtration of water from the river. Both leakage and filtration depend on the quality of the river bed and water level. Let:

$$H_i = \begin{cases} \mu_i C_i S_i & \text{for} \quad 0 \le C_i \le C_{i,\min} \\ \mu_i C_{i,\min} & \text{for} \quad C_i > C_{i,\min} \end{cases}$$

where μ_i is the coefficient of penetration of water through the river bed. Filtration F_i increases with increasing C_i between two critical values of $C_{i,\min}$, when there is no filtration, and $C_{i,\max}$, when it is at a maximum:

$$F_{i,\max} = \begin{cases} 0 & \text{for} \quad 0 \le C_i \le C_{i,\min} \\ \mu_i(C_i - C_{i,\min})S_i & \text{for} \quad C_{i,\min} < C_i < C_{i,\max} \\ \mu_i(C_{i,\max} - C_{i,\min})S_i & \text{for} \quad C_i \ge C_{i,\max} \end{cases}$$

The distribution of water filtering from the river between facies depends on the distance r_{ij} between the ith compartment and the jth facies, as well as on the structure of soil and landscape relief. In particular, this dependence can be described by the function $F_{ij} = F_{i,\max}\chi(r_{ij})$, where $\chi(r_{ij})$ is the decreasing function meeting the condition:

$$\sum_{j=1}^{m} \chi(r_{ij}) = 1$$

Evaporation from the river surface depends on the environmental temperature and can be described by the formula $V_i = V_i^* T^\omega$ or by the relationship $V_i = \mu(\theta)(\rho_V - \rho_2)$, where $\mu(\theta)$ is the function reflecting the impact of the wind, ρ_V is the water vapor pressure at the temperature of the evaporating surface (mb), and ρ_2 is the absolute air humidity at a height of 2 m (hPa).

The volume of overflow is determined by the binary regime of the waterway functioning within a maximum possible water level $C_{i,\max}$, so that:

$$V_i^* = \begin{cases} 0 & \text{for} \quad 0 \le C_i \le C_{i,\max} \\ C_i - C_{i,\max} & \text{for} \quad C_i > C_{i,\max} \end{cases}$$

The distribution of U_i^* between facies depends on the landscape relief, characterized by the matrix of the relief specific run-off $\Psi = \|\Psi_{ij}\|$, it is written as:

$$\sum_{\hat{i}jj=1}^{m} \Psi_{ij} = 1 \qquad \Psi_{ij} \ge 0$$

As a result, $U_{ij} = \Psi_{ij} U_i^*$.

Assuming water for watering from the ith compartment of a waterway is an anthropogenic factor, it should be considered as a free parameter $M_i^* = \sum_{j=1}^{m} M_{ij}$. To take into account a possible heterogeneity of the distribution of M_i^* between facies, take the matrix of the coefficients of the distribution of watering $v = \|v_{ij}\|(v_{ij} \geq 0, \sum_{j=1}^{m} v_{ij} = 1, i = 1, \ldots, n; j = 1, \ldots, m)$, so that $M_{ij} = v_{ij} M_i^*$.

The relationship between the surface water fluxes and ground waters strongly depends on the flux of water infiltrating through the soil layer downward. This flux, called infiltration, with account of only the vertical heterogeneity of the soil can be described in a general form by the equation:

$$\frac{\partial P}{\partial t} = \frac{\partial}{\partial z}\left[p(P)\frac{\partial P}{\partial z} + K_z(P)\right]$$

Bras (1990) gave various solutions to this equation. For practical use the following solution can be recommended:

$$f = f_c + (f_0 - f_c)\exp(-Pl^2 t) \tag{6.8}$$

where $f = (P_i - P_0)P/(\pi t)^{-1}$, f_c is the asymptotic value of the rate of filtration, and f_0 is the initial value of the rate of filtration.

The processes of infiltration and evaporation of ground water depends strongly on the vertical profile of the soil layer. The following soil layers can be selected: saturated and unsaturated. The saturated layer usually covers depths >1 m. The upper unsaturated layer includes soil moisture in the zone of the plants' roots, the intermediate level, and the level of capillary water. The water motion through these layers can be described by the Darsy's law, and the gravitation term $K_z(P)$ in (6.8) is calculated from the equation:

$$K_z(P) = 256.32\delta_s^{-7.28} - 1.27\delta_s^{1.14} \qquad (\text{cm day}^{-1})$$

where δ_s is the volume mass of soil (g cm^{-3}).

Thus, the system of equations (6.4) through (6.7) with the indicated functional descriptions of water fluxes in the region under study at initial values of $W(t_0)$, $G(t_0)$, $C_i(t_0)$, and $\Phi_j(t_0)$ prescribed for a time moment t_0 enables one to calculate the characteristics of the water regime of the whole region for $t \geq t_0$. The initial values are provided by the monitoring system. The regularity of surveys depends on the required accuracy of prognosis and can be realized by planning the monitoring regime. Based on the synthesis of the model and the remote-sensing system the monitoring can be organized practically for any irrigated agro-ecosystem. In this case problems appear in identifying/correlating the airborne measurements with the values of geophysical, ecological, and hydrological parameters. An example of the successful solution of such problems (Vinogradov, 1983) is a determination of the dependence between the coefficient of spectral brightness $\tau_J = \tau_z + (\tau_0 - \tau_z)\exp(-\alpha W^c) + dW^n$, where τ_0 is the coefficient of the dry soil brightness; τ_z is the coefficient of the brightness of soil with moisture content close to a minimum of the field moisture capacity (when there is no free water in the soil); and α, c, d, and n determine the type of the soil (α, d, $n < 1$; $c > 1$; for achromatic loamy soils we have: $\tau_z = 0.09$; $\tau_0 = 0.28$; $\alpha = 0.01$; $c = 2.3$; $n = 0.9$; $d = 0.0001$). Obtaining these estimates is an important problem in the remote sensing of the environment.

Finally, note that the deterministic approach to modeling the water cycle in the zone Ω_L described here cannot be considered as the only possible one. Such an approach gives only average trends in changes of the water cycle components. Their distribution and probabilistic prognosis can be obtained only on the basis of the dynamic–stochastic models of the water balance. In fact, the right-hand parts of (6.4)–(6.7) should be supplemented with random components which reflect oscillations of respective functions and introduce into these equations the factor of the stochastic nature of hydrophysical and hydrological processes. This is especially important in flood modeling. In a general case, it is difficult to evaluate to what degree such a complication of the model will raise its adequacy, but from the formal point of view, this complication leads to a necessity to broaden the theoretical bases of studies of the territorial water balance.

An important aspect of the model of the biospheric water balance as a unit of the GMNSS functioning is its conjunction with the methods of determination of various parameters of the water cycle. Such methods are based on the use of surface, satellite, and airborne measurements. The unit BWBM as the global model makes it easier to understand the role of the oceans and land in the hydrological cycle, to choose the main factors that control it, as well as to trace the dynamics of its interaction with plants, soil, and topographic characteristics of the Earth surface. It is based on an account of the interaction between the elements of the water cycle, as well as natural and anthropogenic factors taken into account through information interfaces with other units of the global model (Krapivin and Kondratyev, 2002).

Consider the block-scheme of the global water exchange and write the respective equations. The basic regularity of the global water exchange is the invariability of water supplies on the Earth over time periods of hundreds of years (i.e., we can reliably write the balance equation $W_E = W_S + W_O$, where W_E, W_S, and W_O are water supplies on the Earth, on land, and in the oceans, respectively). A compartment of the atmosphere is related to the respective region of the water basin. The relationship is valid:

$$\frac{dW_E}{dt} = \frac{dW_S}{dt} + \frac{dW_O}{dt} = 0$$

or $dW_S/dt = -dW_O/dt$. Hence, the trend in changes of water supplies on land is contrary to the similar trend in the oceans.

With the water supply in the atmosphere $W_A = W_{AO} + W_{AS}$, we obtain $W_E = W_A + W_{S1} + W_{O1}$, where W_{AO} and W_{AS} are water supplies in the atmosphere over the oceans and land, respectively; $W_{S1} = W_S - W_{AS}$, $W_{O1} = W_O - W_{AO}$. The balance equation will be:

$$\frac{dW_E}{dt} = \frac{dW_A}{dt} + \frac{dW_{S1}}{dt} + \frac{dW_{O1}}{dt} = 0$$

As is seen, the structure of the trends in the ratios of water supplies is complicated, to analyze it, additional considerations are needed. This complication becomes considerable with further subdivision of the biosphere.

Within the GMNSS, small corrections for the water exchange between the Earth and space are not taken into account. The model of the global water cycle can be based on the method of describing the hydrology of comparatively large territories.

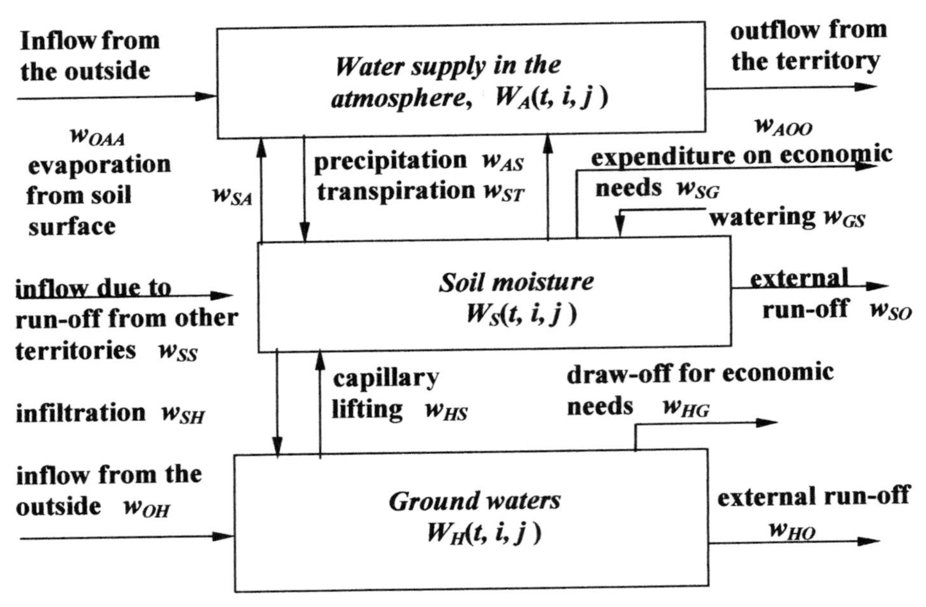

Figure 6.17. Water fluxes across the border of a limited land territory.

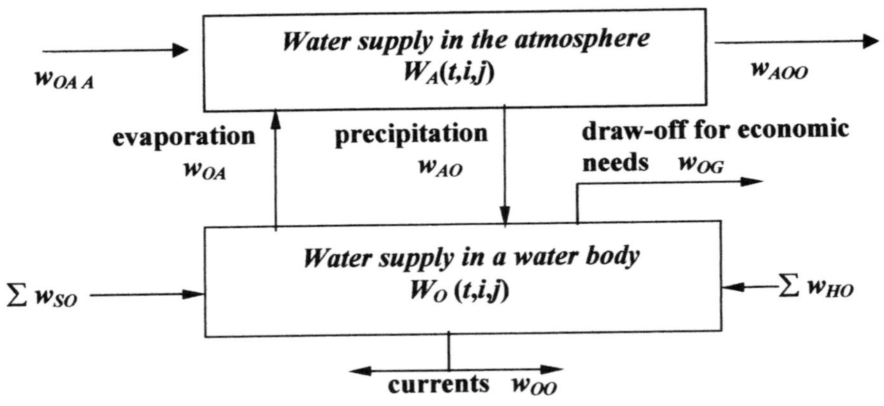

Figure 6.18. Water fluxes across the border of a limited territory and a water body.

In this case the basic unit of such a territory is a compartment Ω_{ij} of the Earth's surface of size $\Delta\varphi_i$ by latitude and $\Delta\lambda_j$ by longitude.

The state of the water component of compartment Ω_{ij} with coordinates (φ_i, λ_j) can be characterized by the magnitude of an equivalent liquid water column over a unit area. Possible water fluxes across the border of Ω_{ij} are shown in Figures 6.17 and 6.18. The intensities of these fluxes depend on the phase state of water, temperature, wind speed, and other geophysical and ecological factors. It is difficult to take into account all the small details of these fluxes within the global model because their interactions have been studied inadequately. Therefore, the degree of detailing chosen here has been orientated toward account of the most important components of their states. Water is considered in liquid, solid, and gas phases. Within the compartment Ω_{ij} there is only one state though in the future with required available information, a vector parameter can be introduced, which would determine the share of precipitation over Ω_{ij} in the form of snow, pellets of snow, granulated snow, pellets of ice, ice rain, hail, rain, drizzle, wet snow, and other forms of precipitation.

The global water balance consists of a mosaic structure of local balances at the level of Ω_{ij}. The proposed description of water fluxes enables one to trace their balance at any of the levels of spatial digitization – region, water basin, continent, ocean, hemisphere, or biosphere. Clearly, the general balance of evaporation and precipitation at the level of the biosphere is maintained. In other cases, on average, with decreasing spatial sizes of the selected unit of the biosphere, one should expect an increasing difference between the precipitation amount and evaporation. In this case the water transport through the atmosphere, with river run-off and sea currents, will serve as an equalizer. Though the quantitative estimates of all these parameters have been well studied, the water cycle dynamics can be described only with the use of the model. As a first approximation, to assess the role of precipitation in the global CO_2 cycle, one can use only the components W_{AU} and W_{AS}. However, with account of the spatially heterogeneous distribution of CO_2, the biosphere should be digitized.

With the notations assumed in Figures 6.17 and 6.18, the balance equations of the water cycle at the level of Ω_{ij} are written as follows:

$$\frac{dW_S(t,i,j)}{dt} = w_{AS} + w_{GS} + w_{SS} + w_{HS} - w_{SO} - w_{SG} - w_{ST} - w_{SA} - w_{SH}$$

$$\frac{dW_H(t,i,j)}{dt} = w_{SH} + w_{OH} - w_{HO} - w_{HG} - w_{HS}$$

$$\frac{dW_O(t,i,j)}{dt} = \sum_{(k,n)\in I_{kn}} [w_{IO}(t,k,n) + w_{HO}(t,k,n)]$$

$$+ w_{AO} + w_O - w_{OA} - w_{OG} - w_{OR} - w_T$$

$$\frac{dW_A(t,i,j)}{dt} = w_{OAA} - w_{AOO} + w_{SA} + \begin{cases} w_V & \text{for water surface} \\ w_{ST} & \text{for land} \end{cases}$$

Detailing of the right-hand parts of these equations with the functional presentations of fluxes in conditions of changing parameters of the environment will determine the level of the qualitative and quantitative reliability of the model. In particular, the model can be simplified by approximating the average value of W_O:

$$\overline{W}_O = \begin{cases} 2,500 + 350\sqrt{t} & \text{for} \quad 0 \le t \le 70 \\ 6,400 - 3,200\exp(-t/62,8) & \text{for} \quad t > 70 \end{cases}$$

where the average depth of the World Ocean is measured in meters, and the age of the ocean t is calculated in millions of years. Variations of the ocean volume can also be approximated using the formula: $\Delta V = \Delta W_O A_O + 59.5(\Delta W_O)^2$, where $A_O = 361.06 \times 10^6\,\text{km}^2$.

6.4.2 Water cycle in the "atmosphere–land" system

The land–atmosphere exchange processes include evaporation of soil moisture from the leaf surface, the stems and trunks of plants, as well as transpiration, precipitation, and evaporation in cases of unstable water accumulations in low ground. The water flow from the soil through the plant is the least studied link in this chain. Regarding the importance of the process of transpiration in the global water cycle, one can judge from available estimates that the process of transpiration takes more water than photosynthesis. For instance, from average assessments, to grow a 20 t yield (wet mass), a plant must extract from the soil about 2,000 t of water, with only 3 t of this being atomic hydrogen bound with atomic carbon in photosynthesis.

The model description of the process of transpiration requires an understanding of the role of physical and physiological factors in this process. A simplified idea about this role is reduced to the following. If the plant roots are in sufficiently wet soil, then the rate of transpiration is a function of temperature, humidity, wind speed, and insolation. Beyond some threshold of soil moisture, when the water supply in soil ceases, the role of these physical factors sharply diminishes, being inferior to the physiological factors – the type of plant, the construction of its roots, the phase of its development, the type of soil, and soil layer thickness. This threshold may occur between 5–50 cm of the precipitable water. At any rate, if for a given type of plant the water is not a limiting factor (i.e., water is not limited), then, as a first approximation, the total growth of plants can be considered proportional to the total potential transpiration for the whole period of growth. The latter is proportional to the amount of incoming solar radiation.

At present about 12% of the total evaporation from the Earth's surface is used by plants in the process of photosynthesis. In this process about 2,250 km^3 of water participate annually with a return coefficient of 0.75. Therefore, the simplest description of transpiration will be: $w_{ST}(t,i,j) = \beta_{ij} W_S(t,i,j)$, where β_{ij} depends on the vegetation productivity. The values of β_{ij} are 0.67 for forests, 0.44 for a meadow steppe, and 0.25 for agricultural crops. However, in real situations W_S is a limiting factor in a more complicated dependence of the impact on transpiration rate through the rate of photosynthesis R_p. In other words, $w_{ST} = k_p R_p$, where k_p is the transpiration coefficient for the plants of the p type. As a first approximation, one can use an approximation $R_p = \varepsilon_p r_p$, where ε_p is the share of solar energy assimilated by the pth type of plants in the process of photosynthesis. The ε_p value depends on the presence of water accessible for plants:

$$\varepsilon_p = \varepsilon_{p,0}[1 - \exp(-\varepsilon_{p,1} W_S(t,i,j))]$$

where $\varepsilon_{p,0}$ is the value of ε_p at a sufficient amount of water; and $\varepsilon_{p,1}$ is the coefficient reflecting a reduction of solar energy assimilated by plants with decreasing accessible water. On average, $\varepsilon_{p,0}$ is reached at $W_S = 10$ mm. Assuming $\varepsilon_p/\varepsilon_{p,0} = 0.9$, we have $\varepsilon_{p,1} = 0.23$. In this case $r_p = 9.6\,\mathrm{kg\,km^{-2}\,day^{-1}}$ of dry substance (or 37 kg of wet phytomass). The coefficient k_p is estimated for each type of plants. The coefficient k_p is equal to 368 for maize, 397 for sugar beet, 435 for wheat, 636 for potatoes, and 462 for cotton (kg H_2O per kg of pure substance).

As one of the models of the process of transpiration we can write $w_{ST} = Y_S(24a + b)$, where Y_S is the specific water return, a is the rate of the ground water lifting (cm hr^{-1}), and b is the average daily change of the ground water level (cm).

Within the GMNSS, to describe the process of atmosphere–land interaction, the fluxes w_{SA}, w_{ST} and w_{AS} are used, whose parametric descriptions serve as the basis for this unit. Information on precipitation w_{AS} is usually included into the information bulletins of hydrometeorological services. The pre-history of the distribution of precipitation in the form of a set of matrices $W_{AS}(\theta) = \|w_{AS}(\theta, i, j)\|$, where $(i,j) \in \Psi$, θ are discrete time moments of the precipitation record from the hydrometeorological service, is used to derive the functional $w_{AS}(t, i, j) = F(W_{AS}(\theta_1), \ldots, W_{AS}(\theta_N), t)$. This is performed with the use of extrapolation, a grouped account of arguments, and evolutionary modeling. Such an approach requires data on the precipitation over the discrete geographical grid $\Delta\varphi \times \Delta\lambda$. However, this database can also be modeled by simulating the global cloud field producing precipitation. The simplest parameterization of clouds consists in prescription of the threshold $W_{A,\max}$, beyond which an excess of atmospheric moisture transforms into water and precipitates. To reduce inevitable errors (mainly caused by overestimation of precipitation), it is expedient to introduce the matrix of threshold $W_{\max} = \|W_{A,\max}(i,j)\|$, $(i,j) \in \Psi$. Then:

$$w_{AS}(t,i,j) = \begin{cases} 0 & \text{for} \quad W_A(t,i,j) \le W_{A,\max} \\ W_{A,\max} - W_A(t,i,j) & \text{for} \quad W_A(t,i,j) > W_{A,\max} \end{cases}$$

If the value $W_{A,\max}(i,j)$ corresponds to a real critical value of the moisture content in the atmosphere over Ω_{ij}, then w_{AS} will be overestimated. It is assumed here that at $W_A(t,i,j) > W_{A,\max}$ the cloud fills the whole cell Ω_{ij}, which does not always correspond to reality. Moreover, the fact that a considerable share of moisture, even with an exceeded critical level, can remain in the cloud and evaporate, is not accounted for. Therefore, to take these special features into account, an adaptive coefficient $\alpha_W < 1$ should be introduced.

Now, divide precipitation into two basic types – solid and liquid. This division can be made by the thermal principle and by seasons. The thermal principle is more preferable due to flexibility at sharp climate changes and the possible shifting of the scale of seasonality. The synoptic division of the periods with different precipitation is justified in various regions of the globe. The average daily temperature in the early period of solid precipitation is below zero, ranging from −4°C to −7°C. On the border of this division precipitation of the mixed type is observed. The relationship between the types of precipitation is described by the formula: $x_T = a - bT$, where

x_T is the share of solid precipitation, T is temperature, and a and b are empirical coefficients. For the Atlantic climatic zone $a = 50$ and $b = 5$.

To parameterize the process of evaporation from land, numerous formulas are used. Here a simple dependence is assumed:

$$w_{SA}(t,i,j) = \rho_{ij}[1 - \exp(-\delta_{ij}/\rho_{ij})]/\sigma_{ij}$$

where δ_{ij} is a maximum possible rate of evaporation in the region and ρ_{ij} is the total amount of moisture getting to the soil per unit time.

The choice of the concrete model of the evaporation process is determined by the character of the applied database. Evaporation from the soil surface substantially depends on the type of vegetation cover. The evaporation in the forest and in the field differs by 30–40%. This is connected with the heterogeneous impacts on the water regime, within various vegetation covers, of such factors as soil freezing, snow-melting intensity, soil structure, radiation budget, and others. To take into account the dependence of the rate of evaporation from land on temperature T_{ij}, properties of vegetation cover, and soil properties, the following formula is used:

$$w_{SA}(t,i,j) = \delta^*(T_{ij}/\overline{T}_{ij})[1 - \lambda \exp(-\lambda_1 A_{ij}/\overline{A}_{ij})] \times [1 - \lambda_2 \exp(-\lambda_3 X_{ij}/\overline{X}_{ij})]$$

where $\overline{T}_{ij}$ is the surface air temperature in the region Ω_{ij}, averaged over the period considered; $\overline{A}_{ij}$ and $\overline{X}_{ij}$ are the average thickness of the soil layer and the density of vegetation cover, respectively; and $\delta^*, \lambda, \lambda_1, \lambda_2, \lambda_3$ are the empirical coefficients.

A detailed analysis of possible models of evaporation from the land surface with account of various types of vegetation cover and changes of climatic parameters is given in the publications by Bras (1990), Chock and Winkler (2000), Karley *et al.* (1993). In particular, there are formulas to calculate evaporation as a function of the height and density of vegetation cover, wind speed, and temperature. For instance, the following dependence is proposed for the rate of complete evaporation:

$$E_T = \begin{cases} (\Delta Q_N + m\gamma L_j)/[\Delta + (1+n)\gamma + I(1-C)] & \text{for} \quad T < 0°\text{C} \\ (\Delta Q_N + \gamma L_j)/[\Delta + \gamma] & \text{for} \quad T \geq 0°\text{C} \end{cases}$$

where Δ is the rate of change of the pressure of saturated vapor as a function of temperature, Q_N is the amount of energy reaching the evaporating surface, γ is the psychometric coefficient ($\approx$0.66 mb K^{-1}), I is the share of complete evaporation due to precipitation caught by foliage, and C is the compensation coefficient due to transpiration. The coefficients m, n, and C are functions of height h and type r_s of vegetation:

$$m = 53 \ln^2(20/h + 2.5) \qquad C = (\Delta + \gamma)/[\Delta + (1+n)\gamma]$$

$$n = r_s[m(1 + U/100)]/250$$

The indicator of the type of vegetation r_s (m s^{-1}) for some types is estimated at: 40 (sunflower and alfalfa), 70 (barley and potatoes), 250 (citrus plants), 130 (cotton), 80 (maize and rice), 50 (sugar beet), 60 (wheat), 400 (tundra), 200 (subtropical meadows), 100 (temperate zone meadows), 100–300 (tropical forests), 200–300 (coniferous forests), and 100–150 (deciduous forests in middle latitudes). Typical values of the parameters n, m, and I are: for grass ecosystems – $n \approx 2.5$, $m \approx 3.5$, $I \approx 0.2r$; for woodlands – $n \approx 30$, $m \approx 5$, $I \approx 0.3r$ (temperate latitudes) and $I \approx 0.15r$ (tropics); where r is precipitation.

Albedo is an important parameter for calculating the solar radiation energy participating in the process of evaporation. The relationship between the height of plants and albedo, as a first approximation, are described by a linear dependence. With the height of plants reaching ~20 m, albedo decreases from 0.25 to 0.1. The albedo values for some types of the Earth's cover are known: heather – 0.14; fern – 0.24; natural pastures – 0.25; shrubs – 0.21; savannah – 0.17; deciduous forests in mid-latitudes – 0.1; coniferous forests and orange groves – 0.16; eucalyptus forests – 0.19; wet tropical forests – 0.13; and waterlogged forests – 0.12. The albedo of agricultural fields varies within 0.15 (sugar cane and fruit trees) and 0.26 (sugar beet, barley, cucumber).

The surface part of the land–atmosphere water exchange is connected with the subdivision of the phase space, at least, into two levels – soil and ground waters. The soil level plays the role of a buffer reservoir between precipitation and ground waters. A simplest parameterization of fluxes between these levels is reduced to linear dependences: $w_{SH}(t,i,j) = \lambda_{ij} W_S(t,i,j)$, $w_{HS}(t,i,j) = \mu_{ij} W_H(t,i,j)$. However, a more strict description of the soil level is dictated by natural heterogeneity of the structure of Ω_{ij}, where small water bodies and land sites with a certain relief can be located. According to the landscape–hydrological principle, to simulate Ω_{ij}, it is necessary to choose facies and sites of water surfaces, which are connected with typifying the floristic background, whose concrete conditions are determined by micro-relief, type and properties of the soil, surface moisture, depth of ground waters, and other factors. It is possible to choose m_{ij} of facies and n_{ij} of water bodies. In this case the soil moisture forms not only due to the fluxes shown in Figures 6.17 and 6.18, but also due to leakage and filtration of water from the water bodies and aqueducts located in Ω_{ij}.

An important factor of the surface part of the water balance is infiltration of precipitation into the soil both during rains and through run-off. The rate of uptake of water into soil w_{SH} is described by the formula: $w_{SH} = k_S l$, where k_S is the coefficient of filtration and l is the hydraulic slope. If we denote as κ the volume mass of the soil, which, on the average, varies within 1.4–1.5 g cm^{-3}, then for k_S it is convenient to use the Azizov formula: $k_S = 256.32\kappa^{-7.28} - 1.27\kappa^{1.14}$ (cm day^{-1}). The parameter l is calculated using the formula $l = (z_0 + z_1 + z_2)/z_0$, where z_0 is the depth of the column of washing-out, z_1 is the capillary pressure, and z_2 is the height of the water layer on the soil surface. At $z_0/z_1 \leq 2$ an approximation $w_{SH} = k_S + t^{-1/2}(0.5 k_S z_1 D)^{1/2}$ is valid, where D is the soil moisture deficit and t is time. Other approximations of the function w_{SH} are known, such as the Horton empirical formula: $w_{SH} = [w_{SH}(t_0, i, j) - k_S]\exp(-\beta t) + k_S$; the Popov formula, $w_{SH} = r \cdot \exp(-rt/D) + k_S$; and the Kostiakov formula, $w_{SH} = k_S + \alpha t^{-n}$, where α, n, and β are calibration parameters and r is the rain intensity.

The interaction between the regions of the assumed grid of the biosphere surface division $\{\Omega_{ij}\}$ is realized through fluxes w_{OAA}, w_{AOO}, w_{SS}, w_{SO}, w_{OH}, and w_{HO}.

6.4.3 Water cycle in the "atmosphere–ocean" system

The processes of transport on the border of the atmosphere–water surface has been well studied. The transport of moisture from the surface of the water body into the atmosphere is one fragment of the complicated physical processes of mass- and energy-exchange across the water–air interface. These processes are functions of many climatic parameters and, to a large extent, are regulated by eddy motions in the surface layer of the atmosphere determined by the wind field.

The possibility to assess the water transport from the water surface into the atmosphere consists of an assessment of the water content of the lower part of the surface layer of the atmosphere, which forms of sea spray and water vapor. The eddy flux of water through the unit surface can be described by the relationship:

$$W_V = -\rho K_W(\partial q/\partial z) = -(\rho w)'q' \approx -\rho < w'q' >$$

where W_V is the vertical eddy flux of water vapor ($\mathrm{g\,cm^{-2}\,s^{-1}}$), K_W is the coefficient of the eddy transport of water vapor ($\mathrm{cm^{-2}\,s^{-1}}$), q is the specific air humidity ($\mathrm{g\,g^{-1}}$), ρ is air density ($\mathrm{g\,cm^{-3}}$), z is the vertical coordinate, w is the vertical wind speed ($\mathrm{cm\,s^{-1}}$), and w' and q are pulsations of w and q values, respectively. Let p be the atmospheric pressure, then we express q through average water vapor pressure e: $q = 0.621e/p$.

Evaporation from the water body surface depends on air temperature and can be described by the function $w_{SA} = w^*T^\omega$, where w^* and w are the empirical parameters. If measurements are made of wind speed θ ($\mathrm{m\,s^{-1}}$), saturated water vapor pressure at a temperature of the evaporating surface E_1, and atmospheric pressure p (mm Hg), then to estimate the rate of evaporation, the Dalton law can be used, $w_{SA} = A(E_1 - e)/p$, and the Shuleikin formula, $w_{SA} = C\theta(E_1 - e)$, where A and C are the parameters related as $A = C\theta/p(C = 0.45 \times 10^{-6}\,\mathrm{g\,cm^{-3}\,mb^{-1}})$. The models by Horton and Rower described above are rather efficient.

6.4.4 Water in the atmosphere

Atmospheric processes of the moisture transport directly connected with the temporal variations of the meteorological elements, play an important role in the global water cycle formation. The global atmospheric circulation can be described by the Monin model (Monin and Krasitsky, 1985):

$$\partial v_\delta/\partial t + V_z\partial v_\delta/\partial z + V_\delta R^{-1}\partial v_\delta/\partial\delta + V_\lambda R^{-1}\sin^{-1}\delta\partial V_\delta/\partial\lambda$$
$$= R^{-1}(V_\lambda)^2 ctg\delta + 2\Omega V_\lambda\cos\delta + (R\rho)^{-1}\partial p/\partial\delta + f_\delta$$
$$\partial v_\lambda/\partial t + V_z\partial v_\lambda/\partial z + V_\lambda R^{-1}\partial v_\lambda/\partial\delta + V_\delta R^{-1}\sin^{-1}\delta\partial V_\lambda/\partial\lambda$$
$$= R^{-1}V_\delta V_\lambda ctg\delta + 2\Omega V_\delta\cos\delta - (R\rho\sin\delta)^{-1}\partial p/\partial\lambda + f_\lambda$$

where Ω is the angular rate of the Earth's rotation; $\delta = \pi/2 - \varphi$ is an addition to latitude; λ is longitude; V_z, V_δ, and V_λ are the components of the velocity of the atmospheric motion; R is the Earth radius; and f_δ and f_λ are the components of acceleration due to friction expressed through the friction stress tensor u_{sl}:

$$\rho f_\delta = \partial u_{\delta z}/\partial z + (R\sin\delta)^{-1}(\partial u_{\delta\delta}/\partial\delta + \partial u_{\delta\lambda}/\partial\lambda) - R^{-1}u_{\lambda\lambda}Ctg\delta$$

$$\rho f_\lambda = \partial u_{\lambda z}/\partial z + (R\sin\delta)^{-1}(\partial u_{\lambda\delta}/\partial\delta + \partial u_{\lambda\lambda}/\partial\lambda) - R^{-1}u_{\lambda\delta}Ctg\delta$$

The velocity field equations are closed by prescribed zero boundary conditions on the Earth's surface, determined by the equation of relief in the form of the prescribed function $z = h(\delta, \lambda)$ and by an added equation of the state of humid air $p = \rho T[r_d + q(r_V - r_d)]$, where $r_d = 0.287$ Joule/g K and $r_V = 0.46$ Joule/g K are the gas constants of dry and water vapour, and $q \approx 3$–4% is the specific humidity. The distribution of temperature $T(\delta, \lambda)$ and function $q(\delta, \lambda)$ can be described by the respective equations of evolution, and the data of the global archive can be used and substituted for tabulated values.

A sufficiently complete description of the general atmospheric circulation models has been given in Nicolis and Nicolis (1995). A simplified description of the atmospheric part of the hydrological cycle is possible by the equation $\partial W_A/\partial t + \nabla Q = E - P$, where W_A is the vertically integrated specific air humidity in this column, and E and P are evapotranspiration and precipitation at the soil level, respectively.

A further simplification of the model of the hydrological cycle is connected with selection of the following three types of prevailing directions: western, eastern, and meridional. For such an approximation, data are used on the amplitude of the wind speed oscillations and the number of days of seasonal flows of moisture transport in the atmosphere. If the mass of water vapor in the air column over the area σ_{ij} is $a = W_A\sigma_{ij}$, then, for instance, for the eastern orientation of the atmospheric circulation the water flux between the adjacent cells of the Earth's surface grid will be $w_{AO} = 2a\theta/d_{ij}$, where θ is the wind speed and d_{ij} is the diameter of Ω_{ij}. Following this scheme, it is easy to re-calculate the moisture supplies at each step of time digitization, since it is unnecessary to solve the problems of numerical integration of partial differential equations. The background information about W_A, σ_{ij}, θ, and d_{ij} is accumulated in the database from different sources. The function W_A is calculated by the balance equation or can be prescribed based on other data. In particular, with changing temperature T and partial pressure of water vapor e, then W_A can be estimated from the relationship: $W_A = meh(1 + \alpha T)^{-1}$, where h is the height of the effective atmospheric layer, and m and α are the proportion coefficients ($m = 0.8$; $\alpha = 1/273$ when measuring W_A in g m^{-2} and T in °C).

One should note that considerable progress has been achieved lately in measurements of the water vapor content in the atmosphere using microwave radiometry, which ensured a reduction of the level of errors from 25% or more to less than 3%. A simultaneous improvement of the methods of calculation of radiation fluxes in the short-wave and long-wave spectral regions gave more reliable estimates of radiation fluxes and RF. However, the problem of the impact of cloud cover dynamics remains far from being resolved. It is important here that developments carried out in 1990 led to the conclusion about a systematic underestimation of the earlier calculated values of solar radiation absorbed by clouds by about 40%, which, naturally, tells upon the reliability of the results of numerical climate modeling in the context of the functioning of "cloud" feedbacks. The reliability can be increased with the use of the results of numerical modeling of the climatic impact of clouds with due regard to different scale processes including direct simulation of the cloud cover dynamics, concentrating especially on the powerful convective cloudiness in the tropics. Use of the present models have given promising results. They provided a sufficiently adequate simulation of the diurnal change of precipitation.

6.4.5 Interaction between global cycles of water and carbon dioxide

This direction of developments demonstrating close connections between the water and carbon cycles and climate has become especially urgent. It has been shown, in particular, that about 60% of carbon supplies on land in North America can be connected with rain enhancement over the North American continent, though it was supposed earlier that the combined impact of SAT increase and fertilization due to the growth of CO_2 concentration in the atmosphere played the main role. New studies of chemical fertilization with due regard to such factors as precipitation, river run-off and water alkalinity have led to the conclusion that the outflow ("export") of carbon characterized by the alkalinity data, has increased with the growing sink and precipitation in the basin of the Mississippi River. The chemical fertilization determines the CO_2 transformation into dissolved bicarbonate and carbonate which is then transported to the ocean by river run-off. The transport of alkaline water masses from land to the ocean determined by river run-off is the main source of seawater alkalinity and, hence, a mechanism for regulation of the level of carbonate concentration in the oceans. This can be important for the formation of the global carbon cycle and functioning of the World Ocean as a carbon sink.

Interesting results (in the context of climate problems) have been obtained from studies of large-scale forest fires taking place in 1997–1998 due to the drought caused by El Niño. According to observational data and numerical modeling results, carbon emissions to the atmosphere caused by forest fires increased by 2.1 ± 0.8 Pg C, which constitutes $66 \pm 24\%$ with respect to the observed anomalous CO_2 concentration growth. On the whole, the results discussed have demonstrated that the variability and intensity of water and energy cycles on inter-annual timescales belong to the most important factors of the carbon cycles formation. For instance, under conditions of persistent forest fires, the region, which had been a sink for a long time, could suddenly become a source of carbon.

In this connection, of importance is a study of energy and water cycles carried out within the Global Energy and Water Cycle Experiment (GEWEX) program. The closure of cycles carried out with analysis of data for the basin of the Mississippi River as an example, has shown that the latter can be balanced within an error of about 15%. Developments concerning the precipitation prediction during the warm season at the south-west of the USA have shown that in this case monsoons play a substantial role. The most important developments planned for 2004–2005 included: an accumulation of the integral base of hydrological data; substantiation of the integral strategy of the global observational system with emphasis on the problems of the water cycle (in this connection, the field observational experiment CEOP (Coordinated Enhanced Observing Period) was planned for related frequent observations); preparation and application of new space-borne remote-sensing instrumentation to measure the water vapor content in the troposphere and lower stratosphere (in particular, with the use of the microwave limb sounder (MLS) carried by the satellite Aura launched in 2004); retrospective analysis of all available observational data (space-derived, in particular) from parameters of water and energy cycles (with the first-priority being long-term global data on precipitation); re-analysis of regional data on climate dynamics for the period 1979–2003; launching of the satellites Cloudsat and CALIPSO in 2005 to retrieve aerosol and cloud cover characteristics (the first of these satellites has been prepared with USA–Canadian cooperation, and the second with USA–French cooperation); the global monitoring of precipitation (the satellite GPM with its complex of instruments is prepared to replace the complex used on the satellite TRMM and SSM/1); "cloud" climatic feedbacks (with emphasis on convective clouds in the tropics); impacts of changes in temperature and hydrological characteristics on concentrations of pollutants and pathogens in the atmosphere near the Earth's surface and in soil water (the main motivation is determined here by the urgent problem of potable water quality); soil moisture monitoring within the studies of droughts; and climate forecasts in the interests of solving problems of water resource control.

A specification of the global model of the CO_2 cycle in the biosphere includes a consideration of the role of the process of its washing-out from the atmosphere due to precipitation. This process has been studied inadequately. Nevertheless, there are direct observational verifications of CO_2 absorption by rain droplets (Egan *et al.*, 1991). In particular, one such verification notes the presence in rain of a considerable amount (up to $15\,mg\,l^{-1}$) of hydrocarbonate ions HCO_3^-. A combined analysis of precipitation amount and variations of the atmospheric CO_2 concentration over the same territory performed using the data of the global observational network, has made it possible to reveal a persistent correlation between these processes. Figure 6.19 shows the curves of changes of average monthly precipitation and concentration of atmospheric CO_2. We see that the dependence between changes of atmospheric CO_2 and precipitation is sufficiently stable. A detailed analysis of this dependence for various latitudinal belts or for other configurations of limited territories reveals similar patterns independent of geophysical coordinates. Here one should point to a high sensitivity of correlation to the duration and type of precipitation. For instance, during a shower the HCO_3^- concentration in precipitation can either double or halve depending on the presence or lack of thunderstorms. Moreover, this ratio depends strongly on the duration of the precipitation period. Observations showed that with an increasing duration of rain the concentration of HCO_3^- decreases. In other words, the interaction between CO_2 concentration and moisture content in the atmosphere is an important component of the global carbon cycle.

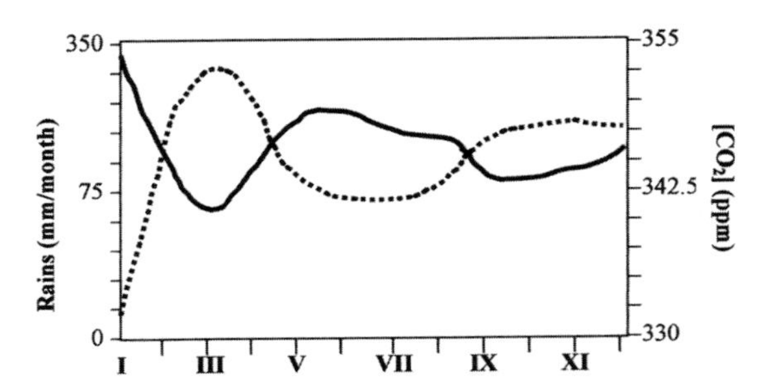

Figure 6.19. Variations of precipitation amounts (solid curve) and CO_2 concentrations in the atmosphere (dashed curve) estimated from integrated data from observatory observations within the framework of the Carbon Cycle Scientific (CCS) Program.

A formalization of the role of rain in the global CO_2 cycle requires a model of CO_2 absorption by water droplets falling at some velocity, u. The most widely used version of such a model is an equation of gas balance on the surface of rain droplets:

$$\frac{dC}{dz} = \frac{3\Delta}{ur^2}\left(1 + 0.3\sqrt{\mathrm{Re}}\sqrt[3]{Sc}\right)(C_A - C^*)$$

where Δ is the coefficient of CO_2 diffusion in the air, r is the droplet's radius, C^* is the balanced concentration of CO_2 in a droplet, C_A is the CO_2 concentration in the atmosphere, z is the altitude, $Sc = \nu/\Delta$ is the Schmidt number, ν is the kinematic viscosity, and $\mathrm{Re} = 2ru/\nu$ is the Reynolds number.

A diversity of the forms of precipitation over the globe complicates the consideration of their role in the global CO_2 cycle. This problem can be solved in two ways. The first way is the formal numerical description of the totality of the processes of precipitation formation. The second way is connected with the use of the present means of global observation of precipitation. In both cases the forms of rain should be clearly classified as functions of meteorological situations. The rain rate can range widely from $1\,\mathrm{mm\,hr^{-1}}$ to $8\,\mathrm{mm\,hr^{-1}}$ and even more. Besides this, there is a certain correlation between the precipitation rate and the size of rain droplets. With the low-intensity rain $r \in [0.1; 0.5]$. A shower can be characterized by the formation of droplets up to $r \sim 6\,\mathrm{mm}$.

Thus, the problem of assessment of the role of precipitation in washing-out CO_2 from the atmosphere is urgent, and to solve it, the global model should take into account a change of hydrological cycles separately over the World Ocean and over land, since these regions of the planet differ in their interaction with the atmosphere.

6.5 THE ROLE OF LAND ECOSYSTEMS AND THE WORLD OCEAN IN PREVENTION OF NATURAL DISASTERS

6.5.1 Land ecosystems and global ecodynamics

The growing global size of the population and the associated increased forcing of human activity on the environment and ecosystems have become not only the main threat to further sustainable development of civilization in the context of global ecological safety but also reflect a dangerous disorder in the normal functioning of various systems of life support (Grigoryev and Kondratyev, 2001). In connection with the key role of the ecosystems in the processes of natural regulation of environmental properties, of principal importance is an analysis of the available data on the global dynamics of ecosystems and an assessment of possible trends.

Global natural and regulated ecosystems play an important role as factors of the environmental dynamics ranging from micro-scales (e.g., soil bacteria) to the whole planet and, on the other hand, are vitally important sources of drinking water, food, timber, paper, and other means of life support.

Global climate change processes are affecting many ecosystems around the globe, and their impact is increasing rapidly causing some shifts in latitudal and altitudal distributions of soil–plant formations. Williams *et al.* (2003) have shown that increasing temperature can result in a significant reduction or complete loss of the core environment of all regionally endemic vertebrates. Mountain ecosystems around the world, such as the Australian Wet Tropics bioregion, are very diverse, often with high levels of restricted endemism, and are therefore important areas of biodiversity. The results presented by Williams *et al.* (2003) suggest that these systems are severely threatened by climate change.

An extreme complexity of the problem discussed is that it is necessary to explain (and, as far as possible, to predict) the dynamics of the interactive NSS (the society should be placed first here since its functioning determines its impact on nature) with its numerous feedbacks, non-linear nature, and "surprises". Unfortunately, the present stage of studies of the "society-nature" system can be considered not more than initial and preliminary. This refers to even a simple description of the present condition of nature (global ecosystems), which results from the observational data deficit with an apparent abundance of some observational means (especially expensive space-borne means). Therefore, the report of the World Resources Institute (USA) (Mock, 2005) is in many respects incomplete, concentrating only on the consideration of five types of ecosystems (the share of land surface is given in parens, except for the Antarctic and Greenland which are occupied by the respective ecosystem): agricultural ecosystems (28%), coastal regions (22% within a 100-km band), forest (22%), freshwater (<1%), and grass (41%) ecosystems. An abandonment of the World Ocean is, of course, a serious, though justified (in view of information deficit) flaw. Special attention should also be paid to the soil ecosystems (Wofsy *et al.*, 2001).

As has been mentioned earlier, ecosystems are very important for the solution of various problems of human life support and regulation of environmental conditions. Analysis of the data on climate for the period 1982–1999, and results of satellite remote sensing of the primary production (NPP) for the same period revealed a mean global growth of land primary production by 6.3% (or by 3.4 Pg C) with considerable spatial heterogeneity manifested through a substantial increase of NPP over 25% of the territory and substantial decrease over 7% of the territory. Some 80% of the contribution to the NPP growth was made by ecosystems in all tropical regions and in high latitudes of the northern hemisphere. Apparently, an increase of NPP in the tropics was connected with decreased cloud amount, which caused an enhanced input of solar radiation, and in other regions – with the combined impact of changes in SAT, precipitation, and solar radiation.

A special problem was analysis of data on the budget of carbon, within the Carbon Sequestration Research Program (CSRP), in studies of the removal of GHGs from the atmosphere on protected territories in 13 states of the USA (Scott, 2005). According to these data, an accumulation of $910\,\mathrm{kg\,C\,ha^{-1}\,yr^{-1}}$ in the upper 20-cm soil layer took place in these territories. Hence, over the whole CSRP territory (5.6 mln ha), 51 mln t C was removed from the atmosphere and accumulated in soil each year. Such detailed analyses over the whole globe will make it possible to understand the real role of land ecosystems in climate stabilization and to evaluate thereby the perspective of growing undesirable phenomena in the environment.

Examples of destructive (and even catastrophic) impacts on ecosystems and their economic consequences are numerous (see for details Grigoryev and Kondratyev, 2001). The collapse of the cod catches in 1990 in Canada made about 30,000 fishermen unemployed, and in the region of Newfoundland brought forth serious economic difficulties in 700 settlements. Material losses in China reaching US$11.2 $\times 10^9$ per year have resulted from a deficit of drinking water due to polluted river and sub-soil waters. In India, commercial forest cutting and the transformation of deforested lands into the agricultural lands have not only changed the traditional way of people's life but also caused a deficit of wood fuel and timber to the detriment of a rural population of 275 million people.

As for estimates of the consequences of global anthropogenic impacts, the situation with water resources is an example: about 28% of the global population have no access to pure drinking water; every year about 5 million people die because of low-quality drinking water and insanitary conditions; about 90% of wastes in developing countries enter rivers, lakes, and coastal regions of the seas, etc. Intensified emissions of CO_2 to the atmosphere have caused considerable changes in the global carbon cycle (Kondratyev and Krapivin, 2004b).

The most important fact is that the levels of impact on the ecosystems have become global in scale. About 75% of marine fish populations have either decreased due to violation of the permissible amounts of catch, or come close to their threshold of survival. An intensive deforestation has almost halved the forested areas, and the construction of various economic infrastructures has caused a fragmentation of forest cover. About 58% of coral reefs are seriously affected by fisheries, tourism, and pollution. Almost 65% of arable lands have partially lost their fertility. The scales of economic usage of ground waters exceed their natural rate of recovery by at least $160 \times 10^9\,\mathrm{m^{-3}\,yr^{-1}}$. In most cases the anthropogenic load on the ecosystems has intensified.

It is well known that the main causes of ecosystem degradation are the growing sizes of populations and, respectively, increased needs for natural resources as well as enhancing loads on the environments. Concrete detailed data characterizing the present global situation can be found in numerous publications (Kondratyev, 1998, 1999, 2000b; Ernst, 2000; Watson *et al.*, 2000). Note that an extremely

	Agro	Coasts	Forests	Water	Grass
Food/fiber	↘ 2	↘ 3	↗ 2	↑ 2	↓ 3
Water quality	↘ 4	↑↓ 3	↘ 3	↘ 4	6
Water amount	↘ 3	6	↘ 3	↘ 3	6
Biodiversity	↘ 4	↘ 3	↘ 4	↘ 5	↘ 3
Carbon supplies	↑↓ 3	6	↘ 3	6	↘ 2
Recreations	6	? 2	6	6	↘ 2
Protection of coasts	6	↘ 4	6	6	6
Wood fuel	6	6	? 3	6	6

Figure 6.20. Expert estimates of the levels of anthropogenic forcings, and their consequences, on ecosystems.

important feature of the growing scales of consumption is their strongest geographical non-uniformity reflecting the socio-economic contrasts in the world.

Now we shall discuss the principal results obtained within the framework of the Pilot Analysis of the Global Ecosystems (PAGES) (*A Guide to World Resources 2000–2001*, 2000). The main difficulties of this analysis are connected with a deficit of available information. The PAGES project is concentrated on the consideration of the three types of indicators of impacts on ecosystem dynamics: (1) anthropogenic loads (increasing size of population, growing level of resources consumption, pollution, overexploitation of natural and controled ecosystems); (2) the spatial extent of ecosystems (size, shapes, localization, geographical distribution); (3) production of economically important products such as agricultural crops, wood, fish, etc. A considerable drawback of each and every one of these indicators is that they do not contain information about the thresholds of an ecosystem's ability to perform its functions of life support.

Figure 6.20 illustrates the presentation of the expert assessments of the state of ecosystems and their functioning as life-supporting systems including the respective trends: increasing ability of ecosystems to perform life-supporting functions (↗), a reduced ability (↘), mixed trends (↑↓), and unknown trends (?). The figures characterize the level of ecosystem functioning as life supporting compared with the situation 20–30 years ago: 1 – excellent, 2 – good, 3 – satisfactory, 4 – weak, 5 – unsatisfactory, and 6 – no estimate. The results of the expert assessments shown in Figure 6.20 testify, on the whole, to considerable anthropogenic forcings on ecosystems (especially during the 20th century) that have led to considerable changes in their ability to perform life-supporting functions. Though in some cases (e.g., food production) ecosystem productivity has substantially increased, providing thereby the required level of production, nevertheless, in other cases (drinking water quality, preservation of biomass, etc.) there was a degradation of ecosystem functioning.

The state of the present agricultural ecosystems responsible for 95% of protein due to cultivated plants and animals, and 99% of calories consumed by humans, is contradictory. On the one hand, the agricultural yields increase, but on the other hand, in most of the countries the quality of agricultural ecosystems worsens. So, for instance, a decrease in the natural soil fertility is recorded on 65% of agricultural land. In spite of this, a wide application of fertilizers, irrigation, and various new technologies (seeds, pesticides, etc.) has more than compensated for the the impact of the deteriorating conditions of agricultural ecosystems, and this trend will remain, apparently, in the near future (for how long it is unclear).

Much more problematic is the situation with fishing. The present state of coastal marine ecosystems is only satisfactory and is at present worsening. In 28% of the areas of water bodies, in the most intensive fishing regions, fish supplies have either decreased due to over-fishing or have only just started to recover. In 47% of water basins the fishing levels are close to maximum permissible biological levels. Unfortunately, the information on the potential supplies of biomass in the ocean is contradictory. The general laws of the distribution of organic life in the World Ocean, and in various seas, have been more or less studied. Phytoplankton as the source of productivity for other trophic levels of the ocean is known to develop mainly in the upper 100-m layer, 65% of the zooplankton biomass and above 90% of nekton being located in the layer descending to 500 m. The benthos animal habitat is mainly within the first 200 m (about 80%). By productivity, the World Ocean is very non-uniform. Less than 25% of its area is occupied by ecosystems with plankton biomasses above 200 $mg\,m^{-3}$, and more than 50% of its area is occupied by unproductive ecosystems ($<50\,mg\,m^{-3}$ of plankton).

The situation in freshwater basins is also rather contradictory. On the one hand, there is deprecatory fishing, but on the other hand, the natural fish supplies are replenished through fish breeding. The scale of fish breeding in ponds is constantly growing. On the whole, an increasing dependence of fish catch on aquacultural production and a decrease of natural fish supplies threaten the population of the developing countries, where aquacultural production is absent.

The water resources of the biosphere have been systematized in detail from the viewpoint of both supplies and quality. The main water fluxes between the oceans and land, as well as volumes of rivers, lakes, sub-soils, and soil waters, have been estimated. On the whole, the global freshwater supplies are not large. Annually, all the rivers bring to the World Ocean about $50 \times 10^3\,m^3$ of water, which is only 1.5 times greater than the volume of the waters of Lake Baikal or the American Great Lakes. Over the continents the water supplies are distributed non-uniformly and not in proportion to the population density (Gleick, 1993).

The construction of dams and other means of regulating the river courses have seriously changed the conditions for use of water resources both for people and for water ecosystems. The level of economic use of river waters has reached, on average, about half the available resources (with this situation strongly differing in various countries). Building of dams and other engineering constructions has led to a considerable fragmentation of about 60% of the largest river systems in the world. For instance, as a result, the time needed for river waters to reach their respective seas has tripled, on average. The regime of inland waters has changed due to deforestation and wetlands drying. This has especially manifested itself in the tropics where the forests are a key factor of the water regime dynamics. The global area of freshwater wetlands has almost halved, which has told upon the water supplies and flood regimes.

The main causes of the ongoing deterioration of the quality of natural waters are chemical pollution of water basins and input of nutrients, as well as (indirectly) a reduced ability of the ecosystems to perform water filtration, and intense soil erosion due to land use (Watson *et al.*, 2000). The pollution by nutrients resulting from the run-off of river waters polluted with fertilizers is a serious problem for all agricultural regions in the world. This kind of pollution causes a eutrophication of waters in lakes and near-shore regions of seas (especially observed in the Mediterranean and Black Seas, in the north-western part of the Gulf of Mexico, and in many large lakes). During the last two decades a dangerous blooming of water basins took place due to the input of nutrients. In many cases the ability of freshwater and marine coastal ecosystems to maintain the required quality of natural waters has been broken. Though, on the whole, in industrial countries the quality of natural inland waters has somewhat improved, the situation in the developing countries is quite different, especially near large cities and in industrial regions. The worsening quality of water tells keenly on poor populations that are without access to pure drinking water.

As has been mentioned above, the level of anthropogenic impacts on water systems has become rather substantial, which is illustrated, for instance, by taking almost half the river run-off for economic (mainly agricultural) needs, which could reach 70% by the year 2025. The transport of freshwater from rivers and other reservoirs to agricultural fields ensures an increase in yields, but in this case various natural ecosystems and water users located down river suffer serious damage. Often the water returns to the rivers but usually after being strongly polluted and, as a rule, unfit for further use. The best example of such processes is the change of the water regime in the Aral Sea basin.

The biodiversity losses during the last century have acquired threatening scales and concern practically all ecosystems (mainly as a result of the loss of habitats). So, for instance, the area of forests has decreased at least by 20% and probably (on global scales) by 50%. Some forest ecosystems (like the dry tropical forests of Central America) have disappeared completely. In many countries up to 50% of mangrove plantations have been lost, the area of wetlands has almost halved, and in some regions the size of territories covered with grass has decreased by 90%. Only the tundra, Arctic, and marine deep-water systems have little changed – though in these cases substantial anthropogenic impacts also occur.

Even when the initial areas of the ecosystems have been preserved, many species are threatened with pollution, over-exploitation, intrusion of "alien" species, and degradation of habitats. From the viewpoint of biodiversity, the freshwater ecosystems have suffered most. For instance, about 20% of the freshwater fish species have been exterminated, and many other species are in a very dangerous condition. Forest, grass, and coastal ecosystems face serious problems of survival. The biodiversity is threatened with frequent diseases of marine organisms, intensive blooming of water bodies, and a considerable reduction of amphibian populations. Apart from the respective losses for medicine (due to a decreasing volume of raw materials for drugs), banks of genes, and eco-tourism, a reduction of biodiversity means also a threat to the ecosystem's productivity, their integrity, and resistance to various external forcings.

Finally, it is clear that there are now numerous signs of anthropogenically caused degradation of ecosystems, their reduced ability to function as systems of life support, and their ability to control environmental properties. The data discussed above show that all the ecosystems considered suffer serious and increasing anthropogenic forcings. Intensive processes of transformation of natural ecosystems take place due to deforestation, pollution (with nutrients, including), dams building, and biological intrusions of alien organisms. Anthropogenic forcings result in considerable transformations of the global biogeochemical cycles, whose normal functioning is a guarantee of ecosystem safety (Krebs *et al.*, 2001).

So far, the negative trends of the state of the global ecosystems do not obviously threaten the high levels of production of various products and services. Food and fiber production has never reached such high levels as today, but dam construction has ensured an unprecedented regulation of river run-off to provide water for agriculture. However, progress in creating material well-being is fraught with a long-term prospect of a reduction of productivity from the global ecosystems and their reduced ability to regulate environmental quality. The use of new technical means and technologies such as synthetic fertilizers, improved means of fishing, and irrigation systems has masked a decrease of bioproductivity of natural ecosystems, ensuring needed growth of food and fiber production. As for the long-term prospects (and it is in this context that sustainable development is defined), one should consider that the reduced functioning of the natural systems of life support may manifest itself, for instance, in the form of a dangerous decrease in biodiversity, reduced quality of drinking water, intensive emissions of greenhouse gases into the atmosphere, and many other negative phenomena.

In light of the World Summit on Sustainable Development (WSSD) in Johannesburg (2002) ("Rio + 10"), of key importance is a substantiation of priorities regarding the problem of the functioning of the "society–nature" interactive system. An acuteness of the situation becomes especially apparent, considering the limited success of the UN Second Conference on Environment and Development (Rio de Janeiro, 1992), a Special UN Session "Rio + 5", and the complete fiasco of the 6-year series of annual conferences of the representatives of the governments signing the UN Framework Convention on Climate Change (FCCC) that served as the basis for the Kyoto Protocol. Speculative exaggerations (reaching sometimes the level of apocalyptic global predictions) characteristic of the concept of "global warming" have diverted attention (and to a considerable extent the financial resources) from the first-priority problem of global ecosystem safety in conditions of an increasing size of the global population and intensifying anthropogenic forcing on natural systems of life support. In this connection the authors of *A Guide to World Resources 2000–2001* (2001) draw attention to the availability of the "ecosystem approach" in that the use of natural resources should be based on a consideration of the ability of the global natural systems of life support to function normally.

The ecosystem approach has the following main features:

- Complex (systematic) character of analysis of interactions in the "society–nature" system.

- Not fragmentary but integral analysis of ecosystem functioning.
- Consideration of all the spatial–temporal diversity of the processes with the indispensable priority of long-term prospects.
- A detailed consideration of interactive processes in the "society–nature" system.
- Justification and accounting of the limits of the normal functioning of ecosystems, and in this connection, permissible anthropogenic loads.

There are differences between traditional and ecosystem approaches to the use of natural resources as applied particularly to forestry. Of course, advantages of the ecosystem approach raise no doubts. The problem is, however, to what extent the conditions of the socio-economic development of a country permit a successive realization of the ecosystem approach. The negative manifestations of the dynamics of various ecosystems reflect great (sometimes insurmountable) difficulties to realize the ecosystem approach. In this connection the participants of the PAGES project have formulated the following general recommendations:

- to stimulate the development of science and observational means;
- to realize and quantitatively estimate the value of functions performed by the ecosystems;
- to stimulate a broad discussion of the goals, scientific foundations, and practical importance of the ecosystem approach; and
- to provide the participation of all needed specialists in the solution of the problems of ecosystem management.

The scientific understanding of the laws of ecosystem functioning and, moreover, a possibility to predict ecosystem dynamics are still in their initial phases. To a great extent this is explained by inadequate complex information about the ecosystems and their functioning. So, for instance, the problem of the global carbon cycle has attracted serious attention since long ago (Kondratyev, 1998, 1999; Kondratyev and Demirchian, 2001; Watson *et al.*, 2000). Nevertheless, not even the system requirements have been formulated yet for complex observational data, which would also make it possible to obtain adequate information about the laws of the carbon cycle (of course, this requires a combined use of both usual and satellite means of observations). In this connection, UN Secretary-General Kofi A. Annan justly noted (*A Guide to World Resources 2000–2001*, 2001): "It is impossible to justify an efficient ecological policy if it is not based on reliable scientific information. Though in many spheres a considerable progress has been reached in obtaining the observational data, there remain considerable gaps in our knowledge. In particular, so far, no complete global assessment of the state of the main ecosystems existing over the globe has been obtained. A response to this necessity was the planned Millennium Ecosystems Assessment – MEA as an important international initiative aimed at mapping the "health" of our planet. This initiative has been backed by many Governments, as well as by UNEP, UNDP, FAO, and UNESCO. I call upon the Governments – UN members to give the needed financial support in assessing the state of ecosystems for the last Millennium and take an active part in the program".

A realization of the MEA program started in 2001 will be only a first step toward an assessment of the state of the global ecosystems, as well as development and accomplishment of measures needed to preserve the ecosystems on the planet in conditions when the need for natural resources grows. There also appears to be an acute problem requiring the preservation not simply of the quality of the environment but the normal functioning of the life-supporting systems on the Earth, without which further development of civilization is impossible.

The land use strategy and its manifestation through environmental changes is one of the important problems of global ecodynamics. Processes on the land surface are still inadequately understood in order to reliably assess their role as climate-forming factors. In this connection, the following questions arise, requiring to be answered:

- What means and methods are needed for the reliable characteristic of land surface changes taking place in the past and present-day?
- What are the main factors of the dynamics of land use and land surface processes?
- How will land use and land surface characteristics change in 50–60 years?
- How does the climate change affect the land use and land surface characteristics and what are the feedbacks between land surface processes and climate?
- What are the possible consequences of the impact of land surface changes on the environment, socio-economic development, and human health?

The respective directions of key developments include:

- Projects of land use and land surface changes. These projects take into account, in particular, changes due to deforestation under conditions of a growing population. According to available estimates, more than 70 mln acres of forested territories and cultivated lands, for instance, in the USA, will be transformed (mainly into the urban regions) during the period up to the year 2025. On the other hand, agricultural and other lands can be transformed into forested areas. This determines the complexity of the dynamics of the carbon budget and a perspective for the creation of the National Land Cover Dataset (NLCD) that characterize the present state of the land surface (mainly satellite observations).
- Land use in the region of the Amazon River. The emphasis on the dynamics of this region is determined by the location there of thick tropical forests intensively transforming due to forest cutting. The UN program is planned to assess the dynamics of global tropical forests with the use of satellite information.
- The state of boreal forests in Russia. The utmost interest in these ecosystems is reflected in some international programs on the specification of sinks and sources of carbon with the use of the satellite systems of NASA and NASDA.

The following directions constitute the main content of development planned for 2004–2005:

- evaluation of agricultural lands on global scales;
- preparation of a new (more detailed) map of the forest biomass resources for the

whole territory of Russia from the data of Landsat (spatial resolution 500 m) and Terra (250 m) satellites, with due regard to results of the ground inventory, which will make it possible to obtain reliable estimates of the role of the Russian forests as a carbon reservoir;
- accumulation of the database on land surface changes in Alaska between 1950–2001 (with annual averaging with the use of 1-km resolved satellite information) to characterize the successive dynamics;
- complex studies of the regional climate with an emphasis on the processes taking place on the land surface connected with the changing hydrological conditions in the south-east of the USA, with subsequent continuation of such developments under conditions of the global tropics; and
- impact of urbanization on the ecosystem dynamics in semi-arid regions of the USA, bearing in mind the solution of problems of water resources and the carbon budget.

6.5.2 Forest ecosystems and the greenhouse effect

During the last years, the problem of the impact of atmospheric carbon dioxide on the global climate has been discussed both by scientists and politicians. Some people believe that humankind will inevitably change the climatic situation on the Earth due to enhanced greenhouse effect, which will change living conditions, probably for the worst. And therefore it is necessary to reduce the industrial emissions of CO_2. Others, agreeing with the consequences of the greenhouse effect, deny the strategy put forward by the Kyoto Protocol and believe that the recommended reduction (quotas) will lead to an aggravation of the economy in many regions of the globe, without solving the problem of the greenhouse effect, but further worsening the global ecological situation. The opponents to the Kyoto strategy think that the greenhouse effect can only be prevented by the correct management of the structure of surface cover and by introducing strict controls over World Ocean pollution (Kondratyev, 1998; Kondratyev and Demirchian, 2001; Kondratyev *et al.*, 2004b). In this connection, the IPCC at the 8th Session in June 1998 in Bonn and at the 14th Session in October 1998 in Vienna prepared a special report on the role of the strategy of using surface cover (forests, in particular) in the global balance of CO_2. This report discusses the problems of interaction between the anthropogenic activity in the field of surface cover reconstruction and the distribution of CO_2 and other greenhouse gases in the biosphere. An assessment is given of various scenarios following from the Kyoto Protocol and concerning the problem of the impact of human society on the surface cover structure in general, and on forested territories, in particular. A brief analysis of this report is given below.

Item 3.1 and Appendix 1 of the Kyoto Protocol foresee a limitation and then a reduction of GHG emissions by 2008–2012. Before this time some problems should be solved to assess the role of the use of the Earth's surface. In particular, among these is the problem of the formalized description of the process of changing Earth cover structure, such as afforestation, forest reconstruction, deforestation and the associated carbon supplies. Understanding of the meteorological processes as

Table 6.16. Carbon supplies in vegetation and in the 1-m soil layer (Watson *et al.*, 2000).

Biome	Area (10^9 ha)	Carbon supplies (Gt C)		
		Vegetation	Soil	Total
Tropical forests	1.76	212	216	428
Temperate zone forests	1.04	59	100	159
Boreal forests	1.37	88	471	559
Tropical savannahs	2.25	66	264	330
Temperate zone meadows	1.25	9	295	304
Deserts and semi-deserts	4.25	8	191	199
Tundra	0.95	6	121	127
Marshes	0.35	15	225	240
Sown areas	1.60	3	128	131
Total	*14.82*	*466*	*2,011*	*2,477*

functions of greenhouse gases refers to one of the key problems of humankind in the first decade of the third millennium. Only an adequate knowledge of the meteorological phenomena on various spatial–temporal scales in conditions of supplies of CO_2 and other GHGs will enable one to make correct and constructive decisions in the field of global environment protection.

The dynamics of the surface ecosystems depends on interactions between biogeochemical cycles, which during the last decade of the 20th century suffered an anthropogenic modification. This especially refers to the cycles of carbon, nitrogen, and water. The surface ecosystems, in which carbon remains in living biomass, decomposing organic matter, and soil, play an important role in the global CO_2 cycle. Carbon exchanges between these reservoirs and the atmosphere take place through photosynthesis, respiration, decomposition, and burning. Human interference into this process takes place through changing the structure of the Earth's cover, pollution of the surface of water basins and soil areas, as well as through direct emissions of CO_2 to the atmosphere (Rojstaczer *et al.*, 2001; Qi *et al.*, 2001).

The role of various ecosystems in the formation of carbon supplies to the biospheric reservoirs determines the rate and direction of changes of the regional meteorological situations and global climate. The accuracy of assessment of the level of these changes depends on the reliability of the data on the surface ecosystem's inventory. The data in Table 6.16 show that a considerable scattering of estimates of carbon supplies in various types of vegetation suggests the conclusion that it is important to classify the surface ecosystems more specifically.

The anthropogenic constituent of the global carbon budget, beginning from the mid-19th century, increases the amplitude of the effect practically on every one of its natural elements. From 1850 to 1998 about 270($\pm$60) Gt C were emitted as CO_2 into the atmosphere due to fuel burning and cement production. About 136($\pm$55) Gt C went to the atmosphere as a result of anthropogenic changes of surface cover. This

Table 6.17. Characteristic of the mean annual CO_2 budget (Watson *et al.*, 2000).

Characteristic	Estimate ($Gt\,C\,yr^{-1}$)	
	1980–1989	1990–1998
1 CO_2 emission due to fossil fuel burning and cement production	5.5 ± 0.5	6.3 ± 0.6
2 CO_2 accumulation in the atmosphere	3.3 ± 0.2	3.3 ± 0.2
3 Assimilation by the oceans	2.0 ± 0.8	2.3 ± 0.8
4 (1) − [(2) + (3)]	0.2 ± 1.0	0.7 ± 1.0
5 CO_2 emission due to changes in the use of land resources	1.7 ± 0.8	1.6 ± 0.8
6 (4) + (5)	1.9 ± 1.3	2.3 ± 1.3

has led to an increase of atmospheric CO_2 of 176($\pm$10) Gt C (i.e., the partial pressure of carbon dioxide in the atmosphere has increased from 285 to 366 ppm (by 28%)). In other words, during 148 years, 48% of emitted carbon remained in the atmosphere and was not assimilated by the land surface or ocean ecosystems (230($\pm$60) Gt C was assimilated).

Some idea about the global carbon budget can be obtained from the data of Table 6.17. This table shows that the rates and trends of carbon accumulation in the surface ecosystems are rather uncertain. However, it is clear that the surface ecosystems are important assimilators of excess CO_2. Understanding the details of such assimilation is only possible through modeling the process of plant growth (i.e., considering the effect of the nutritional elements of soil and other biophysical factors on plant photosynthesis).

According to Table 6.16, the forest ecosystems and associated processes of natural afforestation, forest reconstruction, and deforestation should be studied in detail. The same has been emphasized in Items 3.3 and 3.4 of the Kyoto Protocol, where the necessity to determine national and international strategies of forest management is emphasized. In a forest range, the volume of the reservoir of CO_2 coming from the atmosphere is a function of the density of its canopy, and over a particular time period, a change of this volume is determined by the level and character of the dynamic processes of the transition of a given type of forest to another state. The causes of this transition can be natural, anthropogenic, and/or mixed. Biocenology tries to create a universal theory of such transitions, but so far, there is only a qualitative description of the observed transitions. As mentioned in the Kyoto Protocol, of importance is the correct definition of the notions "afforestation, forest reconstruction, and deforestation". Afforestation means to forest a land area used before (for 20–50 years and longer) for other purposes. Usually this term determines the process of natural succession at the expense of propagation of forest over other territories without human interference. The process of forest reconstruction is defined as planting trees. Deforestation is a substitution of the forest territory for another ecosystem. Thus, two opposite processes are possible in the forest ecosystem dynamics that can be controled by both nature and people. Each of these processes has its versions characterized by special dynamics of the vegetation

Table 6.18. Assessment of the calculated change of mean annual carbon supply for the afforestation/deforestation scenario (Watson *et al.*, 2000).

Region	RF	AF	TR		FR	
			A	B	A	B
Boreal	35	0.4–1.2	0.5	0.1	−18	−185
Temperate	60	1.5–4.5	2.1	1.9	−90	−501
Tropical	120	4–8	13.7	2.6	−1,644	−1,352

Notation: *A* – deforestation; *B* – afforestation; *RF* – change of average carbon supply after deforestation ($t\,C\,ha^{-1}\,yr^{-1}$); *AF* – average rate of CO_2 assimilation at afforestation ($t\,C\,ha^{-1}yr^{-1}$); *TR* – change of area (10^6 ha yr^{-1}) resulting from the deforestation-to-afforestation transition; and *FR* – forecast of changes in carbon supplies ($10^6\,t\,C\,ha^{-1}$) in 2008–2012 after the Food and Agriculture Organization (FAO) scenario.

over a given territory. Of special status is the process of foresting a territory where historically trees had never grown. In this case this territory immediately becomes important to CO_2 dynamics.

Table 6.18 illustrates one impact of the afforestation/deforestation processes on carbon supplies following the FAO scenario (Watson *et al.*, 2000) where the forest is a land area not less than 0.5 ha, with trees more than 5 m high, and the canopy covering more than 10% of the area. Deforestation is determined as a change of the surface cover with a tree canopy covering less than 10% of the area, as well as a change of the class of the forest (e.g., a decrease of productivity). Afforestation is the planting of trees over the area where trees had never grown. Note that "natural broadening" (i.e., propagation of forest over the agricultural territories without human interference) due to the FAO scenario is also referred to as afforestation. Finally, forest reconstruction is a direct planting of trees on the territories earlier covered with forest.

The problem of the correlation of the structure and types of land cover with the greenhouse effect also includes the problems of their control in optimizing the role of vegetation in global processes and achieving stability in NSS dynamics. In particular, Fulé *et al.* (2004) studied the dependence of the fire risk of a forested territory on the state of its canopy (its biomass, density, and composition) and showed that, for instance, in the territory of the national park in North Arizona (USA) the susceptibility of forests to fires during the last 160 years increased from 6% to 33%. This is connected with the change of the structure and composition of forests. During this period the biomass of forest canopy increased by 122% in the lowlands and by 279% in the highlands, which was the cause of the growing fire risk. The conclusion can be drawn that the fire risk can be controled by changing the structure and composition of tree plantations, but it should be borne in mind that the role of this territory in CO_2 assimilation will change. Hence, the GCDC unit of the global model (Figure 6.2) should reflect such changes, and the CON unit should provide a consideration of a possible scenario for reconstruction of the vegetation cover in some totality of pixels of the Earth's surface. It is clear that it is necessary here to take into account the data on water content in the canopy, one of the important parameters in assessing the fire risk. Maki *et al.* (2004) introduced the notion of "vegetation dryness index" (VDI), equal to the ratio of Normalized Differential Vegetation Index (NDVI) to an equivalent water supply to the canopy, and calculated its values for some forest territories of the Far East of Russia from the SPOT/Vegetation data for 1998. It was shown that VDI is an information parameter for detection of the regions with a high potential to catch fire and for determination of the direction of the forest fire propagation.

The technologies available to consider scenarios of the type given in Table 6.18 do not give the possibility to find the scenario which can be recommended for application. The Kyoto Protocol contains too simplified an approach to obtain reliable estimates of CO_2 dynamics as a function of many natural and anthropogenic parameters.

6.6 CORRELATION BETWEEN GLOBAL CLIMATE CHANGE AND NATURAL DISASTERS

6.6.1 Modeling the climate change

Recent decades have witnessed unprecedented attention to climate change problems (this also refers, in particular, to the mass media), which has, of course, stimulated the development of both scientific and applied developments, and, in turn, has ensured the achievement of considerable progress in understanding the causes of the present climate change, the laws of the palaeoclimate, as well as in substantiation of scenarios of possible climate changes in the future (the matter concerns the scenarios but not the forecasts whose abilities should be considered doubtful) (Lal and Yarasawa, 2001). Unfortunately, various speculative exaggerations and apocalyptic prognoses (e.g., complete melting of the Arctic sea ice in the first half of the current century) played a serious role in the growth of the attention given to climate problems. Paradoxically, Presidents and Prime Ministers in various countries entered into discussions on whether the Kyoto Protocol is to be treated as a scientifically sound document (Kondratyev, 2002). This situation is getting even more complicated, in particular, in the absence of sufficiently clear and uniform terminology. Without focusing on the very complicated situation regarding the climate definition, we should remind ourselves that until recently the notion of "climate change" was defined as anthropogenically induced, though one of the main unresolved problems consists in the absence of convincing quantitative estimates of the contribution of anthropogenic factors to global climate formation (nobody doubts, however, that anthropogenic impacts on climate do exist). In the international documents containing analysis of the present ideas of climate the notion of "consensus" was widely used with respect to scientific conclusions as if the development of science were determined not by the difference of views and relevant discussions but by common consent on concrete questions. Apart from the definition, of importance is the problem of vague and uncertain conceptual assessments concerning the various aspects of climate problems.

These (and other) circumstances gain particular importance in the context of the World Summit on Sustainable Development held in September 2002 in Johannesburg (South African). One more unclear definition: the notion of "sustainable development" requires its definition to be agreed upon. This is especially pertinent to the Russian term. At the meeting of the G8 in Genoa (July 2001) by initiative of V. V. Putin, President of Russia, it was decided to hold the World Conference on Climate in Moscow in 2003. In this connection we should remind ourselves that the undeniable success of the Second UN Conference on Environment and Development (Rio-de-Janeiro, 1992) and the special Session of the UN General Assembly "Rio + 5" held five years later (New York, 1997) consisted only in drawing the attention of governments and the public to global change and sustainable development problems. Unfortunately, both these World Forums have been poorly prepared as most clearly evidenced by their failure to elaborate the "Earth Charter" intended to formulate and substantiate priorities. Instead, a very amorphous and declarative "Rio Declaration" document was adopted (Kondratyev, 1998, 1999).

Today, three global ecological problems attract special attention: (1) climate change ("global warming"), (2) the fate of the ozone layer in the stratosphere, and (3) the closed nature of the global biogeochemical cycles (conception of biotic regulation of the environment). A bitter paradox consists in that despite the priority of the third problem and the secondary importance of the other two problems convincingly substantiated in the scientific literature, the documents adopted in Rio de Janeiro in 1992 reflect the absence of a required understanding of the conceptually important fact that of crucial importance is the following sequence of events: socio-economic development (stimulated by population growth) → anthropogenic impact on the biosphere → consequences of such impacts for the environment (climate, ozone, etc.).

Such misunderstanding advanced the "global warming" concept to the foreground, and this resulted in the adoption of the FCCC. This is a rather inadequate and misleading document that treats developing countries unjustly. Its focus – without good reason – is on the anthropogenic origin of the observed global climate change and the recommended (first of all, for developed countries) reductions of GHG emissions, above all, carbon dioxide.

As has been mentioned above, in December 1997, 160 countries participated in the Third Conference of the FCCC signatories held in Kyoto (Japan). The subject of prolonged heated debates was whether it was possible to comply with the required 5% – on average – CO_2 emission reduction by 2008–2012 (with respect to the 1990 level). Naturally, the attitude of the developing countries reflects the fact that their first priority is to raise the living standards of their populations, not to curtail the industry for the sake of CO_2 emissions reduction. However, it was the latter that the USA and other "golden billion" countries laid down as a condition for FCCC ratification. The FCCC history is only one example illustrating the giant (mainly bureaucratic) activity annually devouring hundreds of millions of dollars (instead of investing funds into scientific development). According to Global Ecological Fund (GEF) data as of 30 July 1998, financing of 267 GEF projects constituted US$1.9 bln (Kondratyev, 2002), and at the World Summit in Johannesburg (2002), an agreement was reached to replenish the GEF with US$2.0 bln. It should be remembered that about 10,000 people participated in the Kyoto Conference. The Hague Conference (COP-6) (November, 2000) and the Bonn Workshop (July, 2001) also had large audiences. At the Bonn Workshop were representatives of 178 countries (mainly officers, not specialists).

One can argue that the described situation stems from poorly developed scientific principles regarding global change problems. This conclusion is justified only in part, since as early as 1990 the key aspects of global ecology were discussed in monographs by Gorshkov (1990) and Kondratyev (1990). The authors advanced and substantiated the fundamental concept of the biotic regulation of the environment, and in other publications (see Bibliography) proved the scientific groundlessness of the "greenhouse" hypothesis of global warming, drawing attention to the necessity to study the "atmosphere–ocean–land–ice cover–biosphere" climate system with due regard to the complexity of the feedbacks between its interactive components. The problem of the global observational system was thoroughly analyzed, with emphasis on developments in the field of remote sensing and application of related observational data. Quite unique is the problem of the atmospheric ozone variability.

Now consider the problems of global climate change most vividly reflecting the existing delusions in the field of global ecodynamics. The most important circumstances are as follows:

- Observational data (still inadequate from the viewpoint of their completeness and reliability) do not contain a clear confirmation that global warming is of anthropogenic origin (especially concerning the data of ground observations in the USA, the Arctic, and the results of satellite microwave remote sensing).
- Whereas an enhancement of the atmospheric greenhouse effect caused by a supposed doubling of CO_2 concentration in the atmosphere constitutes about $4\,W\,m^{-2}$, uncertainties connected with consideration of the climate-forming role of atmospheric aerosols and clouds, as well as with the introduction of the so-called "flux correction" in numerical climate modeling, may even reach $100\,W\,m^{-2}$.
- Results of numerical climate modeling that substantiate the hypothesis of the greenhouse global warming and ostensibly agree with the data of observations are not more than an adjustment to observational data.
- Recommendations concerning the levels of GHG emission reductions based on these results are void of any sense (their accomplishment, however, can have far-reaching negative socio-economic consequences).

From the data of numerical modeling carried out by Wigley and Raper (2001) (provided it is realistic), even complete realization of KP recommendations can ensure only a negligible decrease in the annual average global average SAT by not greater than hundredths of a degree.

Serious attention has recently been paid to analysis of uncertainties (incompleteness) in numerical climate modeling. The most serious source of uncertainties is an inadequate consideration of interactive processes in the aerosol–clouds–radiation system (Kondratyev, 2002). There is no doubt that the most complicated aspect of numerical climate modeling is connected with consideration of interactive dynamics of the biosphere, which can be illustrated by two concrete examples that, of course, only partially reflect the complexity of the problem.

To explain the decrease of the amplitude of diurnal change of SAT by about 3–5 K, found from observations during the period 1951–1993 and resulting from a faster increase of minimum, rather than maximum, temperature, it was recommended to take into account the impact of different factors: changes of amount of cloud, water vapor content and tropospheric aerosol, as well as turbulence and soil moisture. Positive trends in the first three of the enumerated factors could lead to a decrease of net radiation in the day-time and to an increase of atmospheric counter-radiation (ACR) at night, whereas changes in the eddy mixing intensity and soil moisture could result in variations of heat and moisture exchange between the Earth's surface and the atmosphere, which should be much more substantial in the day-time than at night.

A strongly pronounced interactivity of processes and insufficient adequacy of their parameterization in climate models substantially complicates the estimates of contributions of various mechanisms on SAT decrease. In this connection, Collatz *et al.* (2000) undertook a numerical modeling of the response of the diurnal change of vegetation covered land to changes of external forcings and the biophysical state of vegetation cover using the SiB2 approximate land biosphere model (Sellers *et al.*, 1996a,b) under given meteorological conditions in accordance with different scenarios which made it possible to simulate the possible impact of the interactive dynamics of vegetation cover on SAT.

Analysis of the numerical modeling results has shown that an ACR increase determines an increase of nocturnal air temperature T_m, over vegetation cover, favoring a SAT decrease, whereas changes in T_m or the growth of $T_m + \text{ACR}$ (it is this that can take place under global warming conditions) promote an increase of both minimum and maximum temperatures, which determines a slight impact of these factors on SAT. This response is mainly determined by the impact of the diurnal change of the aerodynamic stability and radiation budget.

Numerous numerical experiments on climate modeling are based on the use of atmospheric global circulation models (GCM) together with the models of land surface processes (LSM). Results of such numerical experiments depend substantially on the specific interaction between GCM and LSM designed to simulate an exchange with radiation, momentum, and energy between the Earth's surface and the atmosphere. The desire to take into account all land ecosystems heavily complicated the LSM after inclusion of sub-models that consider the processes of photosynthesis, vegetation cover dynamics, and biogeochemical cycles, which radically raised the model's adequacy.

Based on the use of a simple biospheric model SiB as one of the LSM versions, Kim *et al.* (2001) performed various numerical experiments on the assessment of its sensitivity. The results showed that of importance is not only the model's sensitivity with respect to numerous morphological parameters, but also such a factor as sensitivity of transpiration of high vegetation cover to parameters characterizing the vegetation cover resistance. The improved SiB2 model provided consideration of biogeochemical processes that determine the exchange of water vapor, energy, and carbon dioxide between the Earth's surface and the atmosphere. A comparison was made between the results of numerical modeling of processes on a test site (rice field in Thailand (17°03′N, 99°42′E)) with the use of SiB2 models and the modified model SiB2–Paddy (rice field) together with the meso-meteorological model GAME–Tropics (the GAME–monsoon experiment in Asia carried out within the global field experiment GEWEX on the study of energy and water cycles) and the data of meteorological observations in the rainy season (1–6 September 1999).

The comparison revealed good agreement between the results obtained with the use of the two considered models and the data of observations of the diurnal change of radiation budget and latent heat flux calculated with the SiB2 model, except the latent heat flux. Fluxes of latent heat, heat in soils, and the rate of carbon assimilation from the model SiB2–Paddy, agree well with observational data, but the use of SiB2 is connected with substantial systematic errors. With some adjustment of parameters, the SiB2–Paddy model gives sufficiently reliable temperatures for soil, water, and vegetation cover. Results of calculations of the radiation budget, energy and water budgets, latent heat, and the rate of CO_2 assimilation turned out to be adequate. Such results create certain prospects for an adequate consideration of the biosphere as an interactive component of the climate system.

De Rosnay *et al.* (2000) assessed the reliability of parameterization schemes for land surface processes used in GCMs from the viewpoint of agreement between observed and calculated values of annual mean fluxes of energy and moisture depending on a detailed consideration of the soil's vertical structure. The results of calculations revealed a strong dependence of these fluxes on the vertical resolution. The 11-layer scheme of parameterization of heat and moisture transport in soil is adequate with the use of a 1-mm upper layer. Possibilities of realization of the scheme with such a thin upper layer are unclear, but bearing in mind that the horizontal resolution of a GCM constitutes hundreds of kilometers, further efforts are needed to solve such a problem.

An important part of the numerical climate modeling problem is a complex of questions concerning the atmospheric chemistry. For instance, it is well known that the formation of the concentration field for tropospheric ozone (TO) under different conditions (urban, regional, and global distributions) is strongly affected by different short-lived MGCs–ozone precursors, such as nitrogen oxides, methane, numerous organic compounds, hydrogen, and carbon oxide. Each of these MGCs has specific natural (biospheric) and anthropogenic sources.

Since TO is a greenhouse gas, emissions of MGCs mentioned above can affect indirectly the formation of the atmospheric greenhouse effect influencing the TO concentration field. Besides this, MGCs–TO precursors change the hydroxyl concentration field and, hence, the oxidizing ability of the troposphere. In turn, the distribution of hydroxyl concentration in the troposphere controls the lifetime of methane and, thereby, its global-scale concentration.

These factors are responsible for the complicated interactivity of processes that determine both the direct and indirect impact on the atmospheric greenhouse effect formation. Derwent *et al.* (2001) described the global 3-D Lagrangian model which simulates chemical processes taking into account MGC transport. This model was used to simulate the correlated fields of TO and methane concentrations under conditions of emissions to the atmosphere of such short-lived TO precursors as CH_4, CO, NO_x, and hydrogen. In this case the RF of NO_x emissions depends on their location, be it near the surface or in the upper troposphere, in the northern or southern hemispheres. The Global Warming Potential (GWP) was calculated for each of the short-lived MGCs–TO precursors from the data of integration of the response of methane and TO to the RF over the period of 100 years. Introduction of GWP allows an estimate of RF, for instance, for an emission of 1 tonne of any MGC, its equivalent RF for carbon dioxide can be calculated (over the period of 100 years). For the joint impact of methane and TO, the GWP value constituted 23.3.

Analysis of calculation results has shown that the indirect RF due to a change of the content of methane and TO was substantial for all the MGCs considered. While the RF due to changes in methane is mainly determined by the impact of methane emissions, in the case of TO it is controled by all MGC precursors, especially by nitrogen oxides. The TO-induced indirect RF can be so strong that MGCs–TO precursors should be taken into account in assessing possible climate changes and the measures necessary to prevent them.

Despite President G. W. Bush's "anti-Kyoto" statements, the headlines of many American newspapers in January 2001 were dramatic: "Scientists publish frightful predictions of warming: accelerating climate shifting threatens a global disaster in this century" (*The Washington Post*); "The Earth's warming generates a new danger signal" (*The International Herald Tribune*), and others. This stemmed from new climate change scenarios for the 21st century, according to which such changes can be more substantial than expected. According to the data of the Third IPCC Report (IPCC, 2001), an increase of annual mean global mean SAT can reach 5.8°C by 2100 compared with the present-day value (Houghton *et al.*, 2001). In 1996 the respective estimate was only 3.5°C.

As Karl and Gleckler (2001) justly noted, of great importance is the fact that the supposed range of possible SAT increase has turned out to be wider than before. For many specialists in numerical climate modeling it was not surprising, since this sphere is still at an initial stage of development, and besides, the numerical modeling has to be based on a very limited volume of observational data (even the length of the SAT data series constitutes only about 100 years).

Although most scientists believe that the observed global warming has been determined mainly by the growth of GHG concentrations, in some respects the range of estimates of possible climate change has broadened rather than narrowed.

The main uncertainties of climate change assessments include three aspects:

(1) Detection of global warming from observational data.
(2) Attribution of global warming to anthropogenic factors.
(3) Forecast of future climate change.

Karl and Gleckler (2001) believe that the data in IPCC (2001) narrows the range of uncertainties with respect to the first two bullet points, but future climate forecasts have become less definite. According to IPCC (2001), the observed global warming constituted 0.6°C ± 0.2°C (at the 95% level of statistical significance), with "most of the warming observed over the last 50 years being, probably (with a probability between 66–90%) determined by the growth of GHG concentrations". Since one of the main causes of numerical climate modeling uncertainties is still connected with inadequate consideration of the climate-forming role of atmospheric aerosols and clouds, Kiehl and Gent (2004) noted in this connection: "The more we learn about aerosol the better we understand how little we know about it". This especially concerns estimations of the impact of aerosol on clouds and, hence, on climate, which vary widely (for details see Item 7.2 of IPCC (2001)).

Unfortunately, the role of uncertainties in numerical climate modeling has not been adequately evaluated in IPCC (2001). This has prompted criticism from many specialists (Kondratyev and Demirchian, 2001; Essenhigh, 2001; Schrope, 2001).

In the context of global change, the problems connected with assessments of the present and possible future global climate changes are, no doubt, of primary importance (Kondratyev, 2001). Although the global warming concept still prevails in these assessments, as demonstrated in IPCC (2001), it should be considered as no more than an inertia of the earlier speculative ideas whose motivation was far from scientific (convincingly demonstrated by Boehmer-Christiansen (2000)). The contradictory estimates of climate can be vividly illustrated by the radically opposite opinions expressed by the two election candidates for USA Presidency. While A. Gore has long been known as an ardent advocate of the global warming concept and the Kyoto Protocol, G. W. Bush was against (ecological) policies like the Kyoto Protocol, which he considered would lead to a radical increase in the price of oil, oil products to heat houses, natural gas, and electricity. Such an agreement would heavily load the US economy without protecting it against an undesirable climate change. The Kyoto Protocol was seen as inefficient, inadequate, and unjust with respect to America, since it excluded 80% of the world community from participation in implementing the recommendations, including the main centers of population concentrations, such as China and India. According to G. W. Bush's position, of principal importance is the development of new ecologically pure technologies and the use of market mechanisms, including freedom from regulation for the markets of electricity, natural gas, taxation, and "emissions trading" (G. W. Bush believes that natural gas and atomic energy will play an important role in reducing the dangerous dependence of the USA on foreign oil). The USA turned out to be the first country wanting to free its economy from unjustified ecological limitations and raised some proposals deserving attention: to consider atomic energy as an important perspective for future energetics; to search for and study other ways of maintaining the carbon balance (including measures to reconstruct forests as carbon sinks).

It becomes clear that despite a certain progress in the parameterization of the climate-forming processes, the attempts undertaken to compare the numerical climate modeling results with observational data were rather schematic, controversial, and unconvincing. This is connected with the extreme complexity of climate models and an inefficiency of applied modeling technologies. However, gathered experience in modeling the climatic processes provides numerous examples of successful parameterizations of some fragments of the global climate system. In particular, there are several demonstrative estimates of the role of forests in the formation of regional changes of climate characteristics. So, Gedney and Valdes (2000) performed numerical experiments simulating the present ("control") climate and conditions of complete deforestation of the region of the Amazon Basin. For this purpose, a 19-level spectral model of atmospheric general circulation was used. Deforestation should result in the following changes to the basic climate-forming parameters: albedo (13.1% $\rightarrow$ 17.7%); roughness (2.65 m $\rightarrow$ 0.2 m); the share of vegetation cover (0.95 $\rightarrow$ 0.85); leaf index (4.9 $\rightarrow$ 1.9); minimum resistance of vegetation cover (150 s m^{-1} $\rightarrow$ 200 s m^{-1}); and depth of the root zone (1.5 m $\rightarrow$ 1.0 m). All this will also lead to a change in soil type. A consideration of the numerical modeling results revealed deforestation-induced statistically significant changes in the wintertime precipitation in the north-western Atlantic which moved eastwards toward western Europe. Similar variations are connected with changes of large-scale atmospheric circulation in middle and high latitudes. Application of a simple model has confirmed that such variations are due to planetary wave propagation and reveal an interrelationship between processes in the region of deforestation, in the North Atlantic and western Europe, which are independent of the choice of the model, with the level of changes corresponding to estimates of anthropogenic climate change due to growing concentrations of CO_2 and aerosol.

Using the model of global climate CCM1-Oz developed at the National Center for Atmospheric Research (NCAR) (USA), Zhang *et al.* (2001) carried out a much more extensive numerical modeling of the climatic consequences of deforestation in the tropics in conditions of growing greenhouse warming due to a doubled atmospheric CO_2 concentration. Calculations suggested the conclusion that there would be a sharp decrease of evapotranspiration ($\sim$by 180 mm yr^1) and precipitation ($\sim$by 312 mm yr^{-1}), as well as an increase of SAT by 3.0 K in the Amazon Basin. Similar but weaker changes would take place in South-East Asia (a precipitation decrease equal to 172 mm yr^{-1} and a warming of 2.1 K). Still weaker changes would occur in Africa (precipitation increase of 25 mm yr^{-1}). Analysis of the energy balance estimates suggested the conclusion that the climate warming takes place not only due to an enhancement of the greenhouse effect but also as a result of reduced evapotranspiration caused by deforestation. Statistically substantial climate changes due to deforestation in the tropics also appear in the middle latitudes.

The IPCC-1996 report contains the fiercely debated conclusion: "The balance of available evidence supposes a marked impact of man on global climate" (Kondratyev, 1998), as well as the statement that "the anthropogenic signal" manifests itself against a background of natural climatic variability. According to IPCC (2001), "Studies on detection and attribution reveal regularly the presence of anthropogenic signals in the data of climate observations for the last 35–50 years ... Natural impacts could have played a role in the observed warming during the first half of the 20th century, but they cannot explain the warming in the second half of the century". However, the report contains also the following statement: "Reconstruction of the climate for the last 1,000 years testifies to a small probability that the climate warming observed in the second half of the 20th century could have been of totally natural in origin" and further highly uncertain quantitative estimates of the anthropogenic warming are emphasized, especially from the viewpoint of the contributions of various factors of the warming (primarily it refers to atmospheric aerosol). The inconsistency and weakness of the cited statements and conclusions are so apparent that no further comments are needed. Of course, the integral models describing the dynamics of interactions between socio-economic development and nature, and taking into account the functional multi-dimensionality of the NSS, should play the leading role in substantiation of future climate predictions. However, it is still unclear how real the predictions obtained can be with the use of such models of unthinkable complexity and with inadequate input information. One should hope that at least in the foreseeable future the integral models may be the only means to obtain rather conditional scenarios, and the expected consequences of anthropogenic activities will be estimated using the GMNSS within these scenarios giving a prognoses of climate changes for the periods agreeing with geoinformation monitoring systems (GIMS)-technology forecasts (Kondratyev *et al.*, 2002).

6.6.2 Radiative forcing due to aerosols

As has been mentioned, the conclusions of the IPCC (2001) report are based mainly on the results of numerical climate modeling obtained with the use of 40 scenarios of the possible dynamics of GHG emissions in the future. Such scenarios, which depend on prescribed factors, which cannot be reliably predicted, such as changes in the global population size, socio-economic development, and the scientific–technical progress, are very uncertain, which is illustrated, in particular, by the wide range of possible cumulative GHG emissions (in carbon equivalents) of 770–2,540 Gt C by 2100. The respective estimation of the probable range of CO_2 concentrations in the atmosphere in 2100 gave 540–970 ppm. However, these estimates do not take into account the possible measures for reductions of GHG emissions as well as technological progress.

According to IPCC (2001), an expected increase of SAT by 2100 of 1.4–5.8°C will be caused by a high rate of SAT variability, which will exceed the observed rate of SAT increase in the 20th century by a factor of 2–7. The pre-calculated increase in the level of the World Ocean could reach 9–88 cm, such an increase could (due to gigantic thermal inertia of the climate system) continue for several centuries after stabilization of the level of CO_2 concentration (this also refers to SAT, though to a lesser degree). Such a high uncertainty in the estimates of possible future climate changes is connected not only with the inadequacy of the existing information on the role of GHGs but also, to a greater degree, with the role of atmospheric aerosol as a climate-forming factor. This especially concerns the estimates of the so-called indirect aerosol radiative forcing (ARF) connected with the effect of aerosol on the microphysical and optical properties of clouds.

The main source of these uncertainties is the deficit of data on the extremely variable properties of aerosol. Therefore, during the last years, serious efforts have been made to obtain data on aerosol based on the use of various methods of remote sensing (ground, aircraft, satellite) and complex field observations. Bearing in mind the IPCC (2001) Report and several reviews of the results obtained, we shall discuss only some of them. Among the most significant complex observation experiments are ACE-1 (measurements of the main characteristics of aerosols), TARFOX (determination of aerosol radiative forcing), ACE-2 (studies of aerosol in the region of the Pacific Ocean), INDOEX (studies of aerosol in the region of the Indian Ocean), and others. Besides these, there was also the global network of AERONET stations equipped with sun photometers to measure the aerosol optical thickness (AOT) of the atmosphere, as well as lidar soundings made at various points to obtain information on the vertical profiles of aerosol concentration.

Wagner *et al.* (2001) performed comparative calculations of ARF at the levels of the upper boundary of the atmosphere and the surface with the use of observations of the aerosol characteristics for the whole atmospheric thickness and with their vertical profiles taken into account. Input data for calculations were results of lidar soundings (at six wavelengths over the range 355–1,064 nm) as well as measurements with a sun photometer made on 25 March 1999 within the INDOEX program. On this particular day there was multi-layer haze in the troposphere (at altitudes up to 3.5 km), characterized by the presence of strongly absorbing particles. The AOT at the wavelength 530 nm reached 0.57 (with the Angstrom exponent 0.9).

Calculations of ARF with the prescribed ocean albedo of 0.05 and aerosol properties averaged over the atmospheric thickness (the prescribed complex refraction indices are $1.575–0.05i$ and $1.65–0.035i$) gave ARF values ranging between -5 to -12 and -55 to $-81\,\mathrm{W\,m^{-2}}$ at the levels of the upper atmospheric boundary and the surface, respectively. With the vertical profile of the aerosol characteristics taken into account (from the data on the coefficients of backscattering obtained from lidar soundings at six wavelengths and on the coefficient of extinction at wavelengths of 355 nm and 532 nm, the vertical profiles of the efficient radius of particles, number density, and complex index of refraction were retrieved), the respective values of ARF will be equal to -10 and $-60\,\mathrm{W\,m^{-2}}$. The calculated values of ARF efficiency (calculated per unit optical thickness) vary within -11 to $-24\,\mathrm{W\,m^{-2}}$ and -101 to $-154\,\mathrm{W\,m^{-2}}$, respectively.

Since atmospheric aerosol is one of the factors having a direct and indirect effect on climate, on the initiative of NASA and GEWEX, the Global Aerosol Climatology Project (GACP) was born, in 1998. The main goals of which being:

- Retrieval of aerosol characteristics (including their annual course and interannual variations) on global scales from the data of satellite observations of the outgoing short-wave radiation and from conventional observations.
- Numerical modeling of the processes of formation, transformation, and transport of aerosol.

The final goal is to obtain representative data on global climatology of aerosol for the whole period of satellite monitoring, which could be used to study the direct and indirect effects of aerosol on climate. The most important results obtained during the first 3-year phase of GACP are as follows (Mischnenko *et al.*, 2002):

- substantiation of algorithms to retrieve aerosol characteristics from the data of multipurpose measurements of outgoing short-wave radiation;
- analysis of the information content of the existing satellite observations from the viewpoint of aerosol property retrieval;
- comparative analysis of the scenarios of variability of aerosol properties suggested by various groups of specialists; and
- development of methods to use the data of conventional observations to assess the reliability of the aerosol properties retrieved from the data of satellite monitoring.

The results obtained testify to the complex nature of the indirect impact of aerosol on climate and to the need for a complex approach to the solution of this problem. The most important expected result of realization of the first phase of GACP should be a substantiation of the three versions of the global climatology of aerosols including data on AOT and average size of particles, with the use of 1- and 2-channel AVHRR data as well as Total Ozone Mapping Spectrometer (TOMS) data. Also plans have been made to apply the numerical models of the aerosol transport (with chemical transformations taken into account) for theoretical substantiation of aerosol climatology. Bearing in mind the imperfection of the AVHRR calibration, improved devices such as the Moderate-resolution Imaging Spectroradiometer (MODIS) and the Multi-angular Imaging Spectroradiometer (MISR) are supposed to be used to specify the calibration.

As has been mentioned, the estimates of indirect ARF are most complicated. In agreement with the classical Köhler theory (Kokkola *et al.*, 2003), clouds consist of "activated" droplets whose sizes increase spontaneously after reaching a certain critical size (at a certain level of oversaturation of the surrounding water vapor). In this context, clouds are a specific component of the atmosphere with respect to wet non-activated and dry aerosol particles. Sometimes quite a different situation is observed in the atmosphere determined by interactions in the multi-phase system consisting of clouds, aerosols, and the gaseous atmosphere. There is an exchange of mass between the components of this system determined by various physical and chemical processes. So, for instance, the sizes of aerosol particles of natural or anthropogenic origin can grow with increasing relative humidity and, finally, such particles become cloud droplets. Therefore, there is no doubt that clouds should not be considered separately but instead as a component of the system mentioned above.

Another principally important circumstance consists in the exclusive variety of spatial–temporal scales (from molecular to global) connected with the formation and evolution of cloud cover. Quite special is the interaction of clouds and aerosols, affecting substantially both the cloud cover dynamics and aerosol. Analysis of the data of satellite observations has shown that a decrease in the size of cloud droplets caused by aerosol particles functioning as cloud condensation nuclei (CCN) lead to a precipitation reduction, which is manifested over regions of biomass burning, atmospheric pollution, and dust transport in deserts.

Of great importance is also the interaction of clouds with the surrounding atmosphere, especially with water-soluble MGCs such as NH_3, SO_2, HNO_3, organic acids with a small molecular weight, and aldehydes. Though chemical processes with the participation of ice cloud particles remain almost unstudied, the field observations have shown, for instance, that such particles play an important role in removing MGCs from the atmosphere, such as H_2O_2 and NH_3 absorbed by ice crystals. Also it turns out that H_2O_2 reacts with SO_2 much more slowly in ice particles than in water droplets.

Laboratory measurements have led to the conclusion that soot can be an efficient source of condensation nuclei. An increase in the small droplet number density due to CCN depends not only on the CCN number density but also on a number of other factors, such as available moisture, rate of upward fluxes, air temperature, etc., as well as size distribution and chemical composition of aerosol. For many years it was considered that the chemical composition of CCN is determined by the presence of soluble inorganic salts, NaCl or NH_4SO_4, in them. Only inorganic aerosol was taken into account in the initial Köhler theory (Dufour and Defay, 1963), whereas in conditions of the real atmosphere of great importance are organic compounds as elements of aerosols. Based on the use of the modified Köhler theory (Sorjamaa *et al.*, 2004), the important role of organic CCN in cloud formation was demonstrated, manifested through a considerable decrease (in the case of atmospheric pollution up to 110%) in the level of critical oversaturation, especially in the presence of small CCN particles. The share of water-soluble organic compounds (WSOC) varies between 20–70% with respect to total content of WSOC.

Numerical modeling has shown that a consideration of the organic component of CCN leads to an increase in cloud albedo, which is comparable with the Tuomi effect of doubling the CCN concentration (Tuomi, 2004). Another important factor causing changes in cloud properties are the chemical reactions occurring within them (with the participation of MGC and aerosol), which depends on the cloud droplet size distribution, since their chemical compositions are a function of their sizes. Two main compounds participating in reactions in a liquid phase are compounds of sulfur and organic compounds. For instance, oxidation of 80% S(IV) to S(VI) takes place in a liquid phase.

Similar processes including participation of organic compounds remain, however, almost unstudied. This represents one of the remaining uncertainties in the estimates of the contribution of aerosol–cloud interaction as a climate-forming factor. This determines the urgency of laboratory studies of photochemical reactions in clouds with participation of organic compounds, studies of chemical reactions in ice clouds, analysis of the chemical composition of aerosol particles and clouds, and consideration of these and other factors in the models of cloud cover dynamics.

The realization of the Complete Atmospheric Energetics Experiment (CAENEX) has made it possible for the first time to obtain information on ARF and aerosol properties (Kondratyev, 1991, 1993; Kondratyev *et al.*, 1973, 1983; Kondratyev and Binenko, 2000). Airborne observations were used to measure the integrated and spectral characteristics of aerosol. Unfortunately, the results received in the framework of CAENEX had no continuation. However, various observations and calculations of ARF have been intensively continued.

As Boucher (2002) noted, an understanding has been reached now that the aerosol-induced RF manifests itself in the form of direct, semi-direct, and indirect (of the first and second type) forcing. The direct forcing is the impact of the total attenuation of short-wave radiation by aerosol and the thermal emission of aerosol (this impact manifests itself mainly in conditions of a cloudless atmosphere). The semi-direct forcing characterizes the impact of radiation absorption by aerosol in conditions of both clear and cloudy skies on the formation of the vertical profiles of temperature and humidity, which determines the functioning of the feedback that changes the cloud cover dynamics.

The first indirect forcing consists of an increase in the optical thickness and albedo of clouds as a result of decreasing size and increasing number density of cloud droplets (with the constant water content of clouds) under the influence of aerosol particles functioning as CCN. The second type of indirect forcing is connected with the growth of the cloud water content, the altitude of the cloud tops or the increase of cloud lifetimes as a result of decreasing rain rate due to a decrease of the size of cloud droplets.

Direct ARF depends on a set of parameters, which can be classified into three categories:

(1) Biochemical parameters (such as intensity of emissions of primary and secondary aerosols, aerosol lifetime).
(2) Characteristics of the atmosphere (relative humidity, cloud amount, surface albedo).
(3) Microphysical parameters (size distribution, optical properties, hygroscopicity).

Concrete data on the variability of direct ARF and estimates of the respective errors are still limited, which reflects, first of all, the need for extending the programs of field observations of aerosol properties with the use of various observational means including satellites. This will make it possible to carry out a more complete correction of the adequacy of aerosol models and parameterization of the aerosol characteristics in the climate numerical models. The key directions of further developments should include, in particular:

- Specification of the available archive of data on aerosol emissions (especially anthropogenic).
- Extension of observations of the vertical profiles of aerosol characteristics (methods of ground and satellite lidar sounding should play a special role).
- Experimental and theoretical studies of the multi-component aerosol, including obtaining data on hygroscopic and optical properties (single-scattering albedo included).
- Analysis of the contribution of natural and anthropogenic aerosol.
- Improving aerosol models with new data from observations taken into account.

Hansen and Sato (2001) believe that an increase of the mean annual mean global SAT for the last century by more than 0.5°C was caused, at least partially, by anthropogenic forcings on the climate. In this connection, calculations of RF due to the growth of GHG concentrations were made. The obtained estimates testify to a considerable level of RF due to methane, whose contribution reached about half the RF caused by carbon dioxide: $(1.4 \pm 0.2)\,W\,m^{-2}$ (the error is a subjective estimate within 1σ), including $0.5\,W\,m^{-2}$ due to direct and $0.28\,W\,m^{-2}$ due to indirect forcing. These results suggest the need for special attention to be given to the consideration of CH_4 as a GHG and a separate consideration of RF due to methane.

New estimates of RF due to tropospheric ozone give 0.7–0.8 $W\,m^{-2}$ instead of 0.3–0.4 $W\,m^{-2}$ as is supposed in IPCC (2001). Apparently, for this RF the interval 0.4–0.8 $W\,m^{-2}$ can be assumed.

Considerably arbitrary (due to uncertain input data) calculations of RF due to "black" carbon (soot) gave values $(0.8 \pm 0.4)\,W\,m^{-2}$. The RF due to "reflecting" aerosols constituting $(-1.4 \pm 0.5)\,W\,m^{-2}$ with an additional contribution of soil aerosols $(-0.1 \pm 0.2)\,W\,m^{-2}$ playing an important role. A substantial contribution to the total ARF is made by organic aerosol ($-0.3\,W\,m^{-2}$ with an accuracy to the coefficient 2) and ammonium sulfate ($-0.2\,W\,m^{-2}$).

The sum of all positive components of RF reaches $(4.3 \pm 0.6)\,W\,m^{-2}$ (i.e., three times greater than the RF due to CO_2), whereas the sum of the negative component is $(-2.7 \pm 0.9)\,W\,m^{-2}$ and thus the resulting RF is equal to $(1.6 \pm 1.1)\,W\,m^{-2}$ (i.e., close to that only due to CO_2).

Hansen and Sato (2001) performed an analysis of the trends of different GHGs, from which it follows that the rate of the CO_2 concentration growth increased rapidly during the period from the end of World War II to the mid-1970s, reaching a maximum of about 4% yr^{-1}, after which it started decreasing to a comparatively stable level of concentration growth of 1.5 ppm yr^{-1}. During the same period the CH_4 concentration grew rapidly (from 5 to 15 ppm yr^{-1}), slowing down considerably after 1980 (the causes of this are still unknown). The total trend of the rate of RF growth due to 13 chlorofluorocarbon compounds (CFC) was characterized by a rapid intensification until the mid-1980s and a subsequent decrease due to CFC emission reduction in the process of realization of the measures foreseen by the Montreal Protocol. Apparently, before 2010 RF should start decreasing due to CFC. The maximum level of total RF (for the period of 100 years) could reach about 5 $W\,m^{-2}$. With this rate of growth preserved, the level of the equivalent RF corresponding to a doubled CO_2 concentration should be reached by 2050. However, during the last two decades, the rate of the net RF growth decreased to 3 $W\,m^{-2}$. Taking into account a possible reduction of GHG emissions, global warming during several forthcoming decades should constitute $(+0.15 \pm 0.05)$°C during 10 years under conditions of "frozen" CO_2 emissions. The preserved, considerable uncertainty of estimates considered is connected, first of all, with the difficulty of calculations for aerosol-induced RF.

Bearing in mind the importance of comparative estimates of RF due to GHGs, consider the results of the latest calculations of RF due to GHGs discussed in Hauglustaine (2002). According to the estimates obtained, the mean global total RF due to the growth in the atmosphere of the concentration of well-mixed GHGs in the industrial epoch (after 1750) constitutes 2.43 $W\,m^{-2}$, and some of its components are equal to (in W m^{-2}): 1.46 (CO_2); 0.48 (CH_4); 0.15 (N_2O); and 0.34 (CFC) with an error of about 10%.

The RF is characterized by a strong spatial variability. So, for instance, in 1980 the mean global total RF reached 5 W m^{-2} (per 100 years), and the subsequent decrease was connected with a sharp decrease of emissions to the atmosphere of ozone destroying CFCs (in conformity with the respective international agreements) as well as with the retarded growth of CO_2 and CH_4 concentrations in the atmosphere (a decrease of methane concentration continuing for several years was observed). Apart from the GHGs mentioned above, a substantial contribution (up to 8–15%) into the formation of the total RF was also made by tropospheric ozone, for which the initial value of RF was (0.35 ± 0.15) W m^{-2}. During the last few years this value was specified at 0.7 W m^{-2}.

The errors of the estimates of RF due to various GHGs were mainly caused by a lack of information about the spatial–temporal variability of the GHG content in the atmosphere, which depends on the power of the surface sources of GHG emissions and GHG sinks in the free atmosphere. This refers, for instance, to natural emissions of compounds, which contribute most to the variability of the long-lived GHGs and the TO-precursors, such as NO_x and non-methane hydrocarbons (NMHC). The effect of indirect factors and feedbacks on RF can be illustrated using carbon dioxide as an example. Calculations have shown that with CO emissions increasing by 40 Tg, the RF due to TO increases, and this increase is 10–20% greater, if these emissions take place in the tropics and not in middle latitudes. With the increase of NO_x emissions assumed to be equivalent to 1 Tg N, such a difference in the RF variability reaches a factor of 6–8, with the indirect effect of NO_x on RF being negative (i.e., it leads to an ozone-induced decrease of RF). These examples illustrate the critical importance of the consideration of MGC emission geography.

The height at which RF takes place plays a substantial role. In this context, of special importance are processes in the "upper troposphere–lower stratosphere" layer, which necessitates the development of models that take into account chemical processes both in the troposphere and in the stratosphere.

An important factor controling the lifetime of most of GHGs and other MGCs that pollute the atmosphere is the oxidizing capability of the atmosphere determined mainly by the concentration of OH hydroxyl and other oxidants (O_3, HO_2, H_2O_2 or NO_3). Prognostic estimates of the oxidizing capability are extremely uncertain, since it depends on the factors such as content of water vapor and aerosols, the troposphere–stratosphere exchange, intrusion into the troposphere of UV radiation, and the preceding evolution of the oxidizing capability.

A decrease in the level of MGC-induced uncertainty of RF estimates depends on further progress in the development of interactive models of the tropospheric composition considering the complicated interactions between the atmosphere, ocean, and land biosphere, with respect to physical, chemical, and biological processes taking place in these components of the climate system. Of course, the adequacy of such models requires a thorough test with observed data. The problem of adequacy is connected with the complexity of the models, which take into account 50–100 chemical compounds and 100–3,000 photochemical reactions with spatial resolutions of $2^\circ \times 2^\circ$, and consideration of a 20–60-layer atmosphere. One more important aspect is that the RF concept was developed as applied to the analysis of the mean global situation. Therefore, it is necessary to analyze its special features as functions of the scales of spatial–temporal averaging.

In particular, Podgorny and Ramanathan (2001) analyzed the results of studies of microphysical, chemical, and optical properties of aerosols (including data from lidar soundings) obtained in accomplishing the field observation experiment INDOEX in the region of the Indian Ocean. They obtained a complete database on atmospheric aerosol, which can be used to control the reliability of the techniques of numerical modeling of the radiation transfer in the atmosphere (the main goal of INDOEX was to analyze the effect of atmospheric pollution on the climate-forming processes in the tropics of the Indian Ocean). In this regard, calculations have been undertaken of direct RF due to aerosol and clouds and of the respective rate of radiative heating for characteristic conditions of the winter monsoon in the tropics of the Indian Ocean. The main goal of the numerical modeling, based on the use of the Monte-Carlo method, was an analysis of the re-distribution of short-wave radiation between the ocean mixed layer and the atmosphere above it, caused by natural and anthropogenic aerosols.

According to the observation data, the single-scattering albedo $\tilde{\omega}$ at the wavelength 500 nm varied within 0.8–0.9, the aerosol optical thickness of the atmosphere τ_A averaged between 0.1 to 0.8, and the cloud amount f (with the characteristic horizontal scale of several kilometers) constituted 25% (this value is typical for lower clouds over the tropical ocean). In many cases the aerosol layer was far above the lower clouds, which affected substantially the formation of ARF.

In conditions of a cloudy atmosphere, the soot aerosol descending onto the ocean surface from Asia and from the Indian sub-continent affected strongly the ARF (even with a 10% contribution into τ_A). For monthly mean conditions, $\tau_A = 0.4$; $\tilde{\omega} = 0.9$; and $f = 25\%$. The diurnal mean ARF at the surface level R_s was equal to $-25\,\mathrm{W\,m^{-2}}$ and ARF for the atmosphere (i.e., the absorbed solar radiation R_a) varied within 22–25 $\mathrm{W\,m^{-2}}$. The ARF variations at the atmospheric top (R_T) were within 0 and $-3\,\mathrm{W\,m^{-2}}$. The presence of aerosol determines an enhancement of the "cloud" RF by 0.5 or 2.5 $\mathrm{W\,m^{-2}}$, if the aerosol is located mainly either below or above clouds, respectively.

The aerosol-induced rate of radiative heating in the trade wind atmospheric boundary layer reaches 1–1.5 $\mathrm{K\,day^{-1}}$, which can strongly affect the formation of trade wind cumulus clouds. Thus, the main effect of the atmospheric aerosol over the Pacific Ocean (this aerosol is mainly of anthropogenic origin) is manifested through a re-distribution of short-wave radiation between the atmosphere and the mixed layer of the ocean.

Remer *et al.* (2002) discussed the possibilities of retrieving the aerosol-induced RF at the top of a cloud-free atmosphere from satellite observations in connection with the beginning of the functioning of new types of space-borne instruments, including MODIS, which opens up prospects for obtaining various information about the atmospheric aerosol properties. The numerical modeling of the processes of long-range aerosol transport enables one to study the laws of the global distribution of aerosols and in this connection to substantiate the requirements for data from aerosol satellite remote sensing, including retrieval of RF.

Remer *et al.* (2002) analyzed the results of the numerical modeling obtained with the use of a combined model of the tropospheric general circulation and chemical processes in the troposphere, which determine the formation and evolution of the aerosol properties, including the following types of accumulation mode for aerosol: sulfates, organic carbon, black carbon, methanosulphonic acid, sea salts, soil dust, and marine aerosol (fractions of large size dust and sea salt aerosols are considered separately). The total AOT is determined by contributions from the accumulation of two large size modes, though there is also a mode of ultra-small particles, but its contribution into AOT is negligibly small. Since these components of aerosols are considered as internally mixed, this excludes a possibility to evaluate the contribution of individual components to the formation of the total AOT. Naturally, the errors of AOT retrieval from satellite data are at a minimum with the high content of aerosols in the atmosphere.

Remer *et al.* (2002) considered an example of the calculations of monthly mean spatial distribution of smoke ARF resulting from biomass burning in the southern hemisphere, with the use of two models of long-range transport of aerosols. The results obtained show that in 87–97% of cases the values of smoke ARF exceed the level of possible noise (i.e., the prospect for satellite remote sensing of ARF is positive). Basic sources of errors are connected with inadequate data about background aerosol, errors in prescribed optical characteristics of the surface, and some characteristics of aerosol. A combined effect of these three sources of errors forms a total uncertainty of ARF of 1.2–2.2 $\mathrm{W\,m^{-2}}$. However, it is important to note that these estimates of errors do not take into account the contribution of the possible (residual) effects of clouds. The level of errors can be reduced with the combined use of data from satellite and ground observations, as well as results from the numerical modeling of long-range aerosol transport.

Estimates of the aerosol-induced direct RF obtained with several types of aerosol (sulfate soot both due to fossil fuel burning and organic sources) taken into account gave values between -0.25 to $-1.0\,\mathrm{W\,m^{-2}}$, whereas the indirect RF (caused by changes in the size distribution and optical properties of clouds under the influence of aerosol) could vary from 0 to $-1.5\,\mathrm{W\,m^{-2}}$. The possible interval of the values of RF due to greenhouse gases constituted $+2.1$ to $+2.8\,\mathrm{W\,m^{-2}}$ (i.e., the aerosol-induced RF is close in value but opposite in sign to the greenhouse RF). The problem consists, however, in the considerable uncertainty of the ARF values. In this regard, Christopher and Zhang (2002) calculated the RF due to aerosols formed in biomass burning in the tropics, which should be important, since more than 114 Tg of smoke observed in the tropics results from biomass burning.

To calculate the short-wave direct RF, the calibrated hourly data of the GOES-8 scanner was used for the period 20 June–30 August 1998, for the tropics of South America (4°S, 51–65°W). Based on the use of the Mie formulas for calculations of the optical characteristics of individual aerosol particles as well as methods of discrete ordinates of radiation transfer theory, the AOT of smoke formed by biomass burning was calculated at the wavelength 0.67 μm. The results were compared with the AOT from surface observations. Using the AOT values retrieved from the GOES-8 data, fluxes of the outgoing short-wave radiation (OSWR) were calculated (applying the four-flux approximation) and then compared (in some cases) with the OSWR measurements carried out within the framework of the CERES program to study cloud cover and the Earth's Radiation Balance (ERB).

The results obtained by Christopher and Zhang (2002) point to the fact that the AOT values retrieved from the GOES-8 data agree well with the data of surface observations at the AERONET network, with a correlation coefficient of 0.97. The correlation coefficient for CERES data and for the numerical modeling of OSWR was 0.94. From the data for August 1998 the daily averaged values of AOT and ARF in the region under study were, respectively, (0.62 ± 0.39) and $(-45.8 \pm 18.8)\,\text{W m}^{-2}$.

Weaver *et al.* (2002) calculated the short- and long-wave RF due to Saharan aerosol using the results of numerical modeling with a 3-D global model of the aerosol transport and the approximate schemes of parameterization of the short- and long-wave radiation transfer. Calculations enabled one to retrieve (with prescribed results of an objective analysis of the fields of meteorological parameters) the 3-D fields of the concentration of the four sizes of mineral aerosol particles, which served as input information to calculate the radiation fluxes and RF. The results show that dust aerosol causes a decrease in the OSWR (this conclusion agrees with the ERBE satellite measurements). Though a similar agreement is also observed in the case of the outgoing longwave radiation, in this case the errors of an objective analysis of the temperature field caused by neglecting the effect of aerosol on the temperature field is a complicating circumstance. Depending on the level of atmospheric dust content, the values of the shortwave RF at the top of the atmosphere vary within $0 \pm -18\,\text{W m}^{-2}$ (over land).

From observations with the space-borne instruments SeaWiFS over the World Ocean, Chou *et al.* (2002) retrieved the values of AOT τ_A which were used to calculate the ARF with prescribed total atmospheric moisture content (retrieved from the Special Sensor Microwave/Imager (SSM/I) data), monthly mean total content of ozone (from TOMS data), and the field of air temperature (from the data of an objective analysis of the meteorological information carried out by the National Center for Environmental Protection–National Center for Atmospheric Reseach (NCEP–NCAR)).

Analysis of the obtained results revealed mainly a zonal distribution of τ_A, whose basic features of the spatial distribution are determined by the wind field convergence in the lower troposphere. Maximum values of τ_A are found in the equatorial band, in high latitudes of the southern hemisphere in the Arabian Sea, the Gulf of Bengal, as well as in the coastal regions of South-East and eastern Asia. Minimum values of τ_A were recorded in the sub-tropical convergence zones, especially in the southern hemisphere. During most of the year, great values of τ_A are characteristic of the tropical Atlantic Ocean. The dominating trade winds transport the dust aerosol from Africa across the Atlantic Ocean to the coastlines of Central and South America. Great (small) values of AOT are typical of the zones of convergence (divergence). The mean annual value of τ_A at the wavelength 0.865 μm constitutes 0.105.

The mean annual values of ARF over the World Ocean at the level of the upper atmospheric boundary layer (UABL) and at the ocean surface are, respectively, -5.48 and $-5.09\,W\,m^{-2}$. The annual course of ARF is negligible, on average, for the whole World Ocean, but it is substantial in the northern and southern hemispheres. The UABL values of ARF for the period from October to April increase in the northern hemisphere by 33% and in the southern hemisphere (July–January) by 50%. The observed clear-sky UABL value of ARF equal to $-5.48\,W\,m^{-2}$ corresponds to a cooling of about $2.5\,W\,m^{-2}$, if the average cloud conditions are taken into account. Thus, the negative ARF is comparable in value with the positive "greenhouse" forcing (about $4.3\,W\,m^{-2}$) which should happen with a doubled concentration of atmospheric CO_2. A strong ARF was observed in the period of intensive fires in Indonesia in 1997. Over large territories the value of ARF in the period September–October 1997 reached $-108\,W\,m^{-2}$ at the top of the atmosphere and exceeded $25\,W\,m^{-2}$ at the surface level. The presence of aerosol enhanced the heating of the atmosphere due to solar radiation absorption by about $15\,W\,m^{-2}$. The forest fires in Indonesia coincided with the period of an abnormally intensive El Niño event in 1997–1998, when Indonesia was affected by drought and downward air fluxes. The atmospheric heating and surface cooling caused by aerosols originating from forest fires could increase the atmospheric stability and, hence, reduce the atmospheric circulation and enhance drought.

Indirect aerosol forcing is determined by complicated processes of the aerosol–cloud interaction, which require special investigations. Therefore, we shall only briefly consider the latest achievements in this field. By now, one can consider as completed the theory of formation of a single cloud droplet by activation and condensation growth of an aerosol particle on a CCN under condition that CCNs have a simple chemical composition (in particular, they do not contain organic compounds) and they are in equilibrium with chemically active MGCs, such as SO_2, NH_3 and HNO_3. The numerical modeling of the process of cloud droplet formation to simulate the size distribution of cloud droplets is still a difficult problem, especially in the case of mesoscale processes and, moreover, when the atmospheric GCMs require an adequate consideration of various properties of the cloud cover. The main cause of complexity of such a problem is the necessity for an adequate reproduction of the dynamics of the field of oversaturation ($S_{v,w}$) around the CCN particles during their condensation growth until a maximun oversaturation is reached ($S_{v,w}^{max}$) occurring over a short time period. If ($S_{v,w}^{max}$) is known, it means the possibility of an accurate calculation of a minimum size of an aerosol particle, which can be attributed to the category of cloud droplets. Such an approach was realized in detailed microphysical schemes used in the numerical modeling in the case of 1-D models of the pollution tails and 2-D γ-mesoscale models as well as models of large-scale vortices. Another possible (and more economic from the viewpoint of computer time expenditure) approach to a parameterization of the cloud droplets formation as applied to the 3-D numerical modeling consists in the use of the concept of the activation spectrum ($N_{CCN} = f(S_{v,w})$, which contains all the necessary information about the ability of the aerosol population to stimulate the formation of cloud droplets.

Cohard *et al.* (2001) proposed an approximate method of parameterization of the relationship between the form of the spectrum of CCN activation and observed properties of aerosol population (size distribution and solubility) functioning as CCN. In view of considerable differences in the properties of maritime and continental aerosols, they are considered separately. The numerical modeling of the aerosol particle growth is realized within the interactive scheme describing the process of activation of cloud droplets. With respect to each of the two considered aerosol populations (maritime and continental), when considering the variability of solubility, the chemical composition of particles and the log-normal size distribution are assumed to be homogeneous. The reliability of the parameterization scheme provides the use of a large set of aerosol populations with randomized properties.

Estimates of the indirect climatic impact of aerosols through changing the cloud cover properties proceed from suppositions that either the cloud water content remains constant (the Tuomi effect of the cloud albedo increase due to the growth of the small droplets number density) or the water content increases with the growth of the droplet number density (the Albrecht effect of precipitation suppression in the zone of drizzle). On the other hand, both the results of numerical modeling of thermodynamics and cloud cover dynamics and the observed data show that when the number density of cloud droplets grows the water content of clouds may decrease.

In this context Han *et al.* (2002) used the data of satellite observations of clouds in order to analyze the dependence of the cloud water content on the droplet number density. With this aim in view, from the data of observations for January, April, July, and October 1987 the cloud sensitivity (CS) was estimated as a ratio of changes of water content to changes of the total number of droplet in clouds. Analysis of the global database for water clouds (the temperature of the cloud upper boundary is above 273°K, optical thickness $1 \leq \tau \leq 15$) revealed the presence of three regimes of changes of water content depending on changes of aerosols manifested as an increase, approximate stability, or decrease, with the total number density of cloud droplets increasing.

A consideration of the data of satellite observations suggested the conclusion that approximately in one-third of cases (mainly in conditions of the warm atmosphere) the CS is negative (i.e., the cloud water content decreases with increasing number of droplets in clouds) (the regional variability and the annual negative CS agree with the data of other observations). To one-third of cases (with the constant water content) corresponds the relationship $r_{eff} \approx N^{-1/3}$ (r_{eff} is an efficient radius of droplets, N is the droplets number density). In other cases, CS is positive. The results obtained confirm the conclusion that in conditions of increasing cloud droplet number density the size of droplets can decrease and their evaporation directly beneath the cloud bottom can increase, and as a result, the processes in clouds and in the boundary layer are independent in a warm environment. Such a situation is accompanied both by a decrease of water input from the surface and by a decrease of water content in clouds. The main conclusion drawn from the results discussed is that estimates of the negative indirect RF due to the aerosol–cloud interaction obtained using the numerical climate models with the effects of Tuomi and Albrecht taken into account can be overestimated.

One of the manifestations of the indirect climatic impact of aerosols consists in the growth of the number density of small cloud droplets caused by anthropogenic aerosol and by the associated increase of the optical thickness and albedo of clouds, which causes a negative RF (the Tuomi effect) and partial compensation of the positive greenhouse RF. Harshvardan *et al.* (2002) estimated this effect with the combined use of numerical modeling results and data from satellite observations. With this aim in view, the case was considered of an intrusion of the sulfate-aerosol polluted air into the North Atlantic basin (mainly at altitudes of 2–3 km) from the European continent on 2–8 April 1987 in the region 50–55°N, 25–30°W. The numerical modeling was made with the use of the global model of transport with chemical processes taken into account and with the prescribed data from the numerical weather forecast. The evolution of microphysical characteristics of the lower maritime clouds (temperature (260–275 K)) for the period 2–8 April 1987 was retrieved from the AVHRR data in the visible (0.58–0.68 μm) and near-IR (3.55–3.93 μm) regions.

Analysis of the results obtained on days with low and high levels of sulfate pollution permitted the estimation of respective changes in the cloud optical thickness at the wavelength 0.64 μm, number density, and efficient radius of particles. These changes testify to the reality of the Tuomi effect. If with a low level of pollution the number density of droplets varied within 65–95 cm^{-3}, at a maximum level of pollution it increased to 300 cm^{-3}, and the efficient radius of droplets decreased simultaneously from 20 μm to less than 12 μm.

One of the possible indicators of the indirect effect of atmospheric aerosols as CCN causing an increase in the number density of small cloud droplets and, respectively, an increase of cloud albedo, resulting in a climate cooling is the satellite remote-sensing data of changes in the efficient radius of cloud droplets (ERCD). The earlier data revealed the 2-μm difference between the ERCD values over land and over the ocean and the 1-μm difference between the mean hemispherical values, which confirms the validity of the Tuomi hypothesis. An increase in the concentration of small droplets results in a rain rate decrease, which was confirmed by satellite observations.

To further assess the aerosol impact on clouds, Breon *et al.* (2002) undertook an analysis of measurements with the polarized radiometer POLDER carried by the ADEOS satellite (launched in August 1996) performed over the whole period of its functioning from 3 October 1996 to 30 June 1997 (observations could not be continued because of a malfunction of the solar batteries). The results obtained demonstrate the considerable effect of aerosol on the microphysical characteristics of clouds on global scales. Maximum values of ERCD (14 μm) were revealed over distant regions of the tropical oceans, with minimum values (6 μm) over heavily polluted continental regions. High concentrations of small cloud droplets are characteristic of clouds located in the direction of the prevailing winds from the continents.

Some circumstances complicate, however, the final interpretation of the results under consideration:

- In view of specific features of the algorithm for retrieval, the obtained ERCD values are characterized by certain selectivity, not covering clouds of all types (they refer mainly to stratified clouds) and complete cycles of their evolution.
- The POLDER data make it possible to retrieve the content of aerosol in the atmosphere, whereas the Tuomi effect is determined by the concentration of CCN in the atmospheric layer where clouds are formed.

These and other circumstances do not permit us to reliably estimate the influence of the Tuomi effect on RF. Though an analysis of the discussed satellite data cannot be used to distinguish between the contributions of natural and anthropogenic aerosols, one should assume that anthropogenic aerosol plays the prevailing role.

An active discussion of the indirect impact of aerosol on climate (through the effect of CCN on the processes of formation and various, especially optical, characteristics of clouds) has not led so far to sufficiently satisfactory results. Estimates of the RF varied widely from 0.0 to $-4.8\,\mathrm{W\,m^{-2}}$, hence, these estimates are very uncertain. Various observed data used to verify the Tuomi effect have led to rather contradictory results. So, for instance, some studies have shown that this effect manifests itself in the presence of stratified clouds but is absent in the presence of cumulus clouds, though according to data in other publications, the Tuomi effect is also observed in the presence of cumulus clouds.

To resolve such contradictions, Feingold *et al.* (2001) carried out an analysis of the satellite remote-sensing data obtained within the SCAR-B program on studies of the interaction of smoke from forest fires, clouds, and radiation in Brazil. Processing the observed data for two years (in the periods of biomass burning in August–September in 1987 and 1995) covering the region 20°S–5°N, 45°–70°W suggested the conclusion about an expediency of distinguishing between three types of cloud response to smoke aerosols:

- An increase of the cloud droplet number density when the content of aerosol increases, until reaching saturation with a high concentration of aerosol.
- The same evolution as in the case above, but ultimately, the concentration of droplets grows with aerosol content still increasing. The latter is connected with the fact that high concentrations of large particles determines a suppression of oversaturation and prevents an activation of small particles, but opens up the possibility of activation of large particles when the content of smoke aerosols exceeds a certain threshold level.
- The same as in the first case with a subsequent decrease of the droplet number density as the content of smoke aerosol increases, when an intensifying "struggle" for water leads to an evaporation of small droplets. This evaporation results in an unexpected increase in the size of droplets when the content of smoke grows.

A considerable uncertainty of the estimates of the indirect (through changing the properties of cloud cover) effect of aerosol on climate necessitates special efforts to specify the estimates of this effect. An increase of the cloud droplet number density (with the constant cloud water content) is accompanied by a decrease of an efficient radius r_{eff}, which favors an increase of cloud reflectivity. As has been mentioned, this variability was called the first indirect forcing. The second indirect forcing is connected with the fact that with decreasing average sizes of droplets accompanying the growth of the aerosol number density, the efficiency of rain formation decreases, which leads to the formation of clouds with a greater water content and longer lifetime. Observations confirmed that the anthropogenically induced decrease of r_{eff} can considerably change the water content of clouds and reduce precipitation, but the effect on the lifetime of clouds was not observed.

An experimental verification of the reality of the two types of the aerosol climatic impact is hindered by the following:

- Local character of observations of the properties of aerosol and clouds, which prevents us from considering them as globally representative.
- Difficulty with filtering out anthropogenic changes of r_{eff} and water content from the natural changes.

In this regard, Menon *et al.* (2002) performed a numerical modeling of the climatic impact of aerosol based on the combined use of the GSFC climate model as well as models of chemical processes in the atmosphere with the participation of sulfur compounds, sources of organic substances, and sea salts determining the three types of aerosols considered in the model. To determine the cloud droplet number density, a diagnostic approach was used with the data of observations over land and ocean taken into account. Information on the cloud droplet number density thus obtained was used to calculate the variability of r_{eff}, which, in turn, permits the pre-calculation of the impact of aerosol on the optical thickness of clouds and the rate of microphysical processes.

Estimates of the indirect climatic impact of aerosol were obtained from the data on the difference of values caused by the variable characteristics of cloud cover, of the RF at the top of the atmosphere corresponding to present and pre-industrial conditions, with estimates obtained for both types of the aerosol climatic impact. Also, in parameterization of the cloud cover dynamics, calculations were made of the sensitivity of the obtained results to the use of suppositions which determine the vertical distribution of the frequency of occurrence of clouds, rates of auto-conversion, and washing out of aerosol. The content of aerosols in the atmosphere depends on variations of all these parameters.

The numerical modeling results show that the mean global RF due to the indirect impact of aerosol (with the three types of aerosol taken into account) varies from -1.55 to $-4.36\,\mathrm{W\,m^{-2}}$. These results depend markedly on the content of the pre-industrial background aerosol in the atmosphere (when this content is low, the climatic impact of aerosol is stronger) and depends less on the anthropogenic aerosol, whose increase is followed by an enhancement of the indirect impact of aerosol. A substantial dependence of the content of background aerosol determines an insufficient variability of the factors such as the relationship between the size of cloud particles and cloud albedo, as predictors of the aerosol climatic impact, which can be quantitatively estimated from satellite observations. Such a situation reflects the difficulty of a verification of the estimates of the aerosol climatic impact from the observation data.

Analysis of the results of observations of ship smoke emissions into the atmosphere on the lower clouds reveals the possibilities of studying the effect of aerosol on the cloud cover characteristics in real conditions. To estimate the degree of the impact of smoke aerosol on cloud properties, Coakley and Walsh (2002) used the satellite AVHRR data on cloud brightness (reflectivity) bearing in mind that when the lower clouds are above the ship, the clouds catch smoke particles which serve as CCN and favor the formation of new cloud droplets. The growing droplet number density leads to a decrease in their size, which manifests itself, in particular, as an enhancement of the reflected radiation at the wavelength 3.7 μm. Therefore, the AVHRR images of cloud cover at this wavelength can be used to reveal the polluted parts of the cloud cover.

From brightness measurements at the wavelengths of 0.64, 3.7, and 11 μm in July 1999 in the region of the western coastline of the USA, the values of the optical thickness of clouds in the visible were retrieved as well as the efficient radius of droplets and brightness temperature for polluted and clean parts of the cloud cover. Analysis of several hundred cloud cover fragments in the sectors of smoke tails some 30 km long has shown that the resulting changes in the optical thickness turn out to be half that expected, provided the observed changes of droplet radii and assumed stability of the water content of clouds be taken into account. Theoretical estimates of radiation fluxes have led to the conclusion that such a result cannot solely be explained by the effect of radiation absorption by smoke particles. Apparently, the water content of polluted clouds decreases by 15–20% with respect to the original water content. An enhanced absorption of the short-wave radiation by clouds leads to evaporation of a considerable amount of cloud water, which reduces the convection in the atmospheric boundary layer. The latter, in turn, leads to a decrease in the water vapor flux toward the polluted clouds compared with the flux of water vapor to a non-polluted cloud. The processes of absorption, evaporation, as well as contrast between the fluxes of water vapor to polluted and non-polluted parts of the cloud cover result in the loss of water by polluted parts of clouds.

The conclusion drawn from the numerical modeling (with the observed data taken into account), that the mean global climatic ARF is close to the greenhouse RF, emphasizes the necessity to study the role of atmospheric aerosol not only as the climate-forming factor but also as an anthropogenic component, which affects the chemical processes in the atmosphere and, hence, its gas composition. For instance, the substantial effect of the polar stratospheric cloud particles and sulfate aerosol (through heterogeneous chemical reactions on their surfaces) on the formation of the field of ozone concentration in the stratosphere has been identified.

Since the analysis of the effect of aerosol on the chemical reactions in the troposphere is also important, Tie *et al.* (2001) undertook a numerical modeling of the impact of sulfate and carbon ("black") aerosol on tropospheric chemistry. With this aim in view, a modified version of the NCAR global interactive model MOZART of the chemical processes in the atmosphere and atmospheric circulation (transport of MGCs and aerosol) was used. Modification of the earlier meant that apart from gaseous chemical processes, it simulated the processes of formation of sulfate and carbon aerosols as well as heterogeneous reactions on the surface of particles.

A comparison of the results of calculations of the global field of sulfate aerosol concentration and its annual change in the observed data revealed a satisfactory agreement everywhere except the Arctic region. In particular, the vertical profiles of sulfate aerosol concentration in North America agree well with the observed ones (this refers also to the results of calculations of the carbon aerosol distribution near the Earth's surface). The numerical modeling of the processes of formation of sulfate aerosol and of heterogeneous reactions on the surface of its particles has led to the following conclusions:

- The reaction $SO_2 + H_2O_2$ giving sulfate aerosol in cloud droplets reduces the concentration of hydrogen peroxide. In the gas phase reaction $SO_2 + OH$, OH is transformed into HO_2, but the resulting decrease of OH concentration and increase of HO_2 concentration are negligible (<3%).
- The heterogeneous reaction on the surface of sulfate particles with the participation of HO_2 causes a 10% decrease in the concentration of radicals of hydrogen peroxide HO_2 with the coefficient of capture being 0.2. However, this value can be overestimated, which determines an expedience of considering this estimate as an upper limit.
- The reaction on the surface of sulfate particles with the participation of N_2O_5 leads to a decrease of concentrations of nitrogen oxides in middle and high latitudes in winter, reaching 80%. Since the intensity of the ozone formation in winter is low, a decrease of the ozone concentration caused by this reaction does not exceed 10%. However, in summer, a decrease of NO_x concentrations can reach 15%, and that of ozone (in middle and high latitudes) 8–9%.
- The heterogeneous chemical reaction on the surface of sulfate particles with the participation of CH_2O (with the upper limit of the coefficient of capture 0.01) leads to a decrease of CH_2O concentration of 80–90% and a decrease of HO_2 concentration in middle and high latitudes in winter of 8–9%. The obtained estimates should be considered as preliminary, which is explained, first of all, by considerably uncertain input parameters.

Developments aimed at numerical modeling of the climate with the participation of an interactive impact of aerosol represented a new stage in the studies of the indirect climatic impact of aerosol and ARF estimates. This approach is exemplified by the work of Jones *et al.* (2001) in which a numerical modeling was undertaken of the impact of the anthropogenic sulfate aerosol on cloud albedo and an intensity of rain formation (e.g., considering both the first and the second indirect impacts) based on the use of a new version of the global climate model developed at the Hadley Centre (Jenkins *et al.*, 2005). This model uses a new scheme of parameterization of microphysical processes in clouds with an interactive consideration of the sulfur cycle and a parameterization of the effect of sea salt aerosol.

Calculations gave the mean global indirect ARF value of $-1.9\,\mathrm{W\,m^{-2}}$, with the prevailing contribution from the albedo effect. Estimates of contributions of the first and second indirect ARF gave values of, respectively, -1.3 and $-0.5\,\mathrm{W\,m^{-2}}$. The obtained values of the total indirect ARF can be considered as reliable with an accuracy to the coefficient 2. Of course, the global geographic distribution of the indirect ARF is characterized by a high uncertainty in the presence of ARF maxima in the regions of main sources of emissions of sulfur compounds located in the USA, Europe, and China. A comparison of estimates of the ARF due to short-wave and long-wave radiation made by Jones *et al.* (2001) has shown that the contribution of the short-wave ARF (which causes a climate warming) always prevails over that of the long-wave ARF causing a warming. An interactive consideration of the sulfate aerosol climatic impact plays a substantial role.

New ARF estimates with the use of the global climate model developed in the Climate Research Center of the University of Tokyo, with the long-range aerosol transport taken into account but with a priori prescribed sources and sinks of aerosol, were obtained by Takemura (2002). The following types of aerosol were considered: sulfate, sea salt, carbon ("black carbon"), organic carbon, and soil dust. The mean annual mean global values of direct ARF constituted: $0.32\,\mathrm{W\,m^{-2}}$ (anthropogenic sulfate aerosol); $+0.19$ to $-0.05\,\mathrm{W\,m^{-2}}$ (black and organic carbon due to fossil fuel burning); and $+0.15$ and $-0.16\,\mathrm{W\,m^{-2}}$ (black and organic carbon due to biomass burning). Estimation of the total indirect ARF gave $-1.0\,\mathrm{W\,m^{-2}}$.

A consideration of the effect of aerosols on remote sensing results is one of the unsolved problems of satellite monitoring. Therefore, along with the development of the methods of estimation of the aerosol parameters it is necessary to develop algorithms of atmospheric correction for remote sensor data, especially in the coastal regions, where anthropogenic and maritime aerosols get mixed-up (Geogdzhayev *et al.*, 2005). During the last few years (beginning from 1993), as a result of a network of AERONET radiometers, an extensive database has been accumulated on maritime aerosols in the zones of islands (*http://www.aeronet.gsfc.nasa.gov*).

The problem of assessment of the role of atmospheric aerosol as a climate-forming factor is not a new one. An important new stage in understanding and quantitative estimation of this role has become an awareness of its critical importance in the formation of present climate changes. This conclusion is contained, in particular, in the official fundamental documents, such as IPCC (2001). The paradox consists, however, in that with an emphasis on the problem of global warming and the Kyoto Protocol (see Chapter 2) at the World Summit on Sustainable Development (Johannesburg, 29 August–6 September 2002), the problem of the climate-forming contribution of atmospheric aerosol was practically ignored. Meanwhile, the remaining uncertainty of the estimates of the contribution of aerosol to the formation of the climate deprives such an important document as the Kyoto Protocol of scientific grounds.

The resulting conclusion is apparent: in the very near future it is necessary to place an emphasis on the "aerosol and climate" problem bearing in mind both an intensification of the respective theoretical developments and, more importantly, obtaining more adequate information which would permit us to reliably estimate the climatic impact of aerosol.

6.6.3 Global warming, energy, and geopolitics

The multi-dimensionality of the dynamics of civilization development is reflected in a set of problems, which have either been solved or are planned to be studied within various international programs. In 1987 the UN Conference on Environment and Development adopted a resolution to organize an International Panel on Climate Change (IPCC). According to this resolution, certain concepts of the respective studies have been formulated and some preliminary results have been obtained in the following main directions (IPCC, 2001):

- Systematization of the present ideas about possible anthropogenic climate changes.
- Consideration of the consequences of the impact of these changes on the environment and society.
- Discussion of the needed measures to prevent an undesirable development of such impacts.

Publication of the results of the IPCC activity makes it possible to judge the trends of the activity of this organization (Demirchian *et al.*, 2002). Unfortunately, the IPCC emphasizes the idea of the catastrophic character of human impact on global climate, neglecting the opinions of many scientists who show, on a scientific basis, that the role of the anthropogenic factor in the observed changes of climate is more difficult to understand than it is shown in the IPCC reports.

As has been mentioned, the problem of climate change involves several issues, not only scientific but also political. Long-term observations of climate enabled one to select the most important problems, which require a thorough analysis. Nevertheless, other problems should not be ignored, which at the present stage of civilization development secondary. In other words, putting forward some hypotheses on the prospects of the present civilization development should be based on objective laws of nature and data from monitoring, as well as economic and political rules. But the trends of the IPCC activity contradict this concept. According to the IPCC, the rate of growth of atmospheric temperature over a decade constitutes 0.167°C, whereas according to the data of satellite monitoring, this rate ranges between 0.062 and 0.064°C (Demirchian *et al.*, 2002). This contradiction entails various conclusions about the predicted rise of temperature when the atmospheric CO_2 concentration doubles: according to IPCC, by 3°C, and from satellite data, by 0.53°C. A similar contradiction between the IPCC data and results of independent scientific studies is observed with respect to the estimates of temperature changes in the past, when the temperature trends similar to those nowadays were observed. All this reduces the reliability of the results published by IPCC (relevant results are still being actively discussed).

Climate predictions are of political importance. Therefore, a scientifically grounded estimation of the consequences of anthropogenic activity can principally change public opinion and prevent undesirable social consequences. Many scientists in the USA, Europe, Japan, and Russia, where fundamental studies are being carried out on the development of global climate models capable of reliably predicting the NSS dynamics, understand this.

Extremely important is the role of the mass media, which should correctly elucidate the ecological problems, especially global warming. Important decisions have been made at the UN level concerning the prevention of dangerous consequences of global warming for the population. Their realization can markedly hinder the economic development of most countries and further reduce the insufficient levels of their life-supporting abilities. The governments interested in making justified decisions in the field of ecological problems should invest considerable material and financial resources into scientific investigations. The global character of the ecological and climate problems requires studies of the whole surface of the land and oceans, both in space and in time as well as for deep oceans. To finance these studies, considerable investments from national revenues should be made, which is only possible by mutual consent from public institutions driven by a solid political stance. It is impossible to force the ideas of individual groups of the population of the country (or the world), which cause a feeling of danger and raise emotions with respect to such a complicated problem as the problem of future climate.

Since the problems of global warming and economic development are interconnected, one of the most important principles of scientific work in this field is its independence from political situations. On the other hand, the political situation determines the possibility of financing very expensive studies on the climate. At this stage, a situation occurs when political and personal strategies contradict one another. Therefore, for instance, the problem has been brought up (Demirchian *et al.*, 2002) to obtain reliable estimates of the role of the anthropogenic factor in climate changes. The key goal in the solution of this problem is to obtain reliable estimates of the atmospheric temperature increase (ΔT_{2x}) with a doubled concentration of CO_2. It was shown that this problem, as applied to an improvement of climate models, creates a deadlock situation caused by the impossibility of verifying modeling results. One of the ways out of this situation is the adaptive–evolutionary technology of geoinformation monitoring (Kondratyev *et al.*, 2000).

The specified value of ΔT_{2x} in the first overview within 1.6°C–1.7°C enables one to reduce the estimates of the increase in the ocean level and to stop requesting immediate measures to reduce GHG emissions (Houghton *et al.*, 1992). The observed temperature rise by the year 2000 compared with 1850 by about 0.6°C at $\Delta T_{2x} = 1.66 \pm 0.11$°C constitutes 35–36% of ΔT_{2x}. By 2100 it will reach 67–70% (1.0–1.3)°C of ΔT_{2x}. Thus, there are no grounds for the suggested dangerous increase of temperature in the future due to so-called "delayed" heating (Hansen *et al.*, 2000). These data refer to the case of reaching the value of GHG equivalent concentration 502–519 ppm with CO_2 concentration 448–452 ppm by 2100.

Another important direction with numerous discussions and uncertainties is the forecast of possible values of anthropogenic carbon flux. There is a sufficient amount of objective data, consideration of which makes it possible to narrow the range of possible levels of anthropogenic emissions of carbon dioxide and its concentration in the atmosphere. Demirchian *et al.* (2002) established a simple connection between the size of population G and the anthropogenic share of CO_2 in the atmosphere $K_a = 15G(t)$. Such a parameterization substantially simplifies the prediction, since only one assumption on the size of a population is needed. Over the period 1750–2000, this parameterization was characterized by a correlation coefficient 0.994.

Thus, the economic aspect of the problem of global warming is reduced to a search for the strategy of reasonable use of natural resources with constantly improving technologies of energy production and its economic use. The problem of global warming, being a political instrument for various persons and groups from the year 1990, acquires geopolitical importance. An exclusive importance of energy and raw material resources for developed countries determines a necessity for a guaranteed extraction and transportation of these resources. Therefore, at the beginning of the 21st century one should expect substantial changes of economic and political levels of interaction with the global population. The world community is prompted to do this due to the growing intensity of the commodity and financial exchange between countries. Besides this, the level of economic relations has become such that the functioning of an individual country outside these relations becomes impossible. In this context the economic and political relations have become global in scale. For this reason, the solution of the problem of regulating the rules of interaction of countries in this globalized world becomes vitally important. The main source of complicating the relationship is the growing size of the global population, bringing forth a multitude of economic and political problems.

6.7 EVOLUTIONARY TECHNOLOGY FOR PREDICTION OF STRESSFUL CONDITIONS IN THE ENVIRONMENT

The ideas of ecoinformatics stimulate the broadening of the role of computers and numerical modeling in non-traditional spheres of their application, such as ecology, biophysics, and medicine. Model experiments have become traditional, and the term "numerical experiment" is widely used now in many studies, up to experiments with global ecological models (Krapivin, 1993). However, in all these studies the availability of a more or less adequate model is supposed a priori. For concrete operations with the model as a certain series of numerical experiments, a universal computer is needed. As a rule, insurmountable difficulties appear here connected with limited memory and speed, and as a rule, scientists constantly face an insoluble contradiction between the desire to raise the model's accuracy and the computer's capability. Clearly, the development of a model adequate to a real object is eventually impossible. On the one hand, a complete consideration of all parameters of the object leads the scientist to the problem of insurmountable multi-dimensionality. On the other hand, simple models taking into account a small amount of parameters are inadequate to complicated objects and processes of interest to the scientist, especially natural disasters. Besides this, in the spheres such as the physics of the ocean, geophysics, ecology, medicine, sociology, etc., development of an adequate model is principally impossible due to the practical impossibility of obtaining complete information. Also, it is often impossible to define the type of the model required and, hence, the use of developed modeling technologies is unpromising. In particular, attempts of a complex approach to the synthesis of global models have demonstrated that traditional methods of formalization of processes taking place in the environment and society cannot give reliable estimates and forecasts of global change. Nevertheless, a search of new efficient ways to synthesize the global system of environmental control, able to reliably assess the consequences of anthropogenic activity, has led scientists to an idea of self-learning computing systems of the evolutional type (Bukatova *et al.*, 1991). As a result, a new theory of modeling has appeared called the evolutionary informatics.

We preserve the term "model", though its interpretation here has a somewhat different meaning. We will deal with the description of objects changing in time in an unforeseen way and, hence, making removal of information uncertainty at any time moment impossible. Hence, the model in a broad sense should provide a continuous adaptation to the changing behavior and structure of the observed object. So, let a real object A have an unknown algorithm of functioning with only some pre-history $\{Y\}^n$ known, of a final length n. The functioning of A should be simulated using the models created in the system which is functioning in a real time mode. It is clear that a succession of more and more adequate models $\{A_k\}$ should be formed whose reliability is tested by pre-history. Studies by many scientists have resulted in a new approach to the development of self-learning systems based on modeling the mechanisms of evolution. The evolutionary modeling, on the whole, can be presented by a hierarchical 2-level procedure. At the first level, there are two rotational processes called conditionally the process of structural adaptation and the process of usage. At the kth stage of adaptation in the process of functioning of the structural adaptation algorithm, a succession of models $\{A_{s,i}\}, i = 1, \ldots, M_s$, is synthesized, of which the memory is formed of K most effective models $\langle A_s^1, \ldots, A_s^k \rangle$. At the sth stage of usage (following the kth stage of adaptation) the system uses the models stored in memory, going on to choose from them the most efficient ones.

The principal scheme of the ith step of the kth type of the model adaptation is shown in Figure 6.21. Here the unit "object" means that the real object is presented by pre-history $\{Y\}^n$. All other units perform a structural adaptive synthesis of the models (structural adaptation). The function of the ith step begins from the choice of a "mutant" A_{ki}^j, which is either A_{ki}^j or the one the model already used at the preceding $(i-1)$th step of the kth stage of adaptation. In the unit "structure synthesis" the structure $|A_{ki}^j|$ is synthesized from the structure $|A_{ki}|$ using the mode of random changes. The model A_{ki} enters the unit "evolutional choice" where, in correspondence with the criterion F, it is either memorized or forgotten.

The procedure of the evolutional choice of models provides practically unlimited-in-time functioning of the system under conditions of unavoidable information uncertainty. Here, apart from pre-history $\{Y\}^n$, as a rule, not meeting the requirements of statistical homogeneity, scientists have no other information. Clearly, under these conditions, all available information has to be used at most and, in particular, information about the functioning of the stages of adaptation and

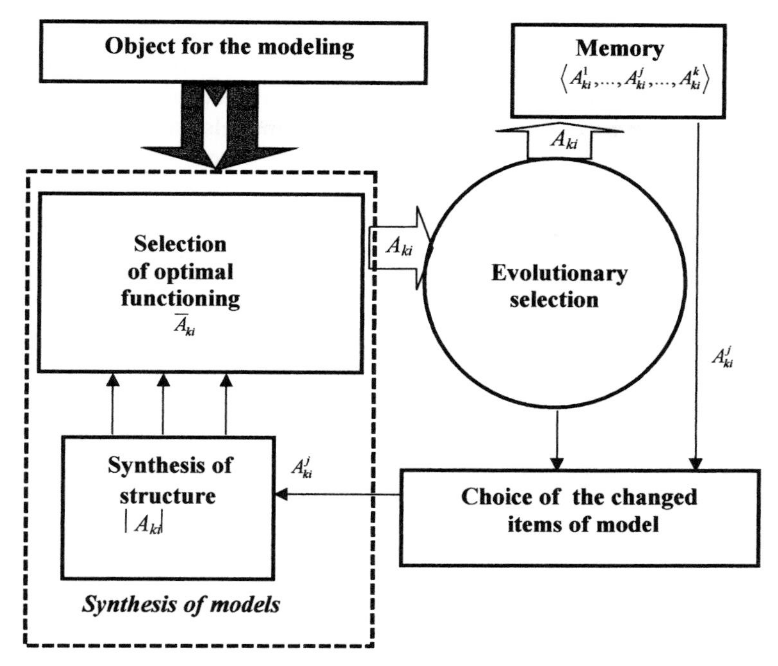

Figure 6.21. The conceptual scheme of the *i*th step of the *k*th type of model adaptation.

usage. In the parametric adaptation algorithm it is used for adaptation of the first-level parameters: characteristic number of each mode of changes, multiplicity of the use of mode of changes, volume of the list of mode of changes, probabilistic distribution of the mode of changes of the list, probabilistic distribution of K-models A^j_k, memory volume at the stage of adaptation, length of pre-history, etc. (Bukatova *et al.*, 1991).

One of the possible ways to raise the efficiency of the evolutionary software tools is the development of the architecture of the specially designed processor implementation of the evolutional systems of data processing including an enlarged functional basis, large-block (multi-processor) construction with homogeneous elemental base, different-type parallelization of computing processes at the level of the system, processors, and elements, alternation of functioning and structure (composition and connections) of the synthesized models, and hierarchical distribution over all levels of control.

Analysis of algorithmic aspects of the evolutional specially designed processors makes it possible to formulate the basic requirements and tasks which should be provided and solved when developing the evolutional hardware. Problems of algorithmic nature appear here inherent to system, processor (unit), and elemental levels of implementation: maximum parallelization, enlargement of base operations, provision of maximum tunability, formation of distributed memory and control, selection of models, synchronization and control, information input and output, etc.

Maximum parallelization can be provided at the expense of parallel hardware organization of computing processes, including an exchange of information between individual processes, execution of base operations by microprocessors (operations on structural changes and assessment of synthesized models), etc. For instance, for hardware implementations to the systems of perceptronic type, parallelization raises the speed of the system by 2–3 orders of magnitude, and parallelism of operations in the base elements adds 102 orders more. So that there are certain reserves here to accelerate the speed, bearing in mind that parallelization of the synthesized model functioning in each microprocessor will further raise this gain in time.

The task of enlarged base operations requires development of microprocessors oriented toward evolutional structural synthesis in a given class of models at a fixed random search procedure. Hardware implementation of such base microprocessors entails difficulties in the development of self-test hardware which can be overcome with the creation of optical super-large integral schemes, which will broaden the functional properties of the elemental base (Bukatova *et al.*, 1991).

Maximum tunability provides an adaptability to changes of input signal characteristics, dynamics of processing conditions (including a change of multiprocessing), input a priori information, and to other factors of the evolutional computation process. In a full volume, such tunability is provided by the current correction of the parameters, transition to other computation procedures and other computing processes, as well as alternation of the composition and structure of connections between the base elements. All this increases the probability of achieving the required state of a specially designed processor, and ensures its reliability.

Parallelization of evolutionary systems of processing at different levels requires the respective distribution by levels of functions of accumulation and storage of information used further to control the respective processes, procedures, and base operations. Thereby, a substantial drawback of the successive (von Neumann) architecture of traditional computers has been overcome, in which both processing and storage of information are separate, leadings to loss of time on address data transfer.

In a general case, the multi-level structure of the distributed memory is hierarchical, when the accumulated information, for instance, about the element functioning, is used in assessing the microprocessor functioning, and the information about the processes is used in assessing the functioning of the specially designed processor.

The process of control in traditional computers is carried out in a centralized, successive address mode, which is rather time consuming, especially in multiprocessor systems. The level distributed control makes it possible to substantially raise the system's speed due to parallelization of the processes of correction (learning). The correlation of distributed memory and control, whose hardware implementation follows the same principles, becomes clear. In particular, the distributed memory enables one to carry out an individual realization of integral corrections at any level of the system. Synchronization and control of base operations and other units of the specially designed processor are an algorithmic and schematic–technical problem. Within the multi-processor implementation of the

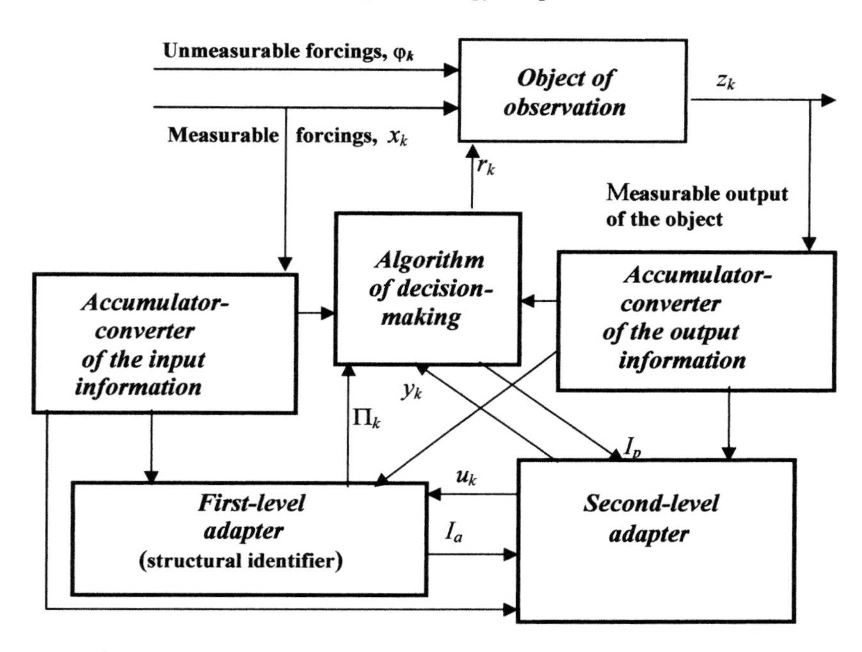

Figure 6.22. Scheme of control with identification of an object on the basis of parametric adaptation.

evolutionary computation process, the asynchronous interaction of parallel fragments of an algorithm is expedient, and, in general, asynchronous schematic technology is needed. One of the perspective ways of hardware implementation of information input–output is the use of optical integral schemes.

Thus, the possibility of the hardware implementation of evolutional modeling technology opens up the possibilities in the field of geoinformation-monitoring systems. The elemental base of micro-electronics makes it possible to raise the level of automation of scientific studies of the environment by including into the monitoring system structure a specialized means of the type of predicting and recognizing processors (Figure 2.3).

In many cases the goal of the data-processing system is to determine or specify the model of the process under study with its simultaneous control. The evolutionary technology provides an achievement of this goal under conditions when other approaches do not work. Equipment of the environmental monitoring system with specialized opto-electronic processors will make it possible to make their functions universal and to broaden their abilities. For instance, when using the method of control with the structural identifier, whose principal scheme is shown in Figure 6.22, the monitoring system can make decisions in the absence of information about the functional connections in the process under study.

The use of evolutionary technology for the prediction of natural disasters and estimations of possible economic losses requires knowledge of only quantitative estimates without an indication of the functional dependences between causes and effects. For instance, the insurance payments in the USA due to natural disasters, in the period 1991–2001 constituted, respectively (US$ bln): 4.7, 23.0, 5.6, 17.0, 8.3, 7.4, 2.6, 10.1, 8.3, 4.3, 24.0. It is difficult to trace any regularity in this succession. The use of evolutional technology makes it possible to continue this succession: 2002 – 37.4, 2003 – 18.7, 2004 – 28.9, and 2005 – 17.2.

6.8 GLOBAL WARMING AND NATURAL DISASTERS

In connection with discussions on the causes of possible climate change, of key importance has recently become an evaluation of uncertainties in the estimates that serve as the basis for conclusions on climate change and the measures to prevent it. Especially important is the problem of estimation of the levels of GHG emissions to the atmosphere connected, primarily, with the solution of the problem of the global carbon cycle. Clearly, without reliable verification of available estimates of emissions, all discussions on the ecological benefit of various measures and related expenditures turn out to be abstract (Nilsson *et al.*, 2002). For instance, how can the fines for neglecting recommendations on GHG emission reductions be substantiated if it is impossible to prove that emissions in 2012 will differ from those in 1990? So far, in discussions on the Kyoto Protocol the problem of the quantitative estimates of uncertainties of the levels of GHG sinks have been ignored (especially concerning the biosphere). However, uncertainties in estimates of total CO_2 fluxes are rather substantial (in Russia they exceed 100%). Calculations of the errors in evaluation of total GHG fluxes gave values ranging within $\sim \pm 5$–25%, whereas the levels of GHG emission reductions foreseen in the Kyoto Protocol average about 5%. The global mean situation is illustrated, for instance, by the fact that uncertainties in the estimates of GHG emissions due to energy production systems nearly equal the uncertainties of the estimates of CO_2 assimilation by the biosphere and land.

In this situation, solution of the problems of uncertainty estimates (first of all, the matter concerns the reliable information of the carbon cycle) and verification is of particular importance. Solution of the verification problem requires an agreement on its mechanisms, which is also important from the financial point of view. The simulation modeling results indicate, for instance, that if the 5.2% confidence level of GHG emission reductions is raised from 50% to 95%, this will entail a 3–4-fold increase of expenditures on respective measures. The main conclusion is that science should be a compass for recommendations of measures in the field of ecological politics. In this respect, the COP-6 failure can become a healing shock (Nilsson *et al.*, 2002).

A serious uncertainty in the estimates of a possible anthropogenic climate change and its determining factors has brought forth a hot discussion on this problem both in the scientific literature and in the mass media. For instance, in his interview to a journalist from *The Scientific American*, Professor R. Lindzen (The Massachusetts Polytechnical Institute, USA) informed that his concern in connection with speculative opinions about climate change originated in 1988, when J. Hansen (Director of the Goddard Institute for Space Studies, New York) declared in public that global climate warming has resulted from the growing concentration of CO_2 emitted to the atmosphere as a result of fossil fuel burning. This statement prompted R. Lindzen to explain that climate science is at an early stage of development and, in particular, there is no consensus whatsoever with respect to the causes of climate change. In early 2001 he made a report at the meeting of the US Cabinet on the climate change problem. The fact that the global mean SAT has increased by about 0.5°C over a period of 100 years and the CO_2 concentration in the atmosphere has grown by about 30%, by no means reflects any cause-and-effect connection between CO_2 concentration increase and temperature rise. Lindzen (2003) believes that the climatic sensitivity (SAT increase with a doubling of CO_2 concentration) estimated at 0.4°C is most reliable, this means that there are no reasons for anxiety about catastrophic changes in global climate in the future.

According to the IPCC mandate, the task set before the IPCC was to prepare an overview of "any climate change in time, both natural and anthropogenic". Pielke (2001a,b, 2002) noted, however, that at least two climate-forming factors have turned out to be either considered unreliably or not considered at all: (1) impact of anthropogenic changes of land surface characteristics on global climate; and (2) biological impacts of the atmospheric CO_2 concentration growth (including the "effect of fertilization") (Tianhong *et al.*, 2003). If both these factors are substantial, then the conclusion suggests that an agreement of the results of global climate numerical modeling is accidental (in fact, the matter concerns mainly the global mean annual mean SAT).

In this regard, Pielke (2002) discussed information that confirms the importance of both climate-forming factors mentioned above and expressed his opinion about the possibilities to test grounds for such a conclusion. For this purpose, data can be used on the impact of anthropogenic changes of land surface characteristics on local, regional, and global climate, illustrating the fact that this impact should be taken into account along with the impact of a doubled CO_2 concentration in the atmosphere (as well as an increase of other GHGs concentrations). The fact is also important that the atmosphere–surface interaction is characterized by various non-linear feedbacks, and therefore the forecast of climate change for the terms longer than a season can turn out to be impossible.

As for possible biological impacts of CO_2 concentration growth, they manifest themselves in the form of a short-term (biophysical), a middle-term (biogeochemical), and a long-term (biogeographical) impact of landscape-forming processes on weather and climate. The biophysical impact includes, for instance, the effect of transpiration on the ratio of latent and sensible heat fluxes as components of the surface heat balance. The biogeochemical impact includes the effect of plant growth (the "effect of fertilization") on the leaf surface area, from which evaporation takes place, on surface albedo, and carbon supply. One of the manifestations of biogeographical impacts is a change of the species composition of vegetation communities with time. Results of numerical modeling indicate that without consideration of biophysical/biogeochemical impacts, estimates of climate change cannot be reliable (Bounoua *et al.*, 2002).

Further development of climate models should contain, in particular, a consideration of the following aspects of climate formation: (1) direct and indirect impacts of landscape dynamics through biophysical, biogeochemical, and biogeographical processes; (2) consideration of anthropogenic changes of land use on different (local, regional, and global) spatial–temporal scales, and (3) estimation of the possibilities of forecasting climate for a term longer than a season, bearing in mind the functioning of numerous non-linear feedbacks that determine the interaction of the atmosphere and surface. These and other unsolved problems confine the importance of IPCC (2001) Report and the US National Report to only estimates of the global climate sensitivity to changes of some climate-forming factors.

During a long period of preparation of three published (1990, 1996, 2001) reports by the IPCC, the completeness of analyses of the climate-forming factors (e.g., apart from GHGs, atmospheric aerosol was considered) and various feedbacks was growing. Despite the fact that numerical climate models have reached a very high level of complexity, they still cannot be considered sufficiently complete from the viewpoint of taking into account all important climate-forming factors. A new step forward in the IPCC (2001) Report was a consideration of the forcing (F) of anthropogenic changes of land use on climate, limited however by the consideration of only the impact of land use dynamics, starting from 1750, on surface albedo. In this connection, the obtained estimates have led to F averaging $-0.2\,\mathrm{W\,m^{-2}}$, with the uncertainty interval 0 to $-0.4\,\mathrm{W\,m^{-2}}$. Thus, such estimates are highly uncertain, and inaccuracies refer even to the sign of F (a more complete consideration of biophysical, biogeochemical, and biogeographical impacts of land use evolution on climate has led to the conclusion that in this case $F > 0$).

In this regard, of importance is an adoption in the IPCC (2001) Report of a new range of possible global warming by the year 2100 (1.4–5.8°C) whose substantiation is based only on numerical modeling data (and therefore will unavoidably change in the future). Also, the problem is that the new range cannot be directly compared with the earlier obtained similar estimates. From the viewpoint of reliability of the future climate change estimates, of importance is the use in the IPCC (2001) Report of the term "projections" instead of "predictions", since the latter means that the factors neglected will not be substantial in the future. Unacceptability of the latter supposition explains why none of the specialists in the field of numerical modeling would state that it is possible to project climate 100 years in advance. This conclusion confirms the results of numerical modeling carried out by Andronova and Schlesinger (2001).

The complexity of the problem of prognostic estimates of climate, especially a selection of the contribution of the anthropogenic component, is illustrated by the preserved contradiction of analysis of the climatic impacts of clouds. Tsushima and Manabe (2001) analyzed the impact of cloud feedback (CF) on the formation of the annual change of global mean SAT with the use of the data from satellite measurements of the Earth's radiation budget, bearing in mind an estimation of the adequacy of the CF consideration in numerical climate models by comparing the calculated and observed change in the global mean SAT. It follows from observational data that the global mean change of SAT is in phase with the change in the northern hemisphere, and its amplitude constitutes 3.3 K (this phase agreement is determined by concentration in the northern hemisphere of continents contributing most to the formation of the amplitude of the annual change of SAT).

Analysis of the data of ERB components for the period February 1985–February 1990 has shown that global mean values of both short-wave and long-wave radiative forcing (SWRF and LWRF) depend weakly on the annual change of global mean SAT (the ERB data refer only to the range 60°N–60°S). Thus, the cloud cover dynamics neither enhance nor attenuate the annual change of SAT. The considered data of SWRF and optical properties of clouds show that not only albedo, but also amounts and the heights of clouds, depend weakly on SAT and, hence, the CF does not affect substantially the annual change in the global mean SAT.

Based on this conclusion, one could deduce that the CF impact on the annual change of global mean SAT or global warming is negligible. However, such a speculative deduction is dangerous in view of a high complexity of the SAT spatial field. Calculations with the use of three climate models which take into interactive account the dynamics of microphysical properties of clouds have led to the conclusion about a slight increase of the cloud top albedo with increasing SAT, which does not agree with observational data. This situation reflects good prospects for comparisons of the estimates of the role of cloud feedback from the data of numerical modeling with observations from the viewpoint of model validation.

6.9 POTENTIAL NATURAL DISASTERS IN THE FUTURE

The dynamics of NSS development indicate that natural disasters in the future can be only more diverse due to new types of catastrophes appearing connected with the broadening of the spheres of application of new technologies. The number of anthropogenic disasters will grow despite an increased efficiency of natural monitoring systems. The problem of interaction between natural and anthropogenic components in NSS dynamics will be of multi-dimensional character in the space of characteristics of the impact of various types of disasters on society. Geological, climatic, ecological, and medical sections of this space will look quite different than they are perceived by the population of the planet under the present conditions of the early 21st century. This will be connected with the change of rhythmic character of catastrophes and transformation of the notions of real and subjective risk.

As Wright and Erickson (2003) noted, natural disasters in the future will be evaluated from the economic point of view. Disasters expected in the future include:

- Impetuous greenhouse effect due to methane positive feedbacks.
- Rapid sea level rise due to polar ice melting.
- Global change of the World Ocean circulation.

These geophysical catastrophes can occur for a lot of reasons, both natural and anthropogenic in origin. For instance, accumulated supplies of methane in deposits, especially in the northern latitudes, in the case of climate warming can become destabilized and cause a sharp increase in its concentration in the atmosphere. So, clathrate compounds buried in permafrost down to 200 m are in an isolated state until the moment of melting. Released methane can enhance the greenhouse effect by 25%. However, as is the case of CO_2, there is no unique solution here, since released methane involved in the global biogeochemical cycle together with other gases, can be transformed and partially absorbed by the ocean. Similar uncertainty appears in assessing the consequences of methane release from deposits in the form of methane hydrates and other gases.

The problem of the consequences of the Arctic ice melting, whose supplies are estimated at 3.8 million cubic kilometers, is reduced to estimation of the World Ocean level rise and enumeration of flooded lands. Apparently, this is possible with a doubling of CO_2 content in the atmosphere. The most pessimistic scenarios show that a World Ocean level rise of 4–7 m due to the Antarctic and Greenland ice melting is possible in the nearest 200 years, with a 5% probability. Even the existing climate models with their imperfection reveal numerous unaccounted factors which can appear with realization of such scenarios and which can include new and enhanced known feedbacks, and stabilize the climate.

The thermohaline circulation of the World Ocean is a key mechanism of the global climate system. Its horizontal constituents are responsible for almost half of the heat exchange between equatorial and polar zones, and the vertical components govern the fluxes of heat, nutritional elements, and gases, including CO_2 between the atmosphere and deep layers of the ocean. Realization of the global warming scenario will apparently lead to a change in the vertical profiles of water temperature and salinity and hence, will change the structure of the ecological system. Wright and Erickson (2003) note that in the discussion of possible changes in the structure of the ocean a large number of questions arise, answers to which with present levels of knowledge are unknown. The role of the time lag of the ice cover dynamics in the northern hemisphere in the formation of mechanisms that change the World Ocean circulation is unclear.

The estimate of the respective variability of atmospheric circulation also remains uncertain. This also confirms the conclusion of Kondratyev (1999) that a study of global dynamics taking into account the geophysical and ecological constituents is only possible with an interdisciplinary approach, with the NSS multi-dimensionality considered. Of course, the use of scenarios of anthropogenic development should remain as one of the approaches to parameterization of some NSS functions.

Consideration of possible natural disasters in the future, with available forecasts of the social development is important for comprehension of their strategy of interaction with nature. A human as a biospheric element will inevitably become an object of biophysical catastrophes with deplorable consequences and, maybe, irreversible results. Therefore, in studies of the global ecodynamics and creation of development scenarios, it is necessary to take into account the NSS multi-dimensionality and remember the mark on the scale of survival, beyond which the existence of humans becomes problematic.

In the future, the character of natural disasters could change due to the impact of new technologies on the environmental processes. In fact, catastrophes for other generations could occur connected with the transformation of mechanisms of biological regulation in ecosystems. Therefore, the theory of catastrophes and risk faces principally new problems of previsions for the directions of civilization development. Modern sociologists and philosophers, when discussing global dynamics, in contrast to specialists in the sciences, instead of the notions of "information society" or "technosphere" and "noosphere", use the concept of "risk society". Protectability of society against natural disasters will diminish because of the broadening spectrum of dangers and their possible scales, despite the development of sciences and industry. Therefore, in the future, the development of the theory of risk will be governed by political decisions. First of all, this will be connected with prevision for new risks, which is possible with the more complete use of the natural–scientific component of the NSS in order to reduce the level of danger and prevent trends toward critical situations in the environment with serious consequences for society. Here we again come to a necessity to realize the scheme in Figure 2.2.

7

The natural catastrophe in the Aral Sea region

7.1 PROBLEMS OF WATER BALANCE IN CENTRAL ASIA

Variations in atmospheric precipitation distributions over vast territories from the steppes of Stavropolye and Kalmykia to the Pamirs and Tien Shan are in many respects determined by large-scale spatial–temporal changes of atmospheric moisture fluxes from the basins of the Caspian and Aral Seas, the Kara–Bogaz–Gol Gulf (KBGG), large reservoirs and accumulators of collector–drainage effluents, saline lands, and other standard evaporators of surface and sub-soil waters on the territory of central Asia. The problem of water balance in central Asia is the subject of numerous publications, since undesirable phenomena are connected with it, such as dust and salt storms, floods, droughts, snow avalanches, landslides and mudflows, as well as other hydrological disasters and man-made catastrophes. Also important is the problem of provision of water for the 129 mln ha of central Asian desert pastures of which only 48% are used. Finally, there is a poorly studied correlation between the central Asia water balance and global changes of the environment. First of all, the rise of the Caspian Sea level and shallowing of the Aral Sea, degradation of many other unique water bodies such as Lake Balkhash and the KBGG if considered as correlated events represent a disaster on a planetary scale, since they entail catastrophic consequences for the population and the environment of vast adjacent territories.

For the regions near the Caspian Sea and central Asia, the areal and hydrological parameters of the basins of the Caspian and Aral Seas, other open water bodies, and conduits are important sources of replenishment of the volumes of atmospheric moisture, the west–eastward transport of which favors an enhancement of the volumes of glaciers and snowfields of the Pamirs and Tien Shan, which, on the one hand, raises and stabilizes the flow of the main rivers in the region, creates favorable conditions to provide ecological safety for the population, prevents dust and salt storms, and resolves the urgent problems of the Aral Sea, near Aral Sea regions, and other regions of central Asia, while on the other hand, leads to snow avalanches, landslides, and mudflows.

From the viewpoint of the ecological safety of a population and prospects of diverse development of the near Caspian and central Asian regions, most favorable were the years 1950–1960. The level of the Caspian Sea varied from -28 to -28.5 m, its area covered between 370 to 374 km^2. Most of this amount fell on the northern shallow sector of the sea. It is expected that with the raised level of the Caspian Sea (3 m from 1978), tidal processes can occur whose waves can propagate over dozens of kilometers from the shoreline.

Preliminary calculations have shown that on the basis of the hydrometeorological situation observed in the near Caspian and central Asian regions, it is possible in the nearest future:

- To compensate the present rise in level of the Caspian Sea and reduce it through evaporation by a volume of $\sim$60 $km^3\,yr^{-1}$ from specially irrigated saline lands and depressions on the eastern coast of the sea.
- To stimulate excess (compared with 1960) precipitation of a total volume of about 110 $km^3\,yr^{-1}$ at rationally chosen points of the regions by forced condensation of vapor and liquid droplet components of the atmosphere and cloudiness. Forced rain can be realized in the regions of the western and southern sectors of the Aral Sea (53 m), Lake Sudochye (50 m), the river bed of the Uzboy, the hollows of Sarakamysh (-38 m), Kazakhlyshor (-28 m), Karashor (-25 m), and others.

The expert level of the idea put forward here is based on the use of geoinformation monitoring system (GIMS)-technology. The following problems appear here:

- Development of a theoretical information model of the formation of atmospheric water fluxes in the near Caspian and central Asian regions, and an assessment of the potential amount of precipitation at local points with different climate scenarios.
- The choice of the sites of saline lands and hollows of the coastal band of the Caspian Sea useful for irrigation using technology that takes into account the hydrological and economic importance of such landscape elements.
- Processing and presentation of the input information in the form of dynamic electronic spatial–temporal thematic maps, as well as producing an archive and database of the experiments (both field and computer).

In connection with the above-mentioned material, the study carried out in this chapter has been aimed at an adaptation of GIMS-technology to conditions of the simulation experiment in the zone of impact of the Aral Sea, and a search for possible ways to change the environmental dynamics in this zone to restore sustainable development characteristics. For this purpose, it was necessary to develop the algorithmic and model means which would make it possible to efficiently control the hydrophysical and hydrological fields of the zone of the impact of the Aral Sea under conditions of its changing level, and assess responses of the aquageosystem to anthropogenic scenarios of intrusion into the elements of its water balance.

7.2 ANALYSIS OF GEOPHYSICAL AND HYDROLOGICAL PROCESSES IN THE ARAL SEA ZONE AND PROBLEMS OF THEIR MODELING

7.2.1 Aral Sea aquageosystem dynamics and the geophysical processes of its water balance formation

The Aral Sea is located in the Turan lowland in the center of the central Asian arid zone at a height of 53 m above World Ocean level, at the boundary with the deserts of Kara-Kum and Lyzyl-Kum. It functions as a gigantic evaporator ($\sim$60 km^3 yr^{-1}). Its basin (area >700,000 km^2) is affected by five countries – Kazakhstan, Kirghizstan, Tadjikistan, Turkmenistan, and Uzbekistan. Before 1960, the Aral Sea was a sufficiently stable water body with an area of 66.459 km^2, with centennial oscillations of its water level within $\pm$3 m and seasonal ones within $\pm$25 cm. It was one of the four largest inland closed saltwater (10 g l^{-1}) bodies on Earth. Before 1960 the maximum width of the sea basin constituted 235 km (from north-west to south-east) and 434 km (from south-west to north-east). The depths of the sea were distributed down to 69 m, so that 88% of them were within 30 m. The sea is deepest in the western sector of the depression. In the southern part of the sea a flat area is located, larger than 11,000 km^2, comprised of the present and old delta of the Amu-Darya River, changing to the south into the sand-hills of the Zaungus Kara-Kum Desert. In the east, the Aral Sea borders the desert plane of Kyzyl-Kum with a general slope toward the sea. Here the mound-ridge sands, with broad dry beds of the old tributaries of the Syr-Darya and Amu-Darya Rivers, are the characteristic relief. In the north and north-west the coastline of the sea is limited by the sand-hills of the foothills of Mugodjar.

The Amu-Darya River has always played an important role for the Aral Sea basin as one of the main components of its water balance. Irrigation measures undertaken in the 1960s have led to the present poor state of this region with the expected propagation of negative processes beyond its boundaries, which has prompted many scientists to search for possible means of their prevention (Krapivin and Phillips, 2001; Schlüter *et al.*, 2005). Based on the modeling system EPIC (Environmental Policies and Institutions for Central Asia), Schlüter *et al.* (2005) created a model of the water regime control in the zone of impact of the Amu-Darya River. This model determines an optimization of the irrigation network with calculated monthly mean control of water fluxes and consideration of water needs for the next 15 years. The model was calibrated against the data of the state of the water system of the Amu-Darya zone recorded in 1994 and 1997. Calculations with this model, not taking into account the totality of direct connections and

Table 7.1. Deviant average air temperatures (°C) by seasons in the Aral Sea region (Kuksa, 1994).

Zone	Spring	Summer	Autumn	Winter
Aral Sea	1.4	0.4	−0.1	0.5
Monsyr	0.6	0.2	1.1	1.3
Kazalinsk	0.9	0.6	1.5	−0.1
Karak	0.8	0.2	0.8	0.3
Chabankazgan	1.1	0.6	0.7	−1.5
Muinak	0.9	0.5	−0.3	−0.6
Chirik-Rabat	0.9	0.5	1.0	0.4
Kungrat	1.4	1.4	1.1	−0.1
Kosbulak	1.3	0.8	0.0	0.0

feedbacks in the system of the water balance of Central Asia, are rather symbolic in nature and thus cannot serve as a guide to making real decisions. Experience from this modeling and ideas expressed by the authors make it possible to proceed to a more complicated model, which is described below.

The climatic conditions of the Aral Sea functioning are determined by its environs characterized above. Temperature oscillations in the sea zone can reach 78°C. The January temperature averages −14°C, sometimes dropping to −33°C. In July the temperature averages 26°C reaching in some years 45°C. In general, the climatic situation in the near Aral Sea zone is variable but not for anthropogenic reasons. For instance, in the period 1951–1960 the inter-annual variability of surface air temperature (SAT) ranged between 4°C and 6°C, and in the period 1971–1980 the winter temperature turned out to be below the norm by 5.5°C. In subsequent years, there was a trend to the transfer of the annual regime of temperature to the continental one. Table 7.1 gives some indicators of deviations of average temperatures from multiyear norms. These values make it possible to prescribe intervals of climatic uncertainty when formulating the synoptic scenarios.

The sum of annual precipitation over the sea oscillates around 100 mm, whereas evaporation is estimated at 1,250 mm yr^{-1} (i.e., every year a layer of 115 cm evaporates from the sea surface). The temperature regime of the sea itself is characterized by water temperature variations from 20–25°C in the summer to −0.7°C in the winter, when a considerable part of sea surface is covered with ice. Since the sea becomes shallow, the heating and cooling of its water masses sometimes reaches the entire water column.

Located among deserts, the sea is constantly wind-affected. In the autumn and winter, north-eastern winds blow, bringing the cold air masses from Siberia, in the spring and summer, the south-western wind brings moisture from the Atlantic Ocean, Mediterranean, and Caspian Seas. Wind-rose and wind speed are important parameters in analysis of the Aral Sea water balance and should be thoroughly taken into account. Figure 7.1 schematically shows wind-roses over the Aral Sea basin. According to estimates of Bortnik and Chistiayeva (1990), the

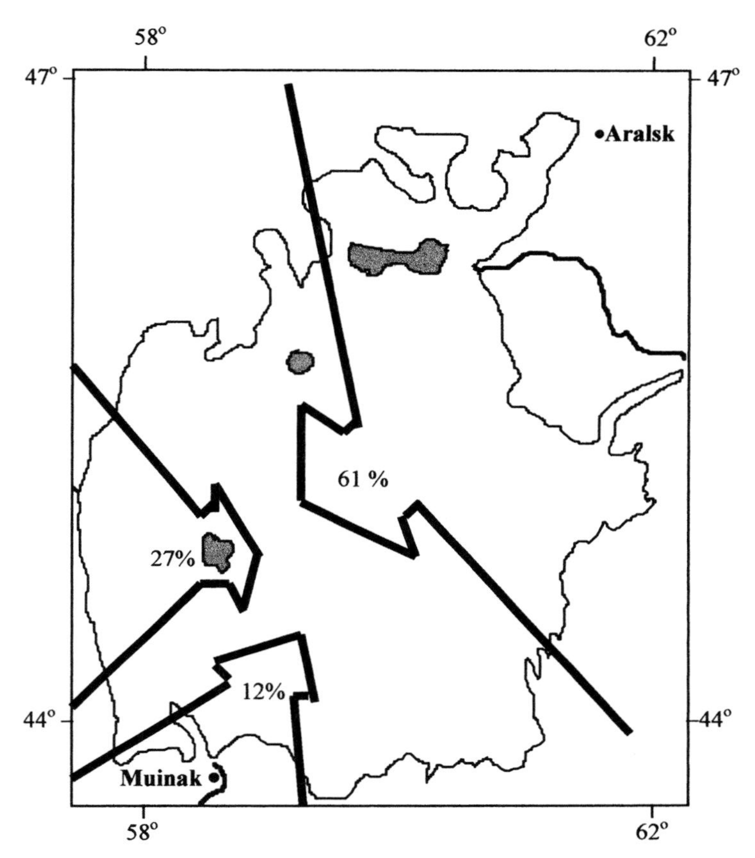

Figure 7.1. Characteristic directions of winds in the zone of the Aral Sea and their repeatability (Lu Hung, 1993; Bortnik and Chistiayeva, 1990).

annual mean wind speed varies depending on the territory within 3–7 $m\,s^{-1}$. The region of the Aral Sea is characterized by a strong variability of wind speed which can steadily reach 30 $m\,s^{-1}$. For instance, on the western coast, such wind speeds are observed annually for more than 50 days, on average, which is very important for the scenario of restoring the Aral Sea level.

The Aral Sea water balance constituents have been discussed elsewhere (Golitsyn, 1995; Berg, 1908; Grigoryev, 1987; Tsytsarin, 1991; Kuksa, 1994; Kosarev, 1975; Bortnik and Chistiayeva, 1990; Aladin and Kuznetsov, 1990; Sumarokova *et al.*, 1991). As far back as 1968, when negative trends in the Aral Sea water balance began to emerge, Kornakov *et al.* (1968) gave a comprehensive analysis of its basic constituents. At that time, due to broadened irrigated lands (by 1980 they increased by 6.5 mln ha), the run-off losses in the delta of the Amu-Darya River reached 9.1 $km^3\,yr^{-1}$, and considering the water intake from irrigation below the city of Nukus, these losses reached 10.7 $km^3\,yr^{-1}$ or 23.3% of the Amu-Darya River run-off. Various experts obtained different estimates, and therefore in modeling the water balance the input values are somewhat uncertain. At any rate, the estimate of the multiyear average (1934–1960) inflow to the delta of the Amu-Darya River is close to 47–49 $km^3 yr^{-1}$. For Syr-Darya it is 15–24 $km^3 yr^{-1}$. The run-off of a further six rivers constituted 43.8 $km^3 yr^{-1}$.

Before 1960, a small portion of the total annual run-off had been spent on irrigation. The irrigated lands were mainly located in the river flood plains, which led to the return of excess irrigation water masses to mother rivers. But in the 1960s, more than 30% of the Amu-Darya River run-off was directed to the Main Turkmenian and other shallower canals and used to fill numerous reservoirs (Rubanov *et al.*, 1987). As a result of the collector–drainage effluents, new unplanned water bodies began to form, whose participation in the water balance of the Aral Sea region manifested itself only via evaporation. Their use in the hydroeconomic balance was limited because of their substantial mineralization (1.5–12.0 $g l^{-1}$). During the period 1975–1988, after which some stabilization was observed, the total area of lakes and reservoirs in the Lower Amu-Darya grew from 483 to 1,256 km^2.

In Kuksa's (1994) opinion, a combination of excess water removal with a drought period and river shallowing between 1974 and 1977 was a stimulus to a catastrophic development of the process of anthropogenic desertification in the Aral Sea region. The river run-off decreased sharply with increasing evaporation from both the sea basin and adjacent territories. By the end of 1980, the total area of dried water bodies in the Amu-Darya delta alone reached 310 km^2. In 1984 the Syr-Darya and Amu-Darya Rivers run-off was estimated at 4 $km^3 yr^{-1}$ and 28 $km^3 yr^{-1}$, respectively. The process of transformation of land surfaces with prevailing conversion of hydromorphous, marshy, and meadow soils into saline and takyr lands had begun. The area of lakes on the delta plains of the Aral Sea region changed from 400,000 ha in 1960 to 120,000 ha in 1970 and almost vanished by the end of the 20th century. The sea coastline retreated from its position in 1960 by dozens of kilometers. For instance, the settlement of Kulanda (a fishing village) is now 35 km from the sea. The characteristic pattern of changes of the elements of the Aral Sea water balance is shown in Tables 7.2 and 7.3 and in Figures 7.2 and 7.3.

The worsening of the ecological situation in the zone of the Aral Sea has led, for instance, in Uzbekistan, to an increase of mother and child mortality. So, in 1994 the mortality of yet unborn children constituted 120 per 100,000 mothers, and newborn mortality was 60 out of 1,000. The climatic situation has markedly changed, which has started to tell on the economy of Kazakhstan and Uzbekistan. Many international organizations have discussed the problem of the Aral Sea. Many experts express their anxiety about the prolonged aimless discussion on the global character of the Aral catastrophe, while concrete decisions should be made saving it or at least stabilizing its level. After all, the developing ecological and human catastrophes connected with the death of the Aral Sea will lead to huge economic and moral losses for the native people.

It is known from the history of the Aral Sea that during the last 10,000 years, it dried out 8 times and, again, was filled. Now we witness its ninth drying-out period. This one differs from the previous ones by its inclusion of powerful anthropogenic

Table 7.2. Dynamics of water inflow to the Aral Sea (Kuksa, 1994).

Year	Amu-Darya	Syr-Darya	*Total*	Year	Amu-Darya	Syr-Darya	*Total*
1959	40.0	18.3	*58.3*	1974	6.2	1.9	*8.1*
1960	37.8	21.0	*58.8*	1975	10.0	0.6	*10.6*
1961	29.2	–	*29.2*	1976	10.3	0.5	*10.8*
1962	29.1	5.7	*34.8*	1977	7.2	0.4	*7.6*
1963	29.9	10.6	*40.5*	1978	18.9	–	*18.9*
1964	36.5	14.9	*51.4*	1979	10.9	2.9	*13.8*
1965	25.2	4.6	*29.8*	1980	8.3	–	*8.3*
1966	33.1	9.5	*42.6*	1981	5.9	–	*5.9*
1967	28.6	8.6	*37.2*	1982	0.04	–	*0.04*
1968	28.9	7.2	*36.1*	1983	2.3	–	*2.3*
1969	55.1	17.5	*72.6*	1984	7.9	–	*7.9*
1970	28.7	9.8	*38.5*	1985	2.4	–	*2.4*
1971	15.3	8.1	*23.4*	1986	0.4	–	*0.4*
1972	15.5	6.9	*22.4*	1987	10.0	–	*10.0*
1973	33.4	8.9	*42.3*	1988	16.0	7.0	*23.0*

Table 7.3. Average multiyear values of the Aral Sea water balance for individual periods. Numertors are volumes of water balance components (km^3), denominators are water layer (cm).

Period	Input		Expenditure, evaporation	Water balance (cm)	Change of level	Imbalance
	River flow	Precipitation				
1911/1960	56.0/84.7	9.1/13.8	66.1/100.0	−1.0/−1.5	0.1	−1.6
1961/1970	43.3/68.5	8.0/12.7	65.4/103.5	−14.1/−22.3	−21.0	−1.3
1971/1980	16.7/29.3	6.3/11.0	55.2/96.8	−32.2/−56.5	−57.6	1.1
1981/1985	2.0/4.1	7.1/14.7	45.9/96.2	−36.8/−77.4	−80.0	2.6
1986/1988	10.8/28.0	6.2/15.4	47.0/116.3	−30.0/−72.9	−65.6	−7.3

factors manifesting itself through human mismanagement. This mismanagement has revealed itself in the unconsidered, drastic, and large-scale changes to the river run-off into the Aral Sea as well as in the absence of any constructive estimates of the consequences of this change. As a result, only 30 years later the regional oasis, the richest in natural resources among the desert sands of the Kara-Kum and Kyzyl-Kum, has been transformed into a lifeless desert. During this period the sea has lost three-quarters of its water volume; its area has contracted by more than 50%. More than 33,000 km^2 of the sea bottom have become exposed and deserted. From this territory, the winds blow off annually more than 100,000 tonnes of salt and fine-dispersed dust with admixtures of various chemicals and poisons, raising them high into the atmosphere. The curves in Figure 7.4 demonstrate the correlative dynamics

(a)

(b)

Figure 7.2. Some fragments of the photos of the Aral Sea zone taken from space at different times. (a) Photo of the Aral Sea zone taken in April 1991 (NOAA–AVHRR). (b) Photo of the Aral Sea taken in 1995 from the Resource-01 satellite using the MSU–SK scanner. (c) The present state of the Aral Sea basin. The contrast indicates the contour of the coastline in 1960 (photo taken from the Resource-01 satellite). (d) Photo from the Terra satellite (EOS–AM1) taken on 5 November 2000, radiometer MODIS, resolution 500 m.

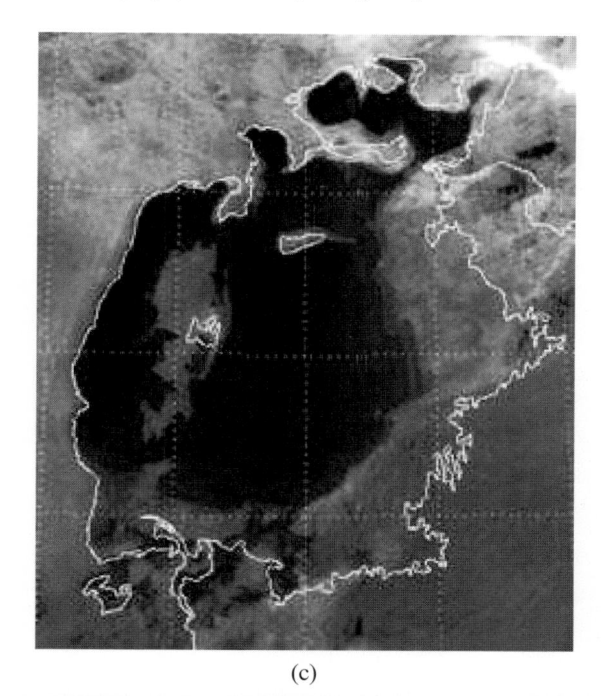

(c)

(d)

Figure 7.3. The state of the Aral Sea on 12 August 2003 from the Aqua satellite (MODIS). Spatial resolution 250 m.

of the main characteristics of the Aral Sea – its level and salinity. Clearly, the development of this dynamic will aggravate the ecological situation over wider areas of adjacent territories.

Many experts seek a solution to the Aral Sea problem, trying to balance the fluxes of water taken for irrigation and returned to the sea (Dukhovny and Stuling, 2001) and to solve the problem, if only partially, of stabilizing the situation. At present, the total volume of return water masses is estimated at 36–40 km^3. Of these, about 50% constitute river waters (18–20 km^3), with 13% being heavily salted. A considerable share of return water is realized through drainage and collector networks, a leakage from which has led to an uncontroled formation of hundreds of water bodies with total volumes exceeding 30 km^3 and wetlands with an areal extent of dozens of thousands of hectares. To stabilize the formed water regime of the Aral Sea basin, thus solving a number of complicated ecological problems, it is

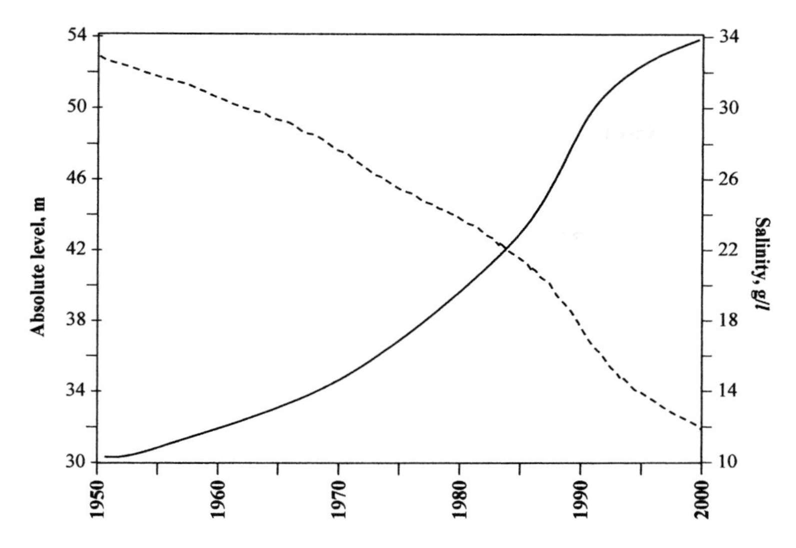

Figure 7.4. Dynamics of the level (solid curve, m) and salinity (dashed curve, $g\,l^{-1}$) of the Aral Sea.

necessary, of course, to develop an intergovernmental system to regulate the return water masses. A broader view of this problem suggests that acceptable solutions can be found, which can lighten the economic load and coordinate the regional water balance with the global position (Borodin *et al.*, 1996; Chukhlantsev *et al.*, 2004; Krapivin and Phillips, 2001; Kondratyev *et al.*, 2002a, 2004d). The project of the annual transfer of 27–60 km^3 of water from Siberian rivers to central Asia, appearing in the 1980s and sometimes discussed even now, turned out to be problematic from the economic and ecological points of view (Micklin, 2002). Nevertheless, this idea has some sense for further theoretical studies of such scenarios. One such scenario is discussed here.

7.2.2 An adaptation of GIMS-technology to the geophysical conditions of the Aral Sea zone

The sphere of geographical information systems (GISs) is the most developed part of the nature monitoring. The GIS combines computer cartography with databases and remote sensing. The GIS elements are the computer network, database, data transmission network, and the system/process of reflecting a real situation as a computer display. Numerous examples of GISs suggest the conclusion that GIS-technology provides the control of the state of the monitored object and serves as an efficient mechanism for the combination of diverse information about the object. However, GIS-technology has serious limitations when the matter concerns complicated problems of nature monitoring, which require the construction of a dynamic image of the medium under conditions of fragmentary data both in space and in time. The main shortcoming of GIS-technology is that it is not oriented toward the versatile forecast of the state of the monitored object. Such problems will be discussed here.

An important step forward in the development of GIS-technology has been made by Kondratyev *et al.* (2002b) where geoinformation monitoring system (GIMS)-technology has been theoretically substantiated and practically applied. This technology removes many shortcomings of GIS-technology and gives a possibility to synthesize the monitoring systems with the functions of a forecast. The key unit of GIMS-technology is a mathematical model of the controled object or process. It is a combination of the empirical and theoretical parts of GIMS-technology that makes it possible to promptly evaluate the current and prognostic changes of the regional environment under study.

The state of natural objects is characterized by diverse parameters, including the type of soil and vegetation, water regime of the territory, salt composition of soils, depth at which ground waters lie, and many others. In principle, the required information on these parameters can be obtained with a different degree of reliability from the data of ground observations, remote sensings, and from data banks of GIS which contain a priori information accumulated over a number of years. The problem a decision maker faces consists in obtaining answers to the following questions:

(1) What devices are better to be used in ground and remote observations?
(2) How to balance the amount of ground measurements and the volume of remote-sensing data with their information content and cost taken into account?
(3) What mathematical models of spatial–temporal changes of the natural object's parameters are expedient to be used for interpolation and extrapolation of the data of *in situ* and remote observations to reduce the volume (amount) of the latter and, respectively, reduce the cost of the work on the whole, as well as predicting the functioning of the observed object?

Any subsystem of the environment is considered as an element of nature interacting through biospheric, climatic, and socio-economic feedbacks with the global nature–society system (NSS). For a concrete object of monitoring, a model is developed which describes such an interaction and functioning on various levels of the spatial–temporal hierarchy of the totality of processes in the environment affecting the state of the object, judging from preliminary estimates. The model incorporates natural and anthropogenic characteristics of a given territory and at its initial state of development it is based on the existing information base. The model's structure is oriented toward an adaptive regime of its use with subsequent episodic corrections of its parameters or its units.

As a result of combination of the system of the environmental data collection, the model of the geoecosystem's functioning on a given territory, the system of computer cartography and means of artificial intelligence, a single GIMS of the territory is synthesized which ensures the prognostic estimates of the consequences of realization of technogenic projects and other estimates of the geoecosystem's functioning. For the Aral Sea zone, a realization of GIMS-technology requires a

Table 7.4. Determination of the identifier A_2 elements.

Type of the surface	Identifier
Open lake water	a
Irrigated territory	b
River sector	c
Waterlogged site	d
Dry river bed	e
Tree–brush vegetation	f
Takyr	g
Saline land	h
Steppe	t
Sea sector	p
Sand	n
Reeds	m
Pasture vegetation	o

selection of characteristic elements of the natural–anthropogenic system which is functioning in this zone. This procedure is realized via a multitude of 2-D matrix structures–identifiers, in the symbolic form describing the geographical configuration of the zone, distribution of soil–plant formations, dislocation of anthropogenic objects, location of characteristic synoptic zones, local topography, and configuration of waterways within the territory.

For equations of the water balance of the Aral Sea region, of importance are elements of land cover, whose impact on evaporation and surface run-off manifests itself via their characteristics. The identifier $A_1 = \|a_{ij,1}\|$ determines the configuration of the territory, which is taken into account in the model of the water balance. Without breaking the community, assume a constant geographic grid of the size $\Delta\varphi$ in latitude φ and $\Delta\lambda$ in longitude λ. Then the identifier A_1 in the GIMS database provides a flexible consideration of the sites of the territory in the Aral Sea zone which will be taken into account:

$$a_{ij,1} = \begin{cases} 1 & \text{the site is included into water balance} \\ 0 & \text{the site is not included into water balance} \end{cases}$$

The identifier $A_2 = \|a_{ij2}\|$ prescribes the spatial distribution of the land cover elements symbolically, referencing them to the water balance components of the territory. Table 7.4 enumerates the characteristic elements of the cover in the zone taken into account by the identifier A_1.

7.2.3 Formation of the database on the elements of the Aral Sea zone environment

In the period 1972–1990, regular sessions of remote monitoring were carried out on the territory of central Asia with the use of microwave, optical, and IR methods of sensing (Borodin and Krapivin, 1998; Krapivin and Phillips, 2001; Chukhlantsev *et al.*, 2004). By synchronous aerospace and ground observations, studies of regional biogeocenoses were carried out. Materials from radar, radio-thermal, photographic, and optoelectronic surveys and measurements have formed the basis of the remote-sensing database which includes information on various characteristics of land cover, hydrometeorological processes, and the atmosphere. In particular, the database contains information about special features of micro- and macro-relief, the type of floristic background, degree of moistening and salting of soils, subsurface anomalies (cavities, ground water lenses, etc.), and atmospheric state. The database contains estimates of dependences of reflecting and emitting properties of the surface at different wavelengths on variations of physico-chemical and geographical parameters of the environmental elements. These data were used in solving the problems of identification on the basis of the algorithms discussed above.

In the qualitative radar studies of reflective properties of extended natural formations, a decoding indicator can be a specific effective scattering surface (ESS). This parameter determines the general background of a radar image of locality and makes it possible to detect, comparatively easily, in the image the sites with anomalous reflective properties. However, it is difficult to use the notion of ESS in quantitative comparisons of radar images of various sites of locality or of the same territory obtained at different time moments. For a qualitative interpretation of radar images, texture features and spectral structures of images were used, determined by the specific local parameters of the respective surface. Both these components have statistical characteristics of their own, the first-order statistics of the spectral structure and texture component being described by multi-dimensional probability density, second-order momentum, and autocorrelation function, which reflect the interaction between the signal's intensity in adjacent elements of the still. In the ultimate case, the spatial radius of correlation of the radar image texture is comparable with the resolution of the measuring device and depends substantially on large-scale variations of relief, plants, biomass, and parameters of other elements of the landscape. Hence, the first-order texture statistics can vary even within one image and one class of objects.

The microwave measurements have shown the presence of typical spectra of radio-brightness temperature. The radio-brightness spectra with positive values of first differences are typical of some types of ice, water bodies with not deeply located dense algae, young (hot) lava flows and fields, concrete surfaces, and some types of dry soils. Monotonically decreasing spectra are characteristic of moistened soils, water bodies, rice checks, and others. Spectra with alternating values of first increments are inherent to multi-layer interfering structures, heterogeneous formations of the type of peat bogs and edges of forest fires.

Polarization and dispersion characteristics of the thermal field turn out to be of significant value to aerospace studies of water bodies, concrete and soil covered aircraft runways, as well as other natural and anthropogenic formations with smooth surfaces. Such formations were used as calibrating microwave reference points.

Synchronous remote and ground measurements of the thermal field radiance of saline lands at wavelengths 1.35, 2.25, and 20 cm have shown that a saline land has

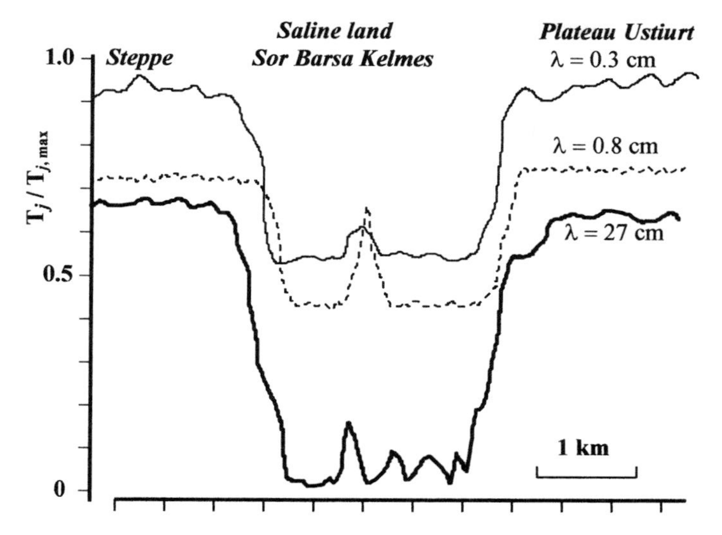

Figure 7.5. Fragment of the record from the IL-18 flying laboratory on the boundary of the Ustiurt Plateau.

vast and stable (from season to season) regions with small variations in radio-brightness temperatures. Along the contour of the saline land a sharp decrease of radio-brightness temperature is observed at wavelengths of 0.8 and 2.25 cm, its minimum value is located in the decimeter interval in the central part of saline land. The Ustiurt Plateau was used as a reference; its comparison with polarization effects makes it possible to carry out a reliable classification of land cover.

The database contains information about radio-brightness contrasts over the whole territory of central Asia. As elements of the territory, the closed systems of water bodies, drainage water accumulators, complexes of man-made and natural lakes, wetted saline lands, and takyrs have been selected. Experience of its formation has shown that only due to remote sensing from a flying laboratory is it possible to operatively estimate the moisture content of the atmosphere along the contour of the territory of the Caspian–Aral system. Episodic trace measurements over inland territories enable one to specify the distribution of land cover and the level of sub-soil water. Examples of records of radio-brightness contrasts from the IL-18 flying laboratory are shown in Figures 7.5 and 7.6.

7.2.4 Specific features of modeling the water balance for the Aral Sea region

The Aral Sea water balance has been calculated by many scientists. However, those calculations did not take into account correlations and estimates averaged over large territories neighboring the Aral Sea hollow. It is apparent that the time dependent character of the climatic situation and variability of the structure of land cover requires a more detailed account, in the water balance equations, of the role of a

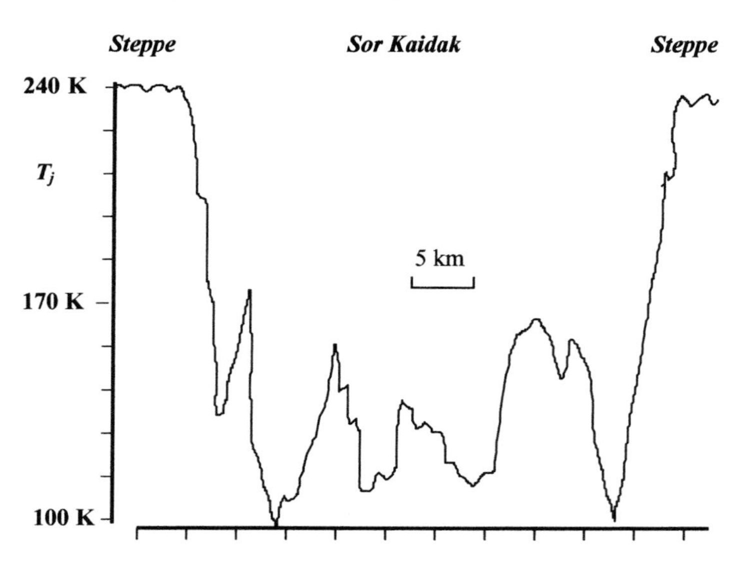

Figure 7.6. Fragment of the record from the IL-18 flying laboratory in the regions of the saline land of Kaidak (wavelength 1.35 cm).

detailed description of climate parameters and morphology of the elements participating in water evaporation.

The model of the water balance of the Aral Sea region can be based on a standard model of the regional balance of moisture for a limited territory, shown schematically in Figure 6.16. Each territory within the Aral Sea zone can have parts of a river network, water body, and land sites. According to the landscape–hydrological principle, to construct a simulation model in the zone of the hydrological system's functioning, it is necessary to select facias. This is connected with the typifying of floristic background, whose concrete form is determined by micro-relief, type and properties of soil, surface moistening, depth of ground water, and other factors. Thus, in general, the territory Ω_L is characterized by m facias, and the input network has n homogeneous sites. With this taken into account, according to the scheme in Figure 6.16, a closed system of balance equations (6.4)–(6.7) is composed, which serves as the basis for calculations of the water balance components in the Aral Sea zone.

7.3 THE ALGORITHMIC PROVISION OF THE SYSTEM MONITORING THE ARAL SEA ZONE

7.3.1 An algorithm to retrieve the dynamic parameters using the differential approximation method

The database of the environmental monitoring system does not always meet the requirement of parametric saturation required by GIMS-technology. Therefore, of interest is an algorithm of parameterization of the functions of the system controled on a given territory, which would not make strict demands on the database. Suppose that in the monitoring regime, measurements are made of N characteristics of the system x_i $(i = 1, \ldots, N)$ at time moments t_s $(s = 1, \ldots, M)$. The formal dependence between $x_i(t)$ will be presented as a system of differential equations with the coefficients $\{a_{ijk}, b_{ij}\}$ known:

$$dx_i/dt = \sum_{k,j=1}^{N} [a_{ijk} x_j(t) x_k(t) + b_{ij} x_j(t)] \tag{7.1}$$

with initial conditions:

$$x_i(0), i = 1, \ldots, N \tag{7.2}$$

The problem of retrieving the $x_i(t)$ values at any time moment in the interval of observations $[0, \mathrm{T}]$ is reduced to a simple Cauchy problem for the system of standard equations. The only obstacle to its solution is an uncertainty of coefficients a_{ijk} and b_{ij}. In this case we follow a traditional method (i.e., we introduce the measure of deviation between calculated $x_i(t_s)$ and measured $\hat{x}_i(t_s)$ values):

$$E = \sum_{s=1}^{M} \left\{ \sum_{i=1}^{N} [x_i(t_s) - \hat{x}_i(t_s)]^2 / N \right\} / M \tag{7.3}$$

where $0 \leq t_1 \leq \cdots \leq t_M \leq T$.

Then a set of coefficients $\{a_{ijk}, b_{ij}\}$ can be determined by solving the following optimization problem:

$$E_0 = \min_{\{a_{ijk}, b_{ij}\}} E \tag{7.4}$$

A search of the minimum function E in (7.4), according to methods described in Bellman and Rous (1971), is reduced to the problem of dynamic programming. Suppose that coefficients $\{a_{ijk}, b_{ij}\}$ are functions of time.

Denote:

$$Y(t) = \begin{Vmatrix} x_1(t) \\ \vdots \\ x_N(t) \\ a_{111}(t) \\ \vdots \\ a_{NNN}(t) \\ b_{11}(t) \\ \vdots \\ b_{NN}(t) \end{Vmatrix} \tag{7.5}$$

Without violating generality, it can be assumed that $a_{ijk} = a_{ikj}$. Then supplementing the system (7.1) instead of Cauchy problem (7.1), (7.2) we obtain:

$$dY/dt = G(Y) \tag{7.6}$$

where the function G has the following components:

$$\left.\begin{aligned} G_i(Y) &= 0 \text{ for } i = N+1, \ldots, N_c, \\ G_i(Y) &= \sum_{k,j=1}^{N} [a_{ijk} x_j(t) x_k(t) + b_{ij} x_j(t)] \text{ for } i = 1, \ldots, N; \end{aligned}\right\} \tag{7.7}$$

with $a_{ijk}(0) = \bar{a}_{ijk}, b_{ij}(0) = \bar{b}_{ij}, N_c = N + N^2 + N^2(N+1)/2$.

Note that using the quasi-linearization method, the solution of a non-linear problem is reduced to the solution of a succession of linear problems. The method is a further development of the known method of Newton–Rafson (Dulnev and Ushakovskaya, 1988) and its generalized version.

Introduce a succession of functions $Y^{(1)}(t), \ldots, Y^{(n)}(t)$ so that $Y^{(1)}(t)$ is a first approximation to the solution of the system (7.6). Then the nth approximation is found by solving the following linear system:

$$dY_i^{(n)}(t)/dt = G_i[Y^{(n-1)}(t)] + \sum_{j=1}^{N_c} \{dG_i[Y^{(n-1)}(t)]/dY\}[Y_j^{(n)} - Y_j^{(n-1)}] \tag{7.8}$$

As shown in Bellman and Dreifus (1965), the iterative process (7.8) is converged following the square law. Solution of (7.8) in a general form is written as:

$$Y^{(n)}(t) = P(t) + \sum_{k=1}^{N_c} C_k H^{(k)}(t) \tag{7.9}$$

where $P(t)$ is the partial solution of the system (7.8), and $H^{(k)}(t)$ is the vector solution of a homogeneous system. To determine $P(t)$, we solve (7.8) under initial conditions $Y_i(0) = 0 (i = 1, \ldots, N_c)$. The functions $H^{(k)}(t)$ are found by solving the Cauchy problem:

$$dY_i^{(n)}(t)/dt = \sum_{j=1}^{N_c} \{dG_i[Y^{(n-1)}(t)]/dY\}[Y_j^{(n)} - Y_j^{(n-1)}](i = 1, \ldots, N_c) \tag{7.10}$$

$$H^{(1)}(0) = \left\| \begin{matrix} 1 \\ 0 \\ \vdots \\ 0 \\ 0 \end{matrix} \right\|, H^{(2)}(0) = \left\| \begin{matrix} 0 \\ 1 \\ \vdots \\ 0 \\ 0 \end{matrix} \right\|, \cdots H^{(N_c)}(0) = \left\| \begin{matrix} 0 \\ 0 \\ \vdots \\ 0 \\ 1 \end{matrix} \right\| \tag{7.11}$$

It follows from (7.8)–(7.11) that the constants C_k are unknown initial conditions of the system of equations (7.7). Therefore, at each iteration in the process of finding a partial solution and solutions of homogenous equations, the constants C_k are found in order to obtain a solution of $x^{(n)}$ which agrees best with observational results in a sense of the method of least squares:

$$E = \min_{\{C_k\}} \sum_{s=1}^{M} \sum_{i=1}^{N} \left[P_i(t_k) + \sum_{k=1}^{N_c} C_k H_i^{(k)}(t_s) - \hat{x}_i(t_s) \right]^2 \tag{7.12}$$

Let

$$\partial E / \partial C_k = 0 \text{ for } k = 1, \ldots, N_c \tag{7.13}$$

It follows from (7.12) and (7.13) that:

$$\sum_{k=1}^{N_c} A_{km} C_k + B_m = 0, m = 1, \ldots, N_c \tag{7.14}$$

where

$$A_{km} = \sum_{s=1}^{M} \sum_{i=1}^{N} H_i^{(k)}(t_s) H_i^{(m)}(t_s), B = \sum_{s=1}^{M} \sum_{i=1}^{N} [P_i(t_s) - \hat{x}_i(t_s)] H_i^{(m)}(t_s)$$

Thus, from each iteration of (7.8) the problem (7.14) should be solved. The rate of convergence of this procedure will depend on successful choice of the initial conditions.

Consider two regimes of the use of this algorithm schematically presented for the Aral Sea basin in Figures 7.7 and 7.8. The distance from the point A_0 to A_i we denote as l_i $(i = 1, \ldots, n)$. Let at points A_i a set of brightness temperatures $T_{\mathrm{B}}^j = (T_{\mathrm{B,A1}}^j, \ldots, T_{\mathrm{B,A}n}^j)$ be fixed at some standard time moments t_j $(j = 1, \ldots, \mathrm{M})$. Then a matrix $\|T_{\mathrm{B,A}i}^j\|$ can be formed within which, using the algorithm of differential approximation, measurements are reduced to a single time moment. Then it only remains to solve an inverse problem on estimation of the geophysical parameters.

7.3.2 Use of the harmonic functions method to process the microwave radiometry data in the case of a closed area

The process of heat propagation in a flat homogeneous medium G with constant thermal–physical properties (ρ is the density, c is the specific heat capacity, K is the coefficient of heat conductivity; $\rho, C, K = \text{const.} > 0$) is described with the equation:

$$\partial T / \partial t = a^2 (\partial^2 T / \partial \varphi^2 + \partial^2 T / \partial \lambda^2) \tag{7.15}$$

where $T = T(\varphi, \lambda t)$ is the temperature at the point $(\varphi, \lambda) \in G$ at a time moment t; and $a^2 = K / \rho c$ is the coefficient of heat conductivity for G. If the process of heat transfer is stationary, then (7.15) becomes the standard Laplace equation:

$$div \cdot grad\ T = \partial^2 T / \partial \varphi^2 + \partial^2 T / \partial \lambda^2 = 0 \tag{7.16}$$

In this case T is a harmonic function of spatial coordinates φ and λ. Together with the temperature field $T(\varphi, \lambda t)$ consider the field of the self-radiation of G in the microwave range, whose intensity in accordance with the Rayleigh–Jeans approximation (Chinlon, 1989) at a local thermodynamic equilibrium is characterized by the brightness temperature $T_J(\varphi, \lambda, \eta, \theta, t)$, where η is the wavelength of the

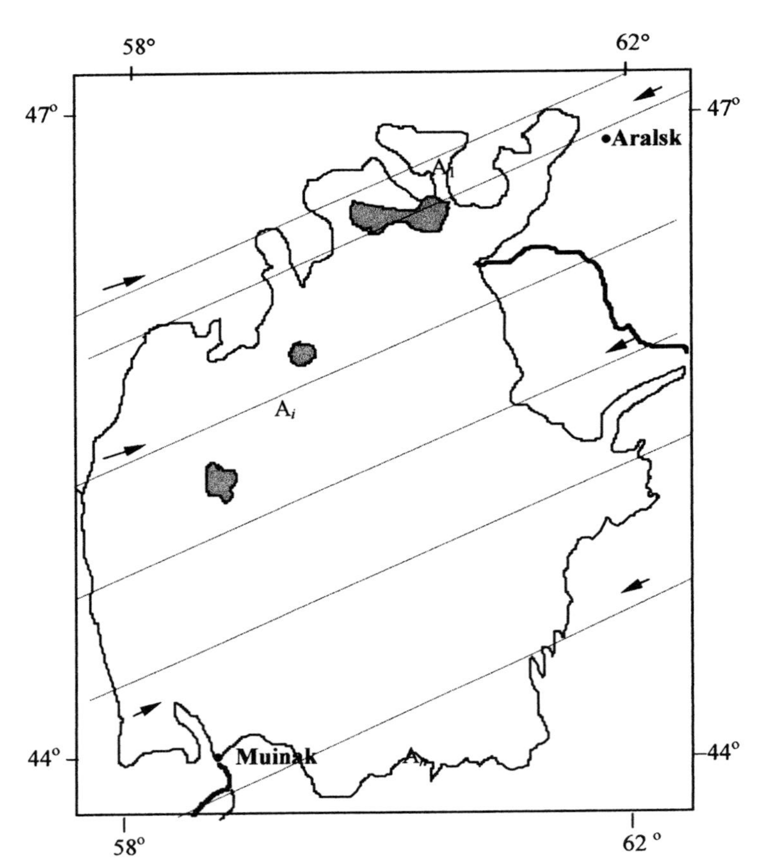

Figure 7.7. Regime of data collection with a random choice of the flying laboratory route parallel to a fixed azimuth.

electromagnetic interval and θ is the observation angle. Assume that for a sufficiently small area V_M of any point $M \in G$ the following condition is satisfied:

$$T_J(\varphi, \lambda, \eta, \theta, t) = A_M + B_M T(\varphi, \lambda t); (\varphi, \lambda) \in V_M; (A_M, B_M = \text{const.}) \qquad (7.17)$$

The form of (7.17) follows from theoretical and experimental estimates of T_J. Thus, for a medium that is homogeneous in depth and limited by a flat surface, the following equation is valid: $T_J = \kappa T_o$, where $\kappa = \kappa(\eta, \theta, \varepsilon)$ is the emissivity coefficient of the medium, ε is the dielectric permeability of the medium, and T_o is the thermodynamic temperature. According to experimental estimates (Shutko, 1987), at wavelengths $\eta \geq$ 5–8 cm, T_j for freshwater practically linearly depends on T_o; the steepness of this dependence constitutes 0.35–0.5°K/°C. An increase of salinity S from 0 to 13–16‰ is followed by a decrease of sensitivity of the radiation field to temperature variations in a wide range of decimeter waves from 10 to 5 cm. In cases of relationships $\eta S \cong 700$; $0 \leq T_o \leq 30$°C; $0 \leq S \leq 180$‰; $0 \leq \theta \leq 25°$ the radiation

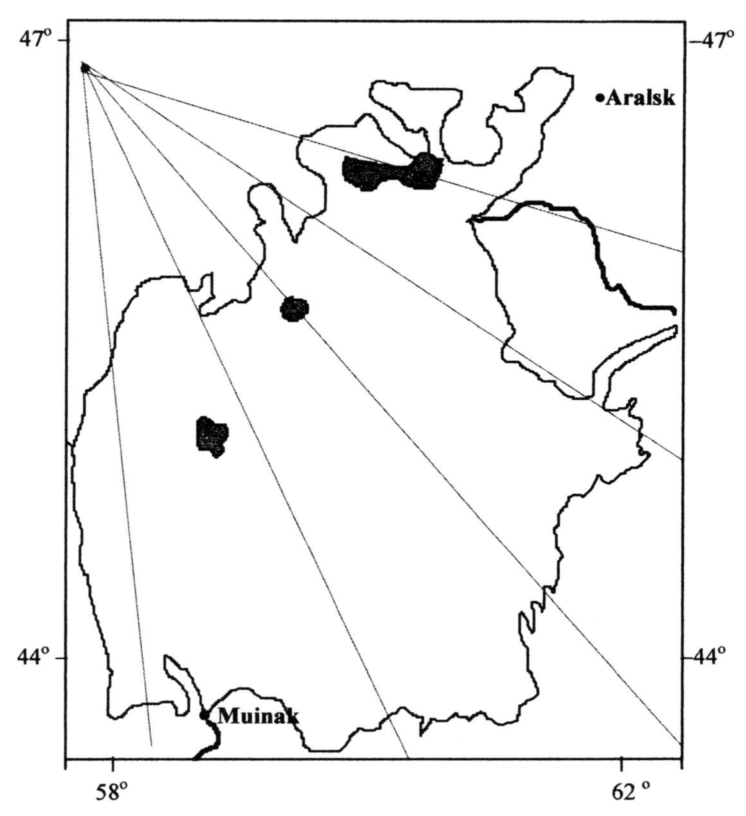

Figure 7.8. Regime of data collection from the flying laboratory with fan-shaped measurements, with azimuthal reference to a fixed object (beacon).

field sensitivity to variations in T_o is at a minimum. It follows from (7.17) that T_J at each point $M \in G$ follows the relationship:

$$T_J(\varphi, \lambda, \eta, \theta, t) = (2\pi)^{-1} \int_0^{2\pi} [A_M + B_M T(\varphi + r\cos a, \lambda + r\sin a, t)da$$

$$= (2\pi)^{-1} \int_0^{2\pi} T_J(\varphi + r\cos a, \lambda + r\sin a, \eta, \theta, t)da$$

for any $r \in (0, r_M)$, from which it follows that T_J is harmonic in G and, hence, satisfies (7.16). A typical boundary value problem for (7.16) is the Dirichlet problem. On the boundary Γ of the medium G, a continuous function $\tilde{T}_J = \tilde{T}_J(u)$ is prescribed, where $u = \varphi + i\lambda$ is the complex coordinate of the point $(\varphi, \lambda) \in \Gamma$. The function T_J should be found harmonic within G and assuming prescribed values of $\tilde{T}_J$ on Γ. This function, according to the complex derivative functions theory is a real part of some analytical function $\Phi(z)$, which is found as the Cauchy integral:

$$\Phi(z) = \frac{1}{2\pi i}\int_\Gamma \frac{\mu(\zeta)}{\zeta - z}\, d\zeta \tag{7.18}$$

with the real density $\mu(\zeta)$, where $\zeta \in \Gamma; z = \varphi + i\lambda$ is a random point in G. Directing z to some point u on the contour Γ and taking into account relationships $\text{Re}\ \varphi(u) = \tilde{T}_J(u)$ and $\text{Im}(d\zeta/(\zeta - u)) = -\cos(r,n)d\sigma/r)$, where r is the distance from ζ to u (the direction is chosen from ζ to u); $d\sigma$ is the length element on Γ; and n is the external normal to Γ. From (7.18) we obtain for $\mu(u)$ the Fredholm integral equation:

$$\mu(u) - \frac{1}{\pi}\int_\Gamma \mu(\varsigma)\frac{\cos(r,n)}{r}\, d\sigma = 2\tilde{T}_J(u)$$

with the continuous core $\cos(r,n)/r$, which can be solved with any right-hand part. Having solved this equation, we find $\varphi(z)$ and, hence:

$$T_J(\varphi, \lambda, \eta, \theta, t) = \text{Re}\ \varphi(z)$$

When G is a circle $|z - z_0| < R$, the solution:

$$T_J(\varphi, \lambda, \eta, \theta, t) = T_J(r, \psi, \eta, \theta, t)(\varphi + i\lambda = z_0 + re^{i\psi}, r < R, 0 \le \psi \le 2\pi)$$

of the Dirichlet problem can be obtained in the form of the Poisson integral:

$$T_J(\varphi, \lambda, \eta, \theta, t) = \frac{1}{2\pi}\int_0^{2\pi} \tilde{T}_J(a)\frac{R^2 - r^2}{R^2 + r^2 - 2Rr\cos(\psi - a)}\, da,$$

where $\tilde{T}_J(a) = \tilde{T}_J(z_0 + \text{Re}^{ia})$; and $(0 \le a \le 2\pi)$.

Without breaking the integrity, we apply the described method together with the method of differential approximation to the procedure of retrieving the data along the route measurements and mapping the territory G at a time moment t^*. Let remote measurements be made in the time interval $[t_0, t_L]$ at a discrete number of points $A_i(i = 1, \dots, N)$ on the boundary Γ. Assume that during the time of measurements Δt, the level of the time-dependence of observational data, is negligibly small (i.e., the whole series of measurements can be divided into $M = |[t_L - t_0]/\Delta t|$ of statistically reliable sites $[t_j, t_{j+1}](j = 1, \dots, M)$, and all measurements can be presented in the form of matrix $\|T_J(i,j)\|$). The method of differential approximation makes it possible to reduce all lines in this matrix to a moment t^* and then, following the method described above to retrieve T_J on the territory G.

7.3.3 The approximation method to solve the inverse problem in the case of identification of geophysical parameters

In the process of monitoring a multitude of data series are formed, use of which is necessary to establish correlations between the parameters of the object under study. Consider a situation that occurs under conditions of radio-physical monitoring. Let at a time moment t_i, at the output of each measuring device (radiometer), the values $Z_{ij}(i = 1, \dots, M; j = 1, \dots, n)$ be fixed so that $Z_{ij} = T_j + \xi_{ij}$. Here T_j is the real value of the jth parameter (radio-brightness temperature at a wavelength λ_j), and ξ_{ij} is the noise constituent. The search for a correlation is reduced to the determination of the dependence:

$$T_j = f_j(X) \tag{7.19}$$

where $X = (x_1, \ldots, x_m)$ are geophysical parameters.

There are a lot of approaches to find the function f. As a rule, the mean-square deviation is used as the criterion of agreement (Borodin and Krapivin, 1998). However, this criterion cannot reflect the dispersive characteristics of the noise constituent in measurements. Therefore, consider the problem from this point of view. Let the function (7.19) be linear, and then we obtain the system $n \geq m$ of equations:

$$\|A_{ij}\| X = T + \Xi \tag{7.20}$$

A solution to (7.20) should be found so that its dispersion is at a minimum. It is assumed that $\Xi = \{\xi_1, \ldots, \xi_n\}$ has a zero average and dispersion $\{\sigma_1^2, \ldots, \sigma_n^2\}$. This solution for $\{x_1^*, \ldots, x_m^*\}$ we call a σ-solution.

Multiply the ith equation of the system (7.19) successively by magnitudes $c_{1i}, \ldots, c_{mi} (i = 1, \ldots, m)$ and let

$$\sum_{i=1}^{n} c_{ji} A_{il} = \sigma_{jl} \tag{7.21}$$

$$\delta_{jl} = \begin{cases} 1 & j = l \\ 0 & j \neq l \end{cases} \quad (l, j = 1, \ldots, m) \tag{7.22}$$

With conditions (7.21) and (7.22) satisfied we obtain:

$$x_1^0 = \sum_{i=1}^{n} c_{1i} T_i \tag{7.23}$$

Similar relationships are written for $x_j^0 (j = 2, \ldots, m)$. Substituting T for Z in (7.23) (i.e., proceeding to the system (7.20)), we have:

$$\tilde{x}_1 = \sum_{i=1}^{n} c_{1i}(T_i + \xi_i) \tag{7.24}$$

From (7.24) we calculate the dispersion:

$$D[\tilde{x}_1] = \sum_{i=1}^{n} c_{1i}^2 \sigma_i^2 \tag{7.25}$$

Since the $\tilde{x}_1$ and x_1^0 averages coincide by definition, to solve the posed problem it is necessary to find a minimum of dispersion (7.25) with conditions (7.22) satisfied. Use the method of uncertain Lagrangian multipliers and form an auxiliary expression:

$$\varphi(c_{11}, \ldots, c_{1k}) = \sum_{i=1}^{n} c_{1i}^2 \sigma_i^2 + \mu_1 \left(\sum_{i=1}^{n} c_{1i} A_{i1} - 1 \right) + \sum_{j=2}^{m} \mu_j \sum_{i=1}^{n} c_{1i} A_{ij} \tag{7.26}$$

Equalizing the first derivatives of the function (7.26) to zero, we obtain:

$$2c_{1k}\sigma_k^2 + \sum_{j=1}^{m} \mu_j A_{kj} = 0 \quad (k = 1, \ldots, n) \tag{7.27}$$

Relationships (7.27) and conditions (7.22) constitute a system $(m+n)$ of equations whose solution makes it possible to determine the sought optimal values of c_{ij}^*. Analysis shows that $D[x_j] = -\mu_j/2$. Values of μ_j we find from the system of equations:

$$\sum_{j=1}^{m} \mu_j \sum_{i=1}^{n} \frac{A_{ij}A_{i1}}{\sigma_i^2} = -2 \quad \sum_{j=1}^{m} \mu_j \sum_{i=1}^{n} \frac{A_{ij}A_{il}}{\sigma_i^2} = 0 \quad l = 2, \ldots, m$$

Quantitative estimates show that the σ-solution is preferable compared with that obtained by the criterion of mean-square deviation. Consider the case $m = 2$ and $n = 3$, where x_1 is the thermodynamic temperature and x_2 is the mineralization degree. From (7.27) we have:

$$c_{1k}^* = \frac{1}{\Delta\sigma_k^2}\left(A_{k1}\sum_{i=1}^{n}\frac{A_{i2}^2}{\sigma_i^2} - A_{k2}\sum_{i=1}^{n}\frac{A_{i1}A_{i2}}{\sigma_i^2}\right) \quad k = 1, \ldots, n \tag{7.28}$$

$$c_{2k}^* = \frac{1}{\Delta\sigma_k^2}\left(A_{k2}\sum_{i=1}^{n}\frac{A_{i1}^2}{\sigma_i^2} - A_{k1}\sum_{i=1}^{n}\frac{A_{i1}A_{i2}}{\sigma_i^2}\right) \quad k = 1, \ldots, n \tag{7.29}$$

where:

$$\Delta = \sum_{i=1}^{n}\frac{A_{i1}^2}{\sigma_i^2}\sum_{i=1}^{n}\frac{A_{i2}^2}{\sigma_i^2} - \left(\sum_{i=1}^{n}\frac{A_{i1}A_{i2}}{\sigma_i^2}\right)^2$$

An optimal estimate of x_j^* is determined from the relationship:

$$x_j^* = \sum_{i=1}^{n} c_{ji}^* Z_i \quad (j = 1, 2)$$

The dispersion of the x_j^* estimate is as follows:

$$D[x_1^*] = \Delta^{-1}\sum_{i=1}^{n}\frac{A_{i2}^2}{\sigma_i^2} \quad D[x_2^*] = \Delta^{-1}\sum_{i=1}^{n}\frac{A_{i1}^2}{\sigma_i^2} \tag{7.30}$$

Compare this estimate with the estimate by the method of least squares. Let:

$$\|A_{ij}\| = \begin{Vmatrix} 1 & 1 \\ 1 & 2 \\ 1 & 3 \end{Vmatrix}$$

Then from the formulas (7.30) we obtain:

$$c_{11}^* = (6\sigma_2^{22} + 2\sigma 3)/\Delta_1 \quad c_{12}^* = (3\sigma_1^2 - \sigma_3^2)/\Delta_1 \quad c_{13}^* = -2(\sigma_1^2 + \sigma_2^2)/\Delta_1$$

$$c_{21}^* = -(2\sigma_2^2 + \sigma_3^2)/\Delta_1 \quad c_{22}^* = (-\sigma_1^2 + \sigma_3^2)/\Delta_1 \quad c_{23}^* = -(\sigma_1^2 + 2\sigma_2^2)/\Delta_1$$

where $\Delta_1 = \sigma_1^2 + 4\sigma_2^2 + \sigma_3^2$.

Then we have:

$$D[x_1^*] = (9\sigma_1^2\sigma_2^2 + 4\sigma_1^2\sigma_3^2 + \sigma_2^2\sigma_3^2)/\Delta_1;$$
$$D[x_2^*] = (\sigma_1^2\sigma_2^2 + \sigma_1^2\sigma_3^2 + \sigma_2^2\sigma_3^2)/\Delta_1$$

Let $\hat{x}_1$ and $\hat{x}_2$ be estimates of the parameters x_1 and x_2, obtained by the method of least squares (i.e., be solutions of the minimization problem):

$$\min_{x_1,x_2}\left(\sum_{i=1}^{n}(T_i + \xi_i - A_{i1}x_1 - A_{i2}x_2)^2\right)^{1/2} = \left(\sum_{i=1}^{n}(T_i + \xi_i - A_{i1}\hat{x}_1 - A_{i2}\hat{x}_2)^2\right)^{1/2}$$

We have:

$$\hat{x}_1 = 4(T_1 + \xi_1)/3 + (T_2 + \xi_2)/3 - 2(T3 + \xi_3) \quad \hat{x}_2 = -(T_1 + \xi_1)/2 + (T_3 + \xi_3)/2$$

$$D[\hat{x}_1] = (16\sigma_1^2 + \sigma_2^2 + 4\sigma_3^2)/9 \qquad D[\hat{x}_2] = (\sigma_1^2 + \sigma_3^2)/4$$

It is seen that $D[\hat{x}_1] \geq D[x_1^*]$ and $D[\hat{x}_2] \geq D[x_2^*]$. Hence, the σ-solution is preferable compared with the estimates obtained by the method of least squares.

7.3.4 Algorithm of randomization for the linear–broken approximation

Measurements of the environmental parameters in the monitoring regime provide sets of series of quantitative characteristics for the system of data processing, which cannot be analyzed because of their stationarity. There are a lot of ways to overcome the time-dependence, which make it possible to remove a contradiction between applicability of statistical methods and the level of the observational data stationarity. One such way consists in the partition of a series of noise-loaded measurements into quasi-stationary parts (Borodin *et al.*, 1996).

Let the results of measurements be presented by a succession of magnitudes Z_{ij}, where $i = 1, \ldots, N$ is the number of time intervals and $j = 1, \ldots, M$ is the number of the measuring devices (information channels). It is assumed that:

$$Z_{ij} = T_{ij} + \xi_{ij} \tag{7.31}$$

where T_{ij} and ξ_{ij} are determinate and stochastic constituents, respectively, with ξ_{ij} having a zero average and dispersion σ_j^2.

The problem of sampling the piecewise constant of a random succession (7.31) is reduced to the classification of this sample by the indicator belonging to samples from distributions with similar averages. To approximate a sample $\{Z_{ij}\}$ by the linear broken randomized function, we perform the following operations. First we find the difference:

$$\Delta Z_{kj} = Z_{k+1,j} - \frac{1}{k}\sum_{l=1}^{k} Z_{lj} = \Delta T_{kj} + \frac{1}{k}\sum_{l=1}^{k}\Delta\xi_{lj}$$

If the magnitudes Z_{kj} and $Z_{k+1,j}$ belong to samples with similar averages, then

$\Delta T_{kj} = 0$. Otherwise $\Delta T_{kj} \neq 0$. Assume that Z_{kj} and $Z_{k+1,j}$ belong to a sample from distributions with similar averages, if:

$$|\Delta Z_{kj}| \leq a_{kj} \tag{7.32}$$

where $a_{kj} = d\sigma_j$, and d is an adaptation coefficient (usually $d = 3 - (1 + 1/k)^{1/2}$). Beginning from $k = 1$ and continuing to successively calculate ΔZ_{kj} and check the condition (7.32), we find the quasi-stationary section of succession $\{Z_{ij}\}$. If the condition (7.32) is not satisfied simultaneously for $Z_{(k+1),j}$ and $Z_{(k+2),j}$, then the element Z_{kj} is considered the last in the sub-multitude, whose elements satisfy the condition of quasi-stationarity. The subsequent sub-multitude of the series $\{Z_{ij}\}$ begins from $Z_{(k+1),j}$ as a first element. The sub-multitude where the average is not a constant value (i.e., the condition (7.32) is never satisfied) is formed as a sub-multitude of random values, whose average changes, following the linear law. In this case, at all stages of the procedure the values $\Delta Z_{(k+m),j} = Z_{(k+m+1),j} - Z_{(k+m),j} (m = 1\text{–}s)$ are calculated. The linear approximation of the section of series $\{Z_{ij}\}$ is constructed between the values $Z_{(k+1),j}$ and Z_{sj}. The equation for the sought straight line is written as:

$$Z - \bar{Z}_{\bar{t}_{sj}} = \overline{\Delta Z}_{sj}(t - \bar{t}_{sj}) \tag{7.33}$$

where t is the time identified by the moment of measurement recording:

$$\bar{t}_{sj} = 0.5(s\text{-}k\text{-}2) \quad \overline{\Delta Z}_{sj} = \frac{1}{N-1}\sum_{i=1}^{N-1} \Delta Z_{(k+1),j} \quad N = 2\bar{t}_{sj} \quad \bar{Z}_{sj} = \frac{1}{N}\sum_{i=1}^{N} Z_{(k+i),j}$$

Checking the stability of the angle of straight line slope (7.33) in the process of its formation is carried out by analyzing the value:

$$\Delta\tilde{Z}_{lj} = \Delta Z_{lj} - \frac{1}{l}\sum_{i=1}^{l} \Delta Z_{(k+1),j}$$

calculated at each step l. Single violations of this stability (i.e., when $|\Delta\tilde{Z}_{lj}| \geq 6\sigma j(1 + l^{-1})^{1/2}$ are considered as accidental releases and either are excluded from consideration or substituted with average values).

7.4 MATHEMATICAL MODELING AND NUMERICAL EXPERIMENT: THE GEOPHYSICAL STUDY OF THE ARAL SEA REGION

7.4.1 Model for structured functional analysis of hydrophysical fields of the Aral Sea

As experience in combating the large-scale change of the geochemical and hydrological situation in the Aral Sea zone has shown, the problem of suspending the process of desertification and ecological degradation in this region cannot be solved without creating a multi-level monitoring system in accordance with functions of a forecast (Golitsyn, 1995; Krapivin and Phillips, 2001). The GIMS-technology

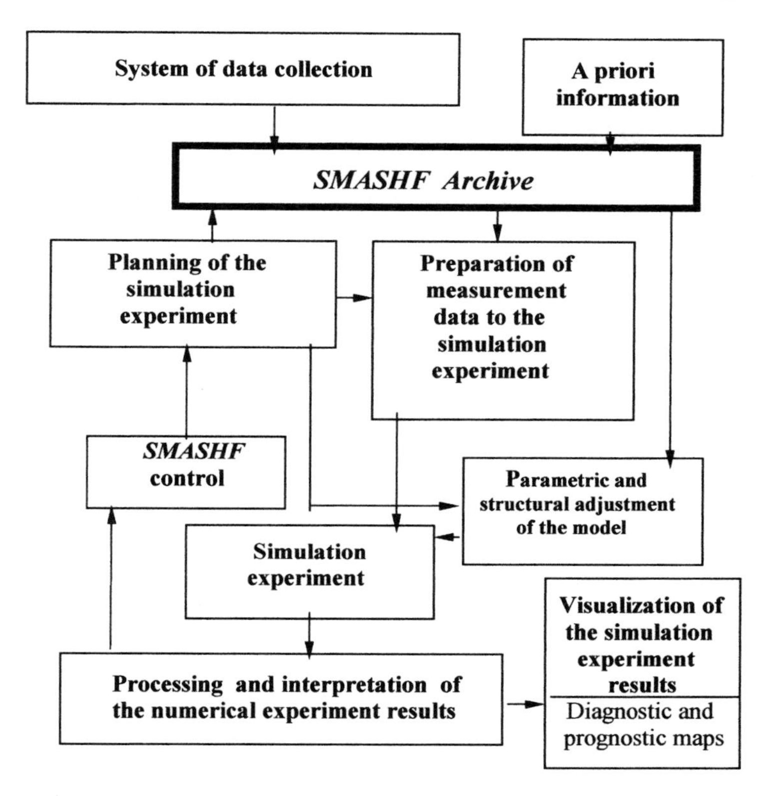

Figure 7.9. Scheme of information fluxes at SMASHF fixing.

suggests a possibility of an adaptive–successive analysis of information about the state of the main hydrophysical fields (temperature τ and salinity S) with correction of the simulation model (by the feedback principle) and control of the processes of collection and processing of the monitoring data. Consider a simulation model of the Aral Sea hydrophysical fields (SMASHF). The SMASHF includes units for information collection, primary processing and accumulation of the monitoring data, simulation of the functioning of the Aral Sea aquageosystem's water regime, prediction of its state, estimation of any discrepancy between measured and predicted states, making decisions on measurement planning, control of hydroeconomic enterprises, and service support when operating with input and output information. The functional structure of the SMASHF is shown schematically in Figure 7.9.

The model of the Aral Sea aquageosystem is a base element of the SMASHF, which provides, at the expense of structural and parametric change, an adaptation of the monitoring regime. The functional detail of the model is shown in Table 7.5. The model describes the hydrophysical processes over Ω_A in the ice-free period. Currents, convective mixing, river run-off, processes of heat and moisture exchange with the atmosphere, and turbulent exchange of heat, salt, and momentum are simulated. At

Table 7.5. The SMASHF units.

Unit	Characteristics of the unit functions
FIC	Formation of initial conditions for the models with regard for multiyear changes in hydrophysical characteristics of the Aral Sea. Choice of steps of space digitization in latitude and longitude. Determination of the step of temporal digitization.
MSC	Model of the seasonal changes of the Aral Sea level.
MSF	Model of the formation of the structure of currents in the Aral Sea.
MTS	Model of the spatial–temporal distribution of water temperature and salinity.
MWD	Model of the water density field formation.
SCT	Simulation of the process of convective mixing of seawater masses.
RCR	Provision of the regime of structural and parametric correction of SMASHF units with modeling results taken into account.

this point, in order to form initial conditions for the end of the winter season t_w a correction is made to the hydrophysical parameters of the aquageosystem obtained in the FIC unit considering multiyear changes in sea level and other indicators of the state of the water basin Ω_A. The calculation grid of depths and initial fields of salinity $S(\varphi, \lambda, z, t)$ and temperature $\tau(\varphi, \lambda, z, t)$ of seawater are corrected. These procedures take into account characteristic estimates of indicated parameters from fragments of multiyear measurements at a moment t_w. The initial field $\tau(\varphi, \lambda, z, 0)$ is prescribed with due regard to water mass geometry at that moment. The field $S(\varphi, \lambda, z, t_w)$ is formed considering a division of Ω_A into areas of water with characteristic homogenous hydrophysical parameters:

$$S(\varphi, \lambda, z, t_w) = \begin{cases} S_1 & \text{for } (\varphi, \lambda) \in \Omega_D \\ S_2 & \text{for } (\varphi, \lambda) \in \Omega_A \backslash \Omega_D \end{cases}$$

where Ω_D is the western deep-water sector of the Aral Sea, $S_1 = (S_D - S_A)/V_D$; $S_2 = (S_P - S_L)/V_P$; V_D and V_P are water volumes in the western deep-water area and the shallow Large Sea, respectively; S_D and S_P are salt supplies in Ω_D and $\Omega_A \backslash \Omega_D$, respectively; and S_A and S_L are salt supplies on dry territories adjacent to Ω_D and $\Omega_A \backslash \Omega_D$, respectively.

The seasonal or intra-annual oscillations of the Aral Sea level are described by the prognostic model MSC which traditionally takes into account the balance constituents of the seawater regime: evaporation, precipitation, and river run-off. The MSC uses multiyear average values of these water balance components as well as information about the present hydrometeorological situation in the Aral Sea zone. The MSC unit simulates fluxes in Ω using the method of Shkodova and Kovalev (Bortnik *et al.*, 1994), which makes it possible to evaluate the horizontal components of velocity $u(\varphi, \lambda, z, t)$ at any point with geographical coordinates $(\varphi, \lambda) \in \Omega_A$ with a random value of depth z. The vertical component of velocity is calculated from the equation of continuity. It is assumed that the motion of water masses is determined by tangential wind stress over Ω_A, and sea surface slopes caused by negative and positive set-ups due to wind unevenness, considering the effect of shores and river run-off. The driving force is balanced by the vertical viscosity and bottom friction. The coefficient of turbulent viscosity is calculated from the formula $v = 0.25aWH/k$, where W is the wind speed, H is the depth, k is the wind coefficient, and a is the proportionality coefficient.

The basic equations of the MSF unit are as follows:

$$u_\varphi = T_\varphi(H - z)/v + (g\rho_o/v) \cdot 0.5(H^2 - z^2) \cdot (\partial\kappa/\partial\varphi)$$

$$u_\lambda = T_\lambda(H - z)/v + (g\rho/v) \cdot 0.5(H^2 - z^2) \cdot (\partial\kappa/\partial\lambda)$$

$$\partial\kappa/\partial\varphi = -3T_\varphi/(2g\rho_o H) - 3v/(g\rho_o H^3) \cdot (\partial\psi/\partial\lambda)$$

$$\partial\kappa/\partial\lambda = -3T_\lambda/(2g\rho_o H) - 3v/(g\rho_o H^3) \cdot (\partial\psi/\partial\varphi)$$

where g is the acceleration of gravity, ρ_o is the water density averaged over Ω, $\boldsymbol{T}(T_\varphi, T_\lambda)$ is the vector of tangential wind stress on the sea surface, κ is the sea level deviation from the surface of the basin, and ψ is the function of complete flows.

The MTS unit simulates the spatial–temporal structure of distribution of salinity S and temperature τ of seawater. The water area Ω_A is divided into compartments $\Omega_i(\cup\Omega_i = \Omega_A)$, which are internally homogeneous both in S and τ. The heat and salt transport between Ω_i is carried out by the currents and differences in gradients. The exchange processes on the boundary with the atmosphere are described by linear relationships taking into account multiyear observational data.

To approximate the vertical variations of the Aral Sea water density $\rho(\varphi, \lambda, z, t)$ the model MWD was used. The SCT unit checks the criterion of the stability of water stratification, and on this basis carries out the convective mixing of water masses. The stratification is considered stable at $\partial\rho_o/\partial z \geq 0$. In this case the convective mixing is absent. With $\partial\rho_o/\partial z < 0$ the stratification is considered unstable, and the process of convective mixing is identified by averaging S and τ between adjacent water layers.

An adaptation of SMASHF is accomplished by the RCR unit based on the data of multiyear measurements from the ship *Otto Schmidt* carried out by the State Oceanographic Institute of Rosgidromet (Bortnik *et al.*, 1994). Measurements were made in the ice-free period over part of Ω_A during 10 days, three times a season at several horizons with an accuracy of 0.01°C for τ and 0.01‰ for S. To correct the obtained measurement data in order to reduce them to daily mean values, corrections from Table 7.6 were used. An exclusion of random fluctuation components in measured values, and their reduction to a form comparable with the modeling results were made by averaging and evaluation of the following characteristics: S value averaged over depth, the depth of the thermocline and the temperature at this depth, and variation and average water temperature above and beneath the thermocline. The parametric adaptation of SMASHF to these values was assessed from the minimum of mean-square deviation. The process of adaptation resulted in schematic maps of distributions of all of the model's components over Ω_A, which

Table 7.6. Corrections of measured values of water temperature (τ, °C) in modeling the Aral Sea hydrophysical fields.

Depth (m)	Time of measurement (hour)			
	7–13	13–18	18–23	23–7
Open areas of the Aral Sea				
0	0	−1.5	0	1.5
5	0	−0.5	0	0.5
≥10	0	0	0	0
Coastal water bodies of the Aral Sea				
0	0	−2.5	0	2.5
5	0	−1.5	0	1.5
10	0	−0.5	0	0.5

Table 7.7. Characteristics of the algorithm of SMASHF adaptation realized in the RCR unit.

Level of adaptation priority	Content of the adaptation process
1	Correction of the vertical constituent of the coefficients of the vertical exchange for τ and $S(\nu_{t,z}, \nu_{S,z})$.
2	Correction of the horizontal constituent of the coefficients of the vertical exchange for τ and $S(\nu_{t,\varphi\lambda}, \nu_{S,\varphi\lambda})$.
3	Change of t_w.
4	Correction of an average value of the sea surface heat balance.
5	Correction of an average value of the sea surface water balance.
6	Correction of $S(\varphi, \lambda, z, t)$.

were used as the initial data for prediction until the next measurement. The SMASHF was adapted using the method of alternative adaptation (Aota *et al.*, 1993). Table 7.7 demonstrates synthesized stages of adaptation and gives a characteristic of this process.

For simulation experiments, two periods of the Aral Sea aquageosystem's functioning have been chosen: 1981 with river run-off to the sea and 1989 without it. The values of the SMASHF parameters are given in Table 7.8. The values of the coefficients of turbulent heat $(\nu_{\tau,\varphi\lambda}, \nu_{\tau,z})$ and salt $(\nu_{S,\varphi\lambda}, \nu_{S,z})$ exchange have been obtained as a result of the model's adaptation. The upper quasi-homogenous layer h was assumed to be 15 m thick for the western deep-water hollow and 10 m for the central zone of the Small Sea.

The results of simulations presented in Figure 7.10 and Table 7.9 show an efficiency of SMASHF as an element of the system of the Aral Sea monitoring in the presence or absence of the Amu-Darya and Syr-Darya River run-off. On average, the SMASHF retrieves the spatial distribution of S and τ with a relative uncertainty

Table 7.8. Estimates of the SMASHF parameters assumed in simulation experiments.

Parameter	Time-dependent estimates of parameters	
	1981	1989
a (g m^{-3})	3.25×10^{-6}	3.25×10^{-6}
k	0.015	0.015
$\Delta\varphi, \Delta\lambda$ (km)	2	1
Δt (minutes)	60	30
W (m s^{-1})	5.5	5.5
$\nu_{T,\varphi\lambda}$ (m^2 s^{-1})	10^3	0.5×10^3
$\nu_{T,\varphi\lambda}$, m^2 s^{-1}	0.5×10^3	0.2×10^3
$\nu_{T,z}, \nu_{S,z}, (z < h)$ (cm^2 s^{-1})	0.5	0.4
$\nu_{T,z}, \nu_{S,z}, (z \geq h)$ (cm^2 s^{-1})	0.5	0.4

Table 7.9. Comparison of field measurements with simulation results for the vertical temperature distribution t (°C) in September–October 1981 and 1989 as an example.

Depth (m)	1981			1989		
	Measured values	Result of adaptation	Forecast	Measured values	Result of adaptation	Forecast
Western deep-water trough						
0	19.6	19.5	20.0	15.3	16.5	16.5
5	19.5	19.5	19.5	17.5	17.5	17.8
10	19.4	19.5	19.5	17.4	17.3	17.8
15	19.3	19.0	19.5	17.5	17.3	17.6
20	19.2	19.0	19.5	17.5	17.3	17.6
25	14.3	16.0	17.0	17.5	17.3	17.2
30	6.7	8.0	9.0	17.5	17.3	17.1
40	4.5	6.0	6.0	15.7	15.5	15.3
50	1.5	3.0	4.0	15.7	15.5	15.3
Central zone						
0	18.3	18.4	18.4	13.5	13.0	13.0
5	18.3	18.4	18.0	13.7	13.0	13.0
10	18.1	18.0	18.0	14.0	13.5	13.5

of up to 10% and gives a prognostic estimate of these distributions for 2 months with a relative error of up to 15%. At other time moments the SMASHF will provide estimates of all hydrophysical parameters over Ω_A with the accuracies indicated above. Thus, the SMASHF controled by the GIMS ensures an acceptable accuracy and permits its inclusion as an element of the structure of the monitoring system of a higher level. The SMASHF is adapted to this function by coordination of the formats of its inputs and outputs with the respective GIMS formats.

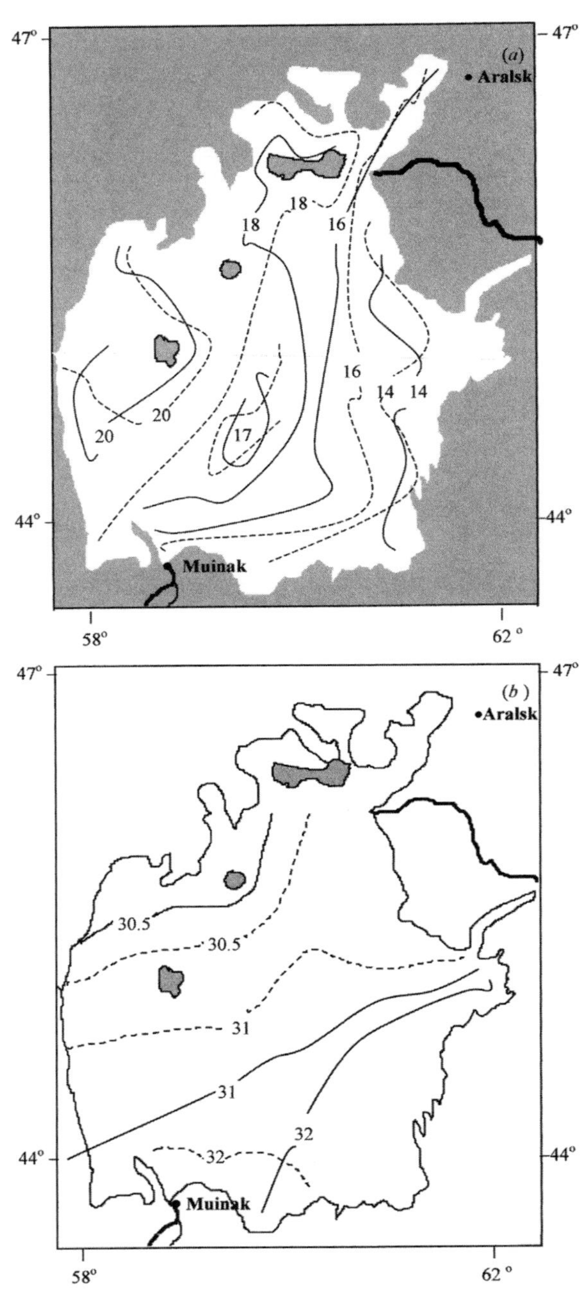

Figure 7.10. (a) Comparison of the results of the forecast (solid curve) of temperature (°C) of the Aral Sea water with field measurements (dashed curves) in October 1981. (b) Comparison of the results of the forecast (solid curve) of salinity (‰) of the Aral Sea water with the field measurements (dashed curves) in October 1989.

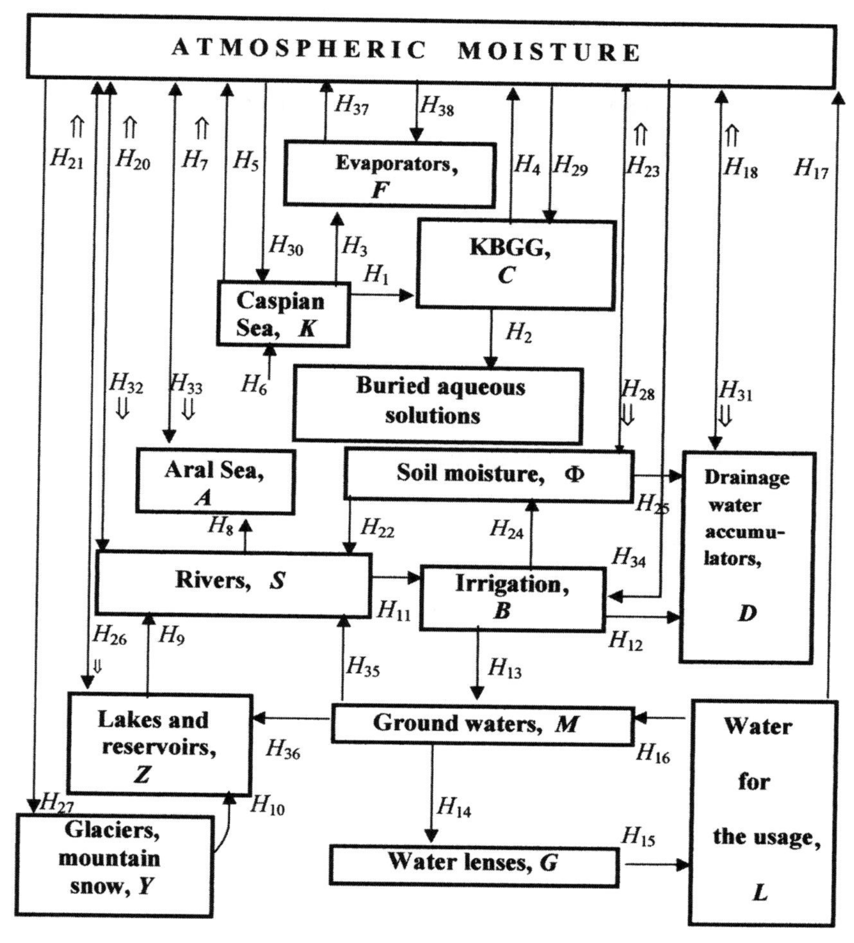

Figure 7.11. The block-scheme of the flux diagram of the Caspian–Aral aquageosystem's water balance realized in SMASHF. Notation is given in Table 7.11.

7.4.2 Model of the regional water balance of the Aral Sea aquageosystem

A conditional scheme of the water balance of the Caspian and Aral Sea's zone of impact is shown in Figure 7.11. To develop a computer analog of the scheme in Figure 7.11, assume a geographical grid of digitizing the territory of the region Ξ into quasi-homogeneous sites Ξ_{ij} of size $\Delta\varphi$ in latitude and $\Delta\lambda$ in longitude, where the magnitudes $\Delta\varphi$ and $\Delta\lambda$ are free parameters chosen in correspondence with the database. Each site Ξ_{ij} is characterized by the type of the floristic background α_{ij} and soil β_{ij}, surface moistening Φ_{ij}, and depth of ground waters M_{ij}. Projecting the grid $\Delta\varphi \times \Delta\lambda$ onto the map of the territory Ξ gives a totality of geophysically heterogeneous objects, which exchange water following the balance relationships

Table 7.10. Totality of the basic matrix identifiers of SMASHF.

Identifier	Description of the identifier
A_1	Selection of the territory Ξ; $A_1 = \|a_{ij}^1\|$, $a_{ij}^1 = 0$ at $(\varphi,\lambda) \notin \Xi$, $a_{ij}^1 = a_1$ at $(\varphi,\lambda) \notin \Xi$ $a_1 = 1$ for the Aral Sea, 2 for the Caspian Sea, 3 for the KBGG, 4 for saline lands, 5 for rivers, 6 for channels, 7 for irrigation systems, 8 for takyrs, etc.
A_k	Statistical mean data on wind speed ($k = 2$), air temperature ($k = 3$), precipitation ($k = 4$), and wind direction ($k = 5$).
A_6	Initial data for all the model's components.
A_7	Relief of the territory Ξ considering the grid $\Delta\varphi \times \Delta\lambda$ (absolute level).
A_8	Classification of X_{ij} as artificial evaporators ($a_{ij}^8 = 1$ for *yes*, $a_{ij}^8 = 0$ for *no*) and as regions with forced precipitation ($a_{ij}^8 = 2$ for *yes*, $a_{ij}^8 = 0$ for *no*).

in accordance with the scheme in Figure 7.11. In each cell Ξ_{ij} with an area σ_{ij} the functioning of the system considered is presented by the sub-multitude of fluxes $\{H_k\}$ selected by matrix identifiers $\{A_s\}$ (Table 7.10), which, according to GIMS-technology, reflect the information structure of the database. In a generalized form, the balance equations of moisture motion in Ξ_{ij} are written as:

$$\sigma_{ij}\left(\frac{\partial W_{ij}}{\partial t} + v_\varphi \frac{\partial W_{ij}}{\partial \varphi} + v_\lambda \frac{\partial W_{ij}}{\partial \lambda}\right) = \sum_{s \in I_{ij}} H_s^{ij} - \sum_{s \in J_{ij}} H_s^{ij} \tag{7.34}$$

$$\sigma_{ij}\frac{dE_{ij}}{dt} = \sum_{s=1}^{38} (\omega_s - \gamma_s) H_s^{ij} \tag{7.35}$$

$$\frac{dL_{ij}}{dt} = H_{15}^{ij} - H_{16}^{ij} - H_{17}^{ij} \tag{7.36}$$

where I_{ij} and J_{ij} are the integral-valued multitudes of indices of fluxes H_k^{ij} corresponding to processes of evaporation and precipitation over Ξ_{ij}; $E = (A, B, C, D, K, S, F, M, Z, Y, G, \Phi)$; and $\omega_i (i = 1$–$5)$ are identifiers of water objects that coordinate the water balance equations with the scheme in Figure 7.11 (e.g., $\omega_1 = 0$ at $E \neq B$ and $\omega_1 = 1$ at $E = B$, etc.). The functions W, E, M, G, Y, and Φ are measured in linear units of the water column, the remaining functions are measured in volume units.

To describe the fluxes $H_k (k = 1$–$38)$ we use the models developed by many experts, parametric dependences, as well as tabulated and graphic functions (Bras, 1990). Evaporation from the soil and water surfaces is parameterized with the various Dalton-type formulas (Hsu *et al.*, 2002; Bras, 1990) depending on the user's desire or in accordance with the presence of the required information in the database. To calculate evaporation from the Aral Sea surface, the formula of Goptarev is used (Bortnik and Chistiayeva, 1990). Evaporation from other water surfaces is described by the RIV-model: $H_s = \mu(\theta)(\rho_w - \rho_2)(s \in I)$, where $\mu(\theta)$ is the indicator of the impact of wind on evaporation, ρ_w is the water vapor pressure at a

Table 7.11. Water flows in Figure 7.10.

Flow	Identifier to the flow
Run-off from the Caspian Sea to the KBGG	H_1
Buried solutions	H_2
Artificial evaporation	H_3
Evaporation from the surface:	
KBGG	H_4
Caspian Sea	H_5
Aral Sea	H_7
accumulators of drainage waters	H_{18}
irrigation waters	H_{19}
Amu-Darya and Syr-Darya Rivers	H_{20}
lakes and reservoirs	H_{21}
soils	H_{23}
artificial evaporators	H_{37}
River run-off:	
to the Caspian Sea	H_6
to the Aral Sea	H_8
Replenishment of Amu-Darya and Syr-Darya Rivers due to lakes and reservoirs	H_9
Melting of glaciers and snowfields	H_{10}
Water expenditure on irrigation	H_{24}
Input of water to accumulators of drainage waters	H_{12}
Leakage from irrigation systems	H_{13}
Water accumulation in lenses	H_{14}
Taking of water from lenses for domestic needs	H_{15}
Return waters	H_{16}, H_{17}
Diversion of flow to irrigation systems	H_{11}
Surface run-off to rivers	H_{22}
Surface run-off from irrigated territories	H_{25}
Precipitation onto:	
lakes and reservoirs	H_{26}
glaciers and snowfields	H_{27}
soil	H_{28}
KBGG	H_{29}
Caspian Sea	H_{30}
Accumulators of drainage waters	H_{31}
Amu-Darya and Syr-Darya Rivers	H_{32}
Aral Sea	H_{33}
irrigation systems	H_{34}
artificial evaporators	H_{38}
Replenishment of water bodies due to ground waters	H_{36}
Replenishment of rivers due to ground waters	H_{35}

temperature of the evaporating surface (gPa), and ρ_2 is the absolute air humidity at a height of 2 m (gPa).

The relationship between surface water fluxes and ground waters is described with the models of Bras, Horton, and Hagen-Poisseville. In general, a consideration of the impact of ground waters in the Aral Sea zone has been studied in detail. In this case the absence of reliable data on the replenishment of the Aral Sea with ground waters makes one introduce the flux H_{35} only. It is assumed that mainly anthropogenic factors affect the depth of ground waters, according to the scheme in Figure 7.11. The level of ground waters is different for different types of soils: in sand deposits it is 2–2.25 m and in clay it is 3.5–4 m. As one of the models of water balance of the ground horizon the following formula was used: $H_{14} = Q_1 + \mu \cdot \Delta M \cdot \sigma / \Delta t + Q_2$, where Q_1 is the overflow from other water horizons due to pressure, Q_2 is the resultant of the horizontal expenditure of the ground horizon, μ is the water yield of the ground horizon, and ΔM is a decrease for the time interval Δt.

The scheme of Figure 7.11 lacks ground water evaporation. It is assumed that this flux, over the whole horizon considered does not affect substantially the water balance, and therefore it is as if distributed among the fluxes H_{14}, H_{15}, H_{35}, and H_{36}. From the available estimates, the amount of ground water evaporation varies within 0.1–1.3 $l\,s^{-1}\,km^{-2}$. According to forecasts made by some experts (Aripov, 1973; Bortnik and Dauletiyarov, 1985), further states of the ground water level will be stable.

The scheme of moisture transfer in the soil–plant–atmosphere system has been assumed according to the algorithm from Kondratyev *et al.* (2002a). The local topography and its impact on the soil moisture dynamics within the assumed geographical grid have been taken into account using the water balance method (Bras, 1990). A comparison of the accuracy of concrete formulas to calculate the coefficients in (7.34)–(7.36) has been made by Krapivin and Phillips (2001) and Brass (1990). The parameterization of H_k fluxes and control of the simulation experiment is carried out using the software characterized in Table 7.12. An exemplary scheme of the model's functioning in the regime of dialogue with the user consists of the following. In the monitoring regime, initial data and varying parameters of the model are corrected with a probable varying time interval Δt, so that at moments $t_i = t_0 + i\Delta t$ a comparison is made between measured and modeled values, and on this basis, either the model is corrected or Δt is changed. The simulated data are identified with geophysical, ecological, and hydrological parameters from respective algorithms described in other chapters.

7.4.3 Model of the Kara–Bogaz–Gol Gulf aquageosystem

The KBGG is one of the unique natural evaporators of Caspian Sea water. Before 1980, the water–salt regime of the Gulf had been determined completely by the volume of inflowing water masses from the Caspian Sea (about 21.5 $km^3\,yr^{-1}$).

Table 7.12. The SMASHF software.

Identifier of the model's unit	Characteristics of the unit
CSII	Calibration and scaling of input information.
IDF	Input data filtering.
R2DD	Reconstruction of 2-D distributions as schematic maps.
EMAM	The Euler model of the atmospheric moisture transfer.
DI	Database interface.
PDF	Parametric descriptions of fluxes H_k $(k = 1, \ldots, 38)$.
SS	Synthesis of scenarios.
CPSD	Cartographic presentation of simulation results.
RSGM	Referencing of the structure $\{\Xi_{ij}\}$ to the geographic map using the identifiers $\{A_i\}$.
CSE	Control of the simulation experiment and forms of visualization of its results.
MAS	Model of the Aral Sea hydrophysical regime.
MKBG	Model of the KBGG hydrophysical and hydrochemical regimes.

The maximum depth of the Gulf has reached 12 m. A combination of climatic and external hydrological conditions over the last 50 years has created a trend toward reducing the Caspian Sea run-off to $6\,\mathrm{km^3\,yr^{-1}}$ in 1939. The run-off volume after negligible oscillations, starting in 1950, has stabilized at a level of $10\,\mathrm{km^3\,yr^{-1}}$. In March 1980, in connection with blocking the strait between the Caspian Sea and the KBGG with a blank dam, their connection became broken and this has led to a new phase in its functioning. In this phase, a trend is clearly seen toward reducing the size of the water surface of the Gulf accompanied by a decrease in depth and accelerated evaporation. As a result, the KBGG basin has started dividing into a western zone, with a depth of about 0.5 m, and an eastern zone 1.25 m deep, with a growth in the total concentration of brines and a substantial change in their chemical composition. In 1984, because of this trend the monitoring of the zone of the KBGG impact has begun; the results obtained have been included into the database mentioned above. It was established that the KBGG hollow has been gradually filling with lagoon deposits, and by 1995 a 40-m salt layer had formed. At present, the KBGG hollow is ready to serve as a powerful evaporator of Caspian waters, if they are transferred naturally. Of course, during this time period additional problems have appeared connected with a possible appearance of undesirable geochemical processes with the transfer of Caspian waters to the KBGG.

Let φ be the latitude and λ the longitude of the geographical point in the KBGG, $(\varphi, \lambda) \in \Omega_g$, where $\Omega_g \subset \Xi$ is the KBGG space. Denote as $v = (v_\varphi, v_\lambda)$ the vector of wind speed, $V = (V_\varphi, V_\lambda)$ the vector of the current velocity, $S_g(\varphi, \lambda, t)$ the concentration of salts in the atmosphere $(\mathrm{mg\,m^{-2}})$, $B_g(\varphi, \lambda, t)$ the concentration of brine $(\mathrm{mg\,m^{-2}})$, and $H_g(\varphi, \lambda, t)$ the concentration of deposited salts $(\mathrm{mg\,m^{-2}})$. Then the kinetic equations taking into account the scheme in

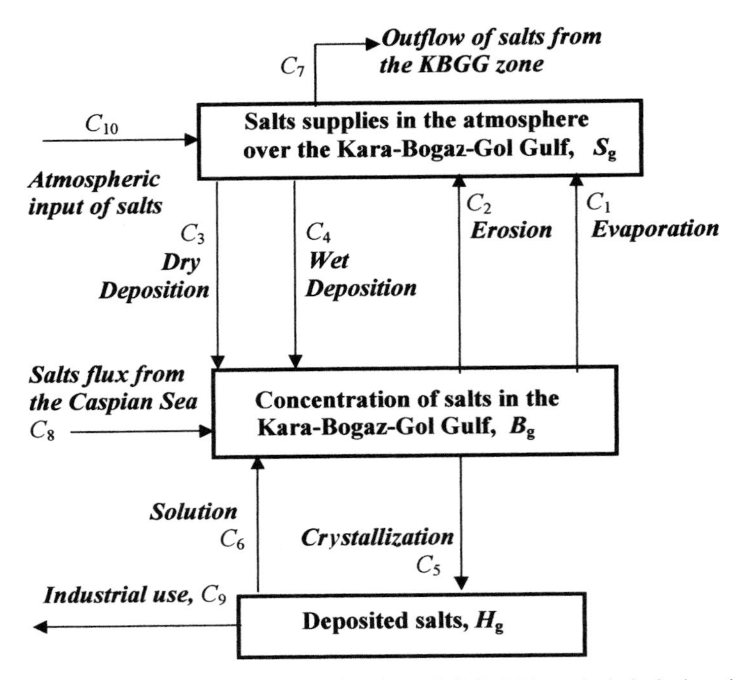

Figure 7.12. The principal scheme of salts in the KBGG (C_8) and their industrial removal (C_9).

Figure 7.12 will be written as:

$$\partial S_g/\partial t + v_\varphi \partial S_g/\partial\varphi + v_\lambda \partial S_g/\partial\lambda = C_1 + C_2 + C_{10} - C_3 - C_4 - C_7$$

$$\partial B_g/\partial t + V_\varphi \partial B_g/\partial\varphi + V_\lambda \partial B_g/\partial\lambda = C_3 + C_4 + C_8 - C_1 - C_2 - C_5 + C_6$$

$$\partial H_g/\partial t = C_5 - C_6 - C_9$$

The fluxes C_i ($i = 1$–10) are presented in Figure 7.12 and are functions of time, temperature, and precipitation. The functional dependences of fluxes C_i on the model's parameters and time t are written as:

$$C_1 = \min\{\lambda_1\theta_1^{\Delta T+\Delta W}, C_1^{\max}\} \qquad C_3 = aS_g \qquad C_4 = bS_gH_{29} \qquad C_{10} = C_{10}(t)$$

$$C_9 = f(t) \qquad C_1^{\max} = 20H_4 \qquad C_7 = \lambda_3 S_g\theta_3^{\Delta\nu} \qquad C_8 = F(t)$$

$$C_2 = \begin{cases} 0 & \text{for wet brine} \\ \lambda_2\theta^{\Delta\nu} & \text{for dry brine} \end{cases} \qquad \tilde{N}_5 = \begin{cases} \beta(B_g - B_{g0}) & \text{for } B_g > B_{g0} \\ 0 & \text{for } B_g \le B_{go} \end{cases}$$

$$\tilde{N}_6 = \begin{cases} 0 & \text{for } H_g = 0 \\ \lambda_4 \exp[\lambda_6\Delta T - \lambda_5 B_g] & \text{for } H_g > 0 \end{cases}$$

where B_o is the concentration of the saturated solution; $\Delta T = T_o - T$; T_o is the

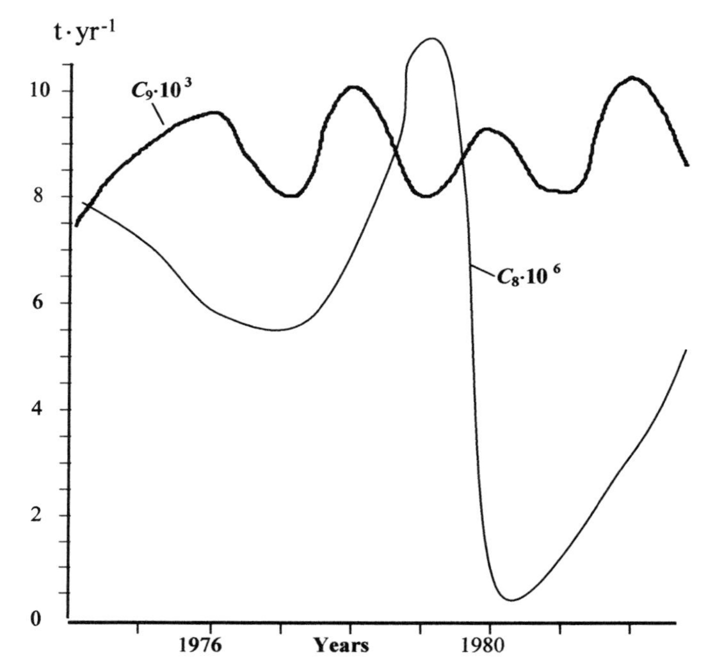

Figure 7.13. The inflow of salts into the KBGG (C_8) and their industrial removal (C_9).

control water temperature; $\Delta v = v_o - v$; v_o is the control wind speed; the functions f and F are given in the form of empirical estimates (Figure 7.13); $C_{10}(t)$ is the scenario describing the pre-history of the inflow of salts from the Caspian Sea; and a, b, β, θ_i, and λ_i are constant coefficients of the model.

The wind speed v and temperature T are prescribed as maps in correspondence with the steps of the spatial grid $(\Delta\varphi, \Delta\lambda)$. The temperature values are calculated from the formula: $T_w = A(t) + B(t)T_a$, where T_a is the air temperature, T_w is the surface water temperature, and the coefficients A and B are determined from:

$$A(t) = \begin{cases} 1.32 \pm 0.04 \text{ for April–September} \\ 3.26 \pm 0.06 \text{ for October–March} \end{cases}$$

$$B(t) = \begin{cases} 0.86 \pm 0.02 \text{ for April–September} \\ 0.88 \pm 0.06 \text{ for October–March} \end{cases}$$

Equations of water balance (7.34) for the territory Ω_g can be written as follows:

$$\sigma_g \partial W/\partial t = E - R + \sum_{j=1}^{N} (H_{4j} - H_{29j}\sigma_{gj})$$

$$\sigma_{gi} \partial c_i/\partial t = H_{1i} - H_{4i} + B_i \sigma_{gi} + K_i + N_i - H_{2i}$$

where σ_g is the area of the Gulf and σ_{gi} is the area of the ith basin of the Gulf in

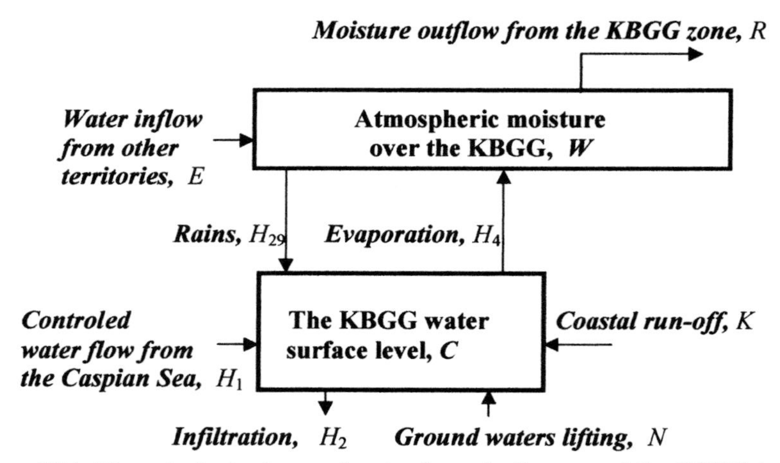

Figure 7.14. The principal scheme of water fluxes in the zone of the KBGG impact.

accordance with the grid of digitization. The remaining symbols are given in Figure 7.11.

7.4.4 Parameterization of the water balance of the Aral Sea region

The set of models described above makes it possible to reduce, into a single system, all fluxes of moisture which can circulate in the territory of the Aral Sea region and move across its boundaries. As seen from the scheme in Figure 7.14, for the numerical experiment on the basis of this system of models, a large data volume is required. This volume can be reduced by taking into account numerous correlations suggested by many authors to describe connections between geophysical parameters and elements of the water balance of the region. A most thorough analysis of such connections has been given in Bortnik and Chistiayeva (1990).

At present, the Aral Sea is a closed water body without run-off, and the equation of its water balance at each stage becomes rather simple:

$$H_8/\sigma + H_{33} + \mathrm{U}_n - H_7 - \mathrm{U}_\Phi/\sigma = \Delta A$$

where U_n is the sub-soil inflow of water, U_Φ is the filtration of seawater into the coastal bottom, and σ is the sea area. In the adaptation of the scheme of total regional water balance (Figure 6.16) to the Aral Sea region conditions, the constituents U_n and U_Φ were included into the fluxes H_{35} and H_{11} to simplify the requirements on the database and as having an insignificant weight compared with other water fluxes.

The role of the underground water expenditure in the water balance of the Aral Sea itself has been discussed elsewhere. From estimates of Chernenko (1981), there is a direct connection between the input of underground waters and the sea level. He calculated the expenditure of underground waters discharged into the sea and obtained the dependence $\mathrm{U}_n\ \sigma = 5.5 + 0.5\,\Delta A$ ($\mathrm{km}^3\ \mathrm{yr}^{-1}$), where ΔA is the change in the Aral Sea level in meters. Hence, for instance, with a 12 m lowering of the level, the inflow of underground waters should reach 11.5 $km^3\,yr^{-1}$, which is not the case. Therefore, this component in the model discussed here has been taken into account in the form of a scenario, which assumes that the flux $U_n = \text{const.}(6.3 - 7.2\,m^3\,s^{-1})$. To calculate a part of the flux H_8, which refers to the Amu-Darya River run-off, we use the equation $\Delta H_8 = 0.52 Y_a - 14.21$, where Y_a is the volume of the run-off at the section of the Kishlak Chatly.

The equation of sea surface evaporation, important for an accurate evaluation of the water balance, has been studied by many authors. So, the above-mentioned formula of Goptarev is written as: $H_7 = K(e_s - e_z)U_z$, where e_s is the maximum partial water vapor pressure (hPa) at a water (or ice) temperature of the sea surface with its salinity taken into account; e_z is the water vapor partial pressure (hPa) in the atmosphere at a height of 2 m; U_z is the wind speed ($m\,s^{-1}$) at a height z (m); $K = 327.5/(Ln^*2 - Ln^*z_o)^2$; z_o is the roughness parameter ($\approx 6 \times 10^{-4}$ m); and $Ln^*z = Lnz + az + (az)^2/(2 \times 2!) + \cdots + (az)^n/(n \times n!)$. The parameter a has been chosen from the condition of the best consideration of the impact of the temperature stratification of the lower atmospheric layer on the rate of evaporation.

7.5 SIMULATION EXPERIMENTS AND FORECAST OF THE WATER BALANCE COMPONENTS OF THE ARAL SEA HOLLOW

7.5.1 Scenario for the potential directions of change in water balance components of the Aral Sea

Figure 7.15 demonstrates schematically the history of the dynamics of the main components of the Aral Sea water balance, and this dynamic is extrapolated under conditions of preserved trends. However, it is clear that the method of extrapolation of the data of previous years cannot be objective and, hence, answer the question about the possibility of the existence of regimes influencing the water balance components which would change these trends. The problem of the control of the Aral Sea water balance remains the subject of numerous studies and discussions. In Mikhailov's (1999) opinion, it is impossible to restore the former Aral Sea; it is only possible to stabilize its water regime at levels close to those nowadays, without large-scale reconstructions of irrigation and drainage systems. In this pessimistic scenario, with an inflow of the Syr-Darya River water 3–5 $km^3\,yr^{-1}$ and regulated water discharge into the Large Sea, and with an additional run-off of 8–10 $km^3\,yr^{-1}$ from the Amu-Darya River waters, the Aral Sea level can be preserved at 31–32 m.

The model of the Aral Sea water balance developed here makes it possible to consider various hypothetical situations of the impact on the water balance of the territory Ξ, to search for means of its positive change with the transfer from the present unsatisfactory condition into a regime of functioning, which may be stable and acceptable to economic and hydro-meteorological criteria. One means of moving away from this critical situation is to reduce the volume of water taken

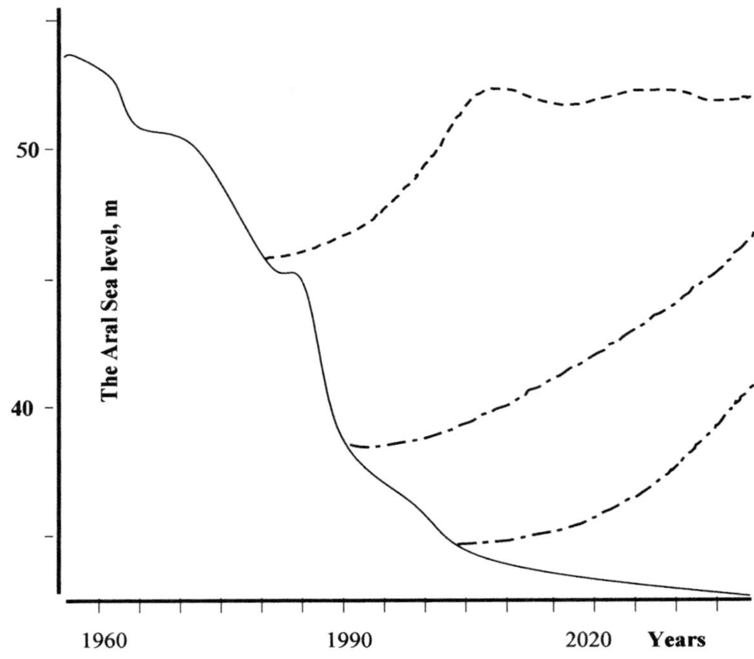

Figure 7.15. Possible dynamics of the Aral Sea levels (in meters with respect to the World Ocean level) as a result of changing the starting time of realization of the scenario in Figure 5.2 (dashed curves). The solid curve corresponds to the natural dynamics of the sea level under conditions of a preservation of the average indicators of natural–anthropogenic parameters.

for irrigation. It is clear that it is impossible to liquidate the Kara-Kum Canal, since many agricultural regions of central Asia are connected with it. Nevertheless, the governments of Kazakhstan and Uzbekistan have discussed the possibility of a partial reduction of cotton plantations in order to return required water volumes to the Aral Sea.

To carry out numerical experiments, we take the region confined to the geographical coordinates [41°,47°]N and [50°,70°]E as the territory Ξ. Let $\Delta\varphi = \Delta\lambda = 10'$. The identifiers $\{A_i\}$ are taken from published data and the electronic database of remote sensing data (Borodin *et al.*, 1996). To form a set of scenarios, consider the hypothetic anthropogenic means of water regime control. The main goal of the computer experiment is to choose a scenario which would best provide a stable transfer of the hydrological regime of the territory Ξ into the state with an increase of the Caspian Sea level by 14 cm yr^{-1} decreasing its level by 1 cm yr^{-1}, as well as achieving a restoration of the main parameters of the Aral Sea at 1960s levels.

Analysis of the data on the dynamics of the Caspian and Aral Sea levels reveals a broken equilibrium in the hydrological regime between them. The hydrometeorological situation established by the end of the 20th century on the territory Ξ cannot be transferred into another equilibrium state without anthropogenic control. The Caspian Sea level can be normalized due to an increase in the runoff into other reservoirs. It can be achieved through a forced withdrawal of water from the Caspian Sea and its transfer to regions of saline land and depressions on the eastern coast of the Caspian Sea at lower levels (−25.7 m). Such elements of the coastal landscape include the sor Dead Kultuk (−27 m), the sor Kaidak (−31 m), the KBGG hollow (−32 m), depressions Karagiye (−132 m), Kaundy (−57 m), Karyn Aryk (−31 m), Chagala-Sor (−30 m), and others. The technology of the transfer of the Caspian Sea waters to these regions of Ξ is not discussed here. Note only that in many cases, open canals are required for the natural motion of water. Of course, additional problems with environmental parameter stability appear here. For instance, for the KBGG it is important not to violate the hydro-chemical processes and bottom relief. For other elements of Ξ one should use Caspian Sea water transfer technologies which would provide the freshening of saline lands and accumulation in the coastal depressions of fresh or weakly mineralized water.

If the indicated procedure of irrigation is realized partially or completely, then Caspian water evaporation will increase. The evaporated moisture is transferred to other territories in accordance with the uncontrolable synoptic situation. From multiyear data on the wind situation in the western part of Ξ, there are time periods with a stable favorable wind-rose. Directions W, NW, and SW are characterized by high repeatability. Hence, the atmospheric transfer of the Caspian Sea water to the Aral Sea hollow is possible in a stable regime. The problem consists in either organization of a forced precipitation of this water or estimation of the natural increase of rain rate. In the model this procedure is referred to the class of the scenario.

7.5.2 Model estimation of the Aral Sea water balance dynamics in the case of a preserved natural–anthropogenic situation in the region

Consider the scenario of a realization of natural trends of the Aral Sea region water balance components. For this purpose, in addition to the accepted estimates of numerous parameters we fix an anthropogenic constituent (flux H_{11}). The volumes of the present and planned irretrievable water consumption have been estimated from the published data and summarized in Table 7.13. Besides this, of importance are the specifications of the values of evaporation from the water surface of the rivers

Table 7.13. Calculated level of irretrievable consumption of water volumes from the Amu-Darya and Syr-Darya River basins (km^3 yr^{-1}).

Zone of the basin	Years			
	1985	1990	2000	2010
Amu-Darya River	33.4	39.0	45.0	45.0
Syr-Darya River	54.0	57.5	61.0	61.0

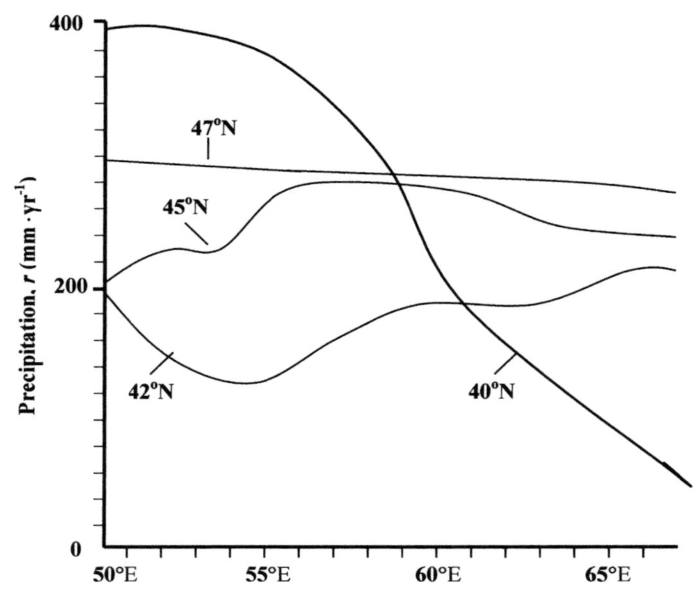

Figure 7.16. Forecast of the distribution of the annual sums of precipitation over the territory of the Aral–Caspian aquageosystem along the latitudinal sections in 2000 from initial data for 1994 adopting the mean statistical wind situation over the last 5 years. The northern hemisphere degrees are shown on the curves.

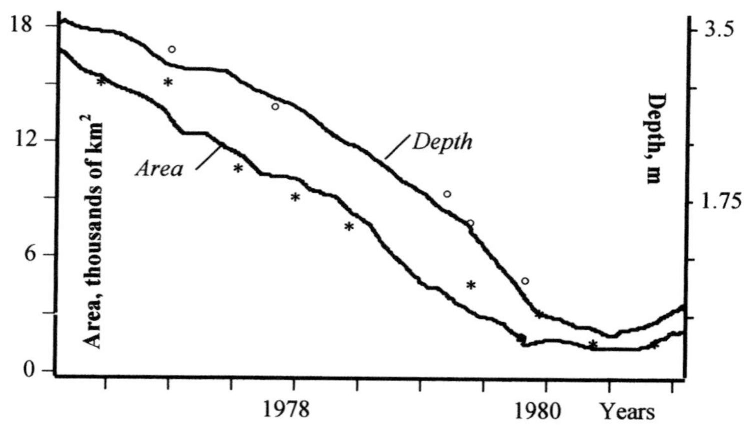

Figure 7.17. Theoretical (solid curves) and measured (o = depth, * = area) estimates of the KBGG aquageosystem's parameters.

and transpiration by hygrophilous plants in the under-flooded zones of the river valleys. To reduce possible uncertainties of the evaporation model's parameters, assume the values $H_{20} = 6\,\text{km}^3\,\text{year}^{-1}$ for Syr-Darya and $H_{20} = 8\,\text{km}^3\,\text{year}^{-1}$ for Amu-Darya. The results of modeling are shown in Figures 7.16, 7.17, and in

Table 7.14. Results of the model estimation of some constituents of the Aral Sea water balance with different wind directions and under conditions of forced precipitation in the Turan lowland (H_8 in $km^3\,yr^{-1}$; H_{33} in $mm^3\,yr^{-1}$).

Time from the beginning of the simulation experiment (years)	Prevailing wind direction								
	NW			W			SW		
	H_8	H_{33}	H_7	H_8	H_{33}	H_7	H_8	H_{33}	H_7
1	38	197	1,010	41	188	998	10	198	1,007
2	44	180	991	37	190	987	12	183	1,011
3	70	160	993	55	171	869	16	160	1,004
4	56	174	968	68	183	901	21	171	1,023
5	48	149	1,001	50	194	977	18	152	1,014
6	51	187	986	44	189	983	14	188	989
7	66	191	999	61	169	1,015	16	190	1,003
8	61	177	956	63	175	994	12	180	1,004
9	59	163	983	52	166	899	9	171	999
10	53	154	979	57	160	908	13	155	991
11	49	142	988	55	159	910	17	143	1,001
12	57	138	985	48	144	1,017	11	140	973
13	52	144	987	54	147	999	8	141	966
14	55	107	1,003	50	133	976	12	110	981

Notation: NW – north-west; W – west; SW – south-west.

Table 7.14. It is assumed that one of the wind directions (W, NW, or SW) is realized during 80 days with repeatability not less than 50%. During the remaining part of the year wind directions are uniformly distributed.

Over the Caspian Sea basin a flux of atmospheric moisture of $1.3\,km^3\,day^{-1}$ is formed. Artificial evaporators add to this flux some $0.2\,km^3\,day^{-1}$. As follows from the calculated results, with persistent westerlies, during the week, forced rains in the region of the Aral Sea equal the annual volume of rains for 1960, and the sea level rises by 0.3 m. During 80 summer days the volume of the Aral Sea is supplemented with $120\,km^3$ of water (i.e., its level has risen by 3.3 m). If the repeatability of W, NW, and SW winds over the territory between the Caspian and Aral Sea is not less than 40% or 50%, with a total duration of 80 days per year or longer, then the Aral Sea level as of 1960 will be reached in 8 or 9 years, respectively. With the wind-rose some 60 days long, the indicated result will be obtained in only 12–15 years. It is assumed in this case that the repeatability of the eastern winds over Ξ does not exceed 15%. The contribution of excess atmospheric moisture from the Caspian Sea to an increase in the river run-off gives about $40\,km^3\,yr^{-1}$, satisfying the relationship $34 < H_7 < 50\,km^3\,yr^{-1}$.

As seen from Figure 7.16, in the distribution of rains over the eastern and central parts of the territory Ξ a stable increase of rains by 8% and 12%, respectively, is observed, which ensures the Aral Sea level dynamics shown in Figure 7.15. The positive balance of moisture transfer on the eastern boundary of Ξ increases by 4%, which stimulates an increase in the river run-off to the Turan lowland. With SW to W and NW winds the amount of rain in the Aral Sea hollow is invariant, the eastern winds turn out to be mainly neutral or in cases give a 4–7% increase in the amount of rain in the zone of the Aral Sea due to the return atmospheric moisture. The indicated invariance of the rain ensures the possibility to regulate the regime of irrigation of arid territories. In particular, with SE winds, an excess evaporation of the Caspian Sea waters can offer a moisture supply for the forced irrigation of the arid steppes in Kalmykia and Stavropol Krai.

Of course, a question of reliability of all these calculations arises. The non-linearity of equations of the models used does not permit us to perform any theoretical estimations of the stability and accuracy of simulation model results. Also, too many factors have not been taken into account. Therefore, as a confirmation of some level of reliability, it is possible to compare simulation (theoretical) calculations and estimates of parameters published in the literature. An example of such a comparison is given in Figure 7.17. It is seen that the model sufficiently reliably reconstructs the history of the dynamics of some characteristics of the KBGG.

Calculations give some hope that the Aral Sea can still be restored or partially stabilized if necessary measures are taken promptly. Here a complex approach is possible including a realization of the considered scenario of irrigation of some territories on the eastern coast of the Caspian Sea and a large-scale reconstruction of irrigation and drainage systems.

7.5.3 Recommendations for the monitoring regime of the Aral Sea aquageosystem

As experience of the State Oceanographic Institute field observations in the Aral Sea region has shown (Bortnik and Chistiayeva, 1990; Bortnik *et al.*, 1994), as well as multiyear remote sensing using the flying laboratories carried out by the Institute of Radioelectronics of the Russian Academy of Sciences and the Institute of Geography of RAS (Borodin *et al.*, 1982), obtaining operational information about the geophysical and hydrometeorological situation requires heavy economic expenditure. Use of GIMS-technology simplifies the problem of organization of a regular monitoring of the indicated region. It is possible due to a coordinated application of the means of observations and mathematical models. As a matter of fact, the water balance of the whole territory, whose impact on the Aral Sea raises no doubts, can be described by a simpler scheme in Figure 7.18. Then, of course, a number of additional problems arise concerning the planning of measurements. They can be solved by creating a specialized information system whose approximate scheme is shown in Figure 7.19. This system is used for monitoring the Aral Sea zone and adjacent territories for an expert comparison of episodic estimations of the water balance components with results of modeling. This comparison can result in either corrections of individual components of the water balance model or additional measurements. On the whole, the scheme in Figure 7.19 can be realized with the use of regular satellite measurements of the areas of different types

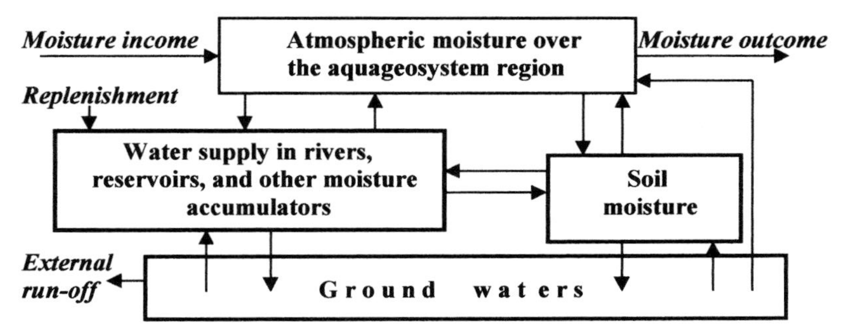

Figure 7.18. The block-scheme of the moisture cycle in the zone of aquageosystem functioning with processes of evaporation, evapotranspiration, leakages, precipitation, and anthropogenic use taken into account.

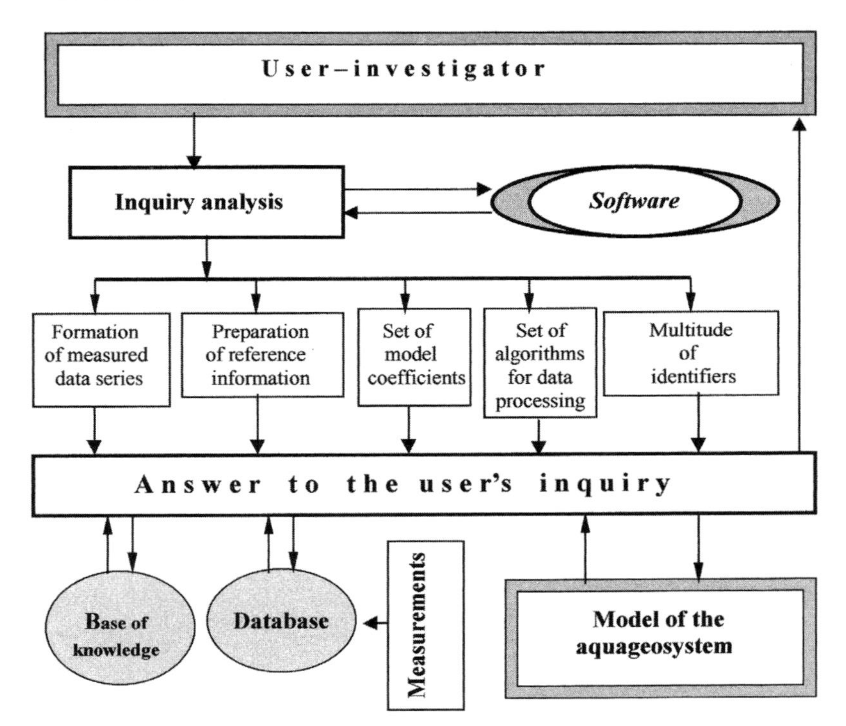

Figure 7.19. An approximate scheme of the dialog regime of the use of algorithmic provision of the hydrophysical experiment.

of land cover (listed in Table 7.4, identifier A_2), temperature, moisture content in the atmosphere, wind speed and direction, as well as salinity of water bodies (Borodin and Krapivin, 1998; Krapivin, 2000a,b; Grankov and Milshin, 1994; Kondratyev, 2000a).

8

Natural disasters as components of global ecodynamics

8.1 PRIORITIES OF GLOBAL ECODYNAMICS

The United Nations Environment Programme's (UNEP) conceptual program "Environmental Change and Humans' Needs: Assessment of Relationships" (Volk, 2003b) contains a substantiation of priorities concerning the environment and goals of development for the millennium. Bearing in mind the preparation of the Fourth Report on the Global Ecological Perspective which should be accomplished in 2007, the emphasis has been placed on analysis of the state of developments in the field of general key problems, such as:

- conceptual and analytical approaches to studies of correlations in the Nature–Society System (NSS);
- detection of correlations between environmental problems and socio-economic development with the ecological politics;
- establishment of correlations between current and possible (future) environmental change;
- substantiation of a means that can be used to study the indicated correlations;
- provision of a complex meeting of commitments in observing the laws of international ecological rights in national programs dedicated to all directions of developments; and
- a revelation of the gaps in scientific developments and estimates.

The enumerated developments should be realized with the following four general requirements taken into account:

(1) Coordination of scientific developments, ecological politics, and their realization.
(2) Taking into account correlations of ecodynamics and socio-economic development.

(3) Consideration of specific processes on various spatial scales.
(4) Reflection of specific processes on different temporal scales.

The fundamental fact for the problem under discussion is a necessity of taking into account the correlation of key problems of ecodynamics, determined by a large number of feedbacks in the NSS, and non-linearity due to which the "threshold effects" and synergism of technologies and ecological politics can occur. Numerous illustrations of the urgency of taking into account various correlations are contained in the problems of global climate change (Kondratyev, 2004b,c). An adequate analysis of the role of feedbacks and NSS non-linearity is seriously complicated by the fragmentary character of available information. In this connection, it is a pity that the concept of biotic regulation of the environment, which could form a conceptual basis for solving the problems of global ecodynamics, has not been acknowledged, so far (Gorshkov *et al.*, 1998; Kondratyev *et al.*, 2003a; Gibson *et al.*, 2005; Kondratyev *et al.*, 2003a,b). It is important that "biospheric mechanisms" serve the basis for the system of life support of manned space vehicles (Grigoryev and Sychev, 2004). Unfortunately, the concept of biotic regulation remains "unnoticed" in the West, which has been illustrated, in particular, by the recent dispute concerning the Gaia concept (Kirchner, 2003; Kondratyev *et al.*, 2004a,b; Lovelock, 2003; Volk, 2003a,b).

The rhetorical opinion of the Earth as an autotrophic self-regulating "superorganism" has been most vividly expressed in the Gaia hypothesis proposed by Lovelock (2003), which is rather schematic and contradictory (Kirchner, 2003; Lenton and Wilkinson, 2003; Volk, 2003a,b). Advocating his concept, Lovelock (2003) reminded us that the adopted Amsterdam Declaration (AD) of 13 July 2001 reads: "The Earth system behaves as a single self-regulating system comprised of physical, chemical, and human components". However, the Declaration does not specify what the Earth system regulates and at what level. On the other hand, the Gaia theory clearly explains that the self-regulation of climate and chemical composition of the atmosphere does take place, and this ensures the Earth's habitability. In Lovelock's (2003) opinion, this view cannot be understood by those who are under the spell of traditional reductionistic conception and cannot perceive that living organisms can not only adapt themselves to changing environmental conditions but also actively change their characteristics in the interests of life support. This truth has long been perceived, though partially, by Vernadsky (1944a,b). As for the AD, on the whole, it deserves a positive response, but it is insufficiently consistent. Lovelock (2003) believes that specialists in the field of climate studies have reached a most complete understanding of the Gaia concept. The opponents to this concept (Volk, 2003b) state that its weak point is the absence of concrete examples illustrating how living organisms affect the environment in order to prevent its degradation. Also, Lovelock's opinion with respect to reductionism cannot be considered substantiated: neither the quantum mechanics nor Darwin's theory can be considered in the context of reductionism.

Although Lovelock states that "the Earth self-regulates its climate and chemistry", this statement needs critical analysis. It is fair that during almost 4 billion years, living beings have changed the chemistry and climate of the planet, and possibly it is this that has determined the habitability of the Earth. However, on the other hand, the present world can be habitable only because the present organisms evolved in the past in accordance with the past change of geological and biological conditions. An adequacy of the Gaia theory can be convincingly substantiated only on the basis of further developments with the use of integral simulation models of the Earth system.

Of course, the latter conclusion can be accepted completely. Efforts of the supporters of the biotic regulation concept substantiated in Russia (Kondratyev, 2001) are directed at development of numerical models of global ecodynamics (Kondratyev *et al.*, 2002a, 2004c,d). It should be added that it is also important to substantiate and realize the global observing system (Kondratyev, 1998).

As has been mentioned elsewhere (Gorshkov *et al.*, 1998), the concept of biological regulation is based on the following conclusions:

(1) The Earth is a unique planet in the Solar System, since on this planet life exists in the form of biota – a totality of all living organisms including humans. Important properties of life include: biological stability of species and their communities as well as a very rigid distribution of energy fluxes absorbed by biota among organisms of different size. Biota is responsible for the formation of environmental properties and their stability in accordance with biotic needs. Only for this reason the existence of biota on Earth has become possible on the basis of the principle of biotic regulation. The maintenance of environmental stability is one of the main goals of all living organisms.
(2) Humans are one biotic species and therefore their important goal is also to maintain global biospheric stability. Otherwise, sustainable development would have been impossible. Humankind left its natural ecological niche long ago and started consuming much more of the biospheric resources than is permitted by the requirements of ecological equilibrium. After the industrial revolution this process of breaking the natural equilibrium has been continuously accelerated under conditions of rapid growth of population size.
(3) Approximate estimates have shown that to provide biospheric stability, not more than 1% of its resources may be used (Gibson *et al.*, 2005). At present, this share is close to 10%. A similar situation exists also with respect to evolution of global biogeochemical cycles of substance. So, for instance, the closedness of the global carbon cycle before the industrial revolution had been near 0.01% (biodiversity had played an important role in establishing biospheric stability). Presently, the closedness has decreased to 0.1% and the threat of a global ecological catastrophe becomes more and more evident. Thus, the biosphere should be considered not as a resource but as a fundamental condition for life support on Earth. At present the main goal should be a restoration of the seriously disturbed biosphere and its maintenance in a state ensuring sustainable development. However, a difficult problem is the presence of many insufficiently studied aspects of biospheric dynamics. In this context, the development of adequate observing systems and further improvement of the numerical modeling methods

play a decisive role. The first of these problems is especially urgent, since an adequate system for global change monitoring not only does not exist but has not even been substantiated conceptually.

(4) A critically important feature of present civilization development consists in inequality between developed and developing countries, which manifests itself first of all through the non-equivalent application of existing resources. The per-capita consumption of the "golden billion" is much higher than that of the remaining part of the world. Therefore, it is necessary to reach an agreement on a new social order based on agreement or cooperation and partnership. The development of adequate systems of life support is only possible on the basis of respective international agreement.

(5) It becomes more and more evident that the present system of market economy does not ensure the transfer from unstable to stable development. The respective efforts undertaken in the spheres such as education, legislation, and management cannot be considered sufficient. Non-governmental and religious organizations can play a more constructive role here.

(6) To stimulate the transition to sustainable development, a complex realization is needed of several measures based on understanding the complexity of the present situation:
   - It is very important to perceive the fact that further development of a consumption society will lead to a global ecological catastrophe and civilization collapse. The only solution to the problem is to reject the traditional paradigm of the society of consumption and to radically change the mode of life based on acceptance of priorities of spiritual values.
   - To overcome the social "North–South" contrasts, prompt measures are needed to help developing countries, based on the accomplishment of the earlier UN recommendations.

The first priority of the UNEP program (Volk, 2003a) defined as interactions between humans and the environment should be based on the concept of biotic regulation. This problem can be solved using the UNEP-developed conceptual model DPSIR (Drivers, Pressures, Impact, Response) which describes the cause-and-effect connections between ecological and socio-economic components of the NSS.

Concrete directions of ecological developments are as follows:

(1) *Natural and anthropogenic changes of the environment*. In this context, of key importance is the system classification and prioritization of correlations between natural and anthropogenic environmental changes (it should be emphasized that the fact that this problem has not been solved yet, is one of the basic reasons for the rather contradictory conclusions about the nature of the present global climate change (Kondratyev *et al.*, 2004a,b)). The UNEP program document justly emphasizes the extreme urgency of the problems of biogeochemical interactions and cycles (even in the case of the carbon cycle we are still far from an adequate understanding of the most substantial mechanisms of its formation)

(Demirchian and Kondratyev, 2004; Kondratyev *et al.*, 2003b; Kondratyev and Krapivin, 2004a).

(2) *Ecological factors and human well-being*. The ecological factors, both positive and negative, demand attention. Positive factors are ecosystems that perform functions of life support (water, food, recreation, etc.), and negative factors are factors of ecological stress (diseases, pests, natural disasters, etc.).
(3) *Anthropogenic impacts on the environment*. In this connection, of special interest is an analysis of demographic dynamics and conditions of socio-economic development.
(4) *Ecological politics and its correlations with ecodynamics*. Bearing in mind the use of technologies, institutional measures, and risk control. Quite special are the problems of interaction between science and ecological politics.

As for concrete factors of ecodynamics, the UNEP program (Volk, 2003a) has justly emphasized the first priority of analysis and assessment of the role of correlations between problems such as biodiversity, climate change, degradation of soil fertility, freshwater, coastal and marine environments, local and regional air quality, ozone layer depletion, stable organic pollutants, and heavy metals.

Questions to be answered are as follows:

- What are the key correlations between various changes of the environment and the factors governing them?
- How are correlations between various anthropogenic impacts realized and how efficiently can they be "disconnected", if needed?
- To what extent does the correlation manifest itself between the ecosystems' production and their functioning?
- What correlations do now exist between the impacts and respective responses and how can they be regrouped or removed in case the interactions in the NSS should be changed?

As for the goals of socio-economic development during the period up to 2015 (compared with conditions in 1990), the basic problems include (Volk, 2003a):

(1) Elimination of extreme poverty and hunger (to halve the number of people with incomes of less that 1 dollar per day).
(2) Realization of universal primary education.
(3) Provision of sexual equality and enhancement of the role of women.
(4) Reduction in children's mortality (children under 5 years) by two-thirds by the year 2015.
(5) Improvement of mothers' health (a three-quarters reduction of mortality in child birth).
(6) Fighting HIV/AIDS, malaria, and other diseases.
(7) Provision of ecodynamics stability (liquidation of the decreasing trends of environmental quality, halving the number of people who have no access to good quality drinking water, improvement of living conditions for, at least, 100 million people living in slums).

(8) Development of a cooperation in the interests of socio-economic evolution (with emphasis on a solution of the problem of the debt of developing countries).

Serious efforts to substantiate the priorities of the environmental studies in the 21st century have been undertaken by National Oceanic and Atmospheric Administration (NOAA) scientists who have worked out a Strategic Plan (SP) for the period 2003–2007 and longer (*New Priorities for the 21st Century*, 2003). The SP determines the priorities for four NOAA directions of development.

(1) Protection, reconstruction, and control of the use of coastal and oceanic resources based on a knowledge of the laws of ecosystem dynamics.

The urgency of these problems is determined by the fact that 55% of the population in the USA live in a 50-mile coastal zone extended for about one-fifth of the total coastline length. Coastal regions develop faster than other regions. Here, in particular, the population size increases by 3,600 people per day, with more than 28 million places of work, and the gross production of goods and services exceeding US$54bln.

(2) Understanding the laws of climate change to raise the ability of society to respond to them (note that two notions are used in the document – *variability* and *change* with respect to climate, without their clear definitions being stated).

In this context, it should be noted that about one-third of the USA Gross Domestic Product (GDP) directly or indirectly depends on weather conditions and climate. The annual and inter-annual climate change similar to that caused by El Niño brought, in 1997–1998, economic losses of about US$25bln, with losses to property of ~US$2.5bln and yields of US$2bln. In the USA, reliable weather forecasts ensure, only in agriculture, a saving of about US$700mln.

(3) Satisfaction of the need of society for information on climate and water resources.

The economic damage caused to the USA by hurricanes, tornados, tsunamis, and other disasters averages about US$11bln. About one-third of the national economy (~US$3 trillion) is "weather sensitive".

(4) Support of national commercial interests by provision of ecologically reliable information to ensure safe and efficient traffic.

Clearly, safe and efficient systems of traffic are one of the key components of an economy. This concerns all types of traffic: ground, sea, and air.

Six directions for development are considered as most substantial:

(1) Creation of a system of integral (complex) observations of the global environment, processing, and analysis of observational data.
(2) Efficient resolution of the problem of propagation of knowledge about the environment.
(3) Accomplishing necessary scientific studies.
(4) Provision of efficient scientific cooperation.

(5) Solution of the problems of national security.
(6) Adequate solution of organizational problems.

The NOAA SP contains the following strategic goals foreseen by the four directions of development mentioned above. The goals of the first direction are:

- to increase the number of stable coastal and marine ecosystems not subject to unfavorable changes;
- to raise the socio-economic value of the marine environment and respective resources (marine food, recreation, tourism, etc.);
- to improve living conditions for coastal and marine flora and fauna;
- to increase the number of species protected from unfavorable impacts;
- to increase the number of cultivated species up to an optimal level; and
- to improve the environmental quality in coastal and marine regions.

These goals can be achieved based on the use of the ecosystem approach and, in this connection, a broadening of the scope of studies of various ecosystems' dynamics within three strategic directions:

(A) Protection, restoration, and control of the use of resources of the coastal and oceanic regions as well as Great Lakes.
(B) Restoration of protected species and their niches.
(C) Broadening and maintenance of stable fisheries.

Of key importance for achieving the goals of the NOAA SP is the development of a complex environmental monitoring system which would provide for obtaining, processing and analysis of data on the interactive ocean–atmosphere–land system on local to global scales.

Assessing the NOAA SP on the whole, one should note that unfortunately, it has been strictly oriented toward national interests. Even in this context, it is impossible to adequately resolve the problems of ecodynamics without considering the interactivity of local, regional, and global processes.

The key problems of global ecodynamics cannot be resolved in isolation from the problems of monitoring and prediction of natural disasters. Attempts of many experts to determine the laws of occurrence of natural disasters and evaluate their role in the environmental system dynamics gives hope that the development of the models such as GMNSS, SiB2, DPSIR, and others will make it possible to obtain, in due time, an efficient instrumentation to resolve these problems. It should be noted here that one should not confine oneself to a consideration of the global temperature and greenhouse gas (GHG) concentration as the only resulting indicators of the 250-year enterprise of the human population, but study global change, considering the biospheric and ecological systems in their interaction with a complex of rapid and slow processes on Earth (Steffen *et al.*, 2004).

Among the most important priorities of ecodynamics are the following.

8.1.1 Inventory of the content of anthropogenic carbon in the ocean

An accomplishment of the field observational programs such as an experiment to study the ocean and atmospheric circulation (WOCE), the joint experiment to study the carbon cycle in the ocean (JGOFS), and the study of the ocean–atmosphere carbon exchange (OACES) have served the basis for obtaining new and more reliable estimates of the anthropogenic CO_2 input to the World Ocean and the CO_2 transfer in the oceans. Results obtained between 1991 and 1998 show that the input of anthropogenic carbon totaled about 117 ± 19 Pg C with a strong spatial–temporal variability. Especially strong contrasts were observed between the North Atlantic, where anthropogenic CO_2 could be observed on the bottom, and the tropical Pacific, where CO_2 was absent at depths exceeding 600 m. Although CO_2 emissions to the atmosphere due to fossil fuel burning take place mainly in the northern hemisphere, about 60% of anthropogenic CO_2 is concentrated in southern hemisphere oceans, a fact that is connected with the effect of transport due to the sub-tropical convergence zone.

8.1.2 Specification of the carbon budget

Observations in the equatorial Pacific Ocean revealed, during the period 1980–1990, drastic changes in the partial CO_2 pressure (pCO_2) in surface waters and CO_2 flux from the atmosphere to the ocean. In 1980 the temporal increase of pCO_2 was slow compared with 1990. By the year 1990 the increasing trend coincided in time with a change of decadal Pacific variability. The latter agrees with the supposition that the main factor of the impact on the CO_2 exchange between the atmosphere and ocean was the natural long-term climatic variability.

One possibility for specifying data on the carbon budget is to use the ocean color observations from satellites with instruments such as SeaWiFS and MODIS calibrated by comparison with the data of marine optical buoys (MOBYs). Calibration ensures the error of retrieving chlorophyll concentrations from seawater does not exceed 6%, which makes it possible to substantially raise the reliability of the estimates of primary production and, respectively, the carbon budget.

For rapid estimates of carbon supplies in the USA forests, a computer algorithm has been developed which has been used to specify the available data of the carbon budget inventory.

New estimates of carbon dynamics, from the data of direct (fluctuation) measurements of the CO_2 exchange between the atmosphere and forest at two sites of old Amazon forests, have led to the conclusion that the annual change of such an exchange is opposite to that obtained earlier. During seven months of the wet season the forest lost carbon, and during five months of the dry season the forest was a sink for carbon. The short dry season strongly limits the processes of respiration (as a result of drying the surface detritus matter), but only weakly affects the process of photosynthesis due to the presence of sufficient moisture in deep soil layers. With the renewal of rains (beginning of the wet season), a large amount of wood remains begins to decay and, respectively, CO_2 is emitted to the atmosphere.

The main goals of developments in 2004–2005 aimed at a deeper understanding of the processes (sources and sinks) of carbon cycles based on accomplishment of the program of carbon budget studies (NACP) in the USA and the program of the ocean carbon cycle and climate studies (OCCC) are as follows: regular observations within the NACP program at networks AmeriFlux and AgriFlux as well as at 500-m masts and ocean platforms (together with the data of satellite remote sensing); broadening of the network of aircraft observations and high masts (their number will increase up to 15); special measurements of the CO_2 exchange between the atmosphere and ocean in the North Atlantic and Pacific Oceans (the northern hemisphere and equatorial belt) to provide the required input information about developments in the field of numerical modeling; analysis of new data from satellite observations with the use of the EOS system; special observations under conditions of different typical landscapes; ship observations of pCO_2 in the Atlantic and Pacific Oceans (the northern hemisphere); studies of correlations between climate, phytoplankton, carbon, and iron content in the Atlantic Ocean; processing of satellite data on the annual change and inter-annual variability of primary productivity of the ocean; numerical modeling of the global carbon cycle; impact of forest ecodynamics on the GHG input to the atmosphere; continuation of the development of the Orbiting Carbon Observatory (OCO) planned to be launched in 2007.

8.1.3 Impact of climate regime shifts on the marine ecosystems

Climate shifts on multi-decadal scales affect the marine and continental ecosystems, which necessitates the monitoring of these shifts to provide the needed information for taking measures to regulate the ecosystems' dynamics. In this connection, of importance are results of analysis of consequences of climate change in the region of the Pacific Ocean in the northern hemisphere, where in 1998 the phase changed from warm to cold, followed by immediate enhancement of phytoplankton productivity and a change of its species composition as well as a considerable increase of the size of populations of salmon and other valuable species of fish in the north-western sector of the Pacific Ocean. In the North Atlantic, the amount of zooplankton and fish changed, especially in the Haine Strait, which is connected with the conversion of North Atlantic oscillations. With these new data taken into account, indicators of ecological changes have been developed, the use of which ensures the calculation of more adequate ecological politics to regulate ecosystem dynamics.

Analysis of the data on a great number of species of plants and animals, corresponding to a wide range of properties of natural ecosystems, has made it possible to reveal changes in the natural habitat and special features of the habitat (migration, blossoming, laying eggs, etc.) of many species that depend on climatic conditions. However, determination of concrete cause-and-effect connections is strongly complicated by the multi-factor character of biological change. According to one of the developments containing data on more than 1,700 species, there has been a considerable poleward shift of habitats reaching, on average, 6.1 km over 10 years during time periods of 16 to 132 years. Taking 279 species as an example, the presence has been demonstrated of spatial–temporal variability determined by climatic trends in the 20th century. Analysis of data for 143 species (from mollusks to mammals and from grass to trees) has demonstrated persistent shifts connected with temperature changes. On the whole, the results obtained reveal the impact of global climate warming on the populations of animals and plants.

8.1.4 Control of natural resources to curb GHG emissions

Problems of regulating the land ecosystem dynamics to curb GHG emissions to the atmosphere and adaptation to climate change can be solved adopting various approaches. For instance, in the case of forests, it is possible to stimulate an accumulation of carbon through an intensification of plant growth by soil fertilization. The results of the 8-year field observational experiment at the site of the broad-leaved forest in Tennessee (USA) have demonstrated the impact of rain on nutrient cycles and the survivability of young trees, however, they have not detected such an impact in the case of big trees, root system development, or litter decay. On the whole, forests demonstrate a resistance to change in rain, but changes of hydrological processes can bring forth a long-term impact on the species composition of trees. It follows from the data of another field experiment carried out in the Wisconsin (USA) that the supposed stimulation of the growth of the aspen forest due to a growing concentration of CO_2 has turned out to be practically compensated for by a simultaneous growth in the tropospheric ozone concentration.

The stimulating impact of the growth of CO_2 concentration in the atmosphere on the development of crops is well known. However, recent studies revealed a negative impact of this trend on the harvest. An artificial doubling of CO_2 concentration over grass covered prairies in the north-eastern part of Colorado (USA) (experiments in an open chamber) has shown that the forage obtained under these conditions is less well digested by animals than that grown under natural CO_2 conditions. Another experiment has shown that an increased CO_2 concentration stimulates the development of five widespread species of weeds better than other vegetation. Hence, there is a danger of a rapid growth of weeds with growing CO_2 concentration. According to the data of one further field experiment, a doubling of CO_2 concentration has involved a much more rapid (by $\sim$90%) consumption of white clover leaves by pests. Thus, the impact of the increased CO_2 concentration on crops can be rather contradictory.

8.1.5 Comparison of the models of ecosystems with observational data

Comparisons of 13 simulation models of forest ecosystems, characterized by various schemes of parameterization of the processes considered and different spatial–temporal resolutions, with the data of field observations in the eastern part of Tennessee (USA) have shown that none of the models has a universal advantage. A good "inter-model" agreement is observed in the case of the water cycle, but considerable differences are observed in the case of the carbon cycle. The best agreement with observations is obtained with the use of data averaged over the results of all models. The models neglecting the key components of forest ecosystems or important processes (like the dynamics of root zones or soil moisture) turned out to be unable to simulate the ecosystem's response to a short-term drought. The best agreement with observational data is observed in the case of models with sufficiently realistic parameterizations of the key processes determining the dynamics of forest ecosystems, with a high temporal resolution (one or several hours). The prognostic potential of the models is reduced in the case of droughts.

The key directions of developments in 2004–2005 determined by the Climate Change Science Program (CCSP) include the following problems: estimation of the climatic impact in the Arctic (the ACIA program) on the environment and human health (e.g., deficit of UV solar radiation); economic infrastructures, etc.; analysis of the data of field observational experiments to assess the impact of an increased CO_2 concentration in the atmosphere on agricultural ecosystems; forecast of localization and ecological impact of weeds with the use of both ground and satellite observational means; prediction of forest fires, evaluation of ecological consequences of the use of chemically dangerous fuels in forests (primarily this concerns the western region of the USA) and carbon accumulation by ecosystems (in this connection, of first priority is the development and application of the Mapped Atmosphere–Plant–Soil System (MAPSS); quantitative description of the impact of climate warming on boreal forests based on the data of observations and numerical modeling results; development of the Long-term Ecological Research (LTER) program monitoring the coastal ecosystem dynamics; quantitative estimates of the impact of varying temperature and precipitation on the development of communities of soil bacteria and micro-fauna in the biologically active soil layer; analysis of the possible impact of climate change on land use and watersheds; quantitative estimates of the impact of consumption of grass cover by cattle on carbon and nitrogen cycles from the data of observations at different pastures; development of models of forest ecosystems to analyze the dynamics of forests under conditions of changing climate; and quantitative estimates of the impact of climate change on ecosystem productivity.

8.1.6 The socio-economic aspect of ecosystem dynamics

Again, with reference to summarizing publications (Kondratyev *et al.*, 2002b, 2003b, 2004d), we enumerate the key directions of developments determined by the CCSP (*Our Changing Planet*, 2004):

- What are the levels, correlations, and significance of the "human dimension" (socio-economic factors) in the development of society and its role in the global change of the environment?
- What are the impacts of the global environmental variability, both present and possible in the future, on economic development; what factors determine the ability of the society to respond to changes that take place; what are the possibilities of providing sustainable development and a reduction of sensitivity of society to the impacts?

- What are possible methods of decision-making in the interests of sustainable development under conditions of NSS complexity and the highly uncertain variability of the global environment?
- What are possible impacts of global environmental changes on human health; what information is needed concerning the ecodynamics and socio-economic factors to assess the respective cumulative risks for health?

The most substantial directions of developments connected with the answers to these questions include the following:

(*i*) *Regional problems of ecodynamics*. In this connection, of primary importance are studies of the problems of energy supply, restoration of ecosystems, human health, and special phenomena such as droughts and forest fires on local and regional scales. In this connection, an integration of physical, chemical, and biological sciences with respective socio-economic aspects taken into account will play a decisive role, which will provide a means to make adequate decisions concerning the ecological politics. Possible approaches to solving such problems can be illustrated by studies of dependences of forest fires on climatic conditions, but the main goal is to forecast anthropogenically caused forest fires.

(*ii*) *Analysis of economic efficiency*. The difficulty of quantitative estimation of the economic efficiency of ecological politics is determined by an ambiguous choice of politics with certain priorities taken into account. In this context, of great importance is an application of ecodynamics simulation models, especially models of changing carbon supplies on land in order to assess the response of carbon supplies in plants and soils to specific land use and changes of land surface characteristics, variations of CO_2 content in the atmosphere, and climate. Solution of such problems will provide a reliable forecast of global "trajectories" of CO_2 emissions to the atmosphere.

(*iii*) *Effect of UV solar radiation on human health*. Intensive developments are being carried out in this direction to study the impact of the lowering level of biologically active UV solar radiation caused by decreasing total ozone content in the atmosphere on the development of skin cancer and cataracts.

Developments planned for 2004–2005 included the following directions:

- creation of centers responsible for decision-making under conditions of uncertainties (DMUU) which should function until the year 2008;
- analysis of domestic energy consumption with the USA, China, and Indonesia as examples, and its perspectives with the current demographic trends taken into account;
- analysis of efficiency of seasonal weather forecasts and prospects for prediction of the inter-annual climate variability;
- substantiation of the possibility to adapt the economic activity to climate change;
- prediction of a possible rise of the ocean's level and substantiation of related measures;

- assessment of the consequences of global change in the context of the problems of risk and plurality of stress factors;
- climate variability and human health; and
- UV solar radiation and health.

Despite the rich content of the USA program on climate studies, one should note that this program:

(1) still lacks the needed systematization (primarily, the program lacks a successive approach to solving the climate problem as an interactive component of global change);
(2) lacks a substantiation of a single approach to solving the problems of the global observation system (the Earth System Science Pathfinder (ESSP) programme of partnership in a complex study of the Earth system does not resolve this problem); and
(3) pays insufficient attention to paleoclimate problems.

These three general opinions determine the prospects for further program developments.

8.2 EVOLUTION OF THE BIOSPHERE AND NATURAL DISASTERS

One of the numerous factors of the development of evolutionary processes in the biosphere–climate system is sudden changes of the environmental characteristics that cause stress for living organisms with possible fatal outcomes. In different periods of evolution the scale and significance of some factors has changed. Two International Council of Scientific Unions (ICSU) programs "Dark nature: rapid natural change and human responses" (started in 2004) and "The role of Holocene environmental catastrophes in human history" (2003–2007) deal with studies of the role of catastrophic processes taking place over 11,500 years, in the development of the current civilization. A longer period of evolution of the planetary system has been considered by Condie (2005) who has analyzed the interaction of different components of the planet for the last 4 million years, selecting those that had affected the history of land ecosystems, oceans, and the atmosphere. The present approach is characterized by a trend toward an increase of natural disasters of anthropogenic origin, such as floods, forest and peat-bog fires, deforestation, desertification, and epidemics. Of course, it is not always possible to distinguish between the causes of a natural disaster. But one thing is apparent, that the present ecodynamics is followed by an increase of extreme situations in the environment. For instance, at the turn of 2000/2001, the ratio of human victims from natural disasters reached 17,400/25,000 despite the attempts of many countries to take preventive measures to protect the population. Maximum economic losses during the last years were recorded in 1995 (US$180bln) and in 2004 (US$220bln). The most dismal years of natural disasters were 1998 and 2004 that resulted in 50,000 and 232,000 deaths, respectively. The contribution of various types of catastrophes to human losses and economic damage varies from year to year, but maximum victims result from earthquakes, floods, winter storms, and tsunamis. For instance, of 700 natural disasters recorded in 1998, winter storms and floods constituted 240 and 170 cases, respectively, and their combined economic damage constituted 85% of all losses. In Europe, in mid-November 1998, more than 215 died from hypothermia.

Significance and scale of critical situations in the environment depend on the state of numerous NSS components. As a rule, elemental phenomena are assessed in the context of the resulting damage for human life and economic activity. It is well known, however, that on the other hand, elemental phenomena are an important factor of ecodynamics on local to global scales. As Lindenmayer *et al.* (2004) note, for many processes that determine the ecosystem dynamics, natural disasters are of key importance as factors that regulate such processes. Therefore, Wright and Erickson (2003) discussed the problem of taking into account the impact of natural disasters within the problems of the numerical modeling of environmental changes. Special attention has been recently paid to forest fires, which, being an important component of ecodynamics, are mainly of anthropogenic origin and bring both material damage and human losses.

Forest fires happening regularly in different global regions act as a factor of ecosystem dynamics, which is manifested through the fire-induced emissions to the atmosphere of GHGs and aerosols. According to available estimates, about 30% of the tropospheric ozone, carbon monoxide, and carbon dioxide contained in the atmosphere is determined by the contribution of forest fires. The aerosol emissions to the atmosphere, connected with forest fires, can substantially affect the microphysical and optical characteristics of cloud cover, leading to climate change. Satellite observations concerning Indonesia have demonstrated, for instance, that the presence of smoke in the atmosphere due to continuing fires has led to a precipitation reduction, which favored the further development of fires. In this context, Ji and Stocker (2002) performed statistical processing of satellite Tropical Rainfall Measuring Mission (TRMM) data (on tropical rains) and the total ozone mapping spectrometer (TOMS) data on the aerosol index (AI) for the period from January 1998 to December 2001 in order to analyze the laws of the annual change, intraseasonal, and inter-annual variability of the quantity of forest fires on global scales. In the period considered there was a clearly expressed annual change of forests in South-East Asia with a maximum in March and in Africa and North and South America, in August. Analysis of the data revealed also an inter-annual variability of forest fires in Indonesia and Central America, correlating with the El Niño/Southern Oscillation (ENSO) cycle in 1998–1999. In 1998, boreal forests were burned down over large territories of Russia and North America. Fires covered an area of about 4.8 mln ha in the boreal forests of Canada and the USA and 2.1 mln ha in Russia.

A distinct correlation is observed between variability of the aerosol content in the global atmosphere and the above-mentioned variation of frequency and intensity of forest fires. An exclusion is the region of south-western Australia where intensive fires recorded from TRMM data were not followed by smoke layer formation (from TOMS data). With the Australian region excluded, the coefficient of correlation between the quantity of fires and AI (from TOMS data) constitutes 0.55. The statistical analysis of the data using calculations of the empirical orthogonal functions (EOF) revealed a contrast between the northern and southern hemispheres, as well as an inter-continental aerosol transfer resulting from fires in Africa and America. The statistical analysis data point to the presence of 25–60-day intra-seasonal variations superimposed onto the annual change of the quantity of fires and aerosol content. A similarity has been detected between the intra-seasonal variability of the quantity of fires and dynamics of the Madden–Julian Oscillation (Kondratyev and Grigoryev, 2004).

As McConnel (2004) justly noted, still it remains enigmatic why some forests become fragmentary, degrade, and loose species diversity, whereas other forests remain in a good condition and even broaden their areal extent. There is no doubt that the growth of population size, further broadening of the spheres of market relations, and intensive development of various economic infrastructures are important factors of the observed deforestation. Factors resisting this process are nature protection measures, which promote forest preservation. Eventually, the forest cover dynamics is determined by a complicated and interactive totality of factors such as biogeophysical processes, the growth of population density, market relations, various forcings (forest fires including), and institutional microstructures.

Forest fires affect the formation of the global carbon cycle. Actually, global scales of forest fires have recently become equivalent in areal extent to the territory of Australia. Almost 40% of the global CO_2 is emitted to the atmosphere; 90% of forest fires are of anthropogenic origin. This means that the inherent balance of natural factors is strongly violated and the laws of natural evolution are subject to a powerful forcing.

Among other natural phenomena affecting the environmental dynamics are volcanic eruptions, dust storms, and thunderstorms. One of the consequences of a large-scale volcanic eruption is heavy atmospheric pollution, which can lead to a climate change in some regions or even a global change in climate. So, in particular, aerosols connected in many respects with emissions of sulfur dioxide during the Pinatubo eruption in 1991 led to a global temperature decrease in the lower atmosphere by about 0.5°C in 1992. Also, it was noted that two years later the level of global ozone content decreased by 4% compared with the previous 12-year period. Additionally, though during the last century there was no drastic climate change due to volcanic eruptions, historically, such occurrences did take place. Some 73,500 years ago, due to the eruption of the Toba Volcano on the island of Sumatra, the global atmospheric temperature decreased by 3–5°C. The area of glaciers increased, and the size of the population on Earth decreased (Oppenheimer, 1996). Thus, some 600 volcanos existing now on Earth could constitute a global threat to the population, since they can seriously change the climate and affect thereby a further NSS development.

The dust storms taking place in many regions of the Earth play a similar role to volcanos regarding environmental change. The dust propagates over great distances negatively affecting the plants and soils and heavily polluting the lower layers of the atmosphere for long time periods. In the regions with a thin vegetative cover and dry climate, dust storms take away the fertile soil layer, damaging the plants. Dust storms occur periodically, for instance, on the plains of the central and western states of the USA during heavy hurricanes. Sometimes they last for several days, raise dust up to 1.5–1.8 km and even to 5–6 km, and transfer it over hundreds and even thousands of kilometers toward the Atlantic Ocean. It is the destruction of the soil–vegetation cover in steppes and partially wooded steppes in the regions of intensive agriculture that the catastrophic phenomena of dust storms are connected with. Along with short-term forcings and apparent damage, dust storms form a regular constituent of global ecodynamics on scales of decades and even centuries.

Thunderstorms play a special role in environmental change. Lightning strikes taking place practically at all latitudes affect the process of photochemical reactions in the atmosphere and are a fire risk. From available estimates, at any moment on the Earth there occurs, on average, 1,800 thunderstorms, each being followed by 200 lightning strikes per hour (or 3.3 strikes per minute). From satellite observations, the global mean frequency of lightning strikes is 22–65 strikes per second. Observations from the Microlab-1 satellite gave 44 ± 5 strikes per second, which corresponds to an occurrence on Earth of 1.4 billion strikes per year. The use of Microlab-1 data has made it possible to draw global maps of the frequency of lightning strikes in different seasons. Analysis of these maps has shown that lightning strikes take place mainly over land, and the ratio between their quantity over land and over the ocean averages 10 : 1. About 78% of lightning strikes fall on the latitudinal band 30°N–70°S. Most intensive is the all-the-year-round regime of lightning strikes in the Congo Basin where the frequency of lightning strikes (in Rwanda) averages 80 strikes $km^{-2}\,yr^{-1}$, which corresponds to conditions in the central region of Florida. The all-the-year-round intensive regime of lightning strikes is characteristic of the central part of the Atlantic Ocean and the western region of the Pacific Ocean, where under the influence of cold air advection over the warm surface of the ocean the atmosphere becomes unstable. Lightning strikes in the eastern sector of the tropical Pacific Ocean and in the Indian Ocean, where the atmosphere is warmer, occur less often. Maximum frequency of strikes in the northern hemisphere occurs in summer, whereas in the tropics a semi-annual cycle of lightning strikes is observed.

In the northern regions, an important constituent of the mechanism of evolution control are severe frosts whose impact on vegetation cover depends on plant hardiness. Jönsson *et al.* (2004), with the Norwegian spruce *Picea abies* as an example, studied the response of boreal forests to temperature variations. It has been shown that sudden changes of temperature cause changes in the wood density, and with an expected climate change the vegetation cover may change, too.

8.3 WILDFIRES AS COMPONENTS OF GLOBAL ECODYNAMICS

8.3.1 Fires and forest ecosystems

Fires play a substantial role in the formation of forest ecosystems. Moreover, it can be stated that they are an integral element of their development. Fire in a forest can be caused either by a lightning strike or anthropogenically. From available estimates, the frequency of lightning strikes in tropical forests reaches $50\,km^{-2}\,yr^{-1}$. In the moderate-zone forests this quantity is less and constitutes $2–5\,km^{-2}\,yr^{-1}$. Of course, not every strike leads to a forest fire. Nevertheless, in different regions of land, lightning strikes lead to 2–30% of forest fires.

During the historical period, geography and quantity of fires have been changing, but their role in the formation of forest landscapes has remained constant. Many species of trees, such as pine, larch, birch, and aspen had a pyrogenic period in their development. Owing to fires these species have preserved their natural habitats and had not been ousted by spruce, cedar, and fir. In the process of evolution, the light-demanding species have become fire resistant due to, for instance, the thick bark and deep root system of the larch and pine, and the birch and aspen's ability to give rich shoots from the roots after a fire. Some species of American pines exhibit a protective reaction. For instance, they open their cones only after a fire, when conditions for seed sprouting are favorable.

Some investigators believe in an objective idea according to which practically all current taiga forests in Russia had been changing due to fires (Furiayev, 1996). Present studies make it possible to confirm this with respect to taiga forests in the historical and pre-historical past. This is confirmed, in particular, by palaeoecological developments carried out in the USA and Finland. Studies of fires using the method of pollen analysis and of traces of burning fixed in stratified lake deposits in eastern Finland, testify to the large scale of fires. During the last 2,100 years, fires have been an integral factor of development of the taiga forests in this region. The repeatability of large-scale fires in eastern Finland varies from 7–110 to 130–180 years. It was noticed that fires became more frequent after 600 AD, when the impact of humans on forests became stronger.

According to the calculations of Kurbatsky (1964), in the early Holocene there had been several thousands of forest fires caused by lightning strikes. Of course, humans had also played a substantial role in the origin of fires in the forest ecosystems (first due to carelessness with fire, and then as a result of loosing control of fires during propagation of the slash-and-burn method used to clear the forests for agriculture). In the 1950s, on all continents about 200,000 forest fires occurred every year, with 3% of these having been caused by lightning strikes. It is not excluded that all forest ecosystems of the planet in the historical and pre-historical past had suffered from fire. There had long been a version of the pyrogenic origin of the present savannas over the vast territory of the Hindustan Peninsula, considered to have propagated in place of burned tropical forests.

Thus, the problem of forest fires has two sides – one negative and one positive. The negative side is connected with the apparent damage to forest ecosystems and economic losses for humans. The positive aspect consists in the evolutionary role of forest fires. For instance, the post-fire renewal of pine forests over vast territories is known to take place easily and rapidly compared with that of clearings during which seeds are destroyed. At the fire sites, where the seeds have not been destroyed, shoots and underwood appear rapidly and simultaneously. Moreover, in different global regions an increase in post-fire productivity of forest ecosystems has been observed. In particular, it was found that in the regions of the Slovenian Karst and Istria at the south-west of the Republic of Slovenia, the fires that occurred affected the species composition of plants (Grigoryev and Kondratyev, 2005). Immediately after a fire the diversity of species and living forms grows even more rapidly. However, in the course of succession to an initial forest composition, this diversity decreases.

Restoration of forest communities depends on the frequency and intensity of fires. In the Siberian taiga forests relatively frequent fires of low intensity cause changes in the age and quality of forest stands. Changes of forest planting are connected with intensive but rare fires. Studies of the post-fire dynamics of forest ecosystems at the far east of Russia show that successions of forest vegetation are connected with fires (Efremov and Sapozhnikov, 1997). At the same time, it was shown that fires are a required element for maintaining biodiversity.

The post-fire regeneration of forest ecosystems depends on many factors (climate, soil, rocks composition, relief, etc.). In the deciduous forests of Siberia, with its permafrost, the shrub and grass-shrub tiers are restored within 4–5 years after a fire, while mosses and lichens grow much more slowly.

Sometimes fires positively affect the soil component of the forest ecosystem and, in particular, promote the preservation and provision of soil with nitrogen, which is known to play an important role in forest ecosystem productivity. This effect has been found by scientists from the Montana University in Missoula in the forests of the western part of Montana (USA) (Newland and De Luca, 2000). Local forests with prevailing yellow pine and Douglas fir develop better in the case of abundant nitrogen-fixing plants and due to this, sufficient content of nitrogen in soil. It was identified that these plants are very important as factors of compensation for nitrogen, as forest ecosystems loose this element after fires. In the absence of fires, the quantity of nitrogen-fixing plants decreases, and subsequently the forest structure changes drastically. On the contrary, periodic fires in the forests of yellow pine and Douglas fir broaden the propagation of nitrogen-fixing plants and favor the forest ecosystem productivity, on the whole.

As has been mentioned above, one of the manifestations of the anthropogenic interference into the forest ecosystem dynamics is forest burning for subsequent agricultural land use. As Krimmer and Lake (2001) have shown, the controled burning of forests maintains their landscape mosaic and promotes the preservation of biodiversity, as well as raising forest ecodynamic productivity. From the observations of Hoffmann (1998), in savannas serrado in Brazil, the productivity of the ecosystem of 3 species of trees, 2 species of shrubs, and 1 species of subshrub increased 4–7 times after burning. Clearly, burning of the forest sites by natives should be attributed to their harmonious co-existence with nature, and planned burning of forests should be carried out on a scientific basis and considered as an element of land use optimization (precisely, forest use). The problem of the balanced use of positive and negative consequences of a forest fire remains unsolved.

8.3.2 Wildfires, dynamics of the biosphere, and climate

Poorly studied positive and negative feedbacks of forest fires makes it impossible to draw the final conclusions about their role in the global ecodynamics. Positive feedbacks mentioned above can be supplemented with facts indicating the opposite post-fire trends in ecosystem dynamics. Numerous results of numerical modeling show that tropical deforestation results in changes of albedo, land surface roughness, leaf area index, and root zone depth, which in turn lead to a decrease of rain rate and relative humidity, but to an increase of surface atmospheric temperature (SAT) and wind speed. Global consequences of such changes will depend on the ratio of scales of reduction of forest areas and measures taken to protect nature. Preliminary results of modeling these consequences, obtained by Hoffmann *et al.* (2003), with the use of the atmospheric general circulation model CCM 3.2, and data of remote sensing, revealed a decrease of rain and relative humidity during the fires in the Amazon Basin. The empirical connection between the scales of desertification in the regions of the Amazon, the Congo, and Indonesia, and the frequency of forest fires established from the NOAA-12 satellite measurements, revealed a dependence of the index of fire risk of a territory on their frequency. This means that forest fires directly affect a climate change, which leads to a change in their frequency of occurrence. This interactivity makes it possible to search a regime of forest regulation which would reduce the level of CO_2 concentration in the atmosphere and exclude an undesirable climate change.

It is supposed that boreal forests are carbon sinks. However, their fires can transform boreal forests into sources of carbon due to direct emissions of carbon during biomass burning and indirect impacts of fires on the thermal and water regimes, as well as the structure and functioning of the ecosystems. The frequency of fires in boreal forests during the last several decades has increased and can increase further under conditions of continuing global climate warming. Thus, this should lead to a shorter time period of ecosystem recovery in the periods between fires, and to an enhancement of emissions of GHGs to the atmosphere. The CO_2 flux from the ground $R_s = H_7^C + H_8^C + H_9^C + H_{15}^C$ (see Figure 6.4) is a second important flux of carbon in boreal forests which determines its role in the formation of the global carbon cycle. The R_s value changes due to different processes:

- fires lead to a partial removal of the vegetation cover and a decrease of soil surface albedo (the latter determines an increase of surface temperature with the rate of decomposition of vegetation remaining stationary);
- fires break the process of accumulation of organic matter in soil and change the balance between the input of detritus and heterotrophic respiration (as a result of a great input of detritus); and
- fires bring forth changes of vegetation succession and composition of its species as well as litter quality.

To better understand the impact of forest fires on the carbon cycle in boreal forests, Wang *et al.* (2003) performed measurements and numerical modeling of R_s for black spruce (*Picea mariana*) with post-fire chronological succession for seven forest fires in the north of the Manitoba Province (Canada) under conditions of good and poor drainage of soils as an example. Studies have been aimed at:

(*i*) a quantitative characteristic of the dependence of R_s on soil temperature for forests of different ages;
(*ii*) study of the post-fire succession dynamics of forest stands; and
(*iii*) estimating the annual CO_2 flux from the soil surface.

The CO_2 flux depends strongly on conditions of drainage and age of the forest stand. There was a positive correlation of CO_2 flux with soil temperature (=0.78), with results of numerical modeling of the flux differing strongly depending on a combination of the level of drainage and age of the forest stand. During the period of the vegetation development season the CO_2 flux from a well-drained soil was much greater than from a poorly drained one. Annual mean values of the CO_2 flux from the soil for different years constituted 244 (the year 1870), 274 (1930), 350 (1964), 413 (1981), 357 (1989), 412 (1995), and 226 $g\,C\,m^{-2}\,yr^{-1}$ (1998), under conditions of well-drained soils and, respectively, 264, 233, 256, 303, 300, 380, and 146 $g\,C\,m^{-2}\,yr^{-1}$ in the case of poorly drained soils. In the winter (from 1 November to 30 April) the values of CO_2 flux varied within 5 to 19% with respect to the annual flux. Apparently, a decrease of the flux at the post-fire sites was determined mainly by the decreasing level of respiration due to tree roots.

In the course of all chronological succession of fires considered, the CO_2 fluxes from soil changed almost two-fold at reaching maximum values before joining of the tree crowns, when the soil was the warmest and an accumulation took place in both ground and sub-ground biomasses. The observed decrease of CO_2 flux in the case of older forest stands can be explained by a lower soil temperature, which is determined by an accumulated heat-isolating organic substance and other factors.

Boreal forests cover territories of about 14 mln km^2 in the circumpolar latitudinal band 50–70°N, which constitutes ~10% of the global land surface. These forests contain disproportionately large amounts of soil carbon because of climatic conditions which are unfavorable for the processes of organic decomposition. The annual mean temperature in boreal forests is close to 0°C with a weakly drained soil explained by prevailing lowlands. Characteristic features of the considered latitudinal band are widespread permafrost and isolation of deep ground horizons from summertime heating because of moss and thin roots which burn in forest fires but restore themselves over decades. This heat insulation favors the preservation of permafrost, which complicates the soil drainage and slows down the processes of decomposition of organic matter stored in deep soil layers beneath the moss layer. Intensification of respiration due to deep layers measured in July and August 1996 reached ~10 $kg\,C\,ha^{-1}\,day^{-1}$. This enhancement of respiration correlates with an increase of the deep layer temperature. Thus, the organic matter of deep layers of the soil is characterized by lability with a low rate of decomposition connected with low temperature.

The data above show that to raise the accuracy of estimates of sinks and sources of CO_2 on land, it is necessary to extend the parametric descriptions of the R_s flux components, apparently not only for forested territories but also for other soil–plant biomes. Smith *et al.* (2003) carried out measurements of CO_2 fluxes under conditions of the mountain ecosystem of the steppe wormwood in the south-eastern part of Wyoming (USA) and found that the difference between CO_2 fluxes measured at different times of day reached 9%, and that even under conditions of homogeneous vegetation cover the temporal variability of CO_2 fluxes between the ecosystem and the atmosphere is substantial.

An important consequence of forest fires is the fire induced changes of atmospheric chemistry due to GHG emissions. For instance, large-scale forest fires in September–November 1997 in Indonesia, Malaysia, and Papua New Guinea resulted in about 130 Tg CO being emitted to the atmosphere, which changed the ozone concentration by 10%. The GHGs emitted to the atmosphere propagated over vast territories, and the total content of CO in December grew by 10–20% in the latitudinal band 30°N–45°S, by 5–10% south of 45°S, and $<5\%$ north of 45°N. On the whole, global emissions of GHGs in biomass burning reached 3,800–4,300 Tg C yr^{-1} with a very small contribution due to fires in boreal forests (23 TgC yr^{-1}, or 0.6%). Emissions of GHGs to the atmosphere due to fires in boreal forests in 1998 constituted 290–383 Tg (total carbon), 828–1,105 Tg (CO_2), 88–128 Tg (CO), and 2.9–4.7 Tg (CH_4). The upper level of these estimates corresponds to 8.9% of the total global emissions of carbon due to biomass burning; 13.8% of global CO emissions due to forest fires, and 12.4% of global emissions of methane due to fires. The contribution of forest fires in Russia constituted 78% of the total emissions (19% is the share from North America). The contribution of peat-bog fires in the far east of Russia to carbon input to the atmosphere in the autumn of 1998 reached 40 Tg C.

A relative impact of fires on the chemical composition of the troposphere was weaker in the northern hemisphere, where the background CO content was greater than in the southern hemisphere. The radiative forcing at the surface level reached 10 W m^{-2} over most of the tropical Indian Ocean, and 150 W m^{-2} in the regions of the forest fires in Indonesia. Thus, forest fires turned out to be an important factor in the impact on the radiative regime in the tropics. About 700 Tg C were emitted to the atmosphere, which constituted 3–4% of the carbon supplies in the peat-bogs of Indonesia. All this testifies to a necessity for a detailed analysis of the consequences of forest fires and a search for a strategy aimed at their control. One of the perspective approaches to this problem is the use of satellite systems with microwave radiometers, which will make it possible to control, on global scales, the moisture content of the forest and to determine thereby the level of their fire risk.

8.3.3 Biomass burning and atmospheric chemistry

Biomass burning is one of the most substantial sources of inputs of many GHGs to the atmosphere. This process takes place both during a forest fire and in burning the wood remains after logging or during agricultural processes. Therefore, the control of the seasonal and inter-annual variability of biomass burning is important for estimation of GHG emissions and calculation of the radiation budget variability.

Duncan *et al.* (2003) proposed a method of control with the use of the satellite scanning radiometer ASTR. Results of measurements between 1979–2000 in south-western Asia, Indonesia, Malaysia, Brazil, Mexico, Central America, Canada, Alaska, and the Asian part of Russia did not reveal any long-term trend of CO emissions in forest fires, but revealed a clear inter-annual variability. The annual volume of CO emission in the indicated regions changed from 2 (the Asian part of Russia and China) to 20 $\mathrm{Tg\,C\,yr^{-1}}$ (Indonesia and Malaysia) with average global total emissions of 437 $\mathrm{Tg\,C\,yr^{-1}}$ (annual values varied from 429 to 565 $\mathrm{Tg\,C\,yr^{-1}}$).

In 1987, in the USA a long-term program was launched to study the spatial–temporal variability of CO concentration in the troposphere. Initial air samples taken in the marine atmospheric boundary layer to analyze the concentration of CO_2 and methane were also used to measure the CO concentration. During the last decade, scientists of the NOAA Climate Modeling and Diagnostics Laboratory carried out an analysis of air samples to estimate the CO concentration from the data of observations over the global network. It follows from the results obtained that the short-term periods of increasing or decreasing CO concentration were superimposed on the long-term decreasing trend of CO concentration in the troposphere observed in the 1990s. It turned out that the instability of standard samples used for calibration determined the systematic underestimation of the observed CO concentration. Novelli *et al.* (2003) showed that the decrease of tropospheric CO is confined mainly to the northern hemisphere, where emissions to the atmosphere due to fossil fuel burning were substantially lower, constituting $1.8 \pm 0.2\,\mathrm{ppb\,yr^{-1}}$. In the southern hemisphere, between 1991 and 2001, this trend was not observed.

The globally averaged CO concentration is characterized by a strong inter-annual variability explained mainly by inter-annual changes of the scale of biomass burning. During the whole period of observations the globally averaged CO concentration was decreasing by about $0.5\,\mathrm{ppb\,yr^{-1}}$; this decrease was mainly concentrated north of 30°N. A strong increase of CO content in the troposphere observed in 1997–1998 was connected with the impact of unusually widespread and intensive forest fires; more that 300 Tg CO were emitted to the troposphere due to fires. In the years of large-scale fires in boreal forests such emissions can cause changes in the dynamics of redox reactions in the troposphere on global scales.

Wet tropical forests in the Amazon Basin play a special role in the global ecodynamics due to the size of their territories (about $5 \times 10^6\,\mathrm{km^2}$) and high biotic activity. Here the all-the-year-round emissions to the atmosphere of natural biogenic aerosol take place both for primary (micro-organisms, pollen, detritus of vegetative origin) and secondary aerosol due to gas-to-particle transformation of organic, nitrogen and sulfur-containing GHGs emitted to the atmosphere by plants. In the dry season, this background aerosol is "suppressed" by smoke aerosol caused by deforestation and burning of biomass remains.

Processes of intensive convection and advection determine the long-range transport of both types of aerosol to higher latitudes, with far-reaching consequences from the viewpoint of affecting the chemical reactions in the atmosphere, as well as the formation of climate and biogeochemical cycles of nutrients. On the other hand, natural aerosol comes to the Amazon Basin from distant regions, including marine aerosol from the Atlantic Ocean and dust aerosol from the Sahara, which can serve as significant nutrients for wet tropical forests. In July 2001, near the city of Balbina in the center of the Amazon region, Graham *et al.* (2003) carried out measurements of aerosol which showed that the atmosphere contains both coarse-scale and fine-scale fractions, which consisted of organic compounds by 70 and 80%, respectively. Coarse-scale aerosols also contained small amounts of soil dust and sea salt particles, while fine-scale aerosols contained amounts of non-marine sulfate. The concentration of coarse-scale aerosol particles averaged $3.9 \pm 1.4\,\mu g\,m^{-3}$ with the ratio of nocturnal to daytime values being 1.9 ± 0.4, whereas the concentration of fine-scale aerosols increased in the daytime (an average of $2.6 \pm 0.8\,\mu g\,m^{-3}$, with the ratio 0.7 ± 0.1).

Biomass burning during the last decades has increased with growing population size. The geographical zones of intensive biomass burning have broadened. The most considerable components of emissions in biomass burning include CO_2, CO, CH_4, NH_3, HCN, methanol, acidic acid, acetol, and others. Knowledge of the volumes of emitted aerosols is important to study the consequences of their impact on the radiation budget and dynamics of ecosystems of land and the World Ocean. Christian *et al.* (2003) found several new components appearing in the burning of African fuels; they are chemically active substances which can take an active part in oxidation reactions. These substances include acetaldehyde, phenol, acetol, glycolaldehyde, methylvinilether, furan, acetone, acetonitrile, propylene-nitrile, and propanenitrile. Emissions of chemically active oxidizing volatile organic compounds in fires in African savannas and in Indonesia constitute 70 and 77% of all emitted aerosol, respectively.

Many uncertainties appear in the studies of the impact of biomass burning on the ozone layer, though there is evidence of this impact. In the period 21 January–14 February 1999, when large amounts of biomass were burned on the western coast of Africa, sonde measurements of the tropospheric ozone content were carried out on board a scientific ship along the route from Norfolk (37°N, 76°W) to Cape Town (34°S, 22°E), and the earlier conclusion was confirmed that the tropospheric ozone and its precursors connected with biomass burning were concentrated in the lower troposphere mainly because of the absence of an intensive convection on land. In the 1,000–799-hPa layer there was a wind-driven transport of ozone and its precursors toward the equator or west. On the other hand, the effect of the upward adiabatic motions connected with the diurnal change of the processes in the west-African planetary atmospheric boundary layer (ABL), could determine the vertical transport of ozone and its precursors to the free atmosphere over the marine ABL. Besides this, lightning strikes in the regions of South Africa, Central Africa, and meso-scale convective systems of the Gulf of Guinea could cause an increase of tropospheric ozone concentration in the middle and upper troposphere of the tropical Atlantic Ocean in the southern hemisphere. At the same time, a decrease of the ozone content called the "ozone paradox" was recorded in the tropical Atlantic Ocean in the southern hemisphere – the explanation of which requires development of a special climate unit for the global model.

8.3.4 Wildfires and the carbon cycle

Wood remains (WR) are an important component of all forest ecosystems, since they affect the nutrient cycle, humus formation, carbon storage, frequency of occurrence of forest fires, water cycle, and are a natural habitat for heterotrophic and autotrophic organisms. Usually the presence of dead trees and WR is left out of account in the estimation of carbon budgets, though the importance of these carbon reservoirs as boreal supplies of carbon has recently attracted attention.

Bond-Lamberty *et al.* (2003) studied the distribution and respiration dynamics of wood remains under conditions of chronological succession in the fires of boreal forests consisting mainly of black spruce, in the northern part of Manitoba Province (Canada). The considered chronological succession included seven forest stands that suffered fires during 1870–1998. Each of the forest stands was a separate well or poorly drained site. The WR biomass varied from 1.4 to 177.6 mg ha^{-1} decreasing, as a rule, with age. Most decayed wood remains were characterized by a higher humidity, lower density, and higher rate of respiration compared with the less decayed remains. Depending on the WR, estimates of the annual emissions of carbon varied between 0.11 and 1.92 mgC ha^{-1} yr^{-1}. This study has shown that in modeling the carbon exchange in the atmosphere–plant–soil system, it is necessary to select a new component that describes the WR (Hoelzemann *et al.*, 2004).

In this regard, the equation of biomass dynamics should contain, at least, three terms that parameterize photosynthesis, respiration, and dying-off. This will make it possible to more adequately reflect the role of the forest ecosystem in the change of CO_2 exchange between the land cover and the atmosphere, by specifying the dependences of the indicated components on climate parameters. In particular, it is known that many boreal ecosystems are sources of carbon in winter and sinks in summer. This process is controled by air temperature and day duration. In summer, in boreal forests during the season of plant vegetation, the amount of carbon absorbed from the atmosphere can reach 80–310 g C m^{-2}, and in winter, carbon emissions can reach 60–90 g C m^{-2}. The inter-annual variability of the seasonal exchange of carbon, from estimates of Suni *et al.* (2003), can constitute from 30 g C m^{-2} in autumn and spring to 80 g C m^{-2} in summer.

8.4 THUNDERSTORMS AS COMPONENTS OF GLOBAL ECODYNAMICS

8.4.1 Thunderstorms and atmospheric chemistry

The level of understanding the photochemical processes in the troposphere that affect its composition depends substantially on the adequacy of ideas about the budget and distribution of nitrogen oxides NO_x ($NO + NO_2$). Due to reactions of the catalytic cycle with the participation of peroxides radicals, NO_x play the decisive role in the processes of tropospheric ozone formation and affects the cycle of HO_x ($OH + HO_2$). One of the most important unclear aspects of the problem under discussion is the insufficient reliability of data on the contribution of lightning strikes to the global budget of NO_x. According to earlier estimates, this contribution varied within 2–20 Tg N yr^{-1}, with most probable values of about 2–6 Tg N yr^{-1} (Kondratyev, 2005a,b).

Crawford *et al.* (2000) carried out an analysis of results of aircraft measurements within the program of the field experiment SONEX on studies of ozone and nitrogen oxides in the troposphere. Results of the analysis revealed the presence, in the upper troposphere, of episodes of NO_x concentration increase determined, apparently, by the impact of lightning strikes. The correlation with specific periods of lightning activity could be found out from analysis of "reverse" trajectories of the motion of air masses as well as from the data of observations at the national network of lightning records. The Lagrangian numerical modeling of "reverse" trajectories has been carried out to trace the evolution of NO_x plumes during a 1–2-day interval between their occurrence and aircraft measurement data. Calculations were made for pre-calculation of expected changes of concentrations of HNO_3, H_2O_2, CH3COOH, HO_2, and OH. Depending on the conditions, the initial concentration of NO_x varied within 1 to 7 ppb. Since the calculated estimates of HNO_3 concentration turned out to be overestimated compared with those observed, the reasons for such differences were analyzed, and it was shown that the observed values of H_2O_2 concentration agree with the supposed removal of H_2O_2 from the atmosphere due to convection. Although it is possible that in the upper troposphere the concentration of CH_3COOH grows due to convection, the numerical modeling has led to the conclusion that this increase can be only of short duration (less than 2 hours), which excludes the possibility to detect this component during aircraft measurement. The possibility of CH_3COOH concentration increase is excluded too in view of a high level of NO concentration.

The reaction between NO and HO_2 in all the cases considered favors a decrease of HO_2 concentration. In some cases the calculated values of OH concentration doubled, but at a maximum level of NO_x concentration the loss of hydroxyl due to the reaction $OH + NO_2$ compensates the formation of hydroxyl due to reactions $NO_2 + NO$. An additional increase of OH by 30–60% can, however, result from the convective transfer of CH_3COOH.

To form an estimate of the contribution of lightning and anthropogenic sources to the formation of NO_x concentration on global scales, Kurz and Grew (2002) undertook a numerical modeling of the frequency and and also global distribution of lightning strikes and the resulting formation of NO_x using the complex model of climate and chemical processes taking place in the atmosphere ECHAM4.L39(DRL)/CHEM(E39/C), whose important feature is a realistic simulation of processes responsible for the formation of penetrating convection.

Calculated estimates of the height of convective clouds, on the whole, agree well with those observed. However, in the mid-latitudes of Western Europe the calculated values were underestimated. The calculated spatial distribution of lightning strikes agrees well with the observed distribution, though the calculated ratio of the density of strikes over land and over the oceans turned out to be underestimated. In the mid-latitudes of Western Europe the density of strikes is underestimated compared with that observed, which can be explained by an inadequate scheme of convection parameterization.

According to numerical modeling results, the NO_x formation is most intensive in the tropics and in mid-latitudes, and is clearly separated (in space) from NO_x emissions due to aviation. The lightning-induced maximum level of NO_x emissions is located at altitudes approximately 5 km below the level of the tropopause, with the level of emissions exceeding 3 times the respective anthropogenic level. Besides this, it is important that maximum aircraft emissions of NO_x tend to be at much higher altitudes and farther north (being concentrated mainly within the North Atlantic corridor 30–60°N at a level of about 200 hPa (i.e., 12 km)) compared with NO_x formed due to lightning strikes.

Zhang *et al.* (2003a) analyzed the possibilities of the numerical modeling of nitrogen oxides formed due to lightning strikes using the 2-D version of the model that simulates an electrization of the storm. It is supposed that the formation of NO is determined by a dissipation of energy whose value is calculated from the value of the electric field before and after a lightning strike. The rate of formation of energy responsible for NO formation is 9.2×10^{16} molecules per Joule. Considering a limited set of chemical reactions in which NO, NO_2, and O_3 participate, a numerical modeling has been carried out of the processes taking place in a small storm with 10 intra-cloud strikes in 2 minutes. The level of dissipation of energy varied between 0.024 and 0.28 GJ. After cessation of lightning strikes the integration continued for a further 18 minutes. Analysis of the modeling results has shown that the mixing ratio of NO formed within a cloud (by order of magnitude) is 10 ppb after the most powerful strikes and 1–2 ppb on the windward side of the thundercloud anvil at the end of the integration interval. These estimates agree with the data of observations. A comparison with results of earlier numerical modeling with the use of a different 2-D model revealed, on the whole, an agreement of estimates, but the results discussed are characterized by greater volumes of energy at high altitudes, which can partially result from a longer integration time and a lack of consideration of strikes in the model in question, as well as from some assumptions made earlier. On the whole, the numerical modeling of lightning formation carried out recently, and its impact on NO formation should be considered more adequate than in previous studies. The necessity of a further improvement to the model has determined the development of a new 3-D model of the processes of formation of electric charges in thunderclouds with a more complete (than earlier) consideration of the chemical processes of nitrogen oxide formation under the influence of lightning strikes (Zhang *et al.*, 2003b).

A totality of the considered GHGs includes in this case NO, NO_2, O_3, CO, CH_4, OH, and HNO_3. The numerical modeling has been carried out for concrete conditions of thunderclouds (storm) observed on 19 July 1981 with 18 intra-cloud strikes during 3 minutes. The numerical modeling has been carried out for a period of 38 minutes, before the thundercloud dissipated. The level of energy dissipated at lightning strikes varied within 0.91 to 2.28 GJ. The maximum level of the NO_x mixing ratio due to lightning strikes reached 35.8 ppb. At the cloud's dissipation, after the cessation of lightning strikes, maximum concentrations of NO and NO_2 (in both cases) constituted about 6.3 ppb and were observed at an altitude of about 4 km. The NO mixing ratio in an anvil reached a maximum of about 2 ppb at an altitude of about 10.5 km. These results agree well with observational data.

Quite surprising was the formation of the NO_2 plume at a concentration of about 0.5 ppb, which reached the surface. In the case of NO there was no plume. On the other hand, NO was transported from the cloud's center to the anvil absent in the case of NO_2, which was determined, probably, by the impact of photolysis. The ratio of concentrations of NO_2/NO decreased with altitude in accordance with observational data. The formation of NO calculated per unit length averaged 2.03×10^{22} molecules per meter. The results obtained show that the short-lived storms determine the formation of the vertical profile of NO_x concentration that differs from the earlier observed C-shape profile.

Mansell *et al.* (2002) carried out a numerical modeling of lightning strikes with the use of a stochastic model of dielectric break and parameterization of electrification mechanisms. This model enables one to simulate a 2-D development of a strike as a stochastic "step-by-step" process. The strike channels propagate over a homogeneous spatial grid, and the direction of propagation (including diagonals) for each step is considered random with the probability of the choice of a certain direction depending on the total electric field. After each step the electric fields are calculated anew with the use of the Poisson equation in order to take into account the impact of the channel's conductivity. The applied parameterization of the process of lightning strike formation provides a realistic 3-D simulation of ramified strikes. The model is able to simulate the formation of different kinds of lightning strikes, including intracloud strikes, and negative and positive "cloud–ground" (CG) strikes.

According to numerical modeling results, the hypothesis that negative strikes appear only when the region of the positive strike is located beneath the center of the negative strike can be considered substantiated. The calculated positive CG strikes were observed only in the parts of the thundercloud where two charged layers located near the Earth had an approximately similar "normal dipole" structure (i.e., positive charges were above negative ones).

Brown *et al.* (2002) studied relationships between CG lightning strokes and stages of the vertical motions development in the region of centers of thunderclouds taking as an example the data of observations of a multi-cell thundercloud center formed on 11 July 1989 in the region of the city of Elgin (North Dakota, USA). Radar observations of reflectance and vertical velocity enabled one to identify some cells within the thundercloud center and trace the evolution of each cell. The evolution process took place in accordance with the model of thundercloud center developed in the late 1940s and in the following succession:

- maturing of upward motions and increasing of the cell's vertical extent at the stage of cumulus cloud development;
- maximum development of vertical motions in the upper part of a cell and downward fluxes with rain from the middle part of the cell; and
- light rain at the stage of the cell's dissipation.

On 11 July there was observed a trend in the formation of the cell clusters, each cluster consisting of both growing (at the stage of maturing) and disintegrating cells. There were no CG strokes when the zone of the storm's convection contained only one cluster of cells. The strokes took place only in the presence of two or more clusters. Except for two cases, lightning strokes were observed in the zone of the storm's convection, as a rule, closer to growing (mature) cells than to dissipating cells. The discussed observational data favor the hypothesis according to which rain falling at the sites of downward fluxes determines the conditions favoring the formation of lightning strikes that reach the Earth. However, it follows from observations that the complicated structure of the electrical field caused by superimposed fields of several cells especially favors an appearance of CG strikes. This situation can promote an earlier formation of CG strikes in the process of new storm cell formation.

8.4.2 The electromagnetic fields of lightning

Although studies of lightning were started in the late 1700s, only in the early 20th century were observations of the electric field carried out to retrieve the spatial distribution of charges in thunderclouds responsible for the formation of lightning. Then, intensive studies began of the distribution of low-frequency electromagnetic waves generated by lightning strikes (atmospherics). Random natural fields received on the Earth's surface in the frequency range from several Hz to MHz are created mainly by thunderstorm sources. Their study for a formalized description in the interests of radar observations has been carried out by Remizov (1985). Radio-noise caused by thunderstorms is disastrous for communication systems just as thunderstorms themselves are disastrous for populations as factors that can lead to fires and destruction.

Chronis and Anagnostou (2003) discussed the preliminary results of the functioning of the experimental network recording lightning strikes (ZEUS) which consists of six receivers located in Western Europe and began in July 2001. The receivers make it possible to record the low-frequency electromagnetic waves (5–15 kHz), generated by lightning, which propagate via the "Earth–ionosphere" waveguide over distances of several thousands of kilometers. Estimates have been obtained of the errors of observations at the ZEUS network (from the viewpoint of reliability of detection, localization, and characteristic of lightning strikes) by comparison with the data of independent observations. Such comparisons have been made for three regions: the eastern coast of the USA/north-western sector of the Atlantic Ocean, Africa, and Spain. Data for comparison came from the results of observations made with the use of the lightning sensor mounted on the TRMM satellite studying precipitation in the tropics, as well as data from the Spanish network for lightning observations. Results of comparisons have shown that the errors of localization (determination of coordinates) of lightning vary within 40 to 400 km at distances up to 5,000 km and farther. Within the territory on which the network is located errors do not exceed 40 km.

Formation of positive charges in the lower parts of clouds revealed by calculations with the use of the so-called tripole model of thunderclouds, was verified by the results of direct balloon soundings and remote sensing from the data of ground measurements of the electric field. However, since the sources of these charges have not been clearly understood, Mo *et al.* (2002) undertook direct aircraft measurements of positive charges in the lower parts of clouds with the use of two aircraft.

Data of observations obtained near New Mexico (USA) on 10 August 1997 demonstrated (at least, in some cases) that near the bottoms of clouds (at an altitude of about 3.4 km) there were located positive charges formed, apparently, under the influence of lightning strikes. The charge recorded in one of the cases at an altitude of about 4 km was $\sim$1.25 C, which agrees with the data of balloon and ground measurements of electric fields. Numerical modeling with electric fields prescribed from the data of aircraft observations near the idealized charge dipole with the instant introduction of positive charge gave results which agree well with observations. The observational data indicate that at a ripe stage of formation of the horizontal distribution of the charge near the cloud bottom this distribution can be very complicated and is characterized by a combination of the contacting regions with opposite charges.

Rakov and Tuni (2003) studied the adequacy of numerical modeling of the lightning electric field at a great distance using the model of the transmission line (TL) and modified model MTLE, taking into account an exponential decrease of current with altitude depending on polar angle (elevation) and the rate of propagation of the opposite charge. The shape of the wave of the latter was approximated by a step function. The same presentation was used for the TL model, whereas in the case of the MTLE model it was supposed that the electric field increases instantly to the level corresponding to the TL and then decreases exponentially. The exponential decrease with altitude (in the case of MTLE) results in a considerable decrease of the electric field intensity about 1 microsecond after reaching a maximum, especially at low values of polar angle and high rates of propagation.

Calculations made by Marshall and Stolzenburg (2002) showed that in the case of positive CG strikes at which Q-flashes occur, the level of energy constitutes about 1×10^{10} J, and the area of strike is $\sim 40 \times 40\,\mathrm{km}^2$. An estimation of total electrostatic energy stored in two stratus clouds of a mesoscale convective cloud system gave values of 5×10^{11} and 2×10^{12} J. These levels of energy are sufficient to provide hundreds or thousands of typical lightning strikes but only 10–100 positive CG strikes with accompanying Q-flashes.

Using Pockels sensors (Miki *et al.*, 2002) in the International Center for Lightning Research and Testing (ICLRT), Florida (USA), Miki *et al.* (2002) carried out measurements of the shape of the electric field wave at horizontal distances between 0.1 to 1.5 m from the lightning channel. The dynamic range of the measuring system varied between $20\,\mathrm{kW\,m^{-2}}$ and $5\,\mathrm{MW\,m^{-2}}$, with the band width in the interval 50 GHz to 1 MHz. Also, electric fields were measured near the bottom of the channel and at distances of 5, 15, and 30 m from the lightning channel. Using Pockels sensors, measurements of electric fields were made for 36 strikes in 9 "trigger" flashes. For 8 of the 36 strikes measurements were also made of horizontal electric fields. According to the results obtained, the shape of the electric field wave looks like an impulse with its front edge determined by the strike leader, and the rear edge determined by the inverse strike. For 6 of the 36 studied, the shapes of the electric field wave were closed V-shapes, whereas the other 30 were characterized by much slower variations in the phase of the reverse stroke than at the leader stage. The vertical electric field reached a maximum in the interval from 176 $kW\,m^{-1}$ to 1.5 $MW\,m^{-1}$ (an average of 577 $kW\,m^{-1}$), and the horizontal electric field – in the range 495 $kW\,m^{-1}$ and 1.2 $MW\,m^{-1}$ (an average of 821 $kW\,m^{-1}$). These values are characterized by a 40% underestimation.

8.5 NATURAL DISASTERS AND SCENARIOS OF GLOBAL ECODYNAMICS

8.5.1 The state of global water resources

One of the components of global ecodynamics is freshwater, whose supplies are an important indicator of humankind's likelihood of survival. The problems of freshwater attract special attention, since on the one hand, water is a key component of ecosystems and, on the other hand, about one-third of the global population is under the threat of a chronic water deficit over several decades. This and other circumstances have determined the UN declaration of the Fresh Water Decade (2005–2015). The following factors characterize the present state of global water resources (*Year-book NEP*, 2004):

- Total volume of water resources on Earth constitutes about 1.4 billion km^3.
- The volume of freshwater resources is about 35 million km^3 or 2.5% of the total volume of water.
- Of these freshwater resources about 24 million km^3 or 68.9% exist as ice and constant snow cover in mountains, the Antarctic, and the Arctic.
- About 8 million km^3 or 30.8% are underground in the form of ground water (in shallow and deep basins of ground waters down to 2,000 m), soil moisture, bog water, and permafrost.
- Freshwater lakes and rivers contain about 105,000 km^3 or 0.3% of the global freshwater supplies.
- Total volume of freshwater supplies that can be used by ecosystems and populations constitutes about 200,000 km^3 (i.e., less than 1% of all freshwater supplies).
- Some 3,011 freshwater biological species are put in the list of species either endangered or irreparable; 1,039 of them are fish. Four of the five existing species of river dolphins and two of the three existing species of sea cows, about 40 species of freshwater tortoises, and more than 400 species of freshwater mollusks are endangered.
- The annual use of ground water is estimated at 600–700 km^3 or about 20% of the global water use; about 1.5 billion people use ground water for drinking.
- From estimates in 2000, 70% of the global freshwater expenditure falls on agriculture.
- The per-capita consumption of water in developed countries is, on average, about 10 times greater than in developing countries. In developed countries

this indicator varies from 500 to 800 l day^{-1}, and in developing countries – from 60 to 150 day^{-1}.

- The industrial consumption of freshwater constitutes about 20% of the global use of freshwater. Between 57% to 69% of the global use of water is applied to produce electrical energy at hydroelectric power stations and nuclear power stations, 30–40% for industrial processes, and 0.5–3% for heat-and-power engineering.

The problems connected with freshwater are numerous and diverse. Tables 8.1 and 8.2 composed by the authors of the *Year-book NEP* (2004) are a summary of the aspects of these problems on which the human life support depends in the context of the UN Millennium Declaration (*UNEP Science Initiative*, 2004).

On the whole, it should be stated that a sufficient provision regarding freshwater is of fundamental importance to achieving the goals of socio-economic development and environmental protection. In this regard, of serious concern are intensified anthropogenic forcings on the environment. For instance, the area of freshwater wetlands that plays an important role in natural water purification and in the formation of the water cycle has almost halved during the last 20 years. Meanwhile, according to economic evaluation of the role of wetland functioning, their losses are equivalent to US$20,000 ha^{-1} yr^{-1}. About 20% of the 10,000 species of freshwater fish are on the verge of extinction. The number of large dams in the world has increased from 5,000 in 1950 to more than 45,000, to which has been associated many negative ecological consequences (*Vital Signs*, 2003). The role of forest ecosystems in regulation of the quality of natural water has changed. For instance, studies by Porvari and Verta (2003) of the content of mercury in the run-off from boreal forest clearings in Finland have shown that the concentration of total mercury constitutes 0.84–24.0 ng l^{-1}, and in surface water run-off an additional 7 to 79 mg C l^{-1} were found.

The geographical distribution of freshwater resources is very irregular: about half the global resources falls on six countries (Brazil, Russia, Canada, Indonesia, China, and Columbia). This irregularity is also characteristic of some countries. For instance, in China, with 7% of the freshwater resources (with its percentage of the global population being 21%), most of the country is arid. Naturally, the countries with water deficit have to use ground water, which leads to a gradual decrease of the ground water level. The data in Table 8.2 characterize the levels of per capita use of ground water in different countries in 2000 (*Vital Signs*, 2003).

Besides this, these data illustrate the socio-economic contrasts existing in the world. Nearly every fifth person in the developing world (1 billion in total) is endangered every day because of the absence of potable water. The main problem is not the absence of water, but unfavorable socio-economic conditions.

Agriculture is known to be the basic consumer of freshwater from rivers, lakes, and underground sources (about 70% on a global scale and up to 90% in many developing countries). Since the broadening of application of irrigation will face the problem of a deficit of freshwater resources in the nearest future, the problem of raising the efficiency of the use of freshwater becomes more and more urgent. In

Table 8.1. Various aspects of vital activity connected (directly or indirectly) with freshwater related problems.

Millennium Declaration	Problems to be resolved
Goal 1 To extirpate extreme poverty and hunger.	Over the period 1990–2015, to halve the share of the population with an income of less than 1 dollar per day and to halve the number of starving people.
Goal 2 To provide universal elementary education.	By the year 2015, to provide the possibility for both boys and girls to be elementarily educated.
Goal 3 To promote equality between men and women and broaden the ability of women.	To liquidate the inequality between boys and girls in their access to elementary and secondary education preferably by 2005 and to all the levels of education by not later than 2015.
Goal 4 To reduce infant mortality.	To reduce by two-thirds the mortality of children under 5 years over the period 1990–2015.
Goal 5 Improve mothers' health.	To reduce by three-quarters mothers' mortality over the period 1990–2015.
Goal 6 To struggle against HIV/AIDS, malaria, and other diseases.	By the year 2015, to stop the propagation of HIV/AIDS, malaria, and other basic diseases and to start reducing the numbers of people affected by them.
Goal 7 To provide ecological stability.	To take into account the principles of sustainable development in national strategies and programs and reduce the loosing trend of ecological resources. By the year 2015, to halve the share of the population without access to pure drinking water, and to substantially improve the lives of the 100 million or so of slum dwellers.
Goal 8 To arrange a global-scale partnership in the interest of development.	To continue the formation of the undisguised, predictable, and non-discriminatory commercial and financial systems based on legal norms. To provide control, development, and reduction in poverty on national and international levels. To satisfy the special needs of the least developed countries having no access to the sea, as well as small island developing countries (SIDC). In cooperation with the private sector, to take measures in order that everybody can use the benefits of new technologies.
Plan of realization of the decisions of the Summit of the Millennium.	By the year 2015, to halve the share of the population without access to safe drinking water partially because of want of money (as foreseen in the Millennium Declaration) and the share of the population without access to basic sanitation. By the year 2005, to deliver plans for a complex control of water resources and efficiency of water use.

Table 8.2. Per capita annual diversion of ground water (m^3 per capita) in different countries.

Country	Level of per capita water diversion
Ethiopia	42
Nigeria	70
Brazil	348
South Africa	354
Indonesia	390
China	491
Russia	527
Germany	574
Bangladesh	578
India	640
France	675
Peru	784
Mexico	791
Spain	893
Egypt	1,011
Australia	1,250
USA	1,932

particular, this refers to micro-irrigation (including drop irrigation) whose scale remains, however, limited.

One area with considerable potential for water saving is connected with food production. For instance, as seen from Table 8.3, production of 10 g of protein in the form of beef needs 5 times more water than in the case of rice, (for 500 food calories this difference reaches 20 times). With an abundant meat diet an average American consumes 5.4 l of water per day, whereas in the case of a vegetarian diet this expenditure is halved.

A serious problem manifests itself in the water supply of cities and water saving. Still more urgent is the problem of the industrial use of freshwater reaching 22% of the global-scale use of freshwater resources (59% in industrial countries and 10% in developing countries).

8.5.2 The energy provision of civilization

Consider the problem of the energy provision of humankind discussed in Chapter 1 in more detail, since the human survival depends on energy provision both on regional and global scales. During the period 1850–1970 the global population size more than tripled, and the energy production increased 12-fold. Since then (by the year 2002) the population has increased by 68%, and fossil fuel consumption by 73% (Starke, 2004). It is important, however, that there was no rigid connection between energy production and economic growth. In the period 1970–1997, measures for

Table 8.3. Water expenditures (in litres) on formation of 10 g of protein and provision of 500 food calories.

Food	10 g of protein	500 calories
Potatoes	67	89
Peanuts	90	210
Onions	118	221
Maize (grain)	130	130
Legumes	132	421
Wheat	135	219
Rice	204	251
Eggs	244	963
Milk	250	758
Poultry	303	1,515
Pork	476	1,225
Beef	100	4,902

energy saving ensured an economic growth with energy expenditure in production reduced by 28%. Data in Table 4.2 illustrate the Gross World Product (GWP) global dynamics, and in Table 4.3 the dynamics of fossil fuel consumption (*Vital Signs*, 2003). The global consumption of fossil fuels in 2002 grew by 1.3% (up to 8,034 mln t of oil equivalent) compared with 0.3% in 2001. The level of fossil fuel consumption has increased by a factor of 4.7 compared with 1950. At present, fossil fuels provide 77% of the global energy consumption.

As for the consumption of various kinds of fossil fuel, the respective trends are rather inhomogeneous. If in 2002 the global mean growth of oil consumption constituted 0.5%, in China it reached 5.7%, the Middle East (2.5%), former USSR countries (1.9%), mainly determined by the increasing export. In the USA (consuming 26% of global oil) the level of consumption has grown negligibly. In some countries (Japan, South Korea, Australia, New Zealand) the level decreased on average by 0.6%, and in the countries of Latin America by 2.6%.

After a short-term but considerable decrease of coal consumption in the late 1990s, in 2002 its global consumption increased (compared with 2001) by 1.9% (2,298 mln t of oil equivalent). In the USA (25% of the global coal consumption) a 0.5% decrease was observed, and in China (23% of the global consumption) a 4.9% increase was recorded (despite the prohibited use of coal in some regions, where smog and acid rain often occurred).

The global mean growth of natural gas consumption constituted 2% (2,207 mln t of oil equivalent), but in the USA (27% of the global consumption) there was a decrease by 3.7% (during the first 10 months of 2002 compared with the respective period in 2001), which was explained mainly by a mild winter. A decrease of natural gas consumption was also observed in some other developed countries, reaching a maximum in Japan (10.4%), but in Norway a strong increase was observed (up to 81%). On the whole, natural gas is characterized by a most rapid growth of consumption (compared with other fossil fuels), providing at present almost 24% of the global energy consumption (compared with 22.5% 10 years ago). The reasons for this growth are many, including the accessibility of natural gas in several countries and its weaker (compared with other fossil fuels) negative influence on the environment.

The levels of energy consumption in various countries differ substantially (Brown, 2004). The population of the richest countries consumes 25 times more energy, on average (per capita), than the population of the poorest countries. For 2.5 billion people in the world, living mainly in Asia and Africa, the basic source of energy is still wood (or other kinds of biomass). Whereas in the USA the per capita level of commercial energy consumption (in weight units of oil equivalent) constitutes $8.1\,t\,yr^{-1}$, in Ethiopia it is only $0.3\,t\,yr^{-1}$. The respective extreme levels of oil consumption in the world are 70.2 and 0.3 barrels per day per thousand people, and electric power – 12,331 and 22 kW per thousand people. Therefore, the level of per capita CO_2 emissions to the atmosphere reaches 19.7 t in the USA and only 0.1 t in Ethiopia.

It should be noted, however, that the rate of growth of energy production (and, respectively, the economy) is highest in Asia. For instance, the scale of economy in China has grown during the period from 1980 by more than 4 times (respectively, the requirement for electric power has grown by 4 times). One of the indicators of development of the Chinese economy can be the growth in the number of personal cars from 5 million in 2000 to 24 million in 2005. The number of "rich" families in India with incomes of more that US$220 per month has grown six-fold during five years, whereas the number of poor families decreased substantially.

According to estimates of the International Energy Agency (2005), in the period 2000–2030, the annual global mean increase of primary energy production should constitute 1.7%, reaching the level of 15,300 mln t of oil equivalent in 2030. The growing need for energy (mainly in developing countries) should be 90% satisfied by fossil fuels. However, even by 2020, about 18% of the global population will be deprived of current sources of energy, such as electricity. Of course, such forecasts should be considered as rather conditional.

Of great importance is substantial progress achieved in the use of alternative energy sources. First of all, this refers to nuclear energy. As is seen from Table 4.10, in the period 2001–2002 the energy production at nuclear power plants (NPPP) increased by about 1.5% (>5,000 MW). After 1993, growth has been most intensive. The quantity of nuclear reactors in the world (at NPPPs) increased to 437 due to the start up in 2002 of new reactors in China (4), South Korea (2), and the Czech Republic (1). In 2002 construction started on six new reactors in India and 26 reactors were in the process of being built (their total power was 20,959 MW). Seven reactors were decomissioned (similar reactors total 106). China plans to build 4 new reactors producing a total power of 20,000 MW during the next several years. In India the total power output from nuclear power plants is to grow by 150% (up to 6,000 MW with perspective growth to 20,000 MW by 2020).

In some countries the situation with nuclear energy is rather difficult. For instance, in Great Britain, the NPPPs are declared non-competitive, and Belgium plans to exclude the use of NPPPs by 2025. At present, in Western Europe no new reactors are being built (except Finland, where parliament has approved the construction of its fifth reactor). Serious difficulties with nuclear energy occur in the USA and Japan (*Vital Signs*, 2004). South Korea plans the set up of eight new reactors during the next 10 years.

As for other alternative kinds of energy, most substantial progress has been achieved in the use of wind power, though an absolute contribution remains small. By the end of 2002 the total production of electric power due to wind generators constituted 32,000 MW, and the global mean growth (compared with the previous year) reached 27% (31% in Western Europe). Compared with 1998, wind power production has tripled (thus, this is the most rapidly developing source of energy). The share of the global wind power in Western Europe is 77% with the main contributions coming from Germany, Spain, and Denmark.

The increasing use of fossil fuels and respective growing CO_2 emissions to the atmosphere have caused anxiety about the possible anthropogenic changes of global climate (Kondratyev and Krapivin, 2003a; Kondratyev *et al.*, 2003a; Kondratyev, 2004b,c). Table 4.4 contains information about respective observed global trends.

Without dwelling upon cause-and-effect connections between temperature and CO_2 concentration (this problem has been discussed in detail by Kondratyev *et al.*, 2004a,b) note only that despite an adoption of the known international documents on the necessity to reduce the GHG emissions (including CO_2) to the atmosphere, the global mean CO_2 concentration continues to grow. This trend will, no doubt, continue in the nearest future. In 2002 alone, 6.44 bln t C were emitted to the atmosphere (a 1% increase compared with 2001). The concentration of CO_2 reached 372.9 ppm, increasing by 18% between 1960 and 2002, and from the beginning of the industrial revolution (1750) by 31%. About 24% of global emissions occur in the USA, whose population constitutes 5% of the global population (per capita CO_2 emissions in the USA exceed by 17 times those observed in India).

One of the important features of the present energy consumption is the growth of energy consumption due to various kinds of transport, primarily, personal cars, whose production reached 40.6 million in 2002, which exceeds five-fold the level in 1950 (*Vital Signs*, 2003). The total quantity of cars exceeds 531 million, increasing annually by 11 million. As a result, in the most developed countries the share of public transport decreased, constituting only 10% in the cities of Western Europe, 7% in Canada, and 2% in the USA.

On the average, about one-third of the energy produced in the world is spent on heating, lighting, air conditioning, and others, with this kind of energy consumption growing rapidly in the presence of strong contrasts between various countries. Domestic energy consumption in the USA and Canada is greater than in the countries of Western Europe by a factor of 2.4, and an average nine-fold difference in the levels of energy consumption is observed between economically developed and developing countries. The share of domestic energy consumption in China is about 40% and in India it reaches 50% (and even greater in many countries of Africa), whereas in developed countries this share does not exceed 15–25%.

According to the data on Human Development Index (HDI) and energy consumption, there is a correlation between these parameters, which, however, does not

Table 8.4. Power consumption and quality of life in different countries.

Country	WI	Per capita power consumption (rank)	Share (%) with respect to power consumption in Sweden
Sweden	1	10	100
Finland	2	6	112
Norway	3	8	104
Austria	5	26	61
Japan	24	19	70
USA	27	4	140
Russia	65	17	71
Kuwait	119	3	162
UAE	173	2	190

manifest itself as direct proportionality (Starke, 2004). For poor people, even a small increase of energy consumption means a substantial increase in their quality of life, but with per capita energy consumption reaching 1 tonne of oil equivalent per year and increasing to 1–3 tonnes per year, the role of additional energy consumption starts decreasing. With a greater increase of energy consumption, its connection with the HDI disappears. The countries with a per capita level of equivalent energy consumption of about 3 tonnes annually are Italy, Greece, and South Africa, whereas in the USA the per capita energy consumption is almost 3 times greater.

To characterize the quality of life, the Wellbeing Index (WI) has been proposed to classify 180 countries, taking into account 87 indicators of the state of humans (health, levels of education and material security, civil rights) and the environment. According to WI data, in this classification first is Sweden and last is the United Arab Emirates, though the level of energy consumption in this country is almost twice as great as in Sweden (Table 8.4). Austria takes a very moderate place in energy consumption but is of a high rank from the viewpoint of wellbeing.

This means that the present state of human society cannot be considered stable from all points of view (including socio-economic and ecological conditions). More and more data indicate that the present structure of consumption results in degradation of the quality of life for many people, manifesting itself through health problems, deterioration of the environment, damage to numerous ecosystems, etc. Despite the continued improvement of technologies and measures on energy saving, under conditions of growing global population size and growing level of per capita consumption (especially in developed countries), prospects of global sustainable development in the 21st century are far from being clear. According to available forecasts, the global population size will increase by more than 40% by the year 2050 (up to 8.9 billion people). Clearly, a possibility of reaching the global mean level of energy consumption as in developed countries (and this level is much lower than in the USA) is completely excluded. This would necessitate raising the energy production more than 8 times over the period 2000–2050. This increase is impossible using fossil fuels (besides which, would follow negative ecological consequences). However, the prospects of intensive development of alternative sources of energy in the 21st century remain unclear.

8.5.3 Assessment of the realization of some scenarios

After the tragedy in Asia caused by the *M* 9 earthquake on 26 December 2004 in the region of the north-western end of Sumatra, the problem of forecasts of extreme natural phenomena has become rather urgent. It is clear that the present geophysical science can only comment on the causes of earthquakes and bring up various hypotheses, trying to explain them by the Earth's crustal shifts. The most complicated problem of present science is the prediction of earthquakes. Despite the existence of special centers recording minute oscillations of the Earth's crust, there is no similar progress in studies of the laws of development of the Earth as a planet. Nevertheless, a certain progress has been achieved in predictions of other kinds of natural disasters due to development of the theory of climate and global ecodynamics. However, all estimates and forecasts are only possible under conditions of certain climatic scenarios and strategies of humankind's development. Therefore, it is important for these scenarios to be developed with NSS's pre-history taken into account (Jolliffe and Stephenson, 2003; Lawrence, 2003).

One of the possible approaches to predicting earthquakes and volcanic eruptions is to use statistics of these natural disasters and the input information for the global model of the nature–society system (GMNSS). The presence of the unit of prediction of random number series using the method of evolutional modeling enables one to determine with some probability the time of occurrence of the next event. As a result, this approach gives the following estimates. During the next 20 years one should expect, on average, 2–3 earthquakes annually of magnitude 7.0 and 4–5 earthquakes of magnitude 5–7. As for volcanic eruptions, the GMNSS predicts five events by the year 2020.

Prediction of other types of natural disasters using the GMNSS becomes possible due to consideration in the model of a totality of feedbacks in the biosphere–climate system. To make such predictions, it is necessary to prescribe a scenario of possible development of the processes of society–environment interaction. A diversity of such scenarios complicates the problem, though here the use of evolutional technology also makes it possible to reveal the most probable trends in this interaction. Consider some situations which can happen in realization of the SRES scenarios (Edmonds *et al.*, 2004). Most pessimistic scenarios A1G MiniCAM and A2ASF lead to an increase of the CO_2 partial pressure in the atmosphere by the year 2020 to 390–410 ppm and up to 520–550 ppm by 2100. As a result, the pH of the upper layer of the oceans decreases, especially in coastal water basins, which leads to changes in the trophic relationships between the ecosystems' components. A characteristic example is the ecosystem of the Peruvian upwelling whose trophic pyramid under standard conditions is characterized by binary space. The curve in Figure 8.1

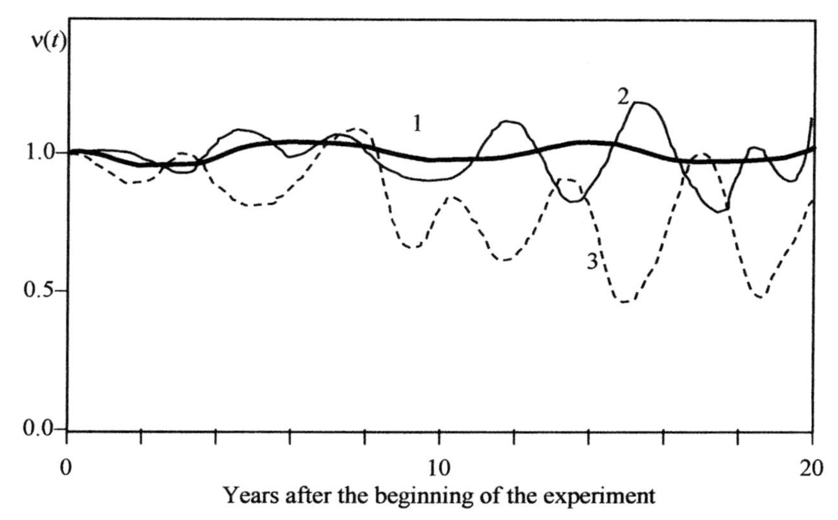

Figure 8.1. Assessment of the Peruvian upwelling ecosystem's survival with different scenarios of global ecodynamics. Notation of scenarios: 1 – A1T-Message, 2 – A1-Aim, 3 – A2-Asf (Edmonds *et al.*, 2004).

shows the state of the survivability of this ecosystem estimated from the criterion:

$$\nu(t) = \sum_{i=1}^{m} B_i(t) \Bigg/ \sum_{i=1}^{m} B_i(t_0),$$

where m is the number of trophic levels, $t_0 = 1999$, and $B_i(t)$ is the total biomass of the ith trophic level over the water body. It is assumed here that minimum concentrations of the elements' biomass not consumed by other levels constitutes 10% of their initial values.

Figure 8.1 demonstrates the response of the system to an increase in temperature of the upper water layer. Calculations show that a temperature increase of 0.4°C is harmless for the system, and with greater changes the system changes its phase state. In the latter case an effect of spatial binarity of the trophic pyramid vanishes, and the system enters the phase of unstable functioning. In the realization of scenarios A1T-Message and B1-Image the climatic situation over the Peruvian upwelling basin does not change substantially, and the ecosystem functions in the regime of binary change of trophic structures in the coastal zone between the El Niño periods (Krapivin, 1996). In the case of scenarios B2-MESSAGE and A1-AIM periods of a long-term upper layer temperature increase appear, which causes some imbalance in energy fluxes in the ecosystem, but on the whole, its stability is preserved. In the third case, when scenarios A1G MiniCAM or A2 ASF are realized, the ecosystem begins to steadily move on to another state characterized by long-term reductions of total biomass. Additional experiments show that considerable water temperature variations change in principle the state of the ecosystem. The phase trajectories of the ecosystem form the quasi-periodic structures of the type of standing waves with a shift of the center of masses toward a decrease of $v(t)$. The ecosystem can survive an increase in temperature by more than 5°C for no longer than 190 days. Oscillations of the concentration of dissolved oxygen decreasing with growing temperature should not exceed $0.2\,\text{ml}\,\text{l}^{-1}$ for more than 100 days, and the rate of vertical advection cannot be less than $0.5 \times 10^{-4}\,\text{cm}\,\text{s}^{-1}$. On the whole, experiments on the assessment of the vitality of the Peruvian upwelling's ecosystem show that with long-term slow changes of external conditions the community re-forms the structure and intensity of energy fluxes between trophic levels. One of the factors of the high stability of the ecosystem is the vertical shift of biomasses of the ecosystem's components, which makes it possible to preserve the phase pattern of the community for a long time even at considerable changes to the environmental parameters, for instance, in the case of scenario A2 ASF.

As seen from studies published elsewhere (Krapivin and Kondratyev, 2002; Kondratyev *et al.*, 2003b, 2004c), a study of global ecodynamics requires development of a mathematical apparatus with simultaneous orientation toward the interdisciplinary needs of biology, geophysics, economics, sociology, climatology, and biocenology. The GMNSS only partially meets these requirements. One of its aspects is a possibility to study the processes of interaction between natural and anthropogenic factors within a wide spectrum of considered direct couplings and feedbacks between NSS components shown schematically in Figure 6.2. The principal non-linearity of parametric ideas of these connections complicates the process of analysis of the laws of global ecodynamics and posts additional problems of evaluation of numerous parameters taking into account their dependence on time and spatial coordinates. Therefore, the reliability of the values obtained and forecasts should be considered depending on the accuracy of the accepted suppositions and scenarios.

One of the possible forecasts that can be made with the use of GMNSS is an estimation of variability of the global water balance components. Based on the IPCC scenario IS92a, which predicted the growth of population size by 2100 up to 11 billion people, we obtain that by the year 2020, increased levels of rain will take place in the north-west of Europe, which will reduce the atmospheric moisture flux from the European continents to America by about $400\,\text{km}^3\,\text{day}^{-1}$. In other regions, changes of moisture cycle will vary within $\pm 7\%$ with a gradual increase in amplitude by 2100. As a result, by the end of the century the regions of the Pacific coastline of the USA, north-eastern India, and south-western China will suffer heavy rain, and the zone of intensive rain in Europe will broaden northward. Hence, floods in these regions will become more frequent. At the same time, the amount of rain on the eastern coast of North America, in the countries of middle Asia and the Middle East will decrease, and the contrast in the transition from rainy to dry seasons in South-East Asia will change. For the European continent, a marked decrease of rain in Greece, Italy, and in the Caucasus will lead to negative consequences. In central Europe the regime of rains will change by no more than 3%.

Table 8.5. Global energy resources.

Non-renewable resources (Terrawatt yr^{-1})	
Usual oil and natural gas	1,000
Non-standard oil and gas except methane clusters	2,000
Methane clusters	20,000
Shale	30,000
Geothermal sources:	
water vapor and hot water	4,000
hot dry rocks	1,000,000
Uranium:	
in reactors with light water	3,000
in breeder reactors	3,000,000
Thermonuclear energy:	
heavy hydrogen–tritium limited with lithium	140,000,000
heavy hydrogen–heavy hydrogen	250,000,000,000
Renewable resources (Terrawatt yr^{-1})	
Hydroenergetics	15
Use of biomass	100
Wind power	2,000
Solar power:	
on land surface	26,000
over the globe	88,000

8.5.4 Conclusion

Lomborg (2001) is right to reject the apocalyptic forecasts of global ecodynamics based on exaggerated misgivings about an insufficiency of natural resources and the state of the environment. Such opinions and estimates are confirmed, in particular, by the data of Holden (2003, table 8.6) and characterizing real and potential global energy resources. Here the energy is expressed (in the case of renewable sources of energy) in Terawatt-years (TW yr^{-1}), which is equivalent to 31 exa-J (1 TW = 31.5 exa-J yr^{-1}).

It should be added that in 2000 the global energy consumption constituted about 15 TW or 15 TW yr^{-1} with the supposed increase to 60 TW yr^{-1} by 2100.

Despite the optimistic data in Table 8.5, the absence of long-term perspectives of the development of the present consumption society illustrated by the values of global ecodynamics (Kondratyev *et al.*, 2003b, 2004d) raises no doubts. Therefore, at the World Summit on Sustainable Development held in Johannesburg in 2002 a necessity was emphasized to accomplish 10-year programs on achieving sustainable development and consumption, including the following recommendations (Starke, 2004):

- developed countries should play a leading role in the provision of stable production and consumption;
- these goals should be achieved based on joint but differentiated responsibilities;

- the problem of stable production and consumption should play a key role;
- special attention should be given to participation of the young generation in the solution of problems of sustainable development;
- application of the principle "polluter pays";
- control of the complete cycle of evolution of product from its production to consumption and waste in order to increase the efficiency of production;
- support to policy that favors the ecologically acceptable production and realization of ecologically adequate services;
- development of more ecologically adequate and efficient methods of energy provision. Exclusion of energetic subsidies;
- support to voluntary initiatives of industry to raise its social and ecological responsibility; and
- study and adoption of experience of ecologically pure production, especially in developing countries as well as in small and medium-sized business.

Although the enumerated recommendations are rather declarative, still they are directed at a necessity to change the paradigm of the socio-economic development (first of all, this refers to developed countries) from the society of consumption to priorities of public and spiritual values. The concrete analysis of the means of such a development requires participation of specialists in the field of the social sciences. Some related ideas have been given in the Introduction to this book in the context of problems of the Earth Charter Initiative (Mukherjee *et al.*, 2004).

9

Interactivity of climate and natural disasters

9.1 GENERAL PROBLEMS OF GLOBAL CLIMATE DYNAMICS

A growing interest in the problems of global climate change determined by its great practical value and contradictory character of assessments of anthropogenic contribution to climate change, dictates, first of all, a necessity for the analysis of the available observational data. Studies of global climate change are connected, first of all, with an accomplishment of the World Climate Research Programme (WCRP) which can be considered as a further development of the Global Atmospheric Research Programme (GARP) started in 1967 in accordance with an agreement between the World Meteorological Organization (WMO) and International Council of Scientific Unions (ICSU) (Kondratyev, 1992; Hague, 2005). One of the goals of the agreement was to study "...the physical processes in the troposphere and stratosphere which are important for understanding the factors that determine the statistical properties of atmospheric general circulation, which can lead to a better understanding of the physical bases of climate".

At a first session of the Joint Committee on GARP planning held in 1967, results of the global numerical climate modeling with the use of a 3-D model with a prescribed doubled CO_2 concentration in the atmosphere were presented. In 1980, the GARP was transformed into the WCRP. Since then the problem of potential global climate change due to the greenhouse gas (GHG) concentration growth has become the subject of not only science but also politics (Boehmer-Christiansen, 2000; Zillman, 2000). Moreover, climate problems have always been connected with economics.

Although during the last years, the anxiety about a potential global ecological crisis has been growing, still many investigators (~70%) of the climate believe that the "emissions trading" will not be an efficient means of GHG emissions reduction, since it is based not on the real reduction of emissions but on the tactics of economic measures. Therefore, a possibility of practical accomplishment of Kyoto Protocol

(KP) recommendations to reduce GHG emissions raises doubts. This assessment agrees with the fact that there are considerable differences in assessments of climate change (especially on regional scales) based on the use of various numerical models (Kerr, 2000). Of no less importance is the conclusion that under conditions of the supposed global climate warming the agricultural productivity in different countries will remain high (though in different regions there will be both "winners" and "losers") and thus of key importance will be an adaptation of agriculture to changing climate (Sirotenko, 2000).

Crichton (2005) discussed the problems arising in modern climatology concerning the survivability of people when global temperatures increase. Among experts no united opinion exists about the future trends in global temperature. Some say temperatures will increase by 1.5°C over the next century, some say 5°C. The majority of experts try to solve this problem (Kondratyev, 2001, 2002). But there exist experts who propose to just see what happens. As Diamand (2005) points out, Crichton's (2005) strategy of "just sitting it out" is a sure path to dirty air, famine, water shortages, war, and death. Diamond (2005) examines why certain ancient societies, such as Easter Islanders and Norse Greenlanders, disappeared, and how many modern societies are on similar paths to destruction due to bad choices made in response to environmental circumstances.

In accordance with the 1987 UN Resolution on the Environment and Development, the ICSU and WMO decided in November 1988 to organize the Intergovernmental Panel on Climate Change (IPCC). Representatives of about 30 countries have reached a conclusion about a necessity not only to analyze the existing ideas of possible anthropogenic climate change but also to consider its probable impacts on the environment and society as well as to discuss measures needed to prevent the undesirable consequences of such impacts. An important result of the IPCC efforts was the development of the Framework Convention in Climate Change (FCCC).

The third Conference of representatives of FCCC signatory powers in Kyoto in December 1997 ended with the signing of the Kyoto Protocol, the most important part of which consists in recommendations of the levels of CO_2 emissions reduction by the countries listed in Appendix 1 of the KP (Table 9.1). Relevant discussions were focused mainly on the economic consequences of emissions reduction and ignored the problem of substantiation of recommended levels of reduction, which causes a serious concern, especially from the viewpoint of the further prospects of development of global biospheric processes.

As for the situation with developing countries, in this connection, the main fact consists in the inability of developed countries to perceive the key significance of the principle of equality (this refers, in particular, to the problem of per-capita emissions). It is easy to predict that the discussed arbitrariness in resolving the problem will remain in the future, though it is apparent that the principles of equality, responsibility, and efficiency formulated in the Climate Convention should serve this basis. The KP recommendations are in apparent contradiction with these principles.

The key aspects of the FCCC were recommendations of reductions of GHG

Table 9.1. The KP recommendations on CO_2 emissions reduction by different countries between the years 2008–2012 (compared with 1990 levels) (Najam and Sagar, 1998).

Level of emissions (%)	Countries	Annual emissions of CO_2 (10^3 t)	Per-capita GDP (US$)	Per capita CO_2 emissions (t)	CO_2 emissions per unit GDP (t/US$$10^3$)
Some countries from the list of Appendix 1					
108	Australia	288,965	16,516	16.91	0.98
94	Canada	462,643	19,705	17.44	0.81
92	Czechia	165,792	3,049	16,00	5.25
92	France	366,536	20,966	6.49	0.31
92	Germany	1,014,155	21,861	12.76	0.59
92	Ireland	30,719	11,349	8.77	0.68
92	Japan	1,155,000	24,205	9.35	0.39
92	Luxemburg	11,343	34,614	30.41	1.10
99	The Netherlands	174,000	18,939	11.22	0.61
100	New Zealand	25,476	12,192	7.61	0.59
101	Norway	35,514	26,387	8.37	0.31
94	Poland	414,930	1,459	10.87	7.04
92	Portugal	42,148	6,774	4.00	0.63
100	Russia	2,388,720	3,897	16.11	4.13
92	England	577,012	16,723	10.08	0.59
93	USA	4,957,022	22,046	19.83	0.90
Some countries not recorded					
	Brazil	197,905	2,877	1.34	0.41
	China	2,340,635	313	2.06	6.60
	India	681,248	347	0.80	2.28
	Indonesia	213,422	613	1.20	1.87
	Iran	187,986	2,139	3.37	1.56
	Mexico	313,826	2,856	3.76	1.27
	South Korea	243,434	5,875	5.68	0.96
	South Africa	294,107	2,758	7.93	2.76

Notation: GDP = Gross Domestic Product.

emissions to the atmosphere in order to prevent undesirable anthropogenic climate changes. As Hulme and Parry (1998) noted, an adoption of the Protocol on GHG emissions reduction at the Third FCCC Conference held in Kyoto in December 1997 serves as a manifestation of the anxiety of the world community about the potential danger for sustainable development of anthropogenic impacts on climate. The logic of GHG emissions reduction proceeds from two assumptions: (1) a reduction of emissions will make it possible to avoid in the future a dangerous climate change; (2) it is cheaper to reduce emissions and provide softer climate changes than to adapt to the changed climate.

However, the problem is that neither scientists nor politicians have adequately studied and estimated the destruction, which can be avoided in accomplishing any scenario of GHG emission reductions, and relative expenditures on adaptation or

Table 9.2. Estimates of the annual mean temperature increase in Great Britain with various scenarios of GHG emission reductions (Spedicato, 1991; Kondratyev *et al.*, 2003a).

Scenario	Annual mean temperature increase (°C)	
	by 2050	by 2100
IS92a	1.39	2.54
KYOTO	1.33	2.39
KYOTO+	1.29	2.26
KYOTO++	1.24	1.79
MINUS-60	0.78	0.11

softening. This is a very complicated problem and therefore it is most difficult to substantiate an adequate strategy.

In this connection, Hulme and Parry (1998) obtained and discussed estimates of climate change for Great Britain based on the use of four scenarios of GHG emission reductions (only CO_2 has been considered) for the period 1990–2100: IS92 (uncontrolled increase of emissions without their reduction); KYOTO (a 5.2% reduction of emissions by the year 2010 compared with 1990 by the "first list" countries (i.e., industrial countries, with subsequently stabilized emissions) (other countries follow the "trajectory" of IS92); KYOTO+ (after 2010, the "first list" countries continue to reduce emissions by 1% per year; for other countries – IS92); and KYOTO++ (the same as KYOTO+ for the "first list" countries; after 2020, other countries reduce emissions by 1% per year). The MINUS-60 scenario of global CO_2 reduction by 60% by the year 2010 has been considered, too.

Data in Table 9.2 illustrate the estimates of respective changes (increase) of the annual mean surface atmospheric temperature (SAT) in Great Britain (in these calculations the impact of atmospheric aerosol was left out of account). As is seen, all discussed measures on CO_2 emission reductions provide only a negligible restraining of climate warming.

McBean (1998) emphasized that though in the context of the FCCC the emphasis has been placed on the problem of GHG emission reductions, clause 4 of the FCCC foresees also a necessity:

(1) to support efforts (scientific, technological, technical, and socio-economic) in the interests of further scientific developments, systematic observations, and creation of the climate data archive, bearing in mind to get a better understanding of the causes of climate change and its socio-economic consequences;
(2) to provide an exchange of the respective scientific, technological, technical, socio-economic, and juridical information to substantiate the strategies for responding to climate change; and

(3) to intensify measures on the development of education, improvement of qualification, and means to inform the population about climate change.

At the intergovernmental meeting of the World Climate Programme (WCP) held in April 1993, a decision was made on the expedience of a complex development on climate problems which was called the Climate Agenda (CA), aimed at providing a more efficient coordination of respective efforts. The CA supported by the WMO, UNEP, ICSU, UNESCO, FAO, and the WHO concentrated on four areas:

(1) Prospects of development of science of climate, and weather forecasting.
(2) Climate services needed to ensure sustainable development.
(3) Study of climatic forcings and development of strategies for responding to the forcings to minimize their consequences.
(4) Adequately planned observations of the climate system.

For the organized realization of the CA, the Interdisciplinary Committee on Climate Agenda (IACCA) was formed in March 1999 in Paris as a result of a substantial reorganization of the Coordination Committee on the CA.

According to the adopted agreement, scientific developments are the sphere of responsibility of WCRP, whereas the IPCC is responsible for negotiations and recommendations of environmental policy. Efforts have been accomplished on the basis of a close cooperation between the WCRP and the IPCC, though not without some problems, first of all, in the underestimation by politicians of the significance of the problem of improving observation means, and in an insufficient understanding by the scientific community of the specific character of information needed by politicians.

As Bolin (1998, 1999) has noted, in the context of the latter, the most important conclusion of the Second IPCC 1995 Report consisted in that "the balance of evidence supposes a marked impact of man on global climate", though on the other hand, it was noted that "our ability to quantitatively estimate the impact of man on global climate is now limited, since the expected signal appears just now at the background of natural variability as well as in view of available uncertainties (in estimates) of key factors". Such conclusions have been mainly supported by the scientific community and stimulated further developments, the results of which have been generalized in the Third IPCC 2001 Report.

In this connection, Bolin (1998, 1999) emphasized that "negotiations between countries that have ratified the FCCC, have shown that substantial political actions are scarcely probable, if the impact of human activity on global climate is not proved". The absence of the required understanding of the impact of climate on humans and ecosystems has determined the resistance of some countries to measures for limitation of the anthropogenic impact on climate. An important aspect of the problem is the groundless statements of the "greens" about the growth of the frequency of extreme natural phenomena such as storms, tropical cyclones, floods, and others. This question requires, however, further thorough analysis (Grigoryev and Kondratyev, 2001).

Table 9.3. The level of CO_2 concentration stabilization, radiative forcing, and change of global mean SAT, with prescribed SAT sensitivity to a doubled CO_2 concentration within 1.5–4.5°C.

Level of CO_2 concentration stabilization (ppm)	Radiative forcing ($W\,m^{-2}$)	Change of global mean SAT		
		Minimum	Middle	Maximum
450	3.6	1.2	2.1	3.6
550	5.0	1.7	2.9	5.2
650	6.2	2.1	3.5	6.4
750	7.1	2.5	4.1	7.4

One of the key unresolved problems is the forecast of development of energy production, increase of CO_2 concentration in the atmosphere, and, respectively, radiative forcing (RF). According to the Special IPCC 1994 Report, various scenarios determine a possible increase of CO_2 emissions by 2100 within 2–60 Gt C (compared with the current value of 7.5 Gt C) (i.e., the obtained estimates differ by a factor of 30).

Data in Table 9.3 illustrate the possible changes in RF at different levels of CO_2 concentration stabilization in the atmosphere (the key unresolved problems consist, however, in what level of stabilization can be considered acceptable). In this connection, special attention should be fixed on further development of global carbon cycle studies, especially bearing in mind the unresolved problem of the "lost sink" which is partially determined, apparently, by the functioning of land ecosystems as CO_2 sinks (on the whole, the role of the biospheric dynamics in the formation of the carbon cycle is far from being clear). Urgently required (also very complicated to achieve) is the modeling of nature–society interaction with long-term, non-linear changes in climate system taken into account. The concept of the risk as applied to the problem of climate change moves to the foreground.

As follows from the above-mentioned, in addition to the large number of publications on key aspects of the climate, the sufficiently problematic task is a regulation at scientific, economic, and political levels of the problem brought forth by the KP. Perhaps, Barenbaum (2002, 2004) and Syvorotkin (2002) were the first to formulate its reconciliatory solution, practically explained by the mechanisms of closing the global carbon cycle, demonstrating its correlation with the cycles of other GHGs and water. Semenov (2004) pointed to the important role of the phase transitions of water, especially between liquid and solid. Masses of ice and permafrost are buffers to temperature changes, taking the role of climate-forming functions. Based on the model calculations, Semenov (2004) analyzed the impact of climate change on the vertical distribution of temperature under conditions of permafrost, and determined laws for the deep seasonal melting of the continental permafrost and the growing greenhouse effect. The main conclusion is that the vertical heat transfer and water phase transitions in the surface layer of the northern latitude lithosphere are climatic stabilizers.

9.2 THE STRATEGY OF CLIMATE STUDIES

As has been mentioned above, the prevailing "anthropogenic" concept of global climate change is based on information on the secular change of the annual mean global mean SAT calculated from the data of observations at land meteorological networks, and ship observations of water temperature in the upper layer of the World Ocean, as well as numerical modeling of SAT change during the last one and a half centuries. Meanwhile, there is no doubt that the information content of the notion of global mean SAT needs further analysis, and being calculated from the data of observations, the SAT values are not only aggravated with errors which are difficult to estimate, but also are insufficiently representative. On the other hand, it is obvious, that reliable results concerning climatic variability can be obtained only by considering the 3-D air temperature fields and other climatic parameters (i.e., by analysis of climate variability on regional scales. In this connection, substantiation of strategies of national and regional studies become especially urgent.

The World Climate Conference (Moscow, 29 September–3 October 2003) has stimulated the publication of a number of summarizing papers on the current studies of global climate dynamics (Demirchian and Kondratyev, 2004; Kondratyev and Krapivin, 2004a,b) with due regard to earlier results (Kondratyev, 1982; Kondratyev and Moskalenko, 1984; Kondratyev and Johannessen, 1993). The related key conclusion consists in establishing the presence of serious uncertainties in results in the fields of both empirical diagnostics and numerical climate modeling. Of exclusive importance becomes an adequate substantiation and efficient realization of further developments in order to reduce the level of existing uncertainties and to assess the reliability of global climate forecasts.

9.2.1 Development of the World Climate Research Programme

Planning of further efforts in the field of global climate studies should be based on the two international programs: the WCRP and the International Geosphere–Biosphere Programme (IGBP). A new program COPES (Coordinated Observations and Prediction of the Earth System) approved by the WCRP Joint Scientific Committee is aimed at substantiation of a new strategy and goals of the WCRP for the period 2005–2015 (*The World Climate Reseach Programme Strategy 2005–2015*, 2004), the beginning of which coincides with the 25th anniversary of the WCRP. As it is known, two main goals of the WCRP consist in assessing climate predictability and the degree of anthropogenic impact on climate. Achievement of these goals should be based on accomplishing the interdisciplinary studies of the climatic "atmosphere–hydrosphere–lithosphere–cryosphere–biosphere" system, influenced by various external factors (including anthropogenic). Such studies should foresee, in particular:

- observations of changes taking place in the climate system;
- a better understanding of the climate variability (mechanisms that determine this variability) on regional scales;

- assessment of substantial trends of regional and global climates;
- development of numerical modeling methods able to simulate the spatial–temporal variability over a wide range of scales and useful for operational predictions;
- study of the climate system's sensitivity to natural and anthropogenic forcings, quantitative assessment of contributions of various forcings;
- analysis of the nature and predictability of the annual change and interannual variability of the climate system on regional and global scales to create the scientific basis for operational predictions of climatic variability in the interests of sustainable development; and
- detection of climate change and recognition of its causes (especially of anthropogenic origin) in the context of the UN FCCC and developments of the IPCC.

The 25th anniversary of the WCRP is one of the stimuli to summarize the main results and substantiate the perspectives of further developments in connection with an appearance of new possibilities and advancements which include the solution of complex interdisciplinary problems such as:

(a) Forecasts of the climate system's dynamics on timescales of weeks–seasons–years–decades–centuries, considering possibilities of climate forecasts for centuries with the use of initial data for observed numerous parameters characterizing the state of the climate system. Solution of the respective problems requires not only radical improvement of the existing means of numerical climate modeling but, primarily, an adequate substantiation of the requirements of observational data and substantial development of the global observational system (these questions have been discussed in detail elsewhere (Kondratyev, 1998; Kondratyev and Cracknell, 1999)).
(b) Development of methods for numerical modeling of global ecodynamics and, in this context, climate dynamics with due regard to interactive chemical processes, the formation of the global carbon cycle, and dynamics of land and ocean ecosystems (Kondratyev and Krapivin, 2003a,b; Kondratyev *et al.*, 2001, 2003b). In other words, the problem of global climate change should be considered as an interactive component of global changes. One should also emphasize the significance of the problems of risk connected with natural and anthropogenic catastrophes (Kondratyev *et al.*, 2002b), as well as unexpected "turns" in the evolution of the global biosphere. The latter are illustrated by a new (demanding further thorough investigation) supposition about an important role of the fine structure of the spectrum of solar radiation (determined by the presence of Fraunhofer lines) in the evolution of the biosphere and, in particular, in the development of immunodeficiency (Kondratyev *et al.*, 2004a,b).
(c) Analysis of significance of an adequate understanding and reliable climate forecasts from the viewpoint of provision of sustainable development (the urgency of this problem has become especially acute in connection with the ratification of the KP by Russia).

These general goals can be achieved only with an efficient coordination of efforts within international programs such as the WCRP, the IGBP, the International Human Dimensions Programme (IHDP; socio-economic aspects of global change), the International Biodiversity Programme (DIVERSITAS), and others.

The main goal of the present stage of the WCRP development is an adequate (well coordinated) solution to the problems of obtaining required observational data, a deeper understanding of the laws of the climate system, and provision of a reliable forecast of its dynamics. This strategy planned for the period 2005–2015 should be realized within the COPES program based on accomplishment of the projects included within the WCRP and directed at the following problems of development (*The World Climate Research Programme Strategy 2005–2015*, 2004):

(1) Determination of the possibility and assessment of expected reliability of the seasonal forecasts of climate for all regions of the globe with the use of available climate models and observational data as well as more adequate information with further development of models and systems of observations.
(2) Further development and test of adequacy of the ensemble forecasts of the natural and anthropogenic variability of climate.
(3) Scientific substantiation and analysis of the adequacy of methods of regional climate forecasts on different temporal scales.
(4) Development of reliably verified models of climate dynamics with chemical processes taken into account.
(5) Realistic numerical modeling of glacial–interglacial cycles with the use of general circulation models.
(6) Shortening of the intervals of uncertainties of the estimates of sensitivity of numerical climate modeling results to external forcings.
(7) Improvement of the numerical modeling of arid climates and analysis of their reliability.
(8) Improvement of numerical modeling of monsoon climates and analysis of their reliability.
(9) Cooperation with the IPCC WG-1 in the interests of preparation of the AR4 Fourth IPCC Report on climate problems.
(10) Analysis of reasons why, how, and where the modes of climate variability change in response to anthropogenic forcings, and affects of such processes on the long-term climatic variability.
(11) Assessment of the predictability of intra-seasonal climate changes based on the use of interactive models.
(12) Determination of spatial–temporal scales of predictability of temperature, salinity, and general circulation of the ocean.
(13) Specification of forecasts of anthropogenic rise of the World Ocean level.
(14) Support of efforts on re-analysis of data on the variability of the climate system's components.

This enumeration is far from being complete, and this clearly follows from information discussed in recent overviews (Kondratyev, 2004b,c,d; Kondratyev *et al.*, 2003b). As has been mentioned, one of the important problems of COPES consists in analysis of the information content of the data of observations of various climatic parameters from the viewpoint of evaluation of contribution of these data to the increase in reliability of climate forecasts on different spatial–temporal scales. Resolving problems connected with this aspect of COPES will require coordinated efforts on collection, accumulation, and re-analysis of the climatological information in order to obtain an internally coordinated (dynamically balanced) structure ensuring an adequate simulation of variability of the climate system in the interests of climate forecasts. Naturally, first steps in this direction should consist in substantiation of an adequate global observational system and detailed requirements to observation data (these problems are far from being resolved). One of the related steps was an accomplishment of the COPES project of coordinated frequented observations in 2002–2004 within the sub-program of the Global Energy and Water Cycle Experiment (GEWEX) as part of the WCRP, the key questions of which were (*The World Climate Research Programme Strategy 2005–2015*, 2004):

- Do variations of the rate of the water cycle take place due to climate change?
- To what degree do local weather changes result from anthropogenic (or natural) global climate change?
- Is it possible to forecast precipitation on spatial–temporal scales determined by requirements connected with solving applied problems?
- How can the total impact of processes determining the water cycle be taken into account (parameterized) in weather forecasts and climate models?

A new stage of the GEWEX is connected with an accomplishment of a complete global description of water and energy cycles, as well as development of improved forecasts of precipitation and water cycle variability (with due regard to natural and anthropogenic climate changes), bearing in mind resolutions of various practical problems, especially those concerning water resources. As Lawford (2004) pointed out, the role of the use of space-borne observational means increases, such as the experimental satellite CloudSat for radar measurements of the vertical structure of clouds and their properties; the system of polar-orbiting environmental satellites (NPOESS); the means of remote sensing of oil moisture and ocean water salinity (SMOS); and a program to study the state of the hydrosphere (Hydros) in accordance with a global survey of changes of soil moisture and conditions of surface permafrost melting and Global Precipitation Measurement (GPM). The Integrated Global Observing Strategy (IGOS) planned for ten years and combined systems of global Earth observations (GEOSS) favor the coordination of such developments. Within the program of the partnership IGOS (IGOS-P) plans are being made to create observational systems with the use of routine (*in situ*) and satellite observational means. In this connection, the problem-oriented special complex field experiments such as CEOP (Koike, 2004) aimed mainly at studies of the global water cycle should play a special role.

The main goal of the Committe on Earth Observation Satellites (CEOS) project, started in October 2002 and completed by the end of 2004, was to assess the impact of the sources and sinks of heat and moisture on land on the global climate formation. The CEOS is considered as a pilot project in accomplishing global observations of the the water cycle within IGOS-P. Tokyo University is the data center for CEOS.

Another important observational component of GEWEX and CLIVAR (programs of climate studies) is the project of multi-disciplinary analysis of the African monsoon (AMMA), which is a continuation of the project CATCH to study tropical atmosphere and water cycle interaction.

The observational programs of special importance include the International Polar Year (IPY) planned for 2007–2008 with coordination carried out by the scientific enterprising group CliC on the problems of climate and the cryosphere. The program THORPEX should be more informative and longer term. This international program of atmospheric studies planned for ten years is carried out under the aegis of the WMO Commission on Atmospheric Sciences and is a part of the World Weather Research Programme (WWRP). Its main goal is to work out methods of weather prediction (especially its manifestations which strongly affect human life and economic activity) with an adequate use of the data of conventional and satellite observations. The THORPEX should culminate in a 1-year global meteorological experiment during the period 2010–2015. The fundamental problem of exclusive importance is development of a full-scale system of global climate observations (GCOS) to obtain a long and homogeneous series of high-quality data on the climate system parameters, bearing in mind the solution of the following set of problems (Steffen *et al.*, 2004).

- Climate monitoring for quantitative estimates of natural climate change in a wide range of spatial–temporal scales and for recognition of anthropogenic climate change.
- Detection and quantitative estimates of various (especially anthropogenic) factors of climate change.
- Diagnostic analysis to get a deeper understanding of the dynamics of the climate system and its components, including an evaluation of contributions of natural climate change.
- Development and analysis of reliability of various hypotheses on the factors of climatic variability on different spatial–temporal scales, as well as assessment of climate predictability.
- Studies of various climate-forming dynamic, physical, chemical, and biological processes with their interactivity taken into account.
- Further improvement of climate models for climate forecasts on timescales from inter-seasonal to inter-annual.

The latter problem is especially urgent: despite serious achievements in the development of methods of numerical climate modeling, the current climate models are still inadequate. This concerns, first of all, a consideration (parameterization) of chemical and biological processes in the climate system. Unresolved fundamental problems of climate theory include:

- interactive numerical modeling of climate and carbon cycle (Kondratyev and Grassl, 1993; Kondratyev and Krapivin, 2004a; Kondratyev *et al.*, 2002a,b, 2003c, 2004b);
- consideration of biospheric dynamics including continental and marine ecosystems (in this context of importance is the concept of biotic regulation of the environment (Kondratyev *et al.*, 2003b,c,d)); and
- parameterization of chemical processes in the troposphere and stratosphere, simulation of paleoclimate dynamics (Lohmann and Sirocko, 2004; Lohmann *et al.*, 2004; Mikolajewicz *et al.*, 2004; Schneider *et al.* 2004; Sirocko *et al.*, 2004), etc.

Of decisive importance is analysis of the adequacy of numerical modeling results by comparing with observational data. The problem of primary importance is the development of approaches to Nature–Society System (NSS) numerical modeling. This will require an accomplishment of an efficient coordination of efforts within four main international programs: WCRP, IGBP, IHDP, and DIVERSITAS. In this connection, one should mention a timely appearance in the IGBP of the program AIMES (analysis, integration, and numerical modeling of the Earth system).

The program of future climate studies is connected, first of all, with the WCRP perspectives based mainly on results obtained in the course of completed programs, such as TOGA (study of Tropical Ocean and Global Atmosphere: 1985–1994); WOCE (World Ocean Circulation Experiment: 1982–2002); and ACSYS (study of the Arctic climate: 1994–2003). Problems of TOGA and WOCE will be continued within the project CLIVAR to study the climate variability on timescales from seasons (with emphasis on the problem of monsoons) to centuries (in this case, detection and assessment of anthropogenic impacts on climate will play a key role), the main goals of which are:

- Description and understanding of physical processes responsible for variability and predictability of the climate on seasonal, inter-annual, decadal, and centennial scales based on the use of available observational data and interactive models of the climate system.
- Use of paleoclimatic information for a better understanding of the present (and possibly future) climate change.
- Development of methods to forecast climate change on scales from seasonal to inter-annual based on the use of global interactive prognostic models.
- Understanding and prediction of the climate system's response to increasing concentrations of minor gas components (MGCs) and aerosol, as well as subsequent comparison of numerical modeling results with observational data with an emphasis on a separation of anthropogenic changes from natural climate "signals".

An accomplishment of the CLIVAR project should answer the following questions, in particular:

- Will El Niño take place next year?
- What type of climate will be in Asia next summer and will the next monsoon bring droughts or floods?
- What will the next winter in Northern Europe be like: "warm and wet" or "cold and dry"?
- What is the global climate warming due to anthropogenic impacts?
- How high could the World Ocean level be in the 21st century?
- Will the global climate warming cause an increase of the frequency of extreme weather phenomena and will these phenomena intensify and become widespread?
- Are sudden climate changes possible?

As has been mentioned, studies of global systems of monsoons are of primary importance for CLIVAR. This has been favored by the successful accomplishment of the field experiment SALLJEX on the studies of the South American jet stream in the troposphere, which was part of a wider project VAMOS to analyze the variability of the system of American monsoons. Also of great importance are studies of the role of processes in the ocean in climate formation, including problems such as variability of the thermohaline circulation in the Atlantic Ocean; dynamics and predictability of the intra-tropical convergence zone in the Atlantic Ocean and its impact on regional climate; atmospheric forcings, their connection with processes in the upper layer of the ocean and feedbacks with SAT, including an assessment of the role of the Kuroshio Current; and upwelling in the Pacific Ocean. In this connection, close attention is devoted to improvement of the observational network in cooperation with GCOS (Global Climate Observation System) and GOOS (Global Ocean Observation System).

An important place in the WCRP is occupied, as before, by the project SPARC, the main goal of which is to study the climatic impact of the stratosphere taking into account interactive chemical, dynamic, and radiative processes that control the atmospheric circulation in the stratosphere and its chemical composition, including ozone content, whose changes determine variations of the UV solar radiation that reaches the Earth's surface (Kondratyev and Varotsos, 2000). A substantial task of SPARC is recognition and quantitative characteristics of the long-term trends of water vapour and ozone content in the stratosphere, as well as temperature. The beginning of the second decade of SPARC functioning has been marked by close attention to such problems as stratosphere–troposphere interaction (parameterization of the stratosphere in climate models), interactive impact of chemical processes on climate, stratospheric indicators of climate change, and various aspects of the problems of assimilation of the data of observations of stratospheric parameters. Development of new models of the stratosphere has opened up the possibilities to reliably simulate the dynamics of the "ozone hole" in the polar stratosphere. By the end of 2004, the estimation of the stratospheric aerosol dynamics was to be completed, and a new initiative begins to study the processes of formation and evolution of polar stratospheric clouds.

The project CliC, being a continuation and development of the program CCSS, is directed at further studies of the impact of processes in the Arctic on the global climate formation. The related problems include: determination of the role of the Arctic Ocean waters and ice cover dynamics as factors of strong changes of Arctic climate; and the study of sensitivity of the Arctic climate system to the growing concentration of GHGs in the atmosphere. The urgency of these and other problems is determined, in particular, by obtaining new data of observations demonstrating substantial changes in the global cryosphere, including: a small extent of multiyear Arctic ice cover (with minimum values in September 2002 and in 2003); intensive melting of Greenland glaciers from the beginning of satellite observations in 1980 (note that according to the latest radio–altimeter data, the coastal parts of glaciers have been melting, whereas in the centre of the ice cover there was an accumulation); destruction of the ice shelf Larsen B in the region of western Antarctic in 2002; accelerated melting of mountain glaciers on all continents. One of the important tasks of CliC is to study the possible additional GHG emissions to the atmosphere resulting from permafrost melting.

An important circumstance is an impossibility of reliable short-range and long-range climate forecasts without an adequate consideration of cryospheric processes (Kondratyev and Johannessen, 1993). This concerns, in particular, the solution of an important problem, such as the forecast of a possible rise in the World Ocean level under conditions of global climate warming. Some other global aspects of CliC include studies of: the thermohaline circulation (freshwater inflow to the North Atlantic); snow and ice cover; reliable methods of taking into account the permafrost and ice cover dynamics in climate models; and changes of solid precipitation with the use of new methods of observations.

The International Programme for Antarctic Buoys (IPAB), using instruments mounted on sea buoys in the southern hemisphere oceans, started in 1995 under the aegis of the WCRP, deserves further support.

9.2.2 Development of the International Geosphere–Biosphere Programme

First of all, it is important to perceive the fact that the discussed climate problems can be resolved only on the basis of a complex study of interactions of all major climate system components and, moreover, the climate problem should be considered as a part of a much more general and extremely difficult problem of study and simulation of NSS dynamics. This approach has been constructively realized in some recent publications (Kondratyev and Krapivin, 2003a,b; Kondratyev *et al.*, 2003a; Kondratyev, 1999, 2001; Kondratyev and Varotsos, 2000). Unfortunately, such developments within the WCRP and related projects are still (in the context under discussion) deprived of the required systematic and constructive character.

As for some aspects of the IGBP (first of all, in the context of COPES), the following problems, in particular, need further consideration and more convincing substantiation:

(1) Adequacy of available observational database from the viewpoint of its completeness and reliability. As an example illustrating the urgency of such

problems, one can refer to the data on mean annual mean global SAT. As Essex and McKitrick (2002) have justly noted, even definition and meaning of this notion have not been clearly substantiated, though it is these data on mean global SAT for the last 1.5 centuries that have formed the basis of the most important conclusions about the nature of the present climate change and various practical recommendations (including the KP). Of no less importance is the fact that the global mean SAT is the product of averaging with the use of rather fragmentary information (especially it refers to the late 19th–early 20th century) consisting of a combination of the data of air temperature observations on land (the problem of filtering out the contribution of the urban "heat islands" from observations at meteorological stations has not been adequately solved as yet) and results of ship observations of water temperature in the upper layer of the ocean (methodically complicated and far from being representative). One of the convincing illustrations of the discrepancy (and insufficient reliability) of available meteorological information is a heated discussion on principally important differences between the data on SAT trends and results of tropospheric temperature retrieval from the data of satellite microwave observations (Kondratyev, 2004c,d; Christy and Spencer, 2003). One more illustration is the unreliable manifestation of the so-called "hockey stick" (a sharp increase of mean global SAT for the last decades on the background of a weak variability during the previous millennium, which excludes, in particular, phenomena such as the "Little Ice Age" which is of principal importance for the characteristic of global warming (McIntyre and McKitrick, 2003)). On the one hand, this emphasizes a necessity to continue analysis of reliability of information on global mean SAT values (there is no doubt that the "orthodox" estimation of global warming $0.6 \pm 0.2°C$ is conditional). On the other hand, there is an acute need of substantiation of a single global system of climate observations in the context of global change.

As for the latter circumstance, at first sight, the situation is quite safe: during the last years, programs have been developed such as GCOS, GOOS, GTOS (global system of land surface observations), IGOS, and others. In fact, a single global observational system is needed in the interests of NSS dynamics monitoring, containing the problem-oriented units based on a substantiation of priorities. The most important aspect of substantiation of such a system is an optimization of the relationship between the use of conventional and satellite observational means. In this connection, there is no doubt that multi-billion expenditures on the creation of the EOS system consisting of three satellites equipped with complexes of the most up-to-date instrumentation have turned out to be unjustified (Goody, 2002).

(2) The role of biospheric dynamics as an interactive component of the climate system. Biotic regulation of the environment still has not been acknowledged (Kondratyev *et al.*, 2003c; Kondratyev *et al.*, 2004d), though the Amsterdam Declaration reflects a substantial progress achieved in this direction. The first step in this direction should be an interactive consideration of the global carbon cycle instead of a priori prescription of arbitrary scenarios of changes of CO_2

concentration in the atmosphere (Kondratyev and Krapivin, 2003a, 2004a; Canadel *et al.*, 2003; Kondratyev *et al.*, 2003a).

(3) With tremendous importance attached to anthropogenic changes of the atmospheric greenhouse effect as the most important climate-forming factors, the absence in the COPES program of a section specially dedicated to these problems seems a paradox. Therefore, contradictory opinions appear about contributions of various components of the greenhouse effect to the formation of global climate. So, for instance, Hansen *et al.* (2002) believe (though one should note that opinions of this group of specialists sometimes suffer sudden changes), that since there is an approximate mutual compensation of RF of different signs due to carbon dioxide and aerosol, of major importance for prognostic RF values is a consideration of the contribution to the atmospheric greenhouse effect enhancement determined by the growth of methane concentration.
(4) In view of the more important role of atmospheric aerosol and clouds in RF formation (and, hence, climate change) compared with MGCs, the respective problems should attract much more serious attention (Kondratyev, 2005a,b). This means, in particular, the necessity of an aerosol–cloud sector in the COPES program. The same refers to the problems of "atmospheric ozone (both stratospheric and tropospheric) and climate" (Braesicke and Pyle, 2004; Kondratyev and Varotsos, 2000).
(5) Under the conditions of growing attention given to problems of risk and natural disasters, of special importance is an analysis of the role of non-linear dynamics of the climate system as a factor of sudden and strong climate change in reaching certain threshold levels of forcing.
(6) Still neglected in "orthodox" climate studies is the important problem of the possible impacts of solar activity on climate via various mechanisms of formation of positive feedbacks with solar activity induced changes of the content of some MGCs in the atmosphere (in particular, nitrogen dioxide) and cloud cover properties (Douglass *et al.*, 2004). In this connection, the work of Solanki *et al.* (2004) deserves attention. They have detected a strong anomaly in solar activity during the last decades, compared with its quieter level during the preceding 11,000 years.

Thus, the main conclusion is that further serious efforts are needed to substantiate a new strategy of global climate studies. It is important, however, that this strategy cannot be realized adequately without more account of the problem of nature–society interaction (Kondratyev *et al.*, 2004b).

9.3 INTERACTIVE COMPONENTS OF THE CLIMATE SYSTEM

9.3.1 Anomalous situations and climate

Climate change manifests itself both on global and regional scales. One of the important features of climate formation, not only on regional but also on global scales, consists in the considerable variability determined by the internal dynamics of the climate system. One of the most substantial factors of internal dynamics is the El Niño/Southern Oscillation (ENSO) event. One of the recent climate warmings due to ENSO started in October–November 2002 and ended in March–April 2003. However, despite the end of ENSO in the period of boreal spring, the ENSO-induced warming has led to regional anomalous rain over a wide range of the Pacific Ocean, including the formation of the zone of increased moisture content along the western coast of South America and the region of moisture deficiency in the eastern part of Australia as well as in the south-western sector of the Pacific Ocean.

The global mean SAT in 2003 turned out to be close to the three maximum values observed during the period from 1880, but below the record level of SAT in 1998. An increase of global mean SAT in 2003 compared with an average value for 1961–1990 constituted 0.46°C. According to the data of satellite thermal sounding, the global mean temperature of the middle troposphere in 2003 was ranked third according to the level of warming compared with the average value for the 1979–1998 period.

The season of hurricanes in 2003 was extremely active in the basin of the Atlantic Ocean, with 16 tropical storms, 7 hurricanes, and 3 powerful hurricanes. Five of these tropical cyclones have caused landslides in north-eastern Mexico. In 2003, Nova Scotia and the Bermudas suffered heavily from hurricanes. A specific feature of the region of the Atlantic Ocean was the formation of five tropical storms in the Gulf of Mexico. Three tropical storms struck outside the usual time period (June–November). One of the storms formed in April, and the other two in December. In the western sector of the Pacific Ocean in the northern hemisphere, the activity of storm formation was less than usual (large-scale storms were absent).

The summer of 2003 in some regions of Western Europe was one of the warmest summer seasons, with heat waves affecting mainly Central and Western Europe. Two anomalous heat waves taking place in June and July–August (especially the second wave) were especially powerful. Droughts accompanying them caused forest fires, which covered a considerable part of the south of France and parts of Portugal in July and August. The summer of 2003 in Western Europe was apparently the hottest period since 1540. The heat wave in France killed 11,000 people. In Germany the summer was hotter than any summer recorded in the 20th century and (except some regions of northern and north-western Germany) the hottest over the whole period of instrumental observations.

The most substantial anomalous situations happening in March 2003 included:

(1) extremely intensive precipitation in the middle part of the Atlantic Ocean, in the south-east and eastern coast of the USA;
(2) extremely low SAT values and unusual snowfall over the European territory of Russia;
(3) 546 tornados in May in the USA, which was unprecedented;
(4) a long-term drought in the west of the USA, where in some regions it was the fourth and fifth year of a considerable deficiency of precipitation;

(5) heavy brush fires in the eastern part of Australia in January and powerful forest fires in the south of California in October;
(6) anomalously intensive precipitation in western Africa and in the Sahel;
(7) return to the normal level of precipitation on the Indian sub-continent in the period of summer monsoon; and
(8) the nearly record extent of the "ozone hole" in the Antarctic reaching a maximum of 28.2 million km^2 in September 2003.

The last years have been marked by an increased interest in studies of the present climate change in high latitudes of the northern and southern hemispheres, mostly determined by the decision to conduct, in 2007–2008, the Third IPY. The major conclusions concerning the Arctic climate diagnostics are concentrated on the analysis of the spatial–temporal variability of polar climate instead of exaggerated attention to unfounded simplification of the situation as manifesting itself through the homogeneous anthropogenic enhancement of climate warming in high latitudes. In this context, of great interest are new results of the paleoclimatic analysis of an ice core from Vostok Station (Vakulenko *et al.*, 2004) which demonstrated a negative correlation between changes of CO_2 concentration in the atmosphere and air temperature.

From the data on the Antarctic discussed in the report by Levinson and Waple (2004) it follows that the last decade in this region was anomalously cold. From the late 1970s to the mid-winter of 1990, the sea ice cover extent round the Antarctic continent was growing.

The monograph of Filatov (2004) dedicated to studies of the climate of Karelia can serve as an example of the informational analysis of the regional features of climate. New developments dedicated to the climate of cities (Mayers, 2004) and analysis of individual long series of meteorological observations (Oganesian, 2004; Alessio *et al.*, 2004; Garcia-Barrón and Pita, 2004) have made an important contribution to studies of regional climate change. A new important stage in comprehending the data of empirical diagnostics of climate was the development and application of interactive models of the climate system and an ensemble approach to numerical climate modeling (Kondratyev, 2004b,d; Palmer *et al.*, 2004).

As has been repeatedly emphasized, the interactive components of the present climate system include a broad spectrum of natural and natural–anthropogenic subsystems and processes, without a complex study of which it is impossible to reliably select the prevailing trends in climate change. In this connection, one should enumerate the most important ones:

- Global water cycle. Effect of "cloud" feedbacks.
- Global carbon cycle. Interaction of water and carbon cycles.
- Land use and land surface changes.
- Present trends in the GHG content in the atmosphere and mechanisms of their control.
- Interaction of climate and land ecosystem productivity.
- Effect of the climate regime shifts on marine ecosystems.

- Control of natural resources to neutralize negative consequences of human activity.
- Socio-economic aspects of ecodynamics and climate and their analysis for optimization of a land use strategy.
- Interactions between processes in the geosphere and biosphere and their dependence on cosmic impacts.

9.3.2 Climate change, forests, and agriculture

Forest and agriculture ecosystems are the environmental components most sensitive to climate change. The former determine many characteristics of biogeochemical cycles of GHGs, and the latter form the human–environmental interaction. Problems appearing here are thoroughly studied within many international programs on the environment and climate. They are especially emphasized in the national programs of the USA. In particular, analysis of consequences with the use of various scenarios of possible changes of global climate has led to the conclusion that in the case of several scenarios, the impact on forestry and agriculture in the USA will be economically favorable. Partially it is connected with the growth of forest productivity (due to the growth of CO_2 concentration) and determined by the ability of forests to adapt to climate change. As for agriculture, according to available prognostic estimates for the period up to 2060, the positive impact of the process of global warming on agriculture in the USA will be less economically favorable than follows from earlier estimates.

Unfortunately, in view of the global and poorly studied character of the correlation between climate change and behavior of vegetation cover (forest ecosystems, in particular) at present there are no reliable estimates of the consequences of climate change on their productivity. Study of the problems appearing here have begun.

9.3.3 Observational data

Analysis of observational data is reduced, as a rule, to the consideration of two categories of information:

(1) SAT changes for the last 1.5 centuries (especially during the last 20–30 years, when an increase of the global mean annual mean SAT was at a maximum).
(2) Paleoclimatic changes. They attract attention from the viewpoint of their comparison with present climatic trends and, to some extent, as an analog to possible climate change in the future (such attempts are being continued, though inadequacy of paleoanalogs for future climate forecasts have been repeatedly and convincingly argued).

By definition, climate is characterized by values of meteorological parameters averaged over 30 years (so, for instance, climate anomalies in 1990 are determined as deviations from averages over the period 1961–1990). An analysis of the spatial–temporal climate variability for individual years is also widely practiced. In particular, the WMO publishes annual surveys of global climate. In these reports attempts are being made to answer the most important questions:

- Does a climate warming take place?
- Is the moisture cycle intensity changing?
- Is the general circulation of the atmosphere and ocean changing?
- Do extreme climate changes (storms, droughts, floods) intensify?
- Is a reliable estimate of the anthropogenic contribution to climate change possible?

The decade of the 1990s, on the whole, was the warmest over the whole period of meteorological observations (beginning from 1860), and the year 1999 was ranked fifth according to its anomalous level of global mean annual mean SAT (+0.33°C) for the period 1860–1999 (it was also ranked fifth for the average SAT anomaly (+0.45°C) in the northern hemisphere – in the southern hemisphere it was only ranked tenth (+0.20°C)).

The band of maximum annual mean SAT extended from the continent of North America eastward across the Atlantic Ocean and the Eurasian continent to the equatorial band of the western sector of the Pacific Ocean. Minimum SAT anomalies were observed in a broad band of the central and north-eastern regions of the Pacific Ocean (including a decrease of SAT). An analysis of observational data revealed the prevailing of positive temperature anomalies in 1999 in many regions of the globe. The most apparent anomalous situations were of both warming and cooling, including:

(1) The cold snap observed in January, which brought a SAT decrease to Norway, Sweden, and some regions of Russia to the levels not observed since the late 19th century.
(2) A temperature decrease in February in Western Europe was followed, in particular, by heavy snowfall in the Alps.
(3) In western Australia the SAT decreased to values below normal, though an extreme warming in early January led to intensive brush fires.
(4) The March temperature in Iceland was at a minimum compared with the last 20 years.
(5) In April, powerful heat waves formed in the northern and central regions of India, and in July and August in the north-eastern and mid-western regions of the USA.
(6) Unusually hot and dry weather was observed in the western part of Russia (SAT anomalies in the central and north-western regions of the European territory of Russia exceeded 5°C).
(7) On the Australian continent, maximum average SAT in November–December turned out to be the lowest since 1950.
(8) The second half of the year was colder (than usual) in central and southern Africa; the Sahel region was colder, wetter, and more cloudy than during preceding years.

(9) The warming in the USA during the last 50 years was weaker than over the rest of the globe, with a weak cooling in the eastern part of the USA.

The land and ocean surface temperature decrease in the tropics in 1999 was determined by the year-long La Niña event. This year was characterized by a great number of destructive meteorological catastrophes, especially floods. In Australia, USA, and Asia there was a multitude of tropical storms; in Europe there were heavy snowfalls, avalanches, and storms; in the USA there were droughts and tornados.

The global mean annual mean SAT value in the late 20th century exceeded by more than 0.6°C the value recorded in the late 19th century (the error of this estimate ±0.2°C corresponds to a 95% confidence level). Analysis of the SAT observational data suggested the conclusion that beginning in 1850, there has been an irregular but substantial trend of climate warming on a global scale. This trend was very weak in the period from the mid-19th century to 1910, and then it increased to 0.1°C during 10 years (from combined data on SAT and sea surface temperature (SST) in the period 1910–1940 and during the last two decades). Two positive episodes of cooling were separated by the interval of a weak cooling, especially in the northern hemisphere. During the periods from 1951–1960 to 1981–1990 the sign of the inter-hemispherical difference of temperatures changed: the northern hemisphere became colder than the southern hemisphere.

As Moron *et al.* (1998) noted, the present global warming was considered by some specialists as being connected with sudden changes in the region of the Pacific Ocean in about 1976 or with a gradual warming of the tropical band of the Pacific Ocean as well as with other regional-scale phenomena. The irregular trend mentioned above and an attempt to recognize the external forcings on it (anthropogenic or natural), complicated by the presence of the internally determined variability of the climate system, have been traditionally based on the interpretation of the trend as red noise.

The existence of such regularities has been well established and ascribed mainly (if not completely) to instability of the interactive "atmosphere–ocean" system in the tropical Pacific Ocean. Periodicities of about 4–6 years and 2–3 years connected with the ENSO event have been detected. Such regularities on the scales of decadal and inter-decadal variability were more difficult to detect in view of the insufficient length of the observation series.

In this connection, Moron et al. (1998) undertook a detailed analysis of all available data on the spatial–temporal variability of the SAT fields of the World Ocean and for its individual regions with the use of a multi-channel singular spectral analysis (MSSA). The main goal of the analysis was to detect the laws of variability and inter-basin relationships between SST on timescales from inter-annual to inter-decadal. The length of observational data series was sufficient for a reliable analysis of SST variability on timescales of 2–15 years, though the statistical reliability of results for longer periods is more difficult to guarantee.

In view of a great interest in SST variability in the Atlantic Ocean, major attention was given to this region. The strongest climatic signal was an irregular long-term SST trend. The use of the MSSA method for data processing for the 20th century revealed the well-known regularities mentioned above: gradual increase of SST in both hemispheres in 1910–1940, with the subsequent increase of SST in the northern hemisphere until the mid-1950s; a lower SST in the southern hemisphere; northern hemisphere ocean cooling in the 1960s until the end of the 1970s; and an initial stability and then increase of SST in both hemispheres in the 1980s with a small weakening of this trend during the last years.

Insufficient lengths of time for the series of instrumental observations makes it impossible to interpret the enumerated global laws as a manifestation of the more or less monotonic increase of SST or as part of long-term centennial oscillations (according to indirect data, oscillations were observed with periods from 65 to 500 years). Possible external factors of variability include: the growth of CO_2 concentration, change of extra-atmospheric insolation, and volcanic eruptions. A new and surprising result was the detection of the fact that a large-scale warming and cooling was preceded by the same SST variability near the southern edge of Greenland and (soon after that) in the central part of the Pacific Ocean in the northern hemisphere. This reflects an important role of high-latitude processes in the North Atlantic and possible interaction (via the atmosphere) with the Pacific Ocean.

On timescales of decadal variability (7–12 years) regular oscillations have not been observed, which are coherent on global scales. In the northern Atlantic, there were 13–15-year and 90-year oscillations. Near Cape Hatteras there were inter-decadal oscillations which propagated along the Gulf Stream to the zone of the North Atlantic, where their phases changed (similar results obtained earlier were rather contradictory). In the context of the search of inter-decadal oscillations of ENSO the data considered did not reveal any substantial maximum of SST variability with periods longer than 10 years in either the Pacific Ocean or the whole World Ocean, but in the Indian Ocean there were observed 20-year SST oscillations, especially regular during the first half of the 20th century (oscillations of this kind have been observed earlier).

Analysis of 7–8-year oscillations revealed the contradictory character of their phase in the sub-tropical and sub-polar cycles of the North Atlantic. As for the inter-annual variability (2–6 years), three dominating periods were recorded: 24–30, 40, and 60–65 months. The first period is a well known quasi-biannual ENSO component which most strongly manifests itself in the tropics of the eastern sector of the Pacific Ocean, with anomalies of constant sign propagating along the western coastline of North and South America (in other oceans such a variability is negligibly small).

An important new result consists in the detection of two clearly differing low-frequency modes of oscillation, combined by the common physical nature of the quasi-four-year mode and characterized by a drastic change of periodicity in 1960 – from $\sim$5 years to almost 4 years. This set of observational data suggests the conclusion that ENSO irregularity occurs due to interaction of the internal instability of the atmosphere–ocean system in the tropical Pacific Ocean with the annual change. Since the mode of quasi-biannual oscillations also exists in other oceans, but does not correlate practically with the index of the southern oscillations, one should assume that a stronger quasi-four-year signal forming here can remotely propagate beyond the Pacific Ocean, whereas the opposite process is practically impossible.

A weak oscillation with periods about 28–30 months is observed in the SST field in the southern hemisphere Atlantic Ocean and agrees with the Hermanito event observed earlier, which is, probably, an ENSO analog. The results discussed are a stage of developments aimed at comparison of numerical modeling results with the data of observations using the 2- and 3-D models of the atmosphere–ocean system.

An important contribution into the idea of SST changes in the past has been made by analysis of the data of temperature observations in boreholes. So, for instance, Bodri and Čermák (1999) noted that while the amplitude of long-term SAT changes in transitions from glacial to inter-glacial periods reached 10–15 K, during the Holocene (the last 10,000–14,000 years) changes of the order of several Kelvin took place on timescales from decades to several centuries. In this connection, an analysis has been made of the data on the vertical profiles of temperature measured at different depths in boreholes in Czechia and maps were drawn of SAT changes in Czechia taking place between 1100–1300 years and the present-day (little climatic optimum), 1400–1500 years, and 1600–1700 years (the main phases of the Little Ice Age).

Huang *et al.* (2000) discussed the results of processing the data on temperature measured at different depths in 616 boreholes in the southern hemisphere, which has opened up possibilities to retrieve the change in global mean temperature over five centuries. The data for 479 holes revealed a global warming of about 1.0 K had taken place over the last five centuries. Only during the 20th century, the warmest one, did the increase of the continents' surface temperature reach 0.5 K (about 80% of climate warming fell within the 19th–20th centuries). This warming over five centuries had been stronger in the northern hemisphere (1.1 K) than in the southern hemisphere (0.8 K). On the whole, the results obtained agree with conclusions drawn on the basis of data on tree rings, though the latter demonstrate a weaker centennial SAT trend, which can be explained by special features of dendroclimatic methods.

Analysis of paleo-information on SAT obtained from the data on oxygen isotopes in Greenland ice cores for the Quaternary period has shown that long-term temperature changes are superimposed by faster changes on timescales from millennia to tens of years (Bowen, 2000). An analysis of the Antarctic ice cores revealed similar changes. In particular, in both polar regions, substantial changes of temperature had taken place in the period of the Holocene.

The data on plankton foraminifera *Neogloboquadrina pachyderma* from the north-eastern region of the Atlantic (west of Ireland) made it possible to retrieve the SST and trace the Heinrich events connected with iceberg outbrakes.

Information on the content of carbon dioxide and methane in air bubbles contained in ice cores reflects an important role of those MGCs in climate formation. For instance, it was shown that during four interglacial periods the temperature in the Antarctic had preceded changes in CO_2 concentration by about 4,000 years.

New results of numerical modeling of the dynamics of the El Niño event caused by variations of orbital parameters, have satisfactorily followed "Milankovich" frequencies during the last 150,000 years as well as a variation on the timescale of a millennium (Morén and Påsse, 2001). However, most surprising was a detection of a climate variation with a period of 1,450 years from different data for different regions of the globe, which had regularly repeated, in particular, in Greenland during the last 110,000 years, including the last glaciation and Holocene (an increase in the amplitude of such changes was observed in the periods of glaciation). These results reflect a radical reorganization of the climate system taking place during comparatively short time periods. The Holocene looked (compared with these changes) like a period of comparatively stable climatic conditions. There is no doubt that in the absence of climatic feedbacks the growth of GHG concentrations in the atmosphere should bring forth a climate warming. However, the real situation turns out to be much more complicated, and to understand it, a reliable detection and quantitative estimate of the role of feedbacks are needed. Otherwise, a reliable forecast of climate change in the future is impossible. Since one of the very important sources of respective information is peat bogs, they should be thoroughly protected.

Having analyzed the data of satellite observations of SST starting from 1982, Strong *et al.* (2000) noted a warming over most of the tropics and in mid-latitudes of the northern hemisphere (with the global mean trend of $+0.005°C$ per year not exceeding the limits of observational errors). Less representative southern hemisphere SST data reflect the existence of the opposite cooling trend (the problem of SST data reliability needs serious attention).

According to the data of Levitus *et al.* (2000), during the last 50 years (1948–1998) the World Ocean has substantially warmed. The upper 300-m layer has warmed most (by 0.31°C, on average), whereas the temperature of the 3-km layer increased by 0.06°C. This increase in the temperature of the upper layer of the ocean had preceded the SAT increase which started in 1970.

Satellite data on the sea ice cover extent are important as an indicator of the global climate dynamics. Gloersen *et al.* (1999) detected a statistically substantial decrease in the global area of sea ice constituting $(-0.01 \pm 0.003) \times 10^6\ km^2$ per 10 years.

Microwave remote-sensing data is quite special, analysis of which has not revealed any substantial changes in the average temperature of the lower troposphere during the last decades. This is also confirmed by the results of aerologic soundings. From the data of Woodcock (1999a,b), the global mean SAT in October 1999 was 0.2°C below the average value for the period between 1979–1999.

Santer *et al.* (2000) have discussed the causes of different trends of SAT and the lower tropospheric temperature. Having analyzed the SAT data for the periods between 1925–1944 and 1978–1999, Delworth and Knutson (2000) came to the conclusion that the main cause of SAT changes was a combined impact of anthropogenic RF and unusually substantial multi-decadal internal variability of the climate system.

An important indicator of climate dynamics can come from satellite data on changes of the balance of the mass of Greenland's glaciers. Results of laser altimetry in northern Greenland, for the period 1994–1999, show that, on the whole, at altitudes above 2 km the ice sheet was balanced, with local changes of different signs. A decrease of the glacier's thickness dominated at low altitudes, exceeding 1 m per year, which is enough to raise the World Ocean level by 0.13 mm per year (this is equivalent to about 7% of the observed rise of the ocean level).

The data of observations of the moisture cycle parameters still remain fragmentary. An exception are publications such as Russo *et al.* (2000), in which an analysis has been made, at the Observatory of Genoa University, of the change of the diurnal sum of precipitation for 1833–1985. A decrease of the quantity of rainy days for the whole period of observations has been revealed, as well as a considerable growth of rain rate starting from 1950. During the last 30 years there was a considerable increase in the number of days with intensive precipitation.

Yu *et al.* (1999) performed an analysis of available climate data on the heat balance of the atmosphere with the use of results of both observations and calculations. The atmospheric radiation budget was found from the data of satellite observations of the fluxes of outgoing short-wave and long-wave radiation and radiation fluxes at the surface level retrieved from satellite data. Quantities of turbulent heat fluxes at the surface level were taken from the data of observations within the COADS program, and the horizontal heat transport was calculated with the use of the respective meteorological information. To minimize random errors, spatial–temporal averaging was made: the zonally averaged components of the atmospheric heat balance components for the latitudinal band 50°N–50°S, as well as values for this latitudinal band, have been considered.

An analysis of the data discussed has shown that it is impossible to close the atmospheric heat balance: an additional 20 W m^{-2} are needed. Attempts to use different versions of the input volumes of information did not help to remove this "imbalance". Since the closing of the water vapor balance with the use of the same data was successful, one can assume that the cause of this "imbalance" is an inadequacy of estimates of the atmospheric radiation budget, manifesting itself though underestimated solar radiation absorbed by the atmosphere.

Having analyzed the completeness and reliability of the available data of climatic observations, Folland *et al.* (2000) have come to the conclusion that the existing volume and quality of data make it impossible to give adequate answers to the questions enumerated above. In this connection, anxiety is aroused because of degradation of the systems of conventional meteorological observations taking place during the last decades, which are of importance also for calibration of the satellite remote-sensing results. Therefore, even calculations of decadal mean values of climate parameters are difficult for some regions of Africa and vast regions of the World Ocean.

Discussing the observed regularities of global climate change with time and its causes, Wallace (1998) noted that the first-priority should be to consider the following problems:

(1) periodic climate change due to variations of extra-atmospheric solar radiation;
(2) quasi-periodic climate variability (its most vivid manifestation – quasi-biennial oscillations in the equatorial stratosphere);

(3) the ENSO event (in view of a wide range of frequencies, this event cannot be considered quasi-periodic);
(4) inter-decadal climate variations, which are to a great extent determined by the internal intra-seasonal and intra-annual variability of the climate system;
(5) climate variability on timescales from inter-decadal to centennial;
(6) analysis of the statistical significance of estimates of unprecedented events and "shifts of regimes" in the light of the time-dependent series of many climatic parameters; and
(7) revealing the phase relationships between climate change on timescales from inter-annual to inter-decadal.

An important fact is that most of the climate variability can be described by a separate consideration of dependences on time and space coordinates. In view of the exceptional complexity of the climate system, with its numerous degrees of freedom and a multitude of feedbacks, highly regular structures and modes of climate evolution can be rather the exception than a rule. "Rough" schemes of parametrization for reconstructing the structure and evolution of climate anomalies without superfluous detail should have a higher degree of stability. Important climatic "signals" considered in solving the problems of detection and forecast of global climate change should be seen "with the naked eye". A much more complicated problem than usually supposed is an assessment of the statistical significance of some quantitative characteristics of climate variability, especially unprecedented events and "shifts of regime" from the data on the time series of limited duration (as a rule, in view of the time-dependence of such series).

Apparently, one of the directions of studies of climatic time dependence and related catastrophic events (of the type of the tropical storm Katrina, which 13 years after Hurricane Andrew was the most powerful in the history of Miami (Florida), causing the USA huge economic damage, completely submerging New Orleans and ruining a large number of constructions in late August 2005) is a search for connections between temperature variations of different scales in different water bodies of the World Ocean. For instance, Chang *et al.* (2000) and Yamagata *et al.* (2004) have shown that there is a stable correlation between changes of water surface temperature in the Indian and Pacific Oceans, which especially strongly manifest themselves in the season of monsoons in the Indian Ocean basin. The ENSO event favoring the propagation of the sub-tropical anticyclone over the western Pacific plays a marked role in stirring up the feedback mechanisms. Studies of appearing correlations are successfully carried out with the use of an interactive ensemble Canadian Global Climate Model (CGCM) developed at the Centre for Ocean–Land–Atmosphere Studies (COLA).

In the Pacific Ocean there are two key regions which play an important role in the variability of the upper water layer temperature. These are the western and central sectors of the northern Pacific Ocean. Changes taking place here affect the climatic situation in many regions of Asia and in more remote areas (Nakamura and Yamagata, 1997). Therefore, studying of complicated climatic situations in the

Pacific region is important for detection of latent dependences between stimulators of global climate change in the future (Timmermann *et al.*, 1999).

9.3.4 Climate-forming factors

Discussing the prospects of developments within the CLIVAR program, Bolin (1999) emphasized that "the IPCC was very careful in its assessments in order to stick to conclusions known from scientific literature, which serve the basis for such assessments. The key fact is that it is necessary to distinguish between something that can be considered real and which remains uncertain. As for the future climate forecasts, many uncertainties still remain. This approach has determined the confidence in the scientific community when making concrete decisions and should be preserved in the future".

Bolin (1999) emphasized that though the IPCC 1995 assessment report contains the statement that global climate change taking place in the 20th century is partially determined by human activity, this conclusion was formulated very carefully. Of principal importance in this context was an evaluation of the probable contribution of random climate variations independent of human impact. Results of recent studies have clarified this question, showing that random variations of global mean SAT on timescales from decades to centuries for the last 600 years were within $\pm 0.2^{\circ}C$ or less (this conclusion, as has been mentioned above, disagrees with data of observations from which it follows that SAT changed in the past within wider limits). In this connection (as Bolin (1999) believes), the skeptics about estimates of the contribution of anthropogenic warming for the last 50–75 years should be asked how they can explain a much stronger global warming observed during the last decades.

In reports of official representatives of several countries at conferences in Kyoto and Buenos Aires and in the mass media, weather and climate anomalies like tropical hurricanes and unusual El Niño events were ascribed to the impacts of global warming. These opinions should, however, be thoroughly tested scientifically, though a possibility of more frequent anomalous events under conditions of global warming is not excluded. Therefore, especially urgent becomes the development of methods of climate forecasts on regional scales, considering first, the period between 2008–2012.

In Bolin's (1999) opinion, to assess the socio-economic consequences of the accomplishment of measures foreseen by the KP on GHG emission reductions, it is very important to develop integral models – a combination of models of climate, carbon cycle, as well as power engineering and socio-economic development, which will require much more time and much effort. In this respect, a difficult problem is to validate such models in order to analyze their reliability. The absence of an adequate validation means that the results of numerical modeling with the use of integral models can be considered only as possible scenarios but not forecasts.

Characterizing the climatic forcings, Hansen *et al.* (1998, 1999) pointed out that they are still not determined with an accuracy sufficient for reliable climate forecasts.

Table 9.4. Greenhouse radiative forcing F for the period after the beginning of the industrial revolution.

Gas	Radiative forcing
CO_2	$F = f(c) - f(c_0)$, where $f(c) = 5.04 \lg(c + 0.0005c^2)$
CH_4	$0.04(m^{1/2} - m_o^{1/2}) - [g(m, n_0) - g(m_0, n_0)]$; $g(m, n) = 0.5 \lg[1 + 0.00002(mn)^{0.75}]$
N_2O	$0.04(n^{1/2} - n_o^{1/2}) - [g(m_0, n) - g(m_0, n_0)]$;
CFC-11	$0.25(x - x_o)$
FC-12	$0.30(y - y_o)$

There is reliable information about the GHG content in the atmosphere, which determines a positive RF, but serious difficulties are connected with assessments of the impacts caused by such factors as atmospheric aerosol, clouds, and land use change, causing a negative RF, which determines a partial compensation of the "greenhouse" climate warming. One of the consequences of this compensation consists in a much more significant role of changes of extra-atmospheric insolation (solar constant (SC)) as a climate-forming factor than was supposed earlier, based on numerical modeling with only GHG contributions taken into account (the "greenhouse" RF due to the growth of CO_2 concentration in the period from the beginning of the industrial revolution to the present-day was estimated at about $1.5\,W\,m^{-2}$).

In connection with these circumstances, Hansen *et al.* (1998, 1999) obtained new estimates of global mean RF. The data in Table 9.4 characterize the results of an analytical approximation (with an error of about 10%) of various components of the "greenhouse" RF (the recent detection of SF_5CF_3 as a substantial GHG shows that the problem of substantiation of GHG priority cannot be considered resolved completely). The GHG concentrations are expressed in ppm (CO_2, c); ppb (CH_4, m); ppb (CFC-11, x; CFC-12, y); CFC – chlorofluoroorganic compounds (freons). The total RF value is $2.3 \pm 0.25\,W\,m^{-2}$. Of interest is the fact that the rate of RF increase from 0.01 to $0.04\,W\,m^{-2}\,yr^{-1}$ over the period 1950–1970 decreased to $0.03\,W\,m^{-2}$ per year during the subsequent 20 years in connection with the decreasing rate of the growth of CO_2 concentration (despite the continuing growth of CO_2 emissions), the reasons for which remaining unclear. A certain contribution was made also by the reduction of the growth of CO_2 concentration, but again for reasons unknown.

The RF due to the growing concentration of tropospheric ozone was estimated at $0.4 \pm 0.15\,W\,m^{-2}$. A drop of stratospheric ozone could result in the RF equal to $-0.2 \pm 0.1\,W\,m^{-2}$. Although these changes of the sign are mutually partly compensated, it does not mean they are insignificant, since variations of the ozone content in the troposphere and stratosphere affect substantially (and in a different way) the formation of the vertical profile of temperature.

As for the RF due to aerosol, its determination is still unreliable due to a lack of adequate information on the real atmospheric aerosol. A numerical modeling, with anthropogenic sulphate, organic, and soil aerosol taken into account (with a latent consideration of soot aerosol by prescribed realistic aerosol absorption), for prescribed global distributions of aerosol optical thickness has made it possible to

Table 9.5. Global mean RF for three types of anthropogenic aerosol.

Type of aerosol	$\varpi = 1$		"More realistic" ϖ	
	ΔF (W m^{-2})	ΔT_s (°C)	ΔF (W m^{-2})	ΔT_s (°C)
Sulphate	−0.28	−0.19	−0.20	−0.11
Organic	−0.41	−0.25	−0.22	−0.08
Dust	−0.53	−0.28	−0.12	−0.09
Total	*−1.22*	*−0.72*	*−0.54*	*−0.28*

evaluate the global distributions of RF and balanced surface temperature, and then to obtain respective global mean values of changes of RF (ΔF) and surface temperature (ΔT_s) for purely scattering aerosol (single scattering albedo $\omega = 1$) and more realistic aerosol (Table 9.5).

Hansen *et al.* (1998) noted that the most reliable value of total RF due to aerosol constitutes -0.4 ± 0.3 W m^{-2} instead of the value -0.54 given in Table 9.5, though in any case this estimate remains very uncertain due to unreliable input data on aerosol properties.

Anthropogenic RF changes due to clouds are undoubtedly more substantial than those due to aerosol, but they are even more uncertain. Such changes (including the impact of aircraft contrails) mainly result from the indirect impact of anthropogenic aerosol, which causes (functioning as condensation nuclei) variations of cloud droplet size distributions and optical properties. Rough estimates give "cloud" RF from -1 to -1.5 W m^{-2}, but it can change by an order of magnitude (depending on the prescribed input parameters). The conditional value $-1^{+0.5}_{-1}$ W m^{-2} can be assumed. Some increase of cloud amount observed in the 20th century may be attributed to an indirect impact of aerosol. Specification of the obtained estimates requires an accomplishment of complex observational programs in different regions of the globe.

The contribution of land use changes into RF variations is connected with the processes of deforestation, desertification, and biomass burning, which affect the surface albedo and roughness as well as evapotranspiration. It is also important that the bare surface albedo changes more strongly being covered with snow than vegetation. Approximate estimates of the Earth's Radiation Balance (ERB) change due to land use evolution is -0.2 ± 0.2 W m^{-2}.

The natural RF due to SC changes during the last century (including an indirect impact on the ozone layer) can be assumed to equal 0.4 ± 0.2 W m^{-2}. Since the total RF constitutes only about 1 W m^{-2}, the contribution of extra-atmospheric insolation variability could play a substantial role. Volcanic eruptions cause RF changes from 0.2 to -0.5 W m^{-2} (these estimates are, however, conditional). For analysis of possible anthropogenic impacts on global climate, of extreme importance are estimates of sensitivity of the climate system to external forcings. Hansen *et al.* (1998) assumed that the change of global mean SAT at a doubled CO_2 concentration should constitute 3 ± 1°C. Since the RF estimates are not reliable enough, it is expedient to use different scenarios of RF change. One of the developments in this sphere is a study of Tett *et al.* (1999).

Crowley (2000) estimated the contribution of various factors to climate formation (SAT change) for the last 1,000 years using the energy balance climate model. According to the results obtained:

(1) Changes of global mean SAT during the last 1,000 years may be explained as a result of the combined impact of the known RFs (in the pre-industrial epoch 41–64% of SAT changes had taken place due to extra-atmospheric insolation and volcanic activity).
(2) Global warming observed in the 20th century was mainly of anthropogenic ("greenhouse") in origin, substantially exceeding the internally caused variability of the climate system.

Unfortunately, the argumentation contained in the study of Crowley (2000) is unconvincing even from the viewpoint of explanation of the centennial change of global mean SAT. For instance, the causes of climate cooling in the late 19th–early 20th century have not been explained. Of course, the model considered cannot simulate changes of regional climate. The role of North Atlantic oscillations in climate formation has not been revealed. On the whole, the fact that possibilities of approximate energy balance models do not exceed the limits of substantiation of rather conditional scenarios of climate but cannot describe the dynamics of the real climate system raises no doubts. This conclusion refers also to results obtained with much more complicated global interactive climate models (see, e.g., Knutson *et al.*, 1999).

Noting that climate models with a low spatial resolution ($\sim 3^\circ$–6° lat.) cannot reliably simulate (and, moreover, forecast) climate changes on regional scales, Mearns *et al.* (1999) showed that to resolve such problems, two approaches can be used:

(1) statistical (with regard to observational data) scaling (reducing to a higher spatial resolution) of the numerical modeling results obtained with the help of low-resolution models; and
(2) expanding such models by including the "nested" regional models with a higher resolution.

In this connection, Mearns *et al.* (1999) undertook a comparison of scenarios of anthropogenic climate change (with a doubled CO_2 concentration) calculated with the use of the National Center for Atmospheric Research (NCAR) "nested" model RegCM2 and semi-empirical method of scaling (Scott Data System, SDS). In both cases the large-scale numerical modeling was carried out with the use of the GCM developed in the Commonwealth Scientific and Industrial Organization (CSIRO).

The results obtained show that the RegCM2 model reveals a stronger spatial variability in the fields of temperature and precipitation than the SDS model, which leads, however, to a greater amplitude of the annual change of temperature than the models RegCM2 or GCM. The diurnal change of temperature turned out to be weaker in the cases of SDS and GCM than RegCM2, and the amplitude of the diurnal change of precipitation varied in the interval between those corresponding to SDS and RegCM2. Calculations with RegCM2 reproduce both an increase and decrease of probability of precipitation with a doubled CO_2 concentration, whereas SDS gave only an increase of precipitation.

One of the causes of the differences mentioned above could be the fact that the semi-empirical model SDS is based on the data that refer only to the surface level 700 gPa, whereas in the other two models the vertical structure of the atmosphere is taken into account. This comparison does not permit, however, drawing the conclusion as to which of the results obtained reflects "correctly" the impact of forcing. To answer this question and establish the cause of these differences, the numerical modeling should be further improved.

One of the important illustrations of highly uncertain theoretical estimates of the causes of climate change is a re-assessment of the role of the "Milankovich mechanism" as the main factor of paleodynamics of climate observed during the last years.

According to the Milankowich theory (Morén and Påsse, 2001), changes of paleoclimate had been determined by latitudinal redistribution of extra-atmospheric insolation and in the annual change as a result of variations of the parameters of the Earth's orbit (especially it refers to glacial–interglacial cycles during the Quaternary period) which include:

(1) inclination of the rotation axis with respect to the orbital plane (fluctuating within 22–24.5° with the current estimate of 23.4°; average periodicity of variations constitutes 41,000 years and affects mainly the high-latitude insolation);
(2) precession of equinoctial points affecting the time of the onset of equinoxes and solstices, which tells mainly on low-latitude insolation (precession is characterized by dual periodicities of 19,000 and 23,000 years); and
(3) eccentricity of the Earth's orbit which changes from being almost circular to strongly elliptical with a periodicity of about 95,800 years (these changes cause a modulation of precession).

Milankovich supposed, in particular, that the summertime low-level insolation in high latitudes is the cause of the onset of glaciations and formations of ice sheets. The resulting increase of surface albedo determines the functioning of a positive feedback which intensifies the impact of insolation decrease. The summertime low-level insolation was observed at a minimum angle of the orbit's inclination, high eccentricity, and at an apogee in the northern hemisphere summer. According to Milankovich's calculations, this configuration took place 185,000, 115,000, and 70,000 years ago.

Although during the last years the concept of Milankovich has been acknowledged (but not completely), its individual aspects have been critically analyzed. This being for two reasons:

(1) many new geological data of analysis of the cores of sea bottom rocks and ice have appeared; and
(2) a considerable progress has been achieved in numerical climate modeling.

An analysis of both these sources of information has shown that an adequate explanation of paleoclimate changes is only possible with regard for not only variations of the orbital parameters but also other climate-forming factors – in particular, variations of the GHG contents in the atmosphere, including most importantly, carbon dioxide.

In this connection, Palutikof *et al.* (1999) performed an analysis of new geological data on paleoclimate changes, with a higher temporal resolution, in the context of the present ideas of global climate dynamics. The data of analysis of ice cores and pollen obtained in 1990 have led to the following two important general conclusions:

(1) Part of the observational data does not confirm the glaciation cycles having been determined by the Milankovich mechanism (this refers especially to data on $\delta^{18}O$ in calcite veins at Devils Hole in Nevada, USA, which testify to opposite phases of the cycles of glaciation and theory of Milankovich).
(2) It follows from other data that this mechanism can explain only slow quasi-periodic variations but not the short-term variability (on timescales from decades to millennia), with respect to which it has turned out that it had happened much more often than was supposed.

The variability of the global mean temperature could reach several degrees during several decades. In particular, a large-scale sudden climate cooling had happened in the Ames interglacial period ($\sim$122,000 years ago), when climatic conditions had been very close to the present ones. A typical example of a short-term climatic variability is the Heinrich and Dansgaard/Oeshger events (Keigwin and Boyle, 2000). Such events could repeat in the future.

Many uncertainties remain also concerning the impacts of present changes of extra-atmospheric insolation on climate. Soon *et al.* (2000) have demonstrated, for instance, the presence of a hyper-sensitivity of the climate system to changes in UV insolation whose effect is intensified by a feedback due to the statistical stability of clouds, the effect of tropical cirrus clouds, and the stratospheric ozone (the "ozone–climate" problem needs special analysis).

Of particular interest is an interactive consideration of the biospheric dynamics as a climate system component. The significance of this problem can be exemplified by estimates of the climatic impact of deforestation in the tropical Amazon basin, obtained by Bunyard (1999). Nature in the Amazon basin (first of all, wet tropical forest, WTF) performs a number of important functions, still not completely taken account of, including the energy input from the tropics to higher latitudes, which, however, is under threat in view of a high rate of WTF destruction.

According to present assessment, every year up to 17 mln ha of tropical forest is removed, with $\sim$6 mln ha falling within Brazil/Amazon. By the end of 1988, 21 mln ha had been deforested, and 10 years later this reached 27.5 mln ha, which exceeds the size of Great Britain.

The WTF destruction is important from the viewpoint of the impact on the global carbon cycle, since there is a danger of transformation of the WTF zone from being a sink to a source of carbon for the atmosphere. Not less substantial are ecological aspects of WTF elimination, in view of the ecological uniqueness of the tropical forests of Central and South America. According to available estimates, deforestation in the Brazil/Amazon alone (over an area of 360 mln ha) will result in the loss of an annual sink of carbon of 0.56 bln t, and on global scales this level could reach 4 bln t C yr^{-1}. Bearing in mind that in 1988, as a result of forest fires, the tropical forests were burned over an area of about 9 mln ha, then due to this source alone the atmosphere gained 1–2 bln t C.

Over the virgin tropical forests, about 75% of the incoming solar radiation is spent on evapotranspiration. Therefore, removal of WTF will result in radical changes of energy exchange and the global atmospheric circulation. Changes of local climate should be even more substantial, especially from the viewpoint of precipitation, which can be reduced by 65%. Of key importance is the fact that the threshold level of WTF elimination that determines the loss by the ecosystem of its self-supporting ability, remains unclear. For instance, if it is 20%, then this threshold has been exceeded already.

Bengtsson (1999) drew attention to the fact that since non-linear processes exhibit the prevailing impact on the climate system's variability, it is impossible to establish any simple connection between external forcings (e.g., the growth of the GHG content or variability of extra-atmospheric insolation) and response of the climate system to such forcings. With the unpredictability of some factors of climate taken into account, the difficulty of distinguishing between anthropogenic and natural variability of the climate becomes apparent and even increases due to the fact that both internally and externally forced modes of climate variability are determined by the same mechanisms and feedbacks.

Although considerable progress has been recently achieved in numerical modeling of the climate system, it refers mainly to the atmosphere. Results of the "ensemble" numerical experiments indicate that the 3-D atmospheric circulation in the tropics is mainly determined by the impact of boundary conditions, whereas in high latitudes the impact of atmospheric dynamics prevails. The simulation of the water cycle in the atmospheric turns out to be rather realistic.

A considerable progress in modeling of the interactive atmosphere–ocean system has made it possible to successfully predict the seasonal and inter-annual variability and, in particular, El Niño events. An adequate consideration of the processes on the land surface has ensured a substantial increase of reliability of hydrological forecasts (including river run-off).

In this context, Bengtsson (1999) discussed the progress made in numerical climate modeling in three areas. A successful accomplishment of the TOGA program has made an important contribution to provision of the transition to operational forecasts of the seasonal and inter-annual variability considering the prescribed SAT changes in the tropics, which has determined the critical importance of reliable SAT data. The second of these directions is connected with numerical modeling of climate change on scales of decades and longer and especially with explanation of the centennial change of global mean annual mean SAT. Apparently, in the case of long-term climatic variability the stochastic forcing can be considered as a zero hypothesis.

A consideration of the impact of low-frequency climatic fluctuations on the level of the Caspian Sea has shown that the long-term variability of the level is connected first of all with SST anomalies in the eastern sector of the tropical Pacific Ocean. It turns out that positive SST anomalies correlate with enhanced precipitation in the basin of the Volga watershed and vice versa. The main cause of variations of the Caspian Sea level is the long-term dynamics of ENSO events, which should be considered chaotic.

An important part of the discussed problem is a study of anthropogenic climate change. Calculations have shown that a doubling of CO_2 concentration should result in the outgoing long-wave radiation at the level of tropopause decreasing by $3.1\,W\,m^{-2}$, and the downward long-wave radiation flux in the stratosphere growing by about $1.3\,W\,m^{-2}$. Thus, the total RF at the tropospheric top level will constitute $4.4\,W\,m^{-2}$. Calculations of the resulting SAT change with the use of 11 climate models revealed a warming within 2.1°C to 4.8°C, as well as an intensification of the global mean precipitation of between 1 to 10%.

According to Bengtsson (1999), before 1980, the centennial change of the global mean SAT was characterized by the prevailing contribution of natural variability with the subsequent increase of the anthropogenic contribution. An important task of subsequent developments is to improve the numerical modeling (first of all, from the viewpoint of a more adequate consideration of various mechanisms of feedbacks) in order to provide reliable forecasts on regional and local scales. Quite an urgent problem facing global modeling consists in the interactive consideration of the biogeochemical cycles.

One of the most important aspects of numerical climate modeling is an assessment of the contribution of anthropogenic climate-forming factors. In this connection, Allen *et al.* (1999) discussed possibilities of recognition, evaluation, and forecast of the contribution of anthropogenic global climate change characterized by the SAT with regard to the available data of observations and numerical modeling. The latter is carried out bearing in mind the internal variability of the climate system, as well as the impact of the greenhouse effect (and respective climate warming) and sulphate aerosol (the effect of climate cooling).

The four global 3-D models of the interactive atmosphere–ocean system predicted an increase of the global mean SAT for the decade 2036–2046 (compared with pre-industrial levels) within 1.1–2.3 K. Calculations of the climate system sensitivity to a doubled CO_2 concentration gave values within 2.5–3.5 K. According to the HadCM2-G5 climate model developed at the Hadley Centre (Great Britain), the global mean "greenhouse" warming for the period 1996–2046 should constitute 1.35 K, and the "sulphate" cooling will constitute 0.35 K, thus the resulting warming will reach 1 K.

With the anthropogenic SAT increase in the 20th century equal to 0.25–0.5 K per 100 years, calculations made with a simple climate model predict a 1–2 K uncertainty of the balanced SAT forecast by the year 2040. Such estimates can, however, be sufficiently reliable only with an adequate consideration of the characteristic time taken for ocean adaptation. A critically important aspect of prognostic estimates is a necessity to take into account the possible sudden non-linear climate changes, which seriously limit the lead time of forecasts.

The most attractive perspective for assessment and forecast of anthropogenic SAT changes is connected with analysis of the spatial–temporal variability of the SAT fields that takes into account the impacts of the greenhouse effect and aerosol. Realization of this approach is seriously complicated, however, by the impossibility of a reliable prescription of aerosol forcing on the SAT field. A serious problem is also the necessity to take into account the climatic impact of changes in the content of stratospheric and tropospheric ozone.

9.3.5 Contradictory nature of climate study results

The problem of anthropogenic global warming is now at the centre of attention of not only specialists but also wide circles of the population. With a deeper understanding of this problem, there appears a feeling of inconsistency of results obtained in this field, especially during the last decade. In this connection, Mahlman (1998) carried out an overview of such results aimed at analysis of the fundamental scientific aspects of the problem under discussion, in which the emphasis is placed on the role of numerical modeling and analysis of observational data in understanding the present climate change. The monograph by Weber (1992) has been dedicated to the same theme.

The main difficulty in understanding the causes of climate change is still connected with the impossibility of an adequate consideration of climatic feedbacks. First of all, this refers to the cloud–radiation feedback, direct and indirect (through the impact on radiative properties of clouds) climatic impact of atmospheric aerosol, as well as the impact of the atmosphere–ocean interaction on climate formation.

The often ignored specific manifestation of the "greenhouse" climate warming is the system's large inertia (the thermal "memory" of deep layers of the ocean spans centuries and even thousands of years). Schlesinger *et al.* (2000) pointed out, for instance, the existence of the global SAT oscillations, with a period of about 65–70 years, from the observational data for the period 1858–1992. It is important that principal differences between numerical modeling (and forecasts) of weather and climate should be taken into account. In the case of numerical climate modeling, it is important to use "tuning" and adjustment because of the difficulty of an adequate consideration of a complicated totality of interactive processes and spatial–temporal scales. In this context, the use of paleoclimatic data plays a substantial role, though they cannot be analogs to a possible climate change in the future.

A serious anxiety has been caused by the inadequacy of global observation systems and degradation of ground observations, especially manifesting themselves in some cases. Mahlman (1998) emphasized that the controversy of the problem of anthropogenic climate change consists in the absence of reliable quantitative estimates of relationships between the contributions of natural and anthropogenic factors of change. This circumstance creates serious difficulties for the practical realization of recommendations contained in the KP.

Such conclusions, discussed in detail before, have received universal recognition illustrated by the recent review of Grassl (2000). In this context, quite surprising is the wide use of the term "climate change" as determining only the anthropogenic change. The substitution of the notion of "climate change" (in its true meaning) with the term "global warming" is also incorrect, since both observational data and the results of numerical modeling indicate a highly inhomogeneous present climate change, far from being reduced to only a SAT increase.

Such terminological misunderstanding is not accidental, however. It is aimed at disinformation for the sake of establishing a false conception of anthropogenic ("greenhouse") global warming, which has been convincingly explained by Boehmer-Christiansen (1997, 2000), who analyzed the political motivation for this concept.

In August 1997, the US Minister of geology, V. Babbitt, addressing about 3,000 participants of the annual Congress of the US Ecological Society, said that they should implement their civil obligation – help to convince the skeptical American public that global warming is both real and dangerous: "We have a scientific consensus but we have not a public consensus". In this connection, Morris (1997) carried out an overview of available scientific information in order to analyze the grounds for this opinion, since many specialists do not share the apocalyptic predictions of anthropogenic global warming. The emphasis in the overview has been placed on the problem of distinguishing between natural and anthropogenic climate change.

In the mid-1970s the forecasts of global cooling due to sulphate aerosol predicted, for instance, that this impact will limit an increase of global mean temperature due to an enhanced greenhouse effect of the atmosphere by less than 2°C even with a 8-fold increase of CO_2 concentration. The warming trend observed in the 1980s has attracted attention to the problem of climate warming. At the height of the summer of 1988, J. Hansen (Hansen *et al.*, 1988; Hansen, 1998) declared to the US Congress, the 99% probability of anthropogenic global warming and its destructive consequences for the ecosystem in the future, suggesting a general consensus had been reached among specialists. Many meteorologists and climatologists did not share these views: respective developments did not permit establishment of reliable cause-and-effect relationships between anthropogenic GHG emissions and observed climate change.

The political motivation behind the support given to the global "greenhouse" warming concept determined the gigantic growth of government financing of developments in the USA between 1990–1995 of US$600mln to US$1.8bln.

In the context of contributions of different factors to the formation of global climate, Morris (1997) emphasized an importance of the combined consideration of contributions of the "greenhouse" warming and solar activity, which during the period of instrumental observations constituted 0.31°C and 0.41°C, respectively. According to these estimates, a doubling of CO_2 concentration will lead to a global warming of 1.26–1.33°C. With such climate changes, there are no grounds to expect any often predicted catastrophic consequences such as the rise of the World Ocean level and increases of epidemics. However, there is no doubt that one should expect an increase in agricultural yield. On the whole, the climate-warming consequences should be positive. Although reliable forecasts of ecological consequences of GHG emissions are still impossible, it is apparent that their reduction will strongly affect the economy. Therefore, the main conclusion which may be drawn is that one should not try to prevent the still unreliably predicted climate change but adapt to it. However, this conclusion has been disputed in numerous publications (Hansen *et al.*, 1998; Houghton, 2000; Hulme *et al.*, 1999; Wigley, 1999).

In view of climate problem complexity, relevant scientific publications usually emphasize a serious uncertainty in available estimates, which favors the possibility of continuing studies, but on the other hand, but is fraught with serious consequences in the context of ecological policy. As Boehmer-Christiansen (1997) noted, "ecological" bureaucrats and experts as well as well-organized representatives of the fuel-energy industry are interested in the proof of the existence of anthropogenic warming and want to obtain needed scientific support. These three groups with common interests support non-governmental ecological organizations and thoroughly consolidate them after the Rio Conference.

The main target of "green" pressure is the governmental hierarchy which, in view of the wide range of its responsibilities, as a rule, either does not want or cannot realize the required measures including ecological taxes, subsidizing development of renewable energy sources, nuclear energy, public transport, energy efficiency, and others. The World Bank (WB) and Global Ecological Fund (GEF) are also lobbied, to support numerous and expensive developments which are controled by many experts and specialists in the spheres of finance. These experts are the most vigorous advocates of measures to prevent the supposed undesirable climate change, and the WB and respective UN organizations use their recommendations more frequently, in particular, those connected with so-called "Joint Implementation".

This system is based on such fundamental documents as the International FCCC, and scientific bases are the IPCC reports. To achieve these goals, the following components should be provided: science that substantiates the threats; the "green movement" that expresses emotion; rhetoric and "principle of precaution"; developments of new technologies aimed at reducing GHG emissions; and bureaucratic bodies that prepare the required plans and develop strategies.

Only after all that do politicians, who should ensure financial support in order to transform plans into concrete actions, start acting. In this context Boehmer-Christiansen (1997) discussed various circumstances concerning "scientific provision". Clearly, the IPCC Reports may be the fundamental documents, if anthropogenic impact on climate is really dangerous. However, the difficult problem is how to determine the criteria of "danger" and who should be responsible for it (considering also the socio-economic and political factors).

Still there are many uncertainties in this problem. In particular, in connection with the discussion of the necessity to reduce GHG emissions the emphasis is placed on decreasing the scale of the use of coal, especially in electric power production (while for natural gas the conditional "factor of CO_2 emissions" constitutes 15, in the case of coal it reaches 25 or more, being very variable). An "attack" on coal was not, however, connected with ecological motives. The question of ecology did not rise at all until the mid-1980s, when the price of oil and gas dropped (especially in 1986). To resolve the appearing economic problems, ecological allies were needed to provide the competitive ability of "pure" energy for the sake of "sustainable development".

It is no mere chance that the IPCC was forethought in 1995, really planned in 1987, and started functioning in 1988. Governments of various countries support the plans for GHG emission reductions not for ecological but for other reasons: enhancement of national nuclear energy (Germany); an increase of the export potential for electric power produced by nuclear power stations and gas (France, Norway); and an increase in the size of a country's financial support, etc. Naturally, the countries where electric power depends on coal, are most skeptical about the "greenhouse" warming.

Having analyzed the role of various international organizations and programs in the problem of global climate change, Boehmer-Christiansen (1997) emphasized: "Climate policy cannot be understood without a deeper analysis of the role of science and scientific understanding of the coalition of non-ecological interests (both commercial and bureaucratic) which serve as a driving force of development of events on international scales. Where this coalition will lead us, remains unclear, so far".

Constructive prospects of resolving this problem are connected with development and application of complex models to assess possible changes of climate and socio-economic development. In the detailed overview of methods and results of numerical modeling of global climate change with regard for dynamics of socio-economic processes, prepared by Parson and Fisher-Vanden (1997), the main aspects of the so-called Integrated Assessments (IA) have been discussed. The main goal of such developments are to substantiate recommendations for people making respective decisions concerning ecological policy.

Four concrete goals include the following problems:

(1) assessment of a possible response to climate change;
(2) analysis of the structure of scientific bases of modeling and characteristic uncertainties of the results obtained;
(3) comparative estimates of possible risks; and
(4) analysis of the achieved scientific progress.

During the last several decades, two approaches prevailed in developments of IA models: (1) use of assessments obtained by interdisciplinary groups of experts; and (2) formal numerical modeling. The former approach is characteristic of IPCC efforts and developments within the Montreal Protocol, whereas the latter approach was realized by individual specialists. The emphasis was placed on the use of IA models to analyze possible impacts of climate change on the development of energy and economy in the context of the problem of CO_2 emissions to the atmosphere.

The general assessment of the results obtained consists in the fact that the IA models cannot be used to substantiate the highly specialized measures in view of the insufficiently detailed character of such models. However, the use of such models is important for assessment of possible uncertainties and, hence, an expediency of making any decisions. In this case it is important that considerable uncertainties concerning the system on the whole, or recommendations on the required ecological policy, can turn out not to be those uncertainties that are most important for understanding the processes responsible for understanding changes taking place in the environment or characterized by most substantial variability.

In considering the data for concrete models, it is possible to range uncertainties by their significance, but this ranging depends on the specific character of the models. Although the use of IA models has made it possible to substantiate some correlations of dynamics of the socio-economic development and variability of the environment, the relevant results should be considered only preliminary. So far, the use of IA models has made only a small contribution to the estimation of comparative risks and to obtaining the answer to the fundamental question, to what degree and in which respects a possible climate change is most substantial.

In view of these circumstances, the results of the use of IA models to substantiate an adequate ecological policy have been, so far, rather limited, and this refers especially to purely didactic estimates based on the use of simple models. The main problems concerning the improvement of IA models include: overcoming insufficient understanding of probable impacts and possibilities of adaptation; poorly substantiated or completely absent descriptions of social and behavioral processes in developing countries; and very limited ideas of a scarcely probable, but radical climate change. Despite these and other unresolved problems, the need for urgency of further development of IA models is obvious.

In conclusion, again, it is emphasized that of basic importance is an improvement of the global observation system. The urgency of this problem has been illustrated in Demirchian *et al.* (2002) where the complicated spatial–temporal SAT variability in high latitudes of the northern hemisphere has been demonstrated as well as the groundlessness of the "greenhouse" warming concept.

9.4 CLIMATE AND GLOBAL URBANIZATION

9.4.1 General features

Cities are one of the most substantial and evident phenomena of present civilization development. Despite an apparent controversy regarding their development, for most of the planetary population they represent a symbol of progress. But nevertheless, "cities do not deserve their epoch, they are not worthy of us". This is the opinion of the outstanding French architect and town builder Le Corbusie (1977), voiced as far back as 1925. One of the major causes of this assessment is the unfavorable ecological situation, another being the continuous growth of the urban population. During the next 20 years most of the population in developing countries will live in cities. This fact, like the problems of large cities, attracts the

Table 9.6. Forecast of the urban population growth (Grigoryev and Kondratyev, 2004a,b).

Level of urbanization (%)				Population size (10^3)		
Years	2000	2015	2030	2000	2015	2030
Global	47.0	53.4	60.3	2,845,049	3,817,292	4,889,393
More developed regions	76.0	79.7	83.5	902,993	968,223	1,009,808
Less developed regions	39.9	48.0	56.2	1,942,056	2,849,069	3,879,585

serious attention of many investigators (Grigoryev and Kondratyev, 2004a; Bulkeley and Betsill, 2003; Wilby, 2003).

The main feature at the turn of the 20th century is the continuing growth of megalopolices. In 1970 the urban population constituted 35% of the planetary population, in 2000 it was already almost 50%. It was supposed that by the year 2003 about 3.3 billion people will live in cities. The prevailing contribution (90%) is made by the cities of developing countries (Grigoryev and Kondratyev, 2004b).

Forecasts of urbanization for the next 30 years show that from the viewpoint of the location of populations, cities head the list. In 2030, the size of the urban population could reach 60% of the global population (Table 9.6).

Serious progress in many spheres of vital activity, which is connected with the growth of cities, is accompanied by heavy negative consequences. They are rather diverse and connected, in particular, with the poverty of many city dwellers, low level of health, high mortality, environmental degradation, as well as shortcomings in the infrastructure and development of municipal services. Of these problems, the most important is the environmental situation and lack of readiness of infrastructures to protect the population against natural disasters. It is this that is the initial cause of other numerous negative aspects of urban life (first of all, city dwellers health and poverty). This hinders the progressive positive development of cities. Of course, specific features of urbanization in the regions of different continents differ drastically. The highest level of urbanization is observed in the countries of North America (77.2%), Latin America (75.3%), and Europe (74.8%). In other regions these indicators are as follows: Oceania and Australia – 70.2%, Asia – 36.7%, and Africa – 37.9%. This statistics is quite natural and reflects the difference between the countries regarding industrial development.

For individual countries, the urbanization phenomenon is characterized by the following indicators: Belgium – 97.3%, Iceland – 92.5%, Uruguay – 91.3%, Great Britain – 89.5%, the Netherlands – 89.4%, Germany – 87.5%, Venezuela – 86.6%, Argentina – 85.9%, Chili – 85.7%, Denmark – 85.3%, Sweden – 83.3%, and Brazil – 81.3%. Analysis of these and other indicators of urbanization shows that there is a dependence of population concentration on the frequency of unfavorable and dangerous natural phenomena on a given territory. In particular, in Iceland, settling is affected by volcanic activity, in the Netherlands by lowland floods. In Chili, vast territories are covered with deserts and mountains, in Uruguay floods happen frequently. In Brazil and Venezuela large areas are covered with tropical forests. Apparently, in other regions, natural disasters also affect the settling of the population. However, this dependence remains poorly studied.

In North America, despite the general high level of population concentration in cities, in the two greatest countries of the continent – the USA and Canada – in 2000 the level of urbanization did not reach 80% (77.2 and 77.1%, respectively). By 2015, the USAs level of urbanization will exceed this threshold and reach 81%. This is explained, at least, by two reasons. One of them is the presence in these countries of vast flat territories within which there is plenty of safe (in natural respect) and comfortable places for settling.

At present, the least urbanized countries are in Africa. In 2000, the level of urbanization exceeded 80% in only four countries. These are Libya (87.6%), Gabon (81.4%), Djibouti (83.3%), and the Western Sahara (95.4%).

In Asia, in the year 2000, the 80% threshold of urban population was exceeded only in 11 countries of 50. Only one large country is on this list – Korea (81.9%). In most densely populated countries the level of urbanization is: China (32.1%), India (28.4%), Indonesia (40.9%), and Pakistan (37.0%). A high level of urbanization is observed in Japan (78.8%), which is explained by the impact of unfavorable natural factors, such as tsunamis, volcanic eruptions, earthquakes, and typhoons.

One of the features of current urbanization is the rise in numbers of cities and urban agglomerations. In North America, by the year 2000 such cities numbered 41 (37 in the USA and 4 in Canada). Among them were 3 cities with populations exceeding 5 million: New York (16,649,000), Los Angeles (13,140,000), and Chicago (6,951,000). Among European countries, Russia and Germany head the list (11 and 13 cities with populations of a million, respectively). In Europe there are only four cities with populations more than 5 millions: Paris (9,264,000), Moscow (9,321,000), Essen (6,541,000), and London (7,640,000). In Africa, there are three such cities: Kinshasa (5,064,000), Cairo (10,552,000), and Lagos (13,427,000). By the year 2015, there will be about 500 mega-cities with populations of more than 1 million.

Thus, an analysis of the general trend in settling of the planetary population carried out by Grigoryev and Kondratyev (2004b) has shown that on the whole, the number of cities containing 1 million people is growing, but the rate of the growth differs in different regions. The global characteristics of the present population dynamics in the cities are as follows:

- on the whole, at present the urbanization is characteristic of all countries in the world. Cities are growing in number, the percentage of the urban population increases. However, the rate of urbanization is slowing;
- the cities containing 1 million people are growing in number. But in the countries with a high level of development the growth of the population in such cities is at a minimum. At present it is slowing down, and in many cities it stopped altogether. In developing, poor countries the growth of populations in these cities is slow; and
- the characteristic feature of the development of urbanization now and in the future is the formation of cities with populations above 5 and even 10 million,

called mega-cities. It is especially interesting to study these cities, since they exhibit features of the future process of urbanization, and, as a matter of fact, features of the future of humankind. They concentrate a multitude of positive aspects from the viewpoint of human activity. At the same time, in such cities every negative aspects of the phenomenon discussed clearly shows itself, especially in countries with developing economies.

9.4.2 Ecodynamics of megalopolises

The 21st century is an epoch of transition of the planetary population to new living conditions within the globalization of many processes. Of course, the elements of globalization of various kinds of human activity have been observed before, even many centuries ago, due to trade, navigation, and military expansions. However, its present features have greatly changed. Globalization of any phenomena has accelerated and is covering larger areas. The range of phenomena incorporated in planetary processes has drastically broadened. The whole world and various kinds of human activity – economy, politics, science, technology, engineering – have turned out to be involved in the process of globalization. Moreover, the whole world has turned out to be not simply in globalization chains, but in a web of chains of events which are propagating across the globe.

Cities, especially large ones, have become the key point of the process of globalization. Globalization in cities is accompanied by many positive phenomena. Among them are, for instance, prompt information on possible acts of terrorism, approaching tropical cyclones, which can threaten the human life, information on global propagation of dangerous infectious diseases, information about new medicines, and many others. Negative aspects of globalization are also known, which vividly and painfully manifest themselves in the cities. Among them are: propagation of drugs, epidemics, criminality, etc. Ecological problems in the cities also contain negative components. These are pollution of the atmosphere and hydrosphere, the high risk of propagation of epidemics resulting from destruction after earthquakes, hurricanes, and floods.

Contrasts in the ecological situation in the cities connected with socio-economic factors, are clearly seen for example with the access of a population to services such as sanitation. People's security in the case of natural disasters depends substantially on these types of services. The best access to such services is provided in the cities of North America (100% access), Europe (99%), and Australia and Oceania (99%). A much lower level of sanitary–hygiene provision is found in the cities of Africa (84%) and Latin America (87%). A low level of access of city dwellers to normal services occurs in cities of Asia (78%), and the worst sanitary conditions occur in India (46%), Nigeria (61%), Ghana and Pakistan (53%).

The cities of the world differ substantially in their waste. For instance, every year, on average, the cities with populations of 1 million or more in Russia dump about 3.5 mln t of solid and concentrated waste. Its main constituents being (in 1,000s of tonnes): wood waste (400), paper (9), textile (8), and glass (3). Most of the waste occurs in cities of highly developed industrial countries, such as the USA and Canada, where per-capita waste is estimated at 2 and 1.5 kg day^{-1}. In the cities of Latin America it is 1.0 kg day^{-1}. Of course, of greater importance is not the amount of waste but the possibilities and practice of its processing. In contrast to the cities of the USA and Canada, where waste is processed practically completely, in most of the cities of South America, some 20–50% of garbage is not removed at all (Gurberlet, 2003).

Large cities concentrate a lot of anthropogenic processes on small territories, the state of which is determined by the ecological, material, sanitary–epidemiological, social, cultural, political, religious, and other specific features of society. All these features define input variables when determining the danger threatening city dwellers in the case of a natural disaster. Clearly, for the optimal development of cities in harmony with the environment, of interest are the development of criteria of ecological risks in cities and their indicators; and development of an efficient system of indicators of ecological situations in cities. At present, there are a large number of such indicators. As a rule, they concern individual elements of the environment. Examples are information systems for the assessment of the quality and condition of the water medium, sanitary–epidemiological situation, quality of atmospheric air, etc. Alimov *et al.* (1999) undertook a multi-criteria assessment of the medium quality and the state of the urban ecosystem of St. Petersburg. Flood (1995) described the system of indicators used by the U.N. having studied more than 100 cities of the world.

Indicators of various phenomena in the cities are an efficient means to analyze and monitor the interaction between the cities and the environment. Among the indicators of the ecological situation in cities, two large groups can be selected. The first group includes indicators of the environment, directly affecting human activity and the ecological situation. First, it concerns the indicators of air and water quality. Such indicators as the areal extent of green plantations are widely used and, to a lesser extent, the characteristics of their state are also used. During the last decades emphasis has been placed on several geoecological characteristics of the urban environment of technogenic origin. Among them are: the level of noise, radioactive pollution, and impact of local electromagnetic fields. The second group includes indicators characterizing the level of poverty, health, special features of the infrastructure, transport, the state of available housing, and quality of environmental control. A detailed analysis of some of these has been made by Grigoryev and Kondratyev (2004a).

In Europe, in 1999, within the framework of conception "Towards sustainable development at a local level: European indicators of sustainability", ten indicators were proposed (Grigoryev and Kondratyev, 2004a,b):

- satisfaction of citizens with the standard of living at a settlement;
- contribution of the cities to global climate change;
- population mobility and passenger traffic;
- availability of public accommodation and services;
- air quality over a settlement;

- transportation of children to and from school;
- control according to principles of sustainable development;
- noise pollution;
- stable land use; and
- output of products that favor sustainable development.

The main apparent shortcoming of these indicators is their inadequacy and some vagueness. Here concrete indicators are absent, such as:

- quality of drinking water;
- rate of mortality (including infant mortality);
- per-capita gross production;
- quantity of poor and homeless people; and
- crime rate.

From the viewpoint of an urban populations security during a natural disaster, the use of such indicators would make it possible to resolve many problems of averting great human losses and reducing economic damage. Of course, all of these problems can be resolved by using a global model that describes the maximum amount of processes and sub-systems of the environment. Such a model has been proposed in Kondratyev *et al.* (2004d). It can be realized only within the framework of the agreement signed by 60 countries on a unification of national systems of observation of the environment into a global system.

9.5 VIEW OF THE FUTURE

9.5.1 Reality and expected changes of the environment

Everything said above makes one re-analyze the observed regularity of transformation of living beings habitats and draws a conclusions about its limits, with regard to survivability. As has been pointed out by many specialists, the frequency of occurrence of natural disasters increases, and this suggests the idea that we are approaching an unknown threshold of permissible anthropogenic impacts on the environment beyond, which human life may be impossible. The main conclusion drawn by scientists is that the accumulated data on various environmental parameters cannot be considered complete and adequate for studies of global ecodynamics. The health of the environment depends on a multitude of correlated factors and processes, whose estimation with a high reliability is impossible with current available knowledge.

Humans introduce noise into processes taking place in nature and do not know how much they change not only the dynamics of these processes but also the laws of their formation. It is enough to note the fact, for instance, that the global primary production changes from year to year, following climatic noise. This change from different estimates varies between 5% and 10%. For instance, the primary production of coniferous forests in the south-east of the USA in July–October averaged over 5 days, changed with amplitude $0.55\,\mathrm{g\,C\,m^{-2}\,day^{-1}}$ about an average level of $6.26\,\mathrm{g\,C\,m^{-2}\,day^{-1}}$. This variability for broad-leaved forests increases moving from middle to high latitudes (10–40°E, 52–70°N). Coniferous forests respond slower to climatic variations in high latitudes compared with low latitudes. Unfortunately, such dependences have been poorly studied, but it is these forests that determine the accuracy of estimates of the sinking of atmospheric carbon to the vegetation cover of the planet, especially Siberian coniferous forests. During the last years attention has grown to studies of climate change in Siberia and in the Far East of Russia as a factor determining the change of the global carbon cycle. In this region the temperature increased by 0.5–0.9°C during the last decade. It has been noted that trends of average temperatures of the atmosphere intensify from the north to the south. Also, a redistribution of precipitation is observed with its increase in the cold period and insignificant reduction on some local territories in the warm season. So, in the basin of the Amur River, during 30 years precipitation, the cold period increased by 35%, and the annual sum of precipitation increased by 12.3%. Irregularity of these changes in space is manifested also in the seawater temperature. For instance, during 100 years the water temperature near Vladivostok increased by 0.64°C, and in Nakhodka it decreased by 0.27°C. Such changes affect the estimate of the CO_2 sink to the World Ocean, especially in coastal upwelling zones. So, along the western coast of the northern sector of the Pacific Ocean during the upwelling season the CO_2 sink constitutes 5% (0.5 Pg C) of the total sink to the Pacific Ocean, whereas the area of this zone constitutes 25% ($0.7 \times 10^6\,km^2$) of the whole shore zone (0–200 m) of the ocean and <2% of the northern part of the Pacific Ocean (14–50°N) where the upwelling season constitutes 30% of the year.

Hales *et al.* (2005) assessed the power of the biological pump that moves excess CO_2 from the atmosphere in the upwelling zone of the Pacific coastline of the USA. This zone covers 25% of the area of the USA shelf. It has been shown that the prevailing of low concentrations of CO_2 in the ocean water in the period of the upwelling season means that this zone becomes a sink for CO_2 for the following reasons:

(1) The rising water masses are rich in nutrients.
(2) The functioning of the oceanic carbonate system changes sharply.
(3) The rising water masses are moderately warmed.

The gas exchange flux H_3^C at the atmosphere–water interface is determined from the formula:

$$H_3^C = (0.79 + 0.0062U_{10}^3)K_{CO2}\Delta P_{CO2}$$

where K_{CO2} is CO_2 solubility ($mol\,m^{-3}\,atm^{-1}$), ΔP_{CO2} is the sea air P_{CO2} difference (atm.), P_{CO2} is the CO_2 partial pressure (atm.), and U_{10} is the hourly mean of the wind speed at $10\,m\,s^{-1}$. The average value of H_3^C in the studied zone was equal to $20\,mmol\,m^{-2}\,day^{-1}$, which is about 15 times greater than the global mean rate of CO_2 assimilation by the World Ocean estimated at $1.3\,mmol\,m^{-2}\,day^{-1}$ ($2\,Pg\,C\,yr^{-1}$). The total CO_2 assimilation for May–August, when the upwelling process on the Pacific coastline of the USA prevails, constituted $2\,mol\,m^{-2}$ (5% of the annual mean sink of CO_2 in this zone estimated at $40 \times 10^{12}\,mol\,yr^{-1}$).

Hales *et al.* (2005) have justly come to the conclusion that similar studies in the zones of upwelling would make it possible to specify the spatial distribution of CO_2 sinks in the World Ocean and to assess thereby the level of danger of anthropogenic emissions of carbon. These studies would also enable one to understand the role of tropical cyclones on the formation of the H_2^C and H_3^C fluxes. As has been mentioned above, hurricanes affect the local rates of gas exchange at the atmosphere–ocean interface through a change of the thermal and physical structure of the upper layer of the ocean. From the estimates of Perrie *et al.* (2004), for instance, Hurricane Gustav happening on 10–12 September 2002 over New England at a speed of $48\,\mathrm{m\,s^{-1}}$ caused a linear increase of local CO_2 flux from the atmosphere to the ocean of up to $2.1\,\mathrm{mmol\,m^{-2}\,hr^{-1}}$.

In general, the impact of hurricanes on fluxes H_2^C and H_3^C (Figure 6.4) has been poorly studied. The water bodies of the North Atlantic where hurricanes are frequent events, creating a zone of upwelling with a decrease of CO_2 partial pressure in water $60\,\mu\mathrm{atm}$, introduce, of course, a substantial change into the ratio H_2^C/H_3^C, but the magnitude of this contribution is unknown.

Discussions on distinguishing between the roles of land biota and that of the World Ocean in stabilization of climate through correlations and feedbacks in the energy system of the planet remain without constructive answers. Of course, one of the principal problems of global ecodynamics is an assessment of the response of vegetative communities to climate change. One can add that changes in vegetation productivity depending on climatic oscillations are characterized by temporal delays specific to the types of vegetation cover. For instance, tropical vegetation responds to climate change with a delay of about 50 years. Besides this, soil–plant formations are determined in many respects by the climatic situation. So, for instance, wet tropical forests survive only on the territories where the dry period lasts for not longer than 4 months. In particular, in the north-west of Brazil, because of the semi-desert character of the climate with a dry period about 8 months long, the tropical forest does not develop. It means, the dry period duration is some threshold magnitude which controls the climate–vegetation relationship. Unfortunately, there are no reliable answers to the question of whether this relationship is reversible (i.e., whether plants can return the the climatic situation through their change). The answer to this question is important in connection with the anthropogenic change of land cover. For instance, during the last decades in Kazakhstan, the areal extent of which ranks ninth in the world, the land covers have changed substantially, which shows itself via changes of the climate situation in the region where the global catastrophe connected with the Aral Sea drying is developing.

The recently observed anomalous climate change is an integral manifestation of the impact of a multitude of factors such as urbanization, deforestation, atmospheric and hydrospheric pollution, decrease of biodiversity, and intrusion of foreign elements to the ecosystems. In many global regions, the following phenomena are observed:

- anomalous temperature maxima and growing quantity of unusually hot days;
- anomalously intensive precipitation;

- decreasing number of cold periods and a decrease of the total time with frosts;
- smoothing the amplitude of the diurnal change of temperature;
- the summertime drying-up of the continents;
- increase of the maximum speed of tropical cyclones;
- increase of average and maximum rains during tropical cyclones.

A sufficiently convincing analysis of the present state of studies in the field of global change has been carried out in the studies of Victor (2001, 2004). The author draws attention to the complexity of the problem of global change and the presence of substantial uncertainties in available scientific ideas, which determine a strong contradiction to the whole problem. This contradiction manifests itself via international decisions, such as the Kyoto Protocol. As an example, we can take the decision of the USA in 2001 to withdraw from the Kyoto Protocol, after which the criticism of the paradigm of global warming used as propaganda by many politicians and scientists has drastically intensified. In the USA, to work out an adequate ecological strategy and in connection with the hopelessness of the search for a compromise over this paradigm, three alternatives have been put forward:

- Moderate precautionary measures including the support of scientific developments, the program of voluntary reduction of GHG emissions to the atmosphere, and rejection of any obligation imposed by the international agreement on these emissions.
- Development of a new international agreement "succeeding" to the Kyoto Protocol, which should foresee realistic measures for the USA and participation of developing countries in GHG emission reductions as well as creation of the global system of "waste trade".
- Stimulating the market of new technologies, which would provide low levels of GHG emissions in the USA and in other countries, especially in developing ones.

These alternatives cover six directions (Kondratyev, 2005b):

- Scientific analysis of the causes and effects of climate change, including measures to support additional developments.
- Adaptation to climate change.
- Strategies of the control of GHG emissions.
- Investments to the development of new technologies.
- Coordination and cooperation of efforts between key developing countries.
- Informing the population.

The 10-year strategic Climate Change Science Program (CCSP) will promote the development of these directions. In this connection, as Victor (2004) and Kondratyev (2005b) note, first of all, it is necessary to overcome some uncertainties in climate science, among which are the following:

- Inadequate and incomplete consideration of climatic feedbacks (especially concerning the role of clouds).

- Insufficiently studied processes of carbon cycle formation and respective feedbacks (including the "fertilization" effect).
- Imperfection of climate models (despite considerable progress achieved in this sphere) and their application limited mainly by consideration of such climatic parameters as temperature (note also that verification of models is of particular importance).
- Leaving out of account a possibility of a sudden sharp and even catastrophic change of climate similar to those in the past.
- Poorly studied socio-economic aspects of the problems of climate change (the main problem here is to obtain quantitative estimates).

9.5.2 The carbon cycle on land

Initiatives enumerated above will, of course, stimulate heated discussions on the problems of substantiation of the ecological policy strategies. It is clear that the problem of climate change should be considered as an interactive part of a broader problem of global changes of the "atmosphere–hydrosphere–cryosphere–biosphere" system. So far, estimates of the contribution of anthropogenic changes in this system are based only inadequate results of numerical modeling. In this context modeling of the global carbon cycle and assessment of its sinks into different biospheric reservoirs are of key importance.

The level of the stabilizing role of the surface part of the global biogeochemical carbon cycle is estimated with an error of not less than 15–20%, which completely depreciates the idea of the Kyoto Protocol. It is quite apparent now that it is not industrial carbon emissions that prevail in the anthropogenic constituent of carbon flux to the atmosphere. Industrialization and urbanization as characteristic indicators of the present ecodynamics, are followed by deforestation, reconstruction of other types of land cover and introduction into them of structural changes that reduce drastically the protective functions of the environment. Cities, being one of the most substantial and vivid phenomena of the present civilization development, expand constantly their areas due to merging of adjacent territories and increase thereby the number of city dwellers. If in 1970 they constituted 35% of the planetary population, in 2005 this indicator exceeded 50% reaching 3.4 billion people. Cities of the developing countries contribute most (90%) to this process. In the spatial respect, the level of urbanization is characterized by the following indicators: North America – 77.2%, Latin America – 75.3%, Europe – 74.8%, Oceania and Australia – 70.2%, Africa – 37.9%, and Asia – 36.7%.

The growth of the size of the population and its concentration on limited territories of megalopolises leads to a necessity to raise the production of food, which will cause an extension of agricultural areas and pastures – to the detriment of forests. All of this reduces the ability of global biota to assimilate CO_2 from the atmosphere. In this connection, Gitz and Ciais (2004) evaluated the global source of CO_2 due to land use and calculated its dynamics from $1.5\,GtC\,yr^{-1}$ in 1950 to $2.4\,GtC\,yr^{-1}$ in 1990, with a predicted level of $4.2\,GtC\,yr^{-1}$ in 2100. The role of various biomes in this dynamic has been assessed (Table 9.7). It should be borne in

Table 9.7. Key biomes and their areas studied by Gitz and Ciais (2004).

Biome	Region											
	North America, Europe, Japan and Australia			Former Soviet Union and Eastern Europe			Asia			Africa, Latin America, and Middle Asia		
	Years											
	1700	1990	2100	1700	1990	2100	1700	1990	2100	1700	1990	2100
Temperate forests	600	530	419	212	145	157	408	397	346	625	602	537
Boreal forests	778	767	756	1,249	1,249	1,249	114	114	114	61	61	61
Tropical forests	52	44	34	–	–	–	455	269	88	1,126	679	52
Temperate crops	–	243	544	–	211	310	–	38	133	–	20	147
Boreal crops	–	11	39	–	–	–	–	–	–	–	–	–
Tropical crops	–	60	96	–	–	–	–	186	499	–	315	1,498
Temperate pasture and grass	511	336	147	722	578	467	513	485	441	70	73	11
Boreal pasture, tundra	513	513	495	514	514	514	43	43	43	–	–	–
Tropical pasture and grass	625	573	547	–	–	–	309	309	177	2,308	2,440	1,885

mind that deforestation leads also to removal of carbon from its circulation ($\sim$90 Mt C yr^{-1}), which is equivalent to an increase of activity of tropical hurricanes. Forest fires make a non-permanent emission of carbon to the atmosphere and then their role becomes equivalent to deforestation process. On the territory of boreal forests the areas of forest fires during the period 1992–2005 varied from 3.0–24.1 mln ha yr^{-1} with carbon emissions 106–209 Tg C yr^{-1}. In the western part of Russia, forest fires during this period gave 11.1–23.1 t C ha^{-1} yr^{-1}, and in North America – 13.0–21.1 t C ha^{-1} yr^{-1}.

The current monitoring network that regularly records the CO_2 concentration in the atmosphere, makes it possible to connect the observed inter-annual changes of this concentration with other natural–anthropogenic phenomena. Murayama *et al.* (2004) discovered a clear correlation between variations of atmospheric CO_2 concentration and such events as ENSO, volcanic eruptions, fuel, and biomass burning. For instance, the ENSO event initiates droughts in tropical latitudes which are followed by reduced photosynthesis and increased areas of forest fires. A volcanic eruption causes an increase of surface temperature, which enhances soil respiration, leading to the growth of CO_2 concentration in the atmosphere. The spatial level of such correlations depends on the power of the event, and oscillations of CO_2 fluxes can range between 0.5 and 1.0 Gt C ,yr^{-1}. However, observations of CO_2 concentration give insufficient information. Calculations of CO_2 dynamics are needed with the use of models, in which the use of detected correlations, as Ito (2005) showed, simplifies the procedure of climate change forecasting.

Ito (2005) repeated an attempt to calculate the character of interaction between the land carbon cycle and global change of the environment using, like Degermendji and Bartsev (2003), a simplified model of land ecosystem "Sim-Cycle" with a spatial resolution of 0.5°. Six IPCC scenarios of CO_2 emissions have been considered. For each scenario, based on the model of atmospheric circulation AGCM, the levels of forecast uncertainties determined by the socio-economic factors have been assessed. Increases in global photosynthesis and carbon supply differed substantially between scenarios, reaching 24 Pg C (23%) and 81 Pg C (37%), respectively. The prognostic uncertainty in estimation of atmospheric CO_2 concentration remained at the same high level.

The model "Sim-Cycle" describes the process of CO_2 exchange between the atmosphere and land ecosystems, reflecting the growth of plants, accumulation of soil carbon, and response of ecosystems to the global change of the environment. The input data of the model "Sim-Cycle" are photosynthetically active radiation (PAR), ambient CO_2 concentration, time, air temperature and humidity, surface temperature, and wind velocity. The general scheme of the model is described by Ito and Oikawa (2002). The model experiments have demonstrated that the land carbon balance can be substantially changed by global variations of the environmental parameters predicted for the 21st century. It is expected that Gross Primary Production (GPP), on average, can increase from 127.9 Pg C yr^{-1} in the first decade of the 21st century to 156–175.8 Pg C yr^{-1} by the year 2090. In a similar way, changes will take place in autotrophic respiration (AR) from 65.8 PgC yr^{-1} to 82.8–94.3 Pg C yr^{-1}, heterotrophic respiration from 58.2 Pg C yr^{-1} to 73.6–81.2 Pg C yr^{-1}, plant biomass from 515 Pg C to 661–720 Pg C, and soil carbon from 1,479 Pg C to 1,293–1,582 Pg C. On the whole, the general conclusion is that in the 21st century, the land ecosystems will play an important role in the global climate–carbon system, though the levels of uncertainties will remain high (2.6–19.4%).

Murayama *et al.* (2004), using the model CASA (the Carnegie–Ames–Stanford Approach), assessed the role of atmospheric circulation variability in the formation of intra-annual oscillations of the observed growth of CO_2 in the northern hemisphere. It has been established that there is a high correlation between indices of the North Atlantic Oscillation (NAO) and Pacific–North America (PNA). It means that in the case of strong variability of atmospheric circulation it should be taken into account when simulating the changes of CO_2 concentration due to its various sources. Consideration of the role of atmospheric transport is important when explaining the inter-annual variation of the observed growth of CO_2 concentration in middle and high latitudes of the northern hemisphere. Here the problem arises of the impossibility of an optimal location of observational stations to detect changes in CO_2 fluxes from biospheric ecosystems in these latitudes. Murayama *et al.* (2004) used the results of observations at two stations: Alert (82°N, 63°W) and Point Barrow (71°N, 157°W).

The number of global models of the carbon cycle has been recently growing (Cox *et al.*, 2000; Delire *et al.*, 2003; Rayner, 2001). The difference between them is determined by the level of consideration of interactive dynamics of various subsystems of the environment and society. The validity of each model consists in a synthesis of new knowledge. The use of new knowledge reduces the uncertainty of the forecast of global climate change by quantities which do not surpass the available global level of uncertainty. Here one should note the Coupled Climate and Carbon Cycle Model Intercomparison Project (C4MIP) which substantially reduces the uncertainty. As Kondratyev *et al.* (2002a, 2003b, 2004d) have noted, a totality of the results obtained makes it possible to develop a complex global model of the NSS which, in the years to come, should provide highly reliable forecasts. A great obstacle in the way is the highly uncertain scenarios of the socio-economic development of humankind. Here of importance is a parameterization of the role of the socio-economic, ethnical, and religious factors in understanding of the level of danger of natural processes, the development of which is increasingly affected by humans. As Kharitonova (2004) notes, the human world view is formed under the influence of numerous factors of spiritual life, including religious dogmas, shaman traditions, and cultural principles. Consideration of all of these aspects in the developed scenarios will enable one to specify means of further NSS dynamics.

9.5.3 Ozone and natural disasters

Among the debatable problems of assessment of climatic trends, the ozone layer is one of the most important. Syvorotkin (2002) considers this problem within the complex study of the correlation between riftogenesis and degassing of the Earth and its impact on the Earth's dynamic shells. The ozone layer destruction is caused by degassing of the Earth's interior mainly due to the income of hydrogen to the upper atmospheric layers. An interesting feature of the process of deep degassing is its irregularity both in time and in space. The main channels of degassing are in the rift zones whose location correlates with the distribution of ozone anomalies.

It is known that the strongest and most frequent destruction of the ozone layer takes place over the Antarctic. Here the rifts of the southern hemisphere are at a minimum distance from each other, providing an intensive purging of the atmosphere by ozone-destroying gases. A particularly strong destruction of the ozone layer is observed in winter, when in the Antarctic atmosphere prevails a stable cyclonic vortex that reduces the intensity of air masses exchange with the mid-latitude atmosphere.

As Syvorotkin (2002) notes, "...in 1995 the ozone problem became a national problem of Russia", over whose territory an ozone anomaly appeared comparable in its parameters with the ozone holes in the Antarctic. So, in March–May 1997, the monthly mean deficit of the total content of ozone from the Kola Peninsula to Kamchatka reached 40%. Over the territory of Russia, five separate centers of ozone anomalies connected with natural sources of ozone-destroying gases have been recorded. Sanctions against Russia, the signatory of the Montreal Protocol, were mainly connected with this phenomenon, though its substantiation can raise doubts, since one of the possible and most probable causes of the origin of this deficit could be the poorly studied phenomenon of degassing (i.e., the formation of present ozone holes under the influence of fluxes of hydrogen, methane, and nitrogen from the Earth's interior).

Sufficiently stable ozone minima in the northern hemisphere are concentrated over Iceland, the Red Sea, and Hawaii. These locations are the most active sites of the rift systems. The centre of a powerful ozone anomaly, where the monthly mean deficit of total ozone content reaches 30%, is located over the zone of the East Pacific Rise. Here, at the ocean bottom, nine sources of hydrogen connected with high seismic activity have been identified.

The total ozone content decreases sharply at the fronts of cyclones, reaching 4–9 DU in amplitude and 1,000 km spatially (Nerushev, 1995). This indicates a correlation between the state of the ozone layer and weather. Hence, there is a clear dependence between natural anomalous events and distribution of the zones of the Earth's degassing. For instance, with emissions of ozone-destroying gases in highlands in the presence of glaciers and snowfields, the landslip–landslide processes, as well as floods, can be activated due to a direct impact of an excess of IR radiation. In particular, ozone layer destruction over the permafrost zones observed presently threatens dangerous technogenic accidents, and the growth of the flux of methane to the atmosphere with a subsequent enhancement of the greenhouse effect. This effect can appear over the region of the Siberian magnetic anomaly, where monthly mean values of total ozone content sometimes reach quantities close to 60 DU (Danilov and Karol, 1991).

The model proposed by Syvorotkin (2002) to describe the role of the processes of the Earth's degassing in the origin of natural disasters, explains sufficiently reliably (but qualitatively) the causes of the origin of many critical situations in the environment due to changes in the total ozone content as a cause of typhoons, cyclones, forest fires, and catastrophic processes in mountains. For instance, there is a mutual connection between forest fires and the state of the ozone layer. A burning forest emits to the atmosphere aerosol which, through the impact on the ozone layer, changes the regional climate, which leads to a change of air temperature and humidity in the forest. Hence, knowledge of the spatial distribution of the ozone content makes it possible to predict the fire risk over a territory.

The hypothesis of dumping of excess fluid pressure through the world rift structures proposed by Syvorotkin (2002) changes drastically the view of the ozone problem. It means that the transport of ozone-destroying gases to the stratosphere determines in many respects the dynamics of many biogeochemical cycles. Hence, the chemical composition of the atmosphere along with such natural phenomena as lightning discharges which provide an additional influx of NO_x ($\sim$2–6 Tg N yr^{-1}), is determined by the processes in the current epoch of civilization development. Therefore, the participants of the Second World Conference on Disaster Reduction (WCDR) held on 18–22 January 2005 in Kobe (Hyogo, Japan) are correct to show anxiety about a necessity to broaden the sphere of scientific cooperation in analysis of the problems appearing in connection with natural disasters. An expectation of and preparation for them requires not only an interaction of social, economic, and administrative systems but also understanding of the causes of their origin (Rizzolio, 2005).

Thus, one of the possible directions of study for mechanisms of natural disaster origins should be a search for connections between concrete types of precursors of natural disasters and the state of the ozone layer over a given territory.

9.5.4 Aerosol and climate

From the viewpoint of global ecodynamics an interest in studies of the role of atmospheric aerosol has recently grown in the context of its impact on climate formation (Kondratyev, 2005b). One of the indicators of air quality is aerosol optical thickness τ, closely connected to the albedo α of the surface–atmosphere system.

Numerous satellite observations have shown that in the presence of aerosol in the atmosphere the amount of cumulus clouds can decrease to 38% and disappear completely in the presence of a thick layer of smoke aerosol (at $\tau \sim 1.3$). In this case the RF decreases drastically from $-28\,\mathrm{W\,m^{-2}}$ to $+8\,\mathrm{W\,m^{-2}}$. The impact of aerosol on the environment manifests itself through a decrease of surface temperature and evaporation from the forest canopy. The presence of smoke aerosol causes a decrease in the size of cloud droplets and leads to a shift of the onset of precipitation from the 1.5-km level above the lower boundary of the clouds. As a result, the cloud layer albedo increases, but the intensity of cloud formation decreases.

Unfortunately, an estimation of the aerosol RF on climate is very uncertain. This is caused by the strong spatial–temporal variability of aerosol properties. The most general estimates of the impact of aerosol on climate are connected with the notion of atmospheric turbidity when the aerosol characteristics are averaged. Various regions are compared using this integral indicator and conclusions are drawn about heavy pollution of the atmosphere. For instance, the coefficient of atmospheric turbidity in South Africa is 2.6–5.8 times lower than the same coefficient for many European countries, where during the last 10 years it decreased by $\sim$10%.

Properties, processes of formation, and consequences of the impact of atmospheric aerosol on climate have been considered in detail in Kondratyev (2005b). One should especially note the role of long-range transport of aerosol. A comparatively long (up to 2–3 weeks) residence time in the atmosphere determines a possibility of long-range transport. The best studied situations of this kind are emissions to the atmosphere of dust aerosol during dust storms in North Africa and subsequent trans-Atlantic transport of particles, though sometimes meridional fluxes of dust aerosol to Western Europe appear. Another, known situation is the long-range transport of dust aerosol to the north-western region of the Pacific Ocean during dust storms in north-western China and in Mongolia.

Understanding of the key role of aerosol in climate formation has stimulated further development of studies of the long-range aerosol transport. Evidence of this is a new program ITCT-Lagrangian-2k4 to study the intercontinental transport and chemical transformation of aerosol. Parrish and Law (2003) briefly characterized the content of this program aimed at studying the long-range (intercontinental) transport and chemical transformation of aerosol and oxidants, as well as their precursors. To obtain the observational data with this aim in view, it would be expedient to use instruments mounted on the "Lagrangian" platform moving together with the investigated air masses. Practically, only "pseudo-Lagrangian" observations are possible with the use of one or more flying laboratories carrying out multiple soundings of certain air masses. This is the main task of the Intercontinental Transport and Chemical Transportation (ITCT) program. The first stage of its accomplishment consisted of aircraft observations in the region of the North Atlantic in the summer of 2004 to study emissions of aerosol precursors and tropospheric ozone in North America, including an analysis of their long-range transport and chemical transformation over the North Atlantic basin, as well as the subsequent impact on the atmosphere of Western Europe.

The key concrete tasks of the ITCT program include:

(1) determination of potentials of photochemical oxidants and formation of aerosol in polluted air masses formed in North America and coming across the Atlantic Ocean to the region of Western Europe;
(2) analysis of atmospheric dynamics responsible for the long-range transport of pollutants from the planetary boundary layer in North America; and
(3) quantitative characteristic of the transport of North American pollutants to the background atmosphere, their subsequent evolution, and climatic implication.

To study the laws of the intercontinental transport of atmospheric pollutants, Stohl *et al.* (2003) performed a numerical modeling for a period of one year (conditions of the year 2000 have been considered) for six passive tracers emitted on different continents (to characterize the levels of emissions, data were taken of the inventory of carbon oxides emissions, CO). Calculations have shown that emissions from the Asian continent most rapidly propagate vertically, whereas the European emissions remain within the lower layers of the troposphere. The European emissions are transferred mainly to the Arctic, where they contribute most to the formation of the Arctic haze. Minor pollution components (MPCs) come from the continent where emissions take place to another continent in the upper troposphere, as a rule, about 4 days later. After this, MPCs from the lower troposphere can arrive, too. With the characteristic residence time of tracers assumed to be 2 days, local tracers turn out to be prevailing components in the atmosphere over all the continents, except Australia, where the share of the "foreign" tracers constitutes about 20% with respect to the whole mass of tracers. With the tracer residence time assumed to be 20 days, even on the continents with a high level of the "home" emissions the share of tracers from other continents exceeds 50%. In connection with the fact that three regions need special attention, further studies should be focused on three directions of developments:

(1) the wintertime accumulation over Indonesia and the Indian Ocean of tracers coming from Asia;
(2) maximum summertime concentration of the Asian tracers in the Middle East;
(3) distribution over the Mediterranean Sea of tracers coming in summer from North America.

Pre-calculations of the spatial distribution of dust (mineral) aerosol is a difficult problem, in view of the episodic nature of its sources and its long-range transport. Based on the use of the data of satellite remote sensing, Luo *et al.* (2003) compared the observed spatial–temporal variability of aerosol distribution with results of numerical modeling, using the "Match" model of aerosol transport in the atmosphere, taking into account chemical reactions which determine the transformation of its properties, as well as the "Dead" model which simulates the processes of formation and transformation of dust aerosol. Results of comparisons have revealed a good agreement but with some differences. To analyze the reasons of these differences, the dependence has been considered of numerical modeling results on variability of various input meteorological parameters as well as on the choice of the scheme of parameterization of the process of "mobilization" of particles (their getting to the atmosphere as a result of saltation). The sensitivity analysis has led to the conclusion that near Australia the difference between calculated spatial distribution of aerosol optical thickness and the observed one is explained by an inadequacy of data on the wind field near the surface and on the sources of aerosol. In the region of Eastern Asia, this difference is mainly determined by unreliable consideration of meteorological conditions. According to the estimates obtained, total emissions of dust aerosol to the atmosphere as a result of dust storms constitutes 1,654 $Tg\,yr^{-1}$. Most powerful sources of dust aerosol emissions are the African deserts whose contribution to the total content of dust aerosol in the atmosphere reaches 73%. Eastern Asia contributes most to the formation of the field of the dust aerosol content over the northern hemisphere Pacific Ocean, whereas in the southern hemisphere the main source of dust aerosol is Australia. The characteristic residence time of dust aerosol in the atmosphere constitutes 6 days.

Important results characterizing the impact of aerosol on the atmosphere in the process of long-range transport have been obtained in Meloni *et al.* (2003). In May–June 1999 on Lampeduza Island (the Mediterranean Sea: 35.5°N, 12.6°E) the second phase of observations was carried out within the PAUR II field experiment to study the photochemical processes in the atmosphere and UV solar radiation. The surface complex of instruments included the Brewer spectrophotometer (Köhler, 1999), 532-nm aerosol lidar, and multi-channel sun radiometer with shadow screen to measure net and scattered radiation fluxes, as well as aerosol optical thickness at wavelengths 415, 500, 615, 671, 868, and 937 nm. At the same time, aircraft (at altitudes up to 4.5 km) measurements of the actinic flux of short-wave radiation were made in the bands of NO_2 and O_3 photodissociation, which enabled one to assess the level of $O(^{\prime}D)$ formation.

Thus, the long-range aerosol transport (especially trans-oceanic) determines various aerosol forcings on the environment and climate on regional and global scales. In this context, of importance is the pollution of the high-latitude atmosphere caused by the long-range transport of anthropogenic aerosol and MGCs. Available data on the long-range transport of aerosol are fragmentary, and therefore it is necessary to carry out further measurements and modeling of global dynamics of atmospheric aerosols. This necessity is dictated by the continuing increase of instability of the global climate system.

One way of decreasing aerosol generation and lies in the use of nanotechnology. Nanotechnology generated products are used in many sectors of human activity, such as aerospace industries, car manufacturing, chemicals, and medicine. Of course, nanoparticles have specific properties and can effect the human body and the environment with unexpected consequences. Nevertheless, nanotechnology opens the new possibility of changing the anthropogenically caused trends in the global ecodynamics.

9.5.5 Natural disasters attack

Non-linear responses to global change in the environment and effects of multiple interacting factors on the climate system play a significant role in the formation of unstable situations in nature within the NSS subsystem which result in the possibility of "surprises" in the form of natural disasters. The main goal of many investigations of global ecodynamics is in the understanding of how and when non-linear changes in the environment can lead to the beginning of natural variability with a high amplitude. In this context, the potential consequences of global change on the NSS development are to be assessed with acceptable reliability. Solution of this problem lies in the use of the global model of the naure–society system (GMNSS) with wide application of existing means for the synthesis of global datasets.

The proposed approach permits us to take into account at least environmental, geographical, socio-economic, and even political aspects of the problem of global ecodynamics. Ideas raised in this book can help us to form effective indices such as the disaster vulnerability index. It is evident that problems arising here have synergistic character and are frequently seen as mutually exclusive. That is why during the past few years, there have been a lot of discussions about the need for indicators of the effectiveness of disaster risk reduction at all levels. Certainly, creation and use of indicators will be to some extent subjective. Also their effectiveness and informational significance depend on the geographical, socio-economic, and cultural circumstances. Nevertheless, the GIMS-technology application to the problem of natural disaster monitoring permits us to consider the series of informational characteristics that can help to forecast disasters and to assess potential losses. The biocomplexity indicator also corresponds to this criterion. Of course, there exist many other indicators such as response time, degree of preparedness, recovery period, degree of efficiency, cost of the disaster reduction system, etc. Therefore, Davis (2003) and Masure (2003) considered the risk management index combined from the risk identification index, risk reduction index, disaster management index, and financial protection index. Finally, assessing progress in disaster risk reduction is a very difficult task. It is especially true for predictions of when and how a disaster will occur. It is possible only to estimate possibilities of a disaster and preparedness for risk minimization. In this context it is proper to mention again the survivability model (Kondratyev *et al.*, 2002a) which opens a possibility to take into consideration not only environmental circumstances but also trends in general quality of life, livelihood, and the local economy.

As it follows from data that characterize the quantity and intensity of natural catastrophes the losses of humanity represented by the victims and economic damage increases with time. Even in the USA, where environmental monitoring has high efficiency, the losses from natural catastrophes are very high. For example, during 2002 alone natural disasters in the USA affected more than 175,000 people and killed more than 500. Insured losses rose to US$44 bln in 2004 in comparison with 2003 (US$15 bln). In general, 2004 and 2005 were thus the most expensive natural disaster years both in insurance history and regarding the enormous collapses of various kinds. For instance, shortly before the end of 2004, south Asia was hit by one of the most destructive natural calamities of recent decades. All the world shuddered with horror when the gigantic tsunami, triggered by the earthquake off the west coast of Sumatra in the Indian Ocean, reached the densely populated coastlines of Sumatra, Thailand, southern India, Sri Lanka, and the Malvides. Somalia, Kenya, and Tanzania in East Africa were also hit. The earthquake of 26 December 2004 in the Indian Ocean corroborated the theory of Syvorotkin (2002) who proposed to consider all natural catastrophes from the point of view of geological processes in the Earth's interior. An application of this theory and search for constructive correlations between environmental parameters to determine the classes of natural disasters with their timely detection are needed as an aim for the international collaboration between scientists from different fields. Such a collaboration is possible within the International Centre of GIMS (Kondratyev *et al.*, 2002a,b, 2003b, 2004d).

During the last decades the NSS components have been characterized by a certain enhancement. Society continues to increase its negative impact on environmental processes. Nature responds to this growing impact to survive. It is enough to list the extreme events that took place in 2004/2005 to understand that nature is still protected. At the present time science cannot explain with a high degree of certainty which processes had anthropogenic origin or which were a product of the natural dynamics of the Earth's system. It is known that 2004 was the fourth warmest year after 1998, 2002, and 2003 since temperature observations began. Apart from 1996, nine of the last ten years were the warmest years since 1861. Global climate change is a reality. For example, the summer of 2003 in Europe was exceptionally hot. This and other extreme events that took place during the last decades were discussed at the 10th World Climate Summit, which took place in Buenos Aires (Argentina) in December 2004 (*http://weblog.greenpeace.org/climatesummit/ipr_02.html*).

The following chronology of extreme events shows that 2004 and 2005 were in particular dominated by atmospheric events and weather related natural disasters:

- Of the 641 events registered in 2004, 85, 150, and 320 were due to geological hazards (75 damaging earthquakes and 10 volcanic eruptions), floods, and windstorms, respectively. The regional distribution of natural hazards is characterized by the Table 9.8.
- In the early hours of 24 February 2004, the towns of Al Hoceima and Ait Kamara in the north of Morocco were shaken by an M 6.4 earthquake. The losses were characterized by the following parameters:

Table 9.8. Statistics of natural catastrophes in 2004.

Region	Number of loss events	Number of fatalities	Economic losses (US$mln)	Insured losses (US$mln)
Africa	35	1,322	444	0
America	185	4,830	68,183	34,585
Asia	245	176,515	72,706	7,887
Australia/Oceania	52	67	343	124
Europe	124	371	3,765	1,218
Worldwide	641	183,105	145,444	43,815

–at least 650 people were killed;
–hundreds of people were injured;
–thousands of buildings collapsed; and
–economic losses were assessed as US$400 mln.

- The period between February–April 2004 was for New Zealand's North Island destructive due to the thunderstorms, rainstorms, and extended precipitation that produced the severest floods for a hundred years. The main damage was to infrastructure installations, agricultural buildings, and machinery. The overall economic loss came to US$200 mln.
- Tropical Storm Catarine (a category 1 storm) that had arisen in the South Atlantic during 27–29 March 2004 reached the state of Catarine in southern Brazil damaging 40,000 buildings and bringing enormous agricultural losses.
- Hurricane Ivan, 7–21 September 2004, rushed across the Caribean and the southern USA leaving dramatic scenes of destruction. For example, 90% of the buildings on Grenada were destroyed. Buildings on the Cayman Islands, oil rigs in the Gulf of Mexico, and citrus harvests in Florida were subjected to destruction. As a result the economic losses reached US$23 bln (including insured losses US$11 bln). Hurricane Ivan was thus one of the costliest storms in insurance history.
- Hurricane Jeanne, 15–19 September 2004, sped over Haiti and the Dominican Republic with a wind speed of 190 km h^{-1} causing floods, landslides, and mudflows that destroyed villages and claimed the lives of more than 1,800 people.
- One of the most devastating natural disasters of recent decades occurred 26 December 2004 in the Indian Ocean. An *M* 9.0 earthquake triggered a tsunami, whose waves caused devastation over coastal areas thousands of kilometers away. The losses from this disaster were not properly assessed until the end of 2005. More that 170,000 people perished and the fate of more than 100,000 people remain unknown.
- Hurricane Katrina in 23 August 2005 was the most destructive and coastliest tropical cyclone in the history of the USA. It resulted in the storms landfall near New Orleans (Louisiana). The hurricane's storm surge destroyed the construc-

Table 9.9. The largest tsunami catastrophes caused by earthquakes since 1700 (Munich Re, 2005b).

Date of event	Earthquake magnitude	Affected region
21 January 1700	9.0	USA, Japan
1 November 1755	8.7	Portugal, Morocco
24–25 November 1833	9.2	Indonesia, Sumatra, India, Sri Lanka
15 June 1896	8.5	Japan, Sanriku
31 January 1906	8.2	Ecuador, Colombia
27 November 1945	8.3	Pakistan, India
1 April 1946	7.5	USA, Hawaii
4 November 1952	8.2	Russia, Kamchatka
9 March 1957	8.3	USA, Hawaii
22 May 1960	9.5	Chile, USA, Hawaii, Japan
28 March 1964	8.4	USA, Alaska, Hawaii, Japan, Chile
26 December 2004	9.0	Indonesia, Sumatra, Thailand, Maldives, India, Sri Lanka, east Africa

tions that protected New Orleans from Lake Pontchartrain, flooding most of the city. The hurricane also damaged the coastal regions of Alabama, Louisiana, and Mississippi. Hurricane Katrina topped Hurricane Andrew as the most expensive natural disaster in US history. Its maximum winds reached $285\,\mathrm{km\,h^{-1}}$.

- Hurricane Rita, the third most intense hurricane in the history of the USA, began in the Atlantic basin practically immediately after Hurricane Katrina and rapidly grew to a category 5 storm with wind speeds of about $280\,\mathrm{km\,h^{-1}}$. This speed was registered along the central Texas coastline. New Orleans was flooded again.

Following to the reviews of natural catastrophes by Munich Re (2005b) during the last 30 years the conclusions can be drawn that a spectrum of problems connected with natural catastrophes certainly broadened and became more complicated. It was observed that the last years were dominated by extreme atmospheric events and weather related natural catastrophes, both in terms of monetary losses generated by them and the number of events. Table 9.9 shows the timely dynamics of the largest tsunami events as a confirmation of this opinion. It has to be emphasized that new kinds of weather risks arose due to the increase in the frequency and intensity of exceptional weather events in the regions where earlier such events were not observed. For example, Hurricane Alex (3 August 2004) intensified to a category 3 storm on the Saffir–Simpson Scale in the region of 40°N – unusually far from the tropics. A situation arose when the hurricanes began to form off the Brazilian coast. Table 9.10 lists the most powerful hurricanes occurring in 2005.

Table 9.10. List of the most powerful hurricanes in 2005.

Date	Hurricane	Category	Highest wind ($km\,h^{-1}$)	Insured losses (US$ bln)
05–13 July 2005	Dennis	4	235	1–2.5
11–21 July 2005	Emily	4	217	0.14
4–18 August 2005	Irene	2	145	4
23–31 August 2005	Katrina	5	280	34.4
1–10 September 2005	Maria	3	185	3.9
6–18 September 2005	Ophelia	1	135	0.8
18–26 September 2005	Rita	5	278	2.5–5
1–05 October 2005	Stan	1	130	3
15–25 October 2005	Wilma	5	280	6–9
27–31 October 2005	Beta	3	185	2.1

An increase in the number and strength of natural disasters during the last years is also characterized by the following events:

- Japan was hit by ten tropical cyclones between June and October 2004. This number represents a record number for Japan throughout the previous century. Typhoons Chaba, Sougda, and Tokage alone were responsible for economic losses exceeding a total of US$14 bln.
- Florida was hit by four hurricanes in the space of a few weeks in 2004. A record number of storms were registered during 2004 in Florida. There had been only three years since 1866 when Florida was struck by three hurricanes.
- Tropical cyclones that occurred in 2004 in the Atlantic and the West Pacific were particularly devastating. Between them, Hurricanes Charley, Frances, Ivan, and Jeanne within the space of only a few weeks resulted, for the Carribean region and Florida, in overall economic losses amounting to over US$50–60 bln, of which US$20–35 bln was borne by the insurance industry (Munich Re, 2005a,b). Hurricane Charley alone cut across Florida on 13–14 August 2004 with gusts of over 280 $km\,h^{-1}$ and caused economic losses of US$20 bln, of which US$7.5 bln came from insurance companies. Hurricanes Frances, Ivan, and Jeanne caused economic losses of US$4.5 bln, 9 bln and 4.5 bln, respectively.
- During the last few days of November 2004 tropical Storm Winnie unleashed torrential rain over the Phillipines and more than 750 people were killed in the flood waters and landslides.
- For Europe the year 2004 was characterized by extreme winter storms and thunderstorms.
- During 2004 a series of tornados caused numerous destruction not only traditionally in the USA but in Europe (Germany, France, Italy, and the UK). For example, in May 2004 alone, 85 tornados passed over the US Midwest generating over US$1 bln in economic losses.

- In the beginning of 2004 Brazil experienced its worst flood catastrophe of the past 15 years. More than 160 people lost their lives.
- Massive floods were observed in 2004 in Haiti, the Dominican Republic, Bangladesh, India, China, and Nepal. The total loss of life reached 5,200.

A comparison of the decades 1950 and 2004 was made by Munich Re (2005a,b). It demonstrated the increase in natural disasters. A comparison of the last ten years with the 1960s reveals a dramatic increase in the number of events (by 2.3 times), economic losses increased by 7 times, and insurance losses by 15.6 times. The above-mentioned trends of severe natural catastrophe frequencies and intensities are unmistakeable. It is evident that observed tendencies in the expansion of negative phenomena in the environment are often caused by ocean warming. But there are many other causes for such processes including:

- Changes in the atmospheric ozone dynamics (Bojkov, 1987; Kondratyev and Varotsos, 2000).
- Emissions of gases by the interiors of the Earth (Dziewonski and Anderson, 1884; Syvorotkin, 2002).
- Changes in the biogeochemical cycles of greenhouse gases (Kondratyev *et al.*, 2004c; Kondratyev, 2005a).
- Variations in the tectonic vortex structures (Leyborne, 1998).

Natural disasters that happened during the 2004–2005 season have shown that the problem of human survivability has become global. Woodworth *et al.* (2005) discovered that the Indonesian tsunami of Boxing Day 2004 was observed around the world. Tsunami amplitudes were measured in meters, and in some cases the waves were large enough to destroy the tide gauge recording equipment (Merrifield *et al.*, 2005). The period of the waves varied between 20 and 45 minutes and amplitudes along the Atlantic coastlines were often tens of centimeters. Tsunami related signals were really recorded in many remote aquatories of the World Ocean. These signals were registered by the US National Tidal and Sea Level Facility (NTSLF) and can be found online at *www.pol.ac.uk/ntlsf*. Some comments are given in Table 9.11.

Globalization of natural disasters has statistical confirmation. Webster *et al.* (2005) examined the number of tropical cyclones and cyclone days as well as tropical cyclone intensity over the past 35 years in an environment of increasing SST (Table 9.12). The following phenomena were noted:

- Tropical ocean SSTs increased by approximately 0.5°C between 1970 and 2004.
- Globally, the annual number of tropical cyclone days reached a peak of 870 days around 1995, decreasing by 25% to 600 days by 2003.
- The observation that increases in North Atlantic hurricane characteristics have occured simultaneously with a statistically significant positive trend in SST has led to speculation that the changes in both fields are a result of global warming.

Zanetti *et al.* (2005) identified 14 tsunamis (Table 9.13) since 1970 that satisfy the selection criteria in terms of the number of victims or the scale of the property

Table 9.11. Sites where the Sumatra tsunami signal of 26 December 2004 was registered (Woodworth *et al.*, 2005).

Site	Comments
Island Signy of the South Okhney group in the south-west Atlantic	A series of oscillations (~30 cm) in sea level took place just after 20:30 on 26 December 2004 under an average value of 6 cm.
Port Stanley	Sea level rose by 9 cm at nearly 23:00, fell 24 cm just after 23:00 and rose 19 cm shortly before midnight. A tsunami signal arrived at Port Stanley after 19–20 hours.
St. Helena	Small tsunami signal was registered at about 02:00 on 27 December with the largest sea level excursion of 3 cm at 06:00.
Gibraltar station	A real tsunami signal did not enter the Mediterranean.
Newlyn (the UK tide guage station to the open ocean)	Some increase in variability (±10 cm) of sea level was recorded at 14:00 on 27 December 2004.

Table 9.12. Change in the number and percentage of hurricanes in categories 4 and 5 for the 15-year periods 1975–1989 and 1990–2004 for different ocean basins (Webster *et al.*, 2005).

Basin	Period			
	1975–1989		1990–2004	
	Number	Percentage	Number	Percentage
East Pacific Ocean	36	25	49	35
West Pacific Ocean	85	25	116	41
North Atlantic	16	20	25	25
Southwestern Pacific	10	12	22	28
North Indian Ocean	1	8	7	25
South Indian Ocean	23	18	50	34

damage caused following an earthquake. We see that in the event of an earthquake with a magnitude of over 7.0, the effects of the resulting tsunami can be observed over entire oceans. The waves often cause damage in places thousands of kilometers from the source.

The coupled nature of these and other processes is a decisive factor of the observed global ecodynamics. The success of relevant studies can be achieved within the framework of new models of thinking based on GIMS-technology (Kondratyev *et al.*, 2002b). Investigations of isolated processes being responsible for global ecodynamics cannot be productive. This truth has been again confirmed by the series of natural disasters that happened during 2004 and 2005.

Table 9.13. List of severe tsunamis since 1970 (Zanetti *et al.*, 2005).

Date	Affected country	Earthquake magnitude	Victims	
			Dead or missing	Injured
17 August 1976	Philippines	7.9	3,739	800
26 May 1983	Japan	7.7	100	na
1 September 1992	Nicaragua	7.0	320	na
11 December 1992	Indonesia	6.8	2,484	na
12 July 1993	Japan	7.8	239	233
8 August 1993	Guam	8.1	0	71
19 January 1994	Indonesia	6.8	6	300
3 June 1994	Indonesia	5.9	235	440
4 October 1994	Japan, Russia	8.2	5	1,500
15 November 1994	Columbia	6.7	74	135
17 July 1998	Papua New Guinea	7.1	2,183	1,000
4 May 2000	Indonesia	7.0	46	246
23 June 2001	Peru, Bolivia, Chile	8.3	145	2,713
26 December 2004	Indonesia, Thailand, Sri Lanka, India ...	9.0	283,000	125,000

Bibliography

Abrahamson, D. E. (1989). *Challenge of Global Warming* (376 pp.). Island Press, Washington, DC.

A Guide to World Resources 2000–2001 (2000). *People and Ecosystems: The Frying Web of Life* (389 pp). World Resources Institute, Washington, DC.

Ahmad, A., Shfiee, M., Hassan, F., and Yaakub, A. (2004). Flood mapping using Radarsat SAR data: A Malaysian experience. *Proceedings of 25th Asian Conference on Remote Sensing, 22–26 November, Chiang Mai,Thailand* (pp. 587–595). AARS, Chiang Mai, Thailand.

Aladin, N. V. and Kuznetsov, L. A. (1990). The present state of the Aral Sea under conditions of growing saltiness. *Proc. Ecological Inst. of the USSR Acad. Sci., Leningrad*, **223**, 123–130 [in Russian].

Alcano, J., Leemans, R., and Kreileman, E. (eds) (2001). *Global Change Scenarios of the 21st Century* (232 pp.). Elsevier, Amsterdam.

Alessio, S., Longhetto, A., and Richiardone, R. (2004). Evolutionary spectral analysis of European climatic series. *Il Nuovo Cimento C.*, **27**, Ser. 2, No. 1, 73–98.

Alimov, A. F., Dmitriyev, V. V., Florinskaya, T. M., Khovanov, N. V., and Chistobayev, A. I. (1999). *Integral Assessment of the Ecological State and Environmental Quality of Urban Territories* (253 pp.). St Petersburg Sci. Centre of RAS, St Petersburg [in Russian].

Allen, M. R., Stott, P. A., Mitchell, J. F. B., Schnur, R., and Delworth, T. L. (2000). Uncertainty in forecasts of anthropogenic climate change. *Nature*, **407**, 617–620.

Alverson, K. (2000). Long-term biogeophysical controls on the carbon cycle and their relevance to human concerns. *International Human Dimensions Programme Update*, **3**, 1–4.

Andronova, N. G. and Schlesinger, M. E. (2001). Objective estimation of the probability density function for climate sensitivity. *J. Geophys. Res.*, **106**(D19), 22605–22611.

Aota, M., Shirasawa, K., Krapivin, V. F., and Mkrtchyan, F. A. (1993). A project of the Okhotsk Sea GIMS. *Proceedings of the Eighth International Symposium on Okhotsk Sea and Sea Ice and ISY/Polar Ice Extent Workshop, 1–5 February, Mombetsu, Japan* (pp. 498–500). Okhotsk Sea and Cold Ocean Research Association, Mombetsu, Japan.

Aripov, S. L. (1973). Aral Sea water balance constituents and their impact on multi-year oscillations of its level. *Water Resources*, **5**, 29–40 [in Russian].

Arsky, Yu. M., Zakharov, Yu. F., and Kalutskov, V. A. (1992). *Ecoinformatics* (520 pp.). Hydrometeoizdat, St Petersburg [in Russian].

Bacastow, R. (1981). Numerical evaluation of the evasion factor. *Carbon Cycle Modelling* (SCOPE-16, pp. 95–101). John Wiley & Sons, New York.

Baibakov, S. N. and Martynov, A. I. (1976). *From the Satellite's Orbit into the Eye of the Typhoon* (176 pp.). Nauka, Moscow [in Russian].

Banerjee, P., Pollitz, F. F., and Bürgmann, R. (2005). The size and duration of the Sumatra–Andaman earthquake from far-field static offsets, *Science*, **308**(5729), 1769–1772.

Barenbaum, A. S. (2002). *Galaxy. Solar System. The Earth: Subordinate Processes and Evolution* (393 pp.). GEOS, Moscow [in Russian].

Barenbaum, A. S. (2004). Mechanism for the formation of gas and oil accumulation. *Annals of Acad. Sci.*, **399**(6), 1–4 [in Russian].

Barnett, V. (2003). *Environmental Statistics* (320 pp.). John Wiley & Sons, London.

Bartsev, S. I., Degermendji, A. G., and Erokhin, D. V. (2003). Global generalized models of carbon dioxide dynamics. *Problems of the Environment and Natural Resources*, **12**, 11–28 [in Russian].

Bazhin, N. M. (2000). Methane emission from a residual layer. *Second International Methane Mitigation Conference, 18–23 June, Novosibirsk* (pp. 231–236). Novosibirsk State University, Novosibirsk, Russia.

Bazilevich, N. I. and Rodin, L. E. (1967). Schematic maps of the productivity and biological cycle of first-priority-type land vegetation. *Proc. All-Union Geograph. Soc.*, **99**(3), 190–194 [in Russian].

Beeby, A. and Brennan, A.-M. (2003). *First Ecology: Ecological Principles and Environmental Issues* (352 pp.). Oxford University Press, Oxford, UK.

Bekoriukov, V. I. and Fedorov, V. V. (1987). An empirical model of the total ozone content over the Southern Hemisphere. *Meteorology and Hydrology*, **3**, 47–53 [in Russian].

Bellman, R. and Dreifus, S. (1965). *Applied Problems of Dynamic Programming* (457 pp.). Nauka, Moscow [in Russian].

Bellman, R. and Rous, R. S. (1971). Method of analysis of a broad class of biological systems. *Cybernetic Problems of Bionics* (pp. 158–169). Mir, Moscow [in Russian].

Bengtsson, L. (1999). *Climate Modelling and Prediction: Achievements and Challenges* (Publication 954, pp. 59–73). World Climate Research Programme, World Meteorological Organization, Geneva.

Berg, L. S. (1908). *The Aral Sea* (580 pp.). Hydrometeoizdat, St Petersburg [in Russian].

Bernard, E. (ed.) (2005). *Developing Tsunami-resilient Communities: The National Tsunami Hazard Mitigation Program* (Vol. VI, 186 pp.). Springer-Verlag, Heidelberg, Germany.

Berner, U. and Hollerbach, A. (2001). Klimasystem Erde – Überschätzen wir das Kohlenoxid? *Stahl und Eisen*, **121**(10), 35–40.

Berz, G. (1999). Catastrophes and climate change: Concerns and possible countermeasures of the insurance industry. *Mitigation and Adaptation Strategies for Global Change*, **4**(3/4), 283–293.

Bgatov, V. I. (1988). *The History of Oxygen of the Terrestrial Atmosphere* (87 pp.). Nedra, Moscow [in Russian].

Bilham, R. (2005). A flying start, then a slow slip. *Science*, **308**, 1126–1127.

Binenko, V. I., Khramov, G. N., and Yakovlev, V. V. (2004). *Extreme Situations in the Modern World and Their Threats to Life* (400 pp.). Scientific Centre for the Ecological Safety of RAS, St Petersburg [in Russian].

Biutner, E. K. (1986). *Planetary Gas Exchange* (240 pp.). Hydrometeoizdat, Leningrad [in Russian].

Björkstrom, A. (1979). A model of CO_2 interaction between atmosphere, ocean, and land biota. *Global Carbon Cycle* (SCOPE-13, pp. 403–458). John Willey & Sons, New York.

Blackmore, P. and Tsokri, E. (2004). Windstorm damage to buildings and structures in the UK during 2002. *Weather*, **59**(12), 336–339.

Blowers, A. and Hinchliffe, S. (2003). *Environmental Responses* (312 pp.). John Wiley & Sons, London.

Bodenbender, J., Wassmann, R., Papen, H., and Rennenberg, H. (1999). Temporal and spatial variation of sulfur–gas transfer between coastal marine sediments and the atmosphere, *Atmos. Env.*, **33**(21), 3487–3502.

Bodri, L. and Čermák, V. (1999). Climate change of last millennium inferred from borehole temperatures: Regional patterns of climate changes in the Czech Republic, Part III. *Glob. Planet. Change*, **21**(4), 225–235.

Boehmer-Christiansen, S. (1997). Who is driving climate change policy? *IEA Stud. Educ.*, **10**, 53–72.

Boehmer-Christiansen, S. (2000). Who determines the policy concerning climate change and how is it determined? *Izv. Russ. Geogr. Soc.*, **132**(3), 6–22 [in Russian].

Bohle, H. (2001). Vulnerability and criticality: Perspectives from social geography. *International Human Dimensions Programme Update*, **2**, 231–239.

Bojkov, P. D. (1987). The 1983 and 1985 anomalies in ozone distribution in perspective. *Monthly Weather Review*, **115**(10), 2187–2201.

Bolin, B. (1998). *The WCRP and IPCC: Research Inputs to IPCC Assessments and Future Needs* (Publication 904, pp. 27–36). World Climate Research Programme, World Meteorological Organization, Geneva.

Bolin, B. (1999). *Global Environmental Change and the Need for International Research Programmes* (Publication 954, pp. 11–14). World Climate Research Programme, World Meteorological Organization, Geneva.

Bond-Lamberty, B., Wang, C., and Gower, S. T. (2003). Annual carbon flux from woody debris for a boreal black spruce fine chronosequence. *J. Geophys. Res.*, **108**(3), WFX1/1-WFX1/10.

Bondur, V. G., Kondratyev, K. Ya., Krapivin, V. F., and Savinykh, V. P. (2005). Problems of monitoring and prediction of natural disasters. *Research of the Earth from Space*, **1**, 3–14 [in Russian].

Borodin, L. F. and Krapivin, V. F. (1998). Remote measurements of the Earth's surface characteristics. *Problems of the Environment and Natural Resources*, **7**, 38–54 [in Russian].

Borodin, L. F., Krapivin, V. F., Krylova, M. S., Kuznetsov, N. T., Kulikov, Yu. N., and Minayeva, E. N. (1982). Multi-purpose flying laboratories to monitor the zones of impact of irrigation systems. *Geography and Natural Resources*, **3**, 31–37 [in Russian].

Borodin, L. F., Krapivin, V. F., and Bui T. L. (1996). Application of GIMS technology to monitor the Aral–Kaspiy aquageosystem. *Problems of the Environment and Natural Resources*, **10**, 46–61 [in Russian].

Borodin, L. F., Krapivin, V. F., Berezin, Yu. V., Levshin, I. P., and Chernikov, A. A. (1998). Ideal and sub-ideal sensors of composite-coded phase-manipulated signals. *Foreign Radioelectronics*, **8**, 15–22 [in Russian].

Bortnik, V. N. and Chistiayeva, S. P. (1990). *The Aral Sea* (195 pp.). Hydrometeoizdat, Leningrad [in Russian].

Bortnik, V. N. and Dauletiyarov, K. Zh. (1985). *Numerical Modelling of Circulation of Aral Sea Waters* (Preprint, 36 pp.). Computer Centre, USSR Academy of Sciences, Moscow [in Russian].

Bortnik, V. N., Lopatina, S. A., and Krapivin, V. F. (1994). A simulation system to study the Aral Sea's hydrophysical fields. *Meteorology and Hydrology*, **9**, 102–106 [in Russian].

Borwein, J., Bailey, D., and Girgensohn, R. (2004). *Experimentation in Mathematics: Computational Paths to Discovery* (29 pp.). A. K. Peters, Natick, MA.

Boucher, O. (2002). Aerosol radiative forcing and related feedbacks: How do we reduce uncertainties? *IGACtivities Newsletter*, **26**, 8–12.

Bounoua, L., Defries, R., Collatz, G. J., Sellers, P., and Khan, H. (2002). Effects of land cover conversion on surface climate. *Clim. Change*, **52**(1–2), 29–64.

Bove, M. And Thráinsson, H. (eds) (2003). *Topics: Annual Review of North American Catastrophes 2002* (50 pp.). American Re, Princeton, NJ.

Bowen, D. Q. (2000). Tracing climate evolution. *Earth Heritage Magazine*, Millennium Issue, 8–9.

Boysen, M. (ed.) (2000). *Biennial Report 1998 and 1999* (130 pp.). Potsdam Institute for Climate Impact Research, Potsdam, Germany.

Bozhinsky, A. N. and Losev, K. S. (1987). *Fundamentals of Avalanche Formation* (280 pp.). Hydrometeoizdat, Leningrad [in Russian].

Braesicke, P. and Pyle, J. A. (2004). Sensitivity of dynamics and ozone to different representations of SSTs in the United Model. *Quart. J. Roy. Meteorol. Soc.*, **130**, Part B, No. 601, 2033–2045.

Bras, R. L. (1990). *Hydrology* (643 pp.). Addison-Wesley, New York.

Braswell, B. H., Schimel, D. S., Privette, J. L., Moore, B., Emery, W. J., Sultzman, E. W., and Hudak, A. T. (1996). Extracting ecological and biophysical information from AVHRR optical data: An integrated algorithm based on inverse modeling. *J. Geophys. Res.*, **101**(D18), 23335–23348.

Braun, R. A., Todd, R. M., and Wallace, N. (1999). A general equilibrium interpretation of damage-contingent securities. *Journal of Risk and Insurance*, **66**(4), 583–595.

Brebbia, C. A. (ed.) (2004). *Risk Analysis* (Vol. IV, 832 pp.). WIT Press, Southampton, UK.

Breon, F.-M., Tanre, D., and Generoso, S. (2002). Aerosol effect on cloud droplet size monitored from satellite. *Science*, **295**(5556), 834–838.

Brown, L. R. (ed) (2004). *State of the World 2004*. Worldwatch Institute, Washington, DC, 245 pp.

Brown, L. R., Flavin, C., French, H., Sampat, P., Matton, A., Dunn, S., Sheehan, M. O., Abramovitz, J. N., Roodman, D. M., Gardner, G., and Masthy, L. (2001). *State of the World 2001* (275 pp.). Earthscan, London.

Brown, R. A., Kaufman, C. A., and MacGorman, D. R. (2002). Cloud-to-ground lightning associated with the evolution of a multicell storm. *J. Geophys. Res.*, **107**(D19), ACL13/1–ACL13/13.

Bulkeley, H. and Betsill, M. M. (2003). *Cities and Climate Change: Urban Sustainability and Global Environmental Governance* (237 pp.). Routledge, London.

Bukatova, I. L. and Makrusev, V. V. (2004). *Theory of Integral–Evolutionary Intellectualization of Social Systems* (126 pp.). Moscow Institute for the National and Cooperative Management Publ., Moscow [in Russian].

Bukatova, I. L., Mikhasev, Yu. I., and Sharov, A. M. (1991). *Evoinformatics: Theory and Practice of Evolutionary Modelling* (206 pp.). Nauka, Moscow [in Russian].

Bunyard, P. (1999). Eradicating the Amazon rainforest will wreak havoc on climate. *Ecologist*, **29**(2), 81–84.

Burt, C. C. (2004). *Extreme Weather: A Guide and Record Book* (304 pp.). W. W. Norton, New York.

Byakola, T. (2000). Technological options and policy measures for methane mitigation in Uganda: Possibilities and limitations. *Second International Methane Mitigation Conference, 18–23 June, Novosibirsk, Russia*, pp. 95–100.

Canadel, I. G., Dickinson, R., Hibbard, K., Raupach, M., and Young, O. (eds) (2003). *Global Carbon Project: The Science Framework and Implementation* (Report No. 1, 69 pp.). Earth System Science Partnership, Canberra.

Carlsson, H., Aspegren, H., Lee, N., and Hilmer, A. (1997). Calcium phosphate in biological phosphorus removal systems. *Water Research*, **31**(5), 1047–1055.

Carpenter, G. (2001). *Natural Hazards:Review of the Year 2000* (17 pp.). The CAT-i Service, London.

Chahine, M. T. (1992). The hydrological cycle and its influence on climate. *Nature (UK)*, **359**(6394), 373–380.

Chang, C.-P., Zhang, Y., and Li, T. (2000). Interannual and interdecadal variations of the East Asian summer monsoon and tropical Pacific SSTs, Part 1: Roles of the subtropical ridge. *J. Climate*, **13**, 4310–4325.

Changnon, S. A. (ed.) (1996). *The Great Flood of 1993: Causes, Impacts, and Responses* (321 pp.). Westview Press, Boulder, CO.

Changnon, S. A. (2000). Flood prediction: Immersed in the quagmire of national flood mitigation strategy. In: D. Sarewitz, R. A. Pielke, Jr, and R. Byerly (eds), *Prediction: Science, Decision Making, and the Future of Nature* (pp. 85–106). Island Press, Washington, DC.

Changnon, S. A. (2001). *Thunderstorms across the Nation: An Atlas of Storms, Hail, and Their Damage in the 20th Century* (93 pp.). Changnon Climatologist and Office of Global Programs, National Oceanic and Atmospheric Administration, Washington.

Chen, J. M., Liu, J., Leblanc, S. G., Lacaze, R., and Roujean, J.-L. (2003). Multi-angular optical remote sensing for assessing vegetation structure and carbon absorption. *Remote Sensing of Environment*, **84**(5), 516–525.

Chernavsky, D. S. (ed.) (2004). *Recognition, Autodiagnostics, Thinking: Synergetics and Human Science* (272 pp.). Radiotekhnika, Moscow [in Russian].

Chernenko, D. S. (1981). Modelling the filtering of artesian water to the Aral Sea hollow. *High School Publ.: Geology and Exploring*, **10**, 82–88 [in Russian].

Chinlon, L. (ed.) (1989). *Optoelectronic Technology and Lightwave Communications Systems* (766 pp.). Van Nostrand Reinhold, New York.

Chock, D. P. and Winkler, S. L. (2000). A trajectory-grid approach for solving the condensation and evaporation equations of aerosols. *Atmospheric Environment*, **34**(18), 2957–2973.

Chou, M.-D., Chan, P.-K., and Wang, M. (2002). Aerosol radiative forcing derived from SeaWiFS-retrieved aerosol optical properties. *J. Atmos. Sci.*, **59**(3), 748–757.

Christensen, O. B. and Christensen, J. H. (2004). Intensification of extreme European summer precipitation in a warmer climate. *Global and Planetary Change*, **44**(1–4), 107–117.

Christian, T. J., Kleiss, B., Yokelson, R. J., Holzinger, R., Crutzen, P. J., Hao, W. M., Saharjo, B. H., and Ward, D. E. (2003). Comprehensive laboratory measurements of biomass-burning emissions, 1: Emissions from Indonesian, African, and other fuels. *J. Geophys. Res.*, **108**(D23), ACH3/1–ACH3/13.

Christopher, S. A. and Zhang, J. (2002). Daytime variation of shortwave direct radiative forcing of biomass burning aerosols from GOES imager. *J. Atmos. Sci.*, **59**(3), Part 2, 681–691.

Christy J.R. and Spencer R.W. (2003). Reliability of satellite data sets. *Science*, **301**(5636), 1046–1047.

Chronis, T. G. and Anagnostou, E. N. (2003). Error analysis for a long-range lightning monitoring network of ground-based receivers in Europe. *J. Geophys. Res.*, **108**(D24), ACL8/1–ACL8/10.

Chukhlantsev, A. A., Golovachev, V. P., Krapivin, V. F., and Shutko, A. M. (2004). A remote sensing-based modelling system to study the Aral–Caspian water regime. *Proceedings of the 25th Asian Conference on Remote Sensing, 22–26 November, Chiang Mai, Thailand* (Vol. 1, pp. 506–511). AARS, Chiang Mai, Thailand.

Clark, W. C. and Dickinson, N. M. (2003). Sustainability science: The emerging research program. *NAS Online*, **10**, 1–5.

Clark, W. C., Crutzen, P. J., and Schellnhuber, H. J. (2005). *Science for Global Sustainability: Toward a New Paradigm* (Working Paper No. 120, 32 pp.). Center for International Development, Cambridge, MA.

Coakley, J. A., Jr and Walsh, C. D. (2002). Limits to the aerosol indirect radiative effect derived from observations of ship tracks. *J. Atmos. Sci.*, **59**(3), Part 2, 668–680.

Coconea, G. (2000). Methane gas emissions from the Romanian natural gas transport system. *Second International Methane Mitigation Conference, 18–23 June, Novosibirsk* (pp. 297–302). Novosibirsk State University, Novosibirsk, Russia.

Coen, J., Mahalingam, S., and Daily, J. (2004). Infrared imagery of crown fire dynamics during FrostFire. *J. Appl. Meteorology*, **43**, 1241–1259.

Cohard, J.-M., Pinty, J.-P., and Suhre, K. (2001). On the parameterization of activation spectra from cloud condensation nuclei microphysical properties. *J. Geophys. Res.*, **105**(D9), 11753–11766.

Collatz, G. J., Berry, J. A., Farquhar, J. A., and Pierce, J. (1990). The relationship between the Rubisco reaction mechanism and models of leaf photosynthesis. *Plant Cell Environment*, **13**, 219–225.

Collatz, G. J., Ball, G. J., Grivet, J. T., and Berry, J. A. (1991). Physiological and environmental regulation of stomatal conductance, photosynthesis and transpiration: A model that includes a laminar boundary layer. *Agricultural and Forest Meteorology*, **54**, 107–136.

Collatz, G. J., Ribas-Carbo, M., and Berry, J. A. (1992). Couples photosynthesis stomatal conductance model for leaves of C_4 plants. *Aust. J. Plant Physiol.*, **19**, 519–538.

Collatz, G. J., Bounoua, L., Los, S. O., Randall, D. A., Fung, I. Y., and Sellers, P. J. (2000). A mechanism for the influence of vegetation on the response of the diurnal temperature range to changing climate. *Geophys. Res. Lett.*, **27**(20), 3381–3384.

Condie, K. C. (2005). *Earth as an Evolving Planetary System* (461 pp.). Elsevier Academic, Burlington, MA.

Cox, P. M., Betts, R. A., Jones, G. D., Spall, S. A., and Totterdell, I. J. (2000). Acceleration of global warming due to carbon-cycle feedbacks in a coupled climate model. *Nature*, **408**, 184–187.

Crawford, J., Davis, D., Olson, J., Chen, G., Liu, S., Fuelberg, H., Hannan, J., Kondo, Y., Anderson, B., Gregory, G. *et al.* (2000). Evolution and chemical consequences of lightning-produced NO_x observed in the North Atlantic upper atmosphere. *J. Geophys. Res.*, **105**(D15), 19795–19809.

Crichton, M. (2005). *State of Fear* (503 pp.). Harper-Collins, London.

Crowley, T. J. (2000). Causes of climate change over the past 1000 years. *Science*, **289**(5477), 270–277.

Csiszar, I., Abuelgasim, A., Li, Z., Jin, J.-Z., Fraser, R., and Hao, W.-M. (2002). Interannual changes of active fire detectability in North America from long-term records of the advanced very high resolution radiometer. *J. Geophys. Res.*, **108**(D2), ACL19/1–ACL19/10.

Danilov, L. D. and Karol, I. L. (1991). *Atmospheric Ozone: Sensation and Reality* (121 pp.). Hydrometeoizdat, Leningrad [in Russian].

Davis, C. A. and Bosart, L. F. (2004). Forecasting the tropical transition of cyclones. *Bull. Amer. Meteorol. Soc.*, **85**(11), 1657–1662.

Davis, I. (2003). The effectiveness of current tools for the identification, measurement, analysis and synthesis of vulnerability and disaster risk. In: O. D. Cardona (ed.), *IDB/IDEA Program on Indicators for Disaster Risk Management* (pp. 1–53). Universidad Nacional de Colombia, Manizales.

De Boer, J. Z. and Sanders, D. T. (2004). *Earthquakes in Human History: The Far-Reaching Effects of Seismic Disruptions* (264 pp.). Princeton University Press, Princeton, NJ.

De Boer, J. Z. and Sanders, D. T. (2005). *Volcanoes in Human History* (317 pp.). Princeton University Press, Princeton, NJ.

Degermendji, A. G. and Bartsev, S. I. (2003). Global small-size models of biospheric dynamics and stability. *Problems of the Environment and Natural Resources*, **7**, 32–34 [in Russian].

Delire, C., Foley, J. A., and Thompson, S. (2003). Evaluating the carbon cycle of a coupled atmosphere–biosphere model. *Global Biogeochemical Cycles*, **17**(1012), doi: 10.1029/2002GB001870.

Del Frate, F., Ferrazzoli, P., and Schiavon, G. (2003). Retrieving soil moisture and agricultural variables by microwave radiometry using neural networks. *Remote Sensing of Environment*, **84**(2), 174–183.

Delgado, J. P. (1998). *Encyclopedia of Underwater and Maritime Archaeology* (135 pp.). New Haven, London.

Delworth, T. L. and Knutson, T. R. (2000). Simulation of early 20th century global warming. *Science*, **287**(5461), 2246–2250.

Dementjeva, T. V. (2000). Emission of gases from peat-bog ecosystems. *Proceedings of Second International Methane Mitigation Conference, 18–23 June, Novosibirsk* (pp. 223–226). Novosibirsk State University, Novosibirsk, Russia.

Demirchian, K. S. and Kondratyev, K. Ya. (1998). Development of energetics and the environment. *Proc. of RAS, Energetics*, **6**, 3–27 [in Russian].

Demirchian, K. S. and Kondratyev, K. Ya. (2004). Global carbon cycle and climate. *Proc. of the Russian Geographical Society*, **136**(1), 16–25 [in Russian].

Demirchian, K. S., Demirchian, K. K., Danilevich, Ya. B., and Kondratyev, K. Ya. (2002). Global warming, energetics, and geopolitics. *Proc. RAS, Energetics*, **3**, 221–235 [in Russian].

De Rosnay, P., Bruen, M., and Polcher, J. (2000). Sensitivity of surface fluxes to the number of layers in the soil model used in GCMs. *Geophys. Res. Lett.*, **27**(20), 3329–3332.

Derwent, R. G., Collins, W. J., Johnson, C. E., and Stevenson, D. S. (2001). Transient behaviour of tropospheric ozone precursors in a global 3D-CTM and their indirect greenhouse effects. *Clim. Change*, **49**(4), 463–487.

Diadin, Yu. A. and Gushchin, A. D. (1998). Gas hydrates. *Soros Educational Journal. Biology. Chemistry. Earth sciences. Physics, Mathematics*, **3**, 55–64 [in Russian].

Diamond, J. (2004). *Collapse: How the World Ends. How Societies Choose to Fail or Succeed* (575 pp.). Golden Penguin Audio Books, New York.

Diamond, J. (2005). *Collapse: How Societies Choose to Fail or Succeed* (575 pp.). Viking, London.

Dilley, M., Chen, R. S., Deichmann, U., Lerner-Lam, A. L., and Arnold, M. (2005). *Natural Disaster Hotspots: A Global Risk Analysis* (132 pp.). World Bank, New York.

Dole, R. M. (2005). The May 2003 extended tornado outbreak. *Bull. Amer. Meteorol. Soc.*, **86**(4), 531–542.

Dong, J., Kaufmann, R. K., Myneni, R. B., Tucker, C. J., Kauppi, P. E., Loski, J., Buermann, W., Alexeyev, V., and Hughes, M. K. (2003). Remote sensing estimates of boreal and temperate forests woody biomass: Carbon pools, sources, and sinks. *Remote Sensing of Environment*, **84**(3), 393–410.

Dore, S. E., Likas, R., Sadler, D. W., and Karl, D. M. (2003). Climate-driven changes to the atmospheric CO_2 sink in the subtropical North Pacific Ocean. *Nature (UK)*, **424**(6950), 754–757.

Douglass, D. H., Blackman, E. G., and Knox, R. S. (2004). Temperature response of Earth to the annual solar irradiance cycle. *Physics Letters A*, **323**(3–4), 315–322.

Dufour, L. and Defay, R. (1963). *Thermodynamics of Clouds* (255 pp.). Academic Press, New York.

Dukhovny, V. A. and Stulina, G. (2001). Strategy of trans-boundary return flow use in the Aral Sea basin. *Desalination*, **139**, 299–304.

Dulnev, G. N. and Ushakovskaya, E. D. (1988). Analysis of the influence exerted by physico-geometric parameters on the temperature field of an object. *J. of Engineering Physics and Thermophysics*, **57**(6), 1487–1492.

Duncan, B. N., Martin, R. V., Staudt, A. C., Yevich, R., and Logan, J. A. (2003). Inter-annual and seasonal variability of biomass burning emissions by satellite observations. *J. Geophys. Res.*, **108**(D2), ACH1/1–ACH1/11.

Dziewonski, M. A. and Anderson, D. L. (1884). Seismic tomography of the Earth's interior. *American Scientist*, **5**, 483–493.

Edmonds, J., Joos, F., Nakicenovic, N., Richels, R. G., and Sarmiento, J. L. (2004). Scenarios, targets, gaps, and costs. In: C. B. Field and M. R. Raupach (eds), *Global Carbon Cycle: Integrating Humans, Climate, and the Natural World* (pp. 77–102). Island Press, Washington, DC.

Edward, B. (2005). *Natural Hazards* (328 pp.). Cambridge University Press, Cambridge, UK.

Efremov, D. F. and Sapozhnikov, A. P. (1997). Far East forest biodiversity and succession dynamics. *Proceedings of the IGBP Siberian Transect Workshop "Spatial–Temporal Dimension of High-Latitude Ecosystem Change", 1–7 September, Krasnoyarsk* (pp. 13–14). V.N. Sukachev Institute of Forest Siberian Branch, Russian Academy of Science.

Egan, W. G., Hogan, A. W., and Zhu, H. (1991). Physical variation of water vapor, and the relation with carbon dioxide. *Geophys. Res. Lett.*, **18**(12), 2245–2248.

Ehhalt, D. H. (1981). Chemical coupling of the nitrogen, sulphur, and carbon cycles in the atmosphere. In: G. E. Likens (ed.), *Some Perspective of the Major Biogeochemical Cycles* (pp. 81–91). Elsevier, Amsterdam.

Ehleringer, J. R., Cerling, T. E., and Dearing, M. D. (eds) (2005). *A History of Atmospheric CO_2 and Its Effects on Plants, Animals, and Ecosystems* (548 pp.). Springer-Verlag, New York.

EPA (2001). *Non-CO_2 Greenhouse Gas Emissions from Developed Countries: 1990–2010* (EPA-430-R-01-007, 79 pp.). US Environmental Protection Agency, Washington, DC.

Erust, W. G. (ed.) (2000). *Earth Systems: Processes and Issues* (566 pp.). Cambridge University Press, Cambridge, UK.

Essenhigh, R. H. (2001). Does CO_2 really drive global warming? *Chem. Innovation*, **31**(5), 44–46.

Essex C. and McKitrick R. (2002). *Taken by Storm. The Troubled Science, Policy and Politics of Global Warming*. Key Porter Books, Toronto, 320 pp.

Feingold, G., Remer, L., Ramaprasad, J., and Kaufmann, Y. J. (2001). Analysis of smoke impact on clouds in Brazilian biomass burning region: An extension of Twomey's approach. *J. Geophys. Res.*, **106**(D19), 22907–22922.

Field, C. B. and Raupach, M. R. (eds) (2004). *Global Carbon Cycle: Integrating Humans, Climate, and the Natural World* (584 pp.). Island Press, Washington, DC.

Field, J. G., Hempel, G., and Summerhayer, C. P. (eds) (2002). *Oceans 2020: Science Trends and the Challenge of Sustainability* (296 pp.). Island Press, Washington, DC.

Filatov, N. N. (2004). *Climate of Karelia: Variability and Impact on Water Objects and Watersheds* (224 pp.). Karel Sci. Centre of RAS, Petrozavodsk, Russia [in Russian].

Fleishman, B. S. (2003). *The Choice Is Yours* (120 pp.). Oecumene, New York.

Flood, J. (1995). Indicators for the implementation and monitoring of agenda. *Habitat*, **1**(5), 13.

Folland, C., Frich, P., Basnett, T., Rayner, N., Parker, D., and Horton, B. (2000). Uncertainties in climate datasets: A challenge for WMO. *WMO Bull.*, **49**(1), 59–68.

Fong, H. and Liang, S. (2003). Retrieving leaf area index with a neural network method: Simulation and validation. *IEEE Trans. Geoscience. Remote Sensing*, **41**(9), 2052–2062.

Fridlingstein, P., Bopp, L., Ciaias, P., Dufresne, J.-L., Fairhead, L., LeTrent, H., Monfray, P., and Orr, J. (2001). Positive feedback between future climate change and the carbon cycle. *Geophys. Res. Lett.*, **28**, 1543–1546.

Fulé, P. Z., Crouse, J. E., Cocke, A. E., Moore, M. M., and Covington, W. W. (2004). Changes in canopy fuels and potential fire behaviour 1880–2040: Grand Canyon, Arizona. *Ecological Modelling*, **175**(3), 231–248.

Fung, Ch. Sh. and Le, D. H. (1997). Control of environmental pollution from industrial and domestic waste in key regions by means of economic development. *Proceedings of the Workshop on Environmental Technology and Management, Ho Chi Minh City, 28–29 May* (pp. 32–42). Institute of Applied Mechanics, Ho Chi Minh City, Vietnam [in Russian].

Furiayev, V. V. (1996). *The Role of Fires in the Process of Forest Formation* (252 pp.). Science, Novosibirsk, Russia [in Russian].

Gale, J. and Freund, P. (2000). Reducing methane emissions to combat global climate change: The role Russia can play. *Proceedings of the Second International Methane Mitigation Conference, 18–23 June, Novosibirsk* (pp. 73–80). Novosibirsk State University, Novosibirsk, Russia.

Garcia-Barrón, L. and Pita, M. F. (2004). Stochastic analysis of time series of temperatures in the south-west of the Iberian Peninsula. *Atmósfera*, **17**(4), 225–244.

Gardner, J. S. (2002). Natural hazards risk in the Kullu District, Himachal Pradesh, India. *Geographical Review*, **92**, 172–177.

Gedney, N. and Valdes, P. J. (2000). The effect of Amazonian deforestation on northern hemisphere circulation and climate. *Geophys. Res. Lett.*, **27**(19), 3053–3056.

Geogdzhayev, I. V., Mishchenko, M. I., Terez, E. I., Terez, G. A., and Gushchin, G. K. (2005). Regional advanced very high resolution radiometer-derived climatology of aerosol optical thickness and size. *J. Geophys. Res.*, **110**(D23205), doi:10.1029/2005JD006170.

Gerstengarbe, F.-W. (2002). *Angewandte Statistic* (PIK Report No. 75, 100 pp.). Potsdam Institute for Climate Research, Potsdam, Germany [in German].

Gibson, R. B., Hassan, S., Holtz, S., Tansey, J., and Whitelaw, G. (2005). *Sustainability Assessment, Criteria, Processes, and Applications* (240 pp.). Earthscan, London.

Gitz, V. and Ciais, P. (2004). Future expansion of agriculture and pasture acts to amplify atmospheric CO_2 levels in response to fossil-fuel and land-use change emissions. *Climatic Change*, **67**(1), 161–184.

Glade, T., Anderson, M. G., and Crozier, M. J. (eds) (2005). *Landslide Hazard and Risk* (608 pp.). John Wiley & Sons, London.

Gleick, P. H. (1993). *A Guide to the World's Fresh Water Resources* (473 pp.). Oxford Scientific, Oxford, UK.

Gloersen, P., Parkinson, C. L., Cavalieri, D. J., Comiso, J. C., and Zwally, H. J. (1999). Spatial distribution of trends and seasonality in the hemispheric ice covers: 1978–1996. *J. Geophys. Res.*, **104**(C9), 20827–20835.

Goldner, J. (2002). *Messages from Space* (132 pp.). Michael Wiese Production, Suite.

Golitsyn, G. S. (1995). Rising of the Caspian Sea level as a problem of diagnosis and forecast of regional climate change. *Physics of the Atmosphere and Ocean*, **31**(3), 385–391 [in Russian].

Golubov, B. N. and Kruchenitsky, G. M. (1999). *Study into the Degassing of the Cavities of Underground Nuclear Explosions as a Factor of Atmospheric Pollution in the Sakha Republic (Yakutia)* (178 pp.). Scientific Council on Biospheric Problems at RAS Presidium, Moscow [in Russian].

Goody, R. (2002). Observing and thinking about the atmosphere. *Annu. Rev. Environ.*, **27**, 1–20.

Gorny, V. I., Salman, A. G., Tronin, A. A., and Shilin, B. V. (1988). The outgoing IR radiation of the Earth as an indicator of seismic activity. *Reports of RAS*, **301**(1), 67–69 [in Russian].

Gorshkov, V. G. (1990). *Energetics of the Biosphere and Environmental Stability* (237 pp.). ARISTI, Moscow [in Russian].

Gorshkov, V. G. (1995). *Physical and Biological Bases of Life Stability: Man, Biota, Environment* (340 pp.). Springer-Verlag, Berlin.

Gorshkov, V. G., Kondratyev, K. Ya., and Losev, K. S. (1998). Global ecodynamics and sustainable development: Natural–scientific aspects and the "human dimension". *Ecology*, **3**, 163–170 [in Russian].

Gorshkov, V. G., Gorshkov, V. V., and Makarieva, A. M. (2000). *Biotic Regulation of the Environment* (364 pp.). Springer/Praxis, Chichester, UK.

Gorshkov, V., Makarieva, A., Mackey, B., and Gorshkov, V. (2002). Biological theory and global change science. *Global Change Newsletter*, **48**, 11–14.

Goudsouzian, A. (2004). *The Hurricane of 1938*, (96 pp.). Commonwealth Editions, Boston.

Graham, B., Guyon, P., Maenhaut, W., Taylor, P. E., Ebert, M., Matthias-Maser, S., Mayol-Bracero, O. L., Gedoi, R. H. M., Artaxo, P., Meixner, F. X. *et al.* (2003). Composition and diurnal variability of the natural Amazonian aerosol. *J. Geophys. Res.*, **108**(D24), AAC5/1–AAC5/16.

Grankov, A. G. and Milshin, A. A. (1994). On the correlation of humidity and moisture content with water surface air temperature. *Research of the Earth from Space*, **10**, 78–81 [in Russian].

Grant, J. A. (2004). Liquid compositions from low-pressure experimental melting of pelitic rock from Morton Pass, Wyoming, USA. *Journal of Metamorphic Geology*, **22**, 65–78.

Grassl, H. (2000). Status and improvements of coupled general circulation models. *Science*, **288**, 1991–1997.

Greenland, D., Goodin, D. G., and Smith, R. C. (2003). *Climate Variability and Ecosystem Response in Long-term Ecological Research Sites* (512 pp.). Oxford University Press, Oxford, UK.

Grigoryev, A. A. (1987). Large-scale changes in the nature of Priaralye from space-borne observations. *Problems of Desert Mastering*, **1**, 16–22 [in Russian].

Grigoryev, A. A. and Kondratyev, K. Ya. (2001). *Ecodynamocs and Geopolitics*, Vol. II: *Ecological Disasters* (688 pp.). St Petersburg Research Centre for Safety, RAS, St Petersburg [in Russian].

Grigoryev, A. A. and Kondratyev, K. Ya. (2004a). Global urbanization, 1: General laws. *Proc. Russian Geographical Society*, **136**(4), 1–8 [in Russian].

Grigoryev, A. A. and Kondratyev, K. Ya. (2004b). Global urbanization, 2: Ecodynamics of large cities. *Proc. Russian Geographical Society*, **136**(5), 1–11 [in Russian].

Grigoryev, A. A. and Kondratyev, K. Ya. (2005). Natural and anthropogenic forest fires: Ecodynamics component and natural disasters. *Proc. Russian Geographical Society*, **137**(1), 3–40 [in Russian].

Grigoryev, A. A. and Sychev, V. N. (2004). Systems of life support for cosmonauts based on biospheric mechanisms. *Herald of RAS*, **74**(8), 675–689 [in Russian].

Grogan, P., Illeris, L., Michelsen, A., and Jonasson, S. E. (2001). Respiration of recently-fixed plant carbon dominates mid-winter ecosystem CO_2 production in sub-Arctic heath tundra. *Climatic Change*, **50**, 129–142.

Grossi, P. and Kunreuther, H. (eds) (2005). *Catastrophe Modeling: A New Approach to Managing Risk* (252 pp.). Springer-Verlag, New York.

Gurjar, B. R. and Leliveld, J. (2005). New directions: Megacities and global change. *Atmospheric Environment*, **39**(2), 391–393.

Gutberlet, J. (2003). Cities, consumption and the generation of waste. *Aviso*, **11**, 12–19.

Gyalistras, D. (2002). *How Uncertain Are Regional Climate Change Scenarios? Examples for Europe and the Alps* (PIK Report No. 75, pp. 85–100). Potsdam Institute for Climate Research, Potsdam, Germany.

Haan, D., Zuo, Y., Gros, V., and Brenninkmeijer, C. A. M. (2001). Photochemical production of carbon monoxide in snow. *J. Atmos. Sci.*, **40**(3), 217–230.

Haque, C. E. (ed.) (2005). *Mitigation of Natural Hazards and Disasters: International Perspectives* (200 pp.). Springer-Verlag, Heidelberg.

Hales, B., Takahashi, T., and Bandstra, L. (2005). Atmospheric CO_2 uptake by a coastal upwelling system. *Global Biogeochemical Cycles*, **19**(GB1009), doi: 10.1029/2004GB002295, 1–11.

Hamill, T. M., Schneider, R. S., Brooks, H. E., Forbes, G. S., Bluestein, H. B., Steinberg, M., Meléndez, D., and Dole, R. M. (2005). The May 2003 extended tornado outbreak. *Bull. Amer. Meteorol. Soc.*, **86**(4), 531–542.

Han, Q., Rossow, W. B., Zeng, J., and Welch, R. (2002). Three different behaviours of liquid water path of water clouds in aerosol–cloud interactions. *J. Atmos. Sci.*, **59**(3), 726–735.

Hansen, J. (1998). Book review of Sir John Houghton's "Global Warming: The Complete Briefing". *J. Atmos. Chem.*, **30**, 409–412.

Hansen, J. and Sato, M. (2001). Trends of measured climate forcing agents. *Proc. Nat. Acad. Sci. USA*, **98**(26), 14778–14783.

Hansen, J., Fung, I., Lacis, A., Rind, D., Lebedeff, S., Ruedy, R., Russell, G., and Stone, P. (1988). Global climate changes as forecast by Goddard Institute for Space Studies' three-dimensional model. *J. Geophys. Res.*, **93**, 9341–9364.

Hansen, J., Ruedy, R., Glascoe, J., and Sato, M. (1999). GISS analysis of surface temperature change. *J. Geophys. Res.*, **104**(D24), 30997–31022.

Hansen, J., Sato, M., Lacis, A., Ruedy, R., Tegen, L., and Matthews, E. (1998). Climate forcing in the industrial era. *Proc. Nat. Acad. Sci. USA*, **95**(22), 12753–12758.

Hansen, J. E., Sato, M., Ruedy, R., Lacis, A., and Oinas, V. (2000). Global warming in the twenty-first century: An alternative scenario. *Proc. Nat. Acad. Sci. USA*, **97**(18), 9875–9880.

Hansen J., Sato M., Nazarenko L., Ruedy R., Laws A., Koch D., Tegen I., Hall T., Shindell D., Santer B., *et al.* (2002). Climate forcings in Goddard Institute for Space Studies SI 2000 simulations. *J. Geophys. Res.*, **107**(D18), ACL2–ACL37.

Hanson, B. (2005). Learning from natural disasters. *Science*, **308**, 1125.

Hardy, J. T. (2003). *Climate Change* (260 pp.). John Wiley & Sons, Washington, DC.

Haritonova V. I. (2004). *Religious Factor in Present Life of the Northern and Siberian People* (39 pp.). Institute of Ethnology and Anthropology, RAS, Moscow.

Harshvardhan, M., Schwartz, S. E., Benkovitz, C. V., and Guo, G. (2002). Aerosol influence on cloud microphysics examined by satellite measurements and chemical transport modeling. *J. Atmos. Sci.*, **59**(2), Part 2, 714–725.

Hasegawa, Y. and Kasagi, N. (2001). The effect of Schmidt number on air–water interface mass transfer. *Proceedings of the Fourth International Conference on Multiphase Flow, New Orleans, 27 May–1 June, University of Nottingham, New Orleans, LA*, pp. 296–292.

Hasegawa, Y. and Kasagi, N. (2005). Turbulent mass transfer mechanism across a contaminated air–water interface. *Proceedings of the Fourth International Symposium on Turbulence and Shear Flow Phenomena (TSFP-4), Williamsburg, VA, 27–29 June*, pp. 971–976.

Hauglistaine, D. (2002). Trace gas radiative forcing and related climate feedbacks: How do we reduce the uncertainties? *IGACtivities Newsletter*, **26**, 20–26.

Heans, K. A. (2001). Assessment of pre-industrial carbon dioxide content in the atmosphere using hydrochemical data. *Proceedings of the First International Conference on Global Warming and the Next Ice Age, 19–24 August, Halifax, Canada* (pp. 140–144). Dalhousie University, Halifax, Canada.

Henderson, V. and Thisse, J. F. (eds) (2004). *Handbook in Regional and Urban Economics*, Vol. 4: *Cities and Geography* (1006 pp.). Elsevier, Amsterdam.

Hide, R., McSharry, P. E., Finlay, C. C., and Peskett, G. D. (2004). Quenching Lorenzian chaos. *International Journal of Bifurcation and Chaos*, **14**(8), 2875–2884.

Hinchliffe, S., Blowers, A., and Freeland, J. (2002). *Understanding Environmental Issues* (216 pp.). John Wiley & Sons, London.

Hoelzemann, J. J., Schultz, M. G., Brasseur, G. P., Granier, C., and Simon, M. (2004). Global wildland fire emission model (GWEM): Evaluating the use of global area burnt satellite data. *J. Geophys. Res.*, **109**(14), D14504/1–D14504/18.

Hoffmann, W. A. (1998). Post-burn reproduction of woody plants in a neotropical savanna: The relative importance of sexual and vegetative reproduction. *J. Appl. Ecology*, **35**(3), 422–433.

Hoffmann, W. A., Orthen, B., and Nascimento, P. K. V. (2003). Comparative fire ecology of tropical savanna and forest trees. *Functional Ecology*, **17**(6), 720–726.

Hoinka, K. P. and de Castro, M. (2005). A renaissance depiction of a tornado. *Bull. Amer. Meteorol. Soc.*, **86**(4), 543–552.

Holdren, J. P. (2003). Environmental change and human condition. *Bull. Amer. Acad. Arts Sci.*, **57**(1), 25–31.

Holweg, E.J. (2000). *Mariner's Guide for Hurricane Awareness in the North Atlantic Basin* (72 pp.). National Oceanic and Atmospheric Administration, Washington, DC.

Honson, B. (2005). Learning from natural disasters. *Science*, **308**(5725), 1125.

Houghton, J. (2000). Global climate and human activities. *Eilss Article*, **1**, 1–13.

Houghton, J., Calander, B. A., and Varney, S. K. (eds) (1992). *Climate Change 1992* (7 pp.). Cambridge University Press, Cambridge, UK.

Houghton, J. T., Ding, Y., Griggs, D. J., Noguer, M., van der Linden, P. J., Dai, X., Masskell, K., and Johnson, C. A. (2001). *Climate Change 2001: The Scientific Basis. Contribution of Working Group I to the Third Assessment Report of the Intergovernmental Panel Group on Climate Change* (881 pp.). Cambridge University Press, Cambridge, UK.

Hsu, S. M., Ni, C.-F., and Hung, P.-F. (2002). Assessment of three infiltration formulas based on model fitting on Richards equation. *Journal of Hydrologic Engineering*, **7**(5), 373–379.

Huang, S., Pollack, H. N., and Shen, P.-Y. (2000). Temperature trends over the past five centuries reconstructed from borehole temperatures. *Nature*, **403**, 756–758.

Hulme, M. and Parry, M. (1998). Adapt or mitigate? Responding to climate change. *Town and Country Planning*, **67**(2), 50–51.

Hulme, M., Barrow, E. M., Arnell, N. W., Harrison, P. A., Johns, T. C., and Downing, T. E. (1999). Relative impacts of human-induced climate change and climate variability. *Nature*, **397**, 689–691.

Iliadis, L. S. (2005). A decision support system applying an integrated fuzzy model for long-term forest fire risk estimation. *Environmental Modelling & Software*, **20**(5), 613–621.

IPCC (2001). *Third Assessment Report*, Vol. 1: *Climate Change 2001. The Scientific Basis* (881 pp.). Cambridge University Press, Cambridge, UK.

Ito, A. (2005). Climate-related uncertainties in projections of the twenty-first century terrestrial carbon budget: Off-line model experiments using IPCC greenhouse-gas scenarios and AO GCM climate projections. *Climate Dynamics*, **44**, 435–448.

Ito, A. and Oikawa, T. (2002). A simulation model of the carbon cycle in land ecosystems (Sim-Cycle): A description based on dry-matter production theory and plot-scale validation. *Ecological Modelling*, **151**, 147–179.

Irion, R. (2001). Fathoming the chemistry of the deep blue sea. *Science*, **293**(5531), 790–793.

Ivanov-Rostovtsev, A. G., Kolotilo, L. G., Tarasiuk, Yu. F., and Sherstiankin, P. P. (2001). *Self-organization and Self-regulation of Natural Systems: Model, Method, and Fundamentals of the D-Self Theory* (216 pp.). Russian Geographical Society, St Petersburg [in Russian].

Ivey, J. P. (2002) *Tropical Storm Allison*. Halff Associates Inc., Allington, TX, 42pp.

Jaeger, C. C., Ortwin, R., Rosa, E. A., and Webler, T. (2001). *Risk, Uncertainty, and Rational Action* (324 pp.). Earthscan, London.

Jagovkina, S. V., Karol, I. L., Zubov, V. A., Lagun, V. E., Reshemikov, A. I., and Rosanov, E. V. (2000a). Estimation of gas deposit leakage into the total methane flux from the West Siberian region. *Proceedings of the Second International Methane Mitigation Conference, 18–23 June, Novosibirsk* (pp. 263–267). Novosibirsk State University, Novosibirsk, Russia.

Jagovkina, S. V., Karol, I. L., Zubov, V. A., Lagun, V. E., Reshemikov, A. I., and Rosanov, E. V. (2000b). Reconstruction of methane fluxes from the west Siberia gas fields by the 3D regional chemical transport model. *Atmospheric Environment*, **34**(29), 5319–5328.

Jarraud, M. (2005). Reducing the risk of nature disasters through early warnings. *WMO Bull.*, **86**(2), 155–156.

Jelle Zeilinga de Boer and Sanders, D. T. (2005). *Earthquakes in Human History* (296 pp.). Princeton University Press, Princeton, NJ.

Jenkins, G., Betts, R., Collins, M., Griggs, D., Lowe, J., and Wood, R. (2005). *Stabilising Climate to Avoid Dangerous Climate Change: A Smmary of Relevant Research at the Hadley Centre* (19 pp.). Met Office Hadley Centre, Exeter, UK.

Ji, Y. and Stocker, E. (2002). Seasonal, intraseasonal, and interannual variability of global land fires and their effects on atmospheric aerosol distribution. *J. Geophys. Res.*, **107**(D23), ACH10/1–ACH10/11.

Johnson, D. E. and Ulyatt, M. I. (2000). Variations in the proportion of methane of total greenhouse gas emissions from US and NZ dairy production systems. *Proceedings of the Second International Methane Mitigation Conference, 18–23 June, Novosibirsk* (pp. 249–254). Novosibirsk State University, Novosibirsk, Russia.

Jolliffe, I. T. and Stephenson, D. B. (2003). *Forecast Verification* (254 pp.). John Wiley & Sons, London.

Jones, A., Roberts, D. L., Woodage, M. J., and Johnson, C. E. (2001). Indirect sulphate aerosol forcing in a climate model with an interactive sulphur cycle. *J. Geophys. Res.*, **106**(D17), 20293–20310.

Jönsson, A. M., Linderson, M.-L., Stjernquist, I., Scglyter, P., and Bärring, L. (2004). Climate change and the effect of temperature backlashes causing frost damage in *Picea abies*. *Global and Planetary Change*, **44**(1–4), 195–207.

Jun, H. B. and Shin, H. S. (1997). Substrates transformation in a biological excess phosphorus removal system. *Water Research*, **31**(4), 893–899.

Kanygin, A. (2004). Praises on catastrophes. *Science at First Hand*, **1**, 29–39 [in Russian].

Karl, T. and Gleckler, P. J. (2001). Tracking changes in AMIP model performance. *Proceedings of the Eighth Scientific Assembly of IAMAS, Innsbruck, 10–18 July* (p. 8). International Association of Meteorology and Atmospheric Sciences, Innsbruck, Germany.

Karley, M. J., Beven, K. J., and Oliver, H. R. (1993). A method for predicting spatial distribution of evaporation using simple meteorological data. *Proceedings of International Symposium "Ech. Proc. Land Surf. Range Space and Time Scales", Yokohama, 13–16 July* (Vol. 212, pp. 619–626). International Association of Hydrological Sciences, Yokohama, Japan.

Karol, I. L. (2000). Impact of transport aircraft flights on the ozonosphere and climate. *Meteorology and Hydrology*, **7**, 17–32 [in Russian].

Kashapov, R. Sh. (2002). On the balance of organic carbon in the natural–economic system of Bashkortostan. *Proceedings of the Russian Geographical Society*, **134**(3b), 39–42 [in Russian].

Kasperson, J. X. and Kasperson, R. E. (eds) (2001). *Global Environmental Risk* (324 pp.). Earthscan, London.

Kasyanova M. A. (2003). *Ecological Risks and Geodynamics* (330 pp.). Sci. World, Moscow [in Russian].

Keeling, R. F. and Visbeck, M. (2001). Antarctic stratification and glacial CO_2. *Nature*, **412**(6847), 605–606.

Keigwin, L. D. and Boyle, E. A. (2000). Detecting Holocene changes in thermohaline circulation. *Proceedings of the National Academy of Sciences*, **97**(4), 1343–1346.

Kendrick, T. D. (1957). *The Lisbon Earthquake* (255 pp.). Lippincott, Philadelphia.

Kerr, R. A. (2000). Dueling models: Future US climate uncertain. *Science*, **288**, 2113.

Khalil, M. A. K., Rasmussen, R. A., Ren, L., Wang, M. X., Shearer, M. J., Dalluge, R. W., and Duan, C.-L. (2000). Methane emissions from rice fields. *Proceedings of the Second International Methane Mitigation Conference, 18–23 June, Novosibirsk* (pp. 13–30). Novosibirsk State University, Novosibirsk, Russia.

Kharitonova, V. I. (2004). *Religious Factor in Present Life of Northern and Siberian Peoples*. Institute for Ethnology and Anthropology, Russian Academy of Sciences, Moscow, 39 pp.

Kharkina, M. A. (2000). Ecological consequences of natural disasters. *Energy*, **1**, 1–6 [in Russian].

Kharkina, M. A. (2003). Volcanic eruptions: Danger or blessing. *Energy*, **9**, 48–53 [in Russian].

Kiehl, J. T. and Gent, P. R. (2004). The Community Climate System Model, Version 2. *J. Clim.*, **17**, 3666–3682.

Kim, W., Arai, T., Kanae, S. Oki, T., and Musiake, K. (2001). Application of the Simple Biosphere Model (SiB2) to a paddy field for a period of growing season in GAME-Tropics. *J. Meteorol. Soc. Japan*, **79**(18), 387–400.

King, C. (2004). *Without Warning: The Great Storm of 1953* (78 pp.). Ian Henry Publications, Romford, UK.

Kirchner, I., Stenchikov, G., Graf, H.-F., Robock, A., and Antuna, J. (1999). Climate model simulation of winter warming and summer cooling following the 1991 Mount Pinatubo volcanic eruption, *J. Geophys. Res.*, **104**, 19039–19055.

Kirchner, J. W. (2003). The Gaia hypothesis: Conjectures and refutations. *Clim. Change*, **58**(1–2), 21–45.

Klyuev V. V. (ed.) (2000). *Safety of Russia: Ecological Diagnostics* (496 pp.). Knowledge, Moscow [in Russian].

Knutson, T. R., Delworth, T. U., Dixon, K. W., and Stouffer, R. J. (1999). Model assessment of regional temperature trends (1949–1997). *J. Geophys. Res.*, **104**(D24), 30981–30996.

Köhler, U. (1999). A comparison of the new Filter Ozonometer Microtops II with Dobson and Brewer Spectrometers at Hohenspeissenberg. *Geophys. Res. Lett.*, **26**(10), 1385–1388.

Koike, T. (2004). The Coordinated Enhanced Observing Period: An initial step for integral global water cycle observation. *WMO Bull.*, **53**(2), 2–8.

Kokkola, H., Romakkaniemi, S., and Laaksonen, A. (2003). Köhler theory for a polydisperse droplet population in the pressure of a soluble trace gas, and an application to stratospheric STS droplet growth. *Atmospheric Chemistry Physics Discussions*, **3**, 3241–3266.

Kondratyev, K. Ya. (1982). *The World Climate Research Programme: State, Perspectives and the Role of Space-borne Observational Means* (Progress in Science and Engineering: Meteorology and Climatology No. 8, 274 pp.). ARISTI, Moscow [in Russian].

Kondratyev, K. Ya. (1990). *Key Problems of Global Ecology* (454 pp.). ARISTI, Moscow [in Russian].

Kondratyev, K. Ya. (1991). Priorities of global ecology. *Proceedings of RAS: Geography*, **6**, 21–30 [in Russian].

Kondratyev, K. Ya. (1992). *Global Climate* (359 pp.). Science, St Petersburg [in Russian].

Kondratyev, K. Ya. (1993). Ecology and politics. *Proceedings of the Russian Geographical Society*, **125**(2), 78–91 [in Russian].

Kondratyev, K. Ya. (1996). Global changes and demographic dynamics. *Proceedings of the Russian Geographical Society*, **128**(3), 1–12 [in Russian].

Kondratyev, K. Ya. (1998). *Multidimensional Global Change* (771 pp.). Wiley/Praxis, Chichester, UK.

Kondratyev, K. Ya. (1999). *Ecodynamics and Geopolitics*, Vol. 1: *Global Problems* (1040 pp.). St Petersburg State University, St Petersburg [in Russian].

Kondratyev, K. Ya. (2000a). Global changes of nature and society on the verge of two millennia. *Proceedings of the Russian Geographical Society*, **5**, 3–19 [in Russian].

Kondratyev, K. Ya. (2000b). Studying the Earth from space: The EOS scientific plan. *Research of the Earth from Space*, **3**, 82–91 [in Russian].

Kondratyev, K. Ya. (2001). Key issues of global change at the end of the second millennium. In: M. K. Tolba (ed.), *Our Fragile World: Challenges and Opportunities for Sustainable Development* (No. 1, pp. 147–165). Eolls, Oxford, UK.

Kondratyev, K. Ya. (2002). Global climate change: Reality, hypotheses, and fiction. *Research of the Earth from Space*, **1**, 3–23 [in Russian].

Kondratyev, K. Ya. (2003). Radiative forcing due to aerosol. *Optics of the Atmosphere and Ocean*, **16**(1), 1–14 [in Russian].

Kondratyev, K. Ya. (2004a). Global climate change: Observational data and numerical modelling results. *Research of the Earth from Space*, **1**, 3–25 [in Russian].

Kondratyev, K. Ya. (2004b). Global climate change: Unsolved problems. *Meteorology and Hydrology*, **4**, 93–102 [in Russian].

Kondratyev, K. Ya. (2004c). Uncertainties in observational data and numerical climate modeling. *Meteorology and Hydrology*, **4**, 103–119 [in Russian].

Kondratyev, K. Ya. (2004d). Priorities of global climatology. *Proceedings of the Russian Geographical Society*, **136**(2), 3–25 [in Russian].

Kondratyev, K.Ya. (2005a). Key aspects of the global climate change problem. *Proceedings of the Russian Geographical Society*, **5**, 92–99 [in Russian].

Kondratyev, K. Ya. (2005b). *Processes of Formation, Properties, and Climatic Impacts of Aerosol* (450 pp.). St Petersburg State University, St Petersburg [in Russian].

Kondratyev, K. Ya. and Binenko, V. I. (2000). On radiative forcing of clouds and aerosols. *Meteorology and Hydrology*, **1**, 33–41 [in Russian].

Kondratyev, K. Ya. and Cracknell, A. P. (1999). *Observing Global Climate Change* (592 pp.). Taylor & Francis, London.

Kondratyev, K. Ya. and Demirchian, K. S. (2001). Global climate and the Kyoto Protocol. *Problems of the Environment and Natural Resources*, **6**, 2–15 [in Russian].

Kondratyev, K. Ya. and Galindo, I. (2001). *Global Change Situations: Today and Tomorrow* (164 pp.). Universidad de Colima, Colima, Mexico.

Kondratyev, K. Ya. and Grassl, S. H. (1993). *Global Climate Change in the Context of Global Ecodynamics* (195 pp.). St Petersburg Research Centre for Ecological Safety of RAS, St Petersburg [in Russian].

Kondratyev, K. Ya. and Grigoryev, Al. A. (2004). Forest fires as a global ecodynamics component. *Optics of the Atmosphere and Ocean*, **136**(4), 279–292 [in Russian].

Kondratyev, K. Ya. and Johannessen, O. (1993). *Arctic and Climate* (140 pp.). Propo, St Petersburg [in Russian].

Kondratyev, K. Ya. and Krapivin, V. F. (2003a). Global change: Real and possible in the future. *Research of the Earth from Space*, **4**, 1–10 [in Russian].

Kondratyev, K. Ya. and Krapivin, V. F. (2003b). Global carbon cycle and climate. *Research of the Earth from Space*, **1**, 3–15 [in Russian].

Kondratyev, K. Ya. and Krapivin, V. F. (2004b). *Modelling the Global Carbon Cycle* (335 pp.). Physics-Mathematics, Moscow [in Russian].

Kondratyev, K. Ya. and Krapivin, V. F. (2004a). Global carbon cycle: State, problems and perspectives. *Research of the Earth from Space*, **3**, 12–21 [in Russian].

Kondratyev, K. Ya. and Moskalenko, N. I. (1984). *The Greenhouse Effect of the Atmosphere and Climate* (Progress in Science and Engineering: Meteorology and Climatology, Vol. 12, 262 pp.). ARISTI, Moscow [in Russian].

Kondratyev, K. Ya. and Varotsos, C. A. (2000). *Atmospheric Ozone Variability: Implicatiions for Climate Change, Human Health, and Ecosystems.* (758 pp.). Springer/Praxis, Chichester, UK.

Kondratyev, K. Ya., Vasilyev, O. B., Ivlev, L. S., Nikolsky, G. A., and Smokty, O. I. (1973). *Impact of Aerosol on Radiation Transfer: Possible Climatic Implications* (266 pp.). LSU, Leningrad [in Russian].

Kondratyev, K. Ya., Grigoryev, A. A., Pokrovsky, O. M., and Shalina, E. V. (1983). *Remote Sounding of Aerosol from Space* (216 pp.). Hydrometeoizdat, Leningrad [in Russian].

Kondratyev, K. Ya., Ortner, J., and Preining, O. (1992). Priorities of global ecology now and in the next century. *Space Policy*, **8**(1), 39–48.

Kondratyev, K. Ya., Moreno-Pena, F., and Galindo, I. (1994). *Global Change: Environment and Society* (47 pp.). Universidad de Colima, Colima, Mexico.

Kondratyev, K. Ya., Moreno-Pena, F., and Galindo, I. (1997). *Sustainable Development and Population Dynamics* (128 pp.). Universidad de Colima, Colima, Mexico.

Kondratyev, K. Ya., Krapivin, V. F., and Pshenin, E. S. (2000). Concept of the regional geoinformation monitoring. *Research of the Earth from Space*, **6**, 1–8 [in Russian].

Kondratyev, K. Ya., Losev, K. S., Ananicheva, M. D., and Chesnokova, I. V. (2001). Some problems of landscape science and ecology in the context of biotic regulation. *Annals of the Russian Geographical Society*, **133**(5), 22–29 [in Russian].

Kondratyev, K. Ya., Grigoryev, Al. A., and Varotsos, C. A. (2002a). *Environmental Disasters: Anthropogenic and Natural* (484 pp.). Springer/Praxis, Chichester, UK.

Kondratyev, K. Ya., Krapivin, V. F., and Phillips, G. W. (2002b). *Global Environmental Change: Modelling and Monitoring*, (319 pp.). Springer-Verlag, Berlin.

Kondratyev, K. Ya., Krapivin, V. F., and Varotsos, C. A. (2003a). *Global Carbon Cycle and Climate Change* (372 pp.). Springer/Praxis, Chichester, UK.

Kondratyev, K. Ya., Krapivin, V. F., and Savinykh, V. P. (2003b). *Perspectives of Civilization Development: Multidimensional Analysis* (574 pp.). Logos, Moscow [in Russian].

Kondratyev, K. Ya., Losev, K. S., Ananicheva, M. D., and Chesnokova, I. V. (2003c). *Natural–Scientific Foundations of Life Stability* (240 pp.). CAGL, Moscow [in Russian].

Kondratyev, K. Ya., Losev, K. S., Ananicheva, M. D., and Chesnokova, I. V. (2003d). Price of ecological service in Russia. *Herald of RAS*, **73**(1), 3–10 [in Russian].

Kondratyev, K. Ya., Losev, K. S., Ananicheva, M. D., and Chesnokova, I. V. (2003e). *Stability of Life on Earth* (152 pp.). Springer/Praxis, Chichester, UK.

Kondratyev, K. Ya., Fedchenko, P. P., and Fedchenko, K. P. (2004a). Solar radiation spectrum and evolution of the biosphere. *Problems of the Environment and Natural Resources*, **11**, 31–51 [in Russian].

Kondratyev, K. Ya., Fedchenko, P. P., and Fedchenko, K. P. (2004b). Fraunhofer lines in the solar spectrum and immunodeficiency problems. *Problems of the Environment and Natural Resources*, **12**, 66–77 [in Russian].

Kondratyev, K. Ya., Krapivin, V. F., and Nitu, C. (2004c). An application of global simulation model to the study of CO_2 greenhouse effect. *Proceedings International Conference "World Energy Systems", 15-17 May* (pp. 467–472). University Oradea, Oradea, Romania.

Kondratyev, K. Ya. , Krapivin, V. F., Savinykh, V. P., and Varotsos, C. A. (2004d). *Global Ecodynamics: A Multidimensional Analysis* (658 pp.). Springer/Praxis, Chichester, UK.

Kornakov, V. I., Borovets, S. A., and Bostandzhoglo, A. A. (1968). *Water Balance and Forecast for a Drop in the Aral Sea Level* (103 pp.). Hydroproject, Tashkent, Uzbekistan [in Russian].

Kosarev, A. N. (1975). *Hydrology of the Caspian and Aral Seas* (271 pp.). MSU, Moscow [in Russian].

Kotelnikov, V. A. (1956). *Theory of Potential Noise Immunity* (156 pp.). Gosenergoizdat, Moscow [in Russian].

Kotlyakov, V. M. (1993). Geographic approach to the theory of disasters. *Proc. of RAS, Ser. Geography*, **5**, 7–17 [in Russian].

Kraabol, A. G. and Stordal, F. (2000). Modelling chemistry in aircraft plumes, 2: The chemical conservation of NO_x to reservoir species under different conditions. *Atmospheric Environment*, **34**(23), 3951–3962.

Kramer, H. J. (1995). *Observation of the Earth and Its Environment* (832 pp.). Springer-Verlag, Berlin.

Krapivin, V. F. (1978). *On the Theory of Complex Systems' Survivability* (248 pp.). Science, Moscow [in Russian].

Krapivin, V. F. (1993). Mathematical model for global ecological investigations. *Ecological Modelling*, **67**(2–4), 103–127.

Krapivin, V. F. (1996). The estimation of the Peruvian current ecosystem by a mathematical model of biosphere. *Ecological Modelling*, **91**(1), 1–14.

Krapivin, V. F. (2000a). Radio-wave ecological monitoring. In: V. V. Klyuev (ed.), *Ecological Diagnostics* (pp. 295–311). Knowledge, Moscow [in Russian].

Krapivin, V. F. (2000b). A simulation model of the biogeochemical cycle of phosphorus in the biosphere. *Problems of the Environment and Natural Resources*, **10**, 26–30 [in Russian].

Krapivin, V. F. and Chukhlantsev, A. A. (2004). Remote UHF radiometric sounding of soil and vegetation in the context of the global environmental change. *Ecological Systems and Devices*, **9**, 37–45 [in Russian].

Krapivin, V. F. and Kondratyev, K. Ya. (2002). *Global Environmental Change: Ecoinformatics* (724 pp.). St Petersburg State University, St Petersburg [in Russian].

Krapivin, V. F. and Mkrtchyan, F. A. (2002). Efficiency of the monitoring systems of detection. *Ecological Systems and Devices*, **6**, 3–5 [in Russian].

Krapivin, V. F. and Nazaryan, N. A. (1997). Mathematical model for investigations of the global sulphur cycle. *Mathematical Modelling*, **9**(8), 36–50 [in Russian].

Krapivin, V. F. and Phillips, G. W. (2001). A remote sensing-based expert system to study the Aral–Caspian aquageosystem water regime. *Remote Sensing of Environment*, **75**, 201–215.

Krapivin, V. F. and Potapov, I. I. (2002). *Methods of Ecoinformatics* (496 pp.). ARISTI, Moscow [in Russian].

Krapivin, V. F. and Vilkova, L. P. (1990). Model estimation of excess CO_2 distribution in biosphere structure. *Ecological Modelling*, **50**, 57–78.

Krapivin, V. F., Svirezhev, Yu. M., and Tarko, A. M. (1982). *Numerical Modelling of Global Biospheric Processes* (272 pp.). Science, Moscow [in Russian].

Krapivin, V. F., Shutko, A. M., Chukhlantsev, A. A., and Potapov, I. I. (2004). Information systems of the ecological monitoring. *Ecological Systems and Devices*, **4**, 3–8 [in Russian].

Krebs, C. J., Boutin, S., and Boonstra, R. (eds.) (2001). *Ecosystem Dynamics of the Boreal Forest* (536 pp.). Oxford Univ. Press, Oxford.

Krimmer, R. W. and Lake, F. K. (2001). The role of indigenous burning in land management. *J. Forest*, **99**(11), 36–41.

Krishnamurti, T. N., Pattniak, S., Stefanova, L., Kumar, T. S. V. V., Mackey, B. P., O'Shay, A. J., and Pasch, R. J. (2005). The hurricane intensity issue. *Monthly Weather Review*, **133**(7), 1885–1912.

Krupchatnikov, V. N. (1998). Simulation of CO_2 exchange processes in the atmosphere–surface biomes system by the climate model ECSSib. *Russ. J. Numer. Anal. Math. Modelling*, **13**(6), 479–492.

Kuksa, V. I. (1994). *The Southern Seas under Conditions of Anthropogenic Stress* (369 pp.). Hydrometeoizdat, St Petersburg [in Russian].

Kurbatsky, N. P. (1964). On the forest fire in the region of impact of the Tungus Meteorite. *Forest Economy*, **2**, 59–61 [in Russian].

Kurz, C. and Grewe, V. (2002). Lightning and thunderstorms, Part 1: Observational data and model results. *Meteorol. Z.*, **11**(6), 379–392.

Kussell, E. and Leibler, S. (2005). Phenotypic diversity, population growth, and information in fluctuating environments. *Science*, **309**, 2075–2078.

Kuznetsov, O. L. and Bolshakov, B. E. (2002). *Sustainable Development: Scientific Basis of Planning Nature–Society–Man System* (615 pp.). Humanistics, St Petersburg [in Russian].

Lal, M. and Yarasawa, H. (2001). Future climate change scenarios for Asia as inferred from selected coupled atmosphere–ocean global climate models. *J. Meteorol. Soc. Japan*, **79**(1), 219–227.

Landsea, C. W., Franklin, J. L., McAdie, C. J., Beven, J. L., II, Gross, J. M., Jarvinen, B. R., Pasch, R. J., Rappaport, E. N., Dunion, J. P., and Dodge, P. P. (2004). A re-analysis of Hurricane Andrew's intensity. *Bull. Amer. Meteorol. Soc.*, **85**(11), 1699–1712.

Langmann, B. (2000). Numerical modeling of regional scale transport and photochemistry directly together with meteorological processes. *Atmos. Env.*, **34**(21), 3585–3598.

Lawford, R. (2004). Earth observations: A renewed opportunity area for GEWEX. *GEWEX News*, **14**(3), 2.

Lawrence, D. P. (2003). *Environmental Impact Assessment: Practical Solutions to Recurrent Problems* (562 pp.). John Wiley & Sons, New York.
Le Corbusie (1977). *Architecture of the 20th Century* (304 pp.). Progress, Moscow [in Russian].
Ledley, T. S., Sundquist, E. T., Schwartz, S. E., Hall, D. K., Fellows, J. D., and Killen, T. L. (1999). Climate change and greenhouse gases. *Earth Observations from Space*, **80**(39), 453.
Legendre, P. and Legendre, L. (1998). *Numerical Ecology* (853 pp.). Elsevier, Amsterdam.
Lenton, T. M. and Wilkinson, D. M. (2003). Developing the Gaia theory: A response to the criticism of Kirchner and Volk. *Clim. Change*, **58**(1–2), 1–12.
Levesque, J. and King, D. J. (2003). Spatial analysis of radiometric fractions from high-resolution multispectral imagery for modeling individual tree crown and forest canopy structure and health. *Remote Sensing of Environment*, **84**(4), 589–602.
Levinson, D. H. and Waple, A. M. (2004). State of the climate in 2003. *Bull. Amer. Meteorol. Soc.*, **85**(6), 1–72.
Levitus, S., Antonov, J. I., Boyer, T. P., and Stephens, C. (2000). Warming of the World Ocean. *Science*, **287**, 2225–2229.
Leyborne, B.A. (1998). Can El Niño be controlled by tectonic vortex structures and explained with surge tectonics? *New Concepts in Global Tectonics*, **4**, 5–6.
Lim, A., Liew, S. C., and Kwoh, L. K. (2004). A new method of active fire detection based on subpixel retrieval of fire temperature using atmospherically corrected MODIS thermal infrared data. *Proceedings of 25th ACRS, 22–26 November, Chiang Mai, Thailand* (pp. 606–609). AARS, Chiang Mai, Thailand.
Lindenmayer, D. B., Foster, D. R., Franklin, J. F., Hunter, M. L., Noss, R. F., Schmiegelow, F. A., and Perry, D. (2004). Salvage harvesting policies after natural disturbance. *Science*, **303**(5662), 1303.
Lindzen, R. S. (2003). The Interaction of waves and convection in the Tropics. *J. Atmos. Sci.*, **60**, 3009–3020.
Liping, G., Erda, L., and Zhongpei, L. (2000). Methane emission flux and mitigation options and its relationship with N_2O emission from paddy soils. *Proceedings of the Second International Methane Mitigation Conference, 18–23 June, Novosibirsk* (pp. 217–222). Novosibirsk State University, Novosibirsk, Russia.
Lisienko, V. G., Druzhinina, O. G., Zobnin, B. B., Rogovich, V. I., and Morozova, V. A. (2002). *Control of Resources: Estimation and Reduction of Ecologic–Economic Damage* (306 pp.). Ural State Technical University, Ekaterinburg, Russia [in Russian].
Löfrstedt, R. and Frewer, L. (eds) (1998). *The Earthscan Reader in Risk and Modern Society* (288 pp.). Earthscan, London.
Lohmann, G. and Sirocko, F. (2004). Paleoclimatic research within DEKLIM. *Pilot Analysis of the Global Ecosystems News*, **112**(12), 6–7.
Lohmann, G., Butzin, M., Dima, M., Grosfeld, K., Knorr, G., Könnecke, L., Romanova, V., Schubert, S., and Zech, S. (2004). Climate transitions: Forcing and feedback mechanisms of glacial–interglacial and recent climate change. *Pilot Analysis of the Global Ecosystems News*, **12**(2), 21–22.
Logofet, D. O. (2002). Matrix population models: Construction, analysis, and interpretation. *Ecological Modelling*, **148**(3), 307–310.
Lomborg, B. (2001). *The Skeptical Environmentalist: Measuring the Real State of the World* (539 pp.). Cambridge University Press, Cambridge, UK.
Lomborg, B. (2005). *Global Crisis, Global Solution* (670 pp.). Cambridge University Press, Cambridge, UK.
Losev, K. S. (2001). *Ecological Problems and Perspectives of Sustainable Development of Russia in the 21st Century* (400 pp.). Kosmoinform, Moscow [in Russian].

Lovelock, J. E. (2003). GAIA and emergence: A response to Kirchner and Volk. *Clim. Change*, **57**(1–2), 1–7.

Lu Hung (1993). Change in the natural conditions of the Aral Sea region under conditions of anthropogenic load (211 pp.). PhD thesis, Moscow State University, Faculty of Geography [in Russian].

Lundlum, D. M. (1963). *Early American Hurricanes 1492–1870* (198 pp.).American Meteorological Society, Boston.

Luo, C., Mahowald, N. M., and del Corral, J. (2003). Sensitivity study of meteorological parameters on mineral aerosol mobilization, transport, and distribution. *J. Geophys. Res.*, **108**(D15), AAC5/11–AAC5/21.

Lutz, W., Sanderson, W. C., and Scherbov, S. (eds) (2004). *The End of World Population Growth in the 21st Century: New Challenges for Human Capital Formation and Sustainable Development* (351 pp.). Earthscan, London.

Mahlman, J. D. (1998). Science and nonscience concerning human-caused climate warming. *Ann. Rev. Energy and Environ.*, **23**, 83–105.

Maki, M., Ishiahra, M., and Tamura, M. (2004). Estimation of leaf water status to monitor the risk of forest fires by using remotely sensed data. *Remote Sensing of Environment*, **90**(4), 441–450.

Maksudova, L. G., Savinykh, V. P., and Tsvetkov, V. Ya. (2000). Integration of environmental sciences in geoinformatics. *Research of the Earth from Space*, **1**, 46–50 [in Russian].

Malinetskiy, G. G. (2002). Scenarios, strategic risks, information technologies. *Information Technologies and Computation Systems*, **4**, 83–108 [in Russian].

Malinetskiy, G. G., Podlazov, A. V., and Kuznetsov, I. V. (2005). On a national system of scientific monitoring. *Herald of RAS*, **75**(7), 592–606 [in Russian].

Malinnikov, V. A., Savinykh, V. P., Sladkopevtsev, S. A., and Tsypina, E. M. (2000). *Geography from Space* (224 pp.). Moscow State University for Geodesy and Cartography, Moscow [in Russian].

Mansell, E. R., MacGorman, D. R., Ziegler, C. L., and Starka, J. M. (2002). Simulated three-dimensional branch lightning in a numerical thunderstorm model. *J. Geophys. Res.*, **107**(D9), ACL2/1–ACL2/14.

Manwell, J. F., McGowan, J. G. and Rogers, A. L. (2002). *Wind Energy Explained* (590 pp.). John Wiley & Sons, New York.

Marakushev, A. A. (1995). The nature of native mineral formation. *Reports of RAS*, **341**(6), 807–812 [in Russian].

Marchuk, G. I. and Kondratyev, K. Ya. (1992). *Priorities of Global Ecology* (264 pp.). Nauka, Moscow [in Russian].

Marshall, T. C. and Stolzenburg, M. (2002). Electrical energy constraints on lightning. *J. Geophys. Res.*, **107**(D7), ACL1/1–ACL1/13.

Mascarenhas, F. C. B., Toda, K., Miguez, M. G., and Inone, K. (2005). *Flood Risk Simulation* (456 pp.). WIT Press, Southampton, UK.

Masure, P. (2003). Variables and indicators of vulnerability and disaster risk for land-use and urban or territorial planning. In: O. D. Cardona (ed.), *BID/IDEA Program on Indicators for Disaster Risk Management* (pp. 1–53). Universidad Nacional de Colombia, Manizales, Colombia. Available at *http://idea.unalmzl.edu.co*

Mayers, J. C. (2004). London's wettest summer and wettest year: 1903. *Weather*, **59**(10), 274–278.

McBean, G. A. (1998). *The Climate Agenda: The Role of the WCRP as the Research Thrust* (WCRP/WMO Publ. 904, pp. 11–18). World Climate Research Programme, World Meteorological Organization, Geneva

McConnel, W. J. (2004). Forest cover change: Tales of the unexpected. *Global Change Newsletter*, **57**, 8–11.

McIntyre, A. D. (1999). The environment and the oil companies. *Marine Pollution Bulletin*, **38**(3), 155–156.

McIntyre , S. and McKitrick, R. (2003). Corrections to the Mann *et al.* (1998). proxy data base and Northern Hemisphere temperature series. *Energy and Environment*, **14**(6), 751–757.

McNulty, S. G. (2002). Hurricane impacts on US forest carbon sequestration. *Environmental Pollution*, **116**, S17–S25.

Meadows, D. H., Meadows, D. L., Randers, J., and Behrens ,W. W. (1972). *The Limits to Growth* (208 pp.). Potomak Associates, New York.

Meadows, M. (2000). United Kingdom methane emissions: Trends, projections, and mitigation options. *Proceedings of the Second International Methane Mitigation Conference, 18–23 June, Novosibirsk* (pp. 37–44). Novosibirsk State University, Novosibirsk, Russia.

Mearns, L. O., Bogardi, L., Giorgi, F., Matyasovsky, I., and Palecki, M. (1999). Comparison of climate change scenarios generated from regional climate model experiments and statistical downscaling. *J. Geophys. Res.*, **104**(D6), 6603–6621.

Meijer, E. W. and Velthoven, P. (1997). The effect of the conversion of nitrogen oxides in aircraft exhaust plumes in global models. *Geophys. Res. Lett.*, **24**(23), 3013–3016.

Meloni, D., Di Sarra, A., Fiocco, G., and Junkermann, W. (2003). Tropospheric aerosols in the Mediterranean, 3: Measurements and modelling of actinic radiation profiles. *J. Geophys. Res.*, **108**(D10), AAC6/1–AAC6/12.

Menon, S., Del Genio, A. D., Koch, D., and Tselioudis, G. (2002). GCM simulations of the aerosol indirect effect: Sensitivity to cloud parameterization and aerosol burden. *J. Atmos. Sci.*, **59**(3), Part 2, 692–743.

Merrifield, M. A., Firing, Y. L., Aarup, T., Agricole, W., Brundrit, G., Chang-Seng, D., Farre, R., Kilonsky, B., Knight, W., Kong, L. *et al.* (2005). Tide gauge observations of the Indian Ocean tsunami, December 26, 2004. *Geophys. Res. Lett.*, **32** L09603, doi: 10.1029/2005GL022610.

Micklin, P. (2002). Water in the Aral Sea basin of Central Asia: Cause of conflict and cooperation? *Eurasian Geography and Economics*, **43**(7), 505–528.

Mikhailov, V. N. (1999). Why did the Aral Sea become shallow? *Soros Educational Collection*, **2**, 85–90 [in Russian].

Miki, M., Rakov, V. A., Rambo, K. J., Schnetzer, G. H., and Uman, M. A. (2002). Electric fields near triggered lightning channels measured with Pockels sensors. *J. Geophys. Res.*, **107**(D16), ACL2/1–ACL2/11.

Mikolajewicz, U., Gröger, M., Maier-Reimer, F., Schurgers, G., Vizcaino, M., and Winguth, A. (2004). Climcyc: Modeling of the Last Glacial cycle: Response of climate and vegetation to insolation forcing between 132–112 ka BP. *Pilot Analysis of the Global Ecosystems News*, **12**(2), 24–25.

Milne, A. (2004). *Doomsday: The Science of Catastrophic Events* (194 pp.). Praeger, Westport, CT.

Mintzer, I. M. (1987). *A Matter of Degrees: The Potential for Controlling the Greenhouse Effect* (Research Report No. 15, 70 pp.). World Resources Institute, Washington, DC.

Mischnenko, M., Penner, J., and Anderson, D. (2002). Global Aerosol Climatology Project. *J. Atmos. Sci.*, **59**(1), 249–250.

Mo, Q., Helsdon, J. H., and Winn, W. P. (2002). Aircraft observations of the creation of lower positive charges in thunderstorms. *J. Geophys. Res.*, **107**(D22), ACL4/1–ACL4/15.

Mock, G. (ed.) (2005). *World Resources 2005. The Wealth of the Poor: Managing Ecosystems to Fight Poverty* (266 pp.). World Resources Institute, Washington, DC.

Monin, A. S. and Krasitsky, V. P. (1985). *Phenomena on the Ocean Surface* (375 pp.). Hydrometeoizdat, Leningrad [in Russian].

Monin, A. S. and Shishkov, Yu. A. (1991). Warming dilemmas in the 20th century. In: A. S. Monin (ed.), *Man and Chaos* (pp. 47–49), Hydrometeoizdat, St Petersburg [in Russian].

Monmonier, M. (1997). *Cartographies of Danger: Mapping Hazards in America* (363 pp.). University of Chicago Press, Chicago.

Morén, L. and Påsse, T. (2001). *Climate and Shoreline in Sweden during Weichsel and the next 150,000 year* (70 pp.). Swedish Nuclear Fuel and Waste Management, Stockholm.

Moron, V., Vautard, R., and Ghil, M. (1998). Trends, interdecadal and interannual oscillations in global sea-surface temperatures. *Climate Dynamics*, **7–8**, 545–569.

Morris, D., Freeland, J., Hinchcliffe, S., and Smith, S. (2003). *Changing Environments* (344 pp.). John Wiley & Sons, London.

Morris, J. (1997). Introduction: Climate change – prevention or adaptation? *International Energy Agency Stud. Educ.*, **10**, 13–37.

Mosier, A. R., Syers, J. K., and Freney, J. R. (2004). *Agriculture and the Nitrogen Cycle* (318 pp.). Island Press, Washington, DC.

Mukherjee, M., Vilela, M., and McDermott, B. (eds) (2004). *Earth Charter Initiative: Annual Report 2004* (28 pp.). ECYI Press, San José, Costa Rica.

Munich Re (1998). *Annual Review of Natural Catastrophes 1997, 1998* (132 pp.). Münchener Ruckversicherungs-Gesellshaft, Munich, Germany. Available at *http://www.munichre.com*

Munich Re (1999). *Topics 2000: Natural Catastrophes – the Current Position* (126 pp.). Münchener Ruckversicherungs-Gesellshaft, Munich, Germany.

Munich Re (2005a). *Annual Report 2004: Advancing Innovation* (220 pp.). Münchener Ruckversicherungs-Gesellshaft, Munich, Germany.

Munich Re (2005b). *Topic Geo. Annual Review: Natural Catastrophes 2004* (60 pp.). Münchener Rückversicherungs-Gesellshaft, Munich, Germany.

Murayama, S., Taguchi, S., and Hiquchi, K. (2004). Interannual variation in the atmospheric CO_2 growth rate: Role of atmospheric transport in the Northern Hemisphere. *Journal of Geophysics Research*, **109**(D02305), doi: 10.1029/2003JD003729, 1–14.

Murnane, R. J. and Liu K.-B. (eds) (2004). *Hurricanes and Typhoons: Past, Present, and Future* (476 pp.). Columbia University Press, New York.

Naidenov, V. I. and Kozhevnikova, I. A. (2003). Why are floods so frequent? *Nature*, **9**, 1–14 [in Russian].

Naidenov, V. I. and Kozhevnikova, I. A. (2005). The law of catastrophic floods. *Herald of RAS*, **75**(1), 46–55 [in Russian].

Najam, A. and Sagar, A. (1998). Avoiding a cop-out: Moving towards systematic decision-making under the Climate Convention. *Clim. Change*, **39**(4), III–IX.

Nakamura, H., Lin, G., and Yamagata, T. (1997). Decadal climate variability in the North Pacific during the recent decades. *Bull. Amer. Meteorol. Soc.*, **78**, 2215–2225.

Negri, A. J., Burkardt, N., Golden, J. H., Halvenson, J. R., Huffman, C. J., Larsen, M. C., McGinley, J. A., Updike, R. G., Verdin, J. P., and Wieczorek, G. F. (2005). The hurricane–flood–landslide continuum. *Bull. Amer. Meteorol. Soc.*, **86**(9), 1241–1247.

Nepomniashchy, N. N. (2002). *Natural Disasters* (456 pp.). Detsky Mir, Moscow [in Russian].

Nerushev, A. F. (1995). The impact of tropical cyclones on the ozonosphere. *Proc. of RAS: Physics of the Atmosphere and Ocean*, **31**(1), 46–52 [in Russian].

New Priorities for the 21st Century (2003). *NOAA's Strategic Plan for FY 2003: FY 2008 and Beyond* (16 pp.). National Oceanic and Atmospheric Administration, Washington, DC.

Newland, J. A. and De Luca, T. H. (2000). Influence of fire on native nitrogen-fixing plants and soil nitrogen status in ponderosa pine–Douglas-fir forests in western Montana. *Can. J. Forest Res.*, **30**(2), 274–282.

Nicolis, C. and Nicolis, G. (1995). From short-scale atmospheric variability to global climate dynamics: Toward a systematic theory of averaging. *J. Atmos. Sci.*, **52**(11), 1903–1913.

Nilsson, S., Jonas, M., and Obersteiner, M. (2002). COP-6: A healing shock? An editorial essay. *Clim. Change*, **52**(1–2), 25–28.

Nishida, K., Nemani, R. R., Glassy, J. M., and Running, S. W. (2003). Development of an evapotranspiration index from Aqua/MODIS for monitoring surface moisture status. *IEEE Trans on Geosci. and Remote Sensing*, **41**(2), 493–501.

Nitu, C., Krapivin, V. F., and Bruno, A. (2000a). *Intelligent Techniques in Ecology* (150 pp.). Printech, Bucharest.

Nitu, C., Krapivin, V. F., and Bruno, A. (2000b). *System Modelling in Ecology* (260 pp.). Printech, Bucharest.

Nitu, C., Krapivin, V. F., and Pruteanu, E. (2004). *Ecoinformatics: Intelligent Systems in Ecology* (411 pp.). Magic Print, Onesti, Bucharest.

Notron, R. (2002). *Early Eighteen-Century Newspaper Reports: A Sourcebook "Natural Catastrophes"* (25 pp.). Available at *http://www.infopt.demon.co.uk/grub/catastro.htm*

Novelli, P. C., Masarie, K. A., Lang, P. M., Hall, B. D., Myers, R. C., and Elkins, J. W. (2003). Reanalysis of tropospheric CO trends: Effects of the 1997–1998 wildfires. *J. Geophys. Res.*, **108**(D15), 4464, doi: 10.1023/2002 JD 003031.

O'Brien, K. R., Kelkar, U., Venema, H., Aandahl, G., Tompkins, H., Javed, A., Bhadwal, S., Barg, S., Nygaard, L., and West, J. (2004). Mapping multiple stressors: Climate change and economic globalization in India. *Global Environmental Change*, **14**(4), 364–370.

Oganecian, V. V. (2004). Change of Moscow climate from 1879 to 2002 in temperature and precipitation extrema. *Meteorology and Hydrology*, **9**, 31–37 [in Russian].

Oppenheimer, C. (1996). Volcanism. *Geography*, **81**(1), 65–81.

Osipov, V. I. (1997). Natural disasters and sustainable development. *Geoecology*, **2**, 5–18 [in Russian].

Osipov, V. I. (2004). The history of natural disasters on the Earth. *Herald of RAS*, **74**(11), 998–1005 [in Russian].

Ougolnitsky, G. A. (1999). *Control of the Ecologic–Economic Systems* (132 pp.). High School Books, Moscow [in Russian].

Our Changing Planet (2004). *The US Climate Change Science Program for Fiscal Years 2004 and 2005* (159 pp.). US Department of Energy, Washington, DC.

Palmer, T. N., Alessandri, A., Anderson, U., Cantelaube, P., Davey, M., Délécluse, P., Déqué, M., Diez, E., Doblas-Reyes, F. J., Feddersen, H. *et al.* (2004). Development of a European multimodel ensemble system for seasonal-to-interannual prediction (DEMETER). *Bull. Amer. Meteorol. Soc.*, **85**(6), 853–872.

Palutikof, J. P., Goodess, C. M., Watkins, S. J., and Burgess, P. E. (1999). Long-term climate change. *Progr. Environ. Sci.*, **1**(1), 89–96.

Parkinson, C. L. (2003). Aqua: An Earth-observing satellite mission to examine water and other climate variables. *IEEE Trans. on Geosci. and Remote Sensing*, **41**(2), 173–183.

Parrish, D. and Law, K. (2003). Intercontinental Transport and Chemical Transformation (ITCT-Lagrangian – 2k4). *J. Geophys. Res.*, **108**(D15), 8–13.

Parson, E. A., and Fisher-Vander, K. (1997). Integrated assessment models of global climate change. *Ann. Rev. Energy Environ.*, **22**, 589–628.

Pauer, J. J. and Auer, M. T. (2000). Nitrification in the water column and sediment of a hypereutrophic lake and adjoining river system. *Water Research*, **34**(4), 1247–1254.

Perrie, W., Zhang, W., Ren, X., and Long, Z. (2004). The role of midlatitude storms on air–sea exchange of CO_2. *Geophys. Res. Lett.*, **31**(L09306), doi: 10.1029/2003GL019212, 1–4.

Pervaniuk, V. S. and Tarko, A. M. (2001). Modelling the global carbon cycle in the atmosphere–ocean system. *Numerical Modelling*, **13**(11), 13–22 [in Russian].

Phelan, J. P. (2003). *Topics: Annual Review of North American Natural Catastrophes* (50 pp.). American Re, Princeton, NJ.

Phelan, J. P. (2004). *Topics: Annual Review of North American Natural Catastrophes* (48 pp.). American Re, Princeton, NJ.

Phothero, J. W. (1979). Maximal oxygen consumption in various animals and plants. *Comp. Biochem. and Physiol.*, **A64**(4), 461–466.

Pielke, R. A., Sr. (2001a). Carbon sequestration: The need for an integrated climate system approach. *Bull. Amer. Meteorol. Soc.*, **82**(11), 20–21.

Pielke, R. A., Sr. (2001b). Earth system modeling: An integrated assessment tool for environmental studies. In: T. Matsuno and H. Kida (eds), *Present and Future of Modeling Global Environmental Change: Toward Integrated Modeling* (pp. 311–337). Terrapub, Tokyo.

Pielke, R. A., Sr. (2002). Overlooked issues in the US national climate and IPCC assessments. *Clim. Change*, **52**(1–2), 1–11.

Pinder, G. F. (2002). *Groundwater Modeling* (248 pp.). John Wiley & Sons, New York.

Pinter, N. (2005). One step forward, two steps back on US floodplains. *Science*, **308**(57), 207–208.

Podgorny, I. A. and Ramanathan, V. (2001). A modeling study of the direct effect of aerosols over the tropical Indian Ocean. *J. Geophys. Res.*, **106**(D20), 24097–24105.

Podlazov, A. V. (2001). The self-organized criticality and analysis of the risk. *Proc. of High School. Applied Non-linear Dynamics*, **9**(1), 49–88 [in Russian].

Pogorelov, A. V. (1998). The stable snow cover regime at the Great Caucasus Ridge. *Data of Glaciological Studies*, **84**, 170–175 [in Russian].

Popovicheva, O. B., Starik, A. M., and Favorsky, O. N. (2000). Problems of the impact of aviation on the gas and aerosol composition of the atmosphere. *Proc. of RAS: Physics of the Atmos. and Ocean*, **2**, 163–176 [in Russian].

Porvari, P. and Verta, M. (2003). Total and methyl mercury concentrations and fluxes from small boreal forest catchments in Finland. *Environmental Pollution*, **123**(2), 181–191.

Posner, R. A. (2004). *Catastrophe: Risk and Response* (332 pp.). Oxford University Press, Oxford, UK.

Qi, F., Guoduong, C., and Masao, M. (2001). The carbon cycle of sandy lands in China and its global significance. *Clim. Change*, **48**(4), 535–549.

Rakov, V. A. and Tuni, W. G. (2003). Lightning electric field intensity at high latitudes: Inferences for production of elves. *J. Geophys. Res.*, **108**(D20), ACL8/1–ACL8/16.

Ramsley, G. (1978). Snow avalanches in Norway. In: S. B. Lavrov and L. G. Nikiforova (eds), *Natural Disasters: Study and Methods of Struggle* (pp. 204–211). Progress, Moscow [in Russian].

Rasch, D., Kubinger, K., Schmidtke, J., and Häusler, J. (2004). The misuse of asterisks in hypothesis testing. *Psychology Science*, **46**(2), 227–242.

Rayner, P. (2001). "Flying Leap" becomes C4MIP. *Res. International Energy Agency*, **4**, 8–12.

Remer, L. A., Kaufman, Y. J., Levin, Z., and Ghan, S. (2002). Model assessment of the ability of MODIS to measure top-of-atmosphere direct radiative forcing from smoke aerosols. *J. Atmos. Sci.*, **59**(3), Part 2, 657–667.

Remizov, L. T. (1985). *Natural Radionoise* (196 pp.). Nauka, Moscow [in Russian].

Richter, C. F. (1969). Earthquakes. *Natural History*, **78**, 37–45.

Ridley, B., Ott, L., Pickering, K., Emmons, L., Montzka, D., Weinheimer, A., Knopp, D., Grahek, F., Li, L., Heymsfield, G. *et al.* (2004). Florida thunderstorms: A faucet of reactive nitrogen to the upper troposphere. *J. Geophys. Res.*, **109**(17), D17305/1–D17305/19.

Rizzolio, D. (2005). Environmental issues in disaster preparation and response. *Division of Early Warning and Assessment/Grid Resource Information Database – Europe Quartely Bulletin*, **7**(1), 1–2.

Robock, A. and Oppenheimer, C. (eds) (2003), *Volcanism and the Earth's Atmosphere* (360 pp.). American Geophysical Union, Washington, DC.

Roch, P. (ed.) (2002). *Environment Switzerland 2002* (239 pp.). International Energy Agency, Zurich, Switzerland.

Roda, F., Avila, A., and Rodrogo, A. (2002). Nitrogen deposition in Mediterranean forests. *Environmental Pollution*, **118**(2), 205–213.

Rojstaczer, S., Sterling, S. M., and Moore, N. J. (2001). Human appropriation of photosynthesis products. *Science*, **294**, 2549–2552.

Ronner, U. (1983). *Biological nitrogen transformations in marine ecosystems with emphasis on denitrification* (165 pp.). Department of Marine Microbiology, University of Goteborg, Goteborg, Sweden.

Roy, A. S. (2004). Remote sensing: A new tool for environmental monitoring and surveillance for disaster control in Ganga–Brahmaputra basin. *Proceedings of 25th Asian Conference on Remote Sensing, 22–26 November, Chiang Mai, Thailand* (pp. 596–599), AARS, Chiang Mai, Thailand.

Rubanov, I. V., Ishniyazov, D. P., Baskakova, M. A., and Chistiakov, P. A. (1987). *Aral Sea Geology* (241 pp.). Fan, Tashkent, Uzbekistan [in Russian].

Ruck, M. (2002). *Natural Catastrophes 2002: Annual Review* (50 pp.). Munich Re Topics, Dresden.

Russo, G., Eva, C., Palau, C., Caneva, A., and Saechini, A. (2000). The recent abrupt increase in the precipitation rate, as seen in an ultra-centennial series of precipitation. *Il Nuovo Cimento C*, **23**(1), 39–51.

Santer, B. D., Wigley, T. M. L., Boyle, J. S., Gaffen, D. J., Hnilo, J. J., Nychka, D., Parker, D. E., and Taylor, K. E. (2000). Statistical significance of trends and trend differences in layer average atmospheric temperature time series. *J. Geophys. Res.*, **105**(D6). 7337–7356.

Satake K. (2005). *Tsunamis: Case Studies and Recent Developments* (343 pp.). Springer-Verlag, Heidelberg, Germany.

Savinykh, V. P. and Tsvetkov, V. Ya. (2001). *Geoinformation Analysis of the Remote Sounding Data* (228 pp.). Geodeoizdat, Moscow [in Russian].

Schlesinger, M. E., Ramankutty, N., and Andronova, N. (2000). Temperature oscillations in the North Atlantic. *Science*, **289**(5479), 547–548.

Schlüter, M., Savitsky, A. G., McKinney, D. C., and Lieth, H. (2005). Optimizing long-term water allocation in the Amudarya river delta: A water management model for ecological impact assessment. *Environmental Modelling & Software*, **20**(5), 529–545.

Schminke, H. U. (2004). *Volcanism* (324 pp.). Springer-Verlag, New York.

Schneider, D. (1995). Global warming is still a hot topic. *Scientific American*, **272**(2), 13–14.

Schneider, R. R., Kim, J.-H., Rimbu, N., Lorenz, S., Lohmann, G., Cubasch, U., Pätzold, J., and Wefewr, G. (2004). GHOST (Global Holocene Spatial and Temporal Climate Variability): Combination of paleotemperature records, statistics, and modeling. *PAGES News*, **12**(2), 25–26.

Schrope, M. (2001). Consensus science or consensus politics? *Nature*, **412**(6843), 112–114.

Schröter, R. (2005). Where was God when the tsunami came? *Guardian Weekly*, 11 January. Available at *http://www.spotlight-online.de/CoCoCMS/generator/viewDocument.php?doc=16170&archive=1*

Scott, K. (2005). *Carbon Sequestration* (28 pp.). US Department of Energy, Washington, DC.

Seidov, D. G. (1987). Numerical models of the ocean circulation. *Earth and Universe*, **5**, 28–34 [in Russian].

Sellers, P. J., Los, S. O., Tucker, C. J., Justice, C. O., Dazlich, D. A., Collatz, G. J., and Randall, D. A. (1996a). A revised land surface parameterization (SiB2) for atmospheric GCMs, Part II: The generation of global fields of terrestrial biophysical parameters from satellite data. *J. of Climate*, **9**(4), 708–737.

Sellers, P. J., Randall, D. A., Collotz, G. J., Berry, J.A., Field, C. B., Dazlich, D. A., Zhang, C., Collelo, G. D., and Bounona L. (1996b). A revised land surface parameterization (SiB2) for atmospheric GCMs, Part 1: Model formulation. *J. of Climate*, **9**(4), 676–705.

Semenov, S. M. (2004). *Greenhouse Gases and the Present Climate of the Earth* (175 pp.). Meteorology and Hydrology Center, Moscow [in Russian].

Semmler, T. and Jacob, D. (2004). Modeling extreme precipitation events: A climate change simulation for Europe. *Global and Planetary Change*, **44**(1–4), 119–127.

Shutko, A. M. (1987). *Microwave Radiometry of Water Surface and Soils* (190 pp.). Nauka, Moscow [in Russian].

Sidorov, A. (1999). First there were volcanoes. *Science and Life*, **2**, 51–55 [in Russian].

Singer, S. F. (1997). *Hot Talk, Cold Science* (111 pp.). Independent Institute, Oakland, CA.

Sinha, P., Hobbs, P. V., Yokelson, R. J., Blake, D. R., Gao, S., and Kirchstetter, T. W. (2004). Emissions from miombo woodland and dambo grassland savanna fires. *J. Geophys. Res.*, **109**(4). D11305/1–D11305/13.

Sirocko, F., Cubash, U., Kaspar, F., von Storch, H., Widmann, M., Litt, T., Kühl, N., Mangini, A., Pachur, H.-J., Claussen, N. *et al.* (2004). Climate change at the very end of a warm stage: First results from the Last Glacial inception at 117,000 yr BP. *International Energy Agency News*, **12**(2), 18–20.

Sirotenko, O. D. (2000). The future of agriculture in Russia in connection with expected climate change. *Problems of Ecological Monitoring and Ecosystems Modelling*, **XVII**, 258–274 [in Russian].

Slovic, P. (2000). *The Perception of Risk* (265 pp.). Earthscan, London.

Smith, W. K., Kelly, R. D., Welker, J. M., Fahnestock, J. T., Reiners, W. A., and Hunt, E. R. (2003). Leaf-to-aircraft measurements of net CO_2 exchange in a sagebrush steppe ecosystem. *J. Geophys. Res.*, **108**(3), ACH15/1–ACH15/9.

Sobrino, J. A., Jiménez-Muñoz, J. C., and Paolini, L. (2004). Land surface temperature retrieval from LANDSAT TM 5. *Remote Sensing of Environment*, **90**(5), 434–440.

Sofronov, M. A. and Vakurov, A. D. (1981). *Fire in a Forest* (128 pp.). Nauka, Novosibirsk, Russia [in Russian].

Solanki, S. K., Usoskin, I. G., Kromer, B., Schorsler, M., and Beer, J. (2004). Unusual activity of the Sun during recent decades compared to the previous 11,000 years. *Nature*, **431**(7012), 1084–1087.

Solozhentsev, E. D. (2004). *Hierarchical Risk Control in Business and Engineering* (416 pp.). Business Press, St Petersburg [in Russian].

Soon, W., Postmentier, E., and Baliunas, S. (2000). Climate hypersensitivity to solar forcing? *Ann. Geophysics*, **18**, 583–588.

Sorjamaa, R., Raatikainen, T., and Laaksonen, A. (2004). The role of surfactants in Köhler theory reconsidered. *Atmospheric Chemistry Physics Discussions*, **4**, 2781–2804.

Sorokhtin, O. G. (2001). The greenhouse effect: Myth and reality. *Herald of RAS*, **1**(1), 8–21 [in Russian].

Sorokhtin, O. G. and Ushakov, S. A. (1996). Drifting of the continents in the geological history of the Earth. In: S. A. Ushakov (ed.), *Life of the Earth: Structure and Evolution of the Lithosphere* (pp. 66–71). Moscow State University, Moscow [in Russian].

Spedicato, E. (1991). A catastrophical scenario for discontinuities in human history. *Journal of New England Antiquities Research Association*, **26**, 1–14.

Starke, L. (ed.) (2004). *State of the World – 2004: Progress towards a Sustainable Society* (246 pp.). Earthscan, London.

Steffen, W. and Tyson, P. (eds.) (2001). *Global Change and the Earth System: A Planet under Pressure* (32 pp.). IGBP Science 4, Stockholm.

Steffen, W., Sanderson, A., Tyson, P. D., Jäger, J., Matson, P. A., Moore, B., III, Oldfield, F., Richardson, K., Schellnhuber, H.-J., Turner, B. L., II, and Wasson, R. J. (2004). *Global Change and the Earth System: A Planet under Pressure* (336 pp.). Springer-Verlag, New York.

Stein, A. F. and Lamb, D. (2000). The sensitivity of sulphur wet deposition to atmospheric oxidants. *Atm. Env.*, **34**(11), 1681–1690.

Stempell, D. (1985). *Weltbevölkerung* (205 pp.). Springer-Verlag, Berlin.

Stephens, B. B. and Keeling, R. F. (2000). The influence of Antarctic sea ice on glacial–interglacial CO_2 variations. *Nature*, **404**(6774), 171–174.

Stevens, C. and Verne, R. (2004). *Renewable Bioresources* (320 pp.). John Wiley & Sons, New York.

Stevens, N. F. and Scott, B. (2002). *Changes in the Ngauruhoe Summit Fumarole Field 1999–2001 from Landsat Data* (Science Report No. 2002/30, 9 pp.). Institute of Geological & Nuclear Science, Lower Hutt, New Zealand.

Stevens, N. F. and Wadge, G. (2003). Towards operational repeat-pass SAR interferometry at active volcanoes. *Natural Hazards*, **00**, 1–30.

Stevens, N. F., Wadge, G., and Williams, C. A. (2001a). Post-emplacement lava subsidence and the accuracy of ERS InSAR digital elevation models of volcanoes. *International Journal of Remote Sensing*, **22**, 818–828.

Stevens, N. F., Wadge, G., Williams, C. A., Morley, J. G., Muller, J.-P., Murray, J. B., and Upton, M. (2001b). Surface movements of emplaced lava flows measured by synthetic aperture radar interferometry. *J. Geophys. Res.*, **106**(B6), 11293–11313.

Stevens, N. F., Glassey, P., and Lyttle, B. S. (2003). *Assessment of Urban Slope Instability in Dunedin, New Zealand, Using Orbital Differential Synthetic Aperture Radar Interferometry* (Client Report 2003/150, 42 pp.). Institute of Geological & Nuclear Science, Dunedin, New Zealand.

Stevens, N. F., Garbeil, H., and Mouginis-Mark, P. J. (2004). NASA EOS Terra ASTER: Volcanic topographic mapping and capability. *Remote Sensing of Environment*, **90**, 405–414.

Stohl, A., Eckhardt, S., Forster, C., James, P., and Spichtinger, N. (2003). On the pathways and timescales of intercontinental air pollution transport. *J. Geophys. Res.*, **108**(D23), ACH6/1–ACH6/17.

Strong, A. E., Kearns, E. J., and Gjovig, K. K. (2000). Sea surface temperature signals from satellite updates. *Geophys. Res. Lett.*, **27**(11), 1667–1670.

Sumarokova, V. V., Babkina, L. P., and Kriventsova, V. E. (1991). The landscape parameterization of the lower Amu-Darya based on decoding of aerospace information. In: E. M. Gluchova (ed.), *Monitoring of the Environment in the Aral Sea Basin* (pp. 200–208). Hydrometeoizdat, Leningrad [in Russian].

Summer, M., Michael, K., Bradshaw, C. J. A., and Hindell, M. A. (2003). Remote sensing of Southern Ocean sea surface temperature: Implications for marine biophysical models. *Remote Sensing of Environment*, **84**(2), 196–213.

Suni, T., Berninger, F., Markkanen, T., Keronen, P., Rannik, Ü., and Vesala, T. (2003). Interannual variability and timing of growing: Season CO_2 exchange in a boreal forest. *J. Geophys. Res.*, **108**(9), ACL2/1–ACL2/8.

Svirezhev, Yu. M. (2002). Simple spatially distributed model of the global carbon cycle and its dynamic properties. *Ecological Modelling*, **155**(1), 53–69.

Syvorotkin, V. L. (2002). *Deep Degassing of the Earth and Global Catastrophes* (250 pp.). Geoinformcenter, Moscow [in Russian].

Szathmáry, E. and Griesemer, J. (2003). *The Principles of Life* (220 pp.). Oxford University Press, Oxford, UK.

Takemura, T. (2002). *A Study of Aerosol Distribution and Optical Properties with a Global Climate Model* (Report No. 16, 113 pp.). CCSR University of Tokyo, Tokyo.

Tareyeva, A. M. and Seliverstov, Yu. G. (2004a). *Byrranga Mountains* (11 pp.). Available at *http://www.geogr.msu.ru* [in Russian].

Tareyeva, A. M. and Seliverstov, Yu. G. (2004b). *Putorana Mountains* (10 pp.). Available at *http://www.geogr.msu.ru* [in Russian].

Tarko, A. M. (2001). Investigation of global biosphere processes with the aid of a global spatial carbon dioxide cycle model. *Sixth International Carbon Dioxide Conference* (Extended Abstract No. 2, pp. 899–902). Tokohu University, Sendai, Japan.

Tarko, A. M. (2005). *Numerical Modelling of Anthropogenic Changes in Global Biospheric Processes* (278 pp.). Physics-Mathematics, Moscow [in Russian].

Tett, S. F. B., Stott, P. A., Allen, M. R., Ingram, W. J., and Mitchell, J. F. B. (1999). Causes of twentieth-century temperature change near the Earth's surface. *Nature*, **399**, 569–572.

The World Climate Research Programme Strategy 2005–2015 (2004). *Coordinated Observation and Prediction of the Earth System (COPES): A Discussion Document* (31 pp.). World Meteorological Organization, Geneva.

Thuong Thuy Vu, Masashi Matsuoka, and Fumio Yamazaki (2004). Lidar signatures to update Japanese building inventory database. *Proceedings of the 25th ACRS, 22–26 November 2004, Chiang Mai (Thailand), Asian Association on Remote Sensing, Chiang Mai, Thailand, 2004*, pp. 624–629.

Tianhong, L., Yanxin, S., and An, X. (2003). Integration of large scale fertilizing models with GIS using minimum unit. *Environmental Modelling*, **18**(3), 221–229.

Tie, X., Brasseur, G., Emmons, L., Horowitz, L., and Kinnison, D. (2001). Effects of aerosol on tropospheric oxidants: A global model study. *J. Geophys. Res.*, **106**(D19), 22931–22964.

Timmermann, A., Oberhuber, J., Bacher, A., Esch, M., Latif, M., and Roeckner, E. (1999). Increased El Niño frequency in a climate model forced by future greenhouse warming. *Nature*, **398**, 684–687.

Timofeyev-Resovsky, N. V. (1961). On some principles of classification of biochorological units. *Proc. of USSR Acad. Sci.*, **27**, 290–311 [in Russian].

Timoshevsky, A., Yeremin, V., and Kalkuta, S. (2003). New method for ecological monitoring based on the method of self-organizing mathematical models. *Ecological Modelling*, **162**(2–3), 1–13.

Titov, V., Rabinovich A. B., Mofield, H. O., Thompson R. E., and González, F. I. (2005). The global reach of the 26 December 2004 Sumatra tsunami. *Science*, **309**, 2045–2048.

Travis, J. (2005). Scientist's fears come true as hurricane floods New Orleans. *News*, **309**, 1656–1659.

Trenberth, K. (2005). Uncertainty in hurricane and global warming. *Science*, **308**(5729), 1753–1754.

Troshkina, E. S. (1992). *The Avalanche Regime of Mountain Territories of the USSR* (196 pp.). ARISTI, Moscow [in Russian].

Tsushima, Y. and Manabe, S. (2001). Influence of cloud feedback on annual variation of global mean surface temperature. *J. Geophys. Res.*, **106**(D19), 22646–22655.

Tuomi, J. (2004). Commentary: Genetic heterogeneity within organisms and the evolution of individuality. *Journal of Evolutionary Biology*, **17**, 1182–1183.

Tuong Thuy, Matsuoka, M. and Yamazaki, F. (2004). Lidar signatures to update Japanese building inventory database. *Proceedings of 25th Asian Conference on Remote Sensing, 22–26 November, Chiang Mai, Thailand* (pp. 624–629). AARS, Chiang Mai, Thailand.

Tupper, A., Carn, S., Davey, J., Kamada, Y., Pott, R., Prata, F., and Tokuno, M. (2004). An evaluation of volcanic cloud detection technique during recent significant eruptions in the western ring of fire. *Remote Sensing of Environment*, **91**(1), 27–46.

Tsytsarin, A. G. (1991). The present state of the elements of the Aral Sea hydrological regime. *Proc. of the State Oceanographic Institute*, **183**, 72–92 [in Russian].

UNEP Science Initiative (2004). *Environmental Change and Human Needs: Assessing Interlinkages. An Input to the Fourth Global Environment Outlook* (Concept Paper GEO-4, 36 pp.). United Nations Environment Programme, Nairobi, Kenya.

Vaganov, E. A., Furiayev, V. V., and Sukhinin, A. I. (1998). Fires in the Siberian taiga. *Nature*, **7**, 51–62 [in Russian].

Vakulenko, N. V., Kotliakov, V. M., Monin, A. S., and Sonechkin, D. M. (2004). Evidence of the leading role of temperature variations with respect to variations of greenhouse gas concentration from data of ice core from the station "Vostok". *Annals of RAS*, **397**(5), 663–667 [in Russian].

Valiela, I. and Bowen, J. L. (2002). Nitrogen sources to watersheds and estuaries: Role of land cover mosaics and losses within watersheds. *Environmental Pollution*, **118**(2), 239–248.

Vansarochana, A. (2004). Synthetic interpolative facies mapping model for landslide susceptibility map. *Proceedings of 25th Asian Conference on Remote Sensing, 22–26 November, Chiang Mai, Thailand* (pp. 612–617). AARS, Chiang Mai, Thailand.

Varotsos, C. A. and Kondratyev, K. Ya. (1998). Dynamics of the total ozone content at middle latitudes of the Northern Hemisphere. *Annals of RAS*, **359**(6), 821–822 [in Russian].

Vernadsky, V. I. (1944a). Some words about the noosphere. *Progress in Modern Biology*, **18**(2), 49–93 [in Russian].

Vernadsky, V. I. (1944b). A few words about the noosphere. *Successes Mod. Biol.*, **18**(2), 49–93 [in Russian].

Victor, D. G. (2001). *The Collapse of the Kyoto Protocol and the Struggle to Slow Global Warming* (178 pp.). Princeton University Press, Princeton, NJ.

Victor, D. G. (2004). *Climate Change: Debating America's Policy Options* (165 pp.). Council of Foreign Relations, New York.

Vilkova, L. P. (2003). Modelling the global carbon cycle. *Problems of the Environment and Natural Resources*, **7**, 34–48 [in Russian].

Villagrán, H. L. (2003). El fenómeno "El Niño" y políticas públicas: Un desafío científico, tecnológico e institucional. *Diálogo Andino*, **22**, 23–34 [in Spanish].

Vinogradov, B. V. (1983). A quantitative presentation of the function of the remote sensing of soil moisture. *Annals of RAS*, **272**(1), 247–250 [in Russian].

Vinogradov, M. E.., Gitelzan, I. I., and Sorokin, J. I. (1970). The vertical structure of a pelagic community in the tropical ocean. *Marine Biology*, **6**(4), 261–268.

Vital Signs, 2003–2004 (2003). *The Trends that Are Shaping Our Future* (149 pp.). Worldwatch Institute/Earthscan, London.

Vladimirov, V. A., Vorobyev, Yu. L., Salov, S. S., Faleyev, M. I., Arkhipova, N. I., Kapustin, M. A., Kashchenko, S. A., Kosiachenko, S. A., Kuznetsov, I. V. *et al.* (2000). *Risk Control: Risk and Sustainable Development* (431 pp.). Nauka, Moscow [in Russian].

Vogel, C. and O'Brien, K. (2004). Vulnerability and global environmental change: Rhetoric and reality. *Aviso* (an international bulletin on global environmental change and human security), **13**, 1–8.

Voitkovsky, K. F. and Korolkov, V. G. (1998). *Water–Snow Fluxes onto the Putorana Plateau* (Data of Glaciological Studies No. 84, pp. 92–94). Moscow State University, Moscow [in Russian].

Volk, T. (2003a). *Gaia's Body: Toward a Physiology of Earth* (296 pp.). MIT Press, Cambridge, MA.

Volk, T. (2003b). Natural selection, Gaia, and inadvertent by-products: A reply to Lenton and Wilkinson's response. *Clim. Change*, **58**(1–2), 13–19.

Von Oosteron, P., Zlatanova, S., and Fendel, E. M. (eds) (2005). *Geo-information for Disaster Management* (1434 pp.). Springer-Verlag, Heidelberg, Germany.

Vorobyev, Yu. L., Gusev, A. V., Zalokhanov, M. Ch., Kuznetsov, I. V., Kulba, V. V., Levashov, V. K., Lvov, D. S., Mayevsky, V. I., Malinetsky, G. G., Makhov, S. A. *et al.* (2003). A project of the scientific monitoring system and crises in present-day Russia. *Herald of RAS*, **3**(4), 71–79 [in Russian].

Wagner, F., Muller, D., and Ansmann, A. A. (2001). Comparison of the radiative impact of aerosols derived from vertically resolved (lidar) and vertically integrated (Sun photometer) measurements: Example of an Indian aerosol plume. *J. Geophys. Res.*, **106**(D19), 22861–22870.

Wainwright, J. and Mulligan, M. (2003). *Environmental Modelling* (412 pp.). John Wiley & Sons, London.

Walker, G. (2003). *Snowball Earth: The Story of the Great Global Catastrophe that Spawned Life as We Know It* (269 pp.). Crown Publishers, New York.

Wallace, J. M. (1998). Observed climatic variability: Spatial structure. In: D. L. T. Anderson and J. Willebrand (eds), *Decadal Climate Variability: Dynamics and Predictability* (NATO ASI Series I 44, pp. 31–81). Springer-Verlag, Berlin.

Wang, C., Bond-Lamberty, B., and Gower, S. T. (2003). Soil surface CO_2 flux in a boreal black spruce fire chronosequence. *J. Geophys. Res.*, **108**(3), WFX5/1–WFX5/8.

Wang, L. K., Vielkind, D., and Mu Hao Wang (1978). Mathematical models of dissolved oxygen concentration in fresh water. *Ecological Modelling*, **5**(2), 115–123.

Wang, Y. and Wu, C.-C. (2004). Current understanding of tropical cyclone structure and intensity changes: A review. *Meteorology and Atmospheric Physics*, **1**, 1–22.

Wange, G. and Archer, D. J. (2003). Evaporation of groundwater from arid playas measured by C-band SAR. *IEEE Trans. on Geosci. and Remote Sensing*, **41**(7), 1641–1650.

Watson, R. T., Noble, I. R., Bolin, B., Ravindranath, N. H., Verardo, D. J., and Dokken, D. J. (eds.) (2000). *Land Use, Land-use Change, and Forestry* (377 pp.). Cambridge University Press, Cambridge, UK.

Wauben, W. M. F., van Velthoven, P. F. J., and Kelder, H. M. (1997). A 3-D chemistry transport model study of changes in Atmospheric ozone due to aircraft emissions. *Atmos. Environm.*, **31**, 1819–1836.

Weaver, C., Ginoux, P., Chou, M.-D., and Joiner, J. (2002). Radiative forcing of Saharan dust: GOCART model simulations compared with ERBE data. *J. Atmos. Sci.*, **59**(3), Part 2, 736–747.

Weber, G. R. (1992). *Global Warming: The Rest of the Story* (188 pp.). Dr Boettiger Verlag, Wiesbaden, Germany.

Webster, P. J., Holland, G. J., Curry, J. A., and Chang, H.-R. (2005). Changes in tropical cyclone number, duration, and intensity in a warming environment. *Science*, **309**, 1844–1846.

Weisenstein, D. K., Ko, M. K. W., Dyominov, I. G., Pitarui, G., Riccardully, L., Visconti, G., and Bekki, S. (1998). The effect of sulfur emissions from HSCT aircraft: A 2D model intercomparison. *J. Geophys. Res.*, **103**(D1), 1527–1547.

White, G. F. (ed.) (1974). *Natural Hazards: Local, National, Global* (437 pp.). University of Colorado, New York.

Widmann, M., Jones, J. M., and von Storch, H. (2004). Reconstruction of large-scale atmospheric circulation and data assimilation in paleoclimatology. *Pilot Analysis of the Global Ecosystems News*, **12**(2), 12–13.

Wigley, T. M. L. (1999). *The Science of Climate Change: Global and US Perspectives* (48 pp.). Pew Center on Global Climate Change, Arlington, VA.

Wigley, T. M. L. and Raper, S. C. B. (2001). Interpretation of high projections for global-mean warming. *Science*, **293**(5529), 451–455.

Wilby, R. L. (2003). Weekly warming. *Weather*, **58**(11), 446–447.

Wilderer, P. A., Schroeder, E. D., and Kopp, H. (eds.) (2005). *Global Sustainability* (266 pp.). Wiley/VCH, Hoboken, N.J.

Williams, S. E., Bolitho, E. E., and Fox, S. (2003). Climate change in Australian tropical rainforests: An impending environmental catastrophe. *Proc. Biol. Sci.*, **270**(1527), 1887–1892.

Wofsy, S. C. (2001). Climate change: Where has all the carbon gone? *Science*, **292**(5525), 2261–2262.

Wofsy, S. C., Goulden, M. L., Munger, J. W., Pyle, E. H., Urbanski, S. P., Hutyra, L., Saleska, S. R., Fitzjarrald, D., and Moore, K. (2001). Factors controlling long- and short-term sequestration of atmospheric CO_2 in a mid-latitude forest. *Science*, **294**(5547), 1688–1691.

Woo, G. (1999). *The Mathematics of Natural Catastrophes* (304 pp.). World Scientific, New York.

Woodcock, A. (1999a). Global warming: A natural event? *Weather*, **54**(5), 162–163.

Woodcock, A. (1999b). Global warming: The debate heats up. *Weather*, **55**(4), 143–144.

Woodbury, P. B., Beloin, R. M., Swaney, D. P., Gollands, B. E., and Weinstein, D. A. (2002). Using the ECLPSS software environment to build a spatially explicit component-based model of ozone effects on forest ecosystems. *Ecological Modelling*, **150**(3), 211–238.

Woodworth, P. L., Blackman, D. L., Foden, P., Holgate, S., Horsburgh, K., Knight, P. J., Smith, D. E., Macleod E. A., and Bradshaw, E. (2005). Evidence for the Indonesian tsunami in British tidal records. *Weather*, **60**(9), 263–267.

Wotawa, G., Kroger, H., and Stohl, A. (2000). Transport of ozone towards the Alps: Results from trajectory analyses and photochemical model studies. *Atmospheric Environment*, **34**(9), 1367–1377.

Wright, E. L. and Erickson, J. D. (2003). Incorporating catastrophes into integrated assessment: Science, impacts, and adaptation. *Climatic Change*, **57**, 265–286.

Xu, Y. and Carmichael, G. R. (1999). An assessment of sulfur deposition pathways in Asia. *Atmospheric Environment*, **33**(21), 3473–3486.

Yakovlev, O. I. (1998). *Cosmic Radiophysics* (432 pp.). Russian Fund for Basic Research, Moscow [in Russian].

Yakovlev, O. I. (2001). *Space Radio Science* (320 pp.). Taylor & Francis, London.

Yamagata, T., Behera, S. K., Luo, J. J., Masson, S., Jury, M. R., and Rao, S. A. (2004). Coupled ocean–atmosphere variability in the tropical Indian Ocean. In: C. Wang, X.-P. Xie, and J. A. Carton (eds), *Earth Climate: The Ocean–Atmosphere Interactions* (pp. 189–212). Springer-Verlag, Berlin.

Yano, Y. and Yamazaki, F. (2004). Building damage detection of the 2003 Bam, Iran earthquake using Quickbird images. *Proceedings of 25th Asian Conference on Remote Sensing, 22–26 November, Chiang Mai, Thailand* (pp. 618–623). AARS, Chiang Mai, Thailand.

Yano Y., Yamazaki, F., Matsuoka, M., and Vu, T. T. (2004). Building damage detection of the 2003 Bam, Iran earthquake using quickbird images. *Proceedings of the 25th ACRS, 22–26 November 2004, Chiang Mai (Thailand), Asian Association on Remote Sensing, Chiang, Thailand, 2004*, pp. 618–623.

Yanovsky, R. G. (1999). *Global Changes and Social Safety* (358 pp.). Academia, Moscow [in Russian].

Yasanov, N. A. (2003). Climate of Phanerozoe and the greenhouse effect. *Herald of the Moscow State University, Ser. 4: Geology*, **6**, 3–11 [in Russian].

Year-book UNEP GEO-2003 (2004). *Nairobi, Kenya* (76 pp.). United Nations Environment Programme, Zurich, Switzerland.

Yu, R., Zhang, M., and Cess, R. D. (1999). Analysis of the atmosphere energy budget: A consistency study of available data sets. *J. Geophys. Res.*, **104**(D8), 9655–9661.

Zagorin, G. K. (1998). Polarization characteristics (Stocks' parameters) of self- and scattered microwave radiation in rain (150 pp.). PhD thesis, Institute of Radioengineering and Electronics of RAS, Moscow [in Russian].

Zalimkhanov, M. Ch. (1981). *The Snow–Avalanche Regime and Perspectives for Mastering the Great Caucasus Ridge* (376 pp.). Rostov-Don State University, Rostov-Don, Russia [in Russian].

Zanetti, A., Schwartz, S., and Enz, R. (2005). Natural catastrophes and man-made disasters. *Sigma*, **1**, 1–40.

Zavarzin, G. A. (2003). Antipode to the noosphere. *Herald of RAS*, **73**(7), 627–636 [in Russian].

Zavarzin, G. A. and Kolotilova, N. N. (2001). *Introduction to Natural History of Microbiology* (256 pp.). Book Yard "University", Moscow [in Russian].

Zerchaninova, I. L. and Potapov, I. I. (2001). Study of the list of technologies of the IPCC in the context of Russian progress. *Problems of the Environment and Natural Resources*, **4**, 3–22 [in Russian].

Zhang, H., Henderson-Sellers, A., and McGuffie, K. (2001). The compounding effects of tropical deforestation and greenhouse warming on climate. *Clim. Change*, **49**, 309–338.

Zhang, X., Helsdon, J. H., Jr., and Farley, R. D. (2003a). Numerical modeling of lightning-produced NO_x using an explicit lightning scheme, 1: Two-dimensional simulation as a "proof of concept". *J. Geophys. Res.*, **108**(D18), 4579, doi: 10.1029/2002 ID003224.

Zhang, X., Helsdon, J. H., Jr., and Farley, R. D. (2003b). Numerical modeling of lightning-produced NO_x using an explicit lightning scheme, 2: Three-dimensional simulation and expanded chemistry. *J. Geophys. Res.*, **108**(D18), 4580, doi: 10.1029/2002 ID003225.

Zhu Xiaoxiang, Liu Ruixia, and Zhang Yeping (2004). Application study on drought monitoring with time-series satellite data. *Proceedings of 25th Asian Conference on Remote Sensing, 22–26 November, Chiang Mai, Thailand* (pp. 600–605). AARS, Chiang Mai, Thailand.

Zillman, J. W. (2000). The challenges ahead. *WMO Bull.*, **49**(1), 8–13.

Zimmerli, P. (2003). *Natural Catastrophes and Reinsurance* (48 pp.). Swiss Re, Zurich.

Index

Printing: Mercedes-Druck, Berlin
Binding: Stein + Lehmann, Berlin